Springer Earth System Sciences

More information about this series at http://www.springer.com/series/10178

Diego Kietzmann · Andrés Folguera
Editors

Opening and Closure of the Neuquén Basin in the Southern Andes

Springer

Editors
Diego Kietzmann
IGeBA - UBA, CONICET
Buenos Aires, Argentina

Andrés Folguera
Instituto de Estudios Andinos - UBA
CONICET
Buenos Aires, Argentina

ISSN 2197-9596 ISSN 2197-960X (electronic)
Springer Earth System Sciences
ISBN 978-3-030-29679-7 ISBN 978-3-030-29680-3 (eBook)
https://doi.org/10.1007/978-3-030-29680-3

This Springer imprint is published by the registered company Springer Nature Switzerland AG
The registered company address is: Gewerbestrasse 11, 6330 Cham, Switzerland

Acknowledgements

The authors acknowledge Jorge Rabassa for the opportunity to show their research through this book.

Book Structure

The Neuquén Basin located in the Southern Central and Northern Patagonian Andes (Fig. 1) contains the largest hydrocarbon reservoirs in Argentina and therefore is characterized by a profound knowledge of the associated sedimentation mechanisms and closure times. During the last 15 years, a considerable amount of new information has been produced that illustrates a complex evolution. Through this, a polyphasic complex structure exhumed the rich sedimentary record that characterizes the basin.

This book is structured through 22 chapters (Fig. 1), the first ten dedicated to the different pulses and sedimentation periods associated with the opening of the Neuquén Basin, and the final twelve dedicated to the pulses linked to its closure, the inception of a foreland basin, and the intermittent destabilization periods of the resulting orogen.

In the first section, the first two chapters are dedicated to the Late Triassic synrift stage of the Neuquén Basin. While Delia et al.'s chapter analyzes the characteristics of the synrift depocenters in the foreland and southern regions of the basin, and Bechis et al.'s paper analyzes the characteristics of the synrift depocenters in the northern part of the basin at the Malargüe fold and thrust belt.

The next three chapters focused on the Early Jurassic stages that corresponded to the first Pacific-marine transgression. Then, Palma et al.'s chapter analyzes this cycle in the northern part of the basin in the Malargue fold and thrust belt, where it responded to a sag stage; while Zavala et al.'s chapter analyzes the particular behavior of this cycle in the southern part of the basin, where this associated with the inversion of the Huincul arch, a structure with perpendicular orientation to the Andes.

The next chapter by Rossel and Carvajal deciphers the life span of the Jurassic volcanic arc on the western slope of the Andes through U/Pb determinations. Additionally, it analyzes its geochemical characteristics showing that this arc coexisted with an attenuated crust and steady subduction after a cycle of arc-trench retraction associated with roll-back in Triassic-Early Jurassic times.

Next chapters link to the renewed synrift stage associated with the Tordillo Formation sedimentation. While Mescua et al.'s chapter describes this cycle in the Malargüe fold and thrust belt, Acevedo et al.'s chapter shifts to the Aconcagua fold

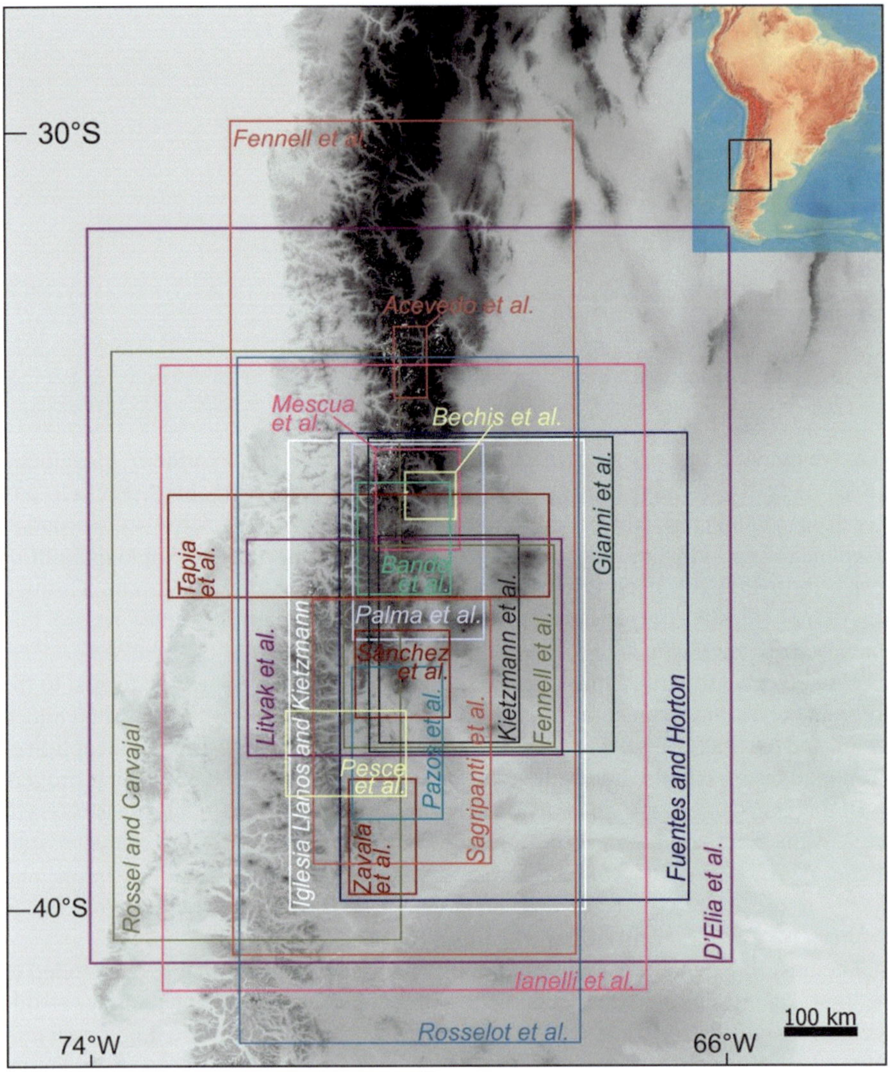

Fig. 1 Neuquén Basin in the Southern Central Andes and the study areas of the different chapters that comprehend this book

and thrust belt region analyzing the particularities of this unit at the northern part of the basin.

The three next chapters dedicate to the second Pacific-derived transgression that flooded the Neuquén Basin. While Iglesia-Llanos et al.'s chapter does paleomagnetic calibration of the Late Jurassic to Early Cretaceous successions; Kietzmann et al.'s chapter analyzes the cyclicity of it as a function of orbital interactions. Finally, Pazos et al.'s chapter analyzes the sequence stratigraphy of the Upper

Mendoza subgroup and its transition to the Huitrin Formation, discussing sources and the inception of tectonic activity through this limit and even deeper into the Early Cretaceous, which traditionally had been linked to sag subsidence (or perhaps dynamic subsidence triggered by trench roll-back of the subducted Pacific plate?).

The second part of this book starts with Tapia et al.'s chapter that constitutes a complete review of the different source areas through U/Pb determinations that characterized basin sag stages and their transition to late Early Cretaceous synorogenic stages. This chapter also deals with the first syn-extensional destabilization deposits of latest Cretaceous age.

The first synorogenic sequence associated with the Andean uplift in the Neuquén Basin corresponding to the Neuquén Formation is analyzed in Fennell et al.'s chapter that associates this to a wedge top position over active-blind contractional structures. The following chapter corresponds to Sánchez et al. and shows through thermochronological data that the westernmost section of the Chos Malal fold and thrust belt in the central Neuquén Basin was exhumed by Late Cretaceous times, while the easternmost section entirely developed in the Late Cenozoic. The following chapter corresponding to Fuentes and Horton analyzes the mechanisms associated with the Late Cretaceous sedimentation, proposing that even though this coexisted with Andean construction to the west, thermal subsidence inherited from the phases of basin extension could have had a persistent control during these first stages of flexural subsidence. Then, Bande et al.'s chapter reveals, through thermochronological data in the northern part of the Neuquén Basin in the Malargüe fold and thrust belt, the importance of Eocene, Miocene, and out-of-sequence early Pliocene contractional exhumation.

Iannelli et al.'s chapter analyzes the Late Cretaceous to Oligocene magmatism geochemically at the latitudes of the Neuquén Basin, relating it to the subduction of a mid-ocean ridge and destabilization of the orogenic wedge during a period of accelerated roll-back. In this sense, based on numerical models, Fennell et al. provide a model for the accelerated roll-back that could have characterized the late Oligocene in the Southern Andes, when the subducted oceanic slab experienced metamorphic transformations above the 410 km depth mantle discontinuity. Rosselot et al.'s chapter compiles available radiometric constraints, La/Yb ratios, and thermochronological data for the late Oligocene to early Miocene volcano-sedimentary depocenters exposed in the High Andes of the Neuquén Basin. The chapter concludes that basal sections (27-22 Ma) of these depocenters coexisted with an attenuated crust associated with an extensional regime, although by 19 Ma these conditions changed through the inception of a contractional regime that controlled the sedimentation. Litvak et al.'s chapter analyzes the tectonic mechanisms associated with the inception of the early Miocene contractional regime, showing an arc expansion toward the San Rafael Block in the foreland zone until the latest Miocene that was associated with a shallowing of a subducted slab at the latitudes of the Malargüe fold and thrust belt. Then, Gianni et al.'s chapter proposes based on tomographic data and geochemistry and morphology of early Miocene magmatism that this shallowing could be related to the subduction of a hot spot that dynamically could have sustained the subducted shallow slab. In this hypothesis, the hot spot would have disrupted the flat slab configuration, provoking the Payenia within-plate volcanic plateau in the latest Pliocene to Quaternary.

The last two chapters are associated with Quaternary deformations in the arc and retroact zones at the latitudes of the Neuquén Basin. First, Sagripanti et al.'s chapter analyzes the correlation between segmented neotectonic belts and asthenospheric circulation patterns revealed by recent magnetotelluric and seismic tomography surveys. Finally, Pesce et al.'s chapter analyzes a Quaternary retro-arc extensional basin and related magmatism in the Loncopué Trough using aeromagnetic and gravimetric data.

This book describes a complex evolution of the Neuquén Basin in the Southern Andes: The Neuquén Basin experienced a Late Triassic synrift stage, such as many other basins in South America. The diachronic character of this extensional process led to the uneven installment of sag conditions: While some Early Jurassic successions correspond to the synrift conditions, others deposited as part of the sag phase. Late Jurassic is again as a period of renewed extension, or a second synrift episode. This second synrift stage could have been a consequence of the retraction of Gondwana respect to the Pacific trench as paleomagnetic data reveal. The latest Jurassic to Early Cretaceous period was linked to renewed sag conditions. However, it seems that subsidence was not homogeneous in this period leading to depocenter abandonment and subsidence focus migration through the Neuquén Basin. We speculate that this occurred by dynamic subsidence imposed by the roll-back of the subducted slab at that time, that most likely coexisted with sag conditions. Late Early Cretaceous linked to the installment of contractional conditions and a first foreland basin at the time when the basin thermally subsided. Since the arc migrated on land during this period, the most likely explanation is the development of a shallow subduction configuration in Late Cretaceous times. This early orogen collapsed partially in the Late Cretaceous presumably in association with the subduction of a mid-ocean ridge and associated asthenospheric window. Latter, a late Paleocene to Eocene neutral tectonic stage installed in the Neuquén Basin. At the end of this period, extension, again associated with a roll-back regime, was related to the development of volcano-sedimentary basins. Late early Miocene was a time of new inception of contractional conditions and development of a foreland basin. Again, this second contractional phase coincides with a substantial inland migration of the volcanic arc that presumably indicates a late Miocene shallow subduction zone, potentially linked to the subduction of a transient hot spot. Late Pliocene to Quaternary conditions finally link to renewed roll-back and associated tearing of the subducted Nazca Plate.

Contents

Opening of the Neuquén Basin in the Western Gondwana margin

The Syn-Rift of the Neuquén Basin (Precuyano and Lower Cuyano Cycle): Review of Structure, Volcanism, Tectono-Stratigraphy and Depositional Scenarios 3
Leandro D'Elia, Andrés Bilmes, Maximiliano Naipauer, Gustavo D. Vergani, Martín Muravchik and Juan R. Franzese

Tectono-Stratigraphic Evolution of the Atuel Depocenter During the Late Triassic to Early Jurassic Rift Stage, Neuquén Basin, West-Central Argentina 23
Florencia Bechis, Laura B. Giambiagi, Maisa A. Tunik, Julieta Suriano, Silvia Lanés and José F. Mescua

Jurassic Uplift Along the Huincul Arch and Its Consequences in the Stratigraphy of the Cuyo and Lotena Groups. Neuquén Basin, Argentina 53
Carlos Zavala, Mariano Arcuri, Mariano Di Meglio, Agustín Zorzano and Germán Otharán

Environmental Controls and Facies Architecture of a Jurassic Carbonate Episode (La Manga Formation), Mendoza Province, Neuquén Basin 75
Ricardo M. Palma, Graciela S. Bressan, Alberto C. Riccardi, José López-Gómez and Javier Martín-Chivelet

Early Andean Magmatism in Southern Central Chile (33°–40° S) 107
Pablo Rossel and Francisco Carvajal

Controls on Deposition of the Tordillo Formation in Southern Mendoza (34°–36° S): Implications for the Kimmeridgian Tectonic Setting of the Neuquén Basin 127
José F. Mescua, Julieta Suriano, Laura Jazmín Schencman, Laura B. Giambiagi, Patricia Sruoga, Elizabeth Balgord and Florencia Bechis

Tectonic Setting of the Tordillo Formation in the Aconcagua Fold-and-Thrust Belt 159
Eliana Acevedo, Eduardo Agustín Rosselot, Federico Martos, Lucas Fennell, Maximiliano Naipauer and Andrés Folguera

Magnetostratigraphy of the Jurassic Through Lower Cretaceous in the Neuquén Basin 175
María Paula Iglesia Llanos and Diego A. Kietzmann

Orbital Controls and High-Resolution Cyclostratigraphy of Late Jurassic–Early Cretaceous in the Neuquén Basin 211
Diego A. Kietzmann, María Paula Iglesia Llanos and Melisa Kohan Martínez

Sedimentology and Sequence Stratigraphy of the Agrio Formation (Late Valanginian–Earliest Barremian) and the Closure of the Mendoza Group to the North of the Huincul High 237
Pablo José Pazos, Marcos Comerio, Diana Elizabeth Fernández, Carolina Gutiérrez, María Candela González Estebenet and Arturo Miguel Heredia

Closure of the Neuquén Basin in the Southern Andes

Middle Jurassic-Late Cretaceous Paleogeography of the Western Margin of the Neuquén Basin (34° 30′–36° S) 269
Felipe Tapia, Marcia Muñoz, Marcelo Farías, Reynaldo Charrier and Daniela Astaburuaga

The Late Cretaceous Orogenic System: Early Inversion of the Neuquén Basin and Associated Synorogenic Deposits (35º–38º S) 303
Lucas Fennell, Pablo Borghi, Federico Martos, Eduardo Agustín Rosselot, Maximiliano Naipauer and Andrés Folguera

Structural and Thermochronological Constraints on the Exhumation of the Chos Malal Fold and Thrust Belt (~37° S) 323
Natalia Sánchez, Martín Turienzo, Isabelle Coutand, Fernando Lebinson, Vanesa Araujo and Luis Dimieri

The Andean Foreland Evolution of the Neuquén Basin: A Discussion 341
Facundo Fuentes and Brian K. Horton

Thermochronological Constraints on the Exhumation of the Malargüe Fold-Thrust Belt, Southern Central Andes 371
Alejandro Bande, Andrés Boll, Facundo Fuentes, Brian K. Horton and Daniel F. Stockli

Late Cretaceous to Oligocene Magmatic Evolution of the Neuquén Basin 397
Sofía B. Iannelli, Lucas Fennell, Lucía Fernández Paz, Vanesa D. Litvak, Alfonso Encinas and Andrés Folguera

The Oligo-Miocene Tectonic Mode Switch: From a Brief Period of Widespread Extension to the Final Closure of the Neuquén Basin 417
Lucas Fennell, Javier Quinteros and Andrés Folguera

Tectonics Associated with the Late Oligocene to Early Miocene Units of the High Andes (Cura-Mallín Formation). A Review of the Geochronological, Thermochronological, and Geochemical Data 431
Eduardo Agustín Rosselot, María Hurley, Lucía Sagripanti, Lucas Fennell, Sofía B. Iannelli, Darío Orts, Alfonso Encinas, Vanesa D. Litvak and Andrés Folguera

The Miocene Magmatism in the Malargüe and Chos Malal Fold and Thrust Belts 449
Vanesa D. Litvak, Sofía B. Iannelli, Lucía Fernández Paz and Andrés Folguera

Plume Subduction Beneath the Neuquén Basin and the Last Mountain Building Stage of the Southern Central Andes 467
Guido M. Gianni, Agustina Pesce, Héctor P. A. García, Marianela Lupari, Sebastián Correa-Otto, Silvina Nacif and Andrés Folguera

Quaternary Deformation in the Neuquén Basin, Explained by the Interaction Between Mantle Dynamics and Tectonics 485
Lucía Sagripanti, Bruno Colavitto, Ana Astort and Andrés Folguera

Pliocene to Quaternary Retroarc Extension in the Neuquén Basin: Geophysical Characterization of the Loncopué Trough 501
Agustina Pesce, Guido M. Gianni, Mario E. Giménez and Andrés Folguera

Contributors

Eliana Acevedo Universidad de Buenos Aires, Facultad de Ciencias Exactas y Naturales, Departamento de Ciencias Geológicas, Buenos Aires, Argentina; CONICET—Universidad de Buenos Aires, Instituto de Estudios Andinos Don Pablo Groeber (IDEAN), Buenos Aires, Argentina

Vanesa Araujo Departamento de Geología, Universidad Nacional del Sur. INGEOSUR-CONICET, Bahía Blanca, Argentina

Mariano Arcuri GCS Argentina SRL, Bahía Blanca, Argentina; Departamento de Geología, Universidad Nacional del Sur, Bahía Blanca, Buenos Aires, Argentina

Daniela Astaburuaga Departamento de Geología, Universidad de Chile, Santiago, Chile

Ana Astort Facultad de Ciencias Exactas y Naturales, Departamento de Ciencias Geológicas, Instituto de Estudios Andinos "Don Pablo Groeber" (IDEAN), CONICET, Universidad de Buenos Aires, Buenos Aires, Argentina

Elizabeth Balgord Weber State University, Ogden, USA

Alejandro Bande Tecpetrol S.A., Buenos Aires, Argentina

Florencia Bechis Instituto de Investigaciones en Diversidad Cultural y Procesos de Cambio (IIDyPCa). Consejo Nacional de Investigaciones Científicas y Técnicas (CONICET). Universidad Nacional de Río Negro, Sede Andina, San Carlos de Bariloche, Argentina

Andrés Bilmes Instituto de Geología y Paleontología (CONICET), Puerto Madryn, Argentina

Andrés Boll Independent Consultant, Buenos Aires, Argentina

Pablo Borghi Wintershall Energía S. A, Buenos Aires, Argentina

Graciela S. Bressan Facultad de Ciencias Exactas y Naturales Departamento de Ciencias Geológicas, IDEAN—CONICET, Universidad de Buenos Aires, La Plata, Argentina

Francisco Carvajal Universidad Andrés Bello, Concepción, Chile

Reynaldo Charrier Escuela de Ciencias de la Tierra, Facultad de Ingeniería, Universidad Andres Bello, Santiago, Chile

Bruno Colavitto Facultad de Ciencias Exactas y Naturales, Departamento de Ciencias Geológicas, Instituto de Estudios Andinos "Don Pablo Groeber" (IDEAN), CONICET, Universidad de Buenos Aires, Buenos Aires, Argentina

Marcos Comerio Centro de Tecnología de Recursos Minerales y Cerámica (CETMIC), CONICET. La Plata, Buenos Aires, Argentina

Sebastián Correa-Otto IGSV, Instituto Geofísico Sismológico Ing. Volponi, Universidad Nacional de San Juan, San Juan, Argentina

Isabelle Coutand Dalhousie University, Halifax, Canada

Leandro D'Elia Centro de Investigaciones Geológicas (CONICET-UNLP), La Plata, Argentina

Mariano Di Meglio GCS Argentina SRL, Bahía Blanca, Argentina;
Departamento de Geología, Universidad Nacional del Sur, Bahía Blanca, Buenos Aires, Argentina

Luis Dimieri Departamento de Geología, Universidad Nacional del Sur. INGEOSUR-CONICET, Bahía Blanca, Argentina

Alfonso Encinas Departamento de Ciencias de la Tierra, Universidad de Concepción, Concepción, Chile

María Candela González Estebenet Universidad de Buenos Aires, Facultad de Ciencias Exactas Y Naturales, Departamento de Ciencias Geológicas, CONICET—Universidad de Buenos Aires, Instituto de Estudios Andinos Don Pablo Groeber (IDEAN), Buenos Aires, Argentina

Marcelo Farías Departamento de Geología, Universidad de Chile, Santiago, Chile

Lucas Fennell Universidad de Buenos Aires, Facultad de Ciencias Exactas Y Naturales, Departamento de Ciencias Geológicas, CONICET—Universidad de Buenos Aires, Instituto de Estudios Andinos Don Pablo Groeber (IDEAN), Buenos Aires, Argentina

Diana Elizabeth Fernández Universidad de Buenos Aires, Facultad de Ciencias Exactas Y Naturales, Departamento de Ciencias Geológicas, CONICET—Universidad de Buenos Aires, Instituto de Estudios Andinos Don Pablo Groeber (IDEAN), Buenos Aires, Argentina

Lucía Fernández Paz Instituto de Estudios Andinos Don Pablo Groeber (IDEAN), Universidad de Buenos Aires-CONICET, Buenos Aires, Argentina

Andrés Folguera Universidad de Buenos Aires, Facultad de Ciencias Exactas Y Naturales, Departamento de Ciencias Geológicas, CONICET—Universidad de Buenos Aires, Instituto de Estudios Andinos Don Pablo Groeber (IDEAN), Buenos Aires, Argentina

Juan R. Franzese Centro de Investigaciones Geológicas (CONICET-UNLP), La Plata, Argentina

Facundo Fuentes YPF S.A., Buenos Aires, Argentina

Héctor P. A. García IGSV, Instituto Geofísico Sismológico Ing. Volponi, Universidad Nacional de San Juan, San Juan, Argentina

Laura B. Giambiagi Instituto Argentino de Nivología, Glaciología y Ciencias Ambientales (IANIGLA). Consejo Nacional de Investigaciones Científicas y Técnicas (CONICET), Centro Científico Tecnológico Mendoza. Av. Ruiz Leal s/n, Parque General San Martín, Mendoza, Argentina

Guido M. Gianni IGSV, Instituto Geofísico Sismológico Ing. Volponi, Universidad Nacional de San Juan, San Juan, Argentina;
CONICET, Instituto de Estudios Andinos Don Pablo Groeber (IDEAN), Universidad de Buenos Aires, Buenos Aires, Argentina

Mario E. Giménez IGSV, Instituto Geofísico Sismológico Ing. Volponi, Universidad Nacional de San Juan, San Juan, Argentina

Carolina Gutiérrez Universidad de Buenos Aires, Facultad de Ciencias Exactas Y Naturales, Departamento de Ciencias Geológicas, CONICET—Universidad de Buenos Aires, Instituto de Estudios Andinos Don Pablo Groeber (IDEAN), Buenos Aires, Argentina

Arturo Miguel Heredia Universidad de Buenos Aires, Facultad de Ciencias Exactas Y Naturales, Departamento de Ciencias Geológicas, CONICET—Universidad de Buenos Aires, Instituto de Estudios Andinos Don Pablo Groeber (IDEAN), Buenos Aires, Argentina

Brian K. Horton Department of Geological Sciences and Institute for Geophysics, Jackson School of Geosciences, University of Texas at Austin, Austin, USA

María Hurley Universidad de Buenos Aires, Facultad de Ciencias Exactas y Naturales, Departamento de Ciencias Geologicas, Buenos Aires, Argentina;
CONICET – Universidad de Buenos Aires, Instituto de Estudios Andinos "Don Pablo Groeber" (IDEAN), Buenos Aires, Argentina

Sofía B. Iannelli Instituto de Estudios Andinos Don Pablo Groeber (IDEAN), Universidad de Buenos Aires-CONICET, Buenos Aires, Argentina;
Universidad de Buenos Aires, Facultad de Ciencias Exactas y Naturales, Departamento de Ciencias Geologicas, Buenos Aires, Argentina

María Paula Iglesia Llanos Facultad de Ciencias Exactas y Naturales, Departamento de Ciencias Geológicas, Universidad de Buenos Aires, Ciudad Autónoma de Buenos Aires, Argentina;
CONICET-Universidad de Buenos Aires, Instituto de Geociencias Básicas, Ambientales y Aplicadas de Buenos Aires (IGeBA), Buenos Aires, Argentina

Diego A. Kietzmann Facultad de Ciencias Exactas y Naturales, Departamento de Ciencias Geológicas, Universidad de Buenos Aires, Ciudad Autónoma de Buenos Aires, Argentina;
CONICET-Universidad de Buenos Aires, Instituto de Geociencias Básicas, Ambientales y Aplicadas de Buenos Aires (IGeBA), Buenos Aires, Argentina

Silvia Lanés Cape Town, South Africa

Fernando Lebinson Departamento de Geología, Universidad Nacional del Sur. INGEOSUR-CONICET, Bahía Blanca, Argentina

Vanesa D. Litvak Departamento de Ciencias Geológicas, Facultad de Ciencias Exactas y Naturales, Instituto de Estudios Andinos "Don Pablo Groeber" (IDEAN), CONICET – Universidad de Buenos Aires, Buenos Aires, Argentina

José López-Gómez Consejo Superior de Investigaciones Cienctíficas, Instituto de Geociencias (CSIC), Universidad Complutense de Madrid, Madrid, Spain

Marianela Lupari IGSV, Instituto Geofísico Sismológico Ing. Volponi, Universidad Nacional de San Juan, San Juan, Argentina

Javier Martín-Chivelet Facultad de Ciencias Geológicas & Instituto de Geociencias (CSIC), Universidad Complutense de Madrid, Madrid, Spain

Melisa Kohan Martínez Facultad de Ciencias Exactas y Naturales, Departamento de Ciencias Geológicas, Universidad de Buenos Aires, Ciudad Autónoma de Buenos Aires, Argentina;
CONICET-Universidad de Buenos Aires, Instituto de Geociencias Básicas, Ambientales y Aplicadas de Buenos Aires (IGeBA), Buenos Aires, Argentina

Federico Martos CONICET—Universidad de Buenos Aires, Instituto de Estudios Andinos Don Pablo Groeber (IDEAN), Buenos Aires, Argentina;
Universidad de Buenos Aires, Facultad de Ciencias Exactas y Naturales, Departamento de Ciencias Geologicas, Buenos Aires, Argentina

José F. Mescua Facultad de Ciencias Exactas y Naturales, Instituto Argentino de Nivología, Glaciología y Ciencias Ambientales (IANIGLA). Consejo Nacional de Investigaciones Científicas y Técnicas (CONICET), Centro Científico Tecnológico Mendoza, Mendoza, Argentina

Marcia Muñoz Escuela de Ciencias de la Tierra, Facultad de Ingeniería, Universidad Andres Bello, Santiago, Chile

Martín Muravchik University of Bergen, Bergen, Norway

Silvina Nacif IGSV, Instituto Geofísico Sismológico Ing. Volponi, Universidad Nacional de San Juan, San Juan, Argentina

Maximiliano Naipauer CONICET—Universidad de Buenos Aires, Instituto de Estudios Andinos Don Pablo Groeber (IDEAN), Buenos Aires, Argentina;
Instituto de Geocronología y Geología Isotópica (CONICET-UBA), CABA, Argentina;
Instituto de Geocronología y Geología Isotópica(INGEIS), Buenos Aires, Argentina

Darío Orts Universidad Nacional de Río Negro, Pcia de Río Negro, Argentina

Germán Otharán GCS Argentina SRL, Bahía Blanca, Argentina;
Departamento de Geología, Universidad Nacional del Sur, Bahía Blanca, Buenos Aires, Argentina

Ricardo M. Palma Facultad de Ciencias Exactas y Naturales Departamento de Ciencias Geológicas, IDEAN—CONICET, Universidad de Buenos Aires, La Plata, Argentina

Lucía Fernández Paz Universidad de Buenos Aires, Facultad de Ciencias Exactas y Naturales, Departamento de Ciencias Geológicas, Buenos Aires, Argentina;
CONICET-Universidad de Buenos Aires, Instituto de Estudios Andinos "Don Pablo Groeber" (IDEAN), Buenos Aires, Argentina

Pablo José Pazos Universidad de Buenos Aires, Facultad de Ciencias Exactas Y Naturales, Departamento de Ciencias Geológicas, CONICET—Universidad de Buenos Aires, Instituto de Estudios Andinos Don Pablo Groeber (IDEAN), Buenos Aires, Argentina

Agustina Pesce IGSV, Instituto Geofísico Sismológico Ing. Volponi, Universidad Nacional de San Juan, San Juan, Argentina

Javier Quinteros GFZ Helmholtz Centre Potsdam, Potsdam, Germany

Alberto C. Riccardi Facultad de Ciencias Naturales y Museo—CONICET, Universidad Nacional de La Plata, La Plata, Argentina

Pablo Rossel Universidad Andrés Bello, Concepción, Chile

Eduardo Agustín Rosselot Universidad de Buenos Aires, Facultad de Ciencias Exactas y Naturales, Departamento de Ciencias Geologicas, Buenos Aires, Argentina;
CONICET – Universidad de Buenos Aires, Instituto de Estudios Andinos "Don Pablo Groeber" (IDEAN), Buenos Aires, Argentina

Lucía Sagripanti Universidad de Buenos Aires, Facultad de Ciencias Exactas Y Naturales, Departamento de Ciencias Geológicas, CONICET—Universidad de Buenos Aires, Instituto de Estudios Andinos Don Pablo Groeber (IDEAN), Buenos Aires, Argentina

Natalia Sánchez Departamento de Geología, Universidad Nacional del Sur. INGEOSUR-CONICET, Bahía Blanca, Argentina

Laura Jazmín Schencman Departamento de Geología, Facultad de Ciencias Exactas y Naturales, Universidad de Buenos Aires, Buenos Aires, Argentina; Instituto de Geociencias Básicas, Aplicadas y Ambientales de Buenos Aires (IGEBA, CONICET-UBA), Buenos Aires, Argentina

Patricia Sruoga CONICET-SEGEMAR, Buenos Aires, Argentina

Daniel F. Stockli Jackson School of Geosciences, University of Texas at Austin, Austin, TX, USA

Julieta Suriano Instituto Argentino de Nivología, Glaciología y Ciencias Ambientales (IANIGLA). Consejo Nacional de Investigaciones Científicas y Técnicas (CONICET), Centro Científico Tecnológico Mendoza, Mendoza, Argentina

Felipe Tapia Departamento de Ciencias Geológicas, Facultad de Ciencias Exactas y Naturales, Universidad de Buenos Aires. Consejo Nacional de Investigaciones Científicas y Técnicas, Instituto de Estudios Andinos "Don Pablo Groeber" (IDEAN), Buenos Aires, Argentina

Maisa A. Tunik Universidad Nacional de Río Negro. CONICET. Instituto de Investigaciones en Paleobiología y Geología. Av. Roca 1242., General Roca, Argentina

Martín Turienzo Departamento de Geología, Universidad Nacional del Sur. INGEOSUR-CONICET, Bahía Blanca, Argentina

Gustavo D. Vergani PLUSPETROL SA, CABA, Argentina

Carlos Zavala GCS Argentina SRL, Bahía Blanca, Argentina;
Departamento de Geología, Universidad Nacional del Sur, Bahía Blanca, Buenos Aires, Argentina

Agustín Zorzano GCS Argentina SRL, Bahía Blanca, Argentina

Opening of the Neuquén Basin in the Western Gondwana margin

The Syn-Rift of the Neuquén Basin (Precuyano and Lower Cuyano Cycle): Review of Structure, Volcanism, Tectono-Stratigraphy and Depositional Scenarios

Leandro D'Elia, Andrés Bilmes, Maximiliano Naipauer, Gustavo D. Vergani, Martín Muravchik and Juan R. Franzese

Abstract The Neuquén Basin is constituted by a system of NW-SE rift clusters formed during the Late Triassic–Early Jurassic at the western margin of Gondwana. The rifting is the result of an extensional tectonic regime triggered by the Gondwana break-up that attained a NE-trending direction ($\lambda 1$) and a magnitude of extension of around 10%. The syn-rift climax occurred synchronously with an important volcanism, shortly after the beginning of crustal extension. The tectono-stratigraphy of the syn-rift indicates that after a short lapse of crustal extension related to mainly epiclastic sedimentation, the rifting was accompanied by widespread volcanism, characterized by sub-alkaline series, with orogenic signature (i.e. Precuyano Cycle). Recent isotope data revealed a purely crustal origin and mantle-derived rocks with high proportion of crustal participation for the magmatic source. The initiation and cessation of the magmatism, as well as the mechanical extension were diachronic through the basin, showing younger ages from North to South. This would have driven the diachronic post-rift initiations that tracked the rift evolution. The combined effect of the ascending global sea level and the diachronic onset of the sag phase determined that the lower Jurassic marine sedimentation (i.e. Lower Cuyano Cycle) had been set either in a syn-rift or post-rift scenarios, depending on its position in the basin. The understanding (in time and space) of the tectono-magmatic-eustacy factors allowed

L. D'Elia (✉) · J. R. Franzese
Centro de Investigaciones Geológicas (CONICET-UNLP), La Plata, Argentina
e-mail: ldelia@cig.museo.unlp.edu.ar

A. Bilmes
Instituto de Geología y Paleontología (CONICET), Puerto Madryn, Argentina

M. Naipauer
Instituto de Geocronología y Geología Isotópica (CONICET-UBA), CABA, Argentina

G. D. Vergani
PLUSPETROL SA, CABA, Argentina

M. Muravchik
University of Bergen, Bergen, Norway

D. Kietzmann and A. Folguera (eds.), *Opening and Closure of the Neuquén Basin in the Southern Andes*, Springer Earth System Sciences,
https://doi.org/10.1007/978-3-030-29680-3_1

defining a conceptual scheme with the following endmembers recorded in the syn-rift megasequence: (i) continental non-volcano-dominated syn-rift, (ii) continental volcano-dominated syn-rift; (iii) marine volcano-dominated syn-rift, and (iv) marine non-volcano-dominated syn-rift depositional scenarios. The principal controls on the syn-rift stratigraphy were the mechanical subsidence, driven by normal faulting, and the volcanism, which depending on the type and magnitude determined stacking patterns of the infill. In extreme situation volcanic processes captured the tectonic structures triggering volcano-tectonic subsidence events and graben-calderas. The relative sea-level change (eustacy + tectonics) and the climate conditions acted as secondary factors, mainly during the final rifting stage. Structural, volcanic and tectono-stratigraphical features suggest that the Neuquén Basin at its initial stages corresponded to a wide rift, triggered by lithospheric extension driven by far-field stresses, controlling magmatism in a context of a "passive" rift.

Keywords Syn-rift cycle · Pangea break-up · Extension · Precuyano and Lower Cuyano Cycles

1 Introduction

This chapter aims to provide a review and new ideas for the syn-rift phase of the Neuquén Basin. However, this will also serve as an introduction to the following chapters of the book, in order to familiarize the reader with the early structural and volcanic evolution of the syn-rift stages in the Neuquén Basin. Then, the proposed goals are as follows:

- To present the developments in the structure of the rift, integrating the latest superficial and subsurface studies at basin scale.
- To characterize the volcanism that accompanied the climax of the rifting.
- To discuss stratigraphic units of the initial stage (i.e. Precuyano and Lower Cuyano Cycles) of the basin from a tectono-sequence stratigraphic approach.
- To document the endmembers of the depositional scenarios (i.e. alluvial/fluvial-dominated, non-marine volcano-dominated and shallow/deep marine-dominated) that characterized the diverse syn-rift infill of the extensional depocentres of the basin.
- To present a model for the early evolution of the Neuquén Basin, taking into account tectonics, magmatism and eustacy, allowing the reader to build up a complete and comprehensive scheme of the rift evolution.

2 The Neuquén Basin: Major Aspects and Evolution

The Neuquén Basin is located on the eastern side of the Andes in Argentina, with isolated outcrops in central Chile between 30° and 40°S. The basin has a roughly triangular shape covering an area of over 120,000 km^2 (Fig. 1; Yrigoyen 1991). The Neuquén Basin experienced a complex tectonic history associated with different mechanisms. Following Howell et al. (2005), the evolution and development of the basin can be considered in three stages: (i) Late Triassic–Early Jurassic: initial extensional tectonic phase that occurred in the palaeo-Pacific margin of Gondwana, associated with volcano-dominated syn-rift (Franzese and Spalletti 2001; Franzese et al.

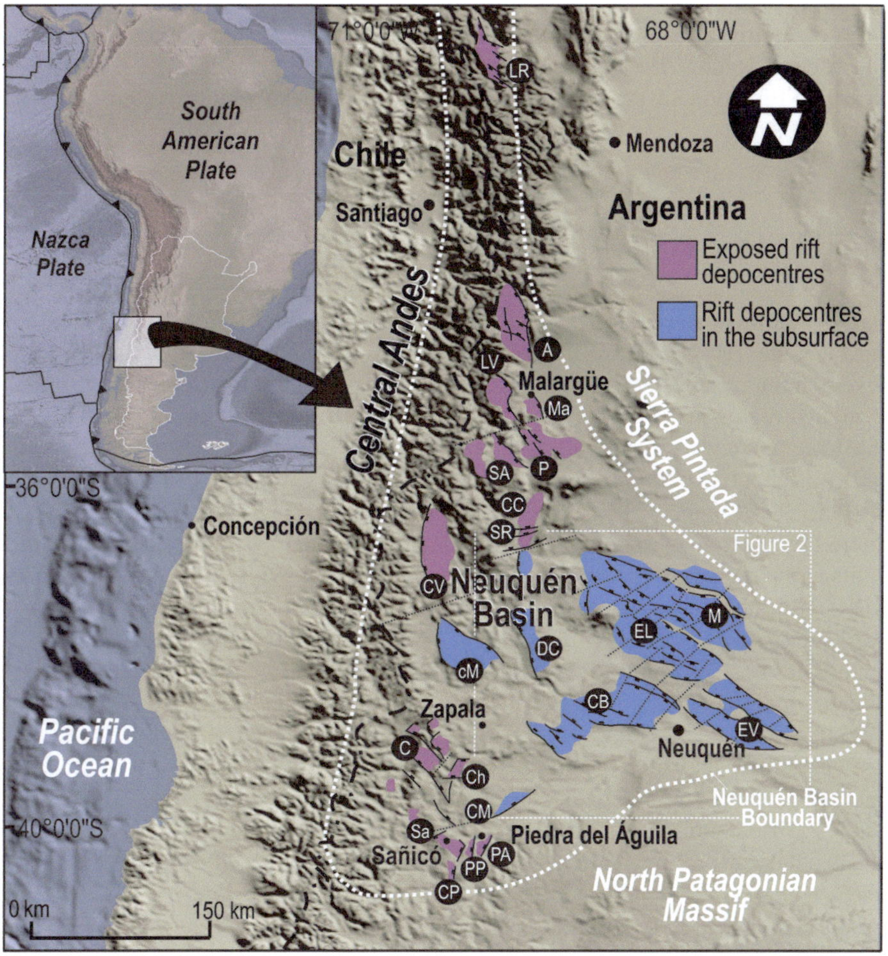

Fig. 1 Distribution of the extensional depocentres in the surface and subsurface of the Neuquén Basin (Modified from Leanza et al. 2013). Black circles identify the main syn-rift depocentres: A: Atuel, C: Catán Lil, CB: Cerro Bandera, CC: Cara Cura, Ch: Chachil, cM: Cerro Mocho, CM. China Muerta, CP: Corral de Piedra, CV: Cordillera del Viento, DC: Dorso de los Chihuidos; EL: Entre Lomas, EV: Estancia Vieja, LR: La Ramada, LV: La Valenciana, M: 25 de Mayo-Medanito, Ma: Malargue, P: Palauco, PA: Piedra del Águila, PP: Piedra Pintada, Sa, Sañicó, SR: Sierra de Reyes

2003; D'Elia et al. 2012a); (ii) Early Jurassic–middle cretaceous: thermal subsidence stage affected by a steeply dipping active subduction associated with a back-arc subsidence during the post-rift phase (Uliana and Legarreta 1993; Legarreta and Uliana 1996); even though these two initial stages are characterized by punctuated episodes of localized inversion (Vergani et al. 1995; Naipauer et al. 2012); and finally (iii) Late Cretaceous-Neogene: the transition to the present-day Andean-type subduction regime, resulting in the contractional tectonic regime that produced the retroarc foreland system. From the onset of this stage, the Andean tectonism produced the uplift of tightly folded outcrops in the western part of the area (i.e. fold and thrust belt region; Ramos 1999; Ramos and Folguera 2005), remaining to the East the Neuquén Embayment in which the majority of the Basin's hydrocarbon fields are located (Fig. 1).

3 Rift Structure

One of the most important expressions of the extensional tectonics occurred during the Late Triassic–Early Jurassic in the western margin of Gondwana was the development of a system of rift clusters (see Sengor 1995) distributed in a roughly triangular region in central-west of Argentina. These clusters were organized in elongate troughs filled by more than 2000 m of volcanic and sedimentary rocks (Legarreta and Gulisano 1989; Manceda and Figueroa 1995; Vergani et al. 1995; Franzese and Spalletti 2001; Franzese et al. 2003; D'Elia et al. 2012a; Fig. 1). Rift troughs are configured by a series of subparallel half-grabens, bounded by planar or listric normal faults with alternating polarity. Historically, the extensional architecture of the basin was determined through a series of structural patterns. The principal rift clusters had been defined with N–S, NW–SE and ENE-WS trending structures, subparallel to the boundaries of adjacent terranes, which seem to be inherited from the pre-Mesozoic western margin of Gondwana (i.e. Sierra Pindada System and North Patagonian Massif; Vergani et al. 1995; Franzese and Spalletti 2001; Fig. 1). Nevertheless, in the last years, 2-D seismic lines and 3-D seismic information combined with borehole data from several oil companies, as well as, information achieved from superficial studies carried out in the exposed depocentres of the basin, have reinterpreted the rift clusters demonstrating an almost unidirectional pattern (Fig. 1)—cf. Vergani (2005) for E-W or NE-SW-trending of the extensional depocentres along the Huincul High area. Thus, based on an accurate tectono-stratigraphic and structural analysis, fault systems were redefined and categorized according to the magnitude with regard to the fault-related thickness infill. The configuration of the Neuquén Basin is nowadays characterized by a series of NW-SE rift clusters that range from 150 to 70 km long and from 40 to 20 km wide, which are divided by interior highs (Fig. 2). The rift clusters consist in a series of, NW–SE-trending, major half-grabens with boundary faults of 15–50 km long, which are prone to NE-dipping direction. The interior highs are formed by the uplifted footwall, defining horst structures. Major interior highs are from 50 to 100 km long and up to 10 km wide and generally parallel to the rift

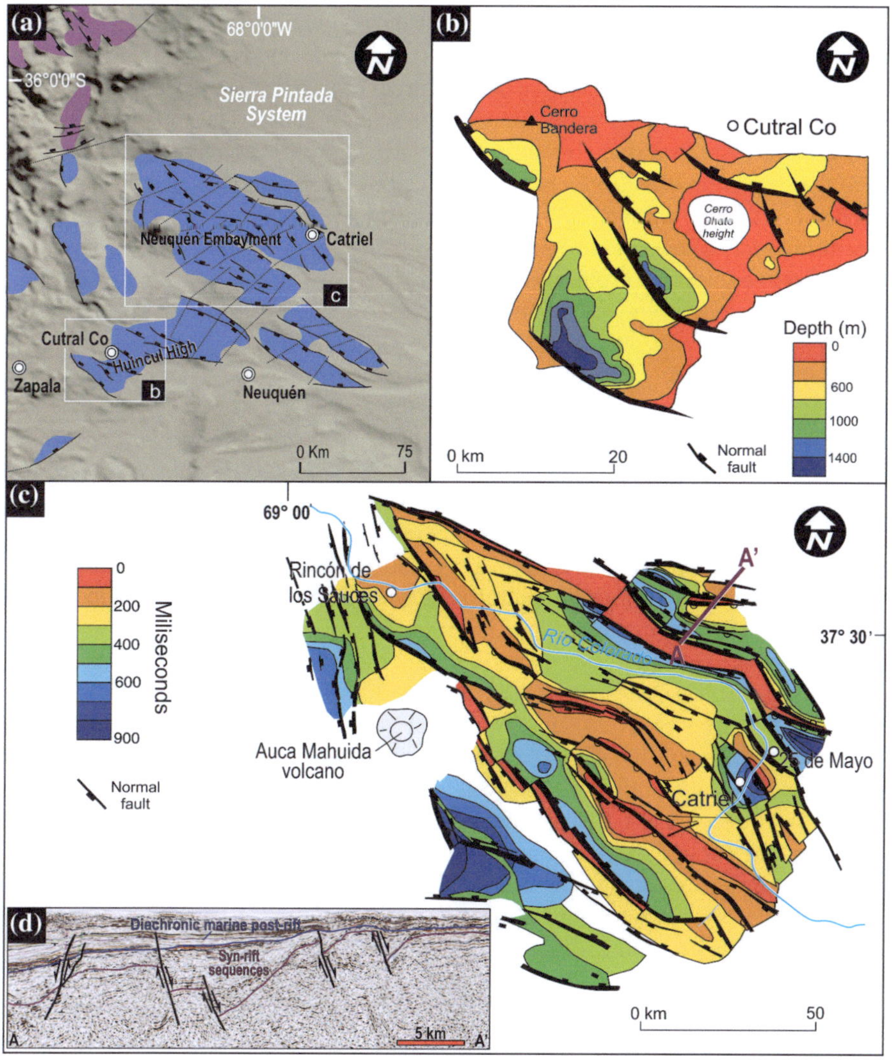

Fig. 2 **a** Distribution of extensional depocentres in the Neuquén embayment and Huincul High areas (see Fig. 1 for location and references). **b** and **c** Maps of the principal grabens and half-grabens of the Neuquén embayment (**b**) and Huincul High (**c**). Colours represent seismic thickness or metres of syn-rift deposits. Modified from Cristallini et al. (2006) and Pángaro et al. (2009), respectively. **d** Structure at the northwestern platform of the Neuquén Basin, based on a three-dimensional (3-D) seismic line. See Fig. 2c for location

margin (e.g. Entre Lomas High, Alto Loma Kauffman High; Figs. 1 and 2), whereas minor highs are from 10 to 30 km long and <5 km wide and are parallel and oblique to the boundaries of the extensional depocentres (Fig. 2). Most normal fault systems terminate in either transfer zones or accommodation zones. The transfer zones are developed as discrete strike-slip and oblique-slip faulting that generally trend parallel to the extension direction whereas the accommodation zones appear as belts of overlapping fault terminations that may separate either systems of uniformly dipping normal faults or adjacent domains of oppositely dipping normal faults, trending parallel, perpendicular, or oblique to the rift structures (Fig. 2). Discrete strike-slip zones are chiefly recognized on the subsurface of the basin (Silvestro and Zubiri 2008; Cristallini et al. 2006, 2009; Pángaro et al. 2009). These structures are narrow zones with NE-trending attitude and have few tens of kilometres in length. They were defined as transfer zones that had facilitated the transfer of strain between extended domains or fragmented depocentres of same polarity, showing right-lateral as well as left-lateral displacements (Fig. 2). The analysis of the transfer zones with regard to the seismic thickness of the depocentres reveals that these structures barely modified the thickness of the infill and generally did not compartmentalized the depocentres within the rift clusters (see Cristallini et al. 2006, 2009). This suggests that transfer zones mainly evolved after the lateral propagation, linkage and evolution of fault zones that configured the major half-grabens of the basin.

Accommodation zones were identified on both subsurface (Silvestro and Zubiri 2008; Cristallini et al. 2006, 2009; Pángaro et al. 2009) and surface studies (Giambiagi et al. 2009; D'Elia et al. 2012b; D'Elia and Martí 2013; Muravchik et al. 2014). Synthetic accommodation zones between similarly dipping systems that connect the hanging wall of one fault to the footwall of another fault, arranged in an *en echelon* pattern, were recognized. These arrangements generally are defined as relay ramps (Silvestro and Zubiri 2008), although faulted-relay ramps also occur (Muravchik et al 2014). To a lesser extent, antithetic accommodation zones that divide the depocentre with opposite polarity with an *en echelon* pattern were also identified (Giambiagi et al. 2009; Silvestro and Zubiri 2008; Cristallini et al. 2006, 2009; Pángaro et al. 2009; Fig. 2). At a basin scale, subsurface and surface studies reveal that different orders of normal faults lie within the half-graben depocentres (Fig. 2).

The major order faults are commonly spaced 2 to 5 km apart, ranging from 5 to 15 km in length (see Silvestro and Zubiri 2008; Cristallini et al. 2006, 2009; Pángaro et al. 2009; Fig. 2), whereas lower-order faults are spaced 0.1–5 km apart, ranging from 2 to 10 km in length (see Giambiagi et al. 2009; Bechis et al. 2009, 2010; D'Elia and Martí 2013). The normal faults are predominantly dip-slip faults. Depending on the order of magnitude, fault displacements range from tens of metres to several hundred metres. The accurate structural analysis of fault orientation to basin scale, allowed to determine two populations, one with azimuth 130°–140°N and the other with azimuth 110°N (Franzese et al. 2006; Silvestro and Zubiri 2008; Cristallini et al. 2006, 2009; Pángaro et al. 2009; Giambiagi et al. 2009; Bechis et al. 2009, 2010; Muravchik et al. 2014; Fig. 2). Based on the structural relationships between them and the stratigraphic control of those upon syn-rift infill, the two fault populations were determined as (quasi) synchronic during the evolution of the diverse rift depocentres.

Unfortunately, few kinematic analyses were carried out on this subject (Giambiagi et al. 2009; Bechis et al. 2009, 2010) (see Chap. "Tectono-stratigraphic evolution of the Atuel depocenter during the Late Triassic to Early Jurassic rift stage, Neuquén basin, west-central Argentina"). To the northern sector of the basin, strain analysis indicated that the intermediate and major axes of the strain ellipsoid (λ2 and λ1) were sub-horizontal, whereas the minor axe (λ3) was sub-vertical, indicating an extensional tectonic regime with a main extension direction (λ1) between NE and ENE trendings. This result is similar to the extension direction inferred from the distribution, orientation and structural arrangement of the half-grabens of the whole Neuquén Basin (e.g. Vergani et al. 1995; Silvestro and Zubiri 2008).

It is important to remark that along the western margin of the Neuquén Basin a series of half-grabens with non-coaxial orientation with regard to the NE-SW extensional regime were described (i.e. Sierra de Reyes-Cara Cura, Sierra de Chacaico, China Muerta, and Sañicó-Piedra del Águila depocentres; Vergani et al. 1995; Franzese et al. 2006; Giambiagi et al. 2009; D'Elia et al. 2015; Fig. 1). These depocentres present E-W or NE-SW trending and are approximately distributed along the 70°S longitude spaced from 70 to 150 km, and bounded, throughout transfer zones by NW-trending extensional depocentres (Fig. 1). The structural analysis carried out in some of these depocentres indicates kinematically heterogeneous sets of fault-slip, determining a bimodal pattern of extensional axes of the strain ellipse during the syn-rift evolution. Based on these outcomes, this non-coaxial depocentre orientation was interpreted as a result of triaxial strain deformation (Giambiagi et al. 2009).

4 Volcanism

Widespread and voluminous volcanism has characterized the Neuquén Basin during its syn-rift evolution. With the exception of some few half-graben depocentres located in the northern part of the basin (e.g. Atuel rift; Fig. 1), the rift system displayed a high volume of volcanism, shortly after the beginning of crustal extension which was asymmetrically distributed in the basin. While most of the half-grabens show that mechanical subsidence was associated with a profuse volcanic activity, in some cases, the rifting initiated with a clastic sedimentary-dominated infill (e.g. Paso Flores, Piedra del Águila, Chihuiu and Llantenes Formations, Franzese and Spalletti 2001; see text below), which was followed by the establishment of the volcanism. Volcanic feeding-systems were characterized by shallow and multiple-injection magmatic chambers and, in the shallower level of the crust, related to the major faults that delimited the depocentres (D'Elia et al. 2012a). Mechanical subsidence was affected by the volcanic activity and, as an endmember situation, volcano-tectonic subsidence episodes with the formation of (graben-) calderas occurred (Muravchik et al. 2008; D'Elia and Martí 2013). The rifting was accompanied by shallow intrusives, lava flows and pyroclastic deposits, which conform up to 70% of the total infill (D'Elia et al. 2012a). While several depocentres are dominated by acid volcanic deposits, in many cases the syn-rift shows a particular spatial-temporal evolution from mainly intermediate volcanic rocks at the base to mainly acidic pyroclastic rocks to the top (Franzese et al. 2006, 2007; D'Elia et al. 2012a; Fig. 3). Trace element data indicate that the lava and pyroclastic products share a co-magmatic trend, dominated by intermediate to acid products (i.e. from andesitic to rhyodacitic compositions; Fig. 3). Rocks belong to the sub-alkaline series, mostly calc-alkaline and transitional, and in a lesser extent tholeiitic, with orogenic signature (Llambías et al. 2007; D'Elia et al. 2012a). Nevertheless, the stratigraphical and compositional features (>50% of acid volcanic products) suggest some differences with typical arc-series. Trace element geochemistry indicates that the magmatic system could be mantle-derived, with shallow mantle origin suggested by rare earth element patterns (D'Elia et al. 2012a). However, recent Hf Isotope analyses carried out in Late Triassic to Early Jurassic detrital zircons (229–192 Ma) at Sierra de Chacaico region, determined low εHft values (−8.1 to −2.1), indicating a crustal origin or a mantle-derived rock with high proportion of crustal participation for the magmatic source (Naipauer et al. 2018).

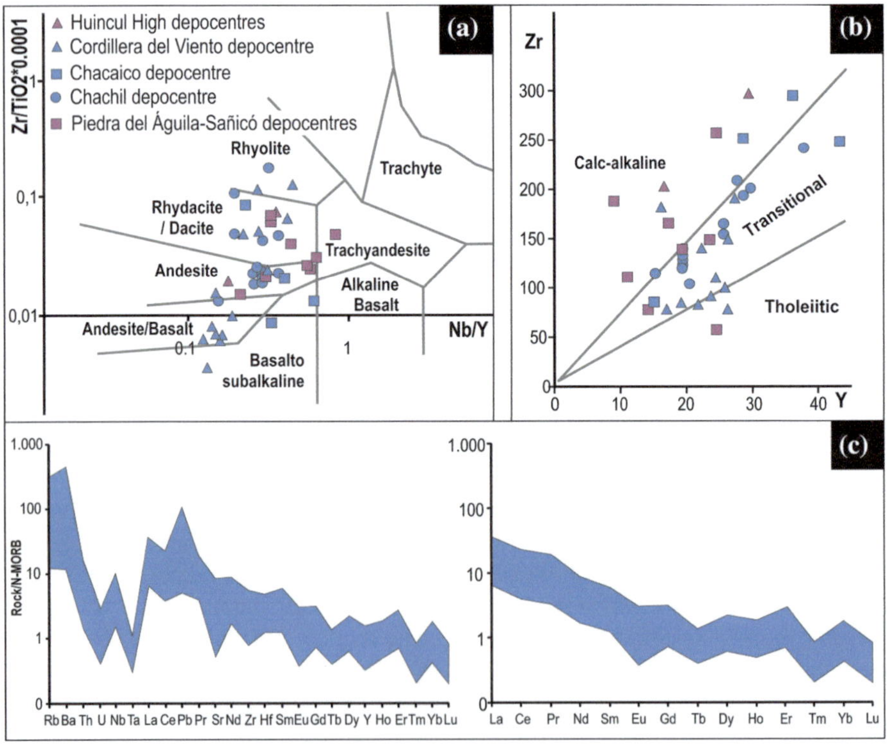

Fig. 3 Geochemistry of the volcanic rocks of the main exposed depocentres of the Neuquén Basin (Modified from D'Elia et al. 2012a). **a** Geochemical discrimination based on immobile elements Zr/TiO_2 versus Nb/Y of Winchester and Floyd (1977). **b** Magma series from Zr versus Y (Barrett and MacLean 1994). **c** Trace elements and Rare Earth Element (REE), N-MOR-B normalized (Sun and McDonough 1989) diagrams

5 Syn-Rift Tectono-Stratigraphy

Based on the pioneering work of Groeber (1946), who identified different major transgressive-regressive marine cycles, basin analysis studies produced a detailed sequence stratigraphy framework of the Neuquén Basin during the 1980s and 1990s (Digregorio and Uliana 1980; Gulisano and Pando 1981; Gulisano et al. 1984; Mitchum and Uliana 1985; Legarreta and Gulisano 1989; Legarreta and Uliana 1996, 1999; Legarreta et al. 1993; Vergani et al. 1995). In these works, the Late Triassic–Middle Jurassic Precuyano and Cuyano cycles were defined as the evidence of the syn-rift phase and the post-rift phase of the basin. The syn-rift deposits were grouped as the Precuyano Cycle (Gulisano et al. 1984), which was considered to be formed by volcanic (lava and pyroclastic flows) and reworked volcaniclastic rocks and to a lesser extent by epiclastic and carbonate deposits. Subsequently, other authors grouped several non-volcanic-dominated syn-rift units into the Precuyano Cycle—see Franzese and Spalletti 2001; Carbone et al. 2011 and Leanza 2009 for reviews of the formal stratigraphy related to the syn-rift deposits—determining it as an Upper Triassic–Lower Jurassic, mainly continental syn-rift (meso) sequence (see Legarreta and Gulisano 1989; Riccardi and Gulisano 1990; Gulisano and Gutiérrez Pleimling 1994). In that scheme, the first generalized marine transgressive deposits (lower Cuyano Cycle) were determined as the onset of the post-rift stage. Nevertheless, deposits of lower Cuyano Cycle were defined as a part of the syn-rift in different sections of the basin, in particular in the central and southern parts of it (see Vergani et al. 1995; Pángaro et al. 2009). In the last years, tectono-stratigraphic analyses combined with geochronologic studies have contributed to determine the tectonically syn-kinematic or post-kinematic conditions of the Upper Triassic–Lower Jurassic infill in diverse places of the basin as well. Nevertheless, these results were not yet included in an integrated tectono-stratigraphic scheme for the whole basin.

Following the scheme of D'Elia et al. (2015), we propose that the diverse types of infill successions of the Precuyano Cycle, which had been grouped in several formal stratigraphic units, correspond to different syn-rift sequences, whereas, depending on the basin location, the Cuyano Cycle infill corresponds to either syn-rift or post-rift sequences (Fig. 4). Therefore, in the present tectono-stratigraphic scheme, the onset of the syn-rift started with isolated and thin half-grabens (Atuel, Malargue, Corral de Piedra rift depocentres; Fig. 1) dominated by epiclastic continental (i.e. Llantenes, Chiuhiu, Paso Flores Formations) and minor marine sedimentation (Arroyo Malo Formation). This initial stage was followed by the rift climax (i.e. phase of high-mechanical subsidence and high-volcanic supply rates) in which the NW-SE rift clusters of the basin were configured. This evolution of the syn-rift depocentres followed a path that is progressively younger from north to south along the basin (Fig. 4). Thereby, the onset of volcanism related to the Precuyano Cycle in the northern part, and in the northeastern boundary of the basin, was constrained at around 210–199 Ma (Pángaro et al. 2002; Barrionuevo et al. 2013; Naipauer et al. 2015), while to the central part of the basin, in the Huincul High region, this onset

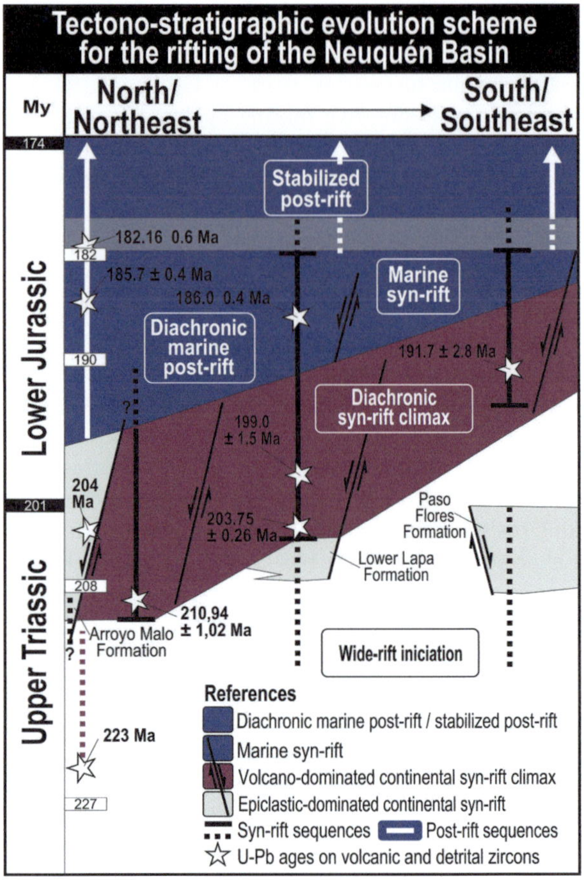

Fig. 4 Tectono-stratigraphic evolution scheme for the rifting of the Neuquén Basin. Geochronological ages taken from Schiuma and Llambías (2007, 2008), Spalletti et al. (2010), Leanza et al. (2013), Barrionuevo et al. 2013, Naipauer et al. (2015, 2016) and Fennell et al. (2017). Tectono-stratigraphy model from north/northeast to south/southeast was devised from Vergani et al. (1995), Spalletti (1999), Lanés et al. (2005), Cristallini et al. (2003, 2009), Franzese et al. (2006, 2007), Giambiagi et al. (2009), Pángaro et al. (2009), Bechis et al. (2010), Muravchik et al. (2014), D'Elia et al. (2015), and Buchanan et al. (2017). Minor and scarce epiclastic deposits within the Precuyano Cycle are not shown

was dated around 203–199 Ma (Schiuma and LLambías 2008), and in the southern margin of the basin it was constrained to ~191 Ma (Spalletti et al. 2010).

The syn-rift climax was followed by the post-rift phase. The almost diachronic ending of the volcanism along the basin, controlled by volcanic syn-rift track evolution, seems to be followed by the post-rift initiation, determining also a diachronism of the post-rift sequences (Fig. 4). To the north of the Neuquén Basin, the syn-rift faults were active until the Sinemurian, positioning the beginning of the post-rift stage by this time (Lanés 2005; Giambiagi et al. 2005, 2008; Lanés et al. 2008). In this region, the structural relief inherited from the syn-rift phase exerted a passive control over the Cuyo Cycle deposits (see Veiga et al. 2013). Towards the north-eastern boundary of the basin, a similar passive control is recorded at the subsurface (Cristallini et al. 2006, 2009). In the central and southern parts of the Neuquén Basin, the tectono-stratigraphic arrangement for the Cuyano Cycle deposits is different (see Chap. "Jurassic uplift along the Huincul arch and its consequences in the stratigraphy of the Cuyo and Lotena groups. Neuquén Basin, Argentina").

The exposed southern syn-rift depocentres located at the foot of the Neuquén Andes (i.e. Chachil-Catán Lil and Sañicó-Piedra del Águila areas; Figs. 1 and 4) show that the lowermost marine deposits of the Cuyo Cycle are synchronous with normal faulting (Muravchik et al. 2014; D'Elia et al. 2015). These deposits were recently dated giving a depositional age of 186.0 ± 0.4 Ma (Leanza et al. 2013). In the central part of the Neuquén Basin, a similar pattern is described (Huincul High region; Vergani et al. 1995; Pángaro et al. 2009). This observation reveals that the stabilized post-rift stage in the central and southern parts of the basin would have begun either during the late Pliensbachian or the early Toarcian. Based on this dataset, we stress that the marine deposits of the first transgression in the whole basin occurred either under thermal subsidence or mechanical subsidence, depending on the position within the basin. Then, the deposits of the Cuyano Cycle could vertically or laterally belong to a different rift stage, that is, syn-rift or post-rift.

This integrated analysis shows a diachronic character of both the syn-rift climax and the post-rift phases along the basin, with a progression from north to south during the temporal evolution. This evolution seems to follow the silicic magmatic evolution of the western Gondwana margin (Ramos 2009 and references therein).

6 Depositional Syn-Rift Scenarios

Basin architecture and depositional scenarios depend upon a complex interaction between the three-dimensional tectonic evolution combined with the climate, volcanism and sea/lake level changes. In this section, the conceptual depositional models will devise in order to characterize the syn-rift sequences of the Neuquén Basin. The conceptual models may be addressed from the combination of two types of endmembers conditions, which are: volcanic syn-rift state versus non-volcanic syn-rift state, and continental versus marine syn-rift conditions (Fig. 5). The following four conceptual models will be addressed within this scheme: (i) *continental non-volcano-dominated syn-rift,* (ii) *continental volcano-dominated syn-rift*; (iii) *marine volcano-dominated syn-rift, and* (iv) *marine volcano-dominated syn-rift depositional scenarios*. It is important to highlight that each conceptual model must not be taken into account as a fixed model, since, despite there is a volcanic, continental dominated condition of the depositional environments within the syn-rift sequences (i.e. Precuyano Cycle s.s.), each of them may have intermediate conditions. In turn, one conceptual model or a combination of the conceptual models can constitute the entire syn-rift infill of a certain depocentre.

The continental non-volcano-dominated syn-rift depositional scenarios are characterized by epiclastic and reworked volcaniclastic deposits with minor pyroclastic deposits (Fig. 5). These facies are organized in several continental environments occurred in isolated extensional depocentres generally during the onset of the rifting (e.g. El Freno, Paso Flores, Piedra del Águila, Llantenes formations; grouped into the Precuyano Cycle) and/or towards the upper part of the syn-rift phase (e.g. Planicie Morada, Remoredo formations; grouped into the Precuyano Cycle), previously

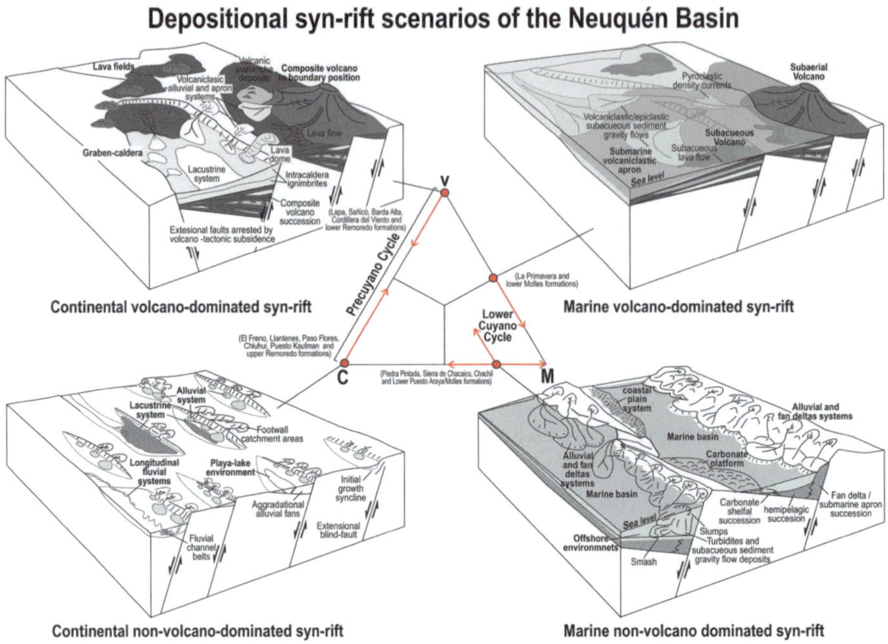

Fig. 5 Depositional scenarios devised in this work for the syn-rift of the Neuquén Basin (i.e. Precuyano and Lower Cuyano Cycle). See text and Fig. 4 for a fully comprehensive scheme. Triangular conceptual graph shows the endmember conditions to the syn-rift, in which: C (continental), M (Marine) and V (Volcanism). Red arrows indicate the prevailing variations that correspond to the different surface and subsurface syn-rift successions recorded along of Neuquén Basin

and/or subsequently to the regional-extended volcanic syn-rift climax. Depending on the displacement of the normal faults and the generation of growth folds that define the footwall relief as well as the gradient of the hanging wall blocks, three different continental sedimentary environments are recorded. Alluvial fans facies associations sourced from incipient footwall catchments occurred as transversal drainage systems, whereas in the centre of the half-grabens either gravelly/sandy bedload-fluvial systems developed as axial drainage system, or ephemeral playa-lakes were interpreted (Spalletti et al. 1999; Buchanan et al. 2017). Mixed-load or bedload-fluvial systems were also determined as the infill of low-evolved depocentres (see Spalletti et al. 1999; D'Elia et al. 2015), characterized as growth synclines, bounded by fault tip monocline or fault breaks. Exceptionally, in evolved rift depocentres deep highstand lakes with progradational, fan deltas/alluvial fan systems were also described (Puesto Kauffman Formation; Pángaro et al. 2002; Barredo et al. 2008). These depositional scenarios were mainly controlled by the tectonic and climate conditions with subordinate volcanic activity (Buchanan et al. 2017).

The continental volcano-dominated syn-rift depositional scenarios (i.e. Precuyano Cycle *s.s.*, Gulisano and Pando 1981; Gulisano et al. 1984) are mainly composed of volcanic (i.e. lava flows and pyroclastic density currents deposits) and resedimented/reworked volcaniclastic facies as well as, and to a lesser extent, carbonate and epiclastic facies (Muravchik et al. 2011; D'Elia et al. 2012b, 2018; Fig. 5). These facies were grouped into the Precuyano Cycle and include deposits of the Sañicó, Ñireco, Lapa, Cordillera del Viento, Medanito, Barda Alta formations and lower Remoredo Formation. Volcanic facies are associated in different sections of the syn-rift which are bounded by unconformity surfaces. Thus, each section corresponds to a higher-order syn-rift sequence characterized by a different type and volume of volcanism, defined by a particular dominated-eruptive style (e.g. effusive volcanism versus explosive volcanism; D'Elia et al. 2018). The volcanism coupled with the activity of extensional faults, determined the types of different volcanic edifices, such as composite volcanoes, graben-calderas and lava fields. During caldera-like events, pyroclastic facies (pyroclastic density currents) constitute up to 90% in volume of the depocentre and were associated with "instantaneous" subsidence occurred along the same normal faults that controlled tectonic subsidence during the previous and subsequent episodes (i.e. volcano-tectonic subsidence; D'Elia and Martí 2013). When volcanic facies are <50% in volume within the extensional depocentres, the decreasing of volcanic input stabilized alluvial/delta-lacustrine environments, which are interbedded with the volcanic facies. The decreasing of the impact of volcanism on the sedimentary system enabled the climate, tectonics or bedrock lithology as main controls on the stratigraphy. Therefore, for the *continental volcano-dominated syn-rift depositional scenarios* the volcanic edifices controlled the stacking patterns of the volcanic units as well as the sedimentary systems (Fig. 5). The landform of the volcanic edifices as well as the styles and scales of the eruptions governed the volcaniclastic input to the basin, setting the main variables of the sedimentary systems, such as provenance, grain size, transport and deposition, and geometry (see D'Elia et al. 2018).

The *Marine volcano-dominated syn-rift depositional scenarios* are mainly developed towards the end of the extensional tectonics only in the western part of the basin, around 37°S (Fig. 5). These facies occurred in the La Primavera Formation, grouped into the Precuyano Cycle (Suárez and De la Cruz 1997). These syn-rift marine scenarios are composed of mainly subaqueous pyroclastic density currents deposits interbedded with subaqueous pyroclastic fall out and sediment gravity flow deposits, with rests of invertebrate fossils and plant remains (i.e. Precuyano Cycle to Leanza 2009). To a lesser extent, basaltic/andesite pillow lava flows and sills with peperite facies were recorded. This facies association may be interpreted as submarine volcaniclastic apron in proximal position, associated with a shallow marine environment, in which volcanic facies could have been related to a subaerial source (Suarez de la Cruz,1997). In this case, the volcanism controlled the input of primary and secondary volcaniclastic materials deposited under wave-base, set in a high relative sea level (Fig. 5).

The *Marine non-volcano-dominated syn-rift depositional scenarios* are formed by epiclastic and minor reworked volcaniclastic facies, deposited in subaqueous marine environments, under sediment gravity flow and fall out processes (Fig. 5). These depositional scenarios are characteristic of the marine half-graben located in the centre (Vergani et al. 1995; Pángaro et al. 2009) and southern part (Muravchik et al. 2014; D'Elia et al. 2015) of the Neuquén Basin, and they are related to the first, basin-scale, marine ingression (Lower Cuyano Cycle) during the Pliensbachian. It is important to remark that these depositional scenarios also correspond to the unique case of marine environments developed in an isolated depocentre during the Upper Triassic at the onset of the basin (i.e. Arroyo Malo Formation in Atuel depocentre; Lanés et al. 2008). Based on facies models carried out from surface studies, these depositional scenarios were defined as longitudinal hangingwall-fed fan deltas or transversal footwall fed fan deltas associated with hemi-pelagic sedimentation (Lanés et al. 2008; Muravchik et al. 2014, Fig. 5). The main controls of this syn-rift scenario were the fault activity, reactivation or activation of new extensional faults that controlled the configuration of the transference of sediments and the relative sea level (eustatic + mechanical tectonic subsidence), setting a positive *A/S* rate (Lanés et al. 2008; Muravchik et al. 2014; D'Elia et al. 2015).

7 Final Remarks: Rift Model, Structure, Volcanism and Sedimentation

The syn-rift of the Neuquén Basin recorded a complex period of extensional tectonics from the Late Triassic to the Early Jurassic. This episode was mainly superimposed to a Late Paleozoic orogenic complex associated with the tectonic configuration of the western margin of Gondwana (Franzese and Spalletti 2001). As a result of the extensional tectonics, a series of rift troughs were distributed over a broad surface conforming a wide rift setting (see Brun 1999; Buck et al. 1999). As pointed out by subsidence (Scivetti 2017) and structural analysis (Bechis et al. 2009, 2010, 2014), the magnitude of extension was around 10% with a roughly NE-SW direction of extension, configuring mainly suborthogonal rifts with participation of oblique fabrics.

During the rifting, the onset and cessation of the magmatism and the mechanical extension were diachronic along the basin, showing younger ages from north to south. The (volcanic) rift climax follows the path evolution of the Mesozoic magmatism of Patagonia (Kay et al. 1989; Ramos 2009), which was related to orogenic or post-orogenic environments (Kay et al. 1989; Ramos and Folguera 2005; Ramos 2009). It is important to remark that the magmatism, which has not typically alkaline-tholeiitic, mafic–felsic bimodal composition like recorded in intracontinental rifts, see Wilson (1989), did not predate the rifting. This indicates that the rifting proceeded in response to lithospheric extension driven by far-field stresses, controlling the magmatism in a context of "'passive' rift (sensu McKenzie 1978; McKenzie and Bickle 1988). The origin of the Neuquén Basin was related to different causes, such as the collapse of Late Paleozoic orogen (Franzese and Spalletti 2001), extensional tectonics related to the Gondwana break-up (Uliana and Biddle 1987; Bechis et al.

2010, 2014) or orogenic-related rift associated with the transition from an arrested-subduction episode to the onset of the Andean margin under tectonic segmentation (D'Elia et al. 2012a). The evidence presented point to the Neuquén Basin like a wide rift mode, in which all features are related to a rift linked in some way with orogenic settings (see Ziegler and Cloetingh 2004) during the Gondwana break-up, as was proposed by Uliana and Biddle (1987) and Bechis et al. (2010, 2014).

The tectono-stratigraphy of the syn-rift indicates that after a short lapse of crustal extension related to mainly epiclastic sedimentation, the rifting was accompanied by widespread, asymmetrically distributed, volcanism of dominantly calc-alkaline signature, which was younger from north to south along the basin (i.e. Precuyano Cycle *s.s.*). This would have driven the diachronic post-rift initiations that tracked the rift evolution. Under this scheme, the combined effect of the ascending global sea level occurred from Early Jurassic (e.g. Hallam 2001) and the diachronic onset of sag phase, determined that the lower Jurassic marine sedimentation has been set in either a syn-rift or post-rift scenarios, depending on the position in the basin. This constitutes a key-issue to evaluate the syn-rift–post-rift transition, since it may determine the misunderstanding that the Cuyano Cycle was defined as a marine sequence in the context of sequence stratigraphy and not as a tectono-sequence.

The tectono-magmatic-eustacy given above allows understanding (in time and space) the diverse depositional environments recorded in the syn-rift megasequence as four depositional scenarios: (i) *continental non-volcano-dominated syn-rift,* (ii) *continental volcano-dominated syn-rift*; (iii) *Marine volcano-dominated syn-rift, and* (iv) *Marine non-volcano-dominated syn-rift depositional scenarios.* The principal controls on the stratigraphy were the mechanical subsidence driven by normal faulting and the magnitude and character of the associated volcanism—that as an extreme situation captured the tectonic structures triggering volcano-tectonic events. In turn, the relative sea-level change (eustacy + tectonics) also acted, although as a secondary factor particularly at the final stage of the rifting.

References

Barredo S, Cristalini E, Zambrano O et al (2008) Análisis tectosedimentario del relleno de edad precuyana y cuyana inferior de la región septentrional del Alto de Kauffman, Cuenca Neuquina. In: Abstracts of the 7° Congreso de Exploración y Desarrollo de Hidrocarburos, IAPG, Mar del Plata, 5–7 Nov 2008

Barrett TJ, MacLean WH (1994) Mass changes in hydrothermal alteration zones associated with VMS deposits of the Noranda area. Explor Min Geol 3:131–160

Barrionuevo M, Arnosio M, Llambías EJ (2013) Nuevos datos geocronológicos en subsuelo y afloramientos del Grupo Choiyoi en el oeste de La Pampa: implicancias estratigráficas. Rev Asoc Geol Argent 70:31–39

Bechis F, Giambiagi LB, Lanés S, García VH, Tunik M (2009) Evidencias de extensión oblicua en los depósitos de sinrift del sector norte de la cuenca Neuquina. Rev Asoc Geol Argent 65:293–310

Bechis F, Giambiagi L, García V, Lanés S, Cristallini E, Tunik M (2010) Kinematic analysis of a transtensional fault system: the Atuel depocenter of the Neuquén basin, southern Central Andes, Argentina. J Struct Geol 32:886–899

Bechis F, Cristallini EO, Giambiagi LB et al (2014) Transtensional tectonics induced by oblique reactivation of previous lithospheric anisotropies during the Late Triassic to Early Jurassic rifting in the Neuquén Basin: INSIGHTS from analog models. J Geodyn 79:1–17

Brun JP (1999) Narrow rifts versus wide rifts: inferences for the mechanics of rifting from laboratory experiments. Philos T Roy Soc A, 695–709

Buchanan AS, Kietzmann DA, Palma RM (2017) Evolución paleoambiental de la Formación Remoredo (Jurásico Inferior) en el depocentro Malargüe, Cuenca Neuquina surmendocina. Rev Asoc Geol Argent 74:163–178

Buck WR, Lavier LL, Poliakov AN (1999) How to make a rift wide. Philos T Roy Soc A, 671–689

Carbone O, Franzese J, Limeres M et al (2011) El Ciclo Precuyano (Triásico Tardío-Jurásico Temprano) en la Cuenca Neuquina. In: Leanza HA, Arregui C, Carbone O et al (eds) Geología y Recursos Naturales de la Provincia del Neuquén. Asociación Geológica Argentina, Buenos Aires, pp 63–75

Cristallini EO, Bottesi G, Gavarrino A et al (2006) Synrift geometry of the Neuquén Basin in the northeastern Neuquén Province, Argentina. In: Kay SM, Ramos VA (eds) Evolution of the Andean Margin: a Tectonic and Magmatic View from the Andes to the Neuquén Basin (35°–39°S Lat). Geol Soc Am SP 407:147–161

Cristallini E, Tomezzoli R, Pando G et al (2009) Controles precuyanos en la estructura de la Cuenca Neuquina. Rev Asoc Geol Argent 65:248–264

D'Elia L, Martí J (2013) Caldera events in a rift depocentre: an example from the Jurassic Neuquén Basin. J Geol Soc London 170:571–584

D'Elia L, Muravchik M, Franzese JR, Bilmes A (2012a) Volcanismo de sin-rift de la Cuenca Neuquina, Argentina: relación con la evolución Triásico-Tardía - Jurásico Temprano del margen Andino. Andean Geol 39:106–132

D'Elia L, Muravchik M, Franzese JR et al (2012b) Tectonostratigraphic analysis of the Late Triassic-Early Jurassic syn-rift sequence of the Neuquén Basin in the Sañicó depocentre, Neuquén Province, Argentina. Andean Geol 39:133–157

D'Elia L, Bilmes A, Franzese JR et al (2015) Early evolution of the southern margin of the Neuquén Basin, Argentina: Tectono-stratigraphic implications for rift evolution and exploration of hydrocarbon plays. J S Am Earth Sci 64:42–57

D'Elia L, Martí J, Muravchik M et al (2018) Impact of volcanism on the sedimentary record of the Neuquén rift Basin, Argentina: towards a cause and effect model. Basin Res 30:311–335

Digregorio JH, Uliana MA (1980) Cuenca Neuquina. In: Turner JCM (ed) Geología Regional Argentina, vol 2. Academia Nacional de Ciencias, Córdoba, pp 985–1032

Fennell LM, Folguera A, Naipauer M et al (2017) Cretaceous deformation of the Southern Central Andes: synorogenic growth strata in the Neuquén Group (35°30′-37°S). Basin Res 29:51–72

Franzese JR, Spalletti LA (2001) Late Triassic–early Jurassic continental extension in southwestern Gondwana: tectonic segmentation and pre-break-up rifting. J S Am Earth Sci 14:257–270

Franzese JR, Spalletti LA, Gómez Pérez I et al (2003) Tectonic and paleoenviromental evolution of Mesozoic sedimentary basins along the Andean foothills of Argentina (32°–54°S). J S Am Earth Sci 16:81–90

Franzese JR, Veiga GD, Schwarz E et al (2006) Tectonostratigraphic evolution of a Mesozoic graben border system: the Chachil depocentre, southern Neuquén Basin, Argentina. J Geol Soc London 163:707–721

Franzese JR, Veiga GD, Muravchik M et al (2007) Estratigrafía de 'sin-rift' (Triásico Superior-Jurásico Inferior) de la Cuenca Neuquina en la sierra de Chacaico, Neuquén, Argentina. Rev Geol Chile 34:49–62

Giambiagi LB, Suriano J, Mescua J (2005) Extensión multiepisódica durante el Jurásico Temprano en el depocentro Atuel de la cuenca Neuquina. Rev Asoc Geol Argent 60:524–534

Giambiagi LB, Bechis F, Lanés S et al (2008) Formación y evolución triásico-jurásica del depocentro Atuel, cuenca Neuquina, provincia de Mendoza, Argentina. Rev Asoc Geol Argent 63:520–533

Giambiagi LB, Tunik M, Barredo S et al (2009) Cinemática de apertura del sector norte de la cuenca neuquina. Rev Asoc Geol Argent 65:278–292

Groeber P (1946) Observaciones geológicas a lo largo del meridiano 70°. Hoja Chos Malal. Rev Soc Geol Argent 1:177–208
Gulisano CA, Gutiérrez Pleimling AR (1994) The Jurassic of the Neuquén Basin: b) Mendoza Province. Guía de Campo. Asociación Geológica Argentina, Buenos Aires
Gulisano C, Pando GA (1981) Estratigrafía y facies de los depósitos jurásicos entre Piedra del Águila y Sañicó, Departamento Collón Curá, Provincia del Neuquén. In: Abstracts of the 8 Congreso Geológico Argentino, Asociación Geológica Argentina, San Luis, 20–26 Sep 1981
Gulisano CA, Gutiérrez Pleimling AR, Digregorio RE (1984) Esquema estratigráfico de la secuencia jurásica del oeste de la provincia del Neuquén. In: Abstracts of the 9 Congreso Geológico Argentino, Asociación Geológica Argentina, San Carlos de Bariloche, 5–9 November 1984
Hallam A (2001) A review of the broad pattern of Jurassic sea-level changes and their possible causes in the light of current knowledge. Palaeogeogr Palaeocl 167:23–37
Howell JA, Schwarz E, Spalletti LA et al (2005) The Neuquén Basin: an overview. In: Veiga GD, Spalletti LA, Howell JA et al (eds) The Neuquén Basin, Argentina: a case study in sequence stratigraphy and basin dynamics. Geol Society, London, SP 252:1–14
Kay SM, Ramos VA, Mpodozis C, Sruoga P (1989) Late Paleozoic to Jurassic silicic magmatism at the Gondwana margin: analogy to the Middle Proterozoic in North America? Geology 17:324–328
Lanés S (2005) Late Triassic to Early Jurassic sedimentation in northern Neuquén Basin, Argentina: tectosedimentary evolution of the first transgression. Geol Acta 3:81–106
Lanés S, Giambiagi LB, Bechis F et al (2008) Late Triassic e early Jurassic successions of the Atuel depocenter: sequence stratigraphy and tectonic controls. Rev Asoc Geol Argent 63:534–548
Leanza HA (2009) Las principales discordancias del Mesozoico de la Cuenca Neuquina según observaciones de superficie. Rev Mus Argent Cs Nat 11:145–184
Leanza HA, Mazzini A, Corfu F et al (2013) The Chachil Limestone (Pliensbachian earliest Toarcian) Neuquén Basin, Argentina: U-Pb age calibration and its significance on the Early Jurassic evolution of southwestern Gondwana. J S Am Earth Sci 42:171–185
Legarreta L, Gulisano CA (1989) Amilisis Estratigráfico de la Cuenca Neuquina (Triásico superior-Terciario inferior), Argentina. In: Chebli GA, Spalletti LA (eds) Cuencas Sedimentarias Argentinas, Universidad de Tucumán, Serie Correlación Geológica, vol 6, pp 221–244
Legarreta L, Uliana MA (1996) The Jurassic succession in west-central Argentina: stratal pattern, sequences and paleogeographic evolution. Palaeogeogr Palaeocl 120:303–330
Legarreta L, Uliana M (1999) El Jurásico y Cretácico de la Cordillera Principal y la Cuenca Neuquina. 1. Facies Sedimentarias. In: Caminos R (ed) Geología Argentina, Instituto de Geología y Recursos Minerales, SEGEMAR, Anales 29, pp 399–416
Legarreta L, Gulisano CA, Uliana MA (1993) Las secuencias sedimentarias jurásico-cretácicas. In: Ramos VA (ed) Geología y Recursos Minerales de Mendoza. Asociación Geológica Argentina, Mendoza, pp 87–114
Llambías EJ, Leanza HA, Carbone O (2007) Evolución tectono-magmática durante el Pérmico al Jurásico temprano en la Cordillera del Viento (37°05′S–37°15′S): Nuevas evidencias geológicas y geoquímicas del inicio de la Cuenca Neuquina. Rev Asoc Geol Argent 62:217–235
Manceda R, Figueroa D (1995) Inversion of the Mesozoic Neuquén Rift in the Malargüe Fold and Thrust Belt, Mendoza, Argentina. In: Tankard AJ, Suárez SR, Welsink HJ (eds) Petroleum basins of South America, AAPG Memoir 62:369–382
McKenzie DP (1978) Some remarks on the development of sedimentary basins. Earth Planet Sci Lett 40:25–32
McKenzie D, Bickle MJ (1988) The volume and composition of melts generated by extension of the lithosphere. J Petrol 29:625–679
Mitchum RM, Uliana MA (1985) Seismic stratigraphy of carbonate depositional sequences, Upper Jurassic–Lower Cretaceous, Neuquén Basin, Argentina. In: Bero BR, Woolverton DG (eds) Seismic stratigraphy: an integrated approach to hydrocarbon exploration, AAPG Memoir 39:255–274

Muravchik M (2008) Sistemas Sedimentarios axiales y transversales a un depocentro de Rift (Ciclo Precuyano) del Suroeste de la Cuenca Neuquina. In: Abstract of the 17 Congreso Geológico Argentino, Asociación Geológica Argentina, San Salvador de Jujuy, 2–6 May 2011
Muravchik M, D'Elia L, Bilmes A, Franzese JR (2011) Syn-eruptive/inter-eruptive relations in the syn-rift deposits of the Precuyano Cycle, Sierra de Chacaico, Neuquén Basin, Argentina. Sediment Geol 238:132–144
Muravchik M, Bilmes A, D'Elia L, Franzese JR (2014) Alluvial fan deposition along a rift depocentre border from the Neuquén Basin, Argentina. Sediment Geol 301:70–89
Naipauer M, Morabito EG, Marques JC et al (2012) Intraplate Late Jurassic deformation and exhumation in western central Argentina: constraints from surface data and U-Pb detrital zircon ages. Tectonophysics 524:59–75
Naipauer M, Tapia F, Mescua J et al (2015) Detrital and volcanic zircon U-Pb ages from southern Mendoza (Argentina): an insight on the source regions in the northern Neuquén Basin. J S Am Earth Sci 64:434–451
Naipauer M., Fennell L, Folguera A et al (2016) Edades U-Pb SHRIMP de volcanitas del Ciclo Precuyano: Controles temporales en la extensión de depocentro Cara Cura-Reyes (36°30′LS), norte de la Cuenca Neuquina. In: Abstracts of the 1 Simposio de Tectónica Sudamericana, Santiago de Chile, 14–16 Nov 2016
Naipauer M, Morabito EG, Manassero M et al (2018) A provenance analysis from the Lower Jurassic Units of the Neuquén Basin. Volcanic Arc or Intraplate Magmatic Input? In: Folguera et al. (eds) The evolution of the Chilean-Argentinean Andes, Springer, pp 191–222
Pángaro F, Corbera R, Carbone O et al (2002) Los reservorios del Precuyano. In: Schiuma M, Hinterwimmer G, Vergani GD (eds) Rocas Reservorio de las Cuencas Producticas Argentinas. Instituto Argentino del Petróleo y del Gas, Buenos Aires, pp 229–254
Pángaro F, Pereira DM, Micucci E (2009) El sinrift de la dorsal de Huincul, Cuenca Neuquina: evolución y control sobre la estratigrafía y estructura del área. Rev Asoc Geol Argent 65:265–277
Ramos VA (1999) Plate tectonic setting of the Andean Cordillera. Episodes 22(3):183–190
Ramos V (2009) Anatomy and global context of the Andes: Main geologic features and the Andean orogenic cycle. In: Kay SM, Ramos V, Dickinson WR (eds) Backbone of the Americas: Shallow Subduction, Plateau Uplift, and Ridge and Terrane Collision. Geological Society of America, vol 204, pp 31–65
Ramos VA, Folguera A (2005) Tectonic evolution of the Andes of Neuquén: constraints derived from the magmatic arc and foreland deformation. In: Veiga GD, Spalletti LA, Howell JA, Schwarz E (eds) The Neuquén Basin, Argentina: a case study in sequence stratigraphy and basin dynamics. Geological Society, London, SP 252:15–35
Riccardi AC, Gulisano C (1990) Unidades limitadas por discontinuidades. Su aplicación al Jurasico Andino. Rev Asoc Geol Argent 45:346–364
Schiuma M, Llambías EJ (2007) New ages on Lower Jurassic volcanism in the Dorsal de Huincul, Neuquén. In: Abstracts of the 3 Simposio Argentino del Jurásico, Mendoza, 2–5 May 2007
Schiuma M, Llambías EJ (2008) New ages and chemical analysis on Lower Jurassic volcanism close to the Huincul High, Neuquén. Rev Asoc Geol Argent 63:644–652
Scivetti N (2017) Análisis comparativo de los controles tectónicos y eustáticos sobre la estratigrafía de Post-Rift (Jurásico inferior-Cretácico inferior) en el Sector Central de la Cuenca Neuquina, Argentina. Ph.D. thesis, Universidad Nacional de la Plata
Sengor AMC (1995) Sedimentation and tectonics of fossil rifts. In: Busby CJ, Ingersoll RV (eds) Tectonics of sedimentary basins, vol 579. Blackwell Science, Oxford, pp 53–118
Silvestro J, Zubiri M (2008) Convergencia oblicua: modelo estructural alternativo para la dorsal neuquina (39°s)—Neuquén. Rev Asoc Geol Argent 63:49–64
Spalletti LA (1999) Cuencas triásicas del oeste argentino: origen y evolución. Acta Geol Hisp 32:29–50
Spalletti LA, Franzese JR, Morel E et al (2010) Consideraciones acerca de la sedimentología, paleobotánica y geocronología de la Formación Piedra del Águila (Jurásico Inferior, Neuquén, República Argentina). Rev Asoc Geol Argent 66:305–313

Suárez M, De la Cruz R (1997) Volcanismo pliniano del Lias durante los inicios de la Cuenca de Neuque'n. Cordillera del Viento, Neuquén, Argentina. In: Abstracts of the 8 Congreso Geológico Chileno, Santiago, 13–17 Oct 1997
Sun SS, McDonough WF (1989) Chemical and isotopic systematics of oceanic basalts: implications for mantle composition and processes. Geol Soc 42(1):313–345
Uliana MT, Biddle KT (1987) Permian to Late Cenozoic evolution of northern Patagonia: main tectonic events, magmatic activity, and depositional trends. In: Mckenzie GD (ed) Gondwana six: structure, tectonics, and geophysics. American Geophysical Union, Washington DC, pp 271–286
Uliana MA, Legarreta L (1993) Hydrocarbons habitat in a Triassic to Cretaceous sub-andean setting: Neuquén basin, Argentina. J Petrol Geol 16:397–420
Veiga GD, Schwarz E, Spalletti LA et al (2013) Anatomy and sequence architecture of the early post-rift in the Neuquén Basin (Argentina): a response to physiography and relative sea-level changes. J Sediment Res 83:746–765
Vergani GD (2005) Control estructural de la sedimentación jurásica (Grupo Cuyo) en la Dorsal de Huincul, Cuenca Neuquina, Argentina. Modelo de falla lístrica rampaplano, invertida. Bol Inf Petrol 1:32–42
Vergani GD, Tankard AJ, Belotti HJ et al (1995) Tectonic evolution and paleogeography of the Neuquén basin, Argentina. In: Tankard AJ, Suárez SR, Welsink HJ (eds) Petroleum basins of South America. AAPG Memoir 62:383–402
Wilson M (1989) Igneous petrogenesis, a global tectonic approach. Unwin Hyman, London, pp 466
Winchester JA, Floyd PA (1977) Geochemical discrimination of different magma series and their differentiation products using immobile elements. Chem Geol 20:325–343
Yrigoyen MR (1991) Hydrocarbon resources of Argentina. Petrotecnia SI/1991:38–54
Ziegler PA, Cloetingh S (2004) Dynamic processes controlling evolution of rifted basins. Earth-Sci Rev 64:1–50

Tectono-Stratigraphic Evolution of the Atuel Depocenter During the Late Triassic to Early Jurassic Rift Stage, Neuquén Basin, West-Central Argentina

Florencia Bechis, Laura B. Giambiagi, Maisa A. Tunik, Julieta Suriano, Silvia Lanés and José F. Mescua

Abstract The Neuquén Basin presents an almost continuous record from the Late Triassic until the Paleocene, making it an excellent case study of the most relevant tectonic stages of southern South America during the Mesozoic. It was initiated in Late Triassic to Early Jurassic times as a continental rift basin in the context of a widespread extensional stage that affected western Gondwana and culminated with the break-up of the supercontinent. The Atuel depocenter is located in the northern sector of the Neuquén Basin. Its synrift and sag units are represented by Upper Triassic to Lower Jurassic siliciclastic marine and continental sedimentary rocks including the oldest marine deposits of the basin, of Late Triassic age. The depocenter infill has been deformed and exhumed during the Andean orogeny, being presently exposed in the northern sector of the Malargüe fold and thrust belt. In this review, we have integrated a large set of stratigraphic, sedimentologic, geochronologic, and structural data in order to unravel the tectono-sedimentary evolution of the Atuel depocenter and to evaluate the main controlling factors of the synrift stage. We analyzed data from the synrift units, such as facies and thickness distribution, sandstone provenance, detrital zircon geochronology data, kinematic data from outcrop-scale normal faults,

F. Bechis (✉)
Instituto de Investigaciones en Diversidad Cultural y Procesos de Cambio (IIDyPCa). Consejo Nacional de Investigaciones Científicas y Técnicas (CONICET). Universidad Nacional de Río Negro, Sede Andina. Mitre 630, San Carlos de Bariloche, Argentina
e-mail: florbechis@gmail.com

L. B. Giambiagi · J. Suriano · J. F. Mescua
Instituto Argentino de Nivología, Glaciología y Ciencias Ambientales (IANIGLA). Consejo Nacional de Investigaciones Científicas y Técnicas (CONICET), Centro Científico Tecnológico Mendoza. Av. Ruiz Leal s/n, Parque General San Martín, Mendoza, Argentina

M. A. Tunik
Universidad Nacional de Río Negro. CONICET. Instituto de Investigaciones en Paleobiología y Geología. Av. Roca 1242., General Roca, Argentina

S. Lanés
Cape Town, South Africa

J. F. Mescua
Facultad de Ciencias Exactas y Naturales, Universidad Nacional de Cuyo, Mendoza, Argentina

D. Kietzmann and A. Folguera (eds.), *Opening and Closure of the Neuquén Basin in the Southern Andes*, Springer Earth System Sciences,
https://doi.org/10.1007/978-3-030-29680-3_2

angular and progressive unconformities, and subsurface information. Reactivation of preexisting NNW-striking anisotropies under a regional NNE extension resulted in an oblique rift setting, which generated a bimodal distribution of NNW- and WNW-striking major normal faults. Reduced strain and stress tensors obtained from the kinematic and dynamic analysis of structural data show a complex heterogeneity that we interpreted as a result of local stress permutations due to the activity of the larger faults and to strain partitioning inside the Atuel depocenter. Sedimentologic and petrographic data revealed a complex evolution with strong lateral variations of the depositional environments during the synrift phase, which lasted from Rhaetian to Pliensbachian times. We identified several stages that were controlled by processes of initiation, propagation, growth, linkage, and deactivation of new and reactivated faults along the depocenter evolution, in combination with sea-level changes related to global eustatic variations. Sandstone provenance data suggest an important basin reorganization by the Toarcian, probably related to the initiation of the sag stage in this depocenter.

Keywords Neuquén Basin · Atuel depocenter · Oblique rift · Kinematic and dynamic analysis · Sedimentary provenance

1 Introduction

The Neuquén Basin is an oil-bearing back-arc extensional basin developed in the southwestern margin of the Gondwana continent and open to the Pacific at its western margin. It hosts an almost continuous record of Late Triassic to Paleocene age (Uliana et al. 1989; Legarreta and Uliana 1999), making it an excellent case study that registers the most relevant tectonic stages of southern South America during the Mesozoic. It began in the Late Triassic to Early Jurassic as a continental rift basin due to a widespread extensional stage at western Gondwana, which gave way to the break-up of the supercontinent and the subsequent opening of the South Atlantic Ocean (Uliana et al. 1989; Vergani et al. 1995; Franzese and Spalletti 2001).

The Neuquén Basin shows a particular triangular shape (Fig. 1), with a western sector that has been deformed by the shortening that has built the Southern Central Andes since the Late Cretaceous (Cobbold and Rosello 2003), obscuring the original distribution of synrift deposits and their structural controls. The southern sector is a wide rift basin, with a more evenly distributed deformation and a more than 4000 m thick succession of volcanic, volcaniclastic, and clastic synrift deposits (Gulisano 1981; Carbone et al. 2011). The northern sector is a narrow and localized rift basin, whose early synrift deposits are continental-marine, clastic-volcanic successions of more than 2000 m (Gulisano and Gutiérrez Pleimling 1994).

The basin has several depocenters, each one showing a particular set of structural and stratigraphic features. Some of their differences can be attributed to strongly varying volumes of volcanic deposits and to a probable diachronic evolution of the early synrift, climax, and sag phases during a eustatic sea-level rise (see

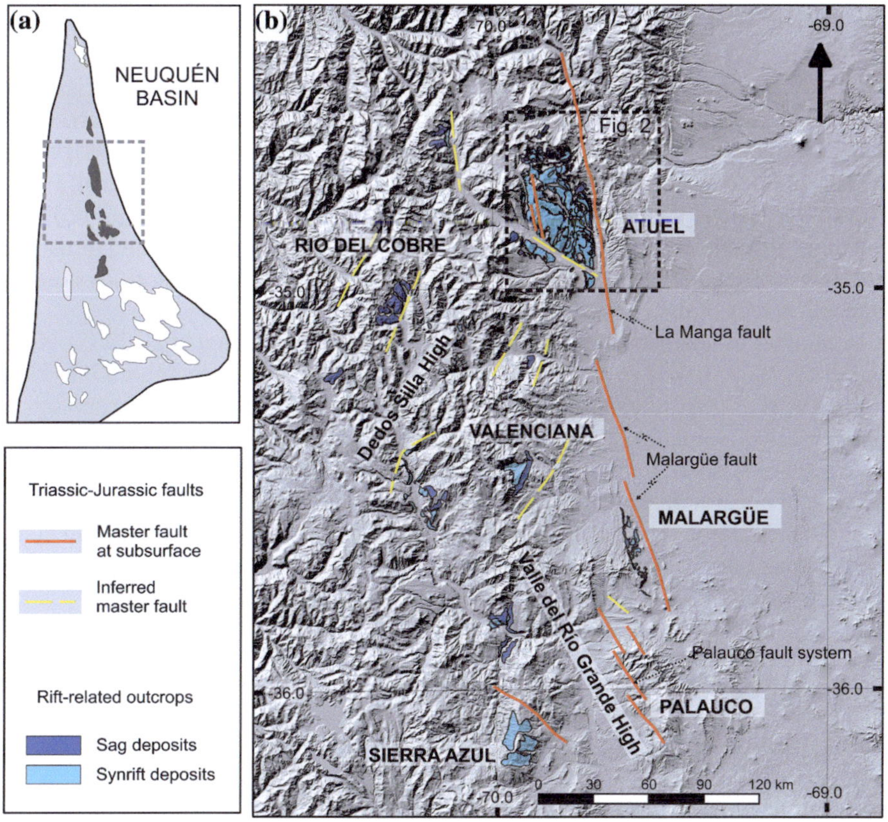

Fig. 1 Regional map showing the distribution of the principal normal faults and the infill of the Late Triassic to Early Jurassic synrift depocenters of the Neuquén Basin in the southern Mendoza province, west-central Argentina (modified from Manceda and Figueroa 1995; Yagupsky et al. 2008; Giambiagi et al. 2009a, b; Bechis et al. 2014; Mescua et al. 2014; Fuentes et al. 2016; Barrionuevo et al. 2019)

Chapter "The Syn-Rift of the Neuquén Basin-Precuyano and Lower Cuyano Cycle: Review of Structure, Volcanism, Tectono-Stratigraphy and Depositional Scenarios"). However, other controlling factors, such as the relative orientation of each depocenter respect to the regional stress field and the role of reactivated structures during the extensional event, have also been proposed (Vergani et al. 1995; Mosquera and Ramos 2006; Giambiagi et al. 2009a; Bechis et al. 2014).

The northern depocenters of the basin are the Yeguas Muertas, Nieves Negras, Atuel, Río del Cobre, Valenciana, Malargüe, Sierra Azul, and Palauco depocenters (Fig. 1; Gulisano and Gutiérrez Pleimling 1994; Manceda and Figueroa 1995; Giambiagi et al. 2009a). Firstly, during the Norian to Early Sinemurian, these depocenters were isolated by structural highs (e.g., Dedos-Silla High, Valle del

Rio Grande High, Manceda and Figueroa 1995), and later became progressively inter-connected during Sinemurian to Late Pliensbachian times.

The Atuel depocenter synrift and sag units are represented by Upper Triassic to Lower Jurassic siliciclastic marine and continental sedimentary rocks (Reijenstein 1967; Volkheimer 1978; Riccardi et al. 1991; Lanés 2005, Lanés et al. 2008), which host the oldest marine deposits of the basin, of Late Triassic age (Riccardi et al. 1997). Those successions have been deformed and exhumed during the Andean orogeny, and they presently crop out in the northern sector of the Malargüe fold and thrust belt (Kozlowski et al. 1993), making the Atuel depocenter one of the best study areas of the early stages of the Neuquén Basin development.

A number of sedimentological (Lanés 2002, 2005; Giambiagi et al. 2005a; Spalletti et al. 2007; Lanés et al. 2008; Tunik et al. 2008, 2011), structural (Giambiagi et al. 2005a, 2008a; Bechis, 2009; Bechis et al. 2009, 2010), and geochronological contributions (Naipauer et al. 2015; Horton et al. 2016) about the Atuel depocenter have been published in the last years. However, an integrated and multidisciplinary analysis of the available information has not been made so far. In this review, we have integrated the large set of available data in order to unravel its tectono-sedimentary evolution and to determine the main controlling factors of the synrift stage.

2 Tectonic Setting

The basement of the Neuquén Basin is affected by structures and rheological contrasts inherited from previous tectonic stages (Mosquera and Ramos 2006; Bechis et al. 2014; Mescua et al. 2016). The early Paleozoic geological evolution of this sector of Gondwana was determined by the accretion of Pampia, Cuyania, Chilenia, and Patagonia allochthonous terranes (Ramos et al. 1986; Ramos 1988, 2008; Astini et al. 1995; Rapalini 2005; Pankhurst et al. 2006; Varela et al. 2011; Tomezzoli 2012; Pángaro and Ramos 2012; see discussion and references therein). These collisions were later followed by the onset of a classical Andean-type margin associated with retroarc basins during the Carboniferous to Early Permian (Limarino and Spalletti 2006), which gave place to the Gondwana orogeny (Keidel 1916), locally known as San Rafael compressional phase (Azcuy and Caminos 1987; Mpodozis and Ramos 1989).

From the Late Permian to the Early Cretaceous, the region underwent extensional conditions, associated with the initial break-up of Gondwana (Charrier 1979; Llambías et al. 1993). The early stages were characterized by a widespread silicic magmatism of Late Permian to Early Triassic age, represented by the upper section of the Choiyoi Group (Llambías 1999; Sato et al. 2015; Giambiagi and Martínez 2008; Kleiman and Japas 2009). During the Early to Middle Triassic, this magmatism was progressively followed by the opening of several NNW-trending narrow continental rift systems, such as the Cuyo, Ischigualasto, and other coeval basins (Charrier 1979; Uliana et al. 1989; Kokogian et al. 1993; Japas et al. 2008; Giambiagi et al. 2011; Espinoza et al. 2018). Younger stages of rifting took place during the Middle to Late

Triassic in the Domeyko Basin (Espinoza et al. 2018) and during the Late Triassic to Early Jurassic in the Neuquén, Ramada, and other coeval basins, which formed new depocenters not geographically connected with the earlier ones (Alvarez and Ramos 1999; Charrier et al. 2007).

Since the late Early Jurassic, the southwestern margin of Gondwana was characterized by active subduction and a well-developed Andean magmatic arc (Mpodozis and Ramos 1989; Charrier et al. 2007), which was probably already active during the Triassic to Early Jurassic extensional stage (Llambías et al. 2007; Schiuma and Llambías 2008; Oliveros et al. 2018) (see Chapter "Early Andean Magmatism in Southern Central Chile, 33°–40° S"). Some authors argued that the regional extensional regime continued until the Early Cretaceous, related to the retreat of the Pacific trench (Mpodozis and Ramos 1989; Ramos and Kay 2006; Ramos 2010). From the Middle Jurassic onwards, the Neuquén Basin was subjected to thermal subsidence (Legarreta and Uliana 1996; Vergani et al. 1995; Howell et al. 2005). Then a second synrift stage is proposed during the Late Jurassic (Vergani et al. 1995; Giambiagi et al. 2003a; Charrier et al. 2007; Mescua et al. 2008) (see Chapter "Controls on Deposition of the Tordillo Formation in Southern Mendoza (34°–36° S): Implications for the Kimmeridgian Tectonic Setting of the Neuquén Basin"). At this time, local episodic shortening phases with selective inversion were registered in the southern sector of the basin during the Jurassic, mainly in the Huincul High (Vergani et al. 1995; Pángaro et al. 2006; Mosquera and Ramos 2006; Silvestro and Zubiri 2008; Cristallini et al. 2009).

Then, Andean uplift started in the Late Cretaceous (Cobbold and Rossello 2003; Zapata and Folguera 2005; Zamora Valcarce et al. 2006; Tunik et al. 2010; García Morabito and Ramos 2012; Mescua et al. 2013; Balgord and Carrapa 2016; Fennell et al. 2017; Gómez et al. 2019), with a renewed pulse of deformation and uplift from the Miocene to Recent (Giambiagi et al. 2003b, 2008b, 2014; Spagnuolo et al. 2012; Orts et al. 2012, Horton et al. 2016). During these shortening phases, the infill of the Neuquén Basin was deformed and exhumed giving way to thick-and thin-skinned fold and thrust belts on the western sector of the basin (Manceda and Figueroa 1995; Uliana et al. 1995; Zapata et al. 1999; Giambiagi et al. 2005b, 2008b; Zapata and Folguera 2005; Zamora Valcarce et al. 2006).

3 The Northern Neuquén Basin

The Late Triassic to Early Jurassic extensional architecture of the northern sector of the Neuquén Basin is made up by a set of NNW and NNE elongated depocenters with highly variable synrift thicknesses, including both continental volcaniclastic deposits and interbedded marine and continental siliciclastic facies (Fig. 1). The structure and infill of most of the depocenters of this sector are outlined in this section, while the main stratigraphic and structural characteristics of the Atuel depocenter are further described and discussed in the following sections.

Deposition of clastic continental synrift deposits (Remoredo Formation) in the **Sierra Azul depocenter** was controlled by movement along its NW-striking and SW-dipping master fault, El Manzano fault (Yagupsky et al. 2008). North of this fault, the basement rocks are covered by Upper Toarcian to Bajocian nearshore and offshore sag deposits (Gulisano and Gutiérrez Pleimling 1994). Toward the east, the **Palauco depocenter** was filled with volcaniclastic and volcanic deposits of the Precuyo Cycle (Gulisano 1981) or Precuyo Mesosequence (Legarreta and Gulisano 1989). Two NNW-striking and ENE-dipping master faults, the Los Cerrillos fault (Barrionuevo et al. 2019) and the Palauco fault (Manceda and Figueroa 1995; Giambiagi et al. 2009a), controlled the space creation within this depocenter. Immediately to the north, the **Malargüe depocenter** was controlled by the NNW-striking, WSW-dipping Malargüe normal fault, interpreted from seismic data (Silvestro and Kraemer 2005; Giambiagi et al. 2009a). Its synrift deposits (Tronquimalal Group; Stipanicic 1979) are clastic and volcaniclastic facies of humid alluvial fans and braided rivers interbedded with lacustrine black shales and sandstones (Spalletti 1997; Artabe et al. 1998; Buchanan et al. 2017) of Late Triassic age based on its paleofloristic content (Artabe et al. 1998). Here, the sag phase is represented by Bajocian nearshore marine deposits (Gulisano and Gutiérrez Pleimling 1994).

The **Valenciana depocenter** was filled with siliciclastic shallow nearshore to offshore marine deposits (Puesto Araya and Tres Esquinas Formations, respectively) with abundant Early Pliensbachian to Aalenian ammonites (Gulisano and Gutiérrez Pleimling 1994). Two tuffs from the lower portion of the Tres Esquinas Formation have been dated in 180.59 ± 0.43 and 181.42 ± 0.24 Ma (U/Pb in zircon; Mazzini et al. 2010). Fuentes et al. (2016) located its probable master fault through the location of the basement-synrift interface derived from well data. This master fault is a NNE-striking, WNW-dipping structure parallel to the Valenciana anticline axis, suggesting either the tectonic inversion of the previous normal fault or a strong control of this structure over the Cenozoic contractional structures.

The **Río Grande depocenter** was developed immediately to the west of the Valenciana depocenter. It was filled with ignimbrites and co-ignimbrite air-fall tuffs interbedded with lacustrine deposits (Lanés and Salani 1998; Lanés and Palma 1998) and overlain by Lower Toarcian–Lower Bajocian marine sediments (Westermann and Riccardi 1982; Damborenea 1987). According to Lanés and Salani (1998), these pyroclastic deposits are synchronic with the Cerro Negro Andesite dated with a crystallization zircon age of 223.3 ± 1.9 Ma (Naipauer et al. 2015). Based on palinspastically restored isopach maps, Manceda and Figueroa (1995) inferred a N to NNE-striking, eastward-facing listric master fault.

The **Los Blancos depocenter**, situated immediately to the south of the Rio del Cobre depocenter, is filled with continental and siliciclastic marine synrift deposits. Their basal portions correspond to sandstones and conglomerates deposited in a gravel braided fluvial system of Hettangian?—Sinemurian age (El Freno Formation; Lanés et al. 2013) in sharp or normal fault contact with basement rocks. Fluvial deposits are overlain by marine siltstone beds deposited in a shallow marine littoral environment with Late Sinemurian? to Early Toarcian ammonites, bivalves and brachiopods (Damborenea 1987) and black shales representing a quick relative sea-level rise during the Toarcian (Gulisano and Gutiérrez Pleimling 1994). This depocenter is inferred to be controlled by the west-dipping, NNE-striking, Los Blancos fault (Mescua et al. 2014; Fuentes et al. 2016).

The **Río Del Cobre depocenter** has been interpreted as controlled by a NNE-trending master fault (Mescua et al. 2014). Its infill is characterized by a succession of 800 m of turbiditic sandstones, limestones, and shales with *Posidonomya alpina* and *Harpocerates* sp. (Gerth 1925), indicating an Early to Middle Jurassic age. Slumps and flute marks suggest a western provenance, consistent with the interpretation of W-sourced proximal submarine fans of the equivalent Nacientes del Teno Formation, cropping out westwards in Chile (Davidson 1988). Based on the fossil content, a correlation with the Puesto Araya and Tres Esquinas Formations of the Atuel depocenter has been proposed (Gulisano and Gutiérrez Pleimling 1994).

4 Stratigraphic Setting for the Atuel Depocenter

The structural basement of the Neuquén Basin in the Atuel depocenter area mainly comprises Permian to Triassic volcanic and plutonic rocks of the Choiyoi Group and a minor amount of Paleozoic metasedimentary rocks of the Las Lagunitas Formation and Carboniferous plutons such as the Carrizalito Tonalite (Fig. 2; Volkheimer 1978; Llambías et al. 1993; Sruoga et al. 2005; Nullo et al. 2005; Tickyj et al. 2009). These units crop out on blocks limited by deep-rooted reverse faults, like in the southernmost sector of the Frontal Cordillera north of the Atuel depocenter, on the Dedos-Silla High, and in the Malargüe and Bardas Blancas anticlines to the south (Figs. 1 and 2; Manceda and Figueroa 1995; Dicarlo and Cristallini 2007; Giambiagi et al. 2012; Mescua et al. 2014). In the Malargüe anticline, zircons from an ignimbrite assigned to the Choiyoi Group yielded a weighted mean U-Pb LA-ICP-MS age of 244.5 ± 2.0 Ma (Fig. 3; Horton et al. 2016). Basement rocks have also been localized in the subsurface to the east of the Atuel depocenter, where boreholes cut the entire sedimentary pile reaching its base (Bechis et al. 2010; Fuentes et al. 2016).

The Late Triassic to Early Jurassic infill of the Atuel depocenter consists of siliciclastic marine and continental sedimentary rocks, included in the Arroyo Malo, El Freno, Puesto Araya, and Tres Esquinas Formations (Fig. 2; Reijeinstein 1967; Stipanicic 1969; Volkheimer 1978; Riccardi et al. 1991, 1997; Gulisano and Gutiérrez Pleimling 1994; Lanés 2005; Lanés et al. 2008; Spalletti et al. 2007). These units have been assigned to the lower Cuyo Group (Stipanicic 1969; Gulisano 1981; Legarreta and Gulisano 1989; Riccardi and Damborenea 1993), although part of this sequence is coetaneous with the Precuyo Cycle or Mesosequence (Gulisano 1981; Legarreta and Gulisano 1989). This stratigraphic section is exposed to the west of the Borbollón–La Manga lineament, a conspicuous NNW-striking structure that extends for more than 50 km along the northeastern sector of the Malargüe fold and thrust belt (Giambiagi et al. 2005a). Exhumation of these deposits was associated with an important denudation, probably favored by the inversion of the Mesozoic normal faults during the Andean shortening (Giambiagi et al. 2012). Conversely, eastwards

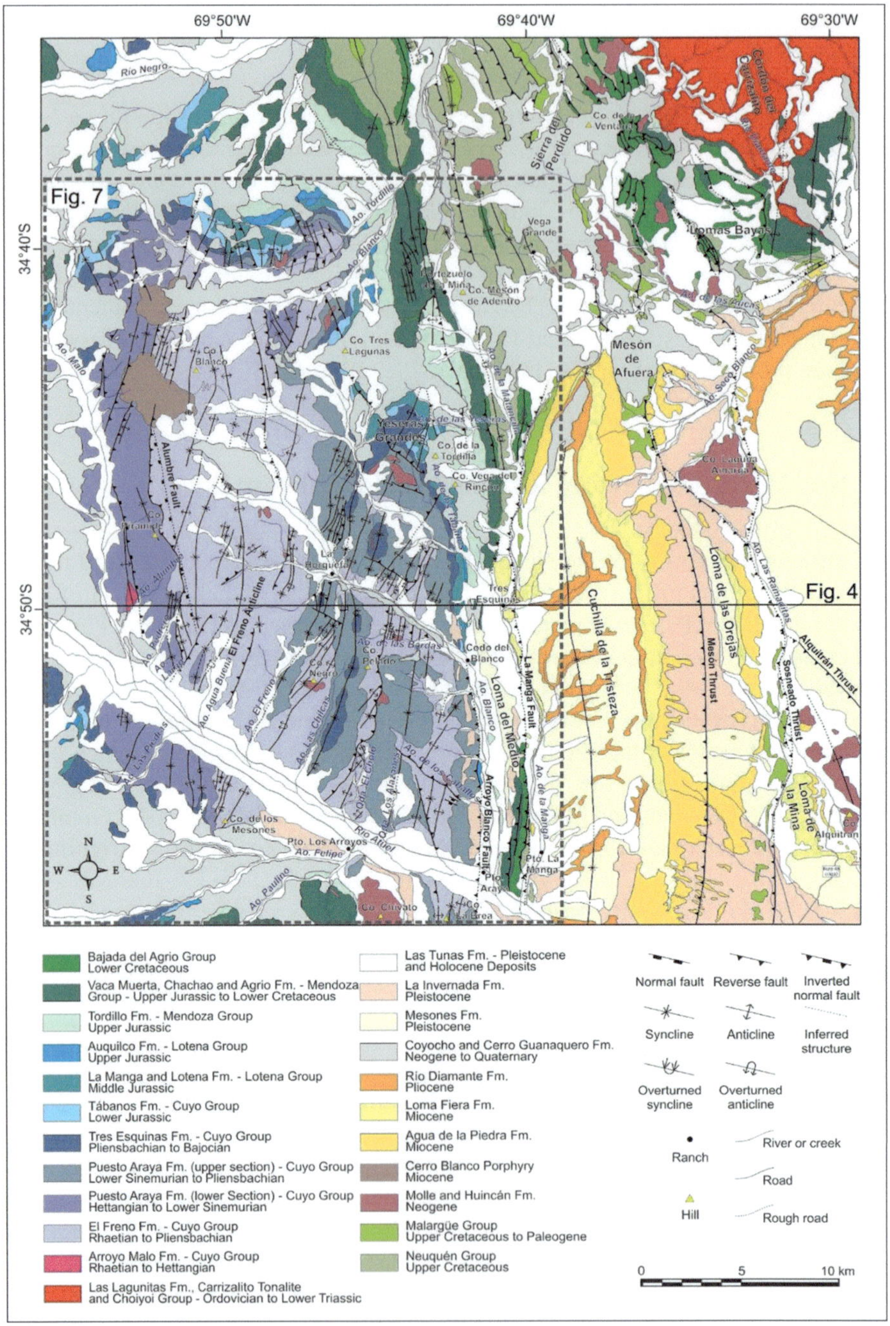

Fig. 2 Geologic map of the studied area, located in the northeastern sector of the Malargüe fold and thrust belt (modified from Giambiagi et al. 2008b; Bechis 2009; and references therein). See location map on Fig. 1

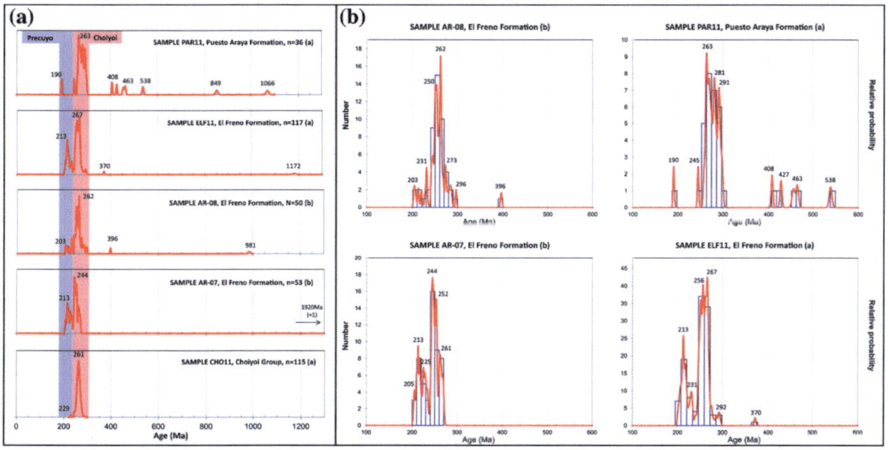

Fig. 3 **A** Probability density plots showing compiled U-Pb LA-ICP-MS geochronologic data, with shaded color bars highlighting major age populations and their probable source units. **B** Age histograms and probability density plots showing the main Phanerozoic peak ages for the El Freno and Puesto Araya Formations. Note that the samples from the older El Freno Formation contain principally 280–200 Ma grains, while the younger Puesto Araya Formation contains mostly 300–260 Ma grains, possibly reflecting progressive exhumation of older Choiyoi basement (Horton et al. 2016). This progressive exhumation is also supported by the higher abundance of older Paleozoic zircons in the Puesto Araya sample. Data obtained from (a) Horton et al. (2016), and (b) Naipauer et al. (2015)

of the Borbollón–La Manga lineament, all these mentioned units are absent at the base of the Mesozoic sedimentary pile, as borehole data evidence (Fig. 4; Bechis et al. 2010; Fuentes et al. 2016).

The Middle Jurassic to Paleogene infill of the Neuquén Basin in this sector consists of a thick succession of evaporitic, calcareous and clastic marine and continental

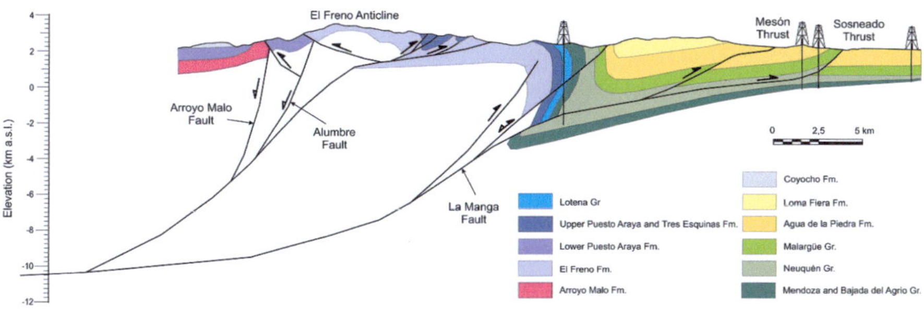

Fig. 4 Vertical cross section showing the present structural configuration of the fold and thrust belt in the Atuel depocenter area (modified from Giambiagi et al. 2012), where the main Mesozoic normal faults have been cut or inverted during the Andean shortening. Lateral thickness variations of the Late Triassic to Early Jurassic infill of the Atuel depocenter are also represented. See location on Fig. 2

sedimentary rocks, assigned to the upper Cuyo, Lotena, Mendoza, Bajada del Agrio, Neuquén, and Malargüe Groups (Fig. 2; Legarreta et al. 1993; Gulisano and Gutiérrez Pleimling 1994; Legarreta and Uliana 1996). The general thickness of this section increases toward the west, and proximal facies deposited at the basin margin crop out in the southern sector of the Frontal Cordillera, at the Cordón del Carrizalito range (Fig. 2).

Finally, the Cenozoic magmatic rocks are represented by Miocene to Quaternary andesitic to basaltic volcanic and subvolcanic rocks of the Molle, Huincán, Coyocho and Cerro Guanaquero Formations, and the Cerro Blanco Porphyry (Fig. 2; Volkheimer 1978; Baldauf 1997; Nullo et al. 2002; Sruoga et al. 2005; Giambiagi et al. 2008b). Thick clastic and volcaniclastic deposits of the Miocene to Holocene synorogenic infill of the Andean foreland basin crop out in the eastern sector, being represented by the Agua de la Piedra, Loma Fiera, Río Diamante, Mesones, La Invernada and Las Tunas Formations (Fig. 2; Yrigoyen 1993; Baldauf 1997; Combina and Nullo 2000; Giambiagi et al. 2008b; Horton et al. 2016).

4.1 Depocenter Infilling During the Late Triassic to Early Jurassic

The stratigraphic setting of the infill of the Atuel depocenter allows differentiating two contrasting areas interpreted as the Arroyo Malo half-graben to the west and the Río Blanco half-graben to the east (Fig. 5; Giambiagi et al. 2008a; Lanés 2005; Lanés et al. 2008).

In the western sector, in the Arroyo Malo half-graben, marine deposits of three coarsening- and thickening-upward cycles of mudstones, sandstones, and conglomerates assigned to the Arroyo Malo Formation and to the lower section of the Puesto Araya Formation, of Rhaetian to late Early Sinemurian age, crop out (Fig. 5; Riccardi et al. 1988, 1997; von Hillebrandt 1989; Riccardi and Iglesia Llanos 1999; Lanés, 2005; Lanés et al. 2008). These successions were interpreted as deposited in shallowing-upward, fluvio-dominated, transverse, normal-fault-controlled, slope-type fan deltas, with the uppermost section representing intermediate shelf to Gilbert-type fan deltas (Lanés 2005).

To the east, in the Rio Blanco half-graben, fining- to coarsening-upward and thinning-upward fluvial successions of lensoidal conglomerates, sandstones, and minor mudstones of the El Freno Formation appear (Fig. 5; Reijeinstein 1967; Artabe et al. 2005; Giambiagi et al. 2005a; Spalletti et al. 2007; Lanés et al. 2008). Interestingly, the geometry and connectivity of the fluvial channels changed along the time from wide ribbons at the base to isolated ribbons surrounded by fines in the middle section, to belts and eventually wider mobile-channel belts at the top (Lanés et al. 2008). These fluvial deposits are sharply overlain by well bedded, fining, and thinning-upward marine sandstones and shales that record a transgressive siliciclastic storm-dominated shelf, evolving from a wave-dominated estuary to a turbidite

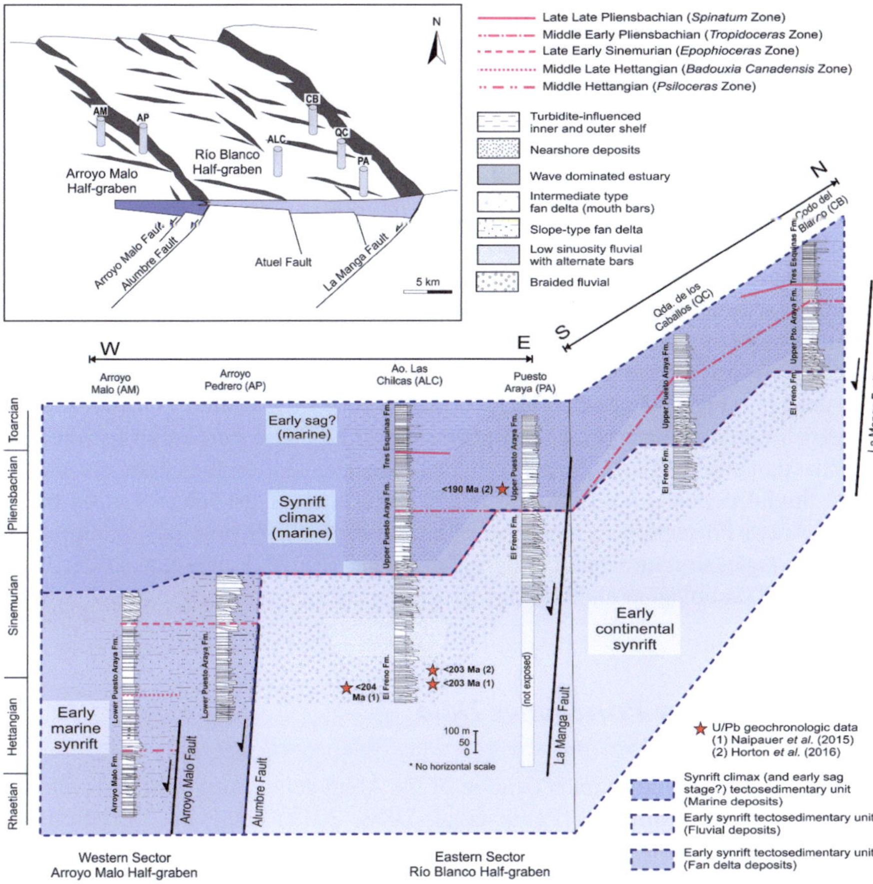

Fig. 5 Stratigraphic setting of the Atuel depocenter infill with biostratigraphic correlation of the vertical sections studied by Lanés (2005) and Lanés et al. (2008). Block diagram shows location of the vertical sections in relation with the main normal faults (modified from Giambiagi et al. 2008a)

outer shelf (Lanés 2005). The basal part of the marine deposits was assigned to the upper section of the Puesto Araya Formation (late Early Sinemurian to Toarcian), transitionally followed by the shales of the Tres Esquinas Formation (Late Pliensbachian–Bajocian), deposited in an anaerobic turbidite marine shelf (Fig. 5; Lanés 2005; Lanés et al. 2008).

The age and correlation of the marine units (Arroyo Malo, Puesto Araya, and Tres Esquinas Formations) were based on a detailed ammonite, bivalve, and brachiopod biostratigraphy (Fig. 5; Riccardi et al. 1991; Riccardi and Damborenea 1993; Lanés 2005; Lanés et al. 2008). Although the base of the Atuel depocenter infill is always below the present exposure level, biostratigraphic data from the outcropping deposits evidence a strong diachronicity for the start of marine sedimentation, varying from Late Triassic in the western sector, up to Early Pliensbachian at its southeastern sector (Fig. 5; Lanés 2005; Lanés et al. 2008). A recent U-Pb (LA-ICP-MS) analysis from detrital zircons of the Puesto Araya Formation in the southeastern sector found a youngest age data of 190.5 ± 3.0 Ma, constraining the maximum sedimentation age at the Early Pliensbachian and confirming previous biostratigraphic data (Fig. 3; Horton et al. 2016). Conversely, age determination of the El Freno Formation was more problematic due to the lack of accurate biostratigraphic markers, though its paleofloristic content yielded a general Early Jurassic age (Herbst 1968; Artabe et al. 2005; Spalletti et al. 2007). Detrital zircons recently dated by U-Pb (LA-ICP-MS) gave youngest populations between 213 and 203 Ma, representing a Late Triassic maximum sedimentation age for the unit (Fig. 3; Naipauer et al. 2015; Horton et al. 2016). Considering that the age of the overlying marine deposits of the upper section of the Puesto Araya Formation varies between the late Early Sinemurian and the Early Pliensbachian (Fig. 5; Riccardi et al. 1991; Lanés 2005), the available biostratigraphic and geochronologic data constrain the age of the El Freno Formation between the late Rhaetian and the Early Pliensbachian. These data support the lateral correlation of the fluvial deposits with the fan-deltaic sequences of the lower section of the Puesto Araya Formation in the western area, proposed by previous stratigraphic, sedimentologic, structural and petrographic studies (Fig. 5; Lanés et al. 2008; Tunik et al. 2008; Giambiagi et al. 2008a).

4.2 Sedimentary Provenance Data

Sandstone petrographic characterization of the Atuel depocenter infill units allowed identifying four petrofacies (Tunik et al. 2008, 2011) using the discrimination diagrams of Dickinson et al. (1983; Fig. 6).

The Petrofacies A characterizes the Arroyo Malo Formation and indicates provenance from a transitional arc (Fig. 6; Tunik et al. 2008). It has low quartz and feldspars content and abundant lithic fragments mainly of felsic and mafic volcanic rocks and subordinated plutonic, sedimentary, and pyroclastic rocks fragments. The low mineralogical maturity of this petrofacies suggests short transport from source areas.

The Petrofacies B includes El Freno Formation and the lower section of Puesto Araya Formation, which show similar petrographic characteristics and provenance from a volcanic arc to a dissected arc (Fig. 6; Tunik et al. 2008). This petrofacies is characterized by a remarkable increase in the quartz and feldspar proportion and a lower lithic fragment content of mainly felsic volcanic rocks. The petrographic data support coetaneous sedimentation of the fluvial deposits of the El Freno Formation to the east and the fan-deltaic deposits of the lower section of the Puesto Araya Formation to the west, as previous studies proposed (Lanés et al. 2008; Giambiagi et al. 2008a).

The Petrofacies C, assigned to the upper Puesto Araya Formation, indicates a dissected arc to a recycled orogen provenance (Fig. 6; Tunik et al. 2008). It shows higher quartz content due to a noticeable increase in polycrystalline quartz and similar abundance of monocrystalline quartz, and subordinated feldspar and felsic lithic

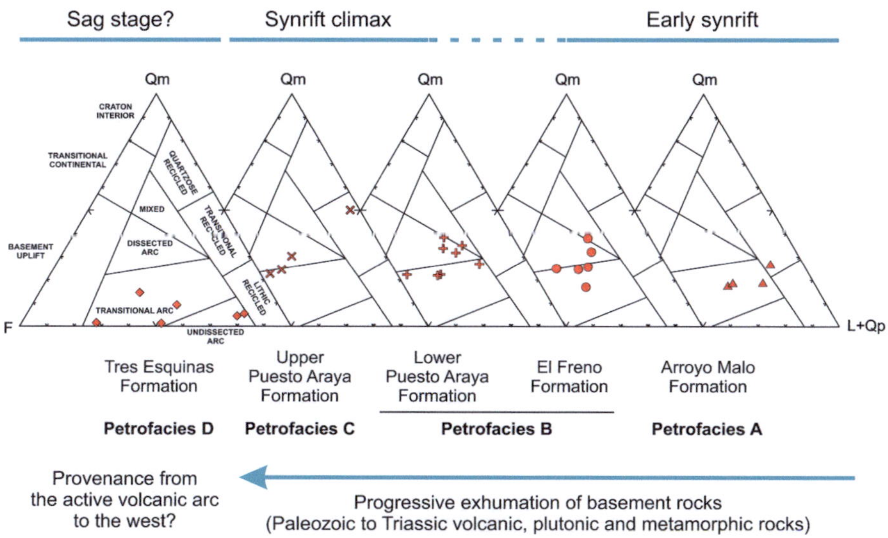

Fig. 6 Detrital sandstone mode for the four petrofacies recognized by Tunik et al. (2008, 2011), plotted in source discrimination diagrams (Dickinson et al. 1983). Qm: monocrystalline quartz, F: total feldspar, L + Qp: total lithics and polycrystalline quartz

fragments. Some of the quartz and feldspars grains show myrmekitic and graphic textures. Therefore, composition and textures point to felsic volcanic and plutonic sources.

Finally, the Petrofacies D is related to the Tres Esquinas Formation and equivalent units located to the west of the Atuel depocenter and revealed provenance from a transitional or undissected arc (Fig. 6; Tunik et al. 2011). This petrofacies composition is very different from the rest with very low quartz and variable feldspar content, with plagioclase more abundant than alkali feldspar. Lithic fragments show higher proportion of basic volcanic rocks respect to felsic textures, while plutonic and metamorphic clasts are uncommon. Volcanic shards found in the matrix and neovolcanic ignimbrites and devitrification lithic clasts indicate active volcanism during deposition of this petrofacies.

5 Structural Setting

In the Atuel depocenter, Mesozoic normal faults have been classified into four orders according to their dimensions and control over synrift subsidence (Bechis et al. 2010). The major normal faults have been inverted or cut by the Andean contractional structures or covered by younger deposits (Giambiagi et al. 2008b; Bechis et al. 2010; Fuentes et al. 2016). Therefore, three lines of evidence have been combined in order to identify the normal faults: (1) controls of the previous extensional structures over the Andean shortening; (2) angular and progressive unconformities within the synrift units, indicating syntectonic sedimentation; and (3) lateral facies and thickness variations of the synrift deposits (Lanés, 2002, 2005; Lanés et al. 2008; Giambiagi et al. 2005a, 2008a; Bechis et al. 2009, 2010). These first- and second-order normal faults are shown in Fig. 7 and described in Sect. 5.1.

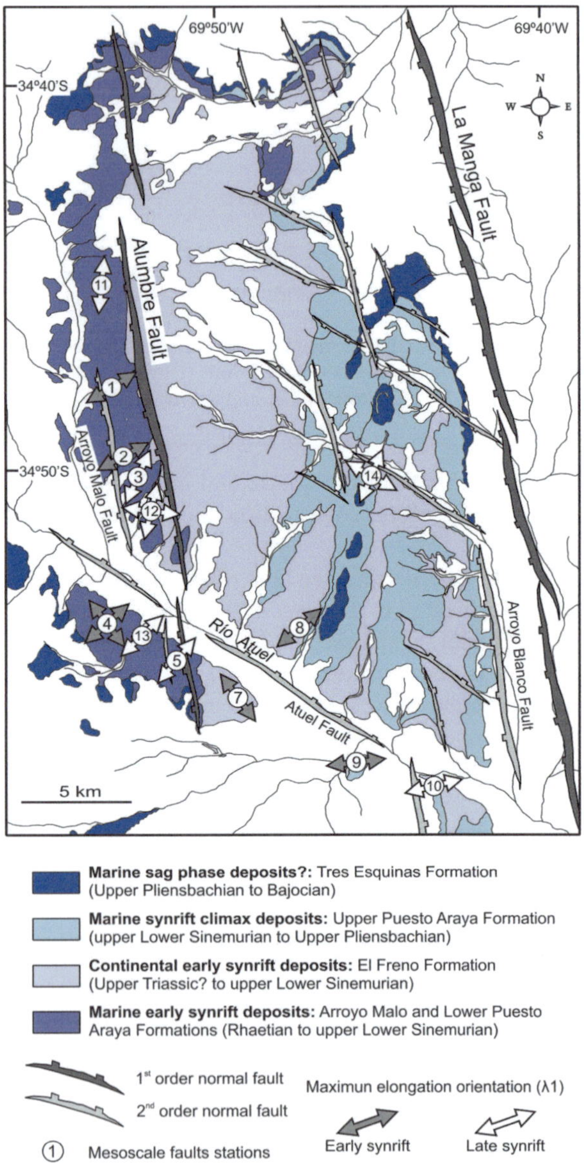

Fig. 7 Map showing the Late Triassic to Early Jurassic extensional architecture of the Atuel depocenter composed of first- and second-order normal faults interpreted by Lanés (2005), Lanés et al. (2008), Giambiagi et al. (2008a) and Bechis et al. (2010). Location of the mesoscale fault stations analyzed in Figs. 8 and 9 is also shown on the map, with double arrows indicating the orientation of the major axis of the obtained strain ellipsoids (λ1). Early synrift data correspond to faults affecting Upper Triassic to Hettangian deposits, while late synrift data were measured in Sinemurian to Pliensbachian deposits. See location of the map on Fig. 2

Fig. 8 **a** Syntectonic progressive unconconformities registered in the fan-deltaic synrift deposits of the hanging-wall of the Alumbre fault (from Giambiagi et al. 2008a). **b** Angular unconformity affecting the fluvial synrift deposits, in the hanging-wall of the La Manga fault (from Bechis et al. 2010). **c** Outcrop-scale faults registered in the southeastern sector of the Río Blanco half-graben (from Bechis et al. 2010). **d** Small-scale faults and related unconformities in the northern sector of the Arroyo Malo half-graben (from Bechis et al. 2010)

On the other hand, mesoscale faults with centimeter- to meter-scale displacements are generally well preserved, and more than 300 outcrop-scale normal faults affecting the synrift strata were recorded (Figs. 8, 9, and 10; Giambiagi et al. 2008a; Bechis et al. 2009, 2010). Field measurements of fault-slip data (fault orientation and direction and sense of slip from kinematic indicators) have been taken on these third and fourth scale faults, and kinematic analysis of these data from previous publications is included in Sect. 5.2 (Kinematic analysis, stations 1–12). Besides, we have also included new data to the kinematic analysis (stations 13 and 14) and we have carried out a new dynamic analysis of the data set, which is presented in Sect. 5.3.

5.1 Geometric Analysis

Two first-order, NNW-striking faults, the La Manga and Alumbre faults, controlled the orientation of the Atuel depocenter, as well as the distribution of sedimentary environments and drainage patterns (Figs. 5 and 7; Lanés 2002, 2005; Lanés et al. 2008; Giambiagi et al. 2008a). The La Manga fault was interpreted as the eastern

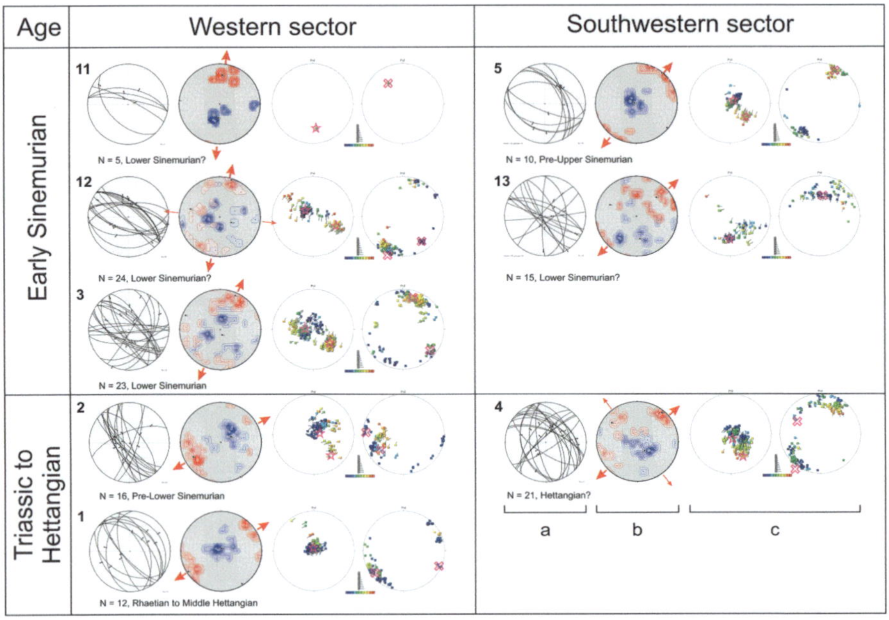

Fig. 9 Kinematic and dynamic analysis of mesoscale faults from the western sector of the Atuel depocenter. Location of measurement stations is shown on Fig. 7. **a** Major circles representing the lower-hemisphere projection of fault planes, with arrows showing the hanging-wall sense of movement. **b** Kinematic diagrams obtained with the FaultKinWin® software (Marrett and Allmendinger 1990; Allmendinger et al. 2012). Distribution of principal incremental shortening axes (*P* axes, blue dots) and extension axes (*T* axes, red dots) for each fault. Black squares represent the obtained axes of the strain ellipsoid ($\lambda 1$, $\lambda 2$ and $\lambda 3$) for every group of faults. Both *P*- and *T*-axis 1% area contours are also shown. Big and small red arrows show the main and secondary extension directions. **c** Diagrams showing the principal orientations of stress axes detected by the multiple inverse method (Yamaji 2000, 2003). The stereograms use lower-hemisphere, equal-area projection. Clusters of symbols with similar colors and similar attitudes on the stereograms represent significant stresses for each set of fault-slip data. Left and right diagrams show results for the maximum ($\sigma 1$) and minimum ($\sigma 3$) stress axes, respectively. Pink stars ($\sigma 1$) and crosses ($\sigma 3$) indicate clusters of the obtained stress axes. Colors indicate the obtained value for the stress ratio $\phi = (\sigma 2 - \sigma 3)/(\sigma 1 - \sigma 3)$. Blue colors for ϕ close to 0 and red for ϕ close to 1. Kinematic data taken from Bechis et al. (2010), except station 13. The number of field data (*N*) and the age of the strata where the faults were observed are also indicated in the figure

border of the depocenter, located along the Borbollón–La Manga lineament, and the Río Blanco half-graben was developed on its hanging-wall. To the east of La Manga fault, there is no record of Upper Triassic to Lower Jurassic synrift deposits neither in outcrops nor in wells, while more than 1500 m of continental synrift strata are recorded on its hanging-wall to the west (Fig. 4; Bechis et al. 2010). Angular unconformities and outcrop-scale normal faults registered on its hanging-wall suggest that this structure remained active until at least Early Pliensbachian times (Fig. 8; Bechis et al. 2010).

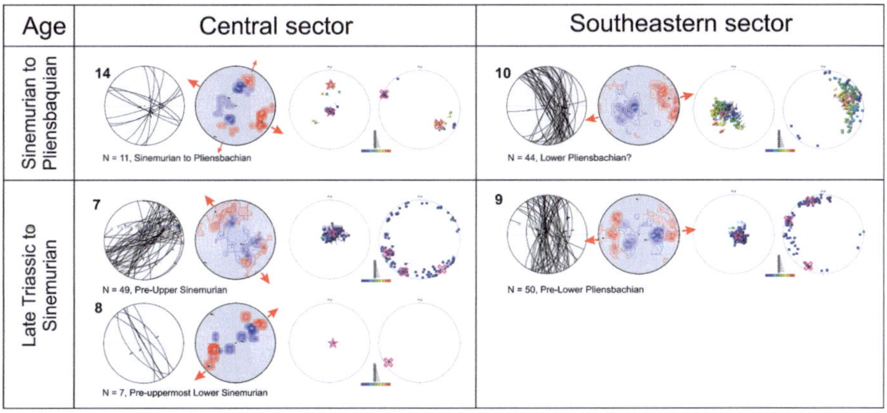

Fig. 10 Kinematic and dynamic analysis of mesoscale faults from the central and eastern sectors of the Atuel depocenter. Location of measurement stations is shown on Fig. 7. Diagram symbology is explained in Fig. 9 caption. Kinematic data taken from Bechis et al. (2010), except station 14

The Alumbre fault was proposed by Lanés et al. (2008) and Giambiagi et al. (2008a) as a WSW-dipping normal fault controlling the coastal line from middle Hettangian to Early Sinemurian times (Figs. 5 and 7). To the west of this fault, the fan-delta deposits of the synrift infill of the Arroyo Alumbre half-graben show a marked wedge geometry, with strata thickening toward the fault and syntectonic angular and progressive unconformities (Fig. 8; Giambiagi et al. 2008a; Bechis et al. 2010). A decrease in the accommodation in the uppermost fan-deltaic deposits on its hanging-wall suggests deactivation of the Alumbre fault by the late Early Sinemurian (Lanés 2005; Lanés et al. 2008).

Second-order faults show a bimodal distribution of NNW and WNW strikes (Fig. 7; Giambiagi et al. 2008a; Bechis et al. 2009, 2010). These structures are associated with synsedimentary deformation, such as abundant slumps within the slope-type fan-deltaic deposits (Lanés 2002), and angular and progressive unconformities within the continental and marine deposits (Bechis et al. 2010). The Arroyo Malo fault, synthetic to the Alumbre fault, was only active during the early synrift stage from the Late Triassic until the middle Hettangian, as sedimentary reactivations observed in the Arroyo Malo section evidence (Figs. 5 and 7; Lanés 2005; Lanés et al. 2008). The Atuel fault is a WNW-striking, inferred structure, which controlled variations in the depth of the fan-deltaic environment in the Arroyo Malo half-graben, with relatively higher depths registered on its hanging-wall to the north (Giambiagi et al. 2008a).

Outcrop-scale third- to fourth-order faults show variable orientation with most faults strikes ranging between NNW and NW, and secondary NNE, NE, and ENE fault populations (Giambiagi et al. 2005a; Bechis et al. 2010).

5.2 Kinematic Analysis

This section presents results from the kinematic analysis carried out by Bechis et al. (2010), together with new data. This analysis consists in obtaining the principal axes of the strain ellipsoid for outcrop-scale faults. Shortening and extension axes and moment tensor solutions were obtained for the different fault populations using the FaultKinWin® software (Marrett and Allmendinger 1990; Allmendinger et al. 2012). The results are shown in Figs. 7, 9 and 10. Data were collected along four sectors within the Atuel depocenter: western (stations 1, 2, 3, 11, and 12), southwestern (stations 4, 5, and 13), southeastern (stations 9 and 10), and central (stations 7, 8 and 14). All data were unfolded to their original position previous to the Andean contractional deformation. The obtained intermediate and major principal axes of the strain ellipsoid ($\lambda 2$ and $\lambda 1$) generally show a subhorizontal position, indicating an extensional tectonic regime, with a subvertical minimum principal axis ($\lambda 3$).

All along and across the studied depocenter, kinematic data indicate a main extensional direction ($\lambda 1$) with a NE to NNE orientation and a secondary extensional direction with a NW to WNW orientation (Figs. 9 and 10). Kinematic data from the western sector indicate a NE extensional direction for the Triassic to Hettangian period and a NNE extensional direction for the Sinemurian period (Fig. 9). Data from the southwestern and central sectors show two clusters of extensional axes, one indicating a principal NE extensional direction and the other one a NW extensional direction (Figs. 9 and 10). A large and representative dataset from station 7, located in the south-central sector, suggests a NW extensional direction. Data from the southeastern sector, close to the La Manga normal fault (stations 9 and 10), show a clear ENE extensional direction, perpendicular to this master fault (Fig. 10).

5.3 Dynamic Analysis

The dynamic analysis consists on paleostress computation from the fault-slip data to obtain the reduced stress tensor, which describes the shape of the stress ellipsoid (orientation of the principal stress axes and the ratio of the principal stress differences $\phi = (\sigma 2 - \sigma 3)/(\sigma 1 - \sigma 3)$). For this analysis, we used the Multiple Inverse Method proposed by Yamaji (2000), which allows the identification of different stress tensors from an heterogeneous fault-slip dataset. We obtained 25 well-constrained reduced paleostress tensor solutions for the Triassic-Hettangian, Sinemurian, and Pliensbachian periods. The results are presented as paired stereoplots showing the orientation of $\sigma 1$ (stars) and $\sigma 3$ (crosses), respectively (Figs. 9 and 10). In all tensors except tensor 13, $\sigma 1$ has a subvertical orientation with a distinctive cluster, and $\sigma 3$ and $\sigma 2$ are subhorizontal, indicating an extensional stress field, in agreement with the orientation of the kinematic axes. Some stations show an azimuthal dispersion of $\sigma 3$ (stations 2, 7, 9, 10, and 13), and others show two or more clusters of $\sigma 3$ (stations 1, 3, and 14).

Most of the tensors have low values of stress ratio ϕ (blue and violet colors in Figs. 9 and 10), suggesting that $\sigma 3$ and $\sigma 2$ are similar in magnitude. Despite that, other tensors (stations 3, 4, 5, 12) show high values of ϕ (orange and red colors) indicating magnitudes of $\sigma 2$ closer to $\sigma 1$ than to $\sigma 3$, which can be interpreted as an extensional/strike-slip regime. The high consistency and wide distribution of the extensional stress states with NNE- to ENE- and NW-oriented $\sigma 3$ and general low values of ϕ indicate that these stress configurations are related to a stress field controlling deformation locally inside the Atuel depocenter.

6 Discussion

6.1 Tectonic Implications for the Kinematic and Dynamic Analyses

Taking into account the orientation and shape of the strain and stress tensors, their distribution and timing, and the relationship between strain and stress axes, we constructed a dynamic model to explain the generation/reactivation of structures, their kinematics, and their evolution through time. Variations between the principal strain and stress axes may indicate that assumptions made for the paleostress analysis are not necessarily satisfied (Kaven et al. 2011). The critical analysis of these variations may shed light into additional assumptions that are required to interpret the stress regimes. Some of these assumptions are (i) mechanically anisotropic material (Jaeger et al. 2007; Hu and Angelier 2004), (ii) preexisting plane of weakness (Marrett and Allmendinger 1990; Nieto-Samaniego 1999; Martinez Días 2002), or (iii) rotations of blocks bounded by major faults (Twiss and Unruh 1998). Clusters of $\sigma 3$ obtained from the dynamic analysis correlate well with the extensional axes obtained from the kinematic analysis (stations 1, 3, 4, 5, 8, 12, and 14), while those stress tensors with a girdle distribution of $\sigma 3$ (stations 2, 7, 9, 10, and 13) differ from the results of the kinematic analysis (Figs. 9 and 10).

During the Late Triassic–late Hettangian period, extensional directions were slightly oblique to the Alumbre fault, in the Arroyo Malo half-graben (stations 1, 2, and 4; Figs. 7, 9 and 11) and perpendicular to the La Manga fault, in the eastern border of the Río Blanco half-graben (station 9; Figs. 7, 10 and 11). The stress ratio values of these tensors indicate close values of $\sigma 3$ and $\sigma 2$. Tensor 4 shows two-orthogonal clusters of $\sigma 3$ (Figs. 9 and 11), which together with low values of stress ratios suggest a $\sigma 2/\sigma 3$ permutation. Inside the Río Blanco half-graben, the kinematic analysis shows that extension was bimodal, with NW and NE directions (stations 7 and 8; Figs. 10 and 11). The NW direction was previously interpreted as probably related to local extension in the hinge zone of Andean anticlines (Bechis et al. 2010). However, the paleostress analysis suggests that the bimodal pattern responds to a local permutation of $\sigma 2$ and $\sigma 3$ during the Mesozoic synrift, in agreement with the values obtained from the stations close to the Alumbre and La Manga faults. We

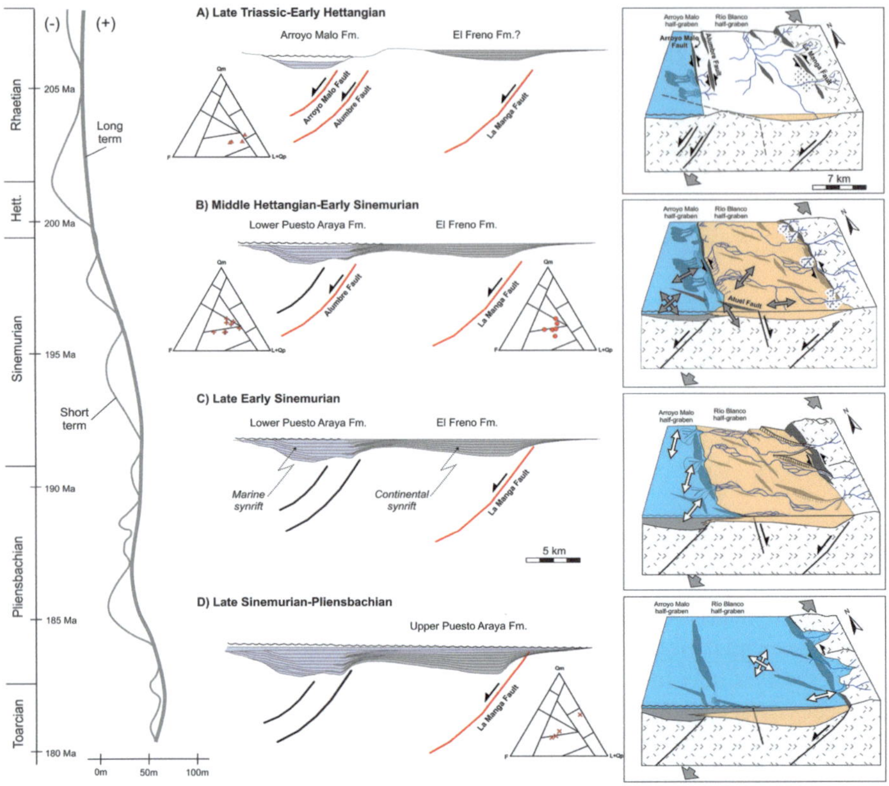

Fig. 11 2D and 3D diagrams showing the main stages of the interpreted tectono-stratigraphic evolution of the Atuel depocenter during the synrift stage (Late Triassic–Pliensbachian). Red colors highlight faults that were active at each stage. Thick gray arrows indicate the interpreted regional direction of extension, while double gray (early synrift) and white (late synrift) arrows show local variations in the extension inside the depocenter. Simplified diagrams showing petrographic provenance data for each unit were also included. The long- and short-term variations of the global eustatic level proposed by Haq et al. (1987) are also shown

interpret that, during the first stage of extension, deformation was focussed on pre-existing NNW-striking faults, such as the La Manga and Alumbre faults. Extensional movement along these structures might have perturbated the local stress field, with magnitudes of $\sigma 2$ close to $\sigma 3$, and local $\sigma 2/\sigma 3$ stress permutations.

During the Sinemurian to Pliensbachian period, the Arroyo Malo half-graben experienced a pure extension with NNE-oriented $\sigma 3$ (stations 3, 11, and 12; Figs. 7 and 9), in concomitance with a progressive deactivation of the Alumbre fault (Fig. 11). This extensional direction is interpreted as reflecting the regional stress regime, in agreement with previous proposals for the regional extension based on kinematic data (Bechis et al. 2010, 2014). On the contrary, in the Río Blanco half-graben, extensional directions are either NW- and NNE-oriented in the central sector (station 14; Figs. 7, 10 and 11) or ENE-oriented near La Manga fault (station 10; Figs. 7, 10 and 11). This is interpreted here as reflecting a strain partitioning between a NNE extension, subparallel to the regional extension direction, and an ENE extension orthogonal to the NNW-striking preexisting La Manga fault.

Our results are consistent with previous kinematic analysis where the Atuel depocenter was interpreted to have been formed by oblique reactivation of NNW-striking anisotropies under a regional NNE extension (Giambiagi et al. 2008a; Bechis et al. 2009, 2010, 2014). The high obliquity between the NNW orientation of the rift system, controlled by the first-order normal faults, and the NNE regional extension would result in a wrench dominated transtension (Tikoff and Teyssier 1994; Teyssier et al. 1995; De Paola et al. 2005; Bechis et al. 2014). This type of transtensional strain produces a complex structural pattern of normal, oblique, and strike-slip faults. Deformation could be either distributed or partitioned, and switching of the main axis of the strain ellipsoid may occur along the evolution of the system, due to finite strain localized along major faults. The results of our kinematic and dynamic analyses show a complex distribution and evolution of the strain and stress inside the Atuel depocenter, which is consistent with predictions from analytical and experimental models of highly oblique rift systems (Tikoff and Teyssier 1994; Bechis et al. 2014). The obtained results suggesting bimodal and/or radial horizontal extension, together with reactivation of NNW-striking anisotropies, also explain the variation in strike of the master faults of the nearby depocenters, between the NNW-trending Atuel, Malargüe and Palauco, and the NNE- to NE-trending Río del Cobre and Valenciana depocenters (Fig. 1).

We infer an important strain localization in the Alumbre and La Manga faults that was responsible for several particular characteristics of the Atuel and Malargüe depocenters, such as: (i) the marked NNW orientation of the depocenters, controlled by the La Manga and Malargüe faults, respectively, (ii) the early marine transgression localized in the hanging-wall of the Alumbre fault, with the only Triassic to Hettangian marine deposits of the entire Neuquén Basin, and (iii) the particular distribution of depositional environments until the end of the Early Sinemurian, as the Alumbre fault limited the marine transgression to the western sector of the depocenter, while the La Manga fault controlled the fluvial sedimentation in the eastern sector.

6.2 Tectono-Stratigraphic Evolution

We integrated the available structural, sedimentologic, petrographic, and geochronologic data into a tectono-stratigraphic model for the evolution of the Atuel depocenter (Fig. 11), whose extensional phase prevailed at least from Rhaetian to Pliensbachian times. A complex interplay between tectonic subsidence related to the activity of the major normal faults and eustatic sea-level changes controlled the evolution of this sector of the Neuquén Basin during the synrift stage.

We interpret the Rhaetian to early Hettangian sediments as deposited during the early synrift phase of the Atuel depocenter, when a paleogeography dominated by early formed fault segments controlled the marked variations in sedimentary fill

(Stage A; Fig. 11). Provenance analysis of the fan-deltaic deposits indicates short transport from local source areas (Tunik et al. 2008), suggesting that the Arroyo Malo and Río Blanco half-grabens were isolated at that time.

During the middle Hettangian to Early Sinemurian (Stage B; Fig. 11), some former fault segments became inactive (i.e., Arroyo Malo fault; Lanés et al. 2008), while other faults gained length and displacement, probably due to the connection of different fault segments (i.e., Alumbre and La Manga faults). The Alumbre fault maintained a nearly fixed position of the coastal line, restricting the marine sedimentation to the western sector of the depocenter. While a braided fluvial system characterizes the synrift infill in the hanging-wall of the La Manga fault during this stage, coetaneous deposits to the west of the Alumbre fault were deposited in a slope-type fan-deltaic environment (Lanés 2002, 2005; Lanés et al. 2008; Tunik et al. 2008).

During the late Early Sinemurian (Stage C; Fig. 11), an eustatic short-term lowstand (Haq et al. 1987) led to fluvial incision on the shelf and fan-deltaic conglomerate deposition in the Arroyo Malo half-graben (Lanés et al. 2008). A subsequent slow sea-level rise increased the accommodation on the hanging-wall of the Alumbre fault, allowing the fan-deltaic mouth bars to prograde (Lanés 2005; Lanés et al. 2008). The restricted accommodation needed for the progradation of the intermediate fan-delta mouth bars suggests deactivation of the Alumbre fault by this stage.

Thereafter, a faster sea-level rise and an important increase of accommodation allowed estuarine deposition in the Río Blanco half-graben (Lanés 2005; Lanés et al. 2008). The decreasing accommodation observed by the end of the Early Sinemurian in the Arroyo Malo half-graben was previously related to the start of the sag phase (Lanés 2005; Lanés et al. 2008). However, the subsequent creation of accommodation in the Río Blanco half-graben, together with the observed outcrop-scale faults affecting Pliensbachian sediments in the southeastern sector of the depocenter (station 10; Fig. 10), suggest that the extensional deformation continued up to this time. We interpret the Late Sinemurian–Pliensbachian (Stage D; Fig. 11) as a period of rift climax associated with important movement along the La Manga master fault (through-going fault zone stage; Gawthorpe and Leeder 2000). During this climax phase, the marine coastal line was controlled by the La Manga and Valenciana master faults.

Provenance data suggest an important basin reorganization by the Toarcian (Tunik et al. 2008, 2011). Petrographic and geochronologic data from Arroyo Malo, El Freno, and Puesto Araya Formations suggest progressive exhumation of the Permo-Triassic magmatic belt during the Hettangian to Pliensbachian period (Figs. 3 and 6; Tunik et al. 2008, 2011; Naipauer et al. 2015; Horton et al. 2016), probably due to normal fault activity. On the contrary, petrographic data of the latest Pliensbachian to Bajocian Tres Esquinas Formation indicate a major change in the source area, as is evidenced by the low quartz content, while a contemporaneous volcanic source, probably located to the west of the basin, was incorporated. This change could be related to the deactivation of the La Manga fault and the initiation of the sag stage in the Atuel depocenter.

7 Concluding Remarks

We compiled, analyzed, and integrated a large set of the previous sedimentologic, petrographic, geochronologic, and structural data from the Atuel depocenter. Geometrical, kinematic and dynamic analysis of structural data allowed us to construct a model explaining the generation and reactivation of structures, their kinematics, and their evolution through time. Reactivation of preexisting NNW-striking anisotropies under a regional NNE extension resulted in an oblique rift setting, which generated a bimodal distribution of NNW- and WNW-striking major normal faults. Concentration of deformation along reactivated NNW-striking faults generated a local stress field superposed to a regional NNE regional extension leading to an in situ stress field with values of $\sigma 2$ close to $\sigma 3$ and local $\sigma 2/\sigma 3$ stress permutations. Contemporaneous NNE- and ENE-oriented extensional directions are probably related to strain partitioning between a NNE extension, subparallel to the regional extension direction, and an ENE extension orthogonal to the NNW-striking preexisting major faults.

Sedimentologic and petrographic data revealed a complex evolution with strong lateral variations of the depositional environments during the synrift stage, mainly controlled by tectonic subsidence related to the activity of the major normal faults and eustatic sea-level changes. The NNW-striking La Manga and Alumbre master faults controlled most of the basin subsidence, the distribution of the sedimentary environments, and the drainage patterns. The Arroyo Malo half-graben developed on the hanging-wall of the Alumbre fault, in the western sector of the depocenter, and beared an early marine synrift infill consisting of fan-deltaic deposits. Eastwards, the La Manga fault controlled subsidence in the Rio Blanco half-graben, whose synrift infill consists of fluvial deposits followed by marine beds deposited in a transgressive siliciclastic storm-dominated shelf.

We identified several stages for the evolution of the Atuel depocenter during the synrift phase, which took place from Rhaetian to Pliensbachian times. These stages were controlled by processes of initiation, propagation, growth, linkage, and deactivation of new and reactivated faults along the depocenter evolution, in combination with eustatic sea-level changes. Provenance data suggest an important basin reorganization by the Toarcian, which could be related to initiation of the sag stage in this depocenter.

Acknowledgements This research was funded by Agencia Nacional de Promoción Científica y Tecnológica (PICT 07-10942, PICT 38295, PICT-2015-1181), Consejo Nacional de Investigaciones Científicas y Técnicas (CONICET PIP 5843), and Universidad de Buenos Aires (UBACYT 855). We wish to give special thanks to Alejandro Celli, Gabriela Da Poian, Victor García, Diego Iaffa, Diego Kietzmann, Darío Orts, Sergio Orts, Marilin Peñalva, Carla Terrizzano, and Daniel Yagupsky for their invaluable help in the field and discussions. We thank Drs. Susana Damborenea, Miguel Manceñido, and Alberto Riccardi, for their comments about the biostratigraphic data. The subsurface information was kindly facilitated by Julián Fantín, Gonzalo Zamora Valcarce, Roberto Varade, and Tomás Zapata, from Repsol-YPF. We also thank Ernesto Cristallini for many fruitful discussions about the topic.

References

Allmendinger RW, Cardozo N, Fisher D (2012) Structural geology algorithms: vectors and tensors in structural geology. Cambridge University Press, Cambridge

Alvarez P, Ramos VA (1999) The Mercedario rift system in the principal Cordillera of Argentina and Chile (32° SL). J S Am Earth Sci 12:17–31

Artabe AE, Morel EM, Spalletti L et al (1998) Paleoambientes sedimentarios y paleoflora asociada en el Triásico tardío de Malargüe, Mendoza. Rev Asoc Geol Argent 53:526–584

Artabe AE, Ganuza DG, Spalletti LA et al (2005) Revisión de la paleoflora del cerro La Brea (Jurásico Temprano), provincia de Mendoza, Argentina. Ameghiniana 42:429–442

Astini RA, Benedetto JL, Vaccari NE (1995) The early Paleozoic evolution of the Argentine Precordillera as a Laurentian rifted, drifted, and collided terrane: A geodynamic model. Geol Soc Am Bull 107:253–273

Azcuy CL, Caminos R (1987) Diastrofismo. In: Archangelsky S (ed) El Sistema Carbonífero en la República Argentina. Academia Nacional de Ciencias, Córdoba, pp 239–251

Baldauf P (1997) Timing of the uplift of the Cordillera Principal, Mendoza Province, Argentina. Master thesis, George Washington University

Balgord EA, Carrapa B (2016) Basin evolution of Upper Cretaceous-Lower Cenozoic strata in the Malargüe fold and thrust belt: northern Neuquén Basin, Argentina. Basin Res 28:183–206

Barrionuevo M, Mescua JF, Giambiagi L et al (2019) Miocene deformation in the orogenic front of the Malargüe fold-and-thrust belt (35°30′–36° S): controls on the migration of magmatic and hydrocarbon fluids. Tectonophysics 766:480–499

Bechis F (2009) Deformación transtensiva de la cuenca Neuquina: análisis a partir de ejemplos de campo y modelos análogos. Ph.D. thesis, Universidad de Buenos Aires

Bechis F, Giambiagi LB, Lanés S et al (2009) Evidencias de extensión oblicua en los depósitos de sinrift del sector norte de la cuenca Neuquina. Rev Asoc Geol Argent 65:293–310

Bechis F, Giambiagi L, García V et al (2010) Kinematic analysis of a transtensional fault system: the Atuel depocenter of the Neuquén basin, southern Central Andes, Argentina. J Struct Geol 32:886–899

Bechis F, Cristallini E, Giambiagi L et al (2014) Transtensional tectonics induced by oblique reactivation of previous lithospheric anisotropies during the Late Triassic to Early Jurassic rifting in the Neuquén basin: insights from analog models. J Geodyn 79:1–17

Buchanan AS, Kietzmann DA, Palma RM (2017) Evolución paleoambiental de la Formación Remoredo (Jurásico inferior) en el depocentro Malargüe, Cuenca Neuquina Surmendocina. Rev Asoc Geol Argent 74:163–178

Carbone O, Franzese J, Limeres M et al (2011) El Ciclo Precuyano (Triásico Tardío-Jurásico Temprano) en la Cuenca Neuquina. In: Leanza HA, Arregui C, Carbone O et al (eds) Geología y Recursos Naturales de la Provincia del Neuquén. Asociación Geológica Argentina, Buenos Aires, pp 63–75

Charrier R (1979) El Triásico en Chile y regiones adyacentes de Argentina: Una reconstrucción paleogeográfica y paleoclimática. Departamento de Geología, Universidad de Chile, Santiago, Comunicaciones 26:1–47

Charrier R, Pinto L, Rodríguez MP (2007) Tectonostratigraphic evolution of the Andean Orogen in Chile. In: Moreno T, Gibbons W (eds) The geology of Chile. The Geological Society, London, pp 21–114

Cobbold PR, Rossello EA (2003) Aptian to recent compressional deformation, foothills of the Neuquén basin Argentina. Mar Petrol Geol 20:429–443

Combina AM, Nullo F (2000) La Formación Loma Fiera (Mioceno superior) y su relación con el volcanismo y el tectonismo neógeno, Mendoza. Rev Asoc Geol Argent 55:201–210

Cristallini EO, Tomezzoli RN, Pando G (2009) Controles precuyanos en la estructura de la cuenca Neuquina. Rev Asoc Geol Argent 65:248–264

Damborenea SE (1987) Early Jurassic bivalvia of Argentina. Part 1: Stratigraphical introduction and Superfamilies Nuculanacea, Arcacea, Mytilacea and Pinnacea. Palaeontogr A 199:23–111

Davidson J (1988) El Jurásico y Cretácico inferior en las nacientes del río Teno (Chile): una revisión. In: Abstracts of the 5° Congreso Geológico Chileno, Santiango de Chile, 8–12 Aug 1988
De Paola N, Holdsworth RE, McCaffrey KJ et al (2005) Partitioned transtension: an alternative to basin inversion models. J Struct Geol 27:607–625
Dicarlo DJ, Cristallini E (2007) Estructura de la margen norte del Río Grande, Bardas Blancas, provincia de Mendoza. Rev Asoc Geol Argentina 62:187–199
Dickinson WR, Beard LS, Brakenridge GR et al (1983) Provenance of North American Phanerozoic sandstones in relation to tectonic setting. Geol Soc Am Bull 94:222–235
Espinoza M, Montecino D, Oliveros V et al (2018) The synrift phase of the early Domeyko Basin (Triassic, northern Chile): sedimentary, volcanic and tectonic interplay in the evolution of an ancient subduction-related rift basin. Basin Res. https://doi.org/10.1111/bre.12305
Fennell LM, Folguera A, Naipauer M et al (2017) Cretaceous deformation of the southern Central Andes: synorogenic growth strata in the Neuquén Group (35° 30′–37° S). Basin Res 29:51–72
Franzese JR, Spalletti LA (2001) Late Triassic–early Jurassic continental extension in southwestern Gondwana: tectonic segmentation and pre-break-up rifting. J S Am Earth Sci 14:257–270
Fuentes F, Horton BK, Starck D et al (2016) Structure and tectonic evolution of hybrid thick-and thin-skinned systems in the Malargüe fold–thrust belt, Neuquén basin, Argentina. Geol Mag 153:1066–1084
García Morabito E, Ramos VA (2012) Andean evolution of the Aluminé fold and thrust belt, Northern Patagonian Andes (38° 30′–40°30′ S). J S Am Earth Sci 38:13–30
Gawthorpe RL, Leeder MR (2000) Tectono-sedimentary evolution of active extensional basins. Basin Res 12:195–218
Gerth E (1925) Estratigrafía y distribución de los sedimentos mesozoicos en los Andes Argentinos. Academia Nacional de Ciencias, vol 9, Cordoba, pp 11–55
Giambiagi LB, Martínez A (2008) Permo-Triassic oblique extension in the Potrerillos-Uspallata area, western Argentina. J S Am Earth Sci 26:252–260
Giambiagi LB, Alvarez P, Godoy E et al (2003a) The control of pre-existing extensional structures on the evolution of the southern sector of the Aconcagua fold and thrust belt, southern Andes. Tectonophysics 369:1–19
Giambiagi LB, Ramos VA, Godoy E et al (2003b) Cenozoic deformation and tectonic style of the Andes, between 33° and 34°. South Latitude. Tectonics 22:1041
Giambiagi LB, Suriano J, Mescua J (2005a) Extensión multiepisódica durante el Jurásico Temprano en el depocentro Atuel de la cuenca Neuquina. Rev Asoc Geol Argent 60:524–534
Giambiagi LB, Álvarez PP, Bechis F et al (2005b) Influencia de las estructuras de rift triásicas – jurásicas sobre el estilo de deformación en las fajas plegadas y corridas Aconcagua y Malargüe. Rev Asoc Geol Argent 60:661–671
Giambiagi LB, Bechis F, Lanés S et al (2008a) Formación y evolución triásico-jurásica del depocentro Atuel, cuenca Neuquina, provincia de Mendoza, Argentina. Rev Asoc Geol Argent 63:520–533
Giambiagi LB, Bechis F, García VH et al (2008b) Temporal and spatial relationships of thick- and thin-skinned deformation in the Malargüe fold and thrust belt, Southern Central Andes. Tectonophysics 459:123–139
Giambiagi LB, Tunik MA, Barredo S et al (2009a) Cinemática de apertura del sector norte de la cuenca Neuquina. Rev Asoc Geol Argent 65:278–292
Giambiagi LB, Ghiglione M, Cristallini E et al (2009b) Kinematic models of basement/cover interactions: insights from the Malargüe fold and thrust belt, Mendoza, Argentina. J Struct Geol 31:1443–1457
Giambiagi LB, Mescua JF, Bechis F et al (2011) Pre-Andean deformation of the Precordillera southern sector, southern Central Andes. Geosphere 7:219–239
Giambiagi LB, Mescua JF, Bechis F et al (2012) Thrust belts of the southern Central Andes: along-strike variations in shortening, topography, crustal geometry, and denudation. Geol Soc Am Bull 124:1339–1351

Giambiagi LB, Tassara A, Mescua JF et al (2014) Evolution of shallow and deep structures along the Maipo-Tunuyán transect (33° 40′ S): from the Pacific coast to the Andean foreland. In: Sepúlveda SA, Giambiagi LB, Moreiras S et al (eds) Geodynamic processes in the Andes of Central Chile and Argentina, The Geological Society, London, SP 399, p 63

Gómez R, Lothari L, Tunik MA et al (2019) Onset of foreland basin deposition in the Neuquén Basin (34°–35°S): new data from sedimentary petrology and U-Pb dating of detrital zircons from the Upper Cretaceous non-marine deposits. J S Am Earth Sci 95:102257

Gulisano CA (1981) El Ciclo Cuyano en el norte de Neuquén y sur de Mendoza. In: Abstracts of the 8 Congreso Geológico Argentino, San Luis, 20–26 Sept 1981

Gulisano CA, Gutiérrez Pleimling AR (1994) The Jurassic of the Neuquén Basin: b) Mendoza Province. Guía de Campo. Asociación Geológica Argentina, Buenos Aires, vol E3, pp 1–103

Haq BU, Hardenbol J, Vail PR (1987) Chronology of fluctuating sea levels since the Triassic. Science 235:1156–1167

Herbst R (1968) Las floras liásicas argentinas con consideraciones estratigráficas. In: Abstracts of the 3° Jornadas Geológicas Argentinas, Buenos Aires, pp 145–162

Horton BK, Fuentes F, Boll A et al (2016) Andean stratigraphic record of the transition from backarc extension to orogenic shortening: a case study from the northern Neuquén basin, Argentina. J S Am Earth Sci 71:17–40

Howell JA, Schwarz E, Spalletti LA et al (2005) The Neuquén Basin: an overview. In: Veiga GD, Spalletti LA, Howell JA et al (eds) The Neuquén Basin: a case study in sequence stratigraphy and basin dynamics. The Geological Society, London, SP 252, 1–14

Hu JC, Angelier J (2004) Stress permutations: three-dimensional distinct element analysis accounts for a common phenomenon in brittle tectonics. J Geoph Res 109:B09403

Jaeger JC, Cook NG, Zimmerman R (2007) Fundamentals of rock mechanics, 4th edn. Wiley-Blackwell, Oxford, p 488

Japas MS, Cortés JM, Pasini M (2008) Tectónica extensional triásica en el sector norte de la cuenca Cuyana: primeros datos cinemáticos. Rev Asoc Geol Argentina 63:213–222

Kaven JO, Maerten F, Pollard DD (2011) Mechanical analysis of fault slip data: Implications for paleostress analysis. J Struct Geol 33:78–91

Keidel J (1916) La geología de las sierras de la provincia de Buenos Aires y sus relaciones con las montañas del Cabo y los Andes. Ministerio de Agricultura de la Nación, Anales Dirección de Geología, Mineralogía y Minería 9:5–77

Kleiman LE, Japas MS (2009) The Choiyoi volcanic province at 34° S–36° S (San Rafael, Mendoza, Argentina): implications for the Late Palaeozoic evolution of the southwestern margin of Gondwana. Tectonophysics 473:283–299

Kokogian DA, Fernández Seveso F, Mosquera A (1993) Las secuencias sedimentarias Triásicas. In: Ramos VA (ed) Geología y recursos naturales de Mendoza. Asociación Geológica Argentina, Buenos Aires, pp 65–78

Kozlowski E, Manceda R, Ramos VA (1993) Estructura. In: Ramos VA (ed) Geología y recursos naturales de Mendoza. Asociación Geológica Argentina, Buenos Aires, pp 235–256

Lanés S (2002) Paleoambientes y Paleogeografía de la primera transgresión en Cuenca Neuquina, Sur de Mendoza. Ph.D. thesis, Universidad de Buenos Aires

Lanés S (2005) Late Triassic to Early Jurassic sedimentation in northern Neuquén Basin, Argentina: tectonosedimentary evolution of the first transgression. Geol Acta 3:81–106

Lanés S, Palma RM (1998) Environmental implications of oncoids and associated sediments from the Remoredo Formation (Lower Jurassic) Mendoza, Argentina. Palaeo 140:357–366

Lanés S, Salani FM (1998) Petrografía, origen y paleoambiente sedimentario de las piroclastitas de la Formación Remoredo (Jurásico Temprano), Argentina (35° 30′ S–70° 15′ W). Rev Geol Chile 25:141–152

Lanés S, Giambiagi LB, Bechis F et al (2008) Late Triassic—early Jurassic successions of the Atuel depocenter: sequence stratigraphy and tectonic controls. Rev Asoc Geol Argentina 63:534–548

Lanés S, Gnaedinger SC, Zavattieri AM et al (2013) Sedimentary paleoenvironment and fossil plants of the El Freno Formation (early Jurassic) in Las Leñas valley, Neuquén basin. Rev Asoc Geol Argentina 70:465–476

Legarreta L, Gulisano CA (1989) Análisis estratigráfico secuencial de la cuenca Neuquina (Triásico superior – Terciario inferior), Argentina. In: Chebli G, Spalletti L (eds) Cuenca Neuquina. Universidad Nacional de Tucumán, San Miguel de Tucuman, Ser. Corr Geol 6, pp 221–243

Legarreta L, Uliana MA (1996) The Jurassic succession in west-central Argentina: stratal patterns, sequences and paleogeographic evolution. Palaeo 120:303–330

Legarreta L, Uliana MA (1999) El Jurásico y Cretácico de la Cordillera Principal y Cuenca Neuquina. In: Caminos R (ed) Geología Argentina. Instituto de Geología y Recursos Minerales, Buenos Aires 29, pp 399–432

Legarreta L, Gulisano CA, Uliana MA (1993) Las secuencias sedimentarias jurásico-cretácicas. In: Ramos VA (ed) Geología y Recursos Naturales de Mendoza. Asociación Geológica Argentina, Buenos Aires, pp 87–114

Limarino CO, Spalletti L (2006) Paleogeography of the Upper Paleozoic basins of southern South America: an overview. J S Am Earth Sci 22:134–155

Llambías EJ (1999) Las rocas ígneas gondwánicas. 1. El magmatismo gondwánico durante el Paleozoico Superior-Triásico. In: Caminos R (ed), Geología Argentina. Instituto de Geología y Recursos Minerales, vol 29. Asociación Geológica Argentina, Buenos Aires, pp 349–363

Llambías EJ, Kleiman LE, Salvarredi JA (1993) El magmatismo Gondwánico. In: Ramos VA (ed.) Geología y Recursos Naturales de Mendoza. Instituto de Geología y Recursos Minerales 29, Asociación Geológica Argentina, Buenos Aires, pp 53–64

Llambías EJ, Leanza HA, Carbone O (2007) Evolución tectono-magmática durante el Pérmico al Jurásico Temprano en la Cordillera del Viento (37° 05′ S–37° 15′ S): Nuevas evidencias geológicas y geoquímicas del inicio de la cuenca Neuquina. Rev Asoc Geol Argent 62:217–235

Manceda R, Figueroa D (1995) Inversion of the Mesozoic Neuquén rift in the Malargüe fold-thrust belt, Mendoza, Argentina. In: Tankard AJ, Suárez R, Welsink HJ (eds) Petroleum basins of South America. AAPG Memoir 62:369–382

Marrett RA, Allmendinger RW (1990) Kinematic analysis of fault-slip data. J Struct Geol 12:973–986

Martinez Días JJ (2002) Stress field variation related to fault interaction in a reverse oblique-slip fault: the Alhama de Murcia fault, Betic Cordillera, Spain. Tectonophys 356:291–305

Mazzini A, Svensen H, Leanza HA et al (2010) Early Jurassic shale chemostratigraphy and U-Pb ages from the Neuquén Basin (Argentina): implications for the Toarcian oceanic anoxic event. Earth Planet Sci Lett 297:633–645

Mescua JF, Giambiagi LB, Bechis F (2008) Evidencias de Tectónica Extensional en el Jurásico Tardío (Kimeridgiano) del suroeste de la Provincia de Mendoza. Rev Asoc Geol Argentina 63:512–519

Mescua JF, Giambiagi LB, Ramos VA (2013) Late Cretaceous Uplift in the Malargüe fold-and-thrust belt (35° S), southern Central Andes of Argentina and Chile. Andean Geol 40:102–116

Mescua JF, Giambiagi LB, Tassara A et al (2014) Influence of pre-Andean history over Cenozoic foreland deformation: structural styles in the Malargüe fold-and-thrust belt at 35 S, Andes of Argentina. Geosphere 10:585–609

Mescua JF, Giambiagi L, Barrionuevo M et al (2016) Basement composition and basin geometry controls on upper- crustal deformation in the Southern Central Andes (30-36°S). Geol Magaz 153:945–961

Mosquera A, Ramos VA (2006) Intraplate deformation in the Neuquén Embayment. In: Kay SM, Ramos VA (eds) Evolution of an Andean margin: a tectonic and magmatic view from the Andes to the Neuquén Basin (35°–39° S latitude). Geol Soc Am, SP 407: 97–123

Mpodozis C, Ramos VA (1989) The Andes of Chile and Argentina. In: Ericksen GE, Cañas MT, Reinemund JA (eds) Geology of the Andes and its relation to hydrocarbon and energy resources. Circum-Pacific Council for Energy and Hydrothermal Resources, Earth Science Series 11, pp 59–90

Naipauer M, Tapia F, Mescua J, Farías M et al (2015) Detrital and volcanic zircon U-Pb ages from southern Mendoza (Argentina): an insight on the source regions in the northern part of the Neuquén Basin. J S Am Earth Sci 64:434–451
Nieto-Samaniego AF (1999) Stress, strain and fault patterns. J Struct Geol 21:1065–1070
Nullo F, Stephens G, Otamendi J et al (2002) El volcanismo del Terciario superior del sur de Mendoza. Rev Asoc Geol Argent 57:119–132
Nullo F, Stephens G, Combina A et al (2005) Hoja geológica 3569-III/3572-IV Malargüe, provincia de Mendoza. Servicio Geológico Minero Argentino. Instituto de Geología y Recursos Minerales, Buenos Aires
Oliveros V, González J, Espinoza M et al (2018) The early stages of the Magmatic Arc in the Southern Central Andes. In: Folguera A, Contreras Reyes E et al (eds) The evolution of the Chilean-Argentinean Andes. Springer, Cham, pp 165–190
Orts DL, Folguera A, Giménez M et al (2012) Variable structural controls through time in the Southern Central Andes (~36ºS). Andean Geol 39:220–241
Pángaro F, Ramos VA (2012) Paleozoic crustal blocks of onshore and offshore central Argentina: new pieces of the southwestern Gondwana collage and their role in the accretion of Patagonia and the evolution of Mesozoic south Atlantic sedimentary basins. Mar Petrol Geol 37:162–183
Pángaro F, Pereira M, Raggio F et al (2006) Tectonic inversion of the Huincul High, Neuquén Basin, Argentina: An Endangered Species. stratigraphic evidences of it's disappearance. In: Abstracts of the 9 Simposio Bolivariano de Exploración Petrolera en las Cuencas Subandinas, Aug 2006
Pankhurst RJ, Rapela CW, Fanning CM et al (2006) Gondwanide continental collision and the origin of Patagonia. Earth Sci Rev 76:235–257
Ramos VA (1988) Late Proterozoic-Early Paleozoic of South America—a collisional history. Episodes 11:168–174
Ramos VA (2008) Patagonia: a Paleozoic continent adrift? J S Am Earth Sci 26:235–251
Ramos VA (2010) The tectonic regime along the Andes: present-day and Mesozoic regimes. Geol J 45:2–25
Ramos VA, Kay SM (2006) Overview of the tectonic evolution of the southern Central Andes of Mendoza and Neuquén (35°–39° S latitude). In: Kay SM, Ramos VA (eds) Evolution of an Andean margin: A tectonic and magmatic view from the Andes to the Neuquén Basin (35°–39° S latitude). Geol Soc Am SP 407:1–18
Ramos VA, Jordan TE, Allmendinger W et al (1986) Paleozoic Terranes of the Central Argentine-Chilean Andes. Tectonics 5:855–880
Rapalini AE (2005) The accretionary history of southern South America from the latest Proterozoic to the Late Palaeozoic: some palaeomagnetic constraints. In: Vaughan APM, Leat PT, Pankhurst RJ (eds) Terrane Processes at the Margins of Gondwana. The Geological Society, London, SP 246, pp 305–328
Reijenstein C (1967) Estratigrafía y tectónica de la zona al Norte del río Atuel, entre los arroyos Blanco y Malo, provincia de Mendoza. Universidad de Buenos Aires, Trabajo final de Licenciatura
Riccardi AC, Damborenea SE (1993) Léxico estratigráfico de la Argentina: Jurásico. Asociación Geológica Argentina, Serie B 21, Buenos Aires, pp 470
Riccardi A, Iglesia Llanos MP (1999) Primer hallazgo de amonites en el Triásico de la Argentina. Rev Asoc Geol Argentina 54:298–300
Riccardi AC, Damborenea S, Manceñido MO et al (1988) Hettangiano y Sinemuriano marinos en Argentina. In: Abstracts of the 5 Congreso Geológico Chileno, Santiago de Chile, 8–12 Aug 1988
Riccardi AC, Damborenea S, Manceñido MO et al (1991) Hettangian and Sinemurian (Lower Jurassic) biostratigraphy of Argentina. J S Am Earth Sci 4:159–170
Riccardi AC, Damborenea SE, Manceñido MO et al (1997) Primer registro de Triásico marino fosilífero de la Argentina. Rev Asoc Geol Argent 52:228–234
Sato AM, Llambías E, Basei M et al (2015) Three stages in the Late Paleozoic to Triassic magmatism of southwestern Gondwana, and the relationships with the volcanogenic events in coeval basins. J S Am Earth Sci 63:48–69

Schiuma M, Llambías EJ (2008) New ages and chemical analysis on Lower Jurassic volcanism close to the Huincul High, Neuquén. Rev Asoc Geol Argent 63:644–652

Silvestro J, Kraemer P (2005) Evolución tecto-sedimentaria de la Cordillera Principal en el sector surmendocino a los 35° 30′ S. Faja Plegada de Malargüe. República Argentina. In: Abstracts of the 6°Congreso de Exploración y desarrollo de hidrocarburos, Mar del Plata, 15–19 November 2005

Silvestro J, Zubiri M (2008) Convergencia oblicua: modelo estructural alternativo para la Dorsal Neuquina (39° S) - Neuquén. Rev Asoc Geol Argent 63:49–64

Spagnuolo M, Litvak V, Folguera A et al (2012) Neogene magmatic expansion and mountain building processes in the southern Central Andes, 36-37° S, Argentina. J Geodyn 53:81–94

Spalletti LA (1997) Sistemas deposicionales fluvio-lacustres en el rift triásico de Malargüe (sur de Mendoza, República Argentina). A Acad Nac Cs Ex Fís Nat 49:109–124

Spalletti LA, Morel EM, Franzese JR et al (2007) Contribución al conocimiento sedimentológico y paleobotánico de la Formación El Freno (Jurásico Temprano) en el valle superior del río Atuel, Mendoza, Argentina. Ameghiniana 44:367–386

Sruoga P, Etcheverría M, Folguera A et al (2005) Hoja Geológica 3569-I Volcán Maipú, Servicio Geológico y Minero Argentino, Boletín 290, 116 p, Buenos Aires

Stipanicic PN (1969) El avance en los conocimientos del Jurásico argentino a partir del esquema de Groeber. Rev Asoc Geol Argent 24:367–388

Stipanicic PN (1979) El Triásico de Valle del río de Los Patos (provincia de San Juan). In: Turner JCM (ed) Geología Regional Argentina. Academia Nacional de Ciencias, Córdoba, pp 695–744

Teyssier C, Tikoff B, Markley M (1995) Oblique plate motion and continental tectonics. Geology 23:447–450

Tickyj H, Fernández MA, Chemale Jr F et al (2009). Granodiorita Pampa de los Avestruces, Cordillera Frontal, Mendoza: un intrusivo sintectónico de edad devónica inferior. In: Abstracts of the 14 Reunión de Tectónica, 19–23 Oct 2015

Tikoff B, Teyssier C (1994) Strain modelling of displacement-field partitioning in transpressional orogens. J Struct Geol 16:1575–1588

Tomezzoli RN (2012) Chilenia y Patagonia: un mismo continente a la deriva? Rev Asoc Geol Argent 69:222–239

Tunik MA, Lanés S, Bechis F et al (2008) Análisis petrográfico preliminar de las areniscas jurásicas tempranas en el depocentro Atuel de la cuenca Neuquina. Rev Asoc Geol Argent 63:714–727

Tunik MA, Folguera A, Naipauer M et al (2010) Early uplift and orogenic deformation in the Neuquén Basin: Constraints on the Andean uplift from U-Pb and Hf isotopic data of detrital zircons. Tectonophysics 489:258–273

Tunik MA, Giambiagi LB; Barredo S et al (2011) Caracterización petrográfica del relleno del rift de la subcuenca Atuel, provincia de Mendoza. In: Abstracts of the 18 Congreso Geológico Argentino, Neuquén, 2–6 May 2011

Twiss RJ, Unruh JR (1998) Analysis of fault slip inversions: Do they constrain stress or strain rate? J Geoph Res 103:12205–12222

Uliana M, Biddle K, Cerdán J (1989) Mesozoic extension and the formation of Argentina sedimentary basins. In: Tankard AJ, Balkwill HR (eds) Extensional tectonics and stratigraphy of the North Atlantic Margin). AAPG Memoir 46: 599–613

Uliana M, Arteaga M, Legarreta L ct al (1995) Inversion structures and hydrocarbon occurrence in Argentina. In: Buchanan J, Buchanan P (eds) Basin inversion. The Geological Society, London, SP88: 211–233

Varela R, Basei MAS, González PD et al (2011) Accretion of Grenvillian terranes to the southwestern border of the Río de la Plata craton, western Argentina. Int J of Earth Sci 100:243–272

Vergani GD, Tankard J, Belotti J et al (1995) Tectonic evolution and paleogeography of the Neuquén Basin, Argentina. In: Tankard AJ, Suárez R, Welsink HJ (eds), Petroleum Basins of South America. AAPG Memoir 62:383–402

Volkheimer W (1978) Descripción Geológica de la Hoja 27b, Cerro Sosneado, provincia de Mendoza. Servicio Geológico Nacional, vol 151, Buenos Aires, p 83 von Hillebrandt A (1989) The Lower Jurassic of the Río Atuel region, Mendoza Province, Argentina. In: Abstracts of the 4° Congreso Argentino de Paleontología y Bioestratigrafía, Mendoza, vol 4, pp 39–43

Westermann GEG, Riccardi AC (1982) Ammonoid fauna from the early Middle Jurassic of Mendoza province, Argentina. J Paleontol 56:11–41

Yagupsky DL, Cristallini EO, Fantín J et al (2008) Oblique half-graben inversion of the Mesozoic Neuquén Rift in the Malargüe Fold and Thrust Belt, Mendoza, Argentina: new insights from analogue models. J Struct Geol 30:839–853

Yamaji A (2000) The multiple inverse method: a new technique to separate stresses from heterogeneous fault-slip data. J Struct Geol 22:441–452

Yamaji A (2003) Are the solution of stress inversion correct? Visualization of their reliability and the separation of stresses from heterogeneous fault-slip data. J Struct Geol 25:241–252

Yrigoyen MR (1993) Los depósitos sinorogénicos terciarios. In: Ramos VA (ed) Geología y Recursos Naturales de Mendoza. Asociación Geológica Argetina, Buenos Aires, pp 123–148

Zamora Valcarce G, Zapata TR, del Pino D et al (2006) Structural evolution and magmatic characteristics of the Agrio fold-and-thrust belt. In: Kay SM, Ramos VA (eds) Evolution of an Andean Margin: a Tectonic and Magmatic View from the Andes to the Neuquén Basin (35°-39°S lat). Geol Soc Am, SP 407, p 125–145

Zapata TR, Folguera A (2005) Tectonic evolution of the Andean Fold and Thrust Belt of the southern Neuquén Basin, Argentina. In: Veiga GD, Spalletti LA, Howell JA, Schwarz E (eds) The Neuquén Basin, Argentina: a case study in sequence stratigraphy and basin dynamics. The Geological Society, London, SP 252, pp 37–56

Zapata TR, Brissón I, Dzelalija F (1999) The structures of the Andean fold and thrust belt in relation to basement control in the Neuquén Basin. Bol Inf Petrol 60:112–121

Jurassic Uplift Along the Huincul Arch and Its Consequences in the Stratigraphy of the Cuyo and Lotena Groups. Neuquén Basin, Argentina

Carlos Zavala, Mariano Arcuri, Mariano Di Meglio, Agustín Zorzano and Germán Otharán

Abstract Marine sedimentation in the Neuquén Basin started in the Early Jurassic with the accumulation of a progradational clastic system: the Cuyo Group. This sedimentary cycle begins with offshore shales of the Los Molles Formation, transitionally overlaid by shelfal to littoral sandstones and conglomerates of the Lajas Formation. According to the current stratigraphic schema, in the southern Neuquén Basin the Cuyo Group ends with continental red beds of the Challacó Formation, a unit interpreted as being the proximal equivalent of the Lajas Formation. Nevertheless, recent studies performed in the Picún Leufú area, complemented with regional stratigraphic evidence, suggest that the accumulation of these stratigraphic units is more likely related to a complex depositional scenario that involves tectonic activity during the Jurassic along the Huincul Arch. The uplift and exposure of Jurassic rocks along the Huincul Arch divided the basin into two sub-basins, each one having a different stratigraphy for the Middle-Late Jurassic time span. The stratigraphic analysis revealed that the red beds traditionally located at the top of the Cuyo Group in fact represent two different stratigraphic units separated by a regional unconformity. The lower unit (1) sharply overlies the Lajas Formation, corresponding to the Challacó Formation *s.s.* This unit was accumulated in a brackish lacustrine basin developed southward of the Huincul Arch (Picún Leufú Sub-basin). The upper unit (2) corresponds to the Bosque Petrificado Formation and is stratigraphically located at the base of the Lotena Group. The Bosque Petrificado Formation was accumulated in a fluvial to marine environment developed in the southern Neuquén Basin. Paleocurrent analysis suggests a sediment supply from the northeast, evidencing a local source of sediments related to an uplifted area. Recent biostratigraphic studies performed in the Bosque Petrificado and Lotena formations revealed an Early Callovian-Early Oxfordian age for these deposits.

C. Zavala (✉) · M. Arcuri · M. Di Meglio · A. Zorzano · G. Otharán
GCS Argentina SRL, Molina Campos 150, Bahía Blanca 8000, Argentina
e-mail: czavala@gcsargentina.com

C. Zavala · M. Arcuri · M. Di Meglio · G. Otharán
Departamento de Geología, Universidad Nacional del Sur, San Juan 670, Bahía Blanca 8000, Buenos Aires, Argentina

D. Kietzmann and A. Folguera (eds.), *Opening and Closure of the Neuquén Basin in the Southern Andes*, Springer Earth System Sciences,
https://doi.org/10.1007/978-3-030-29680-3_3

Keywords Cuyo Group · Lotena Group · Challacó Formation · Synsedimentary tectonics · Huincul Arch

1 Introduction

The Cuyo Group (Groeber 1946; Stipanicic 1969) represents the first fully marine transgressive-regressive depositional cycle recognized in the Neuquén Basin. Depending on the position within the basin, the accumulation of this unit started in the Hettangian–Pliensbachian and extended up to the Middle Callovian.

The internal organization of the Cuyo Group has been described, discussed and analyzed in detail during more than a century. According to the conventional stratigraphy of the Neuquén Basin, the most accepted depositional model suggest that the accumulation of the Cuyo Group started with offshore to prodelta marine shales (Los Molles Formation) transitionally overlaid by shelfal to littoral clastic deposits (Lajas Formation), ending up with continental (fluvial) red beds (Challacó Formation). This classic prograding clastic system has been integrated and adapted to the former sequence stratigraphic framework of the Neuquén Basin.

In recent years, new high-resolution seismic surveys and extensive fieldworks suggested that the stratigraphy of these units is largely more complex and was strongly controlled by synsedimentary tectonics.

This paper provides new stratigraphic evidence for the regional understanding of the origin and evolution of the Jurassic Cuyo and Lotena groups.

2 Geological Framework

Located in west central Argentina, the Neuquén Basin is one of the main oil-producing basins of South America (Fig. 1). The origin of the Neuquén Basin is closely associated with the Mesozoic evolution of the western margin of Gondwana. Basin configuration took place during Late Triassic, over a highly heterogeneous Paleozoic basement resulting from the collision of the Patagonia terrane with the southern margin of Gondwana (Mosquera et al. 2011). The stratigraphic column of the Neuquén Basin comprises a thick (> 7,000 m) sedimentary succession mainly integrated by marine and continental siliciclastic deposits, and minor carbonate and evaporitic units (Fig. 1). Sedimentological and stratigraphic evidences suggest that the Neuquén Basin was a satellite basin with a nearly continuous connection with the Pacific Ocean during the Jurassic and Early Cretaceous.

Sedimentation in the Neuquén Basin started in the Late Triassic as a consequence of a regional rift phase. During this interval, volcanic and volcaniclastic deposits accumulated in irregular and structurally controlled depocenters (Precuyo

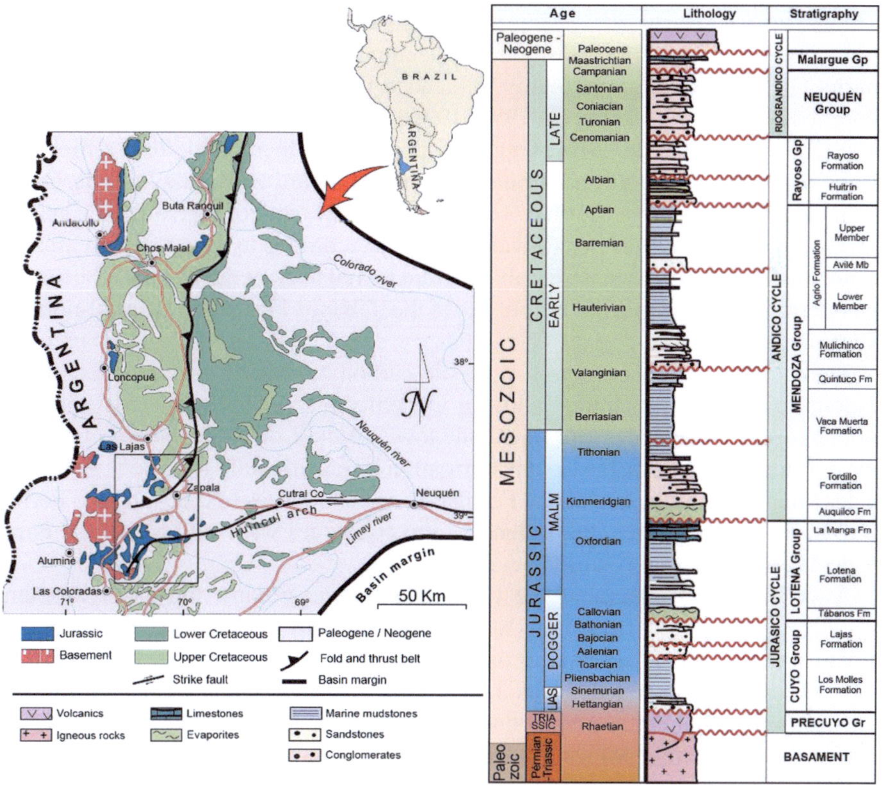

Fig. 1 Location map and generalized stratigraphic column of the Neuquén Basin. Main outcrops of basement, Jurassic and Cretaceous rocks are indicated. The box in the map shows the location of the Jurassic outcrops discussed in this study (modified after Zavala et al. 2006)

Group, Gulisano et al. 1984) (see Chap. "The Syn-Rift of the Neuquén Basin-Precuyano and Lower Cuyano Cycle: Review of Structure, Volcanism, Tectono-Stratigraphy and Depositional Scenarios"). Afterwards, during the Jurassic and Cretaceous the Neuquén Basin was periodically flooded from the paleo-Pacific, resulting in the accumulation of deposits related to several transgressive-regressive cycles. The first two cycles, corresponding to the Cuyo and Lotena groups, comprise the lapse Hettangian-Oxfordian.

The Cuyo Group starts with offshore to prodeltaic marine shales and turbidites of the Los Molles Formation (Weaver 1931). This unit represents a fine-grained clastic succession up to 1000 m thick. Regional data and thickness maps suggest that this unit was accumulated over a tectonically controlled relief. Sediment gravity-flow deposits (muddy, sandy and gravelly) are common in areas associated with lower submarine landscapes, where they form a very thick succession of sandstones and shales related to intrabasinal and extrabasinal turbidity currents (Paim et al. 2011; Zavala and Arcuri 2016). The Los Molles Formation is transitionally overlaid by a thick (up to 750 m) succession of sandstones and conglomerates of the Lajas Formation (Weaver 1931). This unit was accumulated in a complex (littoral to submarine) deltaic environment highly influenced by sandy hyperpycnal flows locally modified by tidal action (Zavala 1993; 1996b; Zavala and Gonzalez 2001; McIlroy et al. 2005; Canale et al. 2014; Gugliotta et al. 2016; Steel et al. 2018). Field evidence suggest a syntectonic accumulation of this unit during Middle Jurassic. One of the most important unconformities recognized in the southern part of the basin is the Intra-Bajocian Unconformity (Zavala 1996a). According to its position within the basin, the Lajas Formation is sharply overlaid by red beds of the Challacó Formation (southern Neuquén Basin) or evaporites of the Tábanos Formation (central-northern Neuquén Basin).

The Lotena Group conforms a clastic-carbonate system accumulated over a regional unconformity (Intra-Callovian Unconformity). This unit starts with conglomerates, marine shales and sandstone lobes of the Lotena Formation. In central and northern basin areas, the Lotena Formation is transitionally followed by shallow-water limestones of the La Manga Formation and evaporites of the Auquilco Formation. In the southern area, the Lotena Group concludes with red beds of the Fortín 1° de Mayo Formation (Gulisano et al. 1984).

After the accumulation of the Cuyo and Lotena groups, the basin was temporarily disconnected from the Pacific Ocean, resulting in the accumulation of fluvial, lacustrine and eolian deposits of the Tordillo Formation and equivalent units. These continental deposits laid over a regional unconformity (Intramalm Unconformity) that defines the boundary between the Lotena and Mendoza groups. Marine sedimentation was reestablished in the Early Tithonian after a catastrophic marine flooding coming from the Pacific Ocean, resulting in the accumulation of offshore black shales of the Vaca Muerta Formation. During the Late Jurassic–Early Cretaceous, the basin was mainly controlled by thermal subsidence (SAG phase). Consequently, the marine deposits of the Mendoza Group and the overlying Huitrín Group display a more regular distribution with an internal stratigraphy controlled by eustatic sea level fluctuations and climatic cycles. Finally, the early uplift of the Andean foothills during the Late Cretaceous (see Chap. "The Late Cretaceous Orogenic System: Early Inversion of the Neuquén Basin and Associated Synorogenic Deposits 35°–38°S") caused the final disconnexion of the Neuquén Basin with the Pacific Ocean. The last results in the final basin infill by a thick interval of continental red beds in the eastern boundary of the basin (Neuquén Group). A simplified schematic stratigraphic column for the central area of the Neuquén Basin is shown in Fig. 1.

3 Materials and Methods

This chapter summarizes more than 30 years of research focused on the understanding of the stratigraphy of the Cuyo Group. During these years, an extensive fieldwork was carried out along the entire basin. More than 15,000 m of stratigraphic sections were described and analyzed in detail along different localities and outcrops of the Early-Middle Jurassic succession of the Neuquén Basin. The stratigraphic sections were measured with a Jacob Staff and positioned by GPS. Over the years, the annotation procedure migrated from the conventional field book towards the digital record using our in-house "LithoHero" software (Iparraguirre et al. 2018). A free version of this software is currently available in Google Play for Android devices. Also, a IOS version will be shortly available. The digital record of field annotations is very important since it supports data integrity, save time, and allows to export all field observations as a .las file, enable data integration with conventional softwares used in the oil industry. Consequently, apart from saving time drawing stratigraphic sections, LithoHero provides a very useful software to integrate outcrop and subsurface data.

Stratigraphic sections were described in detail "bed by bed", with a careful characterization of lithology, sedimentary structures, type of bounding surfaces, geometry, trace fossils and paleocurrent directions. As a complement, stratigraphic sections were conveniently sampled for different paleontologic, palynostratigraphic and petrographic studies. Field observations and interpretations were applied to the construction of a number of cross sections, geologic and paleogeographic maps.

4 The Jurassic of the Neuquén Basin and the Cuyo Group

The current stratigraphic framework for the Jurassic in the Neuquén Basin comes out from the seminal contributions of Weaver (1931), Groeber (1946), Groeber et al. (1953) and Herrero Ducloux (1946). Groeber (1946) termed as "*Jurásico*" to the lower stratigraphic interval of the Neuquén Basin, extended from the Rhaetian up to the Kimmeridgian. This stratigraphic interval comprises three main depositional cycles known as *Cuyano*, *Loteniano* and *Chacayano*. These cycles where later redefined according to the lithostratigraphic nomenclature as the Precuyo, Cuyo and Lotena groups (Stipanicic 1969; Gulisano et al. 1984). These groups have a chronostratigraphic significance, since they represent lithogenetic units bounded by regional unconformities (Gulisano et al. 1984).

The base of the Cuyo Group is defined by the Intraliasic Unconformity, which is characterized by a sharp contact between shelfal and shallow marine clastic/carbonate deposits of the Los Molles Formation on top of different units of the basement (plutonic, volcanic and metamorphic) and the synrift volcanic and volcaniclastic deposits of the Precuyo Cycle (Upper Triassic). According to regional data, the Early Jurassic transgression advanced from north to south during the lapse Hettangian–Pliensbachian (Riccardi et al. 1988).

The Cuyo Group upper boundary was largely more controversial, specially in the southern part of the Neuquén Basin (Picún Leufú sub-basin, Hogg 1993), due to the poor definition of the overlying Lotena Group. Originally, the term Lotena (or "Loteno") Formation was introduced by Weaver (1931) to name the thick-bedded conglomerates and red beds that crop out in the surroundings of the Cerro Lotena locality. These conglomerates are actually considered as the proximal deposits of the Cuyo Group (Lajas Formation, Gulisano et al. 1984). Nevertheless, the term Lotena Formation was widely applied in the past to designate the red beds with some marine intercalations recognized in between the Cuyo and Mendoza groups in the southern part of the Neuquén Basin.

Dellapé et al. (1979) provided an outstanding contribution about the origin and the significance of the red beds attributed to the Lotena Formation. These authors analyzed the classic section outcropped near the bridge of the Ruta Nacional 40 over the Picún Leufú River. According to Dellapé et al. (1979), the red beds recognized on top of the Lajas Formation should be considered as part of the Cuyo Group, while the term "Lotena Formation" should be restricted to the fine-grained marine succession having ammonoids and marine microfossils typical of the Middle Jurassic (Callovian). According to these authors, at this locality the base of the Lotena Formation is defined by a conglomerate that lay on top of a regional unconformity (Intra-Callovian Unconformity, Dellapé et al. 1979), and extends up to the Tithonian shales of the Vaca Muerta Formation. Gulisano et al. (1984) redefined the upper boundary as corresponding to the Intramalm Unconformity (Stipanicic and Rodrigo 1970), located below the fluvial/eolian deposits of the Tordillo/Quebrada del Sapo formations.

A great number of studies have demonstrated that, during the Jurassic, the sedimentation in the southern part of the Neuquén Basin was deeply controlled by coetaneous tectonic activity along the Huincul arch (Vergani et al. 1995; Zavala and Gonzalez 2001; Silvestro and Zubiri 2008; Mosquera et al. 2011; Naipauer et al. 2012). This complex structure (Fig. 1) constitutes an East-West oriented intraplate deformation belt developed along the Permian suture of the Patagonia terrane with Western Gondwana (Mosquera et al. 2011).

Tectonic activity along the Huincul arch during the Middle-Late Jurassic resulted in the growth of localized structural highs that completely disconnected the southern sector of the basin, (known as Picún Leufú sub-basin) from the northern Neuquén embayment.

The stratigraphic consequences of these temporary uplifts are of such magnitude that the Jurassic column at both sides of the Huincul arch is markedly different (Fig. 2). As an example, the Challacó, Fortín 1° de Mayo, Bosque Petrificado and Quebrada del Sapo formations were developed exclusively in the southern area. These units are partially equivalent to the Lajas, Tábanos, La Manga, Barda Negra, Auquilco and Catriel/Sierras Blancas Formations recognized in the central/northern Neuquén Basin. Although in most stratigraphic schemas (e.g., Gugliotta et al. 2016) the Challacó Formation is considered as the fluvial progradational equivalent of the Lajas Formation, field data support a tectonostratigraphic origin for this unit.

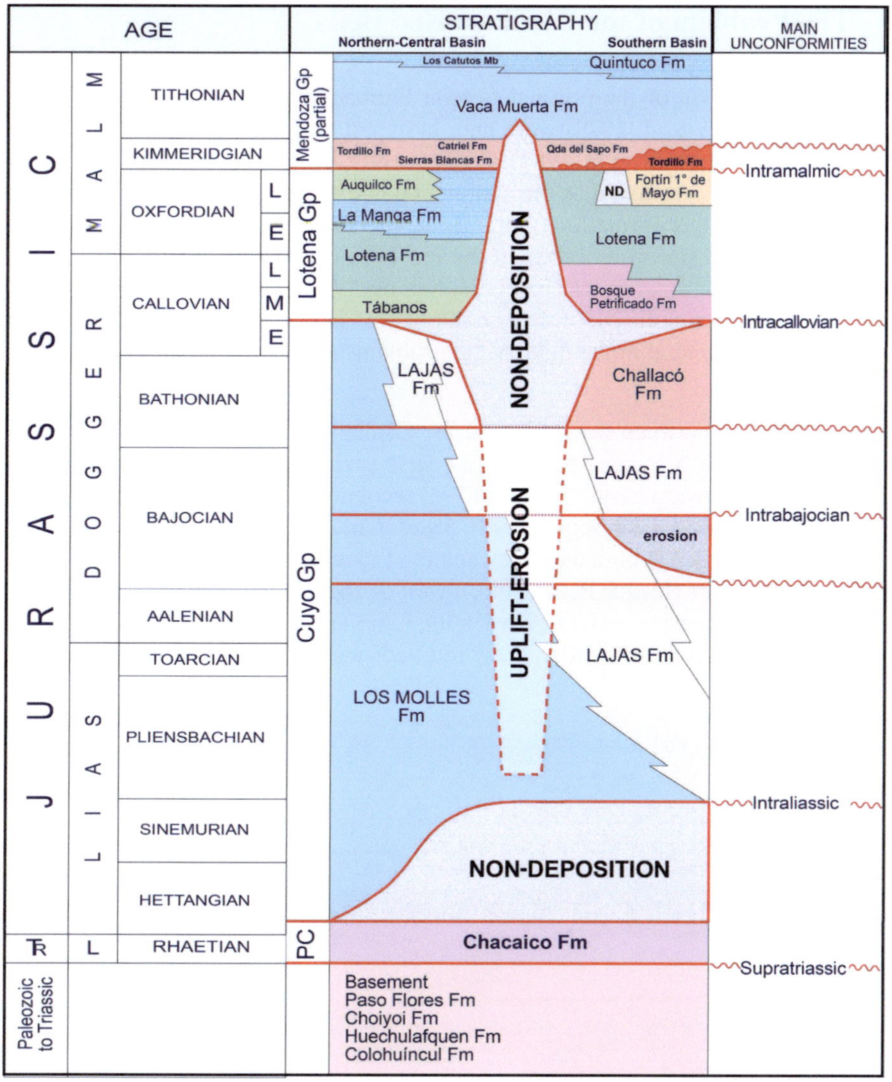

Fig. 2 Chronostratigraphic chart of the Jurassic in the Neuquén Basin. The uplift and local erosion along the Huincul arch occurred during the Late Bajocian-Tithonian temporarily segmented the Neuquén Basin into two different sub-basins, the main Neuquén embayment to the north, and the Picún Leufú sub-basin to the south. Note that the stratigraphy during the Callovian–Kimmeridgian is different at both sides of the Huincul arch

5 The Problem of the Challacó Red Beds

The lower red beds of the original Lotena Formation (re-located in to the Cuyo Group by Dellapé et al. 1979) were later assigned to the Challacó Formation (De Ferrariis 1947; Gulisano et al. 1984). Consequently, the Challacó Formation (Fig. 3) was traditionally considered as the continental equivalent of the prograding deltaic deposits of the Lajas Formation (Gulisano et al. 1984). Most stratigraphic schemas illustrate a lateral transition between the red beds of the Challacó Formation and the shallow marine sandstones and conglomerates of the Lajas Formation (see, e.g., Fig. 3 in Gugliotta et al. 2016). Nevertheless, this lateral transition has never been physically documented in the field. Main problems related to this interpretation are listed below:

- The boundary between the Lajas and the Challacó formations is always a sharp surface commonly located on top of a coarse-grained bed (sometimes displaying a coarsening upward trend). Zavala (1993) recognized a sequence boundary at this level (boundary between sequences JC5 and JC6, Zavala 1993). Veiga (1998) also describes a sequence boundary between the Lajas and Challacó formations.
- At the locality of Bosque Petrificado, a well defined angular unconformity (>40°) is recognized at the boundary between the Lajas and Challacó formations (Zavala 1993, Fig. 4a). At this locality, these red beds were later assigned to the Bosque

Fig. 3 Red beds of the Challacó Formation at the Picún Leufú anticline area. These deposits were traditionally considered as the proximal equivalent of the shallow marine deposits of the Lajas Formation

Fig. 4 **a** Angular unconformity between the Lajas and Bosque Petrificado formations at the Bosque Petrificado locality. **b** Detail of the angular unconformity recognized at Barda Colorada, where Early Jurassic shales of the Los Molles Formation are unconformably overlaid by eolian sandstones of the Neuquén Group. **c** and **d** Composed satellite image of the Cerro Lotena area. The Lajas Formation is composed of near-vertical thick-bedded conglomerates, which are unconformably covered by red beds of the Bosque Petrificado Formation (BP Fm) and offshore marine shales of the Vaca Muerta Formation. Note that the Vaca Muerta shales passively infill a very complex tectonic relief carved by erosion

Petrificado Formation (Zavala and Freije 2002) and were temporarily located in the Lotena Group.

- The red beds of the Challacó Formation (interpreted as accumulated in a fluvial environment, Veiga 1998) are finer-grained than its conceptually related shallow marine equivalent outcropped in the north of Huincul arch. As an example, at the Sierra de la Vaca Muerta locality the Lajas Formation is mainly composed of sandstones and conglomerates, while its apparently equivalent Challacó Formation is dominated by gray to reddish shales with minor channel-fill sandstone deposits.
- The proximal fluvial deposits of the Cuyo Group are well exposed at the Cerro Lotena/Cerro Granito/Bosque Petrificado area, consisting of thick-bedded conglomerates assigned to the Lajas Formation. These deposits are largely coarser-grained than the "proximal" red beds of the Challacó Formation, also outcropped at the same locality.

- The palynological content of the Challacó Formation is commonly characterized by *Botryococcus*, thus suggesting a lacustrine environment for the accumulation of these red beds instead of a fluvial depositional system (Garcia et al. 2006).
- The palynology of the upper part of these "red beds" in the Quebrada del Sapo area indicates marine brackish conditions, exhibiting a palynological association that suggest a Late Callovian age (Olivera et al. 2018).

These observations suggest that the origin of the Challacó Formation is largely more complex than the result of a simple progradation. Zavala and Gonzalez (2001) proposed a tecnostratigraphic significance for this unit. Following these observations, the Challacó Formation was interpreted as accumulated in a lacustrine (endorheic) environment generated as a consequence of an early uplift of the Huincul arch. The early tectonic uplift along this tectonic unit is known for the Late Jurassic–Early Cretaceous, and deeply controlled the stratigraphy of the Neuquén Basin. According to Zavala and Freije (2002), the red beds located between the Lajas and Lotena Formations in reality correspond to two different (partially isopic) stratigraphic units, separated by the Intra-Callovian Unconformity: The Challacó Formation (s.s.) located in the Cuyo Group, and the Bosque Petrificado Formation, located at the base of the Lotena Group. In this chapter, new data are provided focused in the understanding of the physical stratigraphy and significance of these stratigraphic units.

6 Evidence of Early Uplift and Erosion at the Huincul Arch and Their Consequences in the Stratigraphy of the Cuyo Group

Evidence of tectonic uplift and erosion along the Huincul arch during the Middle Jurassic are clearly visible at different outcrops in the Cerro Lotena area. At Barda Colorada locality (Fig. 4b), an angular unconformity is recognized between the Los Molles Formation and the overlying Neuquén Group, evidencing an important uplift and erosion during the lapse Late Jurassic–Early Cretaceous. At a more local scale, in the Cerro Lotena locality (Fig. 4c, d), Bajocian conglomerates of the Lajas Formation are unconformably overlaid by Tithonian shales of the Vaca Muerta Formation (also infilling a structurally controlled relief), thus evidencing a local non-deposition/erosion during the Middle to Late Jurassic (lasting about 20 million years).

Since the subsidence in the Neuquén Basin was continuous at both sides of the Huincul arch during the entire Jurassic, a stratigraphic consequence of this localized uplift and erosion should be a coarse-grained related depositional unit with sediments supplied from the eroded landscapes. The last contrasts with erosional areas, where a time gap is recognized at the unconformity surface. The related depositional units are very important since they can display a complete and detailed record of the time of erosion and rate of uplift (Riba 1976). Recent studies conducted in the Picun Leufú area allowed a clear documentation of the magnitude and timing of these localized tectonic movements. The most interesting stratigraphic interval to characterize this tectonic activity is that corresponding to the time gap registered at main unconformities, represented by the depositional units accumulated between the Bajocian (top of the Lajas Formation) and the Late Callovian (base of the Lotena Formation). This stratigraphic interval is characterized by red beds corresponding to the Challacó and Bosque Petrificado Formations.

7 Stratigraphic Analysis of the Challacó and Bosque Petrificado Formations

A detailed field survey has been conducted in the southern part of the Neuquén Basin trying to unravel the stratigraphic significance of the Challacó and Bosque Petrificado formations. Eight stratigraphic sections were measured in the interval comprised between the Lajas and the Quebrada del Sapo formations. These sections have been described, measured, and sampled in detail, also registering several paleocurrent measurements from tractive sedimentary structures (tabular and trough cross-bedding, ripples and imbrication). The measured sections are distributed along a distance of 12.64 km, and record the thickness and stratigraphic variations recognized along the southern flank of the Picun Leufú anticline. The detailed correlation panel is shown in Fig. 5a. The location of these sections and measured paleocurrent directions in the Bosque Petrificado Formation is shown in Fig. 5b.

In all the analyzed sections, the boundary between the Lajas and Challacó formations is sharp, often placed over a coarsening upward bed recognized at the top of the Lajas Formation. It is considered that this contact represents a sequence boundary having an important tectonostratigraphic significance. The thickness of the Challacó Formation is highly variable (from 10 up to 120 m), since it is often deeply truncated by the Intra-Callovian Unconformity (ICU).

The Challacó Formation is dominantly composed of gray to greenish shales interbedded with lenticular to tabular sandstone and pebbly sandstone bodies. Lenticular sandstone bodies represent channel-fill deposits up to 4 m thick, which can be traced laterally for about 30–50 m, evidencing a ribbon-like geometry. These beds lay on top of a basal erosional surface and internally show large-scale tabular and trough cross-bedding, sigmoidal cross-bedding, parallel lamination and massive bedding. It is common in the presence of fractured rounded clasts. Tabular sandstone bodies are up to 2 m thick and internally show massive bedding, parallel lamination and climbing ripples. Remnants of silicified wood and plant remains are common. These sedimentological evidence suggest that the Challacó Formation was accumulated, in a shallow brackish lake partially affected by tidal influence. Sandstones and pebbly sandstones bodies are here interpreted as sublacustrine hyperpycnal channels and lobe complexes. At present, no relevant biostratigraphic elements have been

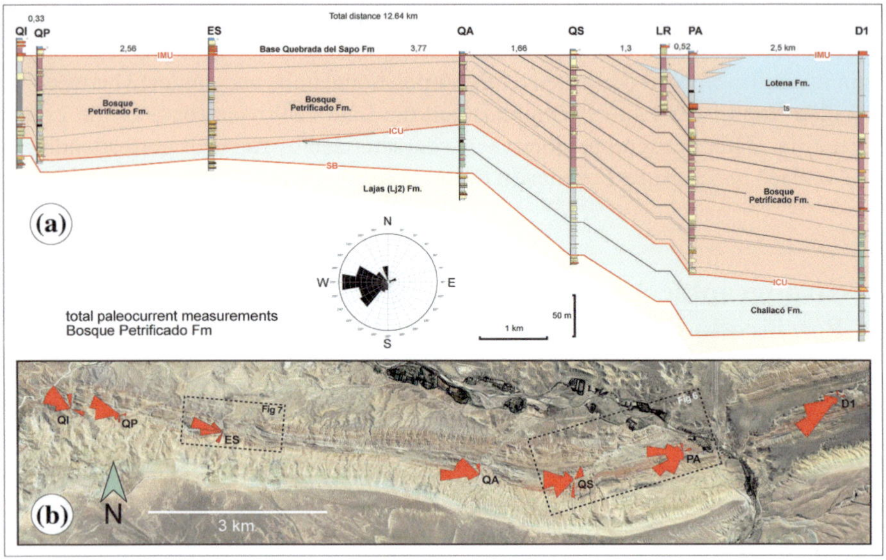

Fig. 5 **a** Stratigraphic cross section along the Picún Leufú anticline. The new proposed boundary between the Cuyo and Lotena Groups in the Picún Leufú area is indicated. The base of the Bosque Petrificado Formation is a regional unconformity, and in places an angular unconformity. **b** Paleocurrent directions measured in the Bosque Petrificado Formation (N: 64). Note that sediments were mainly supplied from tectonically exhumed areas located at the E-NE. D1: Destacamento 1; PA: Puesto Antileo; LR: La Ruta; QS: Quebrada Seca; QA: Quebrada Alvarez; ES: Escuela; QP: Quebrada Painemilla; QI: Quebrada del Indio

recognized in the studied palynological samples of the Challacó Formation. Consequently, according to its stratigraphic position, the Challacó Formation is interpreted as accumulated during the Late Bajocian–Early Callovian.

The contact between the Challacó and Bosque Petrificado formations represents a regional sequence boundary recognized all along the study area, in some places with associated evidence of erosion, angular relationships (Zavala 1993; Zavala and Gonzalez 2001; Zavala and Freije 2002) and tectonic overprint. This unconformity is here correlated with the Intra-Callovian Unconformity (Gulisano et al. 1984), located at the base of the Lotena Group. At different positions along the study area, this unconformity is sealing normal faults affecting the underlying Cuyo Group (Figs. 6 and 7). The existence of a distensive tectonics below the Intra-Callovian Unconformity has been also recognized northward from the Huincul Arch in the Los Catutos and Mallin del Rubio localities (Zavala 2002).

In the study area, the Bosque Petrificado Formation is up to 200 m thick and starts with a regionally continuous conglomerate, pebbly sandstone and sandstone body up to 8 m thick. This basal bed decreases in grain size from east to west, which is coincident with the paleocurrent directions (Fig. 5). Internally, the Bosque Petrificado Formation is composed of red, greenish and dark gray shales interbedded with sandstones, pebbly sandstones and conglomerate beds accumulated in a fluvial to shallow

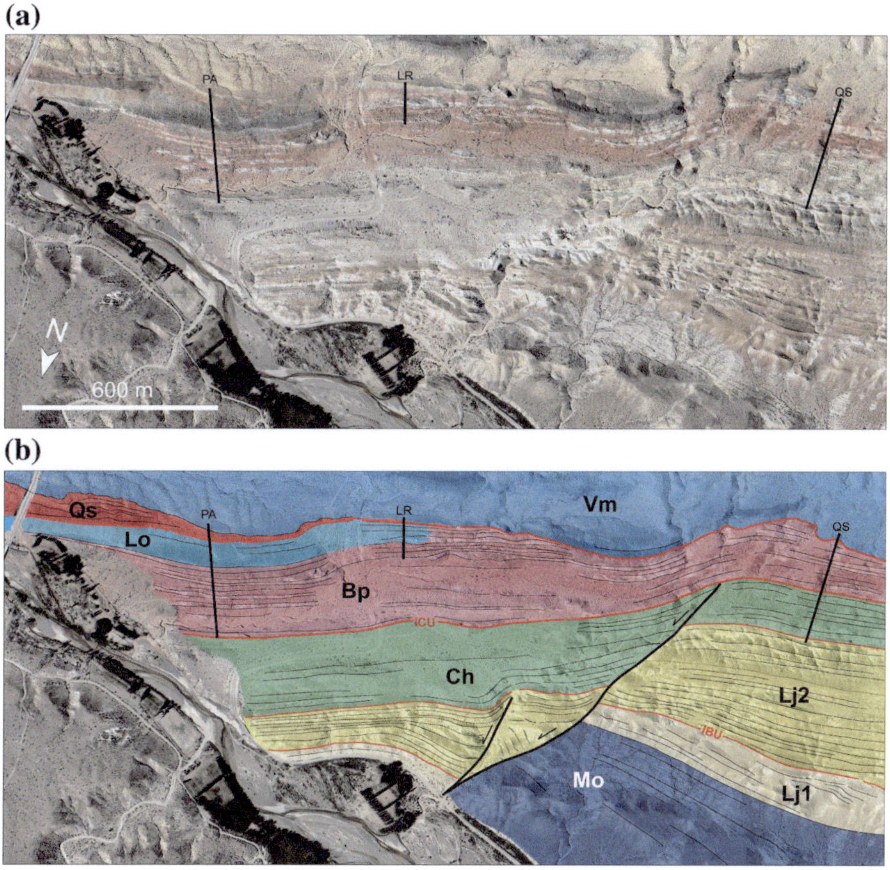

Fig. 6 Linedrawing and interpretation of the Jurassic succession outcropping in the vicinity of the Ruta Nacional 40 bridge over the Picún Leufú River. Note the existence of a series of normal faults affecting the Cuyo Group, which are passively sealed by the Intra-Callovian Unconformity. Stratigraphic evidence supports a genetic relationship between the Bosque Petrificado and Lotena formations. Mo: Los Molles Formation; Lj1: Lajas Formation (1); Lj2: Lajas Formation (2); Ch: Challacó Formation; Bp: Bosque Petrificado Formation; Lo: Lotena Formation; Qs: Quebrada del Sapo Formation; Vm: Vaca Muerta Formation: IBU: Intra-Bajocian Unconformity; ICU: Intra-Callovian Unconformity. PA, LR and QS refer to the sedimentary logs shown in the correlation of Fig. 5 (see Fig. 5 for location)

marine environment. In general, the grain size of these beds is coarser respect to the underlying Challacó Formation. Sandstone and conglomerate beds are tabular to lenticular in shape and internally show large-scale asymptotic cross-bedding. Massive and laminated beds are also common. Dark gray shales are massive or laminated, often showing plant remains and entire leaves with exceptional preservation. In the Picun Leufú area, the upper red beds of the Bosque Petrificado Formation laterally grade up into gray mudstones with ammonoids and other marine fauna that belong

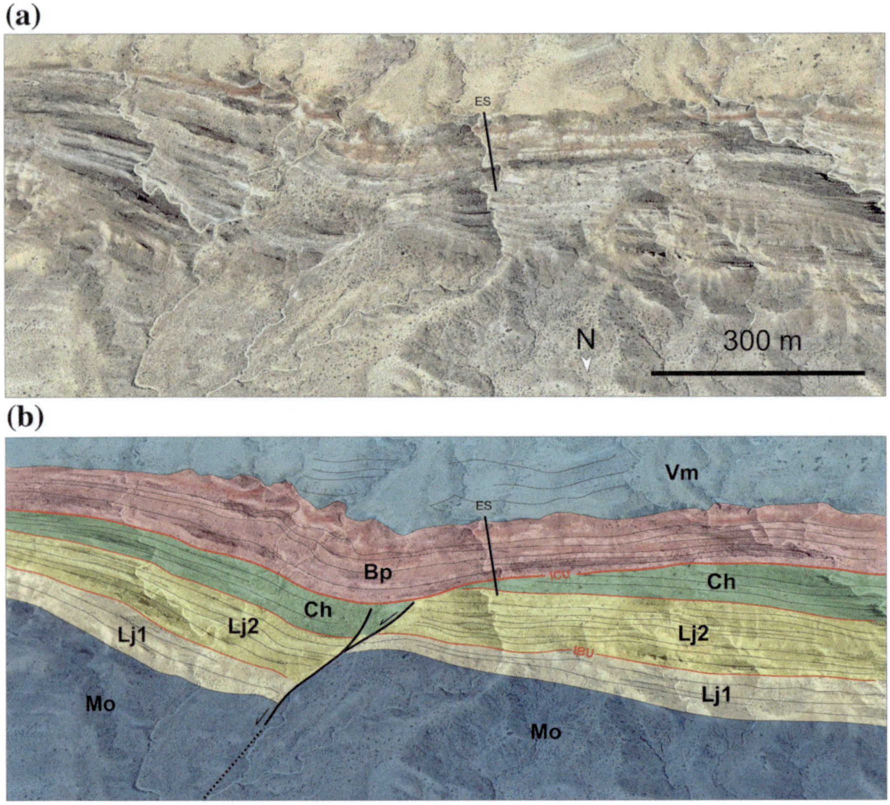

Fig. 7 Linedrawing and interpretation of the Jurassic succession outcropped at the south of the N° 293 School in Picún Leufú. Note that normal faults only affect the Cuyo Group and are sealed by the Intra-Callovian Unconformity. Mo: Los Molles Formation; Lj1: Lajas Formation (1); Lj2: Lajas Formation (2); Ch: Challacó Formation; Bp: Bosque Petrificado Formation; Lo: Lotena Formation; Qs: Quebrada del Sapo Formation; Vm: Vaca Muerta Formation: IBU: Intra-Bajocian Unconformity; ICU: Intra-Callovian Unconformity. ES refers to the sedimentary log shown in the correlation of Fig. 5 (see Fig. 5 for location)

to the Lotena Formation (Fig. 5a). According to their ammonoid fauna, these marine deposits were assigned to the Late Callovian–Early Oxfordian (Garrido and Parent 2013). Recent biostratigraphic studies performed in the basal levels of the Bosque Petrificado Formation at the Quebrada del Sapo locality suggest an age younger than Early Callovian for this unit (Olivera et al. 2018). This age results coherent with the proposed stratigraphic position.

Toward the south, the thickness of the Bosque Petrificado Formation decreases and progressively pinches out, with associated facies change (Fig. 8). At the Estancia Charahuilla section (ECH), this unit is composed of a monotonous succession of reddish shales. The unit is absent at the estancia María Juana section (EMJ), where

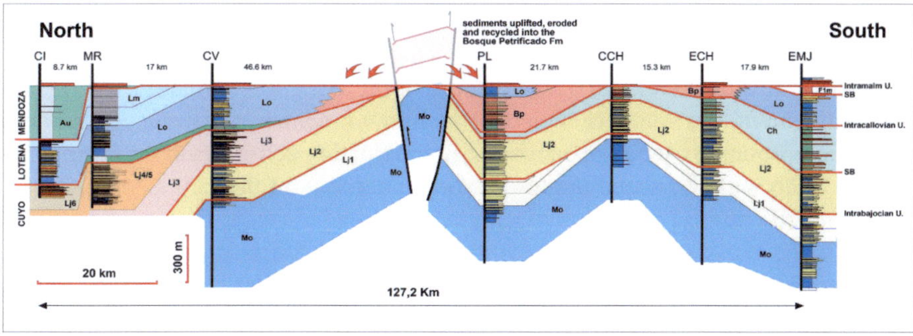

Fig. 8 Regional cross section between the Estancia Maria Juana and the northern part of the Sierra de la Vaca Muerta range (127.2 km). Note that the early uplift and erosion of Lower to Middle Jurassic deposits at the Huincul arch produced the basin segmentation. As a consequence, the southern Picun Leufú sub-basin evolved into a restricted lake (Challacó Formation) followed by a clastic wedge deposit (Bosque Petrificado Formation) resultant from the supply of clastic sediments from the uplifted region. Northward from the Huincul Arch (Sierra de la Vaca Muerta area), the marine sedimentation was not interrupted during the Bathonian–Early Callovian, and sandstone deposits of the Lajas Formation were supplied from the southwest

marine deposits of the Lotena Formation unconformably overlay coarse-grained fluvial deposits of the Challacó Formation (Fig. 8).

A total of 67 paleocurrent measurements were performed in deposits of the Bosque Petrificado Formation. These paleocurrents clearly suggest that clastic materials were supplied from the east and northeast (Fig. 5). This direction is anomalous compared with the main prograding direction from S-SE to N-NW registered for the Cuyo Group in the study area (Zavala 1996a). Consequently, it is interpreted that the conglomerates, sandstones and shales belonging to the Bosque Petrificado Formation represent recycled sedimentary deposits previously accumulated in the Lajas and Los Molles formations. These materials were eroded from tectonically uplifted areas located along the Huincul arch (Fig. 8) and transferred towards the flanks of the structure.

8 Regional Stratigraphic Consequences of the Early Uplift Along the Huincul Arch

The tectonic uplift during the late Jurassic along the Huincul Arch had important consequences for the stratigraphy of the related deposits, since it created a physical barrier that limited the connection between southern and central basin areas. Figure 8 shows a roughly north-south regional cross section along 127.2 km (between the north of the Sierra de la Vaca Muerta and Estancia Maria Juana), flattened at the base of the Upper Jurassic Tordillo Formation. In general, the correlation is relatively simple up to the Bajocian, including the Los Molles, and Lajas (LJ1 and LJ2) Formations

(Fig. 9a). Nevertheless, during the Bathonian-Early Callovian the situation dramatically changed (Fig. 9b), since the basin was partially fragmented into two different depocenters resultant from an early tectonic uplift along the Huincul Arch. Due to this confinement, the southern area evolves into a brackish lacustrine environment (Challacó Formation), while open marine sedimentation continues at the north of the Huincul Arch with the accumulation of shelfal deposits of the Lajas Formation (Lj 3 to Lj 6) (Fig. 9b).

A detailed cross section of the Cuyo Group in the Sierra de la Vaca Muerta is shown in Fig. 10. The sedimentation of the Lajas Formation is characterized by a series of low-angle wedges which are progressively younger toward the north. As documented from paleocurrent analysis, the sediments in the lower sequence (Lajas 2) show a clear provenance from the south-southwest and are mainly composed of

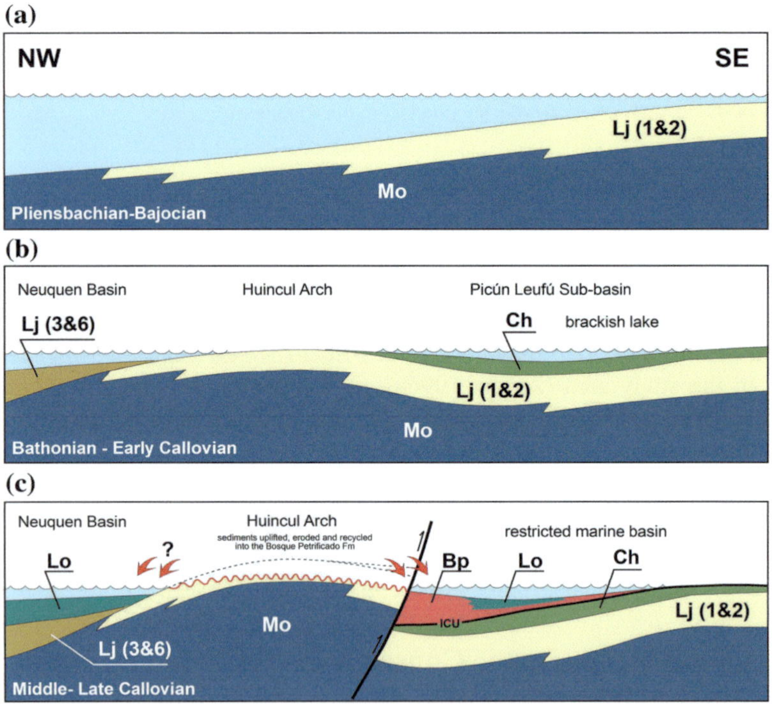

Fig. 9 Chronostratigraphic evolution of the Huincul Arch and its consequences in the stratigraphy. **a** During the Pliensbachian–Bajocian, the Cuyo Group prograded from the southeast. **b** As a consequence of an early tectonic uplift during the Bathonian-Early Callovian, a brackish lake developed in the southern Neuquén Basin, resulting in the accumulation of red beds assigned to the Challacó Formation. Northward from the Huincul Arch (Sierra de la Vaca Muerta) marine sedimentation prevailed, with the accumulation of lithic materials sourced from the southwest. **c** During the Middle-Late Callovian, uplifted and eroded sediments of the Cuyo Group were accumulated in the Bosque Petrificado Formation. These continental deposits laterally grade into marine shales of the Lotena Formation. Eroded sediments accumulated also in the northern flank of the structure, composing coarse grained deposits of the Lotena Group recognized in the subsurface.

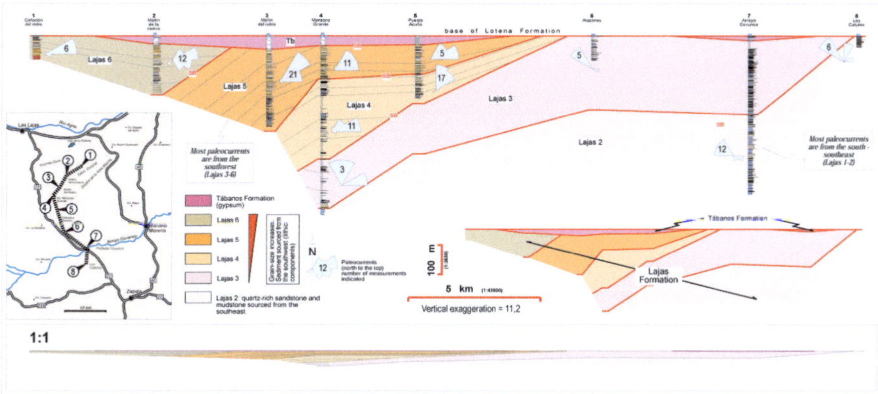

Fig. 10 A 32 km long detailed cross section of the Cuyo Group in Sierra de la Vaca Muerta area (between Los Catutos and Cañadón del Indio). Note that the Lajas Formation is composed of different low-angle wedges that become progressively younger northward. Sequence 2 is quartz-rich and was supplied from the south-southwest. Source area change to the southwest for sequences 3–6. These sequences are coarse-grained and composed of lithic sediments. It is interpreted herein that this change in the source area was a consequence of an early uplift along the Huincul Arch

quartz-rich sandstones. During the next sequence (Lajas 3), the supply area progressively rotates toward the southwest. According to paleocurrent data (Fig. 10), the sediment provenance from the southwest is clear for the younger sequences (Lajas 4–6), which are dominantly composed of lithic elements with an increasing grain size for the younger sequences. This cross section also documents the "out of sequence" accumulation of the Tábanos Formation, since this unit developed above different depositional sequences of the Lajas Formation, evidencing a clear transgressive tendency. For this reason, it is considered that the Tábanos evaporites probably belong to the overlying Lotena Group (see Figs. 2 and 8).

Maximum uplift along the Huincul Arch apparently took place during the Early Callovian, since the main erosional angular unconformity was passively filled with Early (?)–Middle Callovian deposits of the Lotena Group, often laying over a normal-fault distensive system (Figs. 6 and 7). According to these evidences, it is herein interpreted that the Bosque Petrificado, Lotena and Fortin 1° de Mayo Formations recognized in the Picún Leufú Sub-basin are temporarily equivalent to the Tábanos, Lotena, La Manga and Auquilco formations recognized in the central Neuquén Basin (Figs. 2, 8 and 9c). Sediments eroded from exhumed areas accumulated also northward from the Huincul Arch, and are composed of coarse-grained deposits of the Lotena Group recognized in different oil fields (e.g. Aguada Toledo).

9 Consequences for the Stratigraphy of the Picun Leufú Area

Figure 11 shows a conceptual stratigraphic column for the Middle–Upper Jurassic at the Picún Leufú area, with indication of main paleocurrent directions and compressive-distensive tectonic events according to field observations and evidences that emerged from this study. In the study area, the Los Molles, Lajas (Lj 1 and Lj2) and Challacó Formations are bounded by unconformities. Nevertheless, the unconformity between the Los Molles and Lajas Formations was only recognized in certain areas around the southern flank of the Picún Leufú anticline (Freije et al. 2002), being the boundary between these two stratigraphic units generally transitional. On the contrary, the Intra-Bajocian Unconformity (IBU), which establish the boundary betweeen the Lajas 1 and Lajas 2 sequences, is a regional unconformity that locally truncates hundreds of meters of the underlying succession (Zavala 1996a). Compressive tectonic activity progressively increases towards the top of the Lajas 2 sequence, with a maximum activity during the accumulation of the Challacó Formation. Although the main paleocurrents measured in the Challacó Formation suggest a source area located at the southeast, in those areas located closer to the Huincul Arch (Picun Leufú Bridge over Ruta Nacional 40) a local supply from the northeast is also recognized. The accumulation of the Lotena Group started after a distensive tectonic event with a coarse-grained clastic succession supplied from the northeast

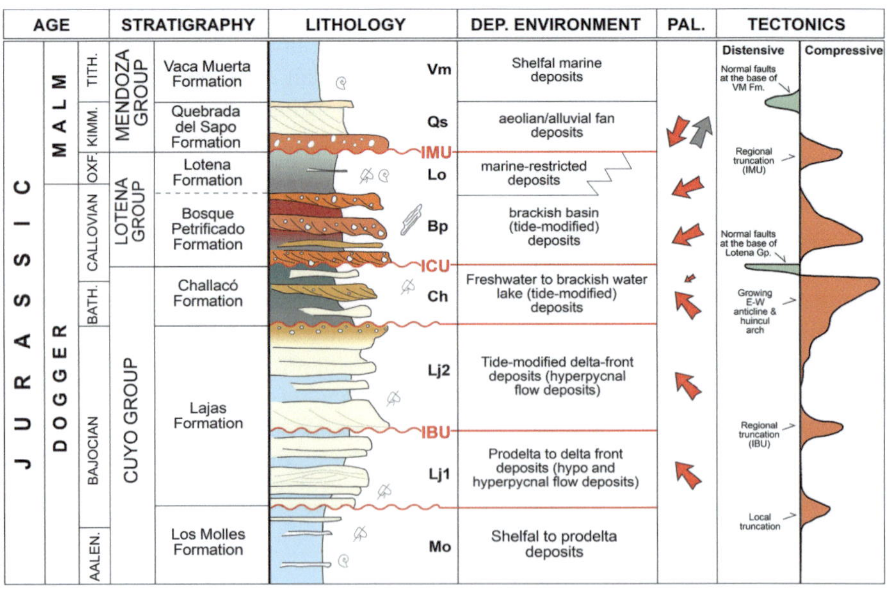

Fig. 11 Simplified stratigraphic column in the Picun Leufú Sub-basin, with indication of the main compressive–distensive tectonic movements registered in the area. The uplift and erosion of Jurassic strata are evidenced by a drastic change in paleocurrents direction (modified from Freije et al. 2002)

(Bosque Petrificado Formation). Compressive tectonic activity also continued during the Middle-Late Callovian and Early Kimmeridgian (Intramalm Unconformity, IMU) since an angular unconformity is commonly recognized at the base of the Quebrada del Sapo Formation.

Finally, it is herein interpreted that the rapid marine flooding registered in the Early Tithonian, which defines the base of the Vaca Muerta Formation, was also tectonically enhanced, since normal fault activity was recognized as well at this boundary (Freije et al. 2002, their Fig. 11a).

10 Conclusions

Stratigraphic and sedimentological evidence suggests a complex scenario for the accumulation of the Jurassic units in the southern part of the Neuquén Basin. Synsedimentary tectonics along the Huincul Arch played a fundamental role in controlling the stratigraphy at both margins of this tectonostratigraphic element. According to these evidences, the Challacó Formation should no longer be considered the continental equivalent of the Lajas Formation, since it corresponds to a depositional unit having a tectonostratigraphic significance, and exclusively developed at the south of the Huincul Arch.

Like in other transgressive depositional cycles in the Neuquén Basin (Tordillo-Vaca Muerta; Mulichinco-Pilmatué; Avilé-Agua de la Mula), the marine flooding of the Lotena Formation in the southern basin is preceded by an initial continental/brackish interval, corresponding to the Bosque Petrificado Formation. A succession equivalent to this basal coarse-grained interval is recognized in the Loncopué area, integrated by conglomerates and coarse-grained sandstones up to 70 m thick (Zavala et al. 2002).

Although recent zircon-based studies on the Quebrada del Sapo Formation support the uplift of the Huincul Arch during the Late Jurassic (Naipauer et al. 2012; Naipauer and Ramos 2016), new studies will be necessary to analyze and discuss these new evidences of an early uplift and erosion during the Late Bajocian-Oxfordian. In the study area, the Challacó and Bosque Petrificado formations can be clearly differentiated because of the existence of an important tectonically enhanced unconformity at their boundary. Nevertheless, in some places these units display isopic facies associations that turn challenging their differentiation. In this way, new regional studies will be required aimed in performing a detailed mapping of these units showing their distribution, facies changes, paleocurrents and biostratigraphy along the Picún Leufú Sub-Basin.

This study is mainly based in outcrops located along the southern flank of the Huincul arch. In this area, the evidence clearly supports a sediment supply from the northeast associated to an early uplift of Jurassic and older deposits. Nevertheless, this ancient uplifted area have probably supplied sediments not only to the south, but also to the depocenters located towards the north. It is probable that part of these recycled coarse-grained sediments were included in the lower section of the

Lotena Formation recognized in the Agua Del Cajón, Loma de la Lata, Aguada Toledo and Sierra Barrosa localities (Zavala et al. 2002). At these positions, the Lotena Formation includes lithic clasts with angular fragments of rounded pebbles that suggest a recycling of Middle Jurassic conglomerates belonging to the Lajas Formation.

Acknowledgements The authors deeply acknowledge to Marcelo Martinez, Daniela Olivera and Ainara Irastorza for their help, support and interesting discussions during fieldwork. We also appreciate the support of the Departamento de Geología (Universidad Nacional del Sur), GCS Argentina SRL and the CONICET for the completion of field activities.

References

Canale N, Ponce J, Carmona N et al (2014) Sedimentology and Ichnology of fluvio-dominated deltas affected by hyperpycnal discharges Lajas Formation (Middle Jurassic), Neuquén Basin, Argentina. Andean Geol 42:114–138

De Ferrariis C (1947) Edad del arco o dorsal antiguo del Neuquén oriental, de acuerdo con la estratigrafía de la zona inmediata. Rev Asoc Geol Argent 2:256–283

Dellapé DA, Pando GA, Uliana MA et al (1979) Foraminíferos y ostrácodos del Jurásico en las inmediaciones del arroyo Picún Leufú y la Ruta 40 (provincia de Neuquén, Argentina), con algunas consideraciones sobre la estratigrafía de la Formación Lotena. In: Abstracts of the 7 Congreso Geológico Argentino, Neuquén, 9–15 Apr 1978

Freije H, Azúa G, González R et al (2002) Actividad Tectónica Sinsedimentaria en el Jurásico del Sur de la Cuenca Neuquina. In: Abstracts of the 5 Congreso de Exploración y Desarrollo de Hidrocarburos, IAPG, Mar del Plata, 29 Oct–2 Nov 2002

Garcia VM, Quattrocchio ME, Zavala C et al (2006) Palinofacies, paleoambientes y paleoclima del Grupo Cuyo (Jurásico Medio) en la Sierra de Chacaico, Cuenca Neuquina, Argentina. Rev Esp Micropal 38:269–288

Garrido AC, Parent H (2013) Estratigrafía y fauna de amonites de los depósitos "Lotenianos" (Caloviano Medio Oxfordiano Inferior?) del anticlinal de Picún Leufú, Cuenca Neuquina, Subcuenca Picún Leufú, Argentina. Bol Inst Fis Geol 83:35–68

Groeber P (1946) Observaciones geológicas a lo largo del Meridiano 70. I- Hoja Chos Malal. Rev Asoc Geol Argent 1:177–208

Groeber P, Stipanicic P, Mingramm A (1953) Jurásico. In: Groeber P, Stipanicic P, Mingramm A (eds) Geografía de la República Argentina, vol 2. Sociedad Argentina de Estudios Geográfico, Buenos Aires, pp 143–347

Gugliotta M, Flint SS, Hodgson DM et al (2016) Recognition criteria, characteristics and implications of the fluvial to marine transition zone in ancient deltaic deposits (Lajas Formation, Argentina). Sedimentology 63:1971–2001

Gulisano CA, Gutierrez Pleimling AR, Digregorio RE (1984) Esquema estratigráfico de la secuencia jurásica al oeste de la provincia del Neuquén. In: Abstracts of the 9 Congreso Geológico Argentino, San Carlos de Bariloche, 5–9 Nov 1984

Herrero Ducloux A (1946) Contribución al conocimiento geológico del Neuquén extrandino. Bol Inf Petrol 23:245–281

Hogg SL (1993) Geology and hydrocarbon potential of the Neuquén, Basin. J Petrol Geol 16:383–396

Iparraguirre J, Arcuri M, Zavala C et al (2018) Lithohero: Creating comprehensive sedimentary logs from cores and outcrops. In: Abstracts of the workshop on deep water sedimentology: progress and perspectives, Bahia Blanca, Apr 2018

McIlroy D, Flint S, Howell JA et al (2005) Sedimentology of the tide-dominated Jurassic Lajas Formation, Neuquén Basin, Argentina. In: Veiga GD, Spalletti LA, Howell JA et al (eds) The Neuquén Basin, Argentina: a case study in sequence stratigraphy and basin dynamics. The Geological Society, London, SP 252, pp 83–107

Mosquera A, Silvestro J, Ramos VA et al (2011) La estructura de la Dorsal de Huincul. In: Leanza HA, Arregui C, Carbone O et al (eds) Geología y Recursos Naturales de la Provincia del Neuquén. Asociación Geológica Argentina, Buenos Aires, pp 385–397

Naipauer M, Morabito EG, Marques JC, Tunik M, Vera EAR, Vujovich GI, Pimentel MP, Ramos VA (2012) Intraplate Late Jurassic deformation and exhumation in western central Argentina: constraints from surface data and U–Pb detrital zircon ages. Tectonophysics 524–525:59–7

Naipauer M, Ramos VA (2016) Changes in source areas at Neuquén Basin: Mesozoic evolution and tectonic setting based on U-Pb ages on zircons. In: Naipauer M, Sagripanti L, Ghiglione M, Orts D (eds) Folguera A. Growth of the Southern Andes, Springer, pp 33–61

Olivera DE, Martínez MA, Zavala C et al (2018) Aportes al conocimiento de la palinología del Grupo Lotena (Jurásico Medio–Tardío) en la Subcuenca de Picún Leufú, Cuenca Neuquina, Argentina. Bol Asoc Latinoam Paleobot Palinolog 18:62R

Paim PSG, Lavina ELC, Faccini UF et al (2011) Fluvial-derived turbidites in the Los Molles Formation (Jurassic of the Neuquén Basin): initiation, transport, and deposition. In: Slatt RM, Zavala C (eds) Sediment transfer from shelf to deep water—revisiting the delivery system. AAPG Studies in Geology 61, pp 95–116

Riba O (1976) Syntectonic unconformities of the Alto Cardener, Spanish Pyrenees: a genetic interpretation. Sediment Geol 15:213–233

Riccardi AC, Damborenea SE, Mancenido MO et al (1988) Hettangiano y Sinemuriano marinos en Argentina. In: Abstracts of the 5 Congreso Geológico Chileno, 8–12 Aug 1988

Silvestro J, Zubiri M (2008) Convergencia Oblicua: Modelo structural alternativo para la Dorsal Neuquina (39°S) – Neuquén. Rev Asoc Geol Argent 63:49–64

Steel E, Simms AR, Steel R et al (2018) Hyperpycnal delivery of sand to the continental shelf: insights from the Jurassic Lajas Formation, Neuquén Basin, Argentina. Sedimentology. https://doi.org/10.1111/sed.12460

Stipanicic P (1969) El avance en los conocimientos del Jurásico argentino a partir del esquema de Groeber. Rev Asoc Geol Argent 24:367–388

Stipanicic PN, Rodrigo F (1970) El diastrofismo jurásico en Argentina y Chile. In: Abstracts of the 4° Jornadas Geológicas Argentinas, Buenos Aires

Veiga GD (1998) Estratigrafía secuencial en series continentales: aplicación a los depósitos de la Formación Challacó, Jurásico de la cuenca neuquina austral (República Argentina). Rev Soc Geol España 11:95–109

Vergani GD, Tankard J, Belotti J et al (1995) Tectonic evolution and paleogeography of the Neuquén Basin, Argentina. In: Tankard AJ, Suárez R, Welsink HJ (eds), Petroleum basins of South America. AAPG Memoir 62:383–402

Weaver C (1931) Paleontology of the Jurassic and Cretaceous of West Central Argentine. Men Univ, Washington, vol. 1, Seattle

Zavala C (1993) Estratigrafía y análisis de facies de la Formación Lajas (Jurásico medio) en el sector suroccidental de la Cuenca Neuquina. Provincia del Neuquén. República Argentina. Ph.D. thesis, Universidad Nacional del Sur

Zavala C (1996a) High-resolution sequence stratigraphy in the Middle Jurassic Cuyo Group, South Neuquén Basin, Argentina. GeoResearch Forum, Advances in Jurassic Research 1–2:295–304

Zavala C (1996b) Sequence stratigraphy in continental to marine transitions; an example from the Middle Jurassic Cuyo Group, South Neuquén Basin, Argentina. GeoResearch Forum, Advances in Jurassic Research 1–2:285–293

Zavala C (2002) El contacto entre los grupos Cuyo y Lotena (Jurásico) en la Sierra de la Vaca Muerta. Cuenca Neuquina, Argentina. In: Abstracts of the 15 Congreso Geológico Argentino, Córdoba, 23–26 Aug 2002

Zavala C, Arcuri M (2016) Intrabasinal and Extrabasinal turbidites: origin and distinctive characteristics. Sediment Geol 337:36–54
Zavala C, Freije H (2002) Cuñas clásticas jurasicas vinculadas a la Dorsal de Huíncul. Un ejemplo del área de Picún Leufú. Cuenca Neuquina, Argentina. In: Abstracts of the 5 Congreso de Exploración y Desarrollo de Hidrocarburos, Mar del Plata, 29 Oct–2 Nov 2002
Zavala C, Gonzalez R (2001) Estratigrafía del Grupo Cuyo (Jurásico inferior-medio) en la Sierra de la Vaca Muerta, Cuenca Neuquina. Bol Inf Petrol 17:52–64
Zavala C, Maretto H, Arcuri M (2002) Las facies clásticas de la Formación Lotena (Jurásico medio) en las áreas de Loncopué y Loma La Lata. Cuenca Neuquina. Argentina. In: Abstracts of the 5 Congreso de Exploración y Desarrollo de Hidrocarburos, Mar del Plata, 29 Oct–2 Nov 2002
Zavala C, Ponce J, Arcuri M et al (2006) Ancient Lacustrine hyperpycnites: a depositional model from a case study in the Rayoso Formation (Cretaceous) of west-central Argentina. J Sediment Res 76:40–58

Environmental Controls and Facies Architecture of a Jurassic Carbonate Episode (La Manga Formation), Mendoza Province, Neuquén Basin

Ricardo M. Palma, Graciela S. Bressan, Alberto C. Riccardi, José López-Gómez and Javier Martín-Chivelet

Abstract La Manga Formation is a vast carbonate system developed in the Neuquén Basin. The age is based in ammonite faunas, ranging from Early Callovian (Bodenbenderi-Proximum Zone) to Middle Oxfordian (Cordatum Standard Zone to Transversarium Standard Zone, and probably to the lower part of the Bifurcatus Standard Zone). A stratigraphical and sedimentological analysis, in the outcrops exposed in the south of Mendoza province, enabled the recognition of five facies associations of a carbonate ramp corresponding to (1) distal outer ramp, (2) proximal outer to distal middle ramp, (3) proximal middle ramp, (4) inner ramp deposits (shoreface, shoal, patch reef, shallow subtidal lagoon and tidal flat) and (5) paleokarstic facies. These facies correspond to homoclinal to distally steepened carbonate ramp. The facies associations are included into three third-order depositional sequences (DS-1, DS-2, DS-3) represented by transgressive and highstand systems tracts with sequence boundaries of regional character. Different controlling factors can be recognised in the deposition of this unit. The abrupt changes of facies, as well as paleokarst and epikarst discontinuity surfaces in the successions provide important evidence in terms of depositional environment and vertical evolution of the carbonate ramp. Facies patterns are variable across the outcrop area and vertically through time because of a combination of ramp morphology, siliciclastic supply, sea level changes and tectonic effects. In the southern sections, siliciclastic influx influenced the deposition

R. M. Palma (✉) · G. S. Bressan
Facultad de Ciencias Exactas y Naturales Departamento de Ciencias Geológicas, IDEAN—CONICET, Universidad de Buenos Aires, La Plata, Argentina
e-mail: palma@gl.fcen.uba.ar

A. C. Riccardi
Facultad de Ciencias Naturales y Museo—CONICET, Universidad Nacional de La Plata, La Plata, Argentina

J. López-Gómez
Consejo Superior de Investigaciones Cienctíficas, Instituto de Geociencias (CSIC), Universidad Complutense de Madrid, Madrid, Spain

J. Martín-Chivelet
Facultad de Ciencias Geológicas & Instituto de Geociencias (CSIC), Universidad Complutense de Madrid, Madrid, Spain

D. Kietzmann and A. Folguera (eds.), *Opening and Closure of the Neuquén Basin in the Southern Andes*, Springer Earth System Sciences,
https://doi.org/10.1007/978-3-030-29680-3_4

of proximal middle ramp facies later overlain by scleractinian patch reefs which grew up throughout progressive stages from aggradational to progradational facies in response to climate controls and nutrient levels influence. In northern outcrops, tectonic controls affected the ramp topography and influenced the development of distal deep marine facies. Shallow subtidal and peritidal cycles indicate a combination of allocyclic and autocyclic processes controlling accommodation space and sediment accumulation.

Keywords Callovian–Oxfordian · Carbonate ramp · Sea level changes · Tectonic controls

1 Introduction

During the Callovian–Oxfordian times in the Neuquén Basin, a widespread carbonate deposition resulted in the settlement of the carbonate ramp of La Manga Formation, which is exceptionally exposed (Fig. 1). It was characterised by a mosaic of shallow-water facies adjacent to deep-water facies (Legarreta 1991; Legarreta and Uliana 1996; Palma et al. 2007, 2009, 2013, 2014; Lo Forte and Palma 2002) that constitutes the main carbonatic unit of the Jurassic of South America.

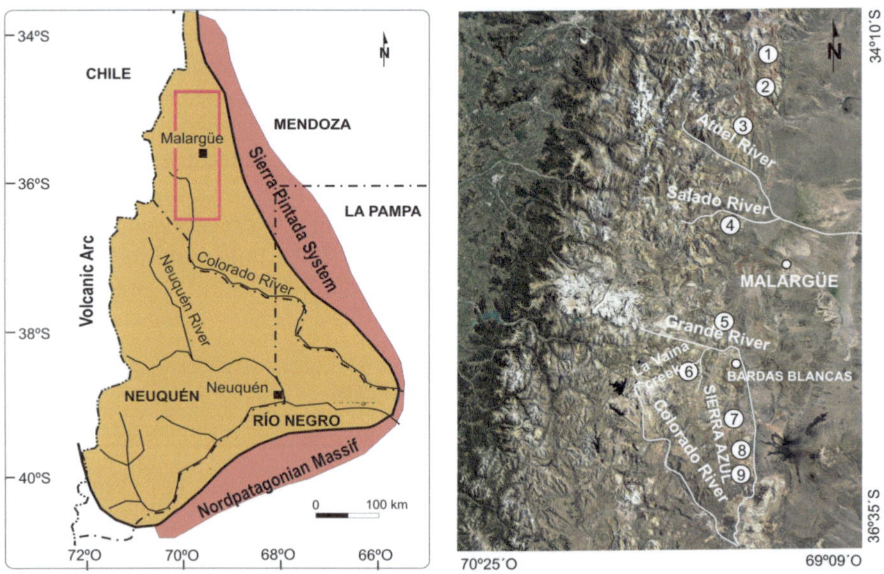

Fig. 1 Location map of the Neuquén Basin and the studied localities: (1) Yeseras Grandes, (2) La Manga creek, (3) Codo del Blanco: (4) Los Blancos creek (1 and 2); (5) Bardas Blancas; (6) La Vaina creek; (7) Coihueco creek; (8) Mechanquil creek; and (9) Mallín Redondo

The development of La Manga Formation has been influenced by autocyclic and allocyclic processes which gave rise to a complex depositional pattern throughout different depocentres (Palma et al. 2007, 2009, 2013, 2014), evidenced by important lateral and vertical shifts of facies from distal outer ramp to intertidal–supratidal deposits, including well-developed reefal deposits, reaching its best development in Bardas Blancas area (Fig. 1).

Studies on the La Manga Formation have been focused mainly on its lithostratigraphy and biostratigraphy (Stipanicic 1965; Leanza 1981; Riccardi 1984, 1992; Palma et al. 2012), sedimentological interpretations (Legarreta 1991; Palma et al. 2005, 2007, 2009, 2011, 2014; Lo Forte and Palma 2002), palaeontology (Morsch 1990; Beresi 2003, 2007; Lazo et al. 2008, Bressan and Palma 2010; Beresi et al. 2017) and diagenetic aspects of the succession (Palma et al. 2007, 2015, 2017). Due to their economic interest, these carbonate ramp sediments have been extensively investigated and many sequence stratigraphic and palaeogeographical data are therefore available (Mitchum and Uliana 1985; Legarreta and Uliana 1996). However, detailed correlation and paleogeographic reconstruction of this ramp are difficult due to the lack of precise biostratigraphic data and the influence of multiple controls in its evolution giving rise to abrupt vertical and lateral shifts of facies through the basin.

Bearing this in mine, this chapter reviewed the state of knowledge of this unit in order to better understand its depositional environments and how eustatic or tectonic controls conditioned its evolution in the southern part of the Mendoza Province.

2 Geological Setting

The deposition of La Manga Formation happened as part of the Jurassic Cycle of Groeber (1946). Years later, Legarreta and Gulisano (1989) agreed generally with the validity of the Groeber cycles and emphasised the importance of eustatic variations in the development of depositional sequences as was suggested by Legarreta and Uliana (1996). In the depositional sequence framework defined by Legarreta and Gulisano (1989), this unit is included in the Lotena Mesosequence, which together with the Precuyo and Cuyo Mesosequences constitute the Lower Supersequence.

The Lotena Mesosequence consists of five depositional sequences that include marine and continental facies (Lotena Formation), carbonate deposits (La Manga Formation) and evaporites (Auquilco Formation), developed from Middle Callovian to Late Oxfordian-Kimmeridgian times (Fig. 2), bounded by two important unconformities (Leanza 2009; Palma et al. 2007). Hiatuses and erosive surfaces separate these formations from the La Vaina creek through Bardas Blancas area and the localities of the northern sector (Yeseras Grandes); some units pinch out or become thinner and only those formations related to dominant transgressive pulses persist.

During Early Callovian–Middle Oxfordian, local and general sea level rise allowed the deposition of carbonate sediments formally assigned to the La Manga Formation, a wide carbonate ramp exceptionally exposed in west-central Argentina. Regional subsidence (thermal sag style) controlled the Pliensbachian–Neocomian

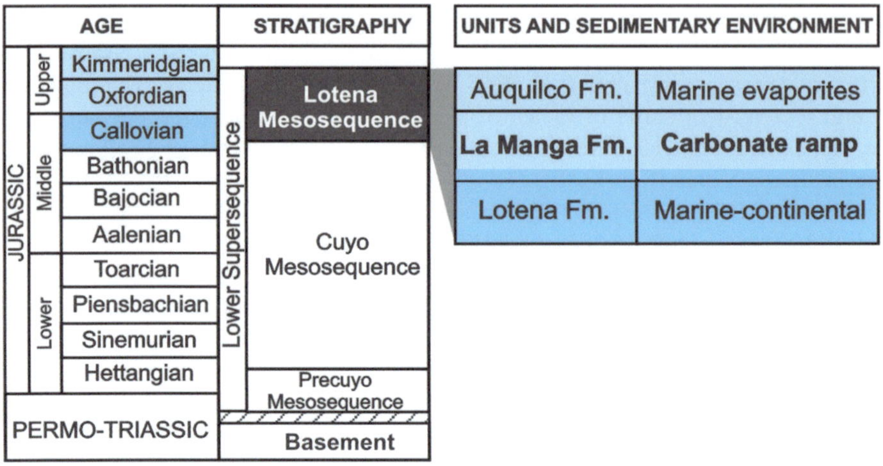

Fig. 2 Stratigraphic chart for the Lower Supersequence and Lotena Mesosequence, in the Neuquén Basin, Mendoza Province (modified from Palma et al. 2009)

period in the basin and also the deposition of the Cuyo, Lotena and Mendoza Groups. This subsidence also included various inversion and fault-controlled stages (Vergani et al. 1995) (see Chap. "Jurassic Uplift along the Huincul Arch and its Consequences in the Stratigraphy of the Cuyo and Lotena Groups. Neuquén Basin, Argentina").

The study area is located in the southern Mendoza province where outcrops of the La Manga Formation are extensive and well exposed. This work examines in detail the lithostratigraphy and sedimentology of the La Manga Formation in different localities (Fig. 1) and provides palaeoenvironmental interpretations of the sedimentary succession and a sequence stratigraphic scheme. Facies were associated into five facies associations briefly summarised in Table 1.

The lithological changes were observed in nine regional areas in southern Mendoza province distributed along a north–south transect throughout 215 km (Fig. 1), where the thickness of the studied sections changes from 23 to 60 m, being smaller in the northward outcrops. The areas reported in this work, from north to south, are: (1) Yeseras Grandes, (2) La Manga creek, (3) Codo del Blanco, (4) Los Blancos creek (1 and 2), (5) Bardas Blancas, (6) La Vaina creek, (7) Coihueco creek, (8) Mechanquil creek and (9) Mallín Redondo. Los Blancos creek 1 and 2 are part of the same succession with a tectonic displacement between them, where Sect. 1 belongs to deeper deposits and the Sect. 2 corresponds to shallower facies. For Bardas Blancas area, a generalised sedimentologic scheme is presented here, but it has to be taken into consideration that this scheme represents four different sections with important lateral changes that were studied in detail by Palma et al. (2007, 2009). Carbonate ramps respond in complex ways to sea level changes and tectonics (Burchette and Wright 1992), and complex facies distribution is inherent to this unit, being largely controlled by different depocentres of the Neuquén Basin (Palma et al. 2007, 2009).

Table 1 Summary of facies associations for Las Lajas Formation proposed in this work

Facies	Description	Components	Interpretation
Distal outer ramp	Mudstone–wackestone to packstone interbedded with shales	Ammonites, radiolarians, sponge spicules, *Chondrites*, *Bositra*. Micritic peloids	Pelagic and hemipelagic suspension settling with thin storm deposits
Distal middle to proximal outer ramp	Massive bioclastic wackestone-packstones	Monotypic accumulations of fragments of branching corals, echinoid spines or *Gryphaea* valves	Reworking by weak bottom currents and storms
	Mounds composed by massive limestones with sponges developed as bodies interbedded with shales. Intermounds of thin bedded limestones	Mounds with sponges and micritic crusts and intermounds with fragments of molluscs, echinoderms, isolated sponges, encrusting serpulids and foraminifers	Deposits controlled by microbial activity, bathimetry, sea-floor morphology and a low sedimentation rate
Proximal middle ramp	Bioclastic packstone and intraclastic–bioclastic packstones. Masive, laminated or graded bioclastic and peloidal wackestones. Frequent amalgamation	Packstones with Gryphaeids, bivalves, gastropods, echinoderms, bryozoans. Intraclasts Wackestones with bivalves and echinoderms fragments, bryozoans, foraminifers, serpulids and intraclasts	Storm deposits above storm wave base
Shoreface (inner ramp)	Fine to mediun grained sandstones interbedded with bioclastic packstone–grainstones. Horizontal laminations, HCS, SCS, wave ripples, tabular cross-stratification, ripple cross-laminations	Intraclasts, scattered oysters, and trace fossils	Storm-dominated shoreface
Shoal (inner ramp)	Oolitic grainstone–packstones. Massive to tabular cross-bedding	Ooids, fragments of molluscs, intraclasts echinoderms, peloids. *Cayeuxia*. Micritised rims	Inner ramp, moderate-high energy shoals
Coral framestones (inner ramp)	Tabular cross-stratified oolitic grainstone–packstone. Massive reefal deposits dominated by scleractinian corals. Intercolony matrix composed by a bioclastic wackestone, packstones, ooids and coated grains	Additional components include bivalves, echinoids, gastropods, dasycladacean and *Cayeuxia*, beside of serpulids, bryozoans, sponges, miliolids ostacods, peloids and intraclasts	Patch reef

(continued)

Table 1 (continued)

Facies	Description	Components	Interpretation
Shallow subtidal lagoon (inner ramp)	Bioturbated wackestones, peloidal-bioclastic wackestone, bioclastic packstones. Oncoidal packstones, floatstone/rudstones. Marls and mudstones Laminated, massive, some herringbone cross-stratification	Oyster and other bivalves, echinoderms, bryozoan, coral fragments, *Cayeuxia* and dasycladacean algae, ostracods, and miliolids. Patch reef. Oncoids	Shallow lagoon. Slightly to moderate energy conditions. Lagoonal channels
Tidal flat (inner ramp)	Mudstone–wackestones Flat-pebble conglomerates, breccias. Planar laminations, massive or normal grading, bioturbation, mud-cracks, tepees	Planar to corrugate stromatolites Ammonites and echinoids.	Restricted intertidal–supratidal setting Occassional storms reworked cemented peritidal facies
Paleokarst	Matrix to clasts supported carbonate breccias. Massive. Dissolution surface, isolated pisolites	Absent	Paleokarst, epikarst

3 Biostratigraphy

The age of the La Manga Formation has been controversial since the pioneer works of Gerth (1925), Jaworski (1925), Groeber (1937), and Groeber et al. (1953). Ammonites were long regarded as Callovian (Groeber 1918, 1933). However, Oxfordian ammonites from the Cordatum-Plicatilis Standard Zones (Lower–Middle Oxfordian) were recognised by Stipanicic (1951) and Stipanicic et al. (1975) in the La Manga Formation type locality (La Manga creek). This unit was later considered Oxfordian in age by Riccardi (1996).

Years later, new detailed sedimentary and stratigraphic studies provided a new record of ammonite faunas from a condensed level and allowed to recognise an Early Callovian to Middle Oxfordian age at the type locality of the La Manga Formation (Palma et al. 2012). This condensed level close to the base of this section has yielded *Rehmannia* sp., *Rehmannia* cf. *paucicostata* (Tornquist) and *Homoeoplanulites* sp., from the Lower Callovian Bodenbenderi-Proximum Zones. The condensed level is overlain by a Lower–Middle Oxfordian succession characterised by *Perisphinctes* (*?Arisphinctes*) sp., *Perisphinctes* (*?Kranaosphinctes*) sp., *Perisphinctes* (*?Kranaosphinctes*) cf. *decurrens*, *Perisphinctes* (*?Antilloceras*) cf. *prophetae*, *Perisphinctes* (*?Otosphinctes*) sp., *Perisphinctes* (*?Subdiscosphinctes*) sp., and *Mirosphinctes* sp., of the Perisphinctes-Araucanites Zone, which was correlated with the upper part of the Cordatum Standard Zone to the Transversarium Standard Zone, and probably to the lower part of the Bifurcatus Standard Zone. Late Callovian and Early Oxfordian ammonites, poorly preserved, are not usually represented because of the existence of a regional hiatus (Riccardi 2008).

4 Facies

The Callovian–Oxfordian in the Neuquén Basin was a time of widespread carbonate deposition that is recorded in the La Manga Formation (Legarreta 1991). On the basis of lithology, sedimentary structures, microfacies and taphonomic attributes, the strata of La Manga Formation allowed a plausible reconstruction of the depositional environments. These deposits include: (1) distal outer ramp, (2) proximal outer to distal middle ramp, (3) proximal middle ramp, (4) a complex record of inner ramp deposits (shoreface, shoal, patch reef, shalow subtidal lagoon and tidal flat) and (5) paleokarstic facies (Palma et al. 2007, 2009, 2011, 2013, 2014, 2017, among others). The terminology for ramps was adopted from Burchette and Wright (1992).

5 Distal Outer Carbonate Ramp

The deep distal outer ramp or basin deposits have been barely discussed, probably because they are represented by only a few, discontinuous outcrops. In this work, they are identified at Los Blancos creek 1 (Fig. 1), where La Manga Formation consists of a cm-scale rhythmic alternation of mudstone–wackestone to packstone carbonates and black shales (Fig. 3a). These elementary cycles show a relatively regular thickness in the order of 5–30 cm, so that they can be regarded as temporarily equivalent units controlled by a regular climate cycle (Palma et al. 2014).

Macrofossils are infrequent but well-preserved ammonites (*Perisphinctes* (*Arisphinctes?*) sp., Perisphinctes-Araucanites Zone) that have served to assign this section to the Early-Middle Oxfordian following the biostratigraphic zones of Riccardi (2008).

The laminated lime-mudstones are black with submillimetre scale of planar to slightly irregular lamination. They commonly occur in very thin layers of 2–8 cm. Calcitised radiolarians and isolated fragments of thin bivalve shells are common.

Dark grey to black laminated radiolarian wackestone to packstone facies with normal grading contain radiolarians (Fig. 3b), sponge spicules and bivalve shells (*Bositra* sp.). Bioclasts show a parallel to subparallel orientation to bedding. *Chondrites* appear scattered. Additional components include micritic peloids, similar to those described by Chafetz (1986) and Kazmierczak et al. (1996), probably related to bacteria-induced precipitation. SEM studies show abundant pyrite framboids, microbial bacterial filaments and coccoid bacteria (Palma et al. 2014). Radiolarians and sponge spicules are common in wackestones while *Bositra* shells are more abundant in packstone facies.

Sharply defined bases and normal grading are characteristic of these skeletal wackestone to packstone facies.

Black shales appear in massive or finely laminated, laterally continuous beds of thicknesses ranging from 2 to 20 cm. Submillimetre laminae are defined by changes in

Fig. 3 **a** Field aspect of the distal outer ramp facies at Los Blancos creek 1 section. Centimetre-scale mudstone–wackestones intercalated with black shales. Scale = 33 cm. **b** Radiolarian wackestone with very few sponge spicules and quartz grains. **c** Photograph showing deposits from distal middle to proximal outer settings. **d** Small sponge mounds from distal middle to proximal outer ramp facies at Bardas Blancas locality. Scale 33 cm. **e** Tabular to lenticular beds showing high degree of fragmentation of gryphaeid shells from proximal middle ramp facies at Bardas Blancas area. **f** Amalgamated shell beds from the proximal middle ramp exhibiting tempestite features at Bardas Blancas area

their contents of clay, tiny pyrite crystals and disseminated organic matter. Subangular to angular silt-size quartz grains may also occur.

5.1 Interpretation

Laminated black shales are interpreted as relatively deep deposits and represent deposition from pelagic and hemipelagic suspension settling on sea floors located below the storm wave base. Their dark colour, general absence of bioturbation and organic content are evidence of oxygen-deficient water conditions during deposition (Kauffman and Sageman 1990; Flügel 2004).

Based on the scarcity of benthic organisms, laminated mudstone facies suggest an oxygen-deficient low-energy environment below storm wave base, which would allow colonisation by low oxygen-tolerant benthic fauna, as revealed by the presence of *Bositra* and *Chondrites* (Duff 1978; Ekdale and Mason 1988; Savrda et al. 1991; Etter 1996; Röhl et al. 2001; Caswell et al. 2009). Fine microlaminations indicate primary sedimentation by suspension settling of pelagic and hemipelagic particles. Intermittent higher energy conditions periodically interrupted periods of low-energy suspension sedimentation evidenced by laminated radiolarian wackestones. Normal gradding suggests a decreasing influx of sediments that may have been deposited by dilute turbidity flows and suspended fine grains (Cook and Mullins 1983; Oschmann 1988; Mulder and Alexander 2001).

Other evidence supporting this interpretation are revealed by their high organic carbon contents and the occurrence of pyrite framboids and fossil microbial remains (Wilkin and Barnes 1997; Schieber and Baird 2001). Furthermore, coccoid bacterial-like textures and micritic peloids could be interpreted to be of bacterial origin (Tribovillard 1998; Perri and Tucker 2007).

The distribution of pyrite framboids (in laminated wackestone levels) indicates periodic development of an anoxic sulphide zone below the sediment surface during early diagenesis (Taylor and Macquaker 2000), where bacterial sulphate reduction plays a key role (Min et al. 2001). In addition, the presence of disarticulated, fragmented and reoriented thin *Bositra* valves reflects weak currents affecting the sea floor. The entire association has been only recognised in Los Blancos creek 1 section.

6 Distal Middle to Proximal Outer Carbonate Ramp

6.1 Bioclastic Wackestones and Wackestone–Packstones

Bioclastic wackestone consists of monotypic concentrations of disperse abraded fragments of branching corals, disarticulated valves of *Gryphaea* sp., or echinoid spines (Bressan and Palma 2010) in beds that can reach 30 cm in thickness (Fig. 3c). Concentrations are bidimensional (one skeletal element in thickness), with coral branches concordant with the bedding plane and slightly oriented in plan view. Orientation is also recognisable for echinoid spines. *Gryphaea* remains appear disarticulated, chaotic in plan view and cross section, but mostly complete. Additionally, small *Gryphaea* accumulations are present in wackestone–packstone deposits (Bressan and Palma 2010), showing disperse packing, disarticulation and valves with different orientations. These deposits have been recognised in Bardas Blancas, Coihueco and Mechanquil creeks, and Mallín Redondo.

6.2 Interpretation

These deposits are interpreted as distal middle to proximal outer ramp settings. Coral branches, echinoid spines and disarticulated *Gryphaea* can be interpreted as the result of redistribution of bioclastic material by the action of weak bottom currents.

Coral fragments suggest transport from shallower waters, from moderate to high-energy environments, and intensive reworking by currents or storms for long periods of time, contributing to fragmentation and abrasion.

Gryphaeids also exhibit evidence of reworking, such as disarticulation and chaotic orientation of the remains, although this could be due bioturbation (Kidwell and Bosence 1991), especially for valves-oriented perpendicular to bedding. Separation and sorting (Kornicker et al. 1963) and axial alignments (Nagle 1967) support current action. Good preservation suggests that this redistribution occurred in low energy environments, in the lower middle to outer ramp.

Small *Gryphaea* accumulations are interpreted as in situ reworked relicts of life assemblages. Frequently gryphaeids formed clumps, and if bioturbators or hydraulic processes would have disturbed the sediment after death of the specimens, they would be disarticulated and scattered short distances (Kidwell et al. 1986).

Considering that these gryphaeids live in low-energy environments (Nori and Lathuilière 2003) and that the shell beds show low dispersion, these deposits would be present below fair-weather wave base, in lower middle ramp or outer ramp environments.

6.3 Sponge Mounds

These mounds are composed of massive, light grey limestones containing sponges (Fig. 3d). Individual mounds vary between 20 cm and 3.5 m high and can reach 15 m wide with an asymmetric shape. Internally, isolated bodies occur, that frequently can appear interbedded and interfingered laterally with dark shales. The intermound facies is a thin bedded, dark grey, fossiliferous limestone containing fragments of molluscs, echinoderms and scattered sponges encrusted by serpulids and foraminifers.

Millimetre-scale micritic crusts were observed in some sponge mounds, so that they contribute more or less significantly toward mound constructions. Micritic crusts in thin sections show occurrence of both clotted and peloidal textures, that can be dense and discontinuous, with relatively smooth upper surfaces. Both textures may be observed within the same crust with gradation from one to the other. Crusts can appear with regular cavities produced by *Lithophaga,* preserved articulated in the borings. In situ precipitated micrite (automicrite) are common.

The main components of these mounds are specimens belonging to the Classes Demospongiae and Hexactinellida similar to those described by Beresi (2003) in Potimalal area. These mounds have been recognised in different localities (Bardas Blancas, La Vaina creek, Coihueco creek, Mallín Redondo) (Fig. 1). In La Vaina creek and Mallín Redondo, reworked hexactinellid bafflestones were found suggesting the presence of sponge mounds in nearby places.

6.4 Interpretation

The sponge mound reflects a lower middle to outer ramp sedimentation environment. The development of these mounds could have been controlled by microbial activity, bathimetry, sea-floor morphology and a low sedimentation rate (Keupp et al. 1990; Pratt 1995). Accordingly, microbial growth, submarine lithification and the presence of encrusting organisms upon microbial laminae and bioerosion testify to a very low sedimentation rate during their emplacement. Although mounds are more common in the base of the unit, their exact ages are difficult to determine owing to the scarcity of biostratigraphic indicators.

From a sequential point of view, the development of these sponge mounds could correspond to a period of rapid increase in accommodation, probably from the beginning of the transgression until the time of maximum flooding is reached. The change in the stacking pattern that consists of an abrupt passage from thin- to thick-bedded massive limestones with an upward enrichment in coral fragments can be accepted as a progradation of the highstand systems tract.

Studies in progress deal with facies succession, microbial fabrics and microencrusters in order to evaluate the palaeoecology and roles of microbial communities in mound construction.

7 Proximal Middle Ramp

These facies are characterised by predominant packstone beds rich in gryphaeid shells followed by wackestones with minor amounts of gryphaeid shells.

Bioclastic packstones and intraclastic–bioclastic packstones are dominated by gryphaeids and minor components that include other bivalves, gastropods, echinoderms and bryozoans. Intraclasts are product of reworked bioclastic or peloidal wackestone facies. Moreover, wackestone facies are massive but locally can show parallel lamination or graded beds. They include both bioclastic and peloidal wackestones. The former have gryphaeid shells as main fossil component but skeletal particles also include bivalve shell fragments, foraminifers and scattered serpulids. Peloidal wackestones consist of peloids, bivalve debris, echinoderm fragments, scattered bryozoans and small intraclasts derived from the peloidal or bioclastic wackestone. Peloids show rounded to elliptical shapes.

Gryphaeid shells appear as tabular and lenticular deposits that vary between 5 and 70 cm in thickness, with complex internal structure and erosive base. Shells show variable degrees of articulation and fragmentation and chaotic orientation in bedding plane and cross section (Fig. 3e). These shell deposits can exhibit lateral and vertical amalgamation reaching thicknesses of more than 5 m (Fig. 3f).

These deposits are well represented at Bardas Blancas area outcropping also in Coihueco and Mechanquil creeks, and Mallín Redondo (Fig. 1), although these facies are absent where only the deepest facies appear (La Manga creek, Los Blancos creek 1, Yeseras Grandes and Codo del Blanco sections).

7.1 Interpretation

These deposits developed in proximal middle ramp settings under high energy conditions (Bressan and Palma 2010). The highly reworked shell deposits with erosive base and vertical and lateral amalgamation can be interpreted as storm beds developed above storm wave base. Highly fragmented material recognised as part of these deposits would be produced above fair water wave base after constant reworking and transported to the middle ramp by storm waves and currents, the same processes that reworked shells in the middle ramp leading to multiple cycles of erosion and deposition before the final burial.

8 Inner Ramp

The inner ramp facies show a more complex record, so it was necessary to subdivide it into shoreface, subtidal shoal, patch reef, shallow lagoon facies and intertidal–supratidal facies.

8.1 Shoreface

This facies is characterised by greyish sandstones fine to medium in size and very subordinate packstone–grainstones where detrital components can reach up to 35% (Palma et al. 2007). Sandstones are composed of quartz, feldspars, volcanic lithoclasts, subrounded carbonate intraclasts and ooids. Some disarticulated and fragmented oysters only appear in the basal layer of this succession.

The fine sandstone beds have 0.20–1.20 m in thickness (Fig. 4a). These are laminated beds that exhibit intensely scoured bases and can appear overlaid by hummocky cross-stratified sandstones or amalgamated packstone–grainstones with HCS. They

Fig. 4 **a** Metre-scale cross-bedded sandstones from storm-dominated shoreface. **b** Metre-scale oolitic grainstone–packstones with tabular cross-bedding underlying the patch reef facies. **c** Oolitic and bioclastic grainstone–packstones. Ooids show concentric lamination despite micritisation. Pore space is occluded by granular calcite cement. **d** Detail of patch reefs composed of scleractinian corals and intercolonial areas. **e** Large coral patch reefs facies. Note different migrating build-ups. Exposed section is approximately 17 m high. **f** Decimetre-scale upward-shallowing cycles from shallow subtidal facies

may pass upward into wave-rippled fine wackestones. Trace fossils, as *Dactyloidites ottoi, Gyrochorte* isp., *Thalassinoides* isp. and *Macaronichnus* isp., are common on the tops of these HCS beds (Palma et al. 2007; Lazo et al. 2008). Swaley cross-stratification on fine-grained sandstone appears in lenticular beds from 20 up to 50 cm in thickness, with sharp or erosive basal contact.

Detrital laminae, both in HCS and SCS consist of a mixture of subangular to subrounded quartz, feldspar, volcanic lithic fragments and allochems, with horizontal laminations or trough and tabular cross-stratification overlain by ripple cross-laminations. Sometimes they appear interbedded with lenticular trough cross-bedded fine sandstones which occur in beds with scoured basal contacts and thicknesses between 40 and 70 cm.

The fine-medium-grained sandstones are characterised by 30–50 cm thick sets of tabular cross-stratified bedding with flat or scoured bases, reaching 14 m in lateral extension and overlain by generated ripple cross-laminations.

These sandstone-dominated deposits are approximately 30 m thick. These facies are better represented in Bardas Blancas area, although similar facies from 1 to 2 m thick have been recognised in the Coihueco and Mechanquil creeks (Fig. 1).

8.2 Interpretation

Fine-grained sandstones with isolated and amalgamated hummocky cross-stratification characterised the lower to middle shoreface, meanwhile sandstones with swaley, trough and tabular cross-stratification indicated a middle–upper shoreface deposits. The predominance of HCS and SCS and local tabular cross-stratification suggest that most of the shoreface was susceptible to combined flow directions (Harms et al. 1982) with dominant storm processes. The swaley cross-stratification and hummocky cross-stratification are indicative of storm-dominated deposition above fair-weather wave base in the upper shoreface and the middle–lower shoreface, respectively (Plint and Norris 1991; Dumas and Arnott 2006).

The presence of trace fossils such as *Dactyloidites ottoi* and *Gyrochorte* isp., records the activity of deposit feeders during fair-weather periods (Pemberton and Frey 1984). These burrows along with *Skolithos* are compatible with the shoreface interpretation (MacEachern and Pemberton 1992). The vertical variation of the sedimentary structures suggests deposition in progressively shallower settings.

From the above observations and interpretations, the succession of these sedimentary facies is characteristic of a typical coarsening-upward sequence accumulated on a storm-dominated shoreface setting.

8.3 Subtidal Shoal

Oolitic grainstone–packstones are arranged in decimetre to metre massive to tabular cross-bedding beds (Fig. 4b). Concentric and some radial laminations in the structure of the ooids are common, although many of them are extensively micritised (Fig. 4c). They have a variety of nuclei, including pellets, mollusc fragments and minor quartz sand grains, with multiple thick oolitic coating and typically interbedded with wackestone–packstone facies. Broken and recoated ooids and oncoids are common.

The main skeletal grains are fragments of oyster, other bivalves, gastropods and echinoderms. Other particles include intraclasts, *Cayeuxia* sp., fragments, miliolid peloids and coprolites varying in abundances. Many shells exhibit micritised rims. Subtidal shoals are best developed at Bardas Blancas area (Fig. 1).

8.4 Interpretation

Concentric ooids with very well-developed coatings suggest that the sediments belong to frequently agitated to intermittently agitaded environments (Strasser 1986). In fact, the presence of intraclasts, fragments of molluscs, echinoderms and algae suggest occasionally action currents or storms. In fact, tabular cross-stratification is associated with migration of megaondules.

The intense micritisation indicates the active action of boring algae and fungi during quiet periods, when the shoal was inactive. Shallow water intermitent high energy conditions are also supported by the presence of oncoids that were able to overturn and roll (Palma et al. 2017).

8.5 Reefs

Reefal core facies is composed of white unstratified limestones dominated by platy, branching and domal scleractinian corals. The coral framework started on tabular cross-stratified oolitic grainstone–packstones. Platy, ramose and dome-shaped scleractinian colonies are recognised (Fig. 4d). Morsch (1990) and Palma et al. (2007, 2009) performed studies about the composition of the coral fauna recognising the presence of *Australoseris* sp., *Actinastrea* sp., *Thamnasteria* sp. and *Garateastrea* sp. Works in progress indicate that the identification of the last three genera is not certain, in part as a consequence of taphonomic artifacts (Hoqui pers. comm.). Preservation can be poor as consequence of the activity of lithophagid bivalves and intense recrystallisation. In situ reef development occurs in small patches with domal structure approximately 20 m across and 18 m in high (Fig. 4e).

The matrix between colonies is composed mainly of bioclastic wackestones and packstone with ooids and coated grains. The bioclastic material consists of abundant fragments of the reef frame builders, bivalves (*Ctenostreon* sp.), echinoids, gastropods (*Harpagodes* sp., and *Natica* sp.), green calcareous algae (dasycladacean) and cyanophytes (*Cayeuxia* sp.), serpulids and bryozoans. Other components are miliolid foraminifers, ostracods, non-skeletal particles such as peloids and intraclasts. At the base of the succession, encrusting and branching sponges are found.

The reef facies are very well exposed in Bardas Blancas area outcropping throughout at least 8 km and reaching 15 m in thickness. Towards the north, they are totally absent, while in the eastern side of the Sierra Azul (Coihueco and Mechanquil area, Fig. 1), they appear poorly preserved, showing high fragmentation.

8.6 Interpretation

In the Bardas Blancas area, the coral morphologies suggest a moderate to high-energy environment (James 1983, cyanophites; Tucker and Wright 1990). The associated fauna represents a benthic community that included suspension feeders, grazing herbivores and scavenger organisms. Bioclasts are highly micritised, due to their exposure on sediment surface for a considerable amount of time. Moreover, grazing gastropods suggest the presence of algal mats, which are a common micrite producer (Bathurst 1976). Such microbialites associated with coral sponges and other microencrusters such as bryozoans, serpulids and foraminifers are related to conditions of high nutrient availability (Leinfelder 1992). There is no evidence of a persistent barrier and they probably formed small patches on the sea floor (cf. James 1983).

Reefal deposits at Coihueco and Mechanquil area can be interpreted as a autoparabiostrome (sensu Kershaw 1994) in base to the presence of corals in growth position and reworked fragments.

8.7 Shallow Subtidal Lagoon Facies

Subtidal facies consist of bioturbated wackestones, peloidal wackestones, bioclastic wackestones, bioclastic packstones, oncoidal packstones, floatstone/rudstones rich in bioclastic debris and isolated patch reefs. These facies are arranged into decimetre-scale upward-shallowing cycles composed of marls, laminated or massive mudstones or bioclastic wackestones and intraclastic wackestone–packstones (Fig. 4f). Bioclastic packstones can exhibit herringbone cross-stratification.

A diverse biota represented by bivalves (mostly oysters), echinoderms, dasycladacean, *Cayeuxia* sp., fragment of corals and small scleractinian coral patch reefs can be recognised. Some fragments of bryozoan colonies, as well as foraminifers and calcispheres are also represented. These beds are usually sheet-like and laterally continuous, commonly resting on erosive bases, ranging from few centimetres up to 30 cm in thickness. Bioturbated wackestones contain horizontal burrows. The matrix is mostly composed of skeletal fragments derived from mechanical breakdown of allochem particles. Micritisation of particles is frequent. Peloidal wackestones are very common and consist of clotted micrite groundmasses locally peloidal, with bioclastic fragments and scattered ooids showing fine radial cortical laminae. With increasing abundance of bioclasts, there is a transition into a bioclastic wackestone rich in a similar fauna to the one mentioned above.

The bioclastic packstone facies are composed of similar bioclasts as mentioned above besides oncoids and delicately branched *Cayeuxia* sp. Oncoids have nuclei consisting of fragments of bivalves, echinoids, microoncoids and *Cayeuxia* sp. Cortex laminae are relatively continuous or display low-angle discontinuities and show a clotted or grumelous texture. Micrite envelopes around bioclasts and bioerosion are common.

Microbial oncoids occur either in packstone or floatstone–rudstones, although they are less frequent in wackestone. They have several shapes, highlighting those spherical, ameboidal and subordinately elliptical (Fig. 5a). Molluscs and echinoid fragments, peloids, and intraclasts occur as oncoid nuclei. Some oncoids, especially the ameboidal or elliptical forms, are characterised by multiple nuclei, represented by small oncoids, bioclasts or intraclasts.

According to the features of the envelopes, different types of oncoids were recognised: (1) micritic laminations, (2) wrinkled laminations and (3) organism-bearing laminations. Laminae follow the shape of nuclei, especially in the inner zone of the cortices. Likewise, frequently discontinuous laminae are observed. Outer laminae contain encrusting organisms dominated by nubeculariids and serpulids. The oncoids are associated with bivalves, echinoderms, foraminifers and frequent intraclasts. The floatstone–rudstones facies is dominated by oncoids which on the bases of their morphology are identified as type C, type R and type L (sensu Flügel 2004). Stromatolites are present in the lagoon facies and record a progression from shallow subtidal to intertidal–supratidal facies.

These deposits are frequent at Bardas Blancas, La Vaina and Mechanquil creeks, and Mallín Redondo.

8.8 Interpretation

Subtidal facies developed under moderately to high energy conditions, characterising lagoonal channels or shoals. Shells and bioclasts usually appear to have been transported and slightly reworked.

A shallow lagoonal protected setting is indicated by the frequent presence of *Cayeuxia* sp. and dasycladacean algae (Schmid and Jonischkeit 1995; Baumgärtner and Reyle 1995).

Tidal influence is clearly visible in the rare wackestone–packstones with herringbone cross-stratification, evidencing the development of tidal channels in a shallow subtidal setting under moderately high energy conditions.

These energy conditions are also suggested by the presence of oncoids that grew under agitated subtidal waters that alternated with periods affected by intermittent currents that reoriented the oncoids parallel to stratification (Fig. 5b). The discontinuous organism-bearing laminations reflect periods of non-agitation and lithification, favouring the growth of encrusting organisms.

Fig. 5 **a** Oncoid with compound nuclei and different growth phase laminae. Encrusters foraminifers and serpulids were incorporated into the laminae. **b** Field aspect from oncoid facies showing variety of size and subspherical shapes of oncoids. Scale = 33 cm. **c** Macroscopic to mesoscopic features of planar to corrugate stromatolites facies from inter-supratidal facies. Scale = 4 cm. **c** Thin section photomicrograph of microbial laminae showing an irregular domal growth shape. **e** Field aspect of intertidal–supratidal cycles, alternating bedding of mudstone–wackestones, microbial laminites, and flat-pebble conglomerates. **f** Detail of mantling breccia composed of chaotic grain-supported to matrix-supported breccias. Scale = 33 cm

8.9 Tidal Flat Facies

Intertidal–supratidal facies are organised in centimetre-scale cycles that show an upward-shallowing succession of mudstones and wackestones, overlaid by horizontal or crinkle microbial laminites (Fig. 5c), flat-pebble conglomerates or breccias beds.

The mudstone facies, from 2 to 5 cm are rich in peloids, showing planar lamination or normal grading on a centimetre-scale. These facies alternate with flat-pebble conglomerate beds or planar microbial laminites that contain pseudomorphs of evaporite crystals.

Wackestone facies appear laminated, weakly bioturbated and, more rarely, with ripple lamination. Centimetric scoured bases show intraclasts, while some tops show mud-cracks.

The internal architecture of the microbial laminites beds is highly variable (Fig. 5d). They show flat-laminated to planar crinkled forms. They form beds from 2 to 80 cm thick and, although some can be traced through tens of metres, most of them pinch out laterally. Crinckly microbial laminites commonly show microtepees, laminar and subcircular fenestrae, mud-cracks, and sheet-cracks. Usually, microbial laminites are overlain by breccia facies or flat-pebble conglomerates. Subangular to angular planar microbial clasts appear as discrete and discontinuous or parallel and continuous to bedding.

Dark to grey massive intraclastic breccias are present as laterally discontinuous beds with scoured base, and as lenses or small channel fills. Beds thicknesses range between 5 and 12 cm. Intraclasts are of microbial carbonate composition, appearing also well-preserved ammonites and regular echinoids. The ammonites have been identified as *Mirosphinctes* sp.

These facies have been recognised in all the sections considered in this work (Fig. 1), except where the deepest facies appear (La Manga creek, Los Blancos creek 1, Yeseras Grandes and Codo del Blanco sections).

8.10 Interpretation

Planar to crinkle microbial beds develop at intertidal–supratidal settings recording the trapping of fine-grained sediments by microbial mats. Studies in recent microbial laminites show that planar to crinkle microbial mats grow frequently above mean low water level or tidal flats which are periodically flooded and exposed (Reid et al. 2011). The frequent tepee structures can be considered according to Assereto and Kendall (1977) as immature tepees. Fenestral pores, sheet-cracks and mud-cracks suggest temporarily emergent conditions or hiatuses indicative of variations in salinity (Husinec and Read 2011).

The origin of the flat-pebble conglomerate facies is related to erosion and redeposition by storms of desiccated microbial mats during the marine flooding on top of the laminated facies (Shinn 1983). Rip-up clasts with random distribution and the presence of abundant ammonites (*Mirosphynctes* sp.), together with regular echinoids (Palma et al. 2011) are interpreted as the result of complete reworking of mud-cracked microbial beds by current activity and storm events, which is also evidenced by scoured bottom and intraclasts.

The intraclastic breccias represent a mix of fragments ripped-up by storm wave action and local autoclastic breccias generated by prolonged subaerial exposure in the upper intertidal–supratidal environment.

According to evidence, this sedimentary succession is interpreted as a restricted very shallow intertidal–supratidal setting with moderate to low energy, affected by high-energy storm events, as suggested by the presence of open marine fauna such as ammonites and nanofossils such as *Cyclagelosphaera* sp. (middle Oxfordian in age, Concheyro, *pers.comm.*).

9 Palaeokarst

Evidence of paleokarst affecting other carbonate facies include mantling breccias, collapse breccias and sinkhole breccias, with chaotic grain-supported to matrix-supported fabrics (Fig. 5e). Clasts of breccias are angular with a fitted clast texture to extreme brecciation showing highly variable textures from silt to cobble size (Fig. 5f). Clasts are polymictic. The petrographical composition of the clasts is consistent with the underlying facies previously described. It reflects derivation from bioclastic packstones, oolitic grainstones and coral reefs and peritidal facies. There are small dissolution cavities in size from 1 to 3 cm below surface, usually indicated by the presence of iron oxides and pisolites with an irregular distribution.

In fact, the base of these breccia deposits, in different areas, is equivalent to the discontinuity mentioned by Legarreta (1991) at the top of reef and grainstone facies seen in the Bardas Blancas section and top of peritidal facies in Yeseras Grandes, La Manga creek, Los Blancos creek 2, La Vaina creek and Mallín Redondo (Fig. 1). The thickness of exposed breccia intervals ranges from 1.20 to 6 m (Fig. 6).

9.1 Interpretation

Paleokarst breccias are interpreted as the result of a main episode of emergence and karstification due to relative sea level fall and the formation of a significant stratigraphic break (148 Ma; Legarreta 1991) which is well developed throughout the area, from the Bardas Blancas section to 60 km southward to the Sierra Azul outcrops. The discontinuity is indicated by the paleokarst and by epikarst features, which are produced by subareal exposure and intense dissolution processes as a consequence of sea level fall. Paleokarst breccias appear throughout the intertidal–supratidal sections, interbedded in some cases, with thin planar microbial laminites. The occurrence of pisolitic grainstones indicates subaereal exposition and vadose diagenesis (Palma et al. 2017).

Sedimentology, morphology and diagenesis let recognise epikarst features in peritidal facies showing truncate shallowing-upward cycles in an inner ramp with low gradient.

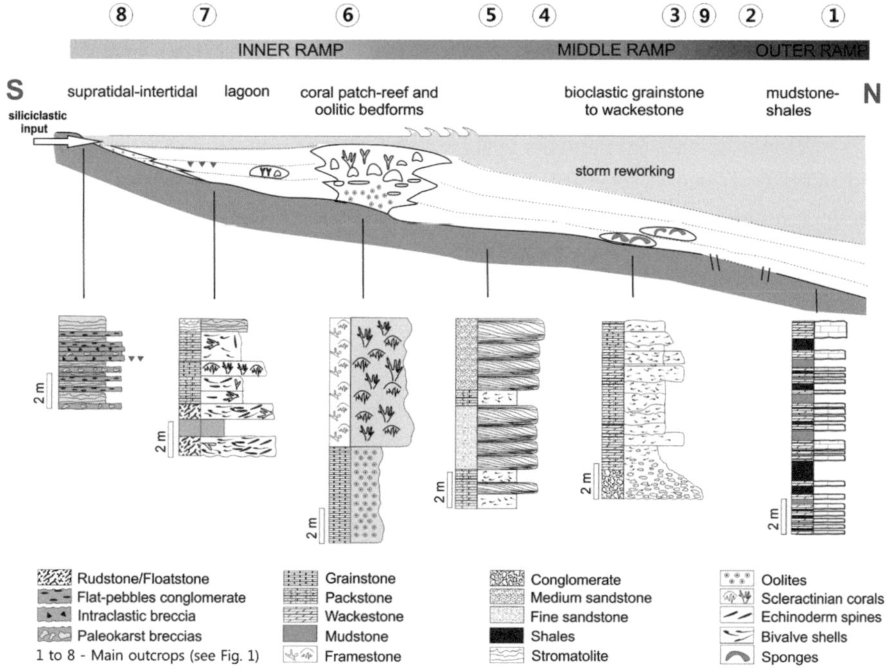

Fig. 6 Schematic reconstructions of the carbonate ramp and synthetic representation of the study sections

Similar paleokarstic features affecting identical facies can be recognised at the Aconcagua Basin, in the northern side of Las Cuevas River near Puente del Inca locality (Argentinian High Andes) (Lo Forte and Palma 2002) but lack of biostratigraphical control does not allow us to correlate this major break over the area.

Moreover, a decrease of accommodation space, influenced by allogenic processes, may be the answer to a sea level fall concomitant with a lower rate of sedimentation.

The exposure of ramp facies represented by significant erosional surfaces is characteristic of a type I sequence boundary (Haq et al. 1987). Polymictic breccia deposits overlain paleokarstic unconformity originated by erosion of ramp areas at lowstand stages (Sarg 1988; Legarreta 1991; Palma et al. 2007, 2016).

10 Controls on Facies Distribution

The described facies are included into three third-order depositional sequences defined by Palma et al. (2007) (DS-1, DS-2, DS-3) mainly represented by transgressive and highstand systems tract stages with sequence boundaries of regional importance. The three third-order depositional sequences (DS-1 to DS-3) are comparable with the 2nd to 4th sequences of Legarreta (1991). These depositional sequences have well development in Bardas Blancas area, where DS-1 shows important erosive surfaces on base and top and consists of a lowstand systems tract (LST-1) and a transgressive systems tract (TST-1). The base is a sequence boundary (SB-1), probably related to the Intra-Callovian Discontinuity (Fig. 7a). For Bardas Blancas area, the LST-1 belongs to alluvial deposits assigned to Lotena Formation (Kietzmann et al.

Fig. 7 **a** Outcrop photograph showing the SB-1, DS-1, and the sharp facies change between DS-1 and DS-2 Bardas Blancas area. **b** Outcrop view of coral reef build-up and sequence boundary type I, overlain by polymictic breccias at Bardas Blancas area. **c** General view of very shallow lagoon facies at Coihueco section. **d** Detail of intertidal–supratidal facies at La Manga creek. Scale = 65 cm. **e** Rhythmic alternation of marls, shales and wackestones from proximal outer ramp at La Manga creek. Scale = 33 cm. **f** Centimetre-scale alternating bedding of mudstones and black shales from distal outer ramp facies at Los Blancos creek 1

2016). Marine deposits of Lotena Formation are overlain by La Manga Formation at La Manga creek, Yeseras Grandes, and La Vaina creek. At Coihueco and Mechanquil creeks, La Manga Formation appears overlain alluvial deposits without formal stratigraphic assignment, meanwhile at Los Blancos creek, Codo del Blanco and Mallín Redondo, the base is not exposed.

The transgressive succession (TST-1) corresponds to La Manga Formation and is characterised by middle ramp packstone–wackestones and proximal outer ramp wackestones, both with gryphaeid shells well developed in Bardas Blancas area. Taphonomic analysis of these beds containing gryphaeid shells (Bressan and Palma 2010) suggests storm events on the middle ramp and weak bottom currents in the proximal outer ramp (Kreisa 1981; Aigner 1982; Handford 1986). Sedimentological and taphonomic features record a decrease in physical reworking evidencing a deepening in the depositional setting. The TST-1 pinches out laterally disappearing northward in Bardas Blancas area. Southward it appears at Coihueco creek, Mechanquil creek and Mallín Redondo sections.

With a sharp and erosive contact, these outer-middle ramp settings are overlain by deposits of the DS-2. This sequence boundary (SB-2) has been caused by a sea level fall.

The DS-2 starts with a progradational succession of inner ramp sandstone facies (3–30 m in thickness) (Fig. 7a). These deposits with scarse fauna and an increase in grain size were interpreted as high-energy shoreface deposits. The packstone–grainstones with tabular cross-stratification represents energetic shallow water depositional conditions, product of migrating megaripples. High energy conditions are also evidenced in the presence of HCS as a result of storm processes. The paucity of bioclasts in this facies suggests that unfavourable conditions existed for the fauna because of the influx of siliciclastic material.

These shallow high-energy shoreface deposits are clearly identified on the southern sections of Bardas Blancas area, reducing their thickness northward. Thin and isolated similar deposits are also preserved in the Coihueco and Mechanquil creeks, and Mallín Redondo (Fig. 1).

These shoreface deposits are included in the lowstand systems tract-2 (Plint and Nummedal 2000) and were interpreted as a lowstand wedge suggesting a significant fall of sea level and a 'type 1' sequence boundary (Legarreta 1991) and correspond to the third sequence of Legarreta (1991) which appears in abrupt contact with the underlying highstand deposits of the preceding sequence (DS-1, Palma et al. 2007).

Meanwhile, basinwards in Bardas Blancas area, the shallowing upward succession of inner ramp deposits is laterally replaced in 8.5 km by the emplacement of sponge mounds of distal middle to proximal outer ramp settings, which constitute the TST-2 (Palma et al. 2003, 2004). Microbialites encrusted by serpulids, bryozoan and foraminifers (Palma et al. 2007) are binding the sponges to one another or grow on the sponge surfaces suggesting conditions of high nutrient availability and a low sedimentation rate (Leinfelder 1992). As sea level continued to rise, oolitic to bioclastic packstone–grainstones constitute the upper part of the TST-2. The ultimate deepening stage of DS-2 is represented by the maximum flooding surface 2 (MFS-2), which represents a regional paraconformity that was buried by downlap deposition of the highstand systems tract 2 (HST-2), described as follows.

The lack of siliciclastic input, the high nutrient levels and the temperature were fundamental in controlling carbonate production and favouring the building of frameworks of scleractinian patch reefs (Fig. 7b) to sea level and keep space with subsequent sea level changes in the shallow and wave-agitated euphotic zone, developing the HST-2. Field evidences allow to accept that the sea level rise could show a deceleration resulting in a thick aggradational and later progradational architecture of the morphology of the ramp, as shown by the growing of the reefs in Bardas Blancas area. As sea level rise slowed down, these patch reefs grew laterally although they never developed as a barrier. Field evidence show that the reef margin was not steep basinwards, possibly due to the absence of a real reef barrier as mentioned previously.

Over a large part of the study area, it is possible to recognise the effects of karstification at the top of reefal or oolitic facies. It could be a consequence of a rapid sea level drop preventing deposition (Palma et al. 2007) and indicates the end of the DS-2 (Fig. 5f).

Paleokarst deposits are represented by mantling breccias, collapse breccias and sinkhole breccias (Palma et al. 2017). Sedimentology, morphology and diagenesis let recognise the epikarst in peritidal facies showing shallowing-upward cycles in an internal ramp with low gradient. A detailed petrographic analysis of this level showed the passage from marine phreatic diagenesis to meteoric-phreatic conditions related to the position of the water table, similar to the examples described by James and Choquette (1984). Fluctuations of the phreatic level during periods of subaerial exposure controlled the development of pendant cementation and drusy calcite cement evidencing the action of meteoric waters (Mazzulo and Mazzulo 1992; Loucks 1999).

The discontinuities produced by subareal exposure and intense dissolution processes represent a stratigraphic break, which is well developed throughout the area, from the Codo del Blanco section to southward, in all localities, to the Sierra Azul outcrops (Coihueco creek, Mechanquil creek and Mallín Redondo).

The DS-3 is less developed than DS-2. Starting on SB-3, it is composed by the TST-3 and the HST-3 consisting in lagoon deposits in Bardas Blancas area (Fig. 7c). The deceleration in the sea level rise and the reduction in the accommodation space resulted in the development of the HST-3 as an open to restricted lagoon with scleractinian patch reefs as well as sporadic and thin storm deposits (Palma et al. 2007, 2009).

The variation of sedimentary structures related to emersion features allowed the recognition of centimetre to decimetre-scale subtidal and intertidal–supratidal cycles (Palma et al. 2013) that exhibit a reduction in their thickness as a result of the deceleration of the sea level rise mentioned above (Fig. 7d). These cycles appear in all the section considered in this work. In the northern outcrops, the transect from Yeseras Grandes to Los Blancos creek 2, passing through Arroyo La Manga, shows the general decrease in cycles thickness towards the south (Palma et al. 2013). Even though individual cycles may not be traceable from one section to another, it is possible to recognise a general shallowing trend in all the studied sections. The upward progressive shallowing tendency is also evidenced by the presence of planar to crinkle microbial laminites and tepee structures, evidencing the reduction in accomodation space.

These cycles would be produced by changes of the depositional environment controlled by fluctuations in sea level (Palma et al. 2011). Their remarkable discontinuity is related to sea level fluctuations (cf. Adams and Grotzinger 1996), even though a local subsidence cannot be ruled out (Hamon and Merzeraud 2008). As Bádenas et al. (2010) suggest the lateral continuity of cycles can be controlled by depositional topography or by local subsidence, where facies distribution results from production rate and accumulation.

These deposits are overlied by karstification. Similar paleokarts features affecting identical facies can be recognized northwards, in the Aconcagua Basin (Lo Forte and Palma 2002), although a lack of biostratigraphical control does not allow us to correlate this facies over the area of this major break.

Although locally in tectonically controlled settings, the lower section from Codo del Blanco, Yeseras Grandes and La Manga creek (Fig. 7e), and deeper deposits from Los Blancos (Los Blancos 1) (Fig. 7f) differs significantly from the other localities. These organic matter facies (outer ramp) would have been deposited in the deepest zones of topographic lows controlled by tilting and differential subsidence of tectonic blocks bounded by normal faults. These faults were probably inherited from previous extensional tectonics of Late Triassic–Early Jurassic age (Giambiagi et al. 2008). It can be accepted that, on a local scale, movement along basement faults could have influenced facies distribution (Palma et al. 2014).

These Middle Oxfordian deep deposits (outer ramp) show marked centimetre-scale rhythmicity, probably controlled by the Earth's orbit parameters. According to biostratigraphic data, the Los Blancos creek 1 section covers the Early to Middle Oxfordian interval (~3.5 Ma). Distinguished elementary cycles give an average duration of each cycle of 20.8 ky. This value is in the range of the precessional cycle of the Earth's orbit. The Blackman–Tuckey spectrum for the thickness series shows a marked peak (above the 95% confidence level) corresponding to 4.5 elementary cycles (~94 ky). This periodicity can be attributed to the high-frequency eccentricity cycle of the Earth's orbit (Palma et al. 2014). Our data thus suggest that the rhythmicity identified in the La Manga Formation can be attributed to environmental changes controlled by regional climate variations. In turn, these variations would have been modulated by the orbital parameters equinoccial precession and eccentricity. Data from other Jurassic deposits are consistent with this rhythmicity and interpretation (e.g., Weedon et al. 1999; Strasser et al. 2004; Husinec and Read 2007).

Along the transect from Yeseras Grandes to Mallín Redondo, facies changes allowed us to reconstruct from distal outer into shallow marine carbonate deposits.

Regional facies distribution reflects the combined processes of subsidence and eustasy which resulted in a homoclinal carbonate ramp to distally steepened ramp. Depositional systems were controlled by their internal dynamics, carbonate production, rate of accommodation space, sea level fluctuations and their interactions with tectonic activity.

Depositional patterns from Codo del Blanco, Yeseras Grandes and Los Blancos creek 1 are consistent with local tectonic activity affecting the ramp topography and hence the depositional environment. Meanwhile, the progradational sandstone deposits at Bardas Blancas area are in response to siliciclastic influx from the south-eastern side of the basin.

The three described depositional sequences (DS-1 to DS-3) are similar to those from the Mesozoic (greenhouse) in which sea level changes appear as low-frequency ups and downs (Heller et al. 1996).

Climatic and tectonic effects on carbonate production were the most probable cause for demise of the La Manga carbonates during the end of Oxfordian and Early Kimmeridgian time (Legarreta et al. 1991; Legarreta and Uliana 1996). A dramatic change resulted in a new basin configuration and thick evaporite deposits (Auquilco Formation) covering the basin (Vergani et al. 1995).

11 Conclusions

An integrated facies analysis of the Middle Callovian–Middle Oxfordian La Manga Formation was carried out to show their facies characteristics and distribution, depositional environments, and controls throughout the northern part of Neuquén Basin. Facies and thickness differences at the studied localities bring out considerable variations in the northern Neuquén Basin, which were controlled by sea level fluctuations and tectonic activity during this time interval.

Analysis of the five facies associations revealed the settlement of a homoclinal carbonate ramp that was shallower in the southern localities (Bardas Blancas, La Vaina, Coihueco and Mechanquil creeks, and Mallín Redondo), while on the northern sections (Yeseras Grandes, La Manga creek, Los Blancos creek, and Codo del Blanco) frequent deep marine (distal outer ramp) facies related to a distally steepened ramp are recognised. These last localities were controlled by reactivated basement faults as well as sea level changes, which are the principal controls in the localities of the south. The five facies are included into three third-order depositional sequences (DS-1, DS-2, DS-3) represented by transgressive and hightstand systems tracts with sequence boundaries of regional importance.

Acknowledgements This work has been supported by different CONICET (PIP), UBA (UBACyT) and FONCyT (PICT) projects directed by R. M. Palma. We want to thank Mr. G. Herrero (Universidad Complutense de Madrid) for technical assistance and J. C. Poblete for his assistance during field work. Also, we would like to thank all the members of the Dirección de Recursos Naturales Renovables of Malargüe as well as the Researcher Group 910198 (Universidad Complutense de Madrid, Comunidad de Madrid, Spain).

References

Adams RD, Grotzinger JP (1996) Lateral continuity of facies and parasequences in Middle Cambrian platform carbonates, Carrara Formation, southeastern California, USA. J Sediment Petrol 66:1079–1090

Aigner T (1982) Calcareous tempestites: storm-dominated stratification in Upper Muschelkalk limestones (Middle Trias, SW Germany). In: Einsele G, Seilacher A (eds) Cyclic and event stratification. Springer, Berlin, pp 180–198

Assereto RL, Kendall CG (1977) Nature, origin and classification of peritidal tepee structures and related breccias. Sedimentology 24:153–210

Bádenas B, Aurell M, Bosence D (2010) Continuity and facies heterogeneities of shallow carbonate ramp cycles (Sinemurian). Lower Jurassic, North-east Spain. Sedimentology 57:1021–1048

Bathurst RGC (1976) Carbonate sediments and their diagenesis developments in sedimentology, 12th edn. Elsevier, New York, p 658

Baumgärtner M, Reyle M (1995) Oberjurassische Rampenentwicklung in der Region von Jabaloyas und Cerezo (Keltiberikum; Spanien). Profil 8:339–361

Beresi MS (2003) Oxfordian sponge association from the Neuquén basin, Mendoza, west central Argentina. J S Am Earth Sci 16:105–112

Beresi MS (2007) Fossil sponges of Argentina: a review. In: Custódio MR, Lôbo-Hajdu G, Hajdu E, Muricy G. (eds) Porifera research: biodiversity, innovation and sustainability, Museu Nacional du Rio de Janeiro, Rio de Janeiro, série livros vol 28, pp 11–21

Beresi MS, Cabaleri NG, Löser H et al (2017) Coral patch reef system and associated facies from southwestern Gondwana: paleoenvironmental evolution of the Oxfordian shallow-marine carbonate platform at Portada Covunco, Neuquén Basin, Argentina. Facies 63:22

Bressan GS, Palma RM (2010) Taphonomic analysis of fossil concentrations from La Manga Formation (Oxfordian), Neuquén Basin, Mendoza Province, Argentina. J Iber Geol 36:55–71

Burchette TP, Wright VP (1992) Carbonate ramp depositional systems. Sediment Geol 79:3–57

Caswell BA, Coe AL, Cohen AS (2009) New range data for marine invertebrate species across the early Toarcian (early Jurassic) mass extinction. J Geol Soc London 166:859–872

Chafetz HS (1986) Marine peloids: a product of bacterially induced precipitation of calcite. J Sediment Petrol 56:812–817

Cook HE, Mullins HT (1983) Basin margin. In: Scholle PA, Bebout DG, Moore CH (eds) Carbonate depositional environments, AAPG Memoir 33:539–618

Duff KL (1978) Bivalvia from the English lower Oxford Clay (Middle Jurassic). Palaeontogr Soc Monogr 132:1–137

Dumas S, Arnott RWC (2006) Origin and hummocky and swaley cross-stratification. The controlling influence of unidirectional current strength and aggradation rate. Geology 34:1073–1076

Ekdale AA, Mason TR (1988) Characteristic trace-fossils associations in oxygen-poor sedimentary environments. Geology 16:720–723

Etter RJ (1996) The effect of wave action, prey type, and foraging time on growth of the predatory snail *Nucella lapillus* (L). J Exp Mar Biol Ecol 196:341–356

Flügel E (2004) Microfacies of carbonate rocks. Analysis, interpretation and application. Springer, Berlin

Gerth E (1925) La fauna neocomiana de la Cordillera Argentina en la parte meridional de la provincia de Mendoza. Academia Nacional de Ciencias, Cordoba, vol 9, pp 57–132

Giambiagi L, Bechis F, Lanés S et al (2008) Formación y evolución Triásico-Jurásica del depocentro Atuel, Cuenca Neuquina, provincia de Mendoza. Rev Asoc Geol Argent 63:520–533

Groeber P (1918) Estratigrafía del Dogger en la República Argentina. Estudio sintético comparativo. Dirección General de Minas, Geología e Hidrogeología, Buenos Aires, vol 18, Serie B, pp 1–81

Groeber P (1933) Confluencia de los ríos Grande y Barrancas (Mendoza y Neuquén). Dirección Nacional de Geología y Minería, Buenos Aires, vol 38, pp 1–72

Groeber P (1937) Descripción geológica de la Hoja 30c Puntilla de Huinc*á*n, provincia de Mendoza. Dirección Nacional de Geología y Minería, Buenos Aires

Groeber P (1946) Observaciones geológicas a lo largo del meridiano 70. I. Hoja Chos Malal. Rev Asoc Geol Argent 1:177–208
Groeber P, Stipanicic PN, Mingramm A (1953) Jurásico. In: Groeber P (ed) Mesozoico, Geografía de la República Argentina. Sociedad Argentina de Estudios Geográficos, Buenos Aires, GAEA 2, pp 143–347
Hamon Y, Merzeraud G (2008) Facies architecture and cyclicity in a mosaic carbonate platform: effects of fault-block tectonics (Lower Lias, Causses platform, south-east France). Sedimentology 55:155–178
Handford CR (1986) Facies and bedding sequences in shelf-storm-deposited carbonates—Fayetteville Shale and Pitkin Limestone (Mississippian), Arkansas. J Sediment Res 56:123–137
Harms JC, Southard JB, Walker RG (1982) Structures and sequences in clastic rocks. Lecture Notes SEPM Short Course, Tulsa 9:249
Haq BU, Hardenbol J, Vail PV (1987) Chronology of fluctuating sea levels since the Triassic. Science 235:1156–1167
Heller PL, Anderson DL, Angevine CL (1996) Is the middle Cretaceous pulse of rapid sea-floor spreading real or necessary? Geology 24:491–494
Husinec A, Read JF (2007) The Late Jurassic Tithonian, a greenhouse phase in the Middle Jurassic-Early Cretaceous "cool" mode: evidence from the cyclic Adriatic platform, Croatia. Sedimentology 54:317–337
Husinec A, Read JF (2011) Microbial laminite versus rooted and burrowed caps on peritidal cycles: salinity control on parasequence development, Early Cretaceous isolated carbonate platform, Croatia. Geol Soc Am Bull 123:1896–1907
James NP (1983) Reefs. In: Scholle PA, Debont DG, Moore CH (eds) Carbonate Depositional Environments, AAPG Memoir vol 33, pp 345–440
James NP, Choquette PW (1984) Diagenesis 9. Limestones—the meteoric diagenetic environment. Geosci Can 11:161–194
Jaworski E (1925) Contribución a la paleontología del Jurásico sudamericano (Revisada por Pablo Groeber). Dir Gral Min Geol Hidrol, Buenos Aires, Sec Geol 4:1–160
Kauffman EG, Sageman BB (1990) Biological sensing of benthic environments in dark shales and related oxygen-restricted facies. In: Ginsburg RN, Beaudoin B (eds) Cretaceous resources, events and rhythms. Kluwer Academic Publishers, Dordrecht, pp 121–139
Kazmierczak J, Coleman ML, Gruszczynski M et al (1996) Cyanobacterial key to the genesis of micritic and peloidal limestones in ancient seas. Acta Palaeont Pol 41:319–338
Kershaw S (1994) Classification and geological significance of biostromes. Facies 31:81–92
Keupp H, Koch R, Leinfelder RR (1990) Steuerungsprozesse der Entwicklung von Oberjura-Spongiolithen Sueddeutschlands: Kenntnisstand, Problem und Perspectiven. Facies 23:141–174
Kidwell SM, Bosence DW (1991) Taphonomy and time-averaging of marine faunas. In: Allison PA, Briggs DEG (eds) Taphonomy: releasing the data locked in the fossil record. Topics in Geobiology 9, pp 115–209
Kidwell SM, Holland SM (1991) Field description of coarse bioclastic fabrics. Palaios 6:426–434
Kidwell SM, Fürsich FT, Aigner T (1986) Conceptual framework for the analysis and classification of shell concentrations. Palaios 1:228–238
Kietzmann DA, Palma RA, Ferreyra TMA (2016) Análisis de facies y asignación estratigráfica de los depósitos fluviales innominados del Jurásico Medio de la Cuenca Neuquina surmendocina. Rev Asoc Geol Argent 73:104–116
Kreisa RR (1981) Storm-generated sedimentary structures in subtidal marine facies with examples from the middle and upper Ordovician of southwestern Virginia. J Sediment Petrol 51:832–848
Kornicker LS, Wise CD, Wise JM (1963) Factors affecting the distribution of opposing mollusk valves. J Sediment Petrol 33:703–712
Lazo DG, Palma RM, Piethé RD (2008) La traza *Dactyloidites ottoi* (Geinitz) en la Formación La Manga, Oxfordiano de Mendoza. Ameghiniana 45:627–632
Leanza HA (1981) The Jurassic-Cretaceous boundary beds in west-central Argentina and their ammonite zones. Neues Jahrb Geol P-A 161:62–92

Leanza HA (2009) Las principales discordancias del Mesozoico de la Cuenca Neuquina según observaciones de superficie. Rev Mus Argent Cs Nat 11:145–184
Legarreta L (1991) Evolution of a Callovian-Oxfordian carbonate margin in the Neuquén Basin, of west-central Argentina: facies, architecture, depositional sequences and global sea-level changes. Sediment Geol 70:209–240
Legarreta L, Gulisano CA (1989) Análisis estratigráfico de la Cuenca Neuquina (Triásico Superior-Terciario Inferior). In: Chebli GA, Spalletti L (eds) Cuencas Sedimentarias Argentinas, Simposio Cuencas Sedimentarias Argentinas, Universidad de Tucumán, Serie Correlación Geológica, vol 6, pp 221–243
Legarreta L, Uliana MA (1996) The Jurassic succession in west-central Argentina: stratal pattern, sequences and paleogeographic evolution. Palaeogeogr Palaeocl 120:303–330
Leinfelder RR (1992) A modern-type Kimmeridgian reef (Ota Limestone); Portugal. Implications for Jurassic reef models. Facies 26:11–34
Lo Forte G, Palma RM (2002) Facies, microfacies and diagenesis of Late Callovian-Early Oxfordian carbonates (La Manga Formation) in the west central Argentinian High Andes. Carbonate Evaporite 17:1–16
Loucks RG (1999) Paleocave Carbonatic Reservoir: origins, burial-depth modifications, spatial complexity and reservoir implications. AAPG Bull 83:1795–1834
MacEachern JA, Pemberton SG (1992) Ichnological aspects of Cretaceous shoreface successions and shoreface variability in the Western interior seaway of North America. In: Pemberton SG (ed) Aplications of ichnology to petroleum exploration, SEPM, Core Workshop 17, Tulsa, pp 57–84
Mazzulo SJ, Mazzulo LJ (1992) Paleokarst and karst associated hydrocarbon reservoir in the Fusselman Formation, west Texas, Permian basin. In: Candelaria MP, Reed CL (eds) Paleokarst, karst related diagenesis and reservoir development: examples from Ordovician-Devonian age strata of West Texas and the mid-continent: Permian Basin Section. SEPM 92, Tulsa, pp 110–120
Min MZ, Lug XZ, Mao SL et al (2001) An excellent fossil wood cell texture with primary uranium minerals at a sandstone-hosted roll-type uranium deposit, NW China. Ore Geol Rev 17:233–239
Mitchum RM Jr, Uliana MA (1985) Seismic stratigraphy of carbonate depositional sequences, Upper Jurassic-Lower Cretaceous, Neuquén Basin, Argentina. AAPG Memoir 39:255–274
Morsch SM (1990) Corales (Scleractinia) de la extremidad sur de la Sierra de la Vaca Muerta, Formación La Manga (Oxfordiano), provincia del Neuquén, Argentina. Ameghiniana 27:19–28
Mulder T, Alexander J (2001) The physical character of subaqueous sedimentary density flows and their deposits. Sedimentology 48:269–299
Nagle JS (1967) Wave and current orientation of shells. J Sediment Petrol 37:1124–1138
Nori L, Lathuilière B (2003) Form and environment of *Gryphaea arcuata*. Lethaia 36:83–96
Oschmann W (1988) Kimmeridge Clay sedimentation—A new cyclicity model. Palaeogeogr Palaeocl 65:217–251
Palma RM, Lo Forte GL, Medhli M et al (2005) High-frecuency cyclicity of the Callovian Calabozo Formation, Neuquén Basin, Argentina. Geol Acta 3:119–132
Palma RM, López-Gómez J, Piethé RD (2007) Oxfordian ramp system (La Manga Formation) in the Bardas Blancas area (Mendoza Province), Neuquén Basin, Argentina: facies and depositional sequences. Sediment Geol 195:113–134
Palma RM, Kietzmann DA, Adamonis S et al (2009) Oxfordian reef architecture of the La Manga Formation, Neuquén Basin, Mendoza Province, Argentina. Sediment Geol 221:127–140
Palma RM, Riccardi AC, Kietzmann DA et al (2011) Depósitos carbonáticos de la Formación La Manga (Caloviano inferior - Oxfordiano medio): evidencias de regresión forzada. Depocentro Atuel, Mendoza, Cuenca Neuquina. In: Abstracts of the 18 Congreso Geológico Argentino, 2–6 May 2018
Palma RM, Kietzmann DA, Martín-Chivelet J et al (2012) New biostratigraphic data from the Callovian-Oxfordian La Manga Formation, Neuquén Basin, Argentina: Evidence from an ammonite condensed level. Rev Paléobiol 11:345–356
Palma RM, Kietzmann DA, Bressan GS et al (2013) Peritidal cyclic sedimentation from La Manga Formation (Oxfordian), Neuquén Basin, Mendoza, Argentina. J S Am Earth Sci 47:1–11

Palma RM, Bressan GS, Kietzmann DA et al (2014) Palaeoenvironmental significance of middle Oxfordian deep marine deposits from La Manga Formation, Neuquén Basin, Argentina. J Iber Geol 40:507–520

Palma RM, Kietzmann DA, Comerio M et al (2015) Oxfordian microbial laminites from La Manga Formation, Neuquén Basin, Argentina: remarkable nanobacteria preservation. J Iber Geol 41:351–363

Palma RM, Bressan GS, López-Gómez J et al (2017) Las facies paleokársticas de la Fm. La Manga (Oxfordiano Medio) en el sur de Mendoza. Rev Asoc Geol Argent 74:40–48

Permberton SG, Frey RW (1984) Ichnology of storm-influenced shallow marine sequence: Cardium Formation (Upper Cretaceus) at Seebe, Alberta. In: Stott DF, Glass DJ (eds.) The Mesozoic of Middle North America. Canadian Society of Petroleum Geologists, vol 9, p 281–304

Perri E, Tucker ME (2007) Bacterial fossils and microbial dolomite in Triassic stromatolites. Geology 35:207–210

Plint AG, Norris B (1991) Anatomy of a ramp margin sequence: facies successions, paleogeography and sediment dispersal patterns in the Muskiki and Marshybank formations, Alberta Foreland Basin. B Can Petrol Geol 39:18–42

Plint AG, Nummedal D (2000) The falling stage systems tract: Recognition and importance in sequence stratigraphic analysis. In: Hunt D, Gawthorpe RLG (eds) Sedimentary responses to forced regressions. The Geological Society, London, SP 172, pp 1–17

Pratt BR (1995) The origin, biota and evolution of deep-water mud-mounds. In: Monty CLV, Bosence DW, Bridges PH, Pratt BR (eds) Carbonate mud-mounds: their origin and evolution. IAS, SP, pp 123–2349

Reid RP, Foster JS, Radtke G et al (2011) Modern Marine Stromatolites of Little Darby Island, Exuma Archipelago, Bahamas: environmental setting, accretion mechanisms and role of Euendoliths. Advances in Stromatolite Geobiology. Lect Note Earth Sci 131:77–89

Röhl H-J, Schmid-Röhl A, Oschmann W et al (2001) The Posidonia Shale (Lower Toarcian) of SW-Germany: an oxygen-depleted ecosystem controlled by sea level and palaeoclimate. Palaeogeogr Palaeocl 165:27–52

Riccardi AC (1984) Las asociaciones de amonitas del Jurásico y Cretácico de Argentina. In: Abstracts of the 9 Congreso Geológico Argentino, San Carlos de Bariloche, 5–9 Nov 1984

Riccardi AC (1992) Biostratigraphy of west-central Argentina. In: Westermann GEG (ed.) The Jurassic of the Circum-Pacific. Cambridge University Press, Cambridge, pp 139–141

Riccardi AC (1996) Heterochronic changes in the Andean Neuquéniceratinae (Ammonoidea, Middle Jurassic). In: Abstractys of the 4 international symposium Cephalopods, Present and Past, 15–17 July 1996

Riccardi AC (2008) The marine Jurassic of Argentina: a biostratigraphic framework. Episodes 31:326–335

Sarg JF (1988) Carbonate sequence stratigraphy. In: Wilgus CK, Hastings BS, Kendall CGStC, Posamentier HW, Ross CA, Van Wagoner JC (eds.) Sea level changes-an integrated approach. SEPM SP 42, pp 155–351

Savrda CE, Bottjer DJ, Seilacher A (1991) Redox-related benthic events. In: Einsele G, Ricken W, Seilacher A (eds) Cycles and events in stratigraphy. Springer, Berlin, pp 524–541

Schieber J, Baird G (2001) On the origin and significance of pyrite spheres in Devonian black shales of North America. J Sediment Res 71:55–166

Schmid DU, Jonischkeit A (1995) The Upper Jurassic Sao Romao limestone (Algarve, Portugal): an isolated carbonate ramp. Profil 8:319–337

Shinn EA (1983) Tidal flats. In: Scholle PA, Bebout DG, Moore CH (eds.) Carbonate depositional environments. AAPG Memoir vol 33, pp 171–210

Stipanicic PN (1951) Sobre la presencia del Oxfordense superior en el Arroyo de La Manga, Mendoza. Rev Asoc Geol Argentina 6:213–240

Stipanicic PN (1965) El Jurásico de la Vega de la Veranada (Neuquén), el Oxfordense y el diastrofismo divesiano (Agassiz-Yaila) en Argentina. Rev Asoc Geol Argent 20:403–478

Stipanicic PN, Westermann GEG, Riccardi AC (1975) The Indo-Pacific ammonite Mayaites in the Oxfordian of the Southern Andes. Ameghiniana 12:281–305

Strasser A (1986) Ooids in Purbeck limestones (lowermost Cretaceous), of the Swiss and French Juras. Sedimentology 33:711–728

Strasser A, Hillgärtner H, Pasquier JB (2004) Cyclostratigraphic timing of sedimentary processes: An example from the Berriasian of the Swiss and French Jura Mountains. In: D'Argenio B, Fischer AG, Premoli Silva I, Weissert I, Ferreri V (eds) Cyclostratigraphy: approaches and case histories. SEPM SP 81, pp 135–151

Taylor KG, Macquaker JHS (2000) Spatial and temporal distribution of authigenic minerals in continental shelf sediments: implications for sequence stratigraphic analysis. In: Glenn CR, Prevot-Lucas L, Lucas J (eds) Marine authigenesis: MICROBIAL TO GLOBal. SEPM, SP 66, p 309–323

Tribovillard N (1998) Bacterially mediated peloids in laminated, organic matter rich, limestones: an unobtrusive presence. Terra Nova 10:126–130

Tucker ME, Wright VP (1990) Carbonate sedimentology. Blackwell, Oxford, p 482

Van Wagoner JC, Posamentier HW, Mitchum RM et al (1988) An overview of the fundamentals of sequence stratigraphy and key definitions. In: Wilgus CK, Hastings BS, Kendall CGStC et al (eds.) Sea level changes—an integrated approach. SEPM, SP 42, pp 39–45

Vergani GD, Tankard AJ, Belotti HJ et al (1995) Tectonic evolution and paleogeography of the Neuquén Basin, Argentina. In: Tankard AJ, Suárez Soruco R, Welsink HJ (eds) Petroleum basins of South America. AAPG Memoir 62, pp 383–402

Weedon GP, Jenkyns HC, Coe AL et al (1999) Astronomical calibration of the Jurassic time scale from cyclostratigraphy in British mudrock formations. Philos T Roy Soc A 357:1787–1813

Wilkin RT, Barnes HI (1997) Formation processes of framboidal pyrite. Geochim Cosmochim Ac 61:323–339

Early Andean Magmatism in Southern Central Chile (33°–40° S)

Pablo Rossel and Francisco Carvajal

Abstract The early Andean arc in Southern Central Chile and Argentina is a main feature of the youngest Andean margin paleogeography, acting as the western edge of the Neuquén Basin sedimentation and sometimes as a barrier to the connection between the open sea and the back-arc basin. Earlier stages of evolution of the magmatic arc are probably related to three pulses of magmatism. The first is related to S-type granites with slightly subduction features and apparently contemporaneous intermediate to highly differentiated volcanic rocks during Late Triassic (~225–220 Ma), that probably represent the first event of magmatism in the area since the Carboniferous. A second is more diverse, with bimodal features, presence of A-type granites and more developed calc-alkaline affinities that took place during the Triassic–Jurassic boundary (210–197 Ma). This event is developed as a paired belt of intrusive outcrops in the present Coastal Cordillera and volcanic deposits in the proximities of the Chile–Argentina border, suggesting an arc—back-arc configuration which is most likely a submerged or slightly emerged immature arc. Finally, the third and definitive instauration and development of the Andean Arc seems to have taken place since late Lower Jurassic, with the development of larger granitic bodies and thick volcanic sequences well exposed in the Coastal Cordillera between 33° and 35° S and close to the Chile–Argentina border between 38° and 40° S. Geochemical and isotopically available data suggest that this early arc was emplaced and evolved through a thin crust with none to little contribution of it during the magma genesis.

Keywords Jurassic · Upper Triassic · Southern Central Andes · Arc magmatism

1 Introduction

Magmatic arcs are among the most striking features in subductive margins, and, in particular, the Andean arc is probably the archetype of subduction-related magmatic systems. With at least 200 Myr of uninterrupted activity and well-exposed outcrops

P. Rossel (✉) · F. Carvajal
Universidad Andrés Bello, Concepción, Chile
e-mail: pablo.rossel@unab.cl

D. Kietzmann and A. Folguera (eds.), *Opening and Closure of the Neuquén Basin in the Southern Andes*, Springer Earth System Sciences,
https://doi.org/10.1007/978-3-030-29680-3_5

(SERNAGEOMIN 2003), the Andean arc is considered one of the main natural laboratories for the understanding of these subduction-related environments. Much of our present understanding of the subduction dynamics and its influence on igneous activity are the result of decades of studies of its products (Buchelt and Téllez 1988; Lucassen et al. 2006; Palacios 1978; Rogers and Hawkesworth 1989; Rossel et al. 2013; Vergara et al. 1995).

Commonly, Andean tectonic evolution, and consequently Andean magmatic evolution, has been divided into two main stages, according to the dominant tectonic conditions ruling the margin. The first is developed mostly under extensional to transtensional conditions, probably induced by the onset of a Mariana-like subduction system in response to roll-back of a cold and dense Phoenix Plate and the incipient breaking of Pangea (Charrier et al. 2007; Grocott and Taylor 2002 and references therein), which took place during Jurassic (and probably part of the Triassic) and Lower Cretaceous. This period was characterized by the effusion of thick piles of mainly mafic to intermediate volcanic material and the intrusion of large co-genetic granitic bodies in the present Coastal Cordillera of Chile and Perú (Buchelt and Téllez 1988; Godoy et al. 2003; González and Niemeyer 2005; Parada et al. 1999 to name a few). To the east of this major paleogeographic domain, extensional conditions persisted since the end of the Paleozoic, leading to the development of extensive NW-SE and N-S oriented extensional basins which worked as counterpart for the record of the Andean arc evolution (Spalleti 2001; Vicente 2006; Charrier et al. 2007) (see Chapter "Controls on Deposition of the Tordillo Formation in Southern Mendoza (34°–36° S): Implications for the Kimmeridgian Tectonic Setting of the Neuquén basin").

However, although the early Andean arc is traceable in Chile almost from its northern to southern edges, most of our understanding of its evolution was the result from the study of rocks that crop out between 18° and 31° S (Kramer et al. 2005; Lucassen et al. 2006; Oliveros et al. 2006, 2007; Rossel et al. 2013, 2015) and to a lesser extent to the south of 42° S (Hervé et al. 2007). This fact introduces an important obstacle, given the scarcity of tectonic and petrogenetical information available for more than 700 km of the arc trace, in a highly segmented margin since the Paleozoic (Mpodozis and Ramos 1989; Charrier et al. 2007). In this context, this chapter intends to compile and summarize the available information regarding the evolution of the early Andean arc in Southern Central Chile (33°–40° S), in order to complement the paradoxically extensive understanding of its paleogeographic counterpart, the Neuquén Basin and evolution of the Andes at these latitudes.

2 Geological Framework

After a period of apparent quiescence or drastic slowdown of subduction in the western margin of Gondwana, induced by the complete amalgamation of the Pangea Supercontinent (Vilas and Valencio 1978) and subsequent collapse of the Gondwanan Orogen during Late Permian and Early Triassic (Llambías et al. 1984; Mpodozis and

Kay 1990; Ramos and Kay 1991), the western margin of South America started a new period of evolution during Jurassic and Lower Cretaceous (Mpodozis and Ramos 1989; Charrier et al. 2007). This new stage was characterized by the development of a well-defined magmatic arc of almost N-S orientation emplaced in the present Coastal Cordillera from Southern Peru to Central Chile (Vicente 2005; Charrier et al. 2007; Mpodozis and Ramos 2008), which apparently shifts to a NW-SE orientation south of 35° S, locating in the actual Main Cordillera of Chile and Argentina at 38–40° S (Charrier et al. 2007; Fig. 1). The arc itself is characterized by large batholitic bodies of Jurassic to Lower Cretaceous ages, that represent the roots of the magmatic system, and related minor subvolcanic bodies and volcanic sequences, that can reach thicknesses over 7000 m in northern Chile (Buchelt and Téllez 1988), which represent the contemporaneous effusive activity. These rocks were emplaced under extensional to transtensional conditions (Schueber and Gonzalez 1999; Grocott and Taylor 2002), although transpressional events are recorded at the end of the Middle Jurassic, especially in Central Chile (Creixell et al. 2011; Ring et al. 2012). Compositionally, the magmatism shows a dominance of mafic and especially intermediate compositions, with mostly medium-K calc-alkaline trends (Kramer et al. 2005; Lucassen et al. 2006; Oliveros et al. 2006; Rossel et al. 2013). Isotopically, these rocks show depleted signatures evidencing small influences of subducted sediments, eroded fore-arc crust or assimilated wall rocks in the storage levels (Kramer et al. 2005; Lucassen et al. 2006; Rossel et al. 2015), which is concordant with the proposed extensional to transtensional setting suggested for the first stage of evolution of the Andean arc.

The configuration of the arc in Central Chile (33–37° S) is similar to that observed in the north (18–31° S) but shows a higher proportion of felsic effusive rocks (Vergara et al. 1995; Vicente 2005), and thicknesses of arc-related lavas and interbedded sediments do not exceed 6000 m (Thiele and Morel 1981; De la Cruz and Suarez 1997). In particular, intrusive bodies exhibit somewhat lower exposures, especially between 34 and 38° S.

Available geogronological data in the area suggest that magmatic activity begun at 225 Ma (Vásquez et al. 2011) with a main pulse of magmatism between 180 and 155 Ma (Creixell et al. 2011; Gana and Tosdal 1996; Godoy and Loske 1988; Ring et al. 2012; Vásquez et al. 2011; Wall et al. 1996).

Scarce available radiometric ages between 155 and 130 Ma in the Coastal Cordillera of Central Chile are coincident with evidence of transpressional tectonics in Middle to Upper Jurassic dykes (Creixell et al. 2011; Ring et al. 2012) and synchronous effusion of thick homogeneous sequences of volcanic material in the backarc during Kimmeridgean to Tithonian (Rossel et al. 2014; Salazar and Stinnesbeck 2016).

No exposures of Jurassic rocks can be observed between 37° and 38° S, and south of these latitudes the arc axis shifts to the actual position in the Chile-Argentina border (Fig. 1). Magmatic evidence of the early Andean subduction system in this area are from volcanic and volcano-sedimentary deposits of the Nacientes del Biobio Formation (38–39° S; Suárez and Emparan 1997) and large batholitic bodies of Jurassic to Lower Cretaceous ages that are exposed almost continuously between

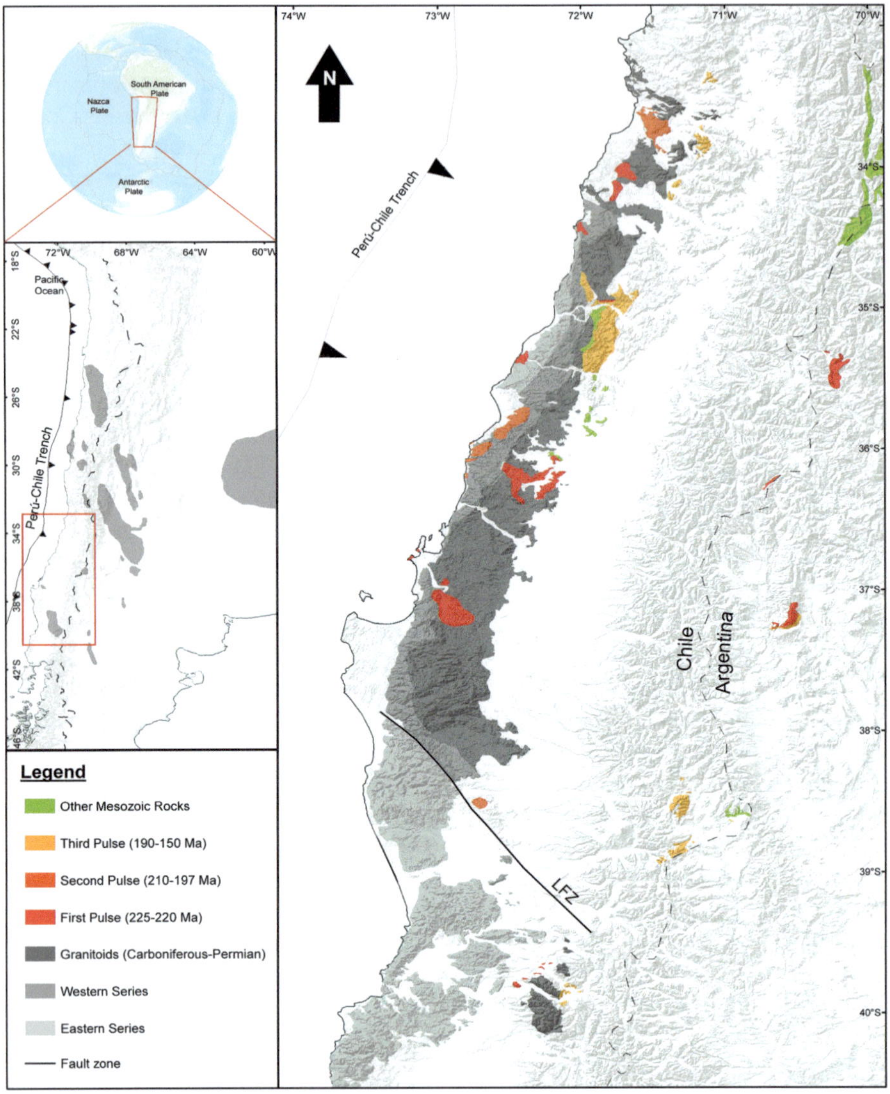

Fig. 1 **a** Location of the Upper Triassic and Jurassic magmatic rocks in Southern Central Chile between 33° and 40° S. Based on available cartography done by the Chilean Geological and Mining Survey; Vásquez et al. (2011), Naipauer et al. (2015) and Llambias et al. (2007). RD.T.F: Rio Damas-Tordillo Formation. C.H.B: Curepto-Hualañe Basin. C.C.A: Cerro Castillo Andesite. P.B: Pocillas Basin. C.T.B: Cajon de Troncóso Beds. S.J.B: Santa Juana Basin. C.D.V: Cordillera del Viento. L.H.F: Llafquentue-Huimpil Formation. N.B.F: Nacientes del Biobío Formation. Pp.B: Panguipulli Basin. **b** Main Triassic extensional depocenters in Chile and Argentina (modified after Spalleti 2001)

38.5° and 40° S (Munizaga et al. 1988; Parada and Munizaga 1978; Rodríguez et al. 1999; Suárez and Emparan 1997).

3 Pre-Andean Magmatism

To the South of 33° S, there are few localities where it is possible to observe direct evidence associated with the early stages of the proto Andean arc (Fig. 1). Those are a series of small epizonal granitic intrusives with an incipient calc-alkaline signature in the coast of Chile between 35° and 37° S (Vásquez and Franz 2009, 2011). These bodies probably represent the initial stages of re-instauration of the Andean arc after the Mid Triassic. This early pulse of magmatism is restricted to a short period between 225 and 220 Ma and is represented by Constitución (35° S) and Hualpén (37° S) granites (Vásquez et al. 2011). According to the previous authors, those granites show high silica content (>68%), biotite as main Fe–Mg phase, and late white mica and tourmaline as distinctive mineral phases, evidencing the presence of water in their genesis. On the other hand, based on geochemical and isotopic data, they can be classified as S-type and apparently, they evolved in a shallow reservoir with significative but less intense contribution of the Paleozoic basement in their genesis than any previous Gondwanian-related granitoids (Parada et al. 1999; Lucassen et al. 2004).

Various volcanic and subvolcanic intermediate to acid rocks, in a close geographic relation to the afore mentioned granitoids, are observed at the base or intercalated within Upper Triassic sediments (Corvalan 1976; Hervé et al. 1976; Abad and Figueroa 2003; Fig. 2).

Northermost evidence of this volcanism through the Coastal Cordillera is at the Crucero de los Sauces Formation (35° S; Corvalan 1976; Thiele and Morel 1981), a Middle–Upper? Triassic unit composed of ignimbrites and rhyolitic tuffs at the base that pass to basaltic and dacitic lavas to the top of the sequence.

According to Thiele and Morel (1981), volcanic rocks were deposited filling paleo depressions developed in the Paleozoic granitoids, underlying the Upper Triassic Marine deposits of the Estero de la Higuera Formation in apparent disconformity.

To the south, on the eastern flank of Coastal Cordillera, in the Cerro Gupo (35.5° S) and Pocillas area (36° S), Triassic volcanic deposits crop out. In the first, igneous rocks were described as tuff levels intercalated in continental sediments that overlie highly deformed and metamorphosed marine deposits (Hervé et al. 1976). The second locality corresponds to the Los Arrayanes Beds, where andesitic to rhyolitic lavas with intercalations of volcanic breccias, lithic tuffs and few clastic sediments are exposed (Lagno 1981). According to the previous author, the sequence overlies in apparent conformity the Upper Triassic Marine deposits of the La Patagua Formation (Muñoz Cristi and Gonzalez Pacheco 1953). Both units were intruded by Lower Jurassic granitoids (Hervé et al. 1976; Lagno 1981).

At the same latitude but in Main Cordillera, Upper Triassic deposits of the Cajon de Troncoso Beds (Cornejo et al. 1982) represent the easternmost evidence of Triassic

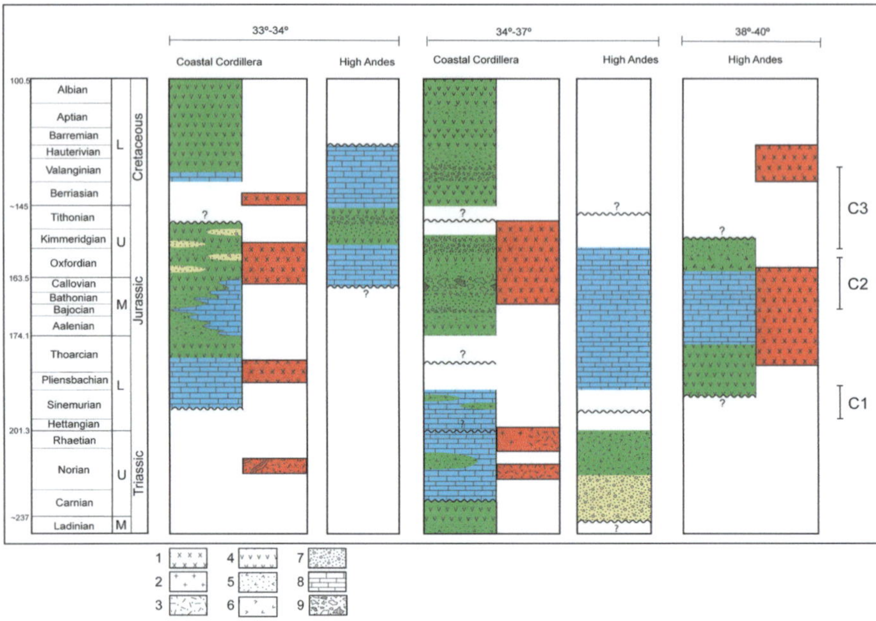

Fig. 2 Chronoestratigraphic chart with Upper Triassic and Jurassic rocks in the studied segment and morphotectonic units. Based on available cartography done by the Chilean Geological and Mining Survey. 1: Mafic and intermediate granitoids, 2: gabroic rocks, 3: felsic granitoids, 4: lavas, 5: volcanoclastic rocks, 6: pyroclastic rocks, 7: continental sedimentary rocks, 8: marine rocks and 9: volcanic breccias. For specific references see the text

volcanism in Chilean Territory. According to these authors, this unit has a basal clastic continental member of ~70 m of thickness and an upper member, composed of tuffs and rhyolitic breccias, with minor intercalations of shales that can reach a maximum thickness of 45 m. The base of the unit is unexposed, and the volcanic deposits are overlaid unconformably by the Middle Jurassic marine rocks of the Nacientes del Teno Formation. Available paleontological data in clastic deposits at the base of this unit suggest an Upper Triassic age.

The southernmost exposures of Triassic volcanic deposits in Southern Central Chile are andesites, dacites and tuffs intercalated in the Upper Triassic Santa Juana Formation (37° S; Abad and Figueroa 2003). Igneous rocks can be recognized forming a narrow belt of outcrops in the eastern margin of this basin. Available fossil flora intercalated in volcanic and clastic deposits suggest a Carnian-Norian age for these rocks (Leppe et al. 2006).

The geographic and chronological relations between intrusive and effusive rocks suggest the existence of synchronous magmatic activity in the area during Late Triassic, but given the scarcity of geochemical and geochronological data for these volcanics, with the exception of the "Andesita Cerro Negro" of 220 Ma. (Naipauer et al. 2015), and the subduction-related Cordillera del Viento Formation (Llambías et al. 2007), both in Argentinian territory, this should be considered only as a working hypothesis.

4 The Transitional to Clearly Subduction Related Arc-Magmatism

After the emplacement of the aforementioned granitic intrusives, a second pulse of magmatism is recognized in the western margin of the Coastal Cordillera of Southern Central Chile. This event was apparently longer than the previous one, covering a life span between 210 and 197 Ma (Vásquez et al. 2011). Compositionally, this event frequently shows bimodal features. Particularly, in the vicinities of the locality of Cobquecura (36° S), it is possible to observe the presence of gabbros and granites with pillows of gabbros, evidencing clearly mingling process (Vásquez and Franz 2009). In contrast to granitic magmas of the preceding pulse, the felsic constituent of this event is more Fe-rich and contains anhydrous phases such as fayalite and hedenbergite or late arfvedsonite that suggest H_2O-poor melt composition and low fO_2 crystallization conditions (Vásquez et al. 2011). Isotopically, magmas of this pulse show a higher input of juvenile asthenospheric sources based on eNd values ranging between 2.8 and −2.0, but high and decoupled values of $^{87}Sr/^{86}Sr$, plotting to the right of the mantle array, and suggesting the existence of a crustal influence in the genesis of the magmas, even in gabbroic rocks (Vásquez et al. 2011). Based on geochemical data, most of the rocks of this pulse can be classified as anorogenic granites even when they show typical subduction-related features (Vásquez et al. 2011). According to these authors, even when many of the tectonic and chemical features of the rocks seem to indicate an origin not related to subduction (i.e., High REE contents, tectonic diagrams, presence of typically anorogenic-related minerals as fayalite and arfvedsonite, the bimodal nature of the magmatic suites, among others), the subduction-related chemistry of the magmas, especially for the mafic ones, cannot be achieved exclusively by the assimilation of the metamorphic basement involved in their genesis and should be considered a genuine and primary feature of the magmas.

No effusive counterpart of this pulse can be recognized in Chilean territory, but at least two localities in western Argentina have volcanic and subvolcanic rocks of this age. The first at 36.5° S is represented by the bimodal volcanic and subvolcanic rocks of the Cara-Cura Reyes Formation (Miranda 1963; Gulisano et al. 1984), dated in 199 Ma by Naipauer et al. 2016 (U–Pb SHRIMP age in zircons). The second is located at 37° S and corresponds to the basalts and basaltic andesites of the Milla Michico Formation in the Cordillera del Viento (Llambías et al. 2007), which, as its intrusive counterpart, shows clear subduction-related affinities but not bimodal compositional features.

5 The Definitive Instauration of the Early Andean Subduction System

After a period of bimodal activity with incipient subduction-related signatures during earlier most Jurassic, a major pulse of magmatic activity can be recognized not only circumscribed to Southern Central Chile but through all the Chilean margin during middle to Late Jurassic (Kramer et al. 2005; Lucassen et al. 2006; Oliveros et al. 2006; Rossel et al. 2013). In this area, the activity is related to thick volcanic successions and genetically related intrusive bodies, mostly of intermediate composition. Between 33 and 34° S, the effusive activity is represented by the mostly Lower–Middle Jurassic Ajial Formation (Vergara et al. 1995). This unit was deposited under alternate marine and continental conditions and consists of a lower sequence of acid lavas and an upper sequence composed of differentiated pyroclastic fall deposits and some ignimbrites, intercalated and grading into sedimentary mostly continental deposits (Thomas 1958; Piraces 1977; Vergara et al. 1995). During Late Middle and Upper Jurassic, volcanic activity in the arc domain was represented by the continental, mostly andesitic deposits of the Horqueta Formation (Piraces 1977). This unit overlies concordantly the marine sediments of the Aalenian to Bajocian? Cerro Calera Formation (or the Ajial Formation where the former is absent; Piraces 1977), and underlies with no clear contact relations (unconformable according to Carter 1963 and Corvalan and Davila 1964; conformable according to Wall et al. 1996) the marine rocks of the Lower Cretaceous Lo Prado Formation. Jurassic volcanic and volcanoclastic deposits related to the Ajial and Horqueta formations can reach a thickness over 6000 m (Vergara et al. 1995), which is in the range but lower than the 7000–8000 m of the volcanic piles observed in northern Chile for the contemporaneous La Negra Formation (Buchelt and Téllez 1988).

Contemporaneous intrusive activity in this area is represented by numerous dioritic, tonalitic and granodioritic bodies and various mafic dike swarms. These bodies intrude the Jurassic volcanic sequences and older Late Paleozoic and Triassic units (Creixell et al. 2011). Available radiometric K–Ar ages in these rocks show a well-defined span of activity between 160 and 145 Ma (Wall et al. 1996), but more accurate methods as U-Pb (zircon and titanite), Ar–Ar (hornblend) and Rb-Sr (whole rock isochrone) ages for the area show no evidence of magmatic activity after 154 Ma (Godoy and Loske 1988; Creixell et al. 2011; Ring et al. 2012).

At the same latitudes, but in main cordillera of Chile and Argentina, a thick pile of continental volcanoclastic sediments and andesitic lava flows of the Rio Damas-Tordillo Formation crops out (Klohn 1960). Based on their actual position, to the east of the main axis of the Jurassic arc plus the geochemical signature of the lavas, showing typical subduction-related features with tholeiitic affinities (Rossel et al. 2014), this unit has been interpreted as the back-arc counterpart of the Upper Jurassic Arc (Charrier et al. 2007). However, a maximum depositional age of 146.4 $\pm$ 4.4 Ma (U–Pb detrital zircons) in clastic sediments from the base of the lava flows in the Rio de las Damas locality (Rossel et al. 2014) suggests that this volcanic activity was diachronous with the main activity in the arc, apparently covering a gap of magmatism and volcanism between the Horqueta and Lo Prado formations.

Between 34 and 35.5° S, the volcanic activity of the Jurassic arc is represented mostly by the Lower to mostly Middle Jurassic deposits of the Altos de Hualmapu Formation. This unit overlies in apparent concordance (Corvalan 1976; Thiele and Morel 1981) to the Hettangian to Sinemurian marine successions of the Rincon de Nuñez Formation (Corvalan 1976), which at its higher levels shows few intercalations of tuffs and epiclastic immature sandstones, suggesting that the arc was active at least since the Late Sinemurian in this area.

Altos de Hualmapu Formation is composed mainly of volcanic breccias, tuff and andesitic lavas, fed by andesitic dykes that intrude both the deposits of the marine Rincon de Nuñez and the Middle Jurassic breccias and lavas. The volcanic deposits are intercalated with marine sandstones with fragments of undifferentiated bivalves at the base, suggesting that the early stages of volcanism occurred under marine conditions or very close to the shoreline. To the top, no evidence of sedimentary rocks is observed but given the absence of hyaloclastites or pillow structures, this volcanism was developed probably under subaerial conditions.

Main intrusive activity in the area is related to calc-alkaline dioritic to granodioritic bodies (Vásquez et al. 2011) that intrude the Rincon de Nuñez and Altos de Hualmapu formations (Corvalan 1976; Thiele and Morel 1981), and can be recognized as isolated outcrops in the Coastal Cordillera at least until 37° S (SERNAGEOMIN 2003). Available K-Ar geochronological data for these rocks indicate a main pulse of activity between 175 and 165 Ma (Gana and Herve 1983), but later U-Pb zircon data suggested that magmatism was active at 155 Ma. (Vásquezet al. 2011), which is highly concordant with the determined life span between 33 and 34° S.

The southernmost exposure of this magmatism in Southern Central Chile is located in the High Cordillera, close to the border between Chile and Argentina between 38 and 40° S. This activity is represented by the volcanic deposits of the Jurassic Nacientes del Biobio Formation and batholitic bodies, mostly of intermediate composition (Suárez and Emparan 1997; Rodríguez et al. 1999).

The volcanic activity of the Nacientes del Biobio Formation can be divided into two main pulses. The first, during Pliensbachian to Aalenian (Suárez and Emparan 1997), corresponds to a series of basaltic to basaltic-andesitic lava flows effused under submarine conditions (Fig. 2) that are part of the Lower Icalma Member (Suárez and Emparan 1997). The base of this unit is not exposed, and no clear stratigraphic relations with overlaying rocks are observed, but according to previous authors, they probably underlie the marine sedimentary deposits of the middle Lolen-Pacunto member of Nacientes del Biobio Formation. The second event of volcanism recorded in Nacientes del Biobio Formation is represented by the intermediate-acid volcanic and pyroclastic deposits that form part of the top of the upper Lonquimay member (Suárez and Emparan 1997; De la Cruz and Suárez 1997). These rocks were deposited over marine regressive clastic and carbonaceous deposits of Callovian to Oxfordian age (Suárez and Emparan 1997). Pyroclastic deposits suggest an origin related to the collapse of an eruptive column, with the presence of a basal surge underflow and overriding ash cloud that was subsequently covered by a high-energy ignimbrite evidenced by the presence of volcanic breccias to the top of the sequence.

Available geochemical data for the lavas of the Icalma and Lonquimay members suggest a subduction-related signature for this volcanism, with small influence of the crust in the genesis of the lavas and with hornblende as the main residual mineral phase during magma fractionation through the Early Jurassic. Traceable but not with a high crustal assimilation during Late Jurassic is suggested by the authors, which could be related to a reduced crustal thickening or the existence of isolated compressive-transpressive events in Late Jurassic, as observed for Central Chile (Creixell et al. 2011; Ring et al. 2012).

Main intrusive bodies in the high Andes of Southern Central Chile are the Galletue Plutonic Group (GPG) of Upper Jurassic to Upper Cretaceous age (Suárez and Emparan 1997) at 38–39° S and the Panguipulli (180–160 Ma), Huechulafquen (170–160 Ma) and Choshuenco (140–130 Ma) plutons in the proximities of the Panguipulli and Riñihue Lakes at 40° S (Parada and Munizaga 1978; Munizaga et al. 1988; Rodríguez et al. 1999). According to Suárez and Emparan (1997), the GPG is composed of monzogranites, quartz diorites and predominantly tonalities and granodiorites. Available K–Ar ages suggest that the magmatic activity in this area is constrained between 148 ± 8 and 73 ± 2 Ma.

To the south, in the Lake District, the Panguipulli Pluton is a northwest–southeast elongated intrusive body, with and exposed area of about 120 km^2, mainly composed of biotite granites and subordinate biotite and hornblende granodiorites and hornblende and biotite tonalities. To the southeast, the Huechulafquen stock can be recognized. It corresponds to a 40 km^2 tonalitic intrusive, with honblende and biotite as its main mafic constituents. Finally, the younger early Andean rocks in this area are the 10 km^2 Lower Cretaceous Choshuenco stock, composed of quartz diorites and hornblende tonalities. No geochemical data are available for these rocks but given the compositional features ranging mainly from quartz diorites to tonalities and subordinate granodiorites and granites, all of them with significative amounts of hydrated minerals as hornblende and biotite as main mafic constituents, a subduction-related origin can be envisaged.

6 Northern Versus Southern Central Early Andean Arc

Geographically, one of the most characteristic features of the early Andean arc in Northern Chile is its location restricted to the present Chilean coast and western edge of the Coastal Cordillera. Additionally, this magmatism is coeval to the development of back-arc activity in the Precordillera and high Andes during Upper Triassic and Upper Jurassic (Rossel et al. 2013; Oliveros et al. 2018). These magmatic belts show a clear N-S to even NNE-SSW orientation south of 26° S. In contrast, early Andean arc exposures in Central and Southern Central Chile show major variations in the location of the magmatic front, with a coastal domain in Central Chile and an Andean domain to the south of this segment (Fig. 1), with apparently, no development of back-arc magmatic activity during the Late Jurassic. Given the lack of continuity presented by the outcrops south of 35° S; it is rather speculative to define if this eastern shift of the volcanic front from north to south is the result of (i) a bending in the Andean axis or (ii) an abrupt arc jump most likely influenced by a major structure (i.e., Lanalhue Fault Zone) or even additional tectonic factors such as (iii) the development of a flat slab or (iv) subduction erosion contrasting rates between northern and southern domains.

Otherwise, as in Northern Chile, the axis of the magmatic arc in the northern edge of the studied area shows an eastern shift from Upper Triassic to Lower Cretaceous, with older rocks cropping out literally in the actual Pacific shoreline and Upper Jurassic to Lower Cretaceous rocks in the eastern flank of the Coastal Cordillera. Contrarily, no clear shifting of the Mesozoic magmatic axis can be observed in southern rocks, which resembles the static behavior of the Patagonian Batholith since Late Jurassic (Rapela et al. 2005).

Stratigraphically, early subduction-related volcanism of Triassic age in northern Chile commonly overlies Upper Triassic clastic deposits (Oliveros et al. 2018) and is typically covered by marine deposit of Rhaetian to Sinemurian, that predates the definitive instauration of the volcanic arc during the Pliensbachian. In Central and Southern Central Chile, these stratigraphic relations are very similar but Late Triassic volcanics as Crucero de Los Sauces and Santa Juana formations, as mentioned before, commonly overlie by unconformity the late Paleozoic crystalline basement.

On the other hand, even though it is rather difficult to locate good outcrops to constraint the existence of disconformities, previous works have suggested the presence of depositional gaps and probably disconformities during early Late Triassic, Early Jurassic and latest Late Jurassic (Fig. 2). Particularly, the two younger disconformities described on the Chilean slope coincide with the C1 and C3 compressional events observed in Argentinian in Patagonia (Navarrete et al. 2016), suggesting that short tectonic compressional episodes through the Jurassic-affected vast sectors including both slopes of the Andes and that Early Mesozoic tectonics is somewhat more complex than was formerly assumed.

As in Northern Chile, the definitive instauration of the volcanic arc in Central Chile took place at the end of Early Jurassic, apparently showing its first peak of activity during the late Early and early Middle Jurassic (Fig. 3), which is in turn contrastingly late compared to Northern Chile. On the other hand, in the southern end of the studied area, the arc activity began quite earlier than in Central Chile (Sinemurian to Pliensbachian), which coincides with the first peak of activity in Northern Chile.

The main pulse of arc activity of the early Andean subduction system in Northern Chile took place since the second half of Middle Jurassic and reached their peak of magmatic production during latest Late Jurassic and earliest Early Cretaceous. On the other hand, the Southern Central Chile arc showed a similar starting point for the second pulse of Jurassic activity during Middle Jurassic, but apparently suffered an abrupt decrease in its production during the second half of the Late Jurassic in the Coastal Cordillera and high Andes region at 38–40° S, which is coincided with the effusion of large amounts of volcanic materials of the Rio Damas Formation in the

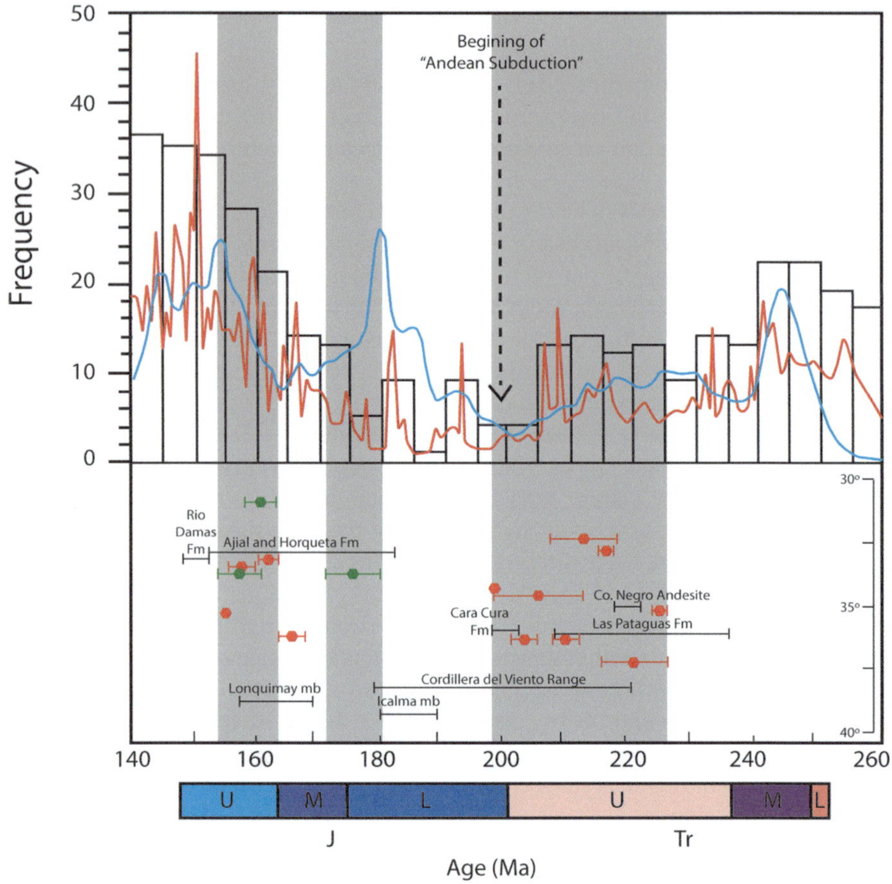

Fig. 3 **a** Frequency distribution histogram for mineral U–Pb and K–Ar (Ar–Ar) radiometric ages from Northern Chile and model distribution of detrital zircón U–Pb ages for Triassic-Jurassic rocks of the Neuquén Basin. Modified after Naipauer et al. (2015). **b** Age distribution of available radiometric U–Pb (red) and Ar–Ar (green), and stratigraphic ages (black) of Triassic and Jurassic igneous rocks of Southern Central Chile. References in the text. Gray bars mark the time span of Patagonian compressive events after Navarrete et al. (2016)

present Chilean-Argentinian boundary during the Kimmeridgian and Tithonian, and as mentioned before with the C3 event of Navarrete et al. (2016) (Fig. 3).

Compositionally, early stages of the Mesozoic magmatism in Southern Central Chile show a transition from anorogenic signatures in the Upper Triassic to clearly calc-alkaline subduction-related compositions in Middle to Upper Jurassic. This compositional variation is systematically coupled with a progressive decrease in silica content, suggesting a progressive decline in the participation of continental crust in the genesis of the magmas during the Mesozoic (Vásquezet al. 2011). The same authors propose that even when the subduction signal is very weak in older

Triassic rocks, this is the result of subduction-related process that probably mark the beginning of the Andean arc in Central Chile and not an inheritance of the Paleozoic crust. Similar behavior can be observed in Northern Chile Triassic and Jurassic igneous rocks (Del Rey et al. 2016; Oliveros et al. 2018). Then, no clear evidence of complete ceasing of the subduction during Triassic could be assumed as proposed in older models (Mpodozis and Kay 1992; Charrier et al. 2007, and references therein), but apparently other driving forces in the genesis of the magmas, as adiabatic melting resulting from the continuous thinning of the crust could have played an important role during the Triassic-Jurassic boundary.

7 Preliminar Geochemical Characterization and Tectonic Implications for the Evolution of the Early Andean Arc in Southern Central Chile

In terms of geochemical compositions, early Andean rocks cropping out in Southern Central Chile show a consistent and uninterrupted sub-alkaline trend that ranges from basaltic to rhyolitic compositions from Late Triassic to Jurassic (Fig. 4a). Same samples plotted in AFM diagrams seem to show a typical curved pattern, suggesting a tholeiitic trend of evolution for their magmas. However, if the samples are filtered according to the above-mentioned three pulses of magmatism, many of the previous characteristics are lost (Fig. 4b). As mentioned before, older Triassic rocks are highly differentiated. Jurassic samples are mostly basaltic to andesitic, and magmas belonging to the Triassic-Jurassic boundary have a bimodal composition. In the same way, if we take separately the different groups, no evidence of tholeiitic fractionation is suggested, although bigger datasets are needed for each period to produce more definitive statements.

Trace elements of all studied rocks, with the exception of three Triassic granitic samples that plot in the WPG field (Fig. 4c), suggest subduction-related affinities (Fig. 4c and d), with marked enrichment in LILE's over HFSE's and the presence of negative Nb–Ta and positive Pb anomalies. REE contents show negative slopes and flat tails suggesting the absence of garnet in the source (Fig. 5). Negative Eu anomalies are a common feature in Triassic and Jurassic rocks, especially in evolved ones, suggesting the presence of plagioclase as an important fractionated phase under apparently low fO2 conditions. This feature suggests that the amount of water available for the genesis of the magmas was restricted, probably as a result of the immaturity of the arc and/or slow rates of convergence, especially during the Triassic (Vilas and Valencio 1978).

Ce/Y, La/Yb and Sr/Y ratios (Fig. 6a, b and c), commonly used as proxies of the crustal thickness (Mantle and Collins 2008; Chapman et al. 2015; Profeta and Ducea 2015), show particularly unexpected behaviors, suggesting an increase in Moho depth from Late Triassic to Late Jurassic, whereas isotopic data of basaltic and intermediate rocks do not clearly reflect a decrease or increment in crustal participation from older

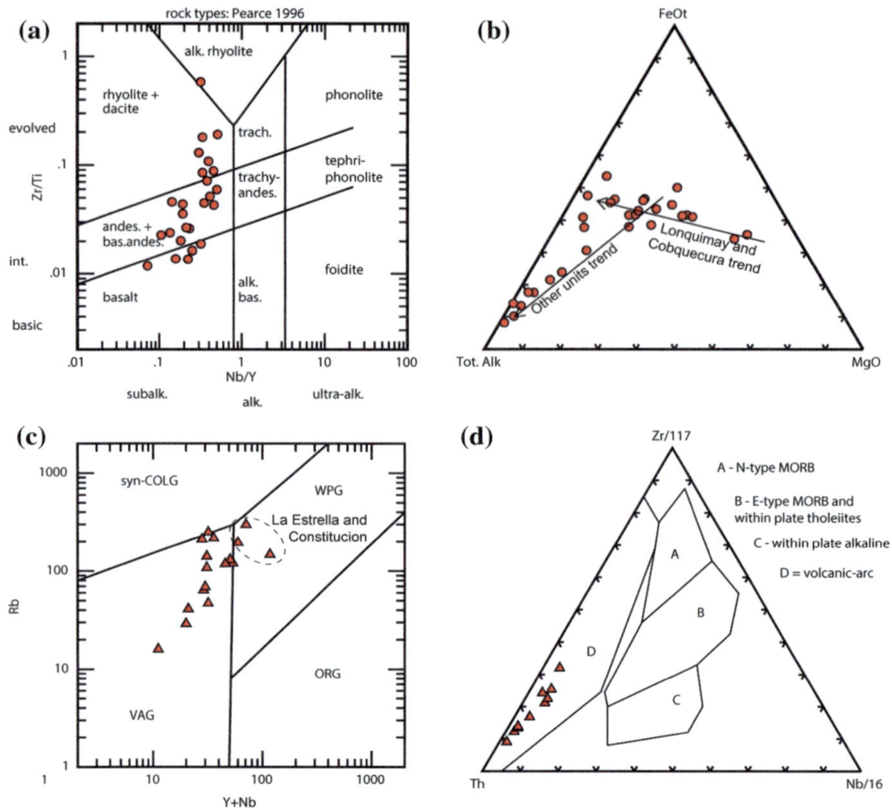

Fig. 4 **a** Nb/Yb versus Zr/Ti diagram for altered rocks (Pearce and Wyman 1996); **b** AFM (Irvine and Baragar 1971); **c** Rb versus Yb + Nb diagram for plutonic and acid volcanic rocks (Pearce et al. 1984); **d** Th–Nb–Zr diagram for basic intermediate and silicic rocks (Wood 1980). Data are from Vergara et al. (1995), Vázquez et al. (2009, 2011) and Oliveros et al. (2018)

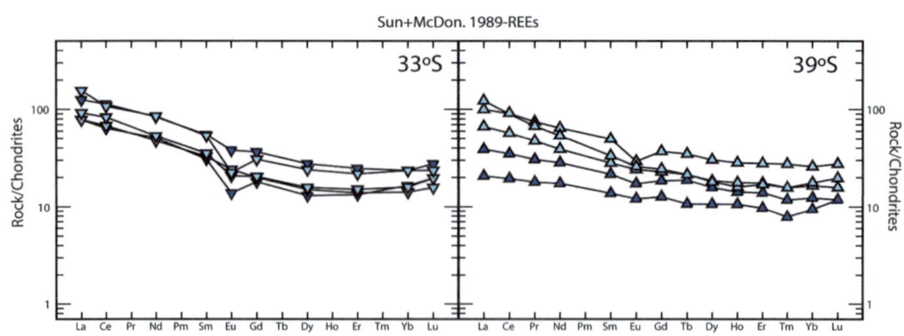

Fig. 5 Chondrite-normalized rare earth spider diagram for volcanic Jurassic rocks from Southern Central Chile. Data are from Vergara et al. (1995) and Oliveros et al. (2018)

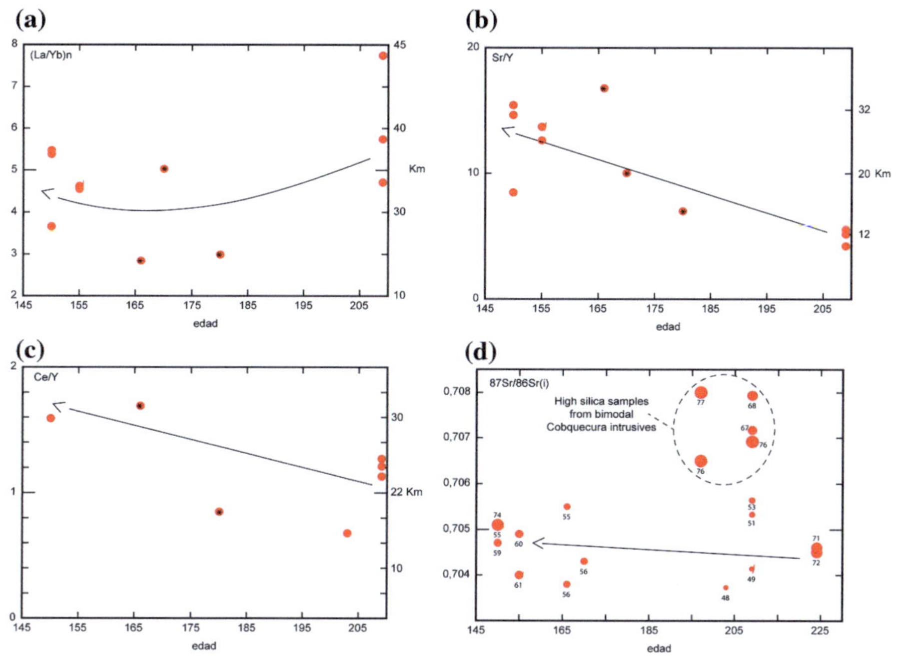

Fig. 6 **a** (La/Yb)n, **b** Sr/Y, **c** Ce/Y and **d** 87Sr/86Sr versus age distribution of Triassic and Jurassic intermediate igneous rocks of Southern Central Chile. Equivalent Moho depth from Mantle and Collins (2008) and Profeta and Ducea (2015). Only (*) samples fit in all the parameters proposed for the previous authors for accurate Moho depth estimations. Data are from Vergara et al. (1995), Vázquez et al. (2009, 2011) and Oliveros et al. (2018)

to younger rocks (Fig. 6d). It is worth noting that the anomalous behavior in REE could probably be related to those only three samples (Fig. 6a and 6c) achieve all the constraints proposed by the original models, and therefore, the available data presented here must be considered only as a preliminary working hypothesis.

Summarizing, the early Mesozoic magmatism in Southern Central Chile was developed under the influence of a subduction regime in a non-thickened crust (Vásquez et al. 2011), probably as the result of continuous extensional conditions ruling the margin since the collapse of the Gondwanan orogen (Charrier et al. 2007). However, it is important to note that the systematic compositional variations exposed before seem to suggest that the influence of thc subduction was gradually higher from Triassic to Late Jurassic times, probably in response to the progressive maturation of the subduction system. In this context, an important detail to take into account is the S-type nature of the older Constitucion and Hualpen granitoids (Vásquez et al. 2011), since this group of rocks has been associated with the subduction initial stages in convergent margins (Chen et al. 2014). The above could indicate that these rocks effectively constitute an older record of the proto-Andean subduction activity in the area, and that previously and unlike to what is observed in Northern Chile; there existed a gap in magmatic production during Early and Middle Triassic in the western Gondwanan margin at these latitudes.

8 Summary and Concluding Remarks

After a period of marked waning of subduction activity coincident with the complete amalgamation and subsequent collapse of the Pangea supercontinent, the first evidence of magmatism in the western margin of Gondwana at these latitudes is the S-type granites of Constitución and Hualpén (225 Ma) and probably related acid and intermediate volcanic rocks of the La Patagua beds, Crucero de Los Sauces and Santa Juana formations. Given the compositional features of these rocks, they could probably represent the first stages of subduction-related magmatism in the area, since the late Paleozoic Granitoids of Coastal Cordillera that took place in inherited NW-SE Paleozoic lineaments along the western margin of Gondwana (Fig. 7a).

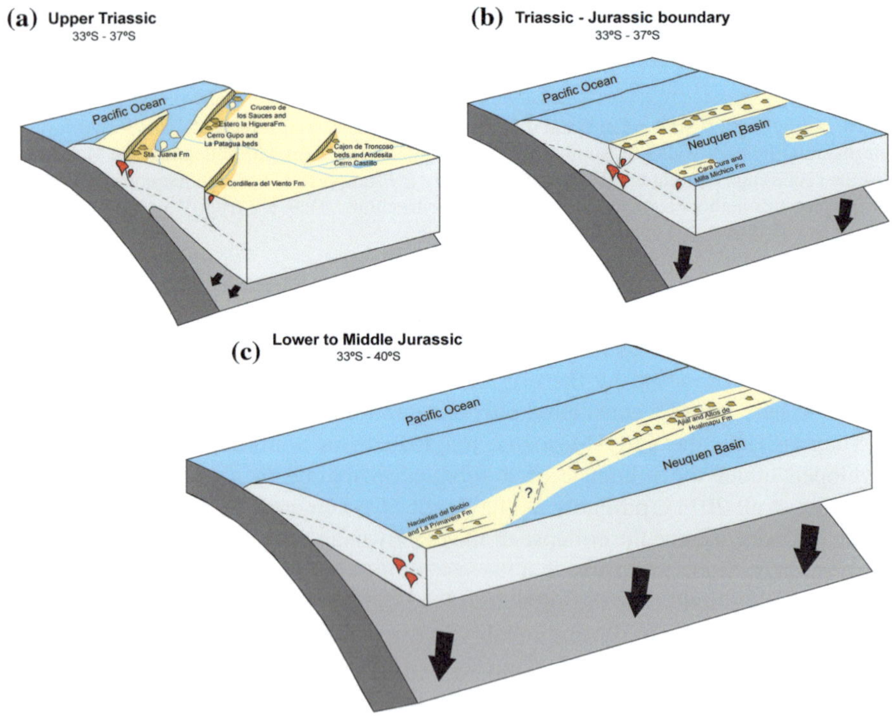

Fig. 7 Paleogeographic and tectonic setting for the magmatism at the Southern Central Chilean margin between Upper Triassic and Middle Jurassic. Location of the schemes corresponds to present-day latitudes

Later during the Triassic–Jurassic boundary (209–198 Ma), a new pulse of magmatism was recorded in the Coastal Cordillera and close to the Chile–Argentina border. Rocks related to this pulse tend to show a bimodal composition and clearer subduction-related signatures, suggesting a strong influence of the fluids released from the subducted slab, but a significative participation of crustal components in the genesis of the magmas, especially in most silicic rocks. The scarcity of stratified units of this age in Chilean territory and the absence of primary volcanic deposits make difficult the construction of paleographic models for this area. However, given the presence of two subparallel magmatic belts with subduction-related signatures and the presence of contemporaneous marine deposits in the Chilean Coastal Cordillera, it is possible to reconstruct an immature arc—back-arc system in which the first was just a narrow and reduced and even not emergent volcanic barrier (Fig. 7b).

Finally, the settlement of a mature mostly subaerial arc since Middle Jurassic is represented by the extrusion of a voluminous pile of differentiated to intermediate volcanic rocks (and associated intrusives) in the position of the present Coastal Cordillera, between 33 and 34° S, and basaltic to andesitic rocks in the Main Cordillera at ~38° S, during a progressive crustal thickening (Fig. 7c).

References

Abad E, Figueroa J (2003) Rocas volcánicas en el Triásico del Biobío: petrografía y marco geológico, cerro Calquinhue, VIII Región; Chile. In: Abstracts of the 10 Congreso Geológico Chileno, Concepción, 6–10 Oct 2003

Buchelt M, Téllez C (1988) The Jurassic La Negra Formation in the area of Antofagasta, northern Chile (lithology, petrography, geochemistry). Lect Notes Earth Sci 17:71–182

Carter WD (1963) Unconformity marking the Jurassic-Cretaceous boundary in the La Ligua area, Aconcagua Province, Chile: U.S. Geological Survey Professional Paper 450-E, pp E61–E63

Cornejo CC, Muñoz BJ, Covacevich CV (1982) Presencia del jurasico sedimentario marino en el Cajon Troncoso, Alta Cordillera de los Andes, VII región, Chile: noticia preliminar. In: Abstracts of the 3 Congreso Geológico Chileno, Concepción, 8–14 Nov 1982

Corvalan J, Davila A (1964) Observaciones geologicas en la Cordillera de la Costa entre los rıos Aconcagua y Mataquito. Soc Geol Chile 9:1–4

Corvalan J 1976 El Triasico y Jurasico de Vichuquen-Tilicura y de Hualañe, Provincia de Curico. Implicaciones paleogeograficas. In: Abstracts of the 1 Congresos Geologico Chileno, Santiago, 2–8 Aug 1976

Chapman JB, Ducea MN, Profeta L et al (2015) Tracking changes in crustal thickness during orogenic evolution with Sr/Y; an example from the Western U.S. Cordillera. Geology 43:919–923

Charrier R, Pinto L, Rodriguez MP (2007) Tectonoestratigraphic evolution of the Andean Orogen in Chile. In: Moreno T, Gibbons W (eds) The geology of Chile. The Geological Society, London, pp 21–144

Chen Y, Shuguang S, Yaoling N et al (2014) Melting of continental crust during subduction initiation: a case of study from the Chaidanuo peraluminous granite in the North Qilian suture zone. Geochim Cosmochim 132:311–336

Creixell C, Parada MA, Morata D et al (2011) Middle-Late Jurassic to Early Cretaceous transtension and transpression during arc building in central Chile: evidence from mafic dike swarms. Andean Geol 38(1):37–63

De La Cruz R, Suárez M (1997) El Jurásico de la cuenca de Neuquén en Lonquimay, Chile; Formación Nacientes del Bío Bío (38°–39°). Rev Geol Chile 24(1):3–24
del Rey A, Deckart K, Arriagada C, Martínez F (2016) Resolving the paradigm of the late Paleozoic-Triassic Chilean magmatism: isotopic approach. Gondwana Res 37:172–181
Grocott J, Taylor GK (2002) Magmatic arc fault systems, deformation partitioning and emplacement of granitic complexes in the Coastal Cordillera, north Chilean Andes (25° 30′ S to 27° 00′ S). J Geol Soc 159(4):425–442
Gana P, Herve F (1983) Geologia del basamento cristalino de la Cordillera de la Costa entre los rios Mataquito y Maule, VII Region. Rev Geol Chile 19:37–56
Gana P, Tosdal RM (1996) Geocronología U-Pb y K-Ar en intrusivos del Paleozoico y Mesozoico de la Cordillera de la Costa, Región de Valparaíso, Chile. Rev Geol Chile 23(2):151–164
Godoy E, Loske W (1988) Tectonismo sinplutónico de dioritas Jurásicas al sur de Valparaíso: datos U-Pb sobre la 'fase Quintay'. Rev Geol Chile 15(2):119–127
Godoy E, Marquardt C, Blanco N (2003) Carta Caldera, Región de Atacama. Serie Geología Básica, vol 76, Servicio Nacional de Geología y Minería, Santiago, 38 p
González G, Niemeyer H (2005) Cartas Antofagasta y Punta Tetas, Región de Antofagasta. Servicio Nacional de Geología y Minería, Carta Geológica de Chile, vol 89, Serie Geología Básica, 35 p
Gulisano C, Gutierrez Pleimling A, Digregorio R (1984) Esquema estratigráfico de la secuencia jurásica del oeste de la provincia del Neuquén. In: Abstracts of the 9 Congreso Geológico Argentino, San Carlos de Bariloche, 5–9 Nov 1984
Hervé F, Thiele R, Parada MA (1976) Observaciones geológicas en el Triásico de Chile central entre las latitudes 35°30′ y 40°00′ sur.. In Abstracts of the 1 Congreso Geologico Chileno, Santiago, 2–8 Aug 1976
Hervé F, Pankhurst RJ, Fanning CM, Calderón M, Yaxley GM (2007) The South Patagonian batholith: 150 my of granite magmatism on a plate margin. Lithos 97 (3–4):373–394
Irvine TN, Baragar WRA (1971) A Guide to the Chemical Classification of the Common Volcanic Rocks. Can J Earth Sci 8 (5):523–548
Klohn C (1960) Geologia de la Cordillera de los Andes de Chile Central, Provincias de Santiago, O'Higgins, Colchagua y Curico. Instituto de Investigaciones Geológicas, Bol 8, 95 p
Kramer W, Siebel W, Romer R et al (2005) Geochemical and isotopic characteristics and evolution of the Jurassic volcanic arc between Arica (18° 30′ S) and Tocopilla (22° S), North Chilean Coastal Cordillera. Chem Erde 65:47–78
Lagno L (1981) Geologia del Cuadrangulo Pocillas, Cauquenes, VII Region, Chile. Degree thesis, Universidad de Chile
Leppe M, Moisan P, Abad E et al (2006) Paleobotánica del Triásico Superior del valle delrío Biobío, Chile: Clase Filicopsida. Rev Geol Chile 33(1):81–107
Llambías E, Caminos R, Rapela C (1984) Las plutonitas y volcanitas del Ciclo Eruptivo Gondwanico. In: Ramos V (ed) Geología y Recursos Naturales de la Provincia de Río Negro. Asociación Geológica Argentina, Buenos Aires, pp 85–117
Llambías EJ, Leanza HA, Carbone O (2007) Evolución tectono-magmática durante el Pérmico al Jurásico temprano en la Cordillera del Viento (37° 05′ S–37° 15′ S): Nuevas evidencias geológicas y geoquímicas del inicio de la Cuenca Neuquina. Rev Asoc Geol Argent 62(2):217–235
Lucassen F, Trumbull R, Franz G et al (2004) Distinguishing crustal recycling and juvenile additions at active continental margins: the Paleozoic to recent compositional evolution of the Chilean Pacific margin (36–41_S). J S Am Earth Sci 17:103–119
Lucassen F, Kramer W, Bartsch V et al (2006) Nd, Pb and Sr isotope composition of juvenilemagmatismin theMesozoic largemagmatic province of northern Chile (18°–27° S): indications for a uniform subarc mantle. Contrib Mineral Petr 152:571–589
Mantle GW, Collins WJ (2008) Quantifying crustal thickness variations in evolving orogens: correlation between arc basalt composition and Moho depth. Geology 36:87–90
Miranda J (1963) Relevamiento geologico de la zona situada entre el bloque oriental de las Sas. De Reyes y Cara Cura, y el extremo sudoeste del Altiplano del Payún (Prov. De Mendoza). YPF, unpublished, pp 46, Buenos Aires

Mpodozis C, Kay SM (1990) Provincias magmáticas ácidas de los Andes Chilenos (28° S-31° S) y evoluciónteectónica de Gondwana. Rev Geol Chile 17:153–180
Mpodozis C, Kay S (1992) Late Paleozoic to Triassic evolution of the Gondwana margin: evidence from Chilean Frontal Cordilleran batholiths (28° S to 31° S). Geol Soc Am Bull 104:999–1014
Mpodozis C, Ramos (2008) Tectónica Jurásica en Argentina y Chile: Extensión, subducción oblicua, rifting, deriva y colisiones?. Rev Asoc Geol Argentina 63(4)
Mpodozis C, Ramos V (1989) The Andes of Chile and Argentina. In: Ericksen GE, Cañas MT, Reinemund JA (eds) Geology of the Andes and its relation to hydrocarbon and energy resources. Circum Pacific Council for Energy and Hydrotermal Resources, Huston, Earth Science Series, vol 11, pp 59–90
Munizaga F, Herve F, Drake R et al (1988) Geochronology of the Lake Region of South Central Chile (39°–42° S). Preliminary results. J S Am Earth Sci 1:309–316
Muñoz Cristi J, Gonzalez Pacheco E (1953) Geologia de los Yacimientos de pirofilita de Arrayan en la provincial de Maule. Inst. De Geol., U. de Chile, vol 3, Santiago, p 26
Naipauer M, Tunik M, Marques JC et al (2015) U-Pb detrital zircon ages of Upper Jurassic continental successions: implications for the provenance and absolute age of the Jurassic-Cretaceous boundary in the Neuquén Basin. The Geological Society, London, SP 399, pp 131–154
Navarrete C, Gianni G, Echaurren A, Kingler FL, Folguera A (2016) Episodic Jurassic to Lower Cretaceous intraplate compression in Central Patagonia during Gondwana breakup. J Geodyn 102:185–201
Oliveros V, Féraud G, Aguirre L et al (2006) The Early Andean Magmatic Province (EAMP): 40Ar/39Ar dating on Mesozoic volcanic and plutonic rocks from the Coastal Cordillera, Northern Chile. J Volcanol Geoth Res 157:311–330
Oliveros V, González J, Espinoza M, Vásquez P, Rossel P, Creixell C, Sepulveda F, Bastías F (2018) The early stages of the magmatic arc in the Southern Central Andes. In: Folguera A et al. (eds) The evolution of the Chilean-Argentinean Andes. Springer Earth System Sciences. Springer, Cham pp 165–190
Oliveros V, Morata D, Aguirre L et al (2007) Jurassic to Early Cretaceous subduction-related magmatism in the Coastal Cordillera of northern Chile (18°30′–24° S): geochemistry and petrogenesis. Rev Geol Chile 34:209–232
Palacios C (1978) The Jurassic Paleovolcanism in Northern Chile. PhD Thesis, Tübingen University
Parada MA, Munizaga F (1978) El batolito de Panguipulli: Caracterizacion del plutonismo jurasicoen la cordillera de los Andes de la Provincia de Valdivia, X región (39° 30′–40° 15′ S). Universidad de Chile, Departamento de Geologia, Comunicaciones 23:12–25
Parada MA, Nystrom J, Levi B (1999) Multiple sources for the Coastal batholith of central Chile (31–34° S): geochemical and Sr–Nd isotopic evidence and tectonic implications. Lithos 46:505–521
Pearce JA, Wyman DA (1996) A users guide to basalt discrimination diagrams, trace element geochemistry of volcanic rocks: applications for massive sulphide exploration. Geological Association of Canada, Short Course Notes 12:79–113
Piraces R (1977) Geologia de la Cordillera de la Costa entre Catapilco y Limache, region de Aconcagua. Ph.D. thesis, Universidad de Chile
Profeta L, Ducea MN (2015) Using trace element proxies for crustal thickness of magmatic arcs: La/Yb. In: Proceedings of the 2015 Goldschmidt Conference, Prague, Aug 2015
Ramos V, Kay S (1991) Triassic rifting and associated basalts in the Cuyo basin, central Argentina. Geol Soc Am SP 265:79–91
Rapela CW, Pankhurst RJ, Fanning CM et al (2005) Pacific subduction coeval with the Karoo mantle plume: the Early Jurasssic Subcordilleran belt of northwestern Patagonia. In: Vaughan APM, Leat PT, Pankhurst RJ (eds) Terrane Processes at the Margins of Gondwana. The Geological Society, London, SP 246, pp 217–239
Ring U, Willner A, Layer P, Richter P (2012) Jurassic to Early Cretaceous postaccretional sinistral transpression in northcentral Chile (latitudes 31°–32° S). Geol Mag 149:202–220

Rodríguez C, Pérez Y, Moreno H et al (1999) Área de Panguipulli-Riñihue, Región de los Lagos: Santiago. Servicio Nacional de Geología y Minería, Mapas Geológicos 10(1):100000

Rogers G, Hawkesworth CJ (1989) A geochemical traverse across the North Chilean Andes: evidence for crust generation from the mantle wedge. Earth Planet Sc Lett 91:271–285

Rossel P, Oliveros V, Ducea MN, Charrier R, Scaillet S, Retamal L, Figueroa O (2013) The Early Andean subduction system as an analog to island arcs: Evidence from across-arc geochemical variations in northern Chile. Lithos 179:211–230

Rossel R, Oliveros V, Mescua J et al (2014) The Upper Jurassic volcanism of the Río Damas-Tordillo Formation (33–35.5 S): Insights on petrogenesis, chronology, provenance and tectonic implications. Andean Geol 41(3):529–557

Rossel P, Oliveros V, Ducea M et al (2015) Across and along arc geochemical variations in altered volcanic rocks: evidence from mineral chemistry of Jurassic lavas in northern Chile, and tectonic implications. Lithos 239:97–113

Salazar C, Stinnesbeck W (2016) Tithonian-Berriasian ammonites from the Baños del Flaco Formation, central Chile. J Syst Palaeontol 14(2):149–182

Scheuber E, Gonzalez G (1999) Tectonics of the Jurassic-Early Cretaceous magmatic arc of the north Chilean Coastal Cordillera (22°–26° S): A story of crustal deformation along a convergent plate boundary. Tectonics, 18(5):895–910

SERNAGEOMIN (2003) Mapa Geológico de Chile: versión digital. Servicio Nacional de Geología y Minería, vol. 4. Publicación Geológica Digital, Santiago

Spalleti L (2001) Evolución de las cuencas sedimentarias. In: Artabe AE, Morel EM, Zamuner AB (eds) El Sistema Triásico en la Argentina. Fundación Museo de la Plata "Francisco Pascascio Moreno", La Plata, Argentina

Suárez M, Emparan C (1997) Hoja Curacautín, Regiones de la Araucanía y del Biobío. Servicio Nacional de Geología y Minería, Carta Geológica de Chile 71, 105 pp, 1:250.000

Thiele R, Morel R (1981) Tectónica triásico-jurásica en la Cordillera de la Costa, al norte y sur del río Mataquito (34° 45′–35° 15′ lat. s), Chile. Rev Geol Chile 13(14):49–61

Thomas H (1958) Geologia de la Cordillera de la Costa entre el Valle de La Ligua y la Cuesta de Barriga: Santiago. Instituto de Investigaciones Geologicas 2, Santiago, 86 p

Vásquez P, Glodny J, Franz G et al (2011) Early Mesozoic Plutonism of the Cordillera de la Costa (34°–37° S), Chile: constraints on the onset of the Andean Orogeny. J Geol 119(2):159–184

Vásquez P, Glodny J, Franz G, Romer RL, Gerdes A (2009) Origin of fayalite granitoids: New insights from the Cobquecura Pluton, Chile, and its metapelitic xenoliths. Lithos 110(1–4):181–198

Vergara M, Levi B, Nystrom J et al (1995) Jurassic and Early Cretaceous island arc volcanism, extension, and subsidence in the Coast Range of central Chile. Geol Soc Am Bull 107:1427–1440

Vicente JC (2005) Dynamic paleogeography of the Jurassic Andean Basin: pattern of transgression and localization of main straits through the magmatic arc. Rev Asoc Geol Argentina 60(1):221–250

Vicente JC (2006) Dynamic Paleogeography of the Jurassic Andean Basin: pattern of regression and general considerations on main features. Rev Asoc Geol Argent 61:408–437

Vilas JF, Valencio DA (1978) Palaeomagnetism and K—Ar age of the Upper Ordovician Alcaparrosa Formation, Argentina. Geophys J Int 55(1):143–154

Wall R, Gana P, Gutiérrez A (1996) Área de San Antonio- Melipilla, Regiones de Valparaíso, Metropolitana y del Libertador General Bernardo O'Higgins. Servicio Nacional de Geología y Minería, Mapas Geológicos 2, Santiago, 20 pp, 1:100.000

Wood DA (1980) The application of a ThHfTa diagram to problems of tectonomagmatic classification and to establishing the nature of crustal contamination of basaltic lavas of the British Tertiary Volcanic Province. Earth Planet Sc Lett 50 (1):11–30

Controls on Deposition of the Tordillo Formation in Southern Mendoza (34°–36° S): Implications for the Kimmeridgian Tectonic Setting of the Neuquén Basin

José F. Mescua, Julieta Suriano, Laura Jazmín Schencman, Laura B. Giambiagi, Patricia Sruoga, Elizabeth Balgord and Florencia Bechis

Abstract The Tordillo Formation is a continental succession of redbeds deposited in the Neuquén Basin during the Kimmeridgian. Based on sedimentological and structural evidence we document an extensional setting between 34° and 36° S, in the northern part of the basin, which contrasts with a transpressional environment in the Huincul arch region (39° S). The tectonic setting of the Neuquén Basin, during the deposition of the Tordillo Formation, was controlled by the combined stresses generated by the subduction system to the west and the break-up of Pangea to the east. We propose that trench roll-back in the Late Jurassic caused regional extension in the back-arc zone. This stress field was locally modified by the clockwise rotation of South America due to the initial break-up of Pangea that ended with the opening of the South Atlantic Ocean. This rotation produced a transpressive stress field in the southern Neuquén Basin in which the structures generated during the Permian

J. F. Mescua (✉) · J. Suriano · L. B. Giambiagi
Instituto Argentino de Nivología, Glaciología y Ciencias Ambientales (IANIGLA), CCT Mendoza, CONICET, Mendoza, Argentina
e-mail: jmescua@mendoza-conicet.gob.ar

J. F. Mescua
Facultad de Ciencias Exactas y Naturales, Universidad Nacional de Cuyo, Mendoza, Argentina

L. J. Schencman
Departamento de Geología, Facultad de Ciencias Exactas y Naturales, Universidad de Buenos Aires, Buenos Aires, Argentina

Instituto de Geociencias Básicas, Aplicadas y Ambientales de Buenos Aires (IGEBA, CONICET-UBA), Buenos Aires, Argentina

P. Sruoga
CONICET-SEGEMAR, Buenos Aires, Argentina

E. Balgord
Weber State University, Ogden, USA

F. Bechis
IIDyPCa, CONICET, Universidad Nacional de Río Negro, San Carlos de Bariloche, Argentina

D. Kietzmann and A. Folguera (eds.), *Opening and Closure of the Neuquén Basin in the Southern Andes*, Springer Earth System Sciences,
https://doi.org/10.1007/978-3-030-29680-3_6

collision of the Patagonia terrane were reactivated as strike-slip and reverse faults leading to variable basin geometries and subsidence patterns recorded by the Tordillo Formation in the north and south.

Keywords Late Jurassic · Extension · Late synrift · Back-arc basin

1 Introduction

The evolution of the Neuquén Basin is characterized by an initial extensional stage that caused the development of isolated depocenters between the latest Triassic and Sinemurian (Legarreta and Uliana 1991, 1996) (see Chaps. "The Syn-Rift of the Neuquén Basin (Precuyano and Lower Cuyano Cycle): Review of Structure, Volcanism, Tectono-Stratigraphy and Depositional Scenarios" and "Tectono-Stratigraphic Evolution of the Atuel Depocenter During the Late Triassic to Early Jurassic Rift Stage, Neuquén Basin, West-Central Argentina"). These depocenters were then connected during the latest synrift stage between the late Sinemurian and the early Toarcian, which in turn was followed by the thermal subsidence stage (Legarreta and Uliana 1991, 1996; Howell et al. 2005). In this classical framework, the extensional phase of the basin ends in the Middle Jurassic and the thermal subsidence lasts for more than 60 million years until the late Cretaceous foreland stage associated with the onset of Andean contraction.

The Tordillo Formation (Stipanicic 1969) is a fluvial dominated continental succession of redbeds (Marchese 1971; Legarreta and Uliana 1999) deposited in an arid climate in the Late Jurassic. Specifically, a Kimmeridgian age was proposed for this unit based on the fossils found in the under- and overlying strata (Groeber 1946; Legarreta and Uliana 1999). Detrital zircon U–Pb ages from sandstones and igneous zircon U–Pb ages from volcanic units indicate that it was deposited around 150–143 Ma (Aguirre et al. 2009; Rossel et al. 2014; Naipauer et al. 2012, 2015a, b; Horton et al. 2016), consistent with biostratigraphic ages but reaching the Tithonian according to the latest International Chronostratigraphic Chart (Cohen et al. 2013).

Thermal subsidence should be the driving mechanism for accomodation during deposition of the Tordillo Formation based on the traditional models for the evolution of the Neuquén Basin (Legarreta and Gulisano 1989; Gulisano and Gutiérrez Pleimling 1995; Vergani et al. 1995; Ramos 1999; Legarreta and Uliana 1996, 1999; Howell et al. 2005). However, other authors have suggested that variable tectonic regimes controled subsidence in the Late Jurassic in the Neuquén Basin, such as a transpressional regime in the southern sector of the basin (Orchuela et al. 1981; Spalletti and Colombo 2005; Mosquera and Ramos 2006; Silvestro and Zubiri 2008) or extensional tectonics in its northwestern sector (Lo Forte 1996; Cegarra and Ramos 1996; Pángaro et al. 1996; Giambiagi et al. 2003; Mescua et al. 2014). Furthermore, in the Chilean slope of the Andes, the coeval clastic and volcanic Río Damas Formation was assigned to an extensional environment (Vergara et al. 1995; Charrier 2007; Charrier et al. 2007) (see Chap. "Early Andean Magmatism in Southern Central Chile, 33°–40°S").

The aim of this chapter is to assess the evidence for extensional control on the deposition of the Tordillo Formation in the northern Neuquén Basin based on analysis of facies variations, sandstone provenance, thickness changes, and subsidence patterns. Using this dataset along with previous studies, we discuss the tectonic setting of the Neuquén Basin during Kimmeridgian times.

2 Geological Setting

During the deposition of the Tordillo Formation, the Neuquén Basin consisted of a narrow N-S trough 90–100 km wide in the north, and a broad section over 300 km wide, the Neuquén embayment, in the south (Fig. 1).

The distribution of facies throughout the basin was dominated by fluvial deposits, along with interbedded alluvial, eolian, and playa lake environments (Fig. 1a; Arregui 1993; Legarreta and Uliana 1996, 1999; Spalletti and Colombo 2005; Spalleti et al. 2008). In many localities, the upper 10–50 m of the Tordillo Formation correspond to green sandstones and mudstones with minor pebble conglomerates, known as the

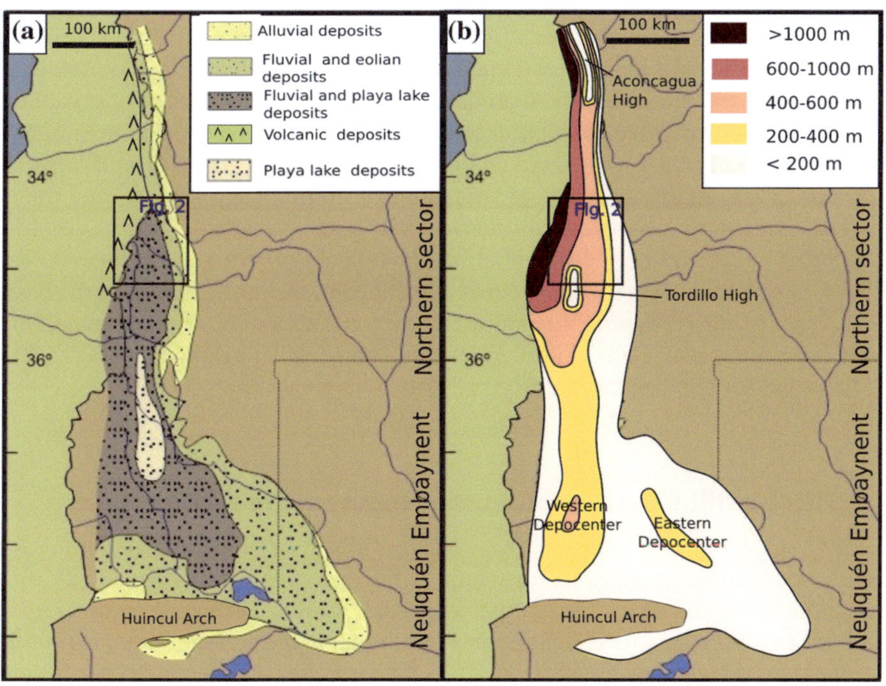

Fig. 1 The Kimmeridgian Neuquén Basin. **a** Facies variations (modified from Legarreta and Uliana 1996). **b** Thickness variations (based on Vergani et al. 1995; Pángaro et al. 1996; Spalletti and Colombo 2005; and this work)

"Faja Verde" (Green Belt) and interpreted as lacustrine or shallow marine deposits (Digregorio 1972; Legarreta and Uliana 1996). Toward the west, the Tordillo Formation interfingers with andesites and basaltic andesites interbedded with conglomerates and volcanoclastic breccias of the Río Damas Formation (Klohn 1960; Charrier 1981; Charrier et al. 2007; Sruoga et al. 2011). The andesites belong to a calcalkaline-subalkaline suite emplaced in an arc-related setting, based on their geochemical signature (Sruoga et al. 2016), although they were emplaced in a retroarc location according to Charrier et al. (2007). Volcanic activity took place throughout the northern portion of the basin in the Late Jurassic reaching the eastern margin in the Aconcagua (Sanguinetti 1989) and Cañada Ancha (López-Gómez et al. 2009) regions (Figs. 1 and 2). The southernmost part of the basin was segmented by the basement uplift of the Huincul arch (Fig. 1; De Ferraris 1947) between the Sinemurian-Toarcian and the Valanginian (Mosquera and Ramos 2006; Silvestro and Zubiri 2008). The E-W orientation of the arch has been attributed to inherited basement structures (Ramos 1977; Mosquera and Ramos 2006).

It is generally assumed that extensional tectonics in the Neuquén Basin were restricted to the initial stages of basin development ending in the late Early Jurassic. Following the synrift stage, subsidence is attributed to thermal processes, including subsidence during deposition of the Tordillo Formation (Legarreta and Uliana 1996, 1999; Howell et al. 2005). This model can explain the depositional features observed in the southern Neuquén embayment, where the Tordillo Formation fills a wide basin with gradual thickness changes and maximum thickness <500 m (Fig. 1b). However, other mechanisms are necessary to produce the subsidence and depositional patterns of this unit in the northern portion of the Neuquén Basin. The development of two separate depocenters in the Neuquén embayment (Fig. 1) has been attributed by Spalletti (2013) to the influence of volcanic load in the magmatic arc, and minor thickness variations associated with normal faults have been interpreted by Cristallini et al. (2009) as the result of differential compaction of Early Jurassic sediments. The narrow trough and depositional patterns developed in the northern part of the basin (Fig. 1) are inconsistent with thermal subsidence and abrupt thickness changes have been interpreted as a result of normal faults controlling deposition (Lo Forte 1996; Cegarra and Ramos 1996; Pángaro et al. 1996; Giambiagi et al. 2003).

3 The Tordillo Formation in the Northern Neuquén Basin

This work focused in the northern sector of the Neuquén Basin, between 34° and 35° S, to assess the depositional controls of the Tordillo Formation in this region (Fig. 2). To present a more complete view of this unit in the northern Neuquén Basin, observations by previous researchers within the study area and also in the Aconcagua region (~33° S) were also taken into account (Fig. 1).

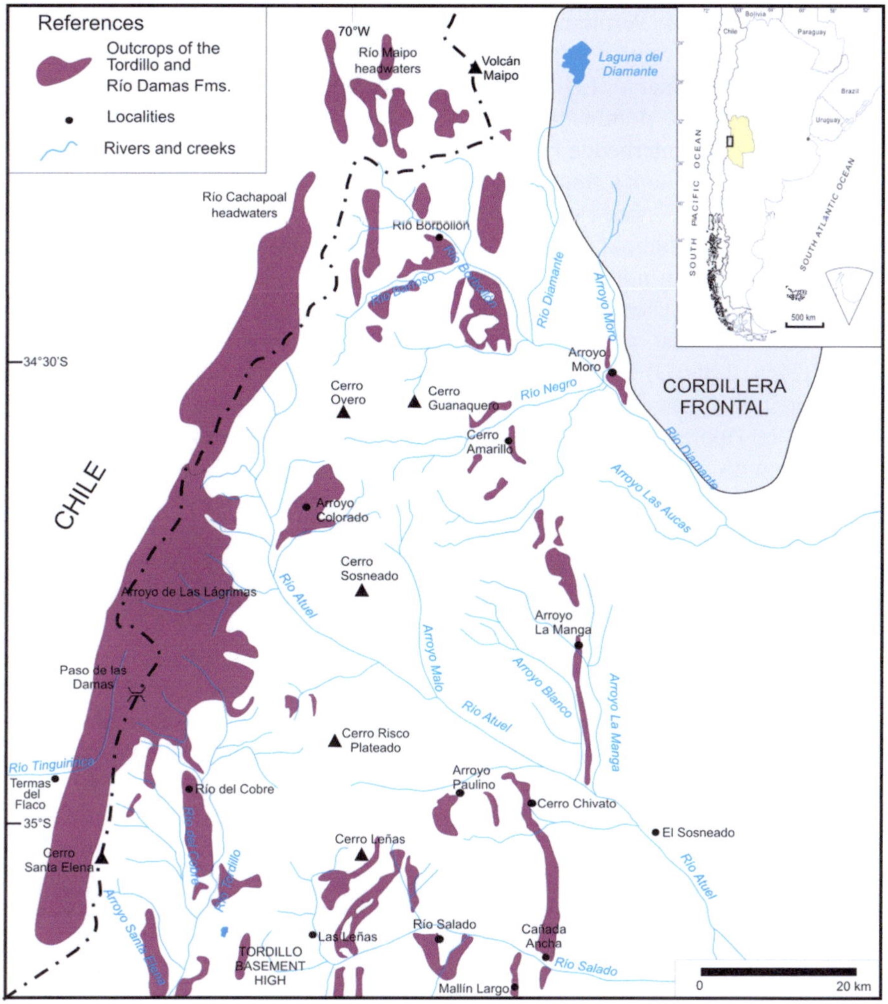

Fig. 2 Outcrops of the Tordillo Formation between 34° and 35° S, with localities mentioned in the text. Based on Sruoga et al. (2005), Nullo et al. (2005), Charrier (1981) and own observations

3.1 Sedimentary Environments

We studied eight profiles between 34° and 35° S to characterize the facies and thickness variations of the Tordillo Formation and to interpret the sedimentary environments in which it was deposited (Fig. 2). These profiles were then compared to observations in other localities.

Throughout most of the study area, the Tordillo Formation overlies gypsum of the Oxfordian Auquilco Formation with a sharp, conformable, or para-conformable contact (Legarreta and Uliana 1999). At the eastern margin of the basin, the Tordillo

Formation overlies the Permian-Triassic rhyolites of the Choiyoi Group (Broens and Pereira 2005). The contact with the marine shales of the overlying Vaca Muerta Formation is conformable (Legarreta and Uliana 1999).

Sandstones are the dominant lithology observed throughout the Tordillo Formation and are locally interbedded with finer lithologies and with minor proportions of conglomerate. Sandstones frequently contain reworked pyroclastic material, especially in the Río del Cobre and Arroyo Moro areas. In Arroyo Moro, multiple tuffs are interbedded throughout the sequence. The westernmost section (Río del Cobre, Figs. 2 and 3) has an important component of andesitic lava flows, indicating the transition to the Río Damas Formation.

Five facies associations were recognized in the eight profiles of the Tordillo Formation (Figs. 2 and 3) in this study. Facies association FA1 (Fig. 4), corresponds to tabular, fine to medium grained sandstones that are massive and horizontally laminated or ripple cross laminated. Thin intercalations of laminated or massive mudstones with mud cracks are common. Occasionally lentiform sandstone beds appear, with 2–4 m of thickness, containing intraclasts and presenting massive or cross-bedded structures. The abundant tabular sand sheets are interpreted as sheet-flood deposits, while mudstones are associated with the waning stage of the flooding. The cross bedded or massive lenses were interpreted as shallow, low sinuosity sandy channels (Hubert and Hyde 1982; Tunbridge 1984).

Facies association FA2 (Fig. 4c) corresponds to amalgamated sandy lenses that are 1–4 m thick and internally massive or planar or trough cross bedded. Some beds have thin conglomeratic layers at the base (channel lags) and intercalated mudstones. This facies association was interpreted as sandy, rapidly migrating, low sinuosity ephemeral rivers in a more proximal setting than FA1 (Tunbridge 1984).

Facies association FA3 (Fig. 4b) comprises fine to medium grained sandy beds with large scale trough cross-bedded foresets of approximately 1.5–4 m of thickness. This facies was interpreted as the result of the migration of eolian dunes (Kocurek and Dott 1981).

Facies association FA4 contains mudstones, that appear massive or horizontally laminated (alternating claystone and siltstone), with abundant mud cracks. Sandy laminas or beds are common intercalations. This association was interpreted as shallow ephemeral lakes (Tunbridge 1984).

Conglomeratic facies association FA5 is dominantly composed of amalgamated lenses of clast-supported pebble to boulder conglomerates, 1–4 m thick that are mostly massive, with rare imbricated or planar cross-bedded sections. Intraclasts are common, and locally gypsum clasts from the underlying Auquilco Formation were found in the Arroyo Moro area. Occasionally, fining upward trends were observed with intercalations of coarse to medium sandstones with planar and trough cross-bedded stratification. This association was interpreted as gravely braided channels dominated by longitudinal bars (Miall 1996).

The transition from the Tordillo Formation to the Río Damas Formation can be observed in the Río del Cobre profile and in an exceptional cross-section along the Arroyo de Las Lágrimas (Fig. 2), which shows the lateral interbedding of volcanic

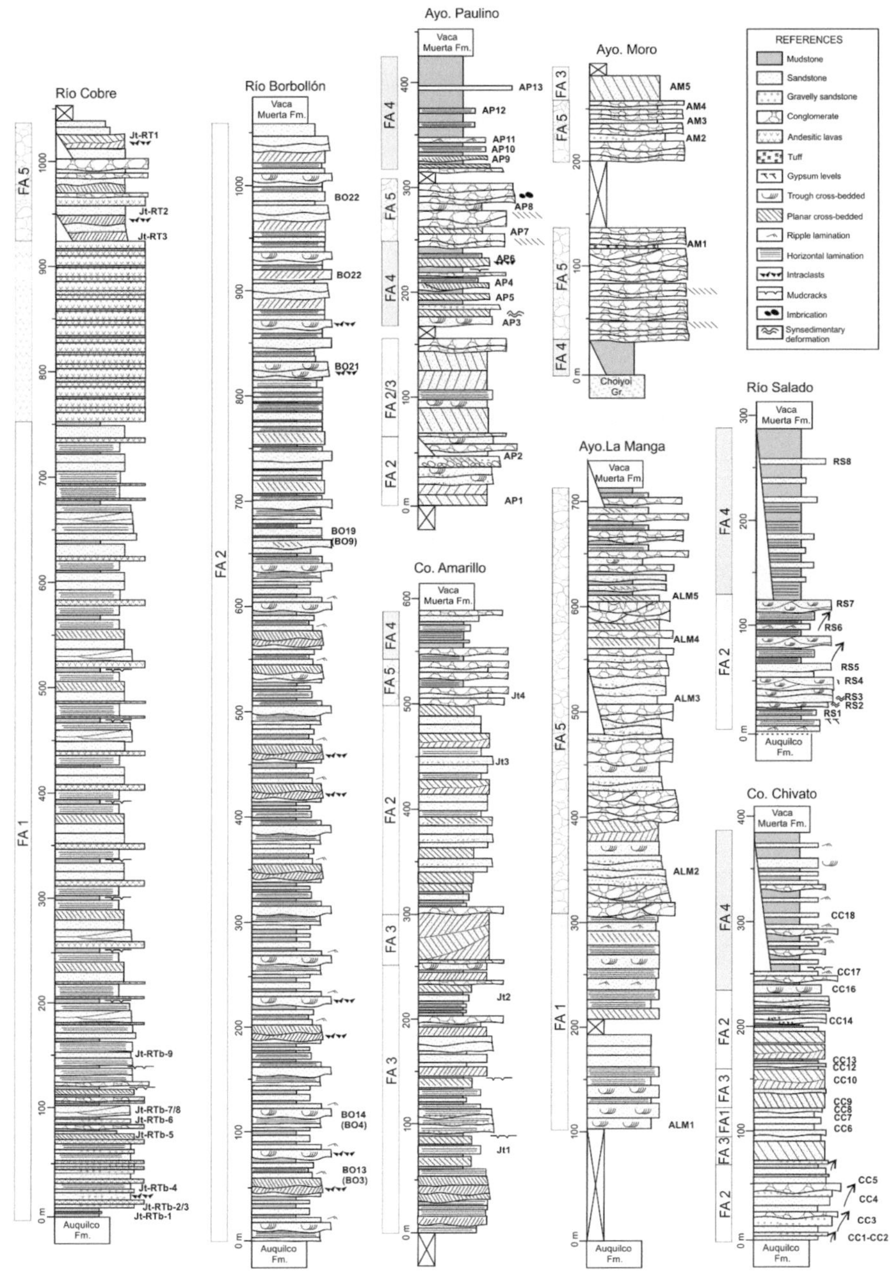

Fig. 3 Sedimentary profiles of the Tordillo Formation between 34° and 35° S. Locations in Fig. 2. Samples used for provenance analysis are shown in the profiles. Cerro Amarillo profile is based on Kim (2003)

Fig. 4 **a** Facies association FA5 at Arroyo Moro corresponding to conglomerates with interbedded red sandstones. White layers contain reworked pyroclastic material. **b** Facies association FA3, eolian cross-bedded sandstones, at Arroyo Moro. **c** Detail of cross-stratification in fine sandstones of facies association FA2 at Río Borbollón. **d** Basaltic andesites of the transition between the Tordillo and Río Damas Formation at Río del Cobre. **e** Facies association FA1 at Río del Cobre. Note dessication cracks filled with sandstone

and sedimentary facies (Fig. 4d and e). To the west, close to the border with Chile, porphyritic andesitic lava flows and coarse polymictic breccias were dominant forming up to ~1000 m thick volcanic piles, which were affected by low-grade metamorphic alteration (chlorite, calcite, epidote and prehnite-pumpellyite; Calderón 2008; Sruoga et al. 2016). A drastic decrease in thickness and grain size, associated with the incorporation of sandstones and conglomerates, was recorded to the east. Minor non-welded ignimbrite deposits were found interbedded in the redbeds. Also, andesitic sills and dykes commonly intrude the sedimentary facies of the Tordillo Formation (Sruoga et al. 2016).

The large thickness of the sequence and the low-grade metamorphic alteration supports a strong structural control during volcanic activity, capable of accommodating large volumes of magma and favoring alteration during burial. Thick lava flows, coarse breccias, and ignimbrites were proximal volcanic products that recorded volcanic activity close to the western margin of the Neuquén Basin and syn-eruptive reworking in a continental environment.

The sedimentary environments represented in the Tordillo Formation in the northern Neuquén Basin can be characterized as an arid alluvial setting, similar to that described by Spalletti and Colombo (2005) for the southern Neuquén Basin.

Our data are consistent with previous work (Legarreta and Uliana 1996) indicating that the facies distribution in the northern section of the basin is characterized by coarse sediments along both margins, with conglomerates restricted to the eastern and western fringes (FA5 in Arroyo Moro and Arroyo La Manga; conglomerates interbedded with lavas of the Río Damas Formation at Arroyo Las Lágrimas and Termas del Flaco, Rossel et al. 2014; Sruoga et al. 2016; Fig. 3). Conglomerates comprise large cobbles and boulders, indicating high relief at the basin boundaries. These coarse facies transition laterally to sandstones deposited in ephemeral sandy fluvial systems toward the center of the basin (FA1 and FA2 in Río Borbollón and Río del Cobre). Playa lake deposits are found locally interbedded with fluvial facies (FA4 in Arroyo Paulino, Cerro Chivato and Río Salado), and are common in the southernmost profiles, suggesting that streams ended in endorrheic depressions to the south.

3.2 Provenance Analysis

Provenance analysis was carried out through petrographic analysis of thirty-five thin sections cut from sandstone samples from the different profiles (Figs. 2 and 3). Point counts were made following the Gazzi-Dickinson method (Gazzi 1966; Dickinson 1970). Table 1 shows the lithotypes counted and the results for each sample. Our data are combined with previous work from the northern Neuquén Basin (Fig. 8; Charrier 1981; Cegarra and Pereyra 1988; Mescua et al. 2008).

The Folk et al. (1970) diagram classifies most sandstones (22/35) as Feldspathic Litharenites (Fig. 5a). Lithic Feldsarenites are also abundant (9/35), especially in the

Table 1 Clast counts of thin sections of the Tordillo Formation

		Lithotypes										
			Qm	Qm (Va)	Qm (Lp)	Qpc	Qpf	Plg	Plg (Lva)	Plg (Lvm)	Plg (Lp)	Fko
Localities	Arroyo La Manga	ALM 1	15	–	–	5	–	46	–	7	–	38
		ALM 2	50	4	–	14	–	33	1	1	–	34
		CAB1	31	–	–	14	–	48	–	–	–	16
		ALM3	47	1	–	10	–	33	–	–	–	35
		ALM4	45	1	1	18	–	24	1	–	–	26
		ALM5	52	–	–	26	–	20	2	–	–	24
	Cerro Chivato	CC1	39	–	1	5	–	20	–	–	1	37
		CC2	46	1	1	13	–	21	–	–	–	48
		CC3	52	1	1	19	–	37	–	–	–	29
		CC4	68	1	–	20	–	29	2	1	–	21
		CC5	56	2	–	19	–	29	5	–	–	22
		CC6	45	–	–	10	–	36	2	–	–	40
		CC7	69	–	–	13	–	32	–	–	–	36
		CC8	50	1	–	8	–	30	1	–	–	31
		CC9	62	–	–	9	–	43	2	–	–	40
		CC10	51	–	–	15	–	41	1	–	–	24
		CC11	48	1	1	22	4	19	2	–	–	7
		CC12	40	4	1	22	–	18	1	–	1	27
		CC13	59	3	4	18	–	32	1	–	2	22
		CC14	59	2	–	14	1	27	2	–	2	30

(continued)

Table 1 (continued)

		Lithotypes										
			Qm	Qm (Va)	Qm (Lp)	Qpc	Qpf	Plg	Plg (Lva)	Plg (Lvm)	Plg (Lp)	Fko
		CC15	54	2	3	20	–	21	–	–	–	26
		CC16	46	3	–	14	1	19	–	–	–	17
		CC17	70	1	–	27	–	22	3	1	–	15
		CC18	89	3	–	29	–	10	1	–	–	21
	Arroyo Moro	AM3	32	–	1	8	–	10	1	–	1	17
		AM4	45	3	–	21	–	12	4	3	–	29
		AM5	34	2	–	11	–	13	3	–	–	22
	Río Salado	RS1	28	–	–	27	–	37	1	–	–	57
		RS2	36	–	–	6	–	36	1	–	–	49
		RS3	56	–	–	10	–	47	–	–	–	64
		RS4	78	–	–	53	–	12	–	–	–	6
		RS5	64	–	–	27	–	55	–	–	–	47
		RS6	52	–	–	17	–	34	–	–	–	56
		RS7	59	1	–	21	–	39	1	–	–	32
		RS8	51	1	–	25	–	52	–	–	–	24

Lithotypes													
Fko (Lva)	Fko (Lm)	Fko (Lp)	Fko (Ls)	Fkm	Fkm (Lp)	Lva	Lvm	Lvb	Lp	Lm	Ls	Lpyro	Px
–	–	–	1	5	–	11	36	12	–	–	1	–	–
–	–	–	–	4	–	27	26	–	4	2	1	–	–

(continued)

Table 1 (continued)

Lithotypes													
Fko (Lva)	Fko (Lm)	Fko (Lp)	Fko (Ls)	Fkm	Fkm (Lp)	Lva	Lvm	Lvb	Lp	Lm	Ls	Lpyro	Px
–	–	–	–	11	–	70	37	–	1	–	–	–	–
–	–	–	–	3	–	48	27	–	1	4	–	–	1
–	–	–	–	–	–	60	15	2	4	3	–	–	–
–	1	–	–	6	–	59	22	4	8	1	1	–	–
–	–	2	–	1	–	66	15	5	–	1	–	–	–
–	–	–	–	1	–	39	26	7	1	–	–	–	–
–	–	–	–	–	–	30	29	6	–	1	2	–	–
–	–	1	–	5	–	36	29	2	1	1	1	–	–
–	–	–	–	–	–	53	15	–	1	1	–	1	–
–	–	–	–	1	–	49	13	1	–	4	–	–	–
–	–	–	–	1	–	40	9	–	–	11	–	–	–
–	–	–	–	–	–	58	19	–	–	1	–	–	–
2	–	–	–	1	–	43	5	–	–	–	–	–	4
3	–	–	–	1	–	56	6	–	–	5	–	–	–
3	–	–	–	–	–	72	10	–	–	11	5	1	–
2	–	3	–	–	–	73	5	1	2	8	1	–	–
–	–	2	–	–	1	63	6	–	–	3	3	2	–
–	–	–	–	3	–	57	10	1	1	4	–	–	–
–	–	–	–	1	–	61	11	2	1	11	2	1	–

(continued)

Table 1 (continued)

Lithotypes													
Fko (Lva)	Fko (Lm)	Fko (Lp)	Fko (Ls)	Fkm	Fkm (Lp)	Lva	Lvm	Lvb	Lp	Lm	Ls	Lpyro	Px
1	–	–	–	–	–	80	6	–	3	12	3	2	–
–	–	–	–	–	–	43	9	3	2	17	1	–	–
–	–	–	–	1	–	27	5	–	3	19	1	–	–
–	–	–	–	3	–	23	4	1	–	3	–	–	–
–	–	–	–	1	–	72	14	3	4	4	1	–	–
–	–	–	–	1	–	74	22	2	–	1	4	2	–
2	–	–	–	1	–	32	9	3	3	–	4	–	–
–	–	–	–	–	–	38	28	5	1	–	1	–	–
–	–	–	–	6	–	17	17	2	–	1	–	–	–
–	–	–	–	3	–	39	14	–	2	2	–	–	–
–	–	–	–	3	–	10	7	1	1	–	–	–	–
–	–	–	–	1	–	42	14	3	–	–	–	–	–
–	–	–	–	–	–	45	13	–	–	5	1	–	–
–	–	–	–	8	–	45	10	–	–	5	–	–	–

Lithotypes					Total	Qm	Qt	F	L	Lt	Lva + FK	Lvm/b + Pg
O	$CaCO_3$–R	Chloritie	Alt	Oo								
11	18	–	3	–	209	8	11	55	34	37	54	101
2	–	16	1	–	149	27	34	36	30	37	66	60
–	–	–	–	–	197	14	20	33	47	54	97	85

(continued)

Table 1 (continued)

Lithotypes					Total	Qm	Qt	F	L	Lt	Lva + FK	Lvm/b + Pg
O	$CaCO_3$–R	Chloritie	Alt	Oo								
1	–	5	2	–	163	23	28	34	38	43	87	60
2	–	–	6	–	155	24	33	26	42	51	88	41
–	–	–	3	–	174	23	35	23	42	54	89	46
4	4	–	1	–	202	21	23	32	45	48	104	40
1	7	–	–	–	212	24	30	34	36	42	89	54
3	4	–	2	–	216	26	35	32	33	42	60	72
2	4	–	6	–	230	32	41	27	32	41	65	61
–	–	–	1	–	205	28	38	27	35	44	82	44
1	3	–	4	–	209	22	27	39	33	38	92	50
–	–	–	–	–	211	33	39	33	28	35	77	41
–	5	–	–	–	204	26	30	31	39	43	91	49
–	4	–	–	–	215	30	34	43	23	28	88	48
–	–	–	–	–	203	25	33	34	33	40	85	47
–	–	–	–	–	206	24	37	15	48	61	85	29
–	–	–	–	–	209	22	32	25	43	54	107	24
–	3	–	2	–	226	30	38	27	35	43	89	38
–	–	–	1	–	214	29	36	30	34	41	91	38
2	–	–	–	–	218	27	37	22	41	50	90	34
–	–	–	–	–	207	24	31	18	51	58	100	25

(continued)

Table 1 (continued)

Lithotypes					Total	Qm	Qt	F	L	Lt	Lva + FK	Lvm/b + Pg
O	$CaCO_3$–R	Chloritie	Alt	Oo								
–	–	–	2	–	216	33	46	19	35	48	62	34
–	–	–	–	–	209	44	58	16	26	40	52	15
–	7	–	1	110	222	32	39	31	30	38	44	15
2	–	–	2	–	220	22	32	23	45	55	109	32
11	–	–	1	–	203	19	25	20	55	61	102	37
6	5	–	2	–	217	14	27	48	25	38	93	49
5	4	–	2	–	212	18	21	43	36	39	88	69
2	–	–	–	–	222	25	30	53	17	21	87	66
–	–	–	–	–	209	37	63	10	27	53	48	26
2	1	–	1	–	219	30	42	49	9	21	60	63
2	4	–	1	–	226	24	32	42	27	35	71	51
–	1	–	1	–	219	28	37	33	29	39	79	52
–	–	–	–	–	221	24	35	38	27	38	78	62

References: Qm: monocrystalline quartz, Qm (Lva): monocrystalline quartz in acidic volcanic fragment, Qm (Lp): monocrystalline quartz in plutonic fragment, Qpc: coarse polycrystalline quartz, Qpf: fine polycrystalline quartz, Plg: plagioclase, Plg (Lva): plagioclase in acidic volcanic fragment, Plg (Lvm): plagioclase in intermediate volcanic fragment, Plg (Lp): plagioclase in plutonicfragment, Fko: potassic feldspar, Fko (Lva): potassic feldspar in acidic volcanic fragment, Fko (Lm): potassic feldspar in metamorphic fragment, Fko (Lp): potassic feldspar in plutonic fragment, Fko (Ls): potassic feldspar in sedimentary fragment, Fkm: microcline, Fkm (Lp): microcline in plutonic fragment, Lva: acidic volcanic fragment, Lvm: intermediate volcanic fragment, Lbv: basic volcanic fragment, Lp: plutonic fragment, Lm: metamorphic fragment, Ls: sedimentary fragment, Lpyro: pyroclastic fragment, Px: pyroxene, O: opaque, CaCO3-R: carbonate replacement, Chlorite: chlorite replacement, Alt: altered fragments, Oo: oolite. QM% = Qm + Qm (Lva) + Qm (Lp), QT% =+ Qm (Lva) + Qm (Lp) + Qpc + Qpf, F% = Lva + Lvm + Lbv + Lp + Lm + Ls + Lpyro, LT% = Lva + Lvm + Lbv + Lp + Lm + Ls + Lpyro + Qpc + Qpf

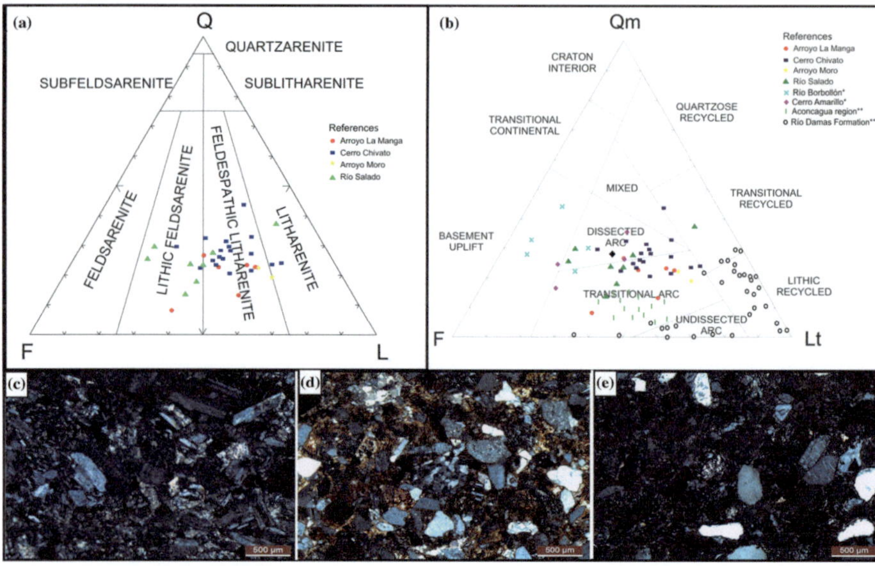

Fig. 5 **a** Sandstone classification of Folk et al. (1970) for the samples studied in this work. **b** Provenance diagram of Dickinson et al. (1983). Localities indicated with * are from Mescua et al. (2008), ** from Cegarra and Pereyra (1988), *** from Charrier (1981). **c** Thin section from the western Río del Cobre profile, composed entirely of plagioclase and fragments of andesites. **d** Thin section from the central Río Salado profile, composed of a mixture of the eastern and western sources: plagioclase and andesite fragments from the retroarc volcanic rocks and acidic volcanic fragments and potassium feldspar from the basement. **e** Thin section from the eastern Arroyo Moro profile, dominated by acidic volcanics from the basement

Río Salado profile. The low maturity of the sandstones suggests that most components are primary (not reworked) and had limited transport before deposition.

The Dickinson et al. (1983) diagram indicates a dissected to transitional arc provenance (Fig. 5b), consistent with the predominance of volcanic lithics and feldspars in the framework components. A more detailed look shows the presence of two populations of volcanic fragments, which can be assigned to various sources. The Kimmeridgian retroarc volcanics, located to the west of the basin (Río Damas Formation, Klohn 1960), are composed exclusively of basic and intermediate rocks, ranging from basaltic andesites to andesites (Charrier et al. 2007). In contrast, the Neuquén Basin basement corresponds to the acidic volcanic and plutonic rocks of the Late Permian-Early Triassic Choiyoi Group, dominated by rhyolites and granites (Llambías et al. 1993), and areally subordinate metamorphic rocks (Azcuy et al. 1999). This basement was cropping out along the eastern margin of the basin.

In order to discriminate these two sources in each sample, we grouped basic and intermediate volcanics with plagioclase fragments corresponding to the retroarc volcanic source and acidic volcanics, potassium feldspar and metamorphic fragments which were supplied by the basement source (Table 1).

Along the eastern margin of the basin (Arroyo Moro, Cerro Chivato), the basement source is clearly dominant (Figs. 5e and 6). This is consistent with observations reported in previous works.

Legarreta (1976) documented the abundance of fragments of acidic volcanic rocks and granites in samples from three localities on the eastern margin of the basin, concluding that the main source of sediments was the basement located to the east. Furthermore, multiple works indicated the presence of granite, rhyolite, and metamorphic clasts in conglomerates along the eastern margin of the basin (Criado Roqué 1950; Volkheimer 1966; Legarreta 1976; Mescua et al. 2008).

In contrast, conglomerates from the western margin of the basin are composed predominantly of basic andesite clasts from the Río Damas Formation (Rossel et al. 2014). While most sandstone samples of the westernmost Río del Cobre profile could not be used for clast counts, because of heavy hydrothermal alteration, three of the samples were unaltered and composed entirely of the retroarc volcanics source (Figs. 5c and 6). Localities closer to the center of the basin show a mixture of both sources (e.g., Río Salado; Figs. 5D and 6). In their study of sandstones from the Aconcagua area (Fig. 1), Cegarra and Pereyra (1988) also pointed out to the presence of both acidic and basic volcanic fragments in their samples. For the Río Damas Formation in the headwaters of the Cachapoal and Maipo rivers (Fig. 1), Charrier (1981) reported that the sandstones were dominated by intermediate volcanic rocks, with variable amounts of plagioclase and quartz (Fig. 5b), and no potassium feldspar. The source area would have been an active volcanic mountain range located to the west (Charrier 1981).

Therefore, provenance studies show that sediments were supplied from both margins of the basin, which is consistent with the distribution of facies. Detrital zircon U–Pb data ages from four samples of the Tordillo Formation (one sample from the Río Salado profile, one from the Cerro Chivato profile, and two samples near Arroyo La Manga and Cañada Ancha) provided ages corresponding to the Choiyoi Group and to the Jurassic arc/retroarc volcanic rocks (Naipauer et al. 2015b), further indicating that sediment was supplied from both sources. Two detrital zircon samples analyzed by Horton et al. (2016) from the confluence of Arroyo La Manga and the Atuel River (Fig. 1) and from the La Valenciana area (35° 30′ S) also contain both the Jurassic arc/retroarc volcanic and basement populations plus a Late Triassic peak likely related to reworking of synrift units of the Neuquén Basin.

3.3 Thickness Variations and Structural Evidence for Extension

In the northern part of the Neuquén Basin, there are abrupt variations in stratigraphic thickness recorded in the Tordillo Formation (Fig. 6). In some cases, these variations coincide with the interpreted master faults of the Early Jurassic rift, such as the La Manga fault of the Atuel depocenter (Giambiagi et al. 2008; Bechis et al. 2010)

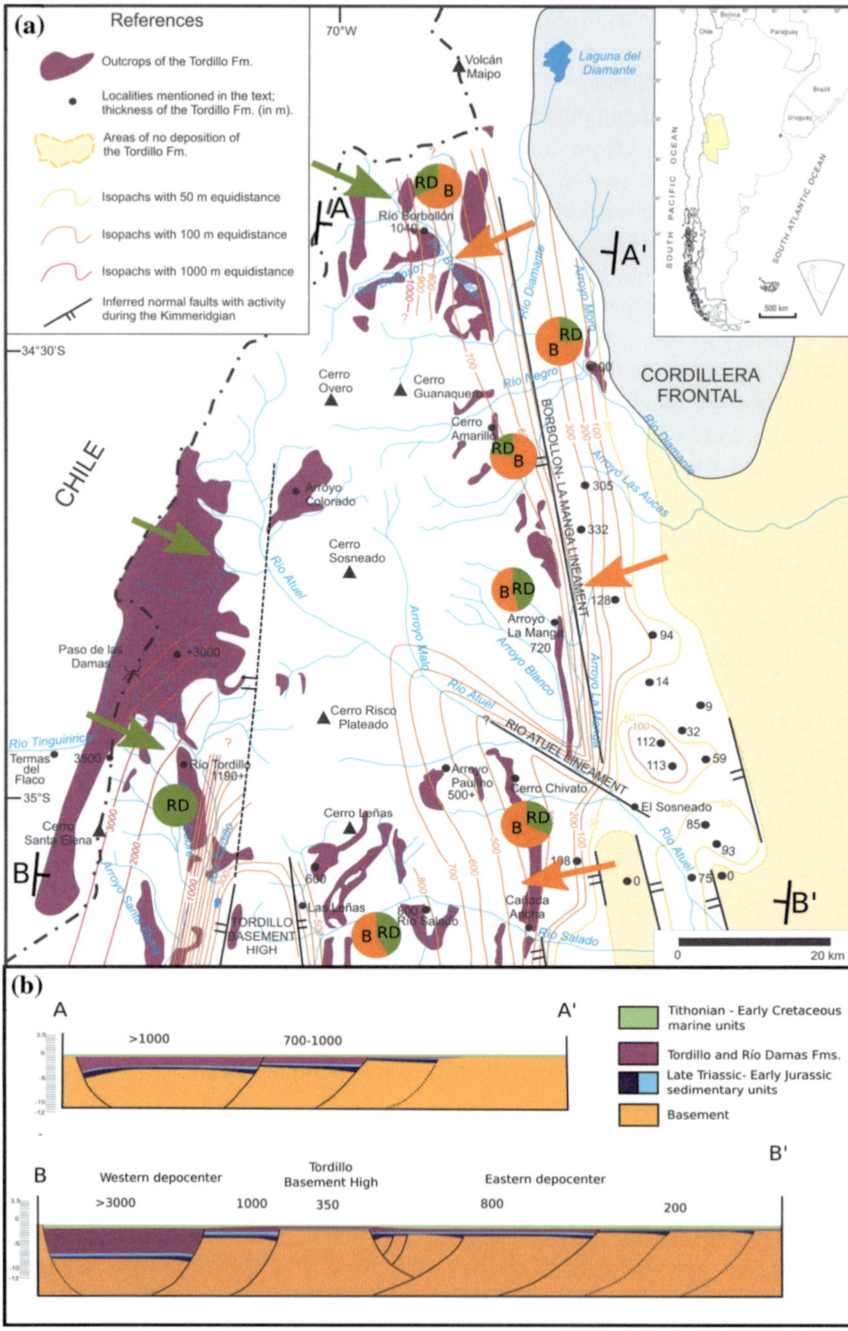
(a)
References
Outcrops of the Tordillo Fm.
Localities mentioned in the text; thickness of the Tordillo Fm. (in m).
Areas of no deposition of the Tordillo Fm.
Isopachs with 50 m equidistance
Isopachs with 100 m equidistance
Isopachs with 1000 m equidistance
Inferred normal faults with activity during the Kimmeridgian
70°W
Volcán Maipo
Laguna del Diamante
SOUTH PACIFIC OCEAN
SOUTH ATLANTIC OCEAN
500 km
A
A'
Río Borbollón 1040
Río Diamante
Arroyo Moro
34°30'S
Cerro Overo
Cerro Guanaquero
Río Negro
CORDILLERA FRONTAL
Cerro Amarillo
BORBOLLON- LA MANGA LINEAMENT
Arroyo Las Aucas
CHILE
Arroyo Colorado
Cerro Sosneado
Río Atuel
305
332
128
Arroyo La Manga 720
Arroyo Malo
Arroyo Blanco
Arroyo La Manga
94
14
9
32
112
113
59
Paso de las Damas
+3000
Cerro Risco Plateado
Río Atuel
RIO ATUEL LINEAMENT
Río Tinguiririca
Termas del Flaco
3500
Río Tordillo 1190+
Arroyo Paulino 500+
Cerro Chivato
El Sosneado
35°S
Cerro Santa Elena
Cerro Leñas
85
93
108
0
75
0
B
B'
600
Las Leñas
TORDILLO BASEMENT HIGH
800 Río Salado
Cañada Ancha
Río Salado
Arroyo Santa Elena
0
20 km
RD
B
(b)
A
A'
>1000
700-1000
Tithonian - Early Cretaceous marine units
Tordillo and Río Damas Fms.
Late Triassic- Early Jurassic sedimentary units
Basement
B
B'
Western depocenter
Tordillo Basement High
Eastern depocenter
>3000
1000
350
800
200

◀**Fig. 6** **a** Thickness map of the Tordillo Formation, with interpreted major normal faults. Red arrows indicate basement sediment source, and green arrows indicate arc/retroarc volcanic source of the Río Damas Formation. Pie charts show aggregate proportion of basement ("B") versus Río Damas ("RD") clasts in all samples counted for each profile. Data for Río Borbollón and Cerro Amarillo are from Mescua et al. (2008). **b** Palispastically restored profiles showing basin geometry across transects A-A′ and B-B′, horizontalized to the top of the Tithonian-Early Cretaceous Vaca Muerta Formation. Numbers indicate thickness of the Tordillo Formation in meters

and the El Cobre fault of the Río del Cobre depocenter (Mescua et al. 2014). In other areas, such as the Sierra de Reyes (37° S), an Early Jurassic depocenter has been found (Giambiagi et al. 2009), but the Tordillo Formation was not deposited (Leanza 2009), indicating that the early rift faults did not reactivate in this sector. Minor normal faults active during deposition of the Tordillo Formation were reported by Vicente and Leanza (2009) in the Aconcagua region, by Mescua et al. (2008) in southern Mendoza province and by Kietzmann and Vennari (2013) in northern Neuquén province (Domuyo region).

The two cross-sections presented in Fig. 6b illustrate the thickness variations and basin geometry observed in the Tordillo Formation. The northern section shows a single asymmetric depocenter, bounded to the east by the La Manga fault. Stratigraphic thickness increases to the west, with over 1000 m in the Río Borbollón area (Fig. 6). The thickness of the Río Damas Formation in Chile ranges between 2000 and 4000 m (Klohn 1960; González and Vergara 1962; Charrier et al. 2007; Rossel et al. 2014). In the southern cross-section (Fig. 6b), the two depocenters are separated by the Tordillo basement high, where a reduced thickness of the Mesozoic units was deposited, indicating that this feature was a topographic high active during much of the basin history (Legarreta and Uliana 1999). The western margin of the basin is not well constrained because of Cenozoic cover; however, we propose that normal faults controlled this margin based on the regional extensional control on the Río Damas Formation (Charrier et al. 2007; Fig. 6), and the westward thickness increase of this unit that suggests an extensional master fault in this direction (Wernicke and Burchfiel 1982; McClay 1990).

3.4 Subsidence Analysis

Extensional control during deposition of the Tordillo Formation is further supported by subsidence analysis (Fig. 7). Rift models (e.g., McKenzie 1978) predict that thermal subsidence during the post-rift stage will decay exponentially during the post-rift (Fig. 7). In contrast, subsidence curves for the northern Neuquén Basin show a dramatic increase in the subsidence rate during deposition of the Tordillo Formation (Fig. 7). In the Aconcagua region, which was a topographic high during initial Neuquén Basin rifting (Álvarez 1996), the fastest rates of subsidence in the Mesozoic are recorded in the Kimmeridgian. These rates then wane during deposition

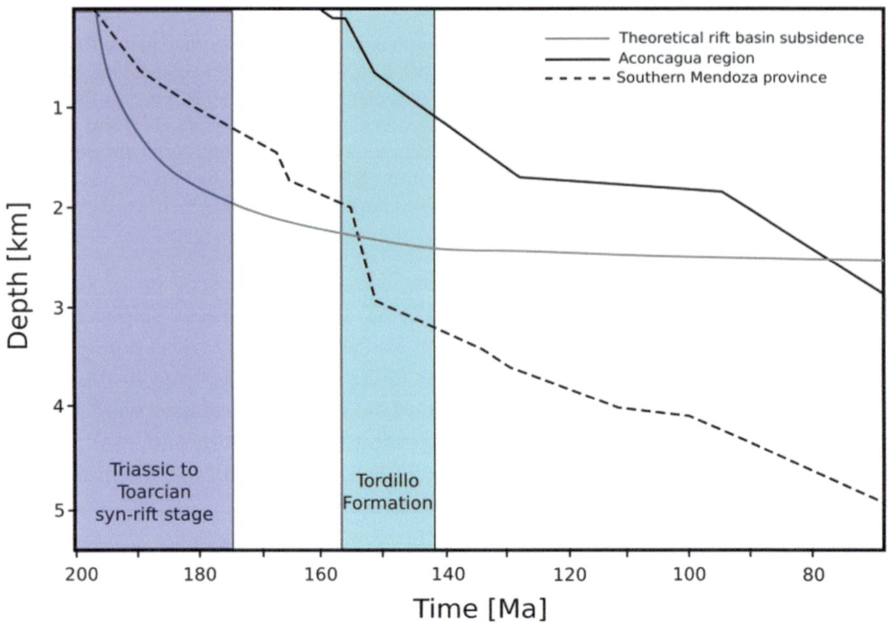

Fig. 7 Decompacted subsidence curves for the northern Neuquén Basin. Aconcagua region based on Balgord (2016). Southern Mendoza corresponds to Arroyo La Yesera from Boll et al. (2014). Theoretical subsidence for a rift basin with a 25 m y-long synrift period and 2 km of synrift subsidence based on the McKenzie (1978) model modified for finite rifting rates (from Cacace et al. 2008)

of the overlying Mendoza Group. We propose that the high subsidence rates during the deposition of the Tordillo Formation are the result of active extension at that time.

4 Discussion

4.1 The Tordillo Formation in the Southern Neuquén Basin

The Tordillo Formation and equivalent units (Quebrada del Sapo, Catriel and Sierras Blancas formations; Legarreta and Uliana 1999) in the Neuquén embayment (Fig. 1) have been studied in detail because of their importance as an oil reservoir (Maretto et al. 2002). Within the embayment, two depocenters were recognized; a western depocenter where fluvial deposits were dominant and there was a maximum thickness of ~500 m, and an eastern depocenter filled by both fluvial and eolian deposits with a maximum thickness of 250 m (Fig. 1; Spalletti and Colombo 2005). Proximal deposits consist of sandy and gravely channel deposits, whereas ephemeral streams fed playa lake environments in the distal sectors of the basin (Spalletti and Colombo

2005). A third depocenter was located south of the Huincul arch (Veiga and Spalletti 2007), with 50 m thick braided fluvial deposit overlain by eolian sandstones. Overall, the Kimmeridgian basin geometry does not seem consistent with normal faulting controlling deposition in the southern part of the basin. The basin was broad and thickness changes of the Tordillo Formation and equivalent units are gradual.

Subsidence curves in the eastern depocenter of the Neuquén embayment (Fig. 1) record increased subsidence with respect to the post-rift sag stage, but with lower rates than during the Early Jurassic synrift stage (Fig. 8a). Cristallini et al. (2009) proposed subsidence linked to differential compaction of the Early Jurassic synrift sediments for the eastern depocenter, whereas Spalletti (2013) suggested that the western depocenter was the result of volcanic load associated with increased activity in the magmatic arc. According to Charrier et al. (2007), the arc was located in the current Cordillera de la Costa, 100 km west of the western depocenter (see Fig. 9), and the Río Damas Formation volcanic units correspond to a retroarc location. If subsidence was dominated by flexure from the loading of the volcanic arc, thickness of the Río Damas Formation should increase toward the west where Cenozoic rocks completely cover the Jurassic units in this region, and therefore this model cannot be tested.

Sandstones from the southern Neuquén Basin correspond to Feldspatic Litharenites and Lithic Feldsarenites in the Folk et al. (1970) diagram, indicating low maturity (Spalletti et al. 2008; Naipauer et al. 2012, 2015a). Sandstone petrography indicates dissected to transtional arc sources based on the high proportion of lithic fragments and feldspar (Fig. 8b). Combining this information with detrital zircon U–Pb ages and paleocurrent data, two main sources were determined: the uplifted basement of the Huincul arch, which provided metamorphic and plutonic fragments, and the arc/retroarc volcanic rocks located west of the basin, which provided mostly intermediate to basic volcanic fragments.

Uplift of the Huincul arch in the southernmost part of the basin (Figs. 1) was recorded in subsidence profiles (Fig. 8a), based on a combination of outcrop mapping and subsurface data from the oil industry, including 3D seismic surveys (Mosquera and Ramos 2006; Silvestro and Zubiri 2008). Early models proposed either strike-slip tectonics (Harding 1974; Orchuela et al. 1981; Orchuela and Ploszkiewicz 1984) or tectonic inversion of Early Jurassic normal faults (Eisner 1991; Uliana et al. 1995; Vergani et al. 1995; Vergani 2005). A model integrating reverse NE-trending faults, dextral NW-trending faults, and oblique-slip E-trending faults (Fig. 9) within oblique NW-directed convergence was proposed by Mosquera and Ramos (2006) and Silvestro and Zubiri (2008), requiring transpression in the southern parts of the Neuquén Basin.

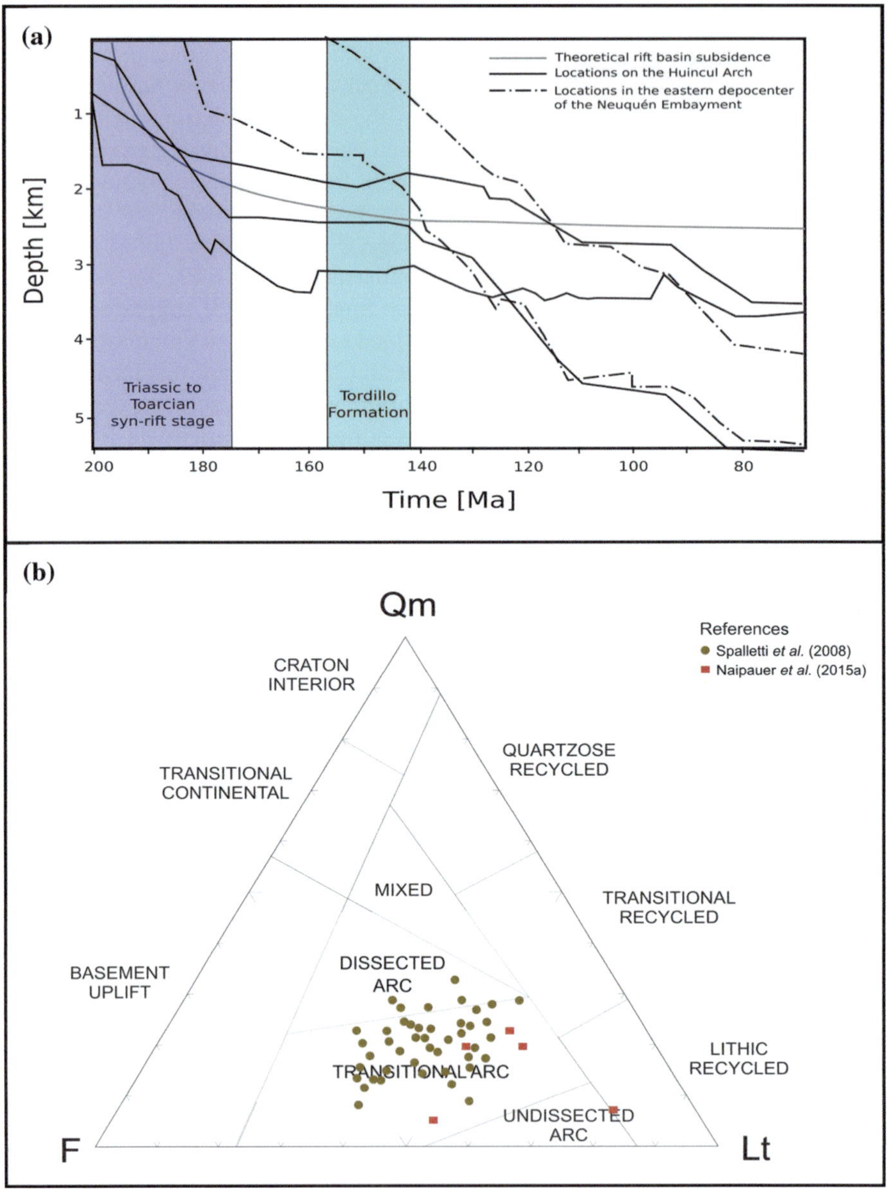

Fig. 8 **a** Subsidence curves for the southern Neuquén Basin. Localities on the Huincul arch are Lindero Atravesado and Las Chivas from Cruz et al. (2002), and Cerro Bandera from Veiga et al. (2002). Locations in the eastern depocenter of the Neuquén embayment are Aguada Pichana and Barreales Colorados, from Veiga (2002). Theoretical subsidence for a rift basin with a 25 m y-long synrift period and 2 km of synrift subsidence based on the McKenzie (1978) model modified for finite rifting rates (from Cacace et al. 2008). **b** Dickinson et al. (1983) diagram for samples from the southern Neuquén Basin from Spalletti et al. (2008) and Naipauer et al. (2015a)

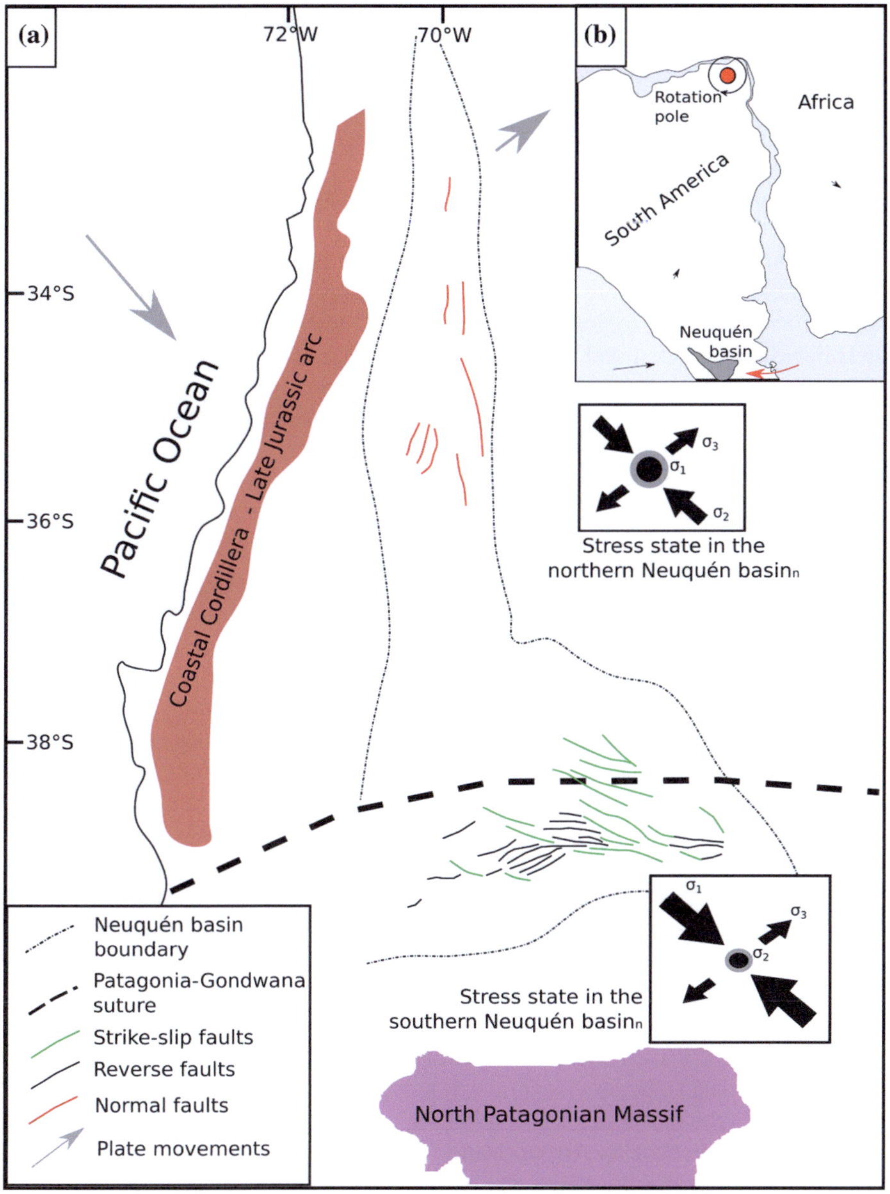

Fig. 9 **a** Main structures in the Neuquén Basin during the Kimmeridgian and stress states in the north and south. **b** Clockwise rotation of South America during opening of the South Atlantic. Based on the 140 Ma reconstruction of Heine et al. (2013). Black arrows are plate movements according to Seton et al. (2012), while red arrow indicates the clockwise rotation movement of southernmost South America

4.2 Unconformities at the Base of the Tordillo Formation and the Proposed Araucanian Orogenic Phase

Based on the unconformity between the Oxfordian Auquilco Formation and the Tordillo Formation in some localities, early studies of the Neuquén Basin suggested an orogenic event denominated "Araucanian phase" between these two units (Burckhardt 1900; Groeber 1918; Stipanicic and Rodrigo 1970; Charrier and Vicente 1972; Davidson and Vicente 1973). However, Groeber et al. (1953) pointed out that the Araucanian movements were restricted to the southern part of the basin and that they did not correspond to a regional orogenic event. Consistent with this view, uplift of the Huincul arch (Figs. 1 and 8) took place during the Kimmeridgian due to tectonic inversion and/or transpressional deformation along a pre-existing structure (Orchuela et al. 1981; Spalletti and Colombo 2005; Mosquera and Ramos 2006; Silvestro and Zubiri 2008), although in the rest of the basin no indication of compressional or transpressional deformation has been reported, suggesting that this tectonic environment was restricted to the Huincul arch area.

The geometry and facies of the northern part of the basin were not consistent with a synorogenic basin, which should have a wedge shape thickening toward the orogen, with proximal facies close to the orogen and distal facies away from it (Jordan 1995; De Celles and Giles 1996). In contrast, the basin presented high relief and coarse facies along both margins, and distal facies in the center (Figs. 1 and 3). Studies in the Aconcagua region of northern Mendoza (Cegarra and Ramos 1996; Pángaro et al. 1996; Giambiagi et al. 2003) proposed an extensional setting for this period based on thickness changes and structural architecture. Evidence of extensional control on the deposition of the Tordillo Formation in southern Mendoza (34°–36° S) were presented by Mescua et al. (2008, 2014). In Chile, regional extension is well documented between 21 and 36ºS (Vergara et al. 1995; Charrier 2007; Charrier et al. 2007). Furthermore, the geochemical studies from the volcanic rocks of the Río Damas Formation indicated no crustal thickening during the Kimmeridgian (La/Yb = 10–14, Sruoga et al. 2016), which suggested that no important orogenic event would have taken place in the Late Jurassic.

The Araucanian unconformity has variable characteristics in different localities (Groeber et al. 1953; Leanza 2009). Most commonly, it is an erosional unconformity (Groeber et al. 1953). In some localities it presents low angularity (e.g., Leanza and Hugo 2001). In the northern part of the basin, the unconformity was observed along the eastern margin, such as Arroyo Moro (Fig. 2) and Mallín Largo (Dajczgewand 2002). In the center of the basin, the gypsum of the Auquilco Formation reached hundreds of meters of thickness (Legarreta and Uliana 1999), suggesting that erosion before the deposition of the Tordillo Formation was minor with respect to the basin margins. This was difficult to reconcile with a regional orogenic phase. We propose that the unconformity was the result of erosion associated with a decrease in base level due to normal faulting in this sector of the basin.

We conclude that the Araucanian unconformity was the result of different processes in different sectors of the basin. In the vicinity of the Huincul arch it was an angular unconformity produced by the uplift of the arch, whereas in the northern sector it was an erosional unconformity with hiatuses of variable duration which were the result of normal faulting and erosion in the footwall of normal faults.

4.3 Tectonic Setting of the Neuquén Basin During the Kimmeridgian

Based on the classical models for the tectonic evolution of the Neuquén Basin, the Tordillo Formation was interpreted to have deposited during the thermal subsidence stage (e.g., Legarreta and Uliana 1991, 1996; Howell et al. 2005). However, the facies, provenance, and thickness variations show a pattern that departs from the gentle downwarping of the basin and slow subsidence typical of post-rift thermal subsidence (Figs. 7 and 8a; McKenzie 1978; Cacace et al. 2006). Structural and sedimentological evidence indicates that deposition in the northern sector was controlled by normal faults. In contrast, the southern sector was characterized by uplift of the Huincul arch in a transpressional or contractional setting. In the following section, we will explore a possible configuration to reconcile these contrasting tectonic settings within the Neuquén Basin (Fig. 9).

The Kimmeridgian subduction system along the western margin of Gondwana was characterized by oblique convergence. The Phoenix oceanic plate had a southeastward convergence respect to southern Gondwana (Seton et al. 2012).

The tectonic plate reconstructions for 160 and 140 Ma by Seton et al. (2012) have shown movement away from the trench for West Gondwana, which would suggest trench roll-back during this period, consistent with the regional extensional setting of Kimmeridgian units along the Pacific margin of Gondwana (Charrier et al. 2007). An active magmatic arc was located in the Cordillera de la Costa in Chile (SERNAGEOMIN 2003), ~80 km west of the western margin of the Neuquén Basin (Fig. 9). The volcanic rocks of the Río Damas Formation corresponded to a retroarc location according to Charrier et al. (2007). Locally, this back-arc volcanism expanded to the east and is interfingered with the sedimentary facies of the Tordillo Formation (Sanguinetti 1989; López-Gómez et al. 2009; Sruoga et al. 2011, 2016).

In this framework, the extensional setting of the northern Neuquén Basin seems to have been governed by the behavior of the subduction dynamics. The uplift of the Huincul arch, however, requires a different process. If the regional tectonic setting determined by plate dynamics was extensional (Charrier et al. 2007), it would be difficult to invoke NW-SE oriented contraction related to subduction as proposed by Mosquera and Ramos (2006), Silvestro and Zubiri (2008), and Mosquera et al. (2011).

Therefore, variable tectonic settings and stress states were required to create the observed sedimentological and structural features observed through the Neuquén Basin in the Late Jurassic (Fig. 9). In the southern sector, a local strike-slip setting was required to produce the structural data presented by Silvestro and Zubiri (2008).

Here, the greatest principal stress, σ_1, had a NW trend based on fault orientations. Open fractures required a NE-trend for σ_3. The intermediate stress σ_2 was vertical. We propose that this stress state was the result of a local permutation of σ_1 and σ_2 with respect to the regional extensional stress, shown in the northern sector (Fig. 9).

During the Kimmeridgian, Gondwana was breaking up. Initial opening of the southernmost South Atlantic Ocean started at 140 Ma, at the end of deposition of the Tordillo Formation (Rabinowitz and LaBrecque 1979; Heine et al. 2013). Rifting between South America and South Africa started before this time, and extension was asymmetric producing a clockwise rotation of South America with a rotation pole in NE Brazil (Fig. 9; Françolin and Szatmari 1987; Heine et al. 2013). We propose that the stress state in the Neuquén Basin was controlled by the combination of trench roll-back along the paleo-Pacific trench and clockwise rotation of South America due to the opening of the South Atlantic. The trench roll-back caused regional extension with horizontal, NE directed minimum stress. Clockwise rotation of South America was greater in the south, driving the North Patagonian Massif northwestward into the southern Neuquén Basin, causing a local permutation of σ_1 and σ_2 and transpression along the Huincul arch (Fig. 9). The orientation of pre-existing structures within the Huincul arch would have led to reactivation as strike-slip faults with local reverse motion.

5 Conclusions

Facies distribution, provenance analysis, thickness variations, and structural data support an extensional setting during deposition of the Tordillo Formation in the northern Neuquén Basin. Extension in the north was coeval with a neutral tectonic setting in the wide central portion of the Neuquén embayment and the transpressional uplift of the Huincul arch in the southernmost section of the basin. We propose that these contrasting tectonic settings are the result of the combination of subduction dynamics and plate rotations caused by the break-up of Pangea. Trench roll-back along the subduction margin would have caused a regional extensional stress state. The break-up of Pangea caused the clockwise rotation of South America around a rotation pole in NE Brazil which pushed the North Patagonian cratonic block toward the NW into the southern margin of the Neuquén Basin leading to a local transpressional stress state along and within the Huincul arch.

Acknowledgements JFM acknowledges funding from the International Association of Sedimentologists (Postgraduate Grant Scheme) in early stages of this project and funding from Agencia de Promoción Científica y Tecnológica (PICT-2015-1181). LG acknowledges the grant from the Agencia de Promoción Científica y Tecnológica (PICT-2011-1079). Field assistance of Maximiliano Viale (IANIGLA) and Jeremías Likerman (UBA) is greatly appreciated.

References

Álvarez PP (1996) Los depósitos triásicos y jurásicos de la alta cordillera de San Juan. In: Ramos VA (ed) Geología de la región del Aconcagua, provincias de San Juan y Mendoza. Subsecretaría de Minería de la Nación, Anales 24, Dirección Nacional del Servicio Geológico, pp 59–137

Arregui C (1993) Análisis estratigráfico paleoambiental de la Formación Tordillo en el subsuelo de la Cuenca Neuquina. In Abstracts of the 12 Congreso Geológico Argentino y 2 Congreso de Exploración de Hidrocarburos, Mendoza, 10–15 October 1993

Azcuy CL, Carrizo H, Caminos R (1999) Carbonífero y Pérmico de las Sierras Pampeanas, Famatina, Precordillera, Cordillera Frontal y bloque de San Rafael. In: Caminos R (ed) Geología Argentina. Instituto de Geología y Recursos Minerales, Anales 29, Servicio Geológico y Minero Argentino, pp 261–318

Balgord E (2016) Triassic to Neogene evolution of the Andean retroarc: Neuquén basin, Argentina. Ph.D. thesis, University of Arizona

Bechis F, Giambiagi L, García V et al (2010) Kinematic analysis of a transtensional fault system: the Atuel depocenter of the Neuquén basin, southern Central Andes, Argentina. J Struct Geol 32:886–899

Boll A, Alonso J, Fuentes F et al (2014) Factores controlantes de las acumulaciones de hidrocarburos en el sector norte de la cuenca Neuquina, entre los ríos Diamante y Salado, provincia de Mendoza, Argentina. In: Abstratcs of the 9 Congreso de Exploración y Desarrollo de Hidrocarburos, Mendoza, 3–7 Nov 2014

Broens S, Pereira M (2005) Evolución estructural de la zona de transición entre las fajas plegadas y corridas de Aconcagua y Malargüe Provincia de Mendoza. Rev Asoc Geol Argent 60(4):685–695

Burckhardt C. (1900) Profils transversaux de la Cordillère Argentino-Chilienne. Stratigraphie et téctonique. Museo de La Plata, Sección Geología y Mineralogía 2, La Plata, pp 1–136

Cacace M, Bayer U, Marotta AM, Lempp C (2006) Driving mechanisms for basin formation and evolution. In: Littke R, Bayer U, Gajewski D, Nelskamp S (eds) Dynamics of complex sedimentary basins. The example of the Central European Basin System. Springer-Verlag, Berlin-Heidelberg, pp 37–66

Cacace M, Bayer U, Marotta AM et al (2008) Driving mechanisms for basin formation and evolution. In: Littke R, Bayer U, Gajewski D, Nelskamp S (eds) Dynamics of complex intracontinental basins. Springer, New York, pp 3–66

Calderón S (2008) Condiciones físicas y químicas del metamorfismo de muy bajo grado de las secuencias mesozoicas en el valle del río Volcán (33° 50′-34° S). Degree thesis, Universidad de Chile

Cegarra MI, Pereyra FX (1988) Estudio de las areniscas de las Formaciones Tordillo y Diamante, Jurásico - Cretácico de la Cuenca Aconcaguina, Provincia de Mendoza. In Abstracts of the II Reunión Argentina Sedimentología, Buenos Aires, 1–5 Aug 1988

Cegarra MI, Ramos VA (1996) La faja plegada y corrida del Aconcagua. In: Ramos VA (ed) Geología de la región del Aconcagua, provincias de San Juan y Mendoza. Subsecretaría de Minería de la Nación, Anales 24, Dirección Nacional del Servicio Geológico, pp 387–422

Charrier R (1981) Geologie der chilenischen Hauptkordillere zwischen 34° und 34° 30′ südlichen Breite und ihre tektonische, magmatische und paläogeopraphische Entwicklung. Berliner Geowissen Abh A 36:1–277

Charrier R (2007) Kimmeridgian backarc extensional reactivation and magmatism in the Northern and Central Chilean Andes (21°–36° LS). In: Abstracts of Geosur 2007, Santiago, 22–25 November 2007

Charrier R, Vicente JC (1972) Liminary and Geosinclinal Andes: major orogenic phases and synchronical evolutions of the Central and Magellan Sectors of the argentine-chilean Andes. Solid Earth Probl Conf, Upper Mantle Project, Buenos Aires, 451–470

Charrier R, Pinto L, Rodríguez MP (2007) Tectonostratigraphic evolution of the Andean orogen in Chile. In: Moreno T, Gibbons W (eds) The Geology of Chile. Geol Soc, London, pp 21–114

Cohen KM, Finney SC, Gibbard PL, Fan J-X (2013-updated) The ICS International Chronostratigraphic Chart. Episodes 36:199–204

Criado Roqué P (1950) Zona comprendida entre los ríos Atuel y Diamante, provincia de Mendoza, escala 1:25000. Yacimientos Petrolíferos Fiscales, Unpublished Internal Report, Buenos Aires

Cristallini E, Tomezzoli R, Pando G, Gazzera C, Martínez JM, Quiroga J, Buhler M, Bechis F, Barredo S, Zambrano O (2009) Controles Precuyanos en la estructura de la cuenca Neuquina. Rev Asoc Geol Argent 65(2):248–264

Cruz C, Boll A, Gómez Omil R et al (2002) Hábitat de hidrocarburos y sistemas de carga Los Molles y Vaca Muerta en el sector central de la cuenca Neuquina, Argentina. In Abstracts of the 5 Congreso de Exploración y Desarrollo de Hidrocarburos, Mar del Plata, 29 Oct–2 Nov 2002

Dajczgewand DM (2002) Faja Plegada y Corrida de Malargüe: estilo de la deformación en Mallín Largo. Degree thesis, Universidad de Buenos Aires

Davidson J, Vicente JC (1973) Características paleogeográficas y estructurales del área fronteriza de las nacientes del Teno (Chile) y Santa Elena (Argentina), Cordillera Principal (35° a 35° 15′ de Latitud Sur). In: Abstracts of the 5 Congreso Geológico Argentino, Cordoba, 22–28 Oct 1973

De Celles P, Giles K, (1996) Foreland basin systems. Basin Res 8:105–123

De Ferraris C (1947) Edad del arco o dorsal antigua del Neuquén Oriental de acuerdo con la estratigrafía de la zona inmediata. Rev Asoc Geol Argent 2(3):256–283

Dickinson WR (1970) Interpreting detrital modes of graywacke and arkose. J Sediment Petrol 40:695–707

Dickinson WR, Beard LS, Brakenridge GR et al (1983) Provenance of North American Phanerozoic sandstones in relation to tectonic setting. Geol Soc Am Bull 94:222–235

Digregorio J (1972) Neuquén. In: Leanza AF (ed) Geología Regional Argentina. Academia Nacional de Ciencias, Córdoba, pp 439–505

Eisner P (1991) Tectonostratigraphic evolution of Neuquén Basin, Argentina. Master thesis, Rice University

Folk RL, Andrews PB, Lewis DW (1970) Detrital sedimentary rock classification and nomenclature for use in New Zealand. NZ J Geol Geophys 13:937–968

Françolin J, Szatmari P (1987) Mecanismo de rifteamento da porçao oriental da margem norte brasileira. Rev Brasil Geociências 17:196–207

Gazzi P (1966) Le arenarie del flysch sopracretaceo dell'Appennino modenese; correlazioni con il flysch di Monghidoro. Miner Petrog Acta 12:69–97

Giambiagi LB, Álvarez P, Godoy E et al (2003) The control of pre-existing extensional structures on the evolution of the southern sector of the Aconcagua fold and thrust belt, southern Andes. Tectonophysics 369:1–19

Giambiagi LB, Bechis F, Lanés S et al (2008) Formación y evolución triásica-jurásica del depocentro Atuel, cuenca Neuquina, Argentina. Rev Asoc Geol Argent 63(4):520–533

Giambiagi LB, Tunik M, Barredo S et al (2009) Cinemática de apertura del sector norte de la cuenca Neuquina. Rev Asoc Geol Argent 65(2):278–292

González O, Vergara M (1962) Reconocimiento Geológico de la Cordillera de los Andes entre los paralelos 35° y 38° Latitud Sur. Publicaciones del Departamento de Geología 24, Universidad de Chile, p 121

Groeber P (1918) Edad y extensión de las estructuras de la Cordillera entre San Juan y Nahuel Huapi. Physis 4:208–240

Groeber P (1946) Observaciones geológicas a lo largo del meridiano 70. 1. Hoja Chos Malal. Rev Asoc Geol Argent 1(3):117–208

Groeber P, Stipanicic P, Mingramm A (1953) Jurásico. In: Groeber P, Stipanicic P, Mingramm A (eds) Geografía de la República Argentina, Sociedad Argentina de Estudios Geográficos GAEA, p 347

Gulisano CA, Gutiérrez Pleimling AR (1995) Guía de Campo: El Jurásico de la Cuenca Neuquina. B) Provincia de Mendoza. Asociación Geológica Argentina, Serie E, vol 3, p 103

Harding TP (1974) Acumulaciones importantes de hidrocarburos originadas por deformaciones causadas por fallas laterales. Petrotecnia 2(17):12–19

Heine C, Zoethout J, Müller RD (2013) Kinematics of the South Atlantic rift. Solid Earth 4:215–253
Horton B, Fuentes F, Boll A et al (2016) Andean stratigraphic record of the transition from backarc extension to orogenic shortening: a case study from the northern Neuquén Basin, Argentina. J S Am Earth Sci 71:17–40
Howell JA, Schwarz E, Spalletti LA et al (2005) The Neuquén basin: an overview. In: Veiga GD, Spaletti LA, Howell JA et al (eds) The Neuquén basin, Argentina: a case study in sequence stratigraphy and basin dynamics. The Geological Society, SP 252, pp 1–14
Hubert JF, Hyde MG (1982) Sheet-flow deposits of graded beds and mudstones on an alluvial sandflat-playa system: Upper Triassic Blomidon redbeds, St Mary's Bay, Nova Scotia. Sedimentology 29:457–474
Jordan T (1995) Retroarc foreland and related basins. In: Busby CJ, Ingersoll R (eds) Tectonics of sedimentary basins. Blackwell Science, Oxford, pp 331–362
Kietzmann D, Vennari V (2013) Sedimentología y estratigrafía de la Formación Vaca Muerta (Tithoniano-Berriasiano) en el área del cerro Domuyo, norte de Neuquén, Argentina. Andean Geol 40(1):41–65
Kim HJ (2003) Geología del área del cerro Amarillo, sur del Arroyo Las Playas, provincia de Mendoza. Degree thesis, Universidad de Buenos Aires
Kocurek G, Dott RH (1981) Distinctions and uses of stratification types in the interpretation of Eolian Sand. J Sediment Petrol 51:579–595
Klohn C (1960) Geología de la Cordillera de los Andes de Chile Central, Provincias de Santiago, O'Higgins, Colchagua y Curicó. Investigaciones Geológicas, Santiago, Bol 8, p 95
Leanza HA (2009) Las principales discordancias del Mesozoico de la cuenca Neuquina según observaciones de superficie. Rev Mus Argent Cs Nat 11(2):145–184
Leanza HA, Hugo C (2001) Hoja Geológica 3969-I Zapala. Servicio Geológico Minero Argentino, vol 275, Instituto de Geología y Recursos Naturales, p 133
Legarreta L (1976) Análisis estratigráfico de la Formación Tordillo (Jurásico superior) en la zona comprendida entre el río Salado y el río Diamante, Departamentos de Malargue y San Rafael, provincia de Mendoza, República Argentina. Degree thesis, Universidad de Buenos Aires
Legarreta L, Gulisano CA (1989) Análisis estratigráfico secuencial de la Cuenca Neuquina (Triásico superior-Terciario inferior). In: Chebli G, Spalletti LA (eds) Cuencas Sedimentarias Argentinas. Universidad de Tucumán, vol 6, Serie Correlación Geológica, p 221–243
Legarreta L, Uliana MA (1991) Jurassic-Cretaceous marine oscillations and geometry of back-arc basin fill, central Argentine Andes. In: McDonald D (ed) Sedimentation, Tectonics and Eustacy. Int Ass Sediment SP 12: 429–450
Legarreta L, Uliana MA (1996) The Jurassic succesion in west-central Argentina: stratal patterns, sequences and paleogeographic evolution. Palaeogeogr Palaeocl 120:303–330
Legarreta L, Uliana MA (1999) El Jurásico y Cretácico de la Cordillera Principal y la cuenca Neuquina: Facies sedimentarias. In: Caminos R (ed) Geología Argentina, vol 29. Servicio Geológico y Minero Argentino, Buenos Aires, pp 399–416
Llambías EJ, Kleiman L, Salvarredi JA (1993) El magmatismo Gondwánico. In: Ramos V (ed) Geología y Recursos Naturales de Mendoza. Asociación Geológica Argentina, Buenos Aires, pp 53–64
López-Gómez J, Martín-Chivelet J, Palma R (2009) Architecture and development of the alluvial sediments of the Tordillo Formation in the Cañada Ancha valley, northern Neuquén basin, Argentina. Sediment Geol 219:180–195
Lo Forte G (1996) El Jurásico de la Alta Cordillera de Mendoza. In: Ramos VA (ed) Geología de la región del Aconcagua, provincias de San Juan y Mendoza. Dirección Nacional del Servicio Geológico, Subsecretaría de Minería de la Nación 24:136–178
Marchese HG (1971) Litoestratigrafia y variaciones faciales de las sedimentitas mesozoicas de la Cuenca Neuquina, Prov. de Neuquén. Rep. Argentina. Rev Asoc Geol Argent 26:343–410
Maretto H, Carbone O, Gazzera C et al (2002) Los reservorios de la Formación Tordillo. In: Schiuma M, Hinterwimmer G, Vergani G (eds) Rocas Reservorio de las Cuencas Productivas Argentinas. Instituto Argentino del Petroleo y el Gas, Buenos Aires, pp 335–358

McClay K (1990) Extensional fault systems in sedimentary basins: a review of analogue model studies. Mar Petrol Geol 7:206–233
McKenzie D (1978) Some remarks on the development of sedimentary basins. Earth Planet Sci Let 40:25–32
Mescua JF, Giambiagi LB, Bechis F (2008) Evidencias de tectónica extensional en el Jurásico superior (Kimmeridgiano) del suroeste de la provincia de Mendoza. Rev Asoc Geol Argent 63(4):512–519
Mescua JF, Giambiagi LB, Tassara A et al (2014) Influence of pre-Andean history over Cenozoic foreland deformation: Structural styles in the Malargüe fold-and-thrust belt at 35° S Andes of Argentina. Geosphere 10(3):585–609
Miall DA (1996) The geology of fluvial deposits, sedimentary facies, basin analysis and petroleum geology. Springer, New York, p 582
Mosquera A, Ramos VA (2006) Intraplate deformation in the Neuquén Embayment. In: Kay SM, Ramos VA (eds) Evolution of an Andean margin: a tectonic and magmatic view from the Andes to the Neuquén Basin (35°-39° S lat). Geol Soc Am, SP 407:97–123
Mosquera A, Silvestro J, Ramos VA et al (2011) La estructura de la dorsal de Huincul. In: Leanza H, Arregui C, Carbone O, Danieli J, Vallés J (eds) Geología y Recursos Naturales de Neuquén. Asocaición Geológica Argentina, Buenos Aires, pp 385–397
Naipauer M, García Morabito E, Marques JC et al (2012) Intraplate Late Jurassic deformation and exhumation in western central Argentina: Constraints from surface data and U-Pb detrital zircon ages. Tectonophysics 524–525:59–75
Naipauer M, Tunik M, Marques JC et al (2015a) U–Pb detrital zircon ages of Upper Jurassic continental successions: implications for the provenance and absolute age of the Jurassic–Cretaceous boundary in the Neuquén Basin. In: Sepúlveda SA, Giambiagi LB, Moreiras et al (eds) Geodynamic processes in the Andes of Central Chile and Argentina. The Geological Society, London, SP 399, pp 131–154
Naipauer M, Tapia F, Mescua JF et al (2015b) Detrital and volcanic zircon U-Pb ages from southern Mendoza (Argentina): an insight on the source regions in the northern part of Neuquén Basin. J South Amer Earth Sci 64(2):434–451
Nullo FE, Stephens GC, Combina AM et al (2005) Hoja Geológica 3569-III Malargüe, Provincia de Mendoza. Instituto de Geología y Recursos Minerales, vol 346, Servicio Geológico Minero Argentino, p 85
Orchuela IA, Ploszkiewicz V, Viñes R (1981) Reinterpretación estructural de la denominada "Dorsal Neuquina". In: Abstracts of the 8 Congreso Geológico Argentino, San Luis, 20–26 September 1981
Pángaro F, Ramos VA, Godoy E (1996) La Faja plegada y corrida de la Cordillera Principal de Argentina y Chile a la latitud del Cerro Palomares (33° 35′ S). In: Abstracts of the 13 Congreso Geológico Argentino y 3 Congreso Exploración de Hidrocarburos, Mendoza, 13–18 Oct 1996
Ramos VA (1977) Estructura. In: Rolleri EO (ed) Geología y recursos naturales del Neuquén. Asociación Geológica Argentina, Buenos Aires, pp 99–118
Ramos VA (1999) Evolución tectónica de la Argentina. In: Caminos R (ed) Geología Argentina, vol 29. Servicio Geológico Minero Argentino, Buenos Aires, pp 715–759
Rabinowitz PD, LaBrecque J (1979) The Mesozoic south Atlantic Ocean and evolution of its continental margins. J Geophys Res 84:5973–6002
Rossel P, Oliveros V, Mescua JF et al (2014) The Upper Jurassic volcanism of the Río Damas-Tordillo Formation (33°–35.5° S): Insights on petrogenesis, chronology, provenance and tectonic implications. Andean Geol 41(3):529–557
Sanguinetti AS (1989) Volcanismo Neojurásico-Neocomiano de la quebrada de Vargas, alta cordillera de Mendoza. Rev Asoc Geol Argent 44:381–393
SERNAGEOMIN (2003) Mapa Geológico de Chile, escala 1:1.000.000. Servicio Nacional de Geología y Minería, Publicación Geológica Digital 4
Seton M, Muller RD, Zahirovic S et al (2012) Global continental and ocean basin reconstructions since 200 Ma. Earth Sci Rev 113:212–270

Silvestro J, Zubiri M (2008) Convergencia oblicua: modelo estructural alternativo para la dorsal Neuquina (39° S) - Neuquén. Rev Asoc Geol Argent 63:49–64
Spalletti LA (2013) Influencia del arco magmático protoandino en la acomodación sedimentaria, la fisiografía y las características de los depósitos del Jurásico superior y Cretácico inferior en la Cuenca Neuquina. Academia Nacional de Ciencias Exactas, Físicas y Naturales 65:28–42
Spalletti LA, Colombo F (2005) From alluvial fan to playa: an upper Jurassic ephemeral fluvial system, Neuquén Basin, Argentina. Gondwana Res 8:363–383
Spalletti LA, Queralt I, Matheos S et al (2008) Sedimentary petrology and geochemistry of siliciclastic rocks from the upper Jurassic Tordillo Formation (Neuquén Basin, western Argentina): Implications for provenance and tectonic setting. J S Am Earth Sci 25:440–463
Sruoga P, Etcheverría M, Folguera A et al (2005) Hoja Geológica 3569-I Volcán Maipo, Provincia de Mendoza. Instituto de Geología y Recursos Minerales, vol, 290Servicio Geológico Minero Argentino, p 92
Sruoga P, Etcheverría M, Cegarra M et al (2011) Engranaje lateral entre las Formaciones Tordillo y Río Damas en la Cordillera Principal de Mendoza (34° 45′ S). In: Abstracts of the 18 Congreso Geológico Argentino, Neuquén, 4–6 May 2011
Sruoga P, Etcheverría M, Cegarra M et al (2016) Hoja Geológica 3569–13, Cerro Risco Plateado, pcia. de Mendoza. Instituto de Geología y Recursos Minerales, vol 420, Servicio Geológico Minero Argentino, p 107
Stipanicic PN (1969) El avance en los conocimientos del Jurásico argentino a partir del esquema de Groeber. Rev Asoc Geol Argent 24(4):367–388
Stipanicic PN, Rodrigo F (1970) El diastrofismo jurásico en Argentina y Chile. In: Abstracts of the 4 Jornadas Geológicas Argentinas, Mendoza, 14–16 Oct 1970
Tunbridge IP (1984) Facies model for a sandy ephemeral stream and clay playa complex; the Middle Devonian Trentishoe Formation of North Devon, U.K. Sedimentology 31:697–715
Uliana M, Arteaga M, Legarreta L et al (1995) Inversion structures and hydrocarbon occurrence in Argentina. In: Buchanan J, Buchanan P (eds) Basin inversion. Geol Soc, London, SP 88, pp 211–233
Vergani G (2005) Control estructural de la sedimentación Jurásica (Grupo Cuyo) en la Dorsal de Huincul, Cuenca Neuquina, Argentina. Modelo de falla lístrica rampa-plano, invertida. Bol Inf Petrol 1:32–42
Veiga R (2002) Migración de hidrocarburos y sistemas petroleros cuyanos en el ámbito central de la cuenca Neuquina, Argentina. In: Abstracts of the 5 Congreso de Exploración y Desarrollo de Hidrocarburos, Mar del Plata, 29 Oct–2 Nov 2002
Veiga GD, Spalletti LA (2007) The Upper Jurassic (Kimmeridgian) fluvial–aeolian systems of the southern Neuquén Basin, Argentina. Gondwana Res 11:286–302
Veiga R, Pángaro F, Fernández M (2002) Modelado bidimensional y migración de hidrocarburos en el ámbito occidental de la dorsal de Huincul, cuenca Neuquina, Argentina. In: Abstracts of the 5 Congreso de Exploración y Desarrollo de Hidrocarburos, Mar del Plata, 29 Oct–2 Nov 2002
Vergani GD, Tankard J, Belotti J et al (1995) Tectonic evolution and paleogeography of the Neuquén Basin, Argentina. In: Tankard AJ, Suárez R, Welsink HJ (eds) Petroleum basins of South America. AAPG Mem 62:383–402
Vergara M, Levi B, Nyström JO, Cancino A (1995) Jurassic and Early Cretaceous island arc volcanism, extension, and subsidence in the Coast Range of central Chile. Geol Soc Am Bull 107(12):1427–1440
Vicente JC, Leanza HA (2009) El frente de corrimiento andino al nivel de los cerros Penitentes y Visera (alta cordillera de Mendoza): aspectos cronológicos y cartográficos. Rev Asoc Geol Argentina 65:97–110
Volkheimer W (1966) Descripción geológica de la Hoja 27b, Cerro Sosneado, provincia de Mendoza. Servicio Geológico Nacional, vol 151, Buenos Aires, p 83
Wernicke B, Burchfiel B (1982) Modes of extensional tectonics. J Struct Geol 4:105–115

Tectonic Setting of the Tordillo Formation in the Aconcagua Fold-and-Thrust Belt

Eliana Acevedo, Eduardo Agustín Rosselot, Federico Martos, Lucas Fennell, Maximiliano Naipauer and Andrés Folguera

Abstract At the northwestern Mendoza province, the Mesozoic infill of the Neuquén Basin is tectonically repeated in the Aconcagua fold-and-thrust belt. Particularly, the Tordillo Formation (commonly associated with the Kimmeridgian) represents a local low stand period of sea level, with mainly alluvial and fluvial sediments. Toward the western sector of the belt, it interfingers with volcanic and volcaniclastic materials and presents a marked increase in the thickness. This unit was studied in two localities at the Blanco River valley, at the undeformed sector and over the second thrust that produces a second repetition of the Upper Jurassic and Lower Cretaceous sequences in the Aconcagua fold-and-thrust belt. This transect exposes facies variations and a significant increase in thickness to the west. Additionally, provenance analysis and paleocurrent directions indicate that the sediment supply was located to the E-SE, and that the underlying units were exhumed at the time of deposition of the Late Jurassic red beds. A consistent thickness increment of the Upper Jurassic deposits to the west through the Aconcagua fold-and-thrust belt suggests that sedimentation was controlled by NNW-directed structures. This is also supported by facies analyses that demonstrate high topographic breaks affecting a smooth west-dipping fluvial ramp toward the volcanic arc. These features support an extensional setting for the deposition of the Tordillo Formation at the latitudes of the Aconcagua fold-and-thrust belt, as other authors have proposed for the Malargüe fold-and-thrust belt to the south. Plate tectonic reconstructions suggest trench rollback during this time previous to the westward migration of the South American plate, which is consistent with the back-arc extension proposed in the previous works.

E. Acevedo (✉) · E. A. Rosselot · F. Martos · L. Fennell · A. Folguera
Universidad de Buenos Aires, Facultad de Ciencias Exactas y Naturales, Departamento de Ciencias Geológicas, Buenos Aires, Argentina
e-mail: elianabacevedo@gmail.com

CONICET—Universidad de Buenos Aires, Instituto de Estudios Andinos Don Pablo Groeber (IDEAN), Buenos Aires, Argentina

M. Naipauer
CONICET—Universidad de Buenos Aires, Instituto de Geocronología y Geología Isotópica (INGEIS), Buenos Aires, Argentina

D. Kietzmann and A. Folguera (eds.), *Opening and Closure of the Neuquén Basin in the Southern Andes*, Springer Earth System Sciences,
https://doi.org/10.1007/978-3-030-29680-3_7

Keywords Tordillo Formation · Río Damas Formation · Back-arc extension · Aconcagua fold-and-thrust belt

1 Introduction

Between the Early Jurassic and the Early Cretaceous, the evolution of the Neuquén Basin is considered as part of a post-rift phase, with a series of transgression-regression cycles controlled by eustatic oscillations, changes in subsidence rates, and localized uplift (Legarreta and Gulisano 1989; Legarreta and Uliana 1991; Vergani et al. 1995; Howell et al. 2005). Through this evolution, the Late Jurassic continentalization is represented by the Tordillo Formation (Stipanicic 1966). In the Aconcagua region (Fig. 1a), it consists mainly of fluvial and alluvial sediments, interbedded with volcanic and volcaniclastic deposits. The volcaniclastic component of the deposits and their thickness have been noted to increase toward the western sector of the basin, where the Tordillo Formation can be correlated with the Río Damas red beds (Yrigoyen 1976; Lo Forte 1996; Sanguinetti 1989; Sanguinetti and Cegarra 1991) (see Chapters "Early Andean Magmatism in Southern Central Chile (33°–40° S)" and "Controls on Deposition of the Tordillo Formation in Southern Mendoza (34°–36° S): implications for the Kimmeridgian tectonic setting of the Neuquén Basin"). Based on the fossil content in the under- and overlying units, a Kimmeridgian age is proposed for the Tordillo Formation. However, U–Pb dating of detrital zircons from the Tordillo and Río Damas Formations is not consistent with this time span,

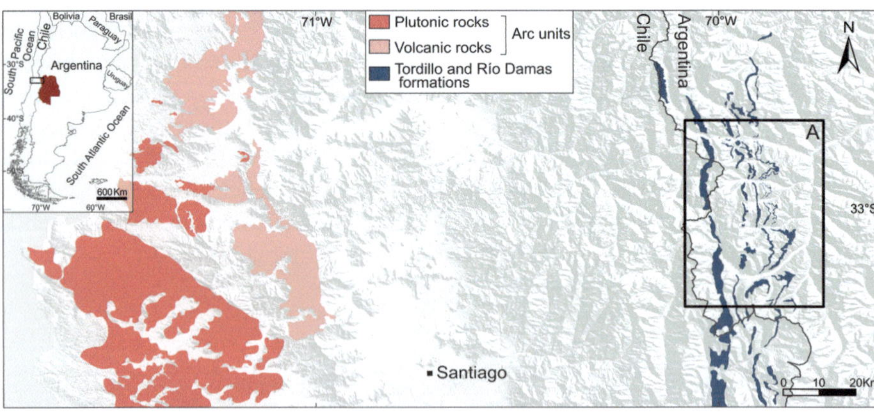

Fig. 1 Digital elevation model with geological units showing the locations of the Late Jurassic retroarc volcano-sedimentary sections and magmatic units on the Pacific coastal zone (the volcanic Ajial and Horqueta formations and associated plutonic rocks interpreted as the Late Jurassic arc roots). The Late Jurassic volcano-sedimentary units on the axial Andean zone correspond to the Río Damas and Tordillo Formations. Based on Gana and Tosdal (1996), Rivano et al. (1993), Ramos (1996), Ramos et al. (2010). The rectangle (A) indicates the studied region shown in Fig. 2, corresponding to the Aconcagua fold-and-thrust belt

yielding ages between 150 and 145 Ma (Naipauer et al. 2012, 2015; Rossel et al. 2014; Naipauer 2016).

The driving mechanism for this Late Jurassic continentalization, represented by the Tordillo Formation, is explained as part of a transgression-regression cycle occurred during a thermal subsidence stage (Legarreta and Gulisano 1989; Legarreta and Uliana 1991). However, other authors have proposed a renewed extensional regime through the Kimmeridgian–Tithonian (?) interrupting sag facies, based on synsedimentay normal faults in the Tordillo red beds, facies distribution, thickness variations, associated volcanic activity, and provenance analysis of the correlative units in Chile (see Chapter "Controls on Deposition of the Tordillo Formation in Southern Mendoza (34°–36° S): implications for the Kimmeridgian Tectonic Setting of the Neuquén Basin") (Charrier et al. 2007; Mescua et al. 2008; Rossel et al. 2014).

Then, the purpose of this chapter is to analyze the tectonic environment associated with the deposition of the Tordillo Formation in the Aconcagua fold-and-thrust belt of the northern Neuquén Basin through the study of its facies changes, thickness variations, and sandstones provenance analysis relating these with the complex structure that characterizes this setting.

2 Geological Setting

The evolution of the Neuquén Basin has been classically divided into three main stages: an initial synrift phase (Late Triassic–Early Jurassic), a thermal subsidence stage (Early Jurassic–Early Cretaceous), and a foreland stage (Late Cretaceous–Cenozoic) (Howell et al. 2005). Syn-extensional and syn-contractional units are fed by a complex register of Paleozoic to Mesozoic units that are exposed in the Aconcagua fold-and-thrust belt and adjacent Frontal Cordillera. The basement of the basin in the Aconcagua region, at northwestern Mendoza province, consists of the upper Paleozoic Alto Tupungato Formation, which is constituted by orthoquartzites and mudstones in fining upward sequences interpreted as turbidites, which are intruded by Permian granites, and the overlying Choiyoi Group composed of acidic to intermediate volcanic rocks of Permian–Triassic age (Polanski 1972; Pérez and Ramos 1996). These units are mainly exposed at the Frontal Cordillera, to the east of the Aconcagua fold-and-thrust belt, uplifted during the Cenozoic contractional stages (Heredia et al. 2012).

The Middle Jurassic to Lower Cretaceous infill of the Neuquén Basin is exposed in the undeformed sector over this basement, being represented by La Manga, Auquilco, Tordillo, and Vaca Muerta formations. The La Manga Formation consists mainly of calcareous rocks unconformably overlying the basement with no development of the syn-rift deposits, which suggests the existence of a paleotopographic high in the area during Late Triassic times, known as the Tigre High (Groeber 1918; Alvarez 1996). Over these, the Auquilco Formation evaporites are interpreted as a sudden desiccation event followed by a continentalization and massive clastic influx represented by the

Tordillo Formation (Legarreta and Uliana 1996). Over these, a marine transgression from the Pacific Ocean allowed the deposition of the Vaca Muerta Formation that consists of a rhythmic alternation of black shales, malrs, and limestones (Legarreta and Uliana 1991, Kietzmann et al. 2014). The Jurassic subduction-related arc was resting on west Gondwana in Central Chile at these latitudes, represented by the magmatic units of the Ajial (Lower to Middle Jurassic) and Horqueta (Middle to Upper Jurassic) Formations that consist of acidic lavas and pyroclastic rocks (Vergara et al. 1995). A plutonic counterpart is also present at the Coastal Cordillera, which consists of an extensive batholith of I-type, calcalkaline diorites, tonalites, granodiorites, and granites whose emplacement occurred between 162 and 156 Ma, representing a fast pulse of voluminous magmas (see Chapter "Early Andean Magmatism in Southern Central Chile, 33°–40° S") (Gana and Tosdal 1996) (Fig. 1).

In the Aconcagua region, the Mesozoic infill of the Neuquén Basin is tectonically repeated by an imbricate fan of thin-skinned structures of the Aconcagua fold-and-thrust belt (Cegarra and Ramos 1996). Between the undeformed sector and the several thrust sheets to the west that integrate this deformational belt, the Tordillo Formation shows a steady increment in thickness and increasing volcaniclastic influence (Fig. 2). At Las Cuevas river valley, the first outcrops of the Tordillo Formation over the undeformed sector and the first thrust sheet present lenses of conglomerates and sandstones in thinning and fining upwards sequences with thicknesses between 30 and 160 m; while in the second and third thrust sheets to the west, it consists of conglomerates and sandstones with intercalations of volcanic and volcaniclastic levels, with variable thickness between 300 and 500 m (Yrigoyen 1972; Sanguinetti 1989; Lo Forte 1996). Then, in the fourth thrust sheet, the Tordillo Formation appears as a volcanic and volcaniclastic sequence, followed by fluvial deposits reaching 1200 m thick (Sanguinetti and Cegarra 1991). In the Chilean slope, this unit is correlated to the Río Damas Formation, with ca. 3000 m of continental successions of alluvial and fluvial deposits intercalated with andesitic lavas and breccias (Klohn 1960).

Despite a retroarc emplacement, the volcanism interfingered in the Tordillo and Río Damas Formations shows a subduction-related affinity, with a source and petrogenetic conditions that cannot be differentiated from the main arc front located to the west (Rossel et al. 2014). Contrastingly, in northern Chile (26°–31° S), the back-arc magmatism is geochemically discriminated from the main arc by their island arc signature, which suggests a development within a highly attenuated crust in an extensional environment (Rossel et al. 2013) (see Chapter "Early Andean Magmatism in Southern Central Chile, 33°–40° S").

Evidence of extension during the sedimentation of the Tordillo Formation in the Aconcagua region is presented by Vicente and Leanza (2009), where they describe Late Jurassic red beds of this unit partially filling a half-graben (Fig. 3a). Similarly, based on facies and thickness variations, Cegarra and Ramos (1996) had proposed that the sedimentation of the Late Jurassic interval was controlled by normal faults in the same area. Furthermore, Vitarella (2000) and Negro (2002) mentioned the presence of normal faults that displace the basement of the Tordillo Formation previous to its deposition.

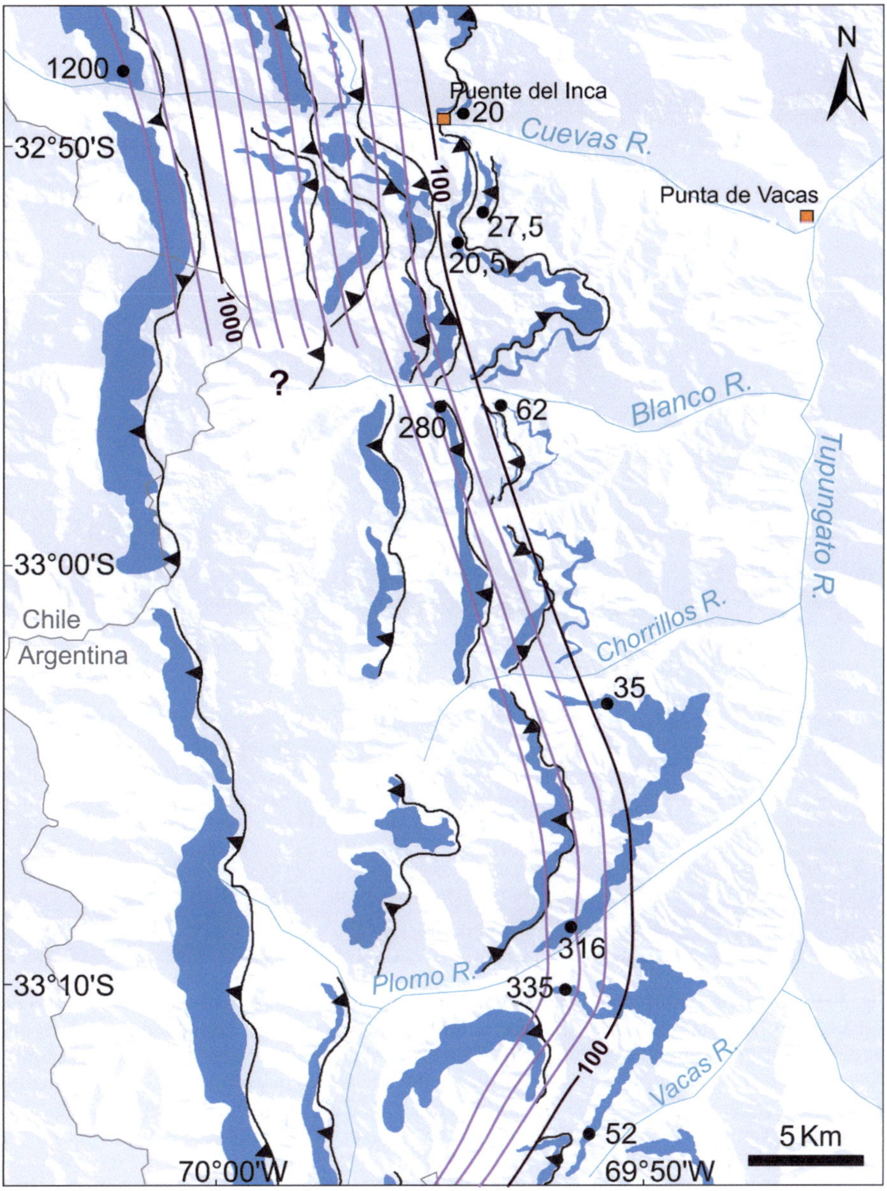

Fig. 2 Outcrops of the Río Damas and Tordillo Formations through the axial Andean zone in the Aconcagua fold-and-thrust belt and structures associated with their emplacement and measured thicknesses. A and B spots indicate the position of the two studied localities. The numbered black dots indicate compiled thicknesses of the Tordillo Formation in meters (Lo Forte 1996; Sanguinetti and Cegarra 1991; Suárez 2000; Rocha 2000; Orts 2001). Note that shortening associated with the Aconcagua fold-and-thrust belt was not restored

Fig. 3 Evidence of extension during the sedimentation of the Tordillo Formation in the Aconcagua and Malargüe fold-and-thrust belts. **a** Displacement of the La Manga Formation limestones and preserved half-graben filled with the Tordillo Formation clastics at the Vargas creek (modified from Vicente and Leanza 2009), **b** Normal faults that loose displacement within the Tordillo Formation at Pincheira creek (35° 25′) (taken from Mescua et al. 2008), and **c** East dipping normal faults affecting the Tordillo Formation at the Colorado creek (also taken from Mescua et al. 2008)

In southern Mendoza (35° S), Mescua et al. (2008) described growth strata in the Tordillo Formation associated with normal faults (Fig. 3b) (see Chapter "Controls on Deposition of the Tordillo Formation in Southern Mendoza (34°–36° S): implications for the Kimmeridgian tectonic setting of the Neuquén basin"). In the region between Vallenar and La Serena in Chile (28° 30′–30° S), the volcanic and volcaniclastic Algarrobal and Mostazal formations and correlatives of the Río Damas and Tordillo formations, respectively, present facies and thickness variations that suggest that they were deposited in extensional sub-basins developed in the back-arc domain (Charrier et al. 2007).

3 The Tordillo Formation in the Aconcagua Fold-and-thrust Belt

The Aconcagua fold-and-thrust belt is a N-S striking thin-skinned belt with the Auquilco evaporites as the main detachment (Cegarra and Ramos 1996). Along the different thrust sheets, the Tordillo Formation is repeated with significant facies and thickness variations. The Blanco River runs across this deformational belt exposing outstanding outcrops of this unit, which were studied in two localities, the undeformed sector and to the west, the second thrust sheet (Fig. 4a) in order to analyze thickness and facies changes.

Fig. 4 **a** South view of the Blanco River valley. A and B point to the locations of the Tordillo sections (A at the undeformed sector of the belt and B at the second thrust sheet). Ptch: Choiyoi Group, Jlm: La Manga Formation, Ja: Auquilco Formation, Jt: Tordillo Formation, Jkvm: Vaca Muerta Formation, Ka: Agrio Formation, Kd: Diamante Formation, **b** Volcanic clasts of the Sec. A, acid (left) and basic (right), **c** Normally graded conglomerate lenses in Sec. A, **d** Tabular levels of conglomerates and sandstones from FA 2 in Sec. B, **e** Interbedded tabular sandstones and mudstones from FA 4 in Sec. B, **f** Through cross-bedding from FA 2 in Sec. B, and **g** Limestone block in a matrix-supported conglomerate in Sec. B

3.1 Sedimentological Analysis

The stratigraphic sections of the Tordillo Formation were logged in two sectors through the analysis of its thickness, texture, sedimentary structures, geometry, spatial relationships, clast composition, and paleocurrents (Fig. 5). The Sec. A was logged from the top of the Auquilco evaporites to the base of the Vaca Muerta shales in the undeformed sector of the Aconcagua belt. The contacts between the different formations were covered and could not be determined with precision. On the other hand, the Sec. B was performed over the contact with the evaporites and ends with the first black shales in the second thrust sheet.

Section A (62 m) consists of a single facies association (FA) (Fig. 5). FA 1 comprises rounded, granule conglomerates that occur in lenticular bodies between 30 and 50 cm thickness, with erosional bases. Internally, these lenses present normal gradation, appear massive, or with horizontal stratification (Fig. 4c). Fragments of acid volcanic rocks and quartz represent the main clastic composition, with basic volcanic rocks and limestones subordinated (Fig. 4b). Clast composition does not present major variations in this section. At 23 m from the base, crops out a 1-m-thick-lenticular andesitic breccia, whose clasts consist mainly of andesites, with sandstones and limestones subordinated. The matrix is volcanic, conformed by plagioclase and amphibole phenocrysts in a hyalopilitic groundmass.

Section B (280 m) was subdivided into four facies associations that represent recurrent facies combinations (Fig. 5). FA 2 consists of clast-supported granule conglomerates interbedded with pebbly and finer sandstones (Fig. 3d). Massive, normal gradation, planar, and through cross-bedding are common structures (Fig. 3f). It presents a tabular geometry and a fining upward tendency. FA 3 consists of interbedded rounded clast and matrix-supported conglomerates and sandstones. Sandstones present cross-bedding and horizontal bedding. Conglomerates are massive or inverse graded. The deposits present a tabular geometry and a coarsening upward tendency. FA 4 consists of interbedded tabular levels of massive or laminated mudstones and horizontally laminated sandstones (Fig. 4e). FA 5 comprises tabular levels of massive and laminated mudstones with lenticular to lentiform sandstones or conglomerates intercalated. Sandstones present through cross-bedding and horizontal lamination. Conglomerates are massive or normally graded.

Clasts consist mainly of acid volcanic and limestone fragments. At 62 m from the base, coarse conglomerate levels present mudstone and sandstone clasts similar to the subjacent deposits and are interpreted as intraclasts. From this level, intraclasts appear in almost every conglomerate strata. At 90 m from the base, a coarse conglomerate strata presents blocks of limestones and acid volcanites up to 50 cm (Fig. 4 g). Paleocurrents in Section B vary between 290° and 335°, suggesting that the sediment source was located to the SE.

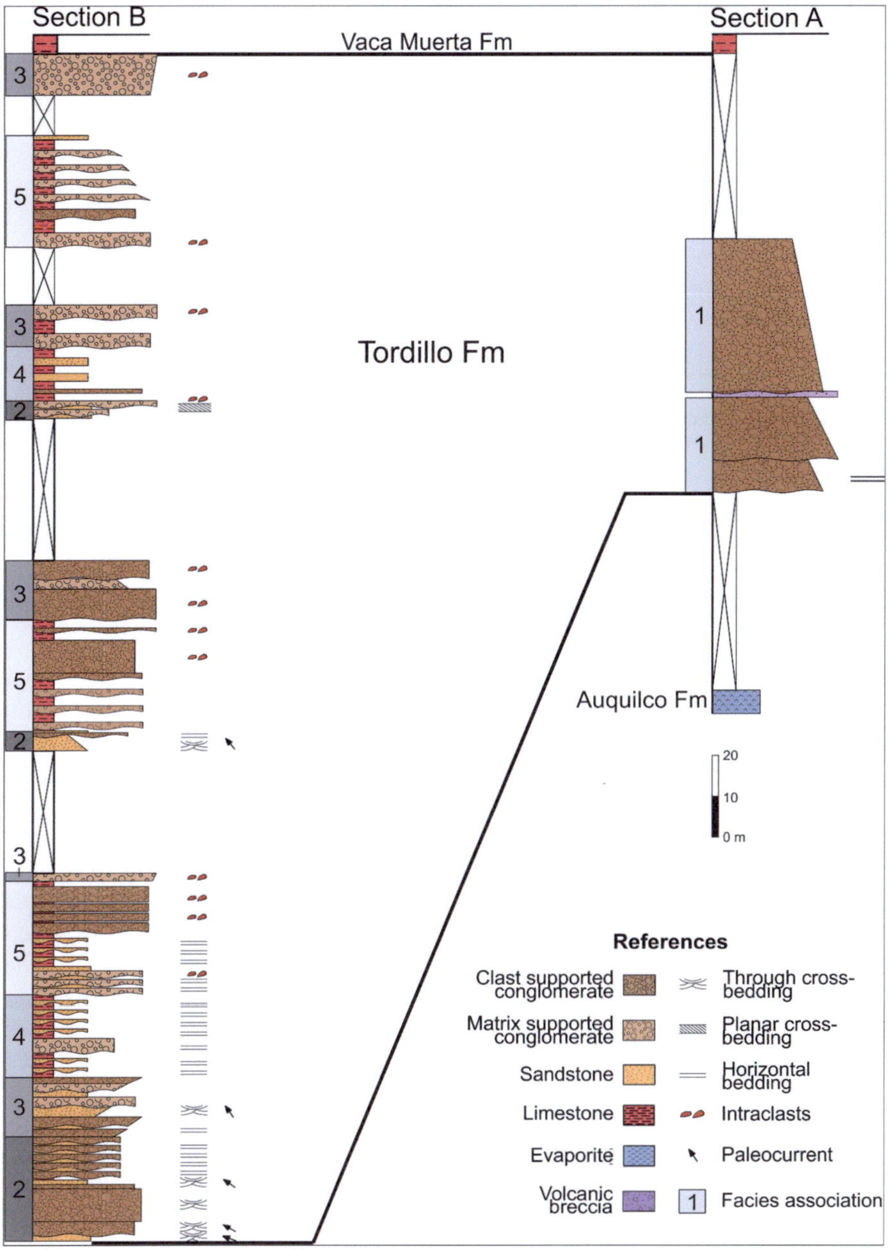

Fig. 5 Sedimentary sections of the Tordillo Formation logged at the Blanco River. Section locations are indicated in Fig. 2

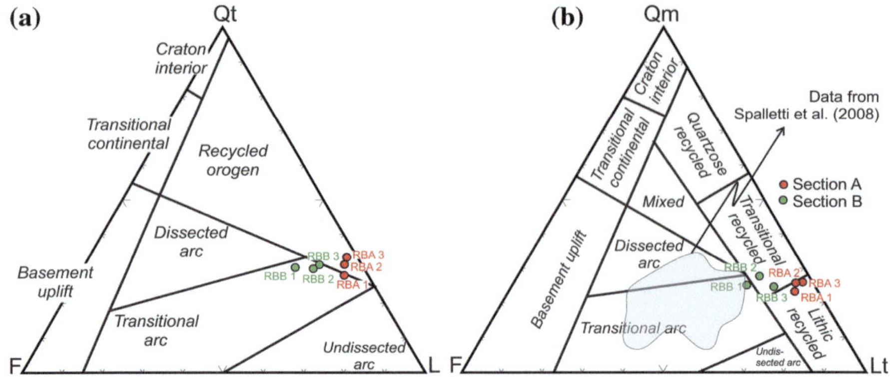

Fig. 6 Provenance data from sandstone petrography of the Tordillo Formation using the Dickinson et al. (1983) diagrams: **a** total quartz, feldspar, and lithic fragments (Qt, F, and L) and **b** monocrystalline quartz, feldspar, and total lithic fragment (Qm, F, and Lt). Data from Spalletti et al. (2008) is shown in order to differentiate the provenance in the Aconcagua region from the southern Neuquén basin

3.2 Petrographic Analysis

Medium- to coarse-grained sandstones samples were collected for petrographic analysis. To determine the provenance of the deposits, thin sections were made from three samples of each stratigraphic section. Grains were counted using the Gazzi-Dickinson method (Ingersoll et al. 1984). All samples are classified as litharenites according to Folk et al. (1970). Point count data of the sandy matrix from Sec. A and sandstones from Sec. B are plotted on QmFLt and QtFL ternary diagrams, indicating a recycled-orogen provenance trending to transitional arc (Dickinson et al. 1983) (Fig. 6). Data from Spalletti et al. (2008) were plotted in order to compare the provenance of the Aconcagua region with the southern sector of the basin, where the diagrams show a pronounced arc provenance (Fig. 6b). In both sections, lithic fragments are dominant and are discriminated in acid and basic volcanic rocks, granitoids, polycrystalline quartz, limestones, and fine sandstones, attributed to various sources. In both sections, acid volcanic fragments are the most abundant fraction.

3.3 Thickness Variations

In the Aconcagua fold-and-thrust belt, a westward thickness increment in the Tordillo Formation can be observed, across the several thrusts sheets (Cegarra and Ramos 1996). In particular, at the Blanco River valley, the second thrust sheet exposes a section (Sec. B, 280 m) that exceeds in more than four times the thickness of the undeformed sector section (Sec. A, 62 m). In comparison, in the fourth thrust sheet at the Las Cuevas River valley, this unit consists of 1200 m of lavas and volcaniclastic

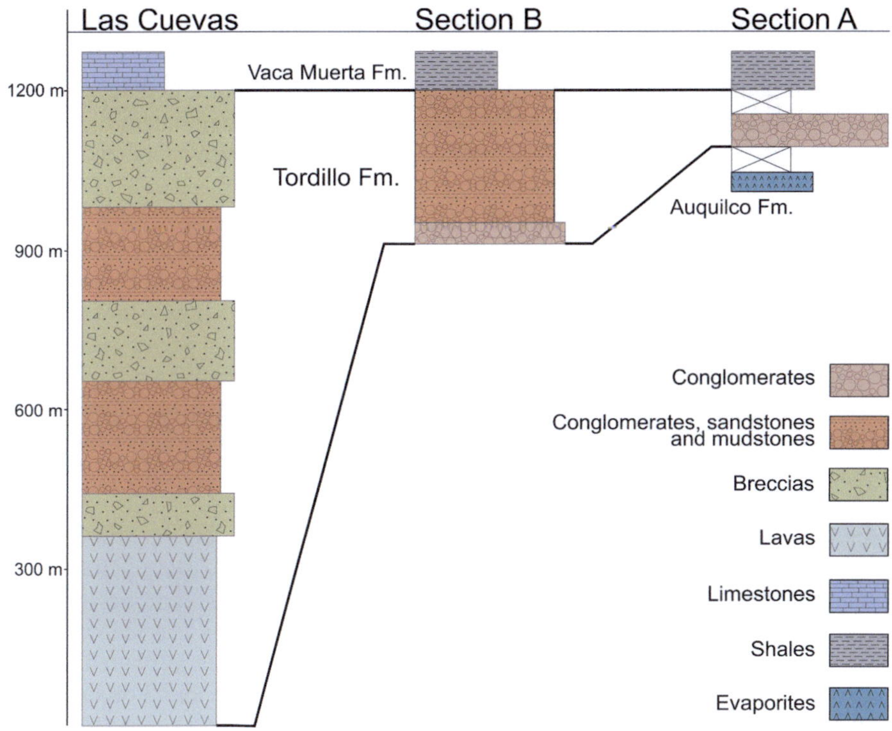

Fig. 7 Schematic sections of the Tordillo Formation across the Blanco River (Secs. A and B) and Las Cuevas River valleys, showing their compositional and thickness lateral variations. Las Cuevas section was taken from Sanguinetti and Cegarra (1991)

deposits that grade to alluvial and fluvial deposits (Sanguinetti and Cegarra 1991) (Fig. 7).

In order to qualitatively visualize the basin geometry during the deposition of the Tordillo Formation through the Aconcagua fold-and-thrust belt, thickness data from various authors was compiled (Lo Forte 1996; Sanguinetti and Cegarra 1991; Suárez 2000; Rocha 2000; Orts 2001). As a result, a thickness map shows a westward constant increase in thickness (Fig. 2).

4 Discussion

The Tordillo stratigraphic sections analyzed at the Blanco River valley are interpreted according to Miall (1977, 2014). The FA 1 from Sec. A represents in-channel deposits identified as longitudinal bars. In Section B, the FA 2 is interpreted as in-channel deposits recognized as channel lag, gravy longitudinal bars, and sandy transversal bars; the FA 3 represents an alternation of debris flows and sandy transversal bars; the

FA 4 is interpreted as floodplain deposits, where the sand levels represent a planar bed flow and the mudstones the waning of the flood; and in the FA 5, the mudstones and laminated sandstones represent the floodplain deposits, and the sand and conglomerates lenses are interpreted as channels, where the conglomerate channels contain channel-lag and longitudinal bar deposits and the sandy channels contain transversal bars.

The stratigraphic Sec. A is interpreted as a braided fluvial system, while the Sec. B presents alternations between braided (FA 2 and FA 3) and anastomosed (FA 4 and FA 5) fluvial systems. These facies variations observed between the analyzed sections indicate differences in the depositional environment and the accommodation space. The conglomeratic Sec. A does not present fine deposits that would correspond to preserved alluvial plains, which suggests a reduced accommodation space. On the other hand, the alternation of FAs in Sec. B is interpreted as variations in the subsidence rate, through which when increasing leads to the deposition of the finer deposits of the anastomosed system and when decreasing allows the progradation of the clastic wedge represented by the braided fluvial systems. The abundance of intraclasts could represent a constant migration of channels. This variations in the accommodation space suggests a higher subsidence rate in the site of the Sec. B during its deposition.

In the petrographic analysis of the Tordillo Formation sands and matrix of the conglomeratic levels, lithic fragments appear as dominant and are attributed to various sources. Acid volcanics and granitoids clasts are attributed to the Choiyoi Group, while basic volcanic fragments represent a supply from the Rio Damas Formation located to the west or from the basic volcanism of the Choiyoi Group. Polycrystalline quartz fragments are assigned to the Alto Tupungato turbidites deposits. Limestone clasts and blocks are attributed to the Oxfordian carbonate deposits of La Manga Formation. The fine sandstone fragments are interpreted as intraclasts based on the similar clast composition.

In both sections, acid volcanic fragments are the most abundant fraction, indicating that the main sediment source corresponds to the Choiyoi Group. The limestone fragments are practically not represented in the sandy fraction, but are abundant in the conglomerates of Sec. B, pointing out to a sediment source close to these deposits. Paleocurrents measured in Sec. B indicate that this source was located to de E-SE. This suggests that the mentioned units were exposed in the eastern margin of the basin presumably at a half-graben shoulder of the depocenter, explaining the increase in limestone fragments and the size of the clasts.

As known, thermal subsidence during the post-rift period decreases exponentially (Royden and Keen 1980). However, subsidence analysis in the Aconcagua region shows rapid tectonic subsidence between 160 and 125 Ma that includes the time of the deposition of the Tordillo Formation and is interpreted as a renewed rifting stage (Balgord 2016) (see Chapter "Controls on Deposition of the Tordillo Formation in Southern Mendoza (34°–36° S): implications for the Kimmeridgian tectonic setting of the Neuquén Basin"). Furthermore, the abrupt thickness changes across the basin are not consistent with a thermal subsidence stage.

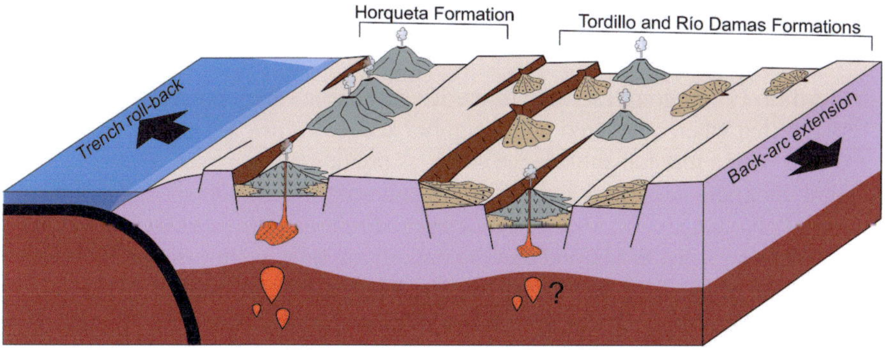

Fig. 8 Inferred extensional Kimmeridgian stage between 32° 30′ S and 33°30′ S. Trench roll back at a time of continental eastward retraction would have favored extensional conditions in the back-arc, where the Tordillo and Río Damas Formations were deposited. The arc front during this time is represented by the Horqueta formation presently exposed on the Coastal Cordillera. An eastern volcanic belt interfingers with the back-arc sedimentation (Rossel et al. 2014)

Additionally, given that the sedimentation of the Tordillo clastics represents a local low stand occurred during a global high stand (Legarreta and Uliana 1996; Haq 2018), an extensional control in the sedimentation could explain the development of depocenters isostatically uplifted, with differential subsidence rates between the different half-grabens (Fig. 8). This could explain the facies, provenance, and thickness variations described at the Blanco River, with Sec. A located on the rift shoulders, where low to none accommodation space is generated. Under this scheme, Sec. B would be located in the hanging wall of a normal fault that would have exposed the underlying units and where the subsidence rate would have been comparatively higher.

Finally, these inferred extensional conditions are in agreement with presently available geodynamical models that propose that during the Kimmeridgian-Tithonian (?), the subduction system presented oblique convergence due to the SE movement of the Phoenix oceanic plate relative to the southern Gondwana trench (Seton et al. 2012; Müller et al. 2016), which is a favorable condition for transtensional reactivation of previous crustal anisotropies. Additionally, these plate reconstructions indicate that west Gondwana moved away from the trench, suggesting that trench rollback could have also favored extensional conditions in the back-arc (see Chapter "Controls on Deposition of the Tordillo Formation in Southern Mendoza (34°–36° S): implications for the Kimmeridgian tectonic setting of the Neuquén Basin").

5 Conclusions

The Tordillo Formation described in the Blanco River valley presents several characteristics that when compared with neighbor sections allow suggesting that it is not consistent with a thermal subsidence stage, as previously proposed. Facies distribution and thickness variations show a significant accommodation space increment to the west. Provenance analysis and paleocurrent directions indicate that sediment supply was located to the E-SE, and that the underlying units were exhumed at the time of deposition of the Tordillo Formation. The thickness of this unit increases rapidly to the west, passing in 15 km from a few meters to more than 1000 m, suggesting that the sedimentation could have been regulated by a steep topography controlled by NNW structures. These features support an extensional setting for the deposition of the Tordillo Formation in agreement with other proposals for the Malargüe fold-and-thrust belt to the south. In addition, available plate tectonic reconstructions suggest trench rollback and highly oblique convergence, at a time of eastward retraction of Gondwana during the Kimmeridgian, which is favorable to back-arc extension at that time.

Acknowledgements This study has been funded by PICT-2016-12252, PIP 11220150100426 and UBACYT 20020150100166BA. This is the R-314 contribution of the Instituto de Estudios Andinos "Don Pablo Groeber".

References

Alvarez P (1996) Los depósitos triásicos y jurásicos de la Alta Cordillera de San Juan. In: Ramos VA (ed) Geología de la región del Aconcagua, provincias de San Juan y Mendoza. Dirección Nacional del Servicio Geológico, Subsecretaría de Minería de la Nación, Buenos Aires, pp 59–137

Balgord EA (2016) Triassic to Neogene Evolution of the Andean Retroarc: Neuquén Basin, Argentina. Ph.D. thesis, University of Arizona

Cegarra MI, Ramos VA (1996) La faja plegada y corrida del Aconcagua. In: Ramos VA (ed) Geología de la región del Aconcagua, provincias de San Juan y Mendoza. Dirección Nacional del Servicio Geológico, Subsecretaría de Minería de la Nación, Buenos Aires, pp 387–422

Charrier R, Pinto L, Rodríguez MP (2007) Tectonostratigraphic evolution of the Andean Orogen in Chile. In: Moreno T, Gibbons W (eds) The geology of Chile. Geol Soc Spec Pub, London, pp 21–114

Dickinson WR, Beard LS, Brakenridge GR et al (1983) Provenance of North American Phanerozoic sandstones in relation to tectonic setting. Geol Soc Am Bull 94(2):222–235

Folk RL, Andrews PB, Lewis DW (1970) Detrital sedimentary rock classification and nomenclature for use in New Zeland. J Geol Geophys 4(13):937–968

Gana P, Tosdal RM (1996) Geocronología U-Pb y K-Ar en intrusivos del Paleozoico y Mesozoico de la Cordillera de la Costa, Región de Valparaíso, Chile. Andean Geol 23(2):151–164

Groeber P (1918) Edad y extensión de las estructuras de la Cordillera entre San Juan y Nahuel Huapi. Physis 4:208–240

Haq BU (2018) Jurassic sea-level variations: a reappraisal. Geol Soc Am Today 28(1):4–10

Heredia N, Farias P, García-Sansegundo J et al (2012) The basement of the Andean Frontal Cordillera in the Cordón del Plata (Mendoza, Argentina): geodynamic evolution. Andean Geol 39(2):242–257

Howell JA, Schwarz E, Spalletti LA et al (2005) The Neuquén Basin: an overview. In: Veiga GD, Spalletti LA, Howell JA et al (eds) The Neuquén Basin: a case study in sequence stratigraphy and basin dynamics. Geol Soc, London (252), pp 1–14

Ingersoll RV, Bullard TF, Ford RL et al (1984) The effect of grain size on detrital modes: a test of the Gazzi-Dickinson point-counting method. J Sediment Res 54(1):103–116

Kietzmann DA, Palma RM, Riccardi AC et al (2014) Sedimentology and sequence stratigraphy of a Tithonian-Valanginian carbonate ramp (Vaca Muerta Formation): A misunderstood exceptional source rock in the Southern Mendoza area of the Neuquén Basin, Argentina. Sediment Geol 302:64–86

Klohn C (1960) Geología de la Cordillera de los Andes de Chile Central, Provincias de Santiago, O'Higgins, Colchagua y Curicó. Investigaciones Geológicas, Bol 8, Santiago, p 95

Legarreta L, Gulisano C (1989) Análisis estratigráfico secuencial de la Cuenca Neuquina (Triásico superior-Terciario inferior). In: Chelbi G, Spalletti L (eds) Cuencas sedimentarias argentinas. Universidad de Tucumán, Serie Correlación Geológica, vol 6, p 221–243

Legarreta L, Uliana MA (1991) Jurassic-Cretaceous marine oscillations and geometry of back-arc basin fill, central Argentine Andes. In: McDonald D (ed) Sedimentation, Tectonics and Eustacy. Int Ass Sediment SP, vol 12, p 429–450

Legarreta L, Uliana MA (1996) The Jurassic succession in west-central Argentina: stratal patterns, sequences and paleogeographic evolution. Palaeogeogr Palaeocl 120(3–4):303–330

Lo Forte GL (1996) Los depósitos jurásicos de la Alta Cordillera de Mendoza. In: Ramos, VA (ed) Geología de la región del Aconcagua, provincias de San Juan y Mendoza. Dirección Nacional del Servicio Geológico, Subsecretaría de Minería de la Nación, Buenos Aires, pp 179–230

Mescua JF, Giambiagi LB, Bechis F (2008) Evidencias de tectónica extensional en el Jurásico Tardío (Kimeridgiano) del suroeste de la provincia de Mendoza. Rev Asoc Geol Argentina 63(4):512–519

Miall AD (1977) Lithofacies types and vertical profile models in braided river deposits: a summary. In: Miall AD (ed) Fluvial sedimentology. Canadian Soc Petrol Geol Mem 5, pp 597–604

Miall A (2014) The facies and architecture of fluvial systems. In Fluvial Depositional Systems. Pringer Geology. Springer, Cham, pp 9–68

Müller MD, Seton M, Sahirovic S, et al (2016) Ocean basin evolution and global-scale plate reorganization events since Pangea breakup. Ann Rev Earth Planetary Sci 44:107–138

Naipauer M, García Morabito E, Marques JC et al (2012) Intraplate Late Jurassic deformation and exhumation in western central Argentina: constraints from surface data and U-Pb detrital zircon ages. Tectonophysics 524–525(1):59–75

Naipauer M, Tapia F, Mescua J (2015) Detrital and volcanic zircon U-Pb ages from southern Mendoza (Argentina): An insight on the source regions in the northern part of the Neuquén Basin. J S Am Earth Sci 64:434–451

Naipauer (2016) Edades U-Pb en circones detríticos de la Formación Tordillo en la Cuenca Neuquina, centro-oeste de Argentina: implicancias en la edad absoluta del límite Jurásico-Cretácico. Acad Nac Cs Ex Fís y Nat, Anales 68:73–84

Negro C (2002) Geología del Cordón Chorrillos, Alto Tupungato, Mendoza. Degree thesis, Universidad de Buenos Aires

Orts S (2001) Geología del sector comprendido entre el río Plomo y Vacas, provincia de Mendoza. Degree thesis, Universidad de Buenos Aires

Pérez DJ, Ramos VA (1996) El basamento prejurásico. In: Ramos VA, Aguirre-Urreta M, Alvarez P (eds) Geología de la Región del Aconcagua, Provincias de San Juan y Mendoza. Dirección Nacional del Servicio Geológico, Subsecretaría de Minería de la Nación, Buenos Aires, pp 27–58

Polanski J (1972) Descripción geológica de la Hoja 24 ab (Cerro Tupungato), provincia de Mendoza, Dirección Nacional de Geología y Minería, Buenos Aires, Bol 128, p 110

Ramos VA (1996) Geología de la región del Aconcagua: provincias de San Juan y Mendoza, República Argentina. Dirección Nacional del Servicio Geológico, Subsecretaría de Minería de la Nación
Ramos VA, Aguirre-Urreta M, Alvarez P (2010) Cerro Tupungato - Hoja Geológica 3369-III. Serv Geológico Min Argentino, Buenos Aires, p 133
Rivano S, Sepúlveda P, Boric R, Espiñe et al (1993) Hojas Quillota y Portillo, V Región. Servicio Nacional de Geología y Minería. Carta Geológica de Chile 73, Santiago
Rocha E (2000) Geología del sector medio del Río Plomo, Mendoza. Degree thesis, Universidad de Buenos Aires
Rossel P, Oliveros V, Ducea (2013) The Early Andean subduction system as an analog to island arcs: evidence from across-arc geochemical variations in northern Chile. Lithos 179:211–230
Rossel P, Oliveros V, Mescua J (2014) The Upper Jurassic volcanism of the Río Damas-Tordillo Formation (33–35.5 S): insights on petrogenesis, chronology, provenance and tectonic implications. Andean Geol 41(3):529–557
Royden L, Keen CE (1980) Rifting process and thermal evolution of the continental margin of eastern Canada determined from subsidence curves. Earth Planet Sci Lett 51(2):343–361
Sanguinetti AS (1989) Volcanismo neojurásico-neocomiano de la Quebrada de Vargas, Alta Cordillera de Mendoza. Rev Asoc Geol Argent 44(1–4):381–393
Sanguinetti AS, Cegarra MI (1991) El volcanismo Jurásico superior al este de Las Cuevas, Cordillera Principal de Mendoza, Argentina. In: Abstracts of the 6 Congreso Geológico Chileno, Viña del Mar, pp 5–9
Seton M, Muller RD, Zahirovic S et al (2012) Global continental and ocean basin reconstructions since 200 Ma. Earth Sci Rev 113:212–270
Spalletti LA, Queralt I, Matheos SD et al (2008) Sedimentary petrology and geochemistry of siliciclastic rocks from the upper Jurassic Tordillo Formation (Neuquén Basin, western Argentina): implications for provenance and tectonic setting. J S Am Earth Sci 25(4):440–463
Stipanicic PN (1966) El Jurásico de la Vega de la Veranada (Neuquén), el Oxfordense y el diastrofismo divesiano (Agassiz-Yaila) en Argentina. Rev Asoc Geol Argent 20(4):403–478
Suárez F (2000) Geología entre los arroyos Potrero Escondido y Chorrillos, Mendoza. Degree thesis, Universidad de Buenos Aires
Vergani GD, Tankard J, Belotti J (1995) Tectonic evolution and paleogeography of the Neuquén Basin, Argentina. In: Tankard AJ, Suárez R., Welsink HJ (eds) Petroleum Basins of South America. AAPG Mem 62:383–402
Vergara M, Levi B, Nyström JO et al (1995) Jurassic and Early Cretaceous island arc volcanism, extension, and subsidence in the Coast Range of central Chile. Geol Soc Am Bull 107(12):1427–1440
Vicente JC, Leanza HA (2009) El frente de corrimiento Andino al nivel de los cerros Penitentes y Visera (Alta Cordillera de Mendoza): Aspectos cronológicos y cartográficos. Rev Asoc Geol Argent 65:97–110
Vitarella G (2000) Geología del valle superior del arroyo Chorrillos. Degree thesis, Universidad de Buenos Aires
Yrigoyen M (1976) Observaciones geológicas alrededor del Aconcagua. In: Abstracts of the 1 Congreso Geológico Chileno, pp 2–7

Magnetostratigraphy of the Jurassic Through Lower Cretaceous in the Neuquén Basin

María Paula Iglesia Llanos and Diego A. Kietzmann

Abstract The first magnetostratigraphic scales for the Jurassic through Early Cretaceous from the Southern Hemisphere have been constructed over the last decades from marine sections in the Neuquén Basin. Paleomagnetic sites were tied to ammonite zones in order to achieve well-refined ages of studied sections. Diverse field tests for the paleomagnetic stability proved the primary origin of isolated magnetizations. In the case of Upper Jurassic–Lower Cretaceous studies, magnetostratigraphic and biostratigraphic data were combined with cyclostratigraphy. Finally, polarities were tied to Andean ammonite zones and from their correlation with the standard zones, calibrated to the GTS2016 (Geomagnetic Polarity Time Scale 2016). For the Early Jurassic, a composite magnetostratigraphic scale was derived out of five sections spanning the Hettangian–Toarcian. The magnetostratigraphic scale portrays 16 reverse (Jr1–Jr16) and 16 normal (Jn1–Jn16) polarity zones that encompass at least 19 ammonite zones. A major difference between both scales rises in the Hettangian involving the Jr1–Jr3 polarity zones. For the Middle Jurassic, the resultant magnetostratigraphy obtained in the Lajas Formation outlines a dominantly reverse polarity pattern. According to the correlation with the GTS2016, the studied section is assigned to the Lower-uppermost Middle Bathonian (Chrons M41 through M39). For the Late Jurassic–Early Cretaceous, the magnetostratigraphic scale obtained in the Vaca Muerta Formation comprises Subchrons M22r.2r through M15r, spanning the *V. andesensis* (Lower Tithonian)–*S. damesi* Zones (Upper Berriasian). The use of diverse chronostratigraphic tools such as biostratigraphy, magnetostratigraphy and cyclostratigraphy, enabled to determine with unprecedented precision the position of the Jurassic–Cretaceous boundary, as well as to assess durations of ammonite zones

M. P. Iglesia Llanos (✉) · D. A. Kietzmann
Facultad de Ciencias Exactas y Naturales, Departamento de Ciencias Geológicas, Universidad de Buenos Aires, Ciudad Universitaria, Pabellón II, Intendente Güiraldes 2160, C1428EHA Ciudad Autónoma de Buenos Aires, Argentina
e-mail: mpiglesia@gl.fcen.uba.ar

CONICET—Universidad de Buenos Aires, Instituto de Geociencias Básicas, Ambientales y Aplicadas de Buenos Aires (IGeBA), Buenos Aires, Argentina

D. Kietzmann and A. Folguera (eds.), *Opening and Closure of the Neuquén Basin in the Southern Andes*, Springer Earth System Sciences,
https://doi.org/10.1007/978-3-030-29680-3_8

Keywords Jurassic · Early Cretaceous · Paleomagnetism · Magnetostratigraphy · Ammonite zones · Cyclostratigraphy · Polarity zones · Jurassic–Cretaceous boundary

1 Introduction

Dating and correlation of marine fossil-bearing Mesozoic sections in the Neuquén Basin have been relatively straightforward due to the abundant content of particularly, ammonites, since duration of a biozone would be of ~1 myr (Westermann 1988). Nonetheless, dating and correlation can become a challenging task when such diagnostic fossils are not found, as in the case of transitional/continental sections or during times of high faunal endemism.

Magnetostratigraphy can be an efficient tool in helping refining the age of the stratigraphic column. Only a few of such studies were performed in Mesozoic sections of the Neuquén Basin, that involve the Lower Jurassic (Iglesia Llanos 1997; Iglesia Llanos and Riccardi 2000; Iglesia Llanos et al. 2006; Iglesia Llanos 2012), Middle Jurassic (Iglesia Llanos et al. in preparation) and Upper Jurassic–Lower Cretaceous (Iglesia Llanos et al. 2017; Kohan Martínez et al. 2018). In this chapter, we present an updates integration of all the magnetostratigraphic data available for this time span in the Neuquén Basin.

The sampled Lower Jurassic sections crop out along the Atuel River (southern Mendoza) and near Chos Malal (northern Neuquén). The entire succession bears ammonites and is thus well-dated, except for some barren intervals. The oldest Lower Jurassic sampled section is Arroyo Malo (Hettangian–Sinemurian) to the westernmost part of the Rio Atuel. Here, ammonites enabled to date most of the sampling horizons besides those of the lower and the upper part. To the east and stratigraphically upwards, Las Chilcas (Sinemurian–lowermost Pliensbachian) contains sampling levels tied to ammonite zones only in the upper part of the section. More to the east at Puesto Araya the sampled Pliensbachian section bears ammonites only in the middle part. To the south in the Neuquén province at Rajapalo–Chacay Melehue (Pliensbachian–Toarcian), the sampling horizons tied to ammonites were those from the lower and upper parts of the section due to the extended coeval extrusion of basaltic flows in the middle portion, particularly affecting the *Tenuicostatum* Zone.

The paleomagnetic study performed in the Middle Jurassic was carried out in the Lajas Formation cropping out along Covunco creek in central Neuquén. The Lajas Formation at this locality is interpreted to represent a river-dominated delta. It bears neither diagnostic fossils nor radiometric ages, and the age of this succession is assigned to the Upper Bajocian–Middle Bathonian based on sequence stratigraphy.

The study of the Upper Jurassic–Lower Cretaceous has been performed in the ammonite-bearing Vaca Muerta Formation, in a section that crops out along the Arroyo Loncoche (southern Mendoza). Ages in this succession are, for the most part, well precised, and the use of cyclostratigraphy, enabled to assess the duration of ammonite zones (Kietzmann et al. 2015, 2018) (see Chap. "Orbital Controls and High-Resolution Cyclostratigraphy of Late Jurassic–Early Cretaceous in the Neuquén Basin"). However, the Jurassic–Cretaceous boundary was not precisely determined yet, compelling the introduction of a magnetostratigraphy tied to the ammonite zones. The definition of Jurassic–Cretaceous boundary represents a drawback at a global scale, since the Cretaceous is the only system/period of the Phanerozoic that so far, has not been defined by a basal boundary stratotype. This is mainly due to the lack of significant evolutionary or physical/chemical event on either at regional or global scale within the commonly used boundary interval, coupled with the difficulty of correlating any biostratigraphic datum due to the pronounced provincialism of marine fauna (Ogg and Hinnov 2012; Ogg et al. 2016). After long years of debate, the Berriasian Working Group has concluded the boundary is marked by the explosion of *Calpionella alpina*, which is placed in the middle of Subchron M19n.2n.

Magnetostratigraphic data are independent of lithology, facies or sections locations. This is due to the fact that reversals of the Earth's magnetic field occur at a global scale and are recorded at the same time in any point of the world. Thereby, magnetostratigraphy constitutes a relevant tool to date and correlate sections/intervals, as well as placing the J-K boundary. It enables correlation of rock strata of diverse depositional and faunal realms as well as the assignment of geologic ages to anomalies of marine magnetic intensities. For Middle Jurassic to present, magnetic anomalies of the ocean floor with their calibrations to biostratigraphy serve as a template against which magnetic polarities isolated, either on-land or in deep-sea sections, can be determined (Ogg and Hinnov 2012). Accordingly, magnetostratigraphy affords the employment of the reference Geomagnetic Polarity Time Scale (GTS) to the high-resolution correlation of marine magnetic anomalies, and the use of the calibrated reference pattern of polarity changes to correlate polarity zones among sections (Ogg and Hinnov 2012). From present to the Kimmeridgian (Chron M25r), polarities from the GTS are directly derived from sea-surface marine magnetic anomalies, whereas from the Kimmeridgian to the Bajocian (Chrons M27 to M44), polarities are modeled from deep-tow surveys. In the latter case, modeling of magnetic polarities is rather complicated and the uncertainty in this time interval is increased by the fact that very high rate of reversals takes place at this time. Because of this, the recent GTS2016 (Ogg et al. 2016) shows two magnetostratigraphic scales for the Middle Jurassic, one obtained from deep-tow surveys and the other from sections on-land (southern Spain). For pre-Bajocian times (170 My), no marine anomalies are available, and therefore, the GTS has been constructed from sections on-land located in Europe and North America, with polarities tied to biostratigraphic zones (Ogg and Hinnov 2012; Ogg et al. 2016).

Polarity successions in the Neuquén Basin were tied to the GTS with ammonite biostratigraphy determined by A.C. Riccardi, following correlations in Riccardi (2008, 2015). Once polarities were tied to ammonite zones, they were correlated with the GPTS standing on their correspondence with the standard zones. Henceforth, a precise dating of particular sampling intervals and correlation among sections has been attained.

2 Lower Jurassic: Geology and Paleomagnetic Sampling

2.1 Hettangian-Sinemurian

Studied sections of this age crop out in Arroyo Malo and Las Chilcas, located along the northern margin of the Atuel River (Fig. 1). Successions are made up of 2 km thick of fine-grained sedimentary and volcanic rocks that become coarser and younger to the east. Deposits in this area are part of the synrift facies that resulted from fault-controlled deltas. To the west, the Arroyo Malo section is made up (Fig. 2) of 1 km thick of fine-grained sedimentary rocks mainly and subordinate volcanic rocks. From the base, the succession comprises the Arroyo Malo Formation, El Freno Formation and El Cholo (=Puesto Araya) Formation (Riccardi et al. 1991, 2004). The Arroyo Malo Formation bears ammonites from the Upper Triassic (Riccardi and Iglesia Llanos 1999), the El Cholo Formation contains six ammonite zones spanning the Lower Hettangian–Lower Sinemurian: *K. bayoensis, D. reissi, S. peruvianus, B. Canadensis, "Vermiceras" and Coronociceras-Arnioceras* (Fig. 2). At least three coeval sills of alkaline composition were recognized and sampled, although polarities were not considered for the magnetostratigraphy. In the El Cholo Formation, flattened ammonites and concretions as well as sedimentary structures prove that the

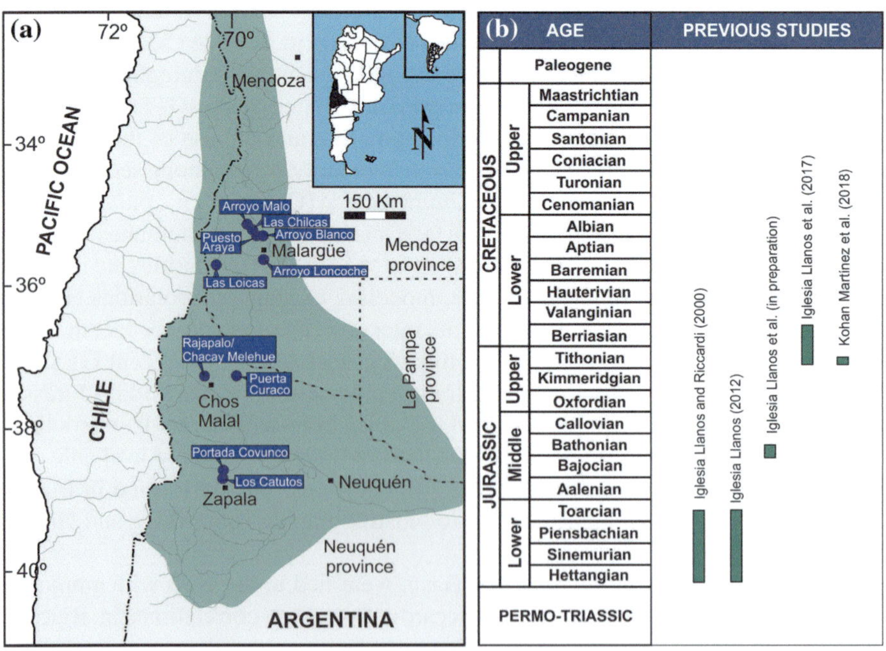

Fig. 1 Sketch map of the Neuquén Basin. Sampled sections are: Lower Jurassic: Arroyo Malo, Las Chilcas, Puesto Araya, Arroyo Blanco, Rajapalo–Chacay Melehue, Middle Jurassic: Portada Covunco, Upper Jurassic–Lower Cretaceous: Arroyo Loncoche, Los Catutos

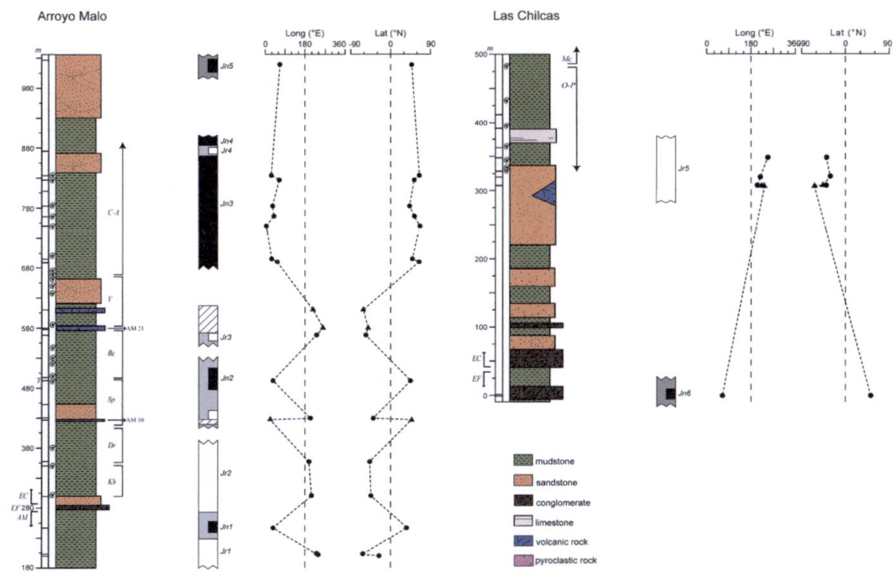

Fig. 2 Litholog, paleomagnetic and ammonite sites, ammonite zones, magnetostratigraphy and virtual geomagnetic poles (VGP) from Arroyo Malo and Las Chilcas (Hettangian–Sinemurian) sections. In the magnetoestratigraphic scale, black: normal polarity (*Jn*), white: reverse polarity (*Jr*). Half-column within gray intervals corresponds to polarities isolated from a single site. *AM:* Arroyo Malo Formation, *EF:* El Freno Formation, *EC:* El Cholo Formation. Ammonite zones: *Kb* (*K. bayoensis*), *Dr* (*D. reissi*), *Sp* (*S. peruvianus*), *Bc* (*B. Canadensis*), *V* ("*Vermiceras*"), *C-A* (*Coroniceras–Arnioceras*), *O-P* (*Ortechioceras–Paltechioceras*), *Mc* (*M. chilcaense*)

sequence was subjected to considerable compaction. Measurements of the axis of fossils indicate compaction values of around 66% with respect to the same undeformed ammonites from northern Chile (Iglesia Llanos and Riccardi 2000).

Sampling horizons or sites in these sections were distributed according to lithologies and biostratigraphic levels, with a mean separation well within the range of ammonite biozones. Two hand samples on average were obtained at each site that was oriented with Brunton and sun compasses. In Arroyo Malo (Fig. 2), 25 sites were distributed over c. 1200 m thick, and in Las Chilcas, nine sites over 500 m thick.

2.2 *Pliensbachian–Toarcian*

Sections of this age crop out along the Atuel River in southern Mendoza (Puesto Araya and Arroyo Blanco) and northern Neuquén (Rajapalo, Chacay Melehue) (Fig. 1). From oldest to youngest, the first section crops out in Puesto Araya and is made up from base to top, by c. 400 m thick of coarse-grained sandstones and conglomerates that belong to the El Freno Formation and finer-grained sandstones of El Cholo Formation (Fig. 3). The latter unit bears ammonites (e.g., Volkheimer

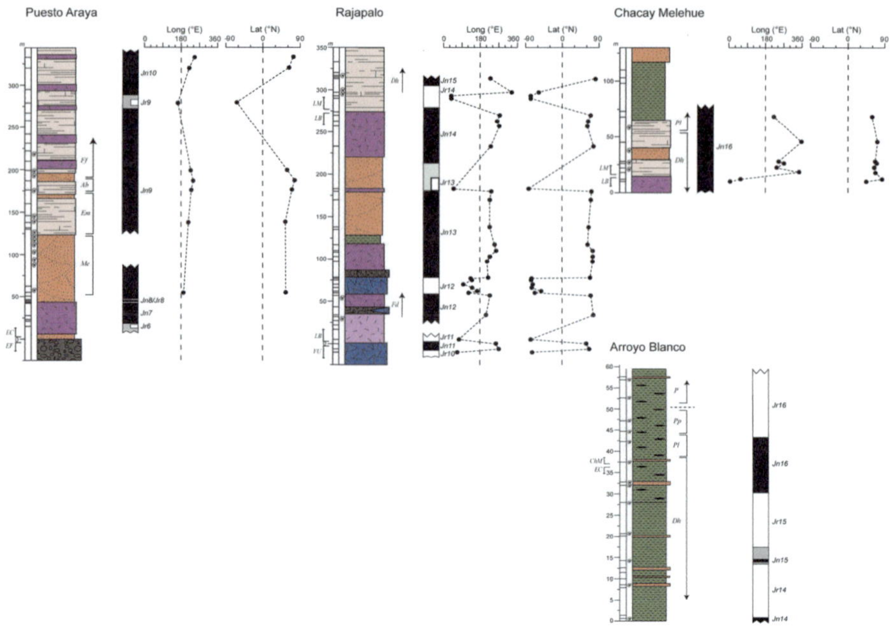

Fig. 3 Litholog, paleomagnetic and ammonite sites, ammonite zones, magnetostratigraphy and virtual geomagnetic poles (VGP) from Puesto Araya, Rajapalo–Chacay Melehue (Pliensbachian–Toarcian) and Arroyo Blanco (polarities only). *VU* volcanic unit, *LB* Lista Blanca Fm., *LM* Los Molles Fm., *ChM* China Muerta Fm. Ammonite zones: *Me* (*M. externum*), *Em* (*E. meridianus*), *Ab* (*A. behrendseni*), *Ff* (*F. fannini*), *Fd* (*F. disciforme*), *T* (*Tenuicostatum*), *Dh* (*D. hoelderi*), *Pl* (*P. largaense*), *Pp* (*P. pacificum*), *Cc* (*C. chilensis*), *P* (*Phymatoceras*). Symbols as in Fig. 2

1978; Riccardi 1983; Hillebrandt 1987) that are assigned to *M. externum*, *E. meridianus*, *A. behrendseni* and *F. fannini* Zones (Early to early Late Pliensbachian). The Rajapalo–Chacay Melehue section (Fig. 3) is 500 m thick and is composed of a volcanic basement, on top of which overlay (Damborenea 1987; Gulisano and Gutiérrez Pleimling 1995) the Lista Blanca (= La Primavera) and Los Molles formations. The succession is made up of volcanic, sedimentary and pyroclastic rocks that correspond to off-shore facies. Ammonites belong to the *F. disciforme*, *D. hoelderi* and *P. largaense* Zones that indicate Late Pliensbachian–Early Toarcian ages (Iglesia Llanos 1997, 2012; Iglesia Llanos and Riccardi 2000). The last section is located in the area of the Atuel River and crops out at Arroyo Blanco. The section is 55 m thick and is made up (Fig. 3) by the El Cholo and China Muerta Formations. Ammonites correspond to the *D. hoelderi*, *P. largaense*, *P. pacificum* and *Plymatoceras* Zones (Early to Late Toarcian).

Paleomagnetic horizons were distributed in such a manner that comprised all represented ammonite zones (Fig. 3). At least two hand samples were collected and oriented with Brunton and sun compasses. In Puesto Araya, 28 sites were placed over c. 380 m thick, in Rajapalo 41 sites over 325 m thick, in Chacay Melehue 12 sites over 125 m thick, and in Arroyo Blanco 20 sites over c. 60 m thick (in Iglesia Llanos 2012).

2.3 Paleomagnetic Analysis and Field Tests

Hand samples were drilled to obtain at least two oriented cores. From each core, two standard specimens were sliced making up a total of 1400 specimens.

Demagnetization procedures were performed at the Laboratorio de Paleomagnetismo D. Valencio, IGEBA, University of Buenos Aires. Demagnetization methods involved increasing alternating fields (AF) and high temperatures (TH), carried out using a 2G AF and a TD 48 Schonstedt oven, respectively. Residual remanent magnetizations were measured mostly in a 2G cryogenic although Spinner magnetometers—Schonstedt, Digico and AGICO JR6 were also employed.

The overall paleomagnetic analysis showed the systematic occurrence of two magnetic components. The softer component was erased at either <250 °C or <15 mT, bearing mostly northern declinations with negative inclinations (Fig. 4a). The other component was isolated between 250 and 550 °C or 15 and 50 mT. When corrected with the paleohorizontal, directions yielded NE (SW) declinations with negative (positive) inclinations (Fig. 4b). The latter is interpreted to represent the characteristic remanent magnetization (ChRM) that showed two main paleomagnetic behaviors: (a) straight trajectories to the origin and (b) curved trajectories defining great circles. Finally, less than 5% of the specimens revealed unstable or unreliable behaviors and were discarded. Components of specimens' type (a) were calculated with principal component analysis—PCA—(Kirschvink 1980) using IAPD (Torsvik 1992) and SuperIAPD (Torsvik et al. 2000) paleomagnetic programs. Magnetic directions with this behavior served as reference directions for the calculation of remagnetization circles. Final mean directions (fmd) for this type of behavior were calculated using Fisherian statistics (Fisher 1953). Specimens' type (b) with curved trajectories was analyzed using remagnetization circles (Halls, 1976; McFadden and McElhinny 1988). Fmd were calculated with the CIRDI program (Mena 1994) that combines remagnetization circles with reference directions. For the purpose of these studies, we considered sites with $N \geq 3$, $\alpha_{95} = \leq 15°$ and $K \geq 10$.

Fmd in the Hettangian–Sinemurian sections were corrected to restore magnetic vectors to the paleohorizontal (inclination shallowing effect, King 1955). The tectonic and compaction-corrected mean direction from the ChRM—reverse polarity—is Decl $= 230°$, $58°$, $\alpha_{95} = 4.5°$, $K = 41$, $N = 25$ (Fig. 4d). In Arroyo Malo and Las Chilcas, the ChRM passed the statistical fold test (McFadden 1990) indicating that the magnetization is pretectonic acquired before the Andean deformation took place. The characteristic magnetization also passed the statistical reversal test as Rc (McFadden and McElhinny 1990) in Arroyo Malo, and the same test was "indeterminate" (Ro) in Las Chilcas due to the small number in the normal directions. In addition, it passed the baked contact test performed in one of the sills in Arroyo

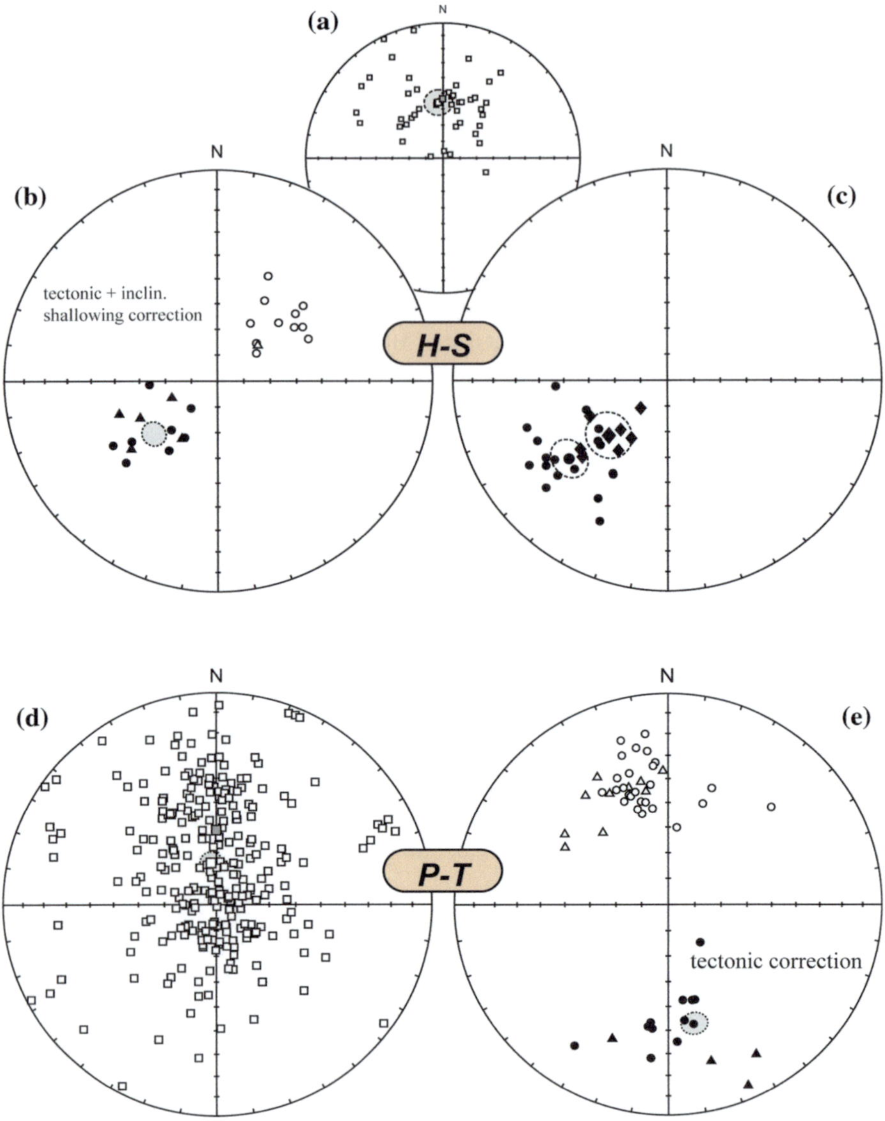

Fig. 4 Stereoplots with Lower Jurassic final mean directions. **H-S**: Hettangian–Sinemurian sections with **a** secondary in situ directions that coincides (Dec = 355°, Inc = −56°, α_{95} = 8°) with the present-day local field, **b** ChRM with tectonic and shallowing inclination corrections and **c** ChRM with tectonic correction only (in reverse polarity) shows two statically distinguishable groups, one with high inclinations in sills (diamonds) and the other with lower inclinations in sedimentary rocks (circles). Symbols: full (open) symbols = lower (upper) hemisphere; circles (triangles) = directions at Arroyo Malo (Las Chilcas); gray square = direction of present-day local magnetic field; gray circle = 95% confidence circle of ChRM. **P-T**: Pliensbachian–Toarcian sections with **d** secondary in situ component coincides with (Dec = 353, Inc = −74, α_{95} = 4.5°) the expected present-day local field, **e** ChRM with tectonic correction. Symbols: circles (triangles) = directions at Rajapalo–Chacay Melehue (Puesto Araya)

Malo, which indicates that the intrusion of the sills was unable to fully reset the ChRM (Iglesia Llanos 1997, 2012; Iglesia Llanos and Riccardi 2000).

Fmd in the Pliensbachian–Toarcian once corrected to the paleohorizontal is (Fig. 4e)—reverse polarity—Decl = 168.5°, Incl = 41, $\alpha_{95} = 4.5°$, $K = 19.1$, $N = 52$, remarkably different from that obtained in the Hettangian–Sinemurian (Fig. 4b) sections (Iglesia Llanos and Riccardi 2000, Iglesia Llanos 2012, Iglesia Llanos and Prezzi 2013). In Rajapalo–Chacay Melehue, the ChRM passed both the statistical fold test (McFadden 1990) and reversal test (McFadden and McElhinny 1990) as Rc. In Puesto Araya, the same reversal test was indeterminate due to the scarce number of data. In Rajapalo–Chacay Melehue, the ChRM passed the relevant conglomerate test, which demonstrates that the clasts' host rocks were not remagnetized after the formation of the conglomerate. Therefore, based on the field tests for the paleomagnetic stability, paleomagnetic behaviors and microscopical observations, the ChRM, isolated in all five sections of Early Jurassic age, is interpreted as primary corresponding to the moment of the deposition of sediments or extrusion of the lava flows/sills (Iglesia Llanos 1997, 2012; Iglesia Llanos and Riccardi 2000).

Based on demagnetization curves and in the case of the Pliensbachian–Toarcian, microscopic studies performed in basaltic lava flows, it is interpreted that the ChRM was carried by titanomagnetite and secondly, titanohematite.

The soft component, on the other hand, shows a mean in situ direction that is coincident with the dipolar local field (Fig. 4a, d), and consequently, it is interpreted as a remagnetization of the dipolar field.

2.4 Magnetostratigraphy

The first magnetostratigraphic scale for the Lower Jurassic in the Southern Hemisphere was constructed in 1994 and progressively refined over the years (e.g., Iglesia Llanos and Vizán 1995; Iglesia Llanos 1997, 2009, 2012; Iglesia Llanos and Riccardi 2000).

Sampling sites with normal (*Jn*) and reverse (*Jr*) polarities were tied to ammonite zones determined by Riccardi (in Iglesia Llanos 1997, 2012) and correlated with the standard zones according to Riccardi (2008) (Fig. 5). Polarities isolated from a single site are shown in Figs. 2 and 3 with half-column within a gray interval, in order to remark boundaries uncertainties. A small number of sites produced no polarity due to the occurrence of the sills such as in the *S. peruvianus* and "*Vermiceras*" Zones in the Hettangian.

Polarity zones were tied to ammonite zones in each section in order to construct a composite magnetostratigraphic scale (Fig. 6), that was correlated with the GTS2016 (Ogg et al. 2016). The correlation is, in general, very consistent considering the important intervals with sparse or null paleomagnetic/biostratigraphic sampling, except for the Hettangian. Despite the fact that paleomagnetic sampling sites include all ammonite zones, the correlation with the GPTS needs some caution, like the case of polarity zones *Jn5* and *Jn6* placed amid a rather extended unsampled interval. All

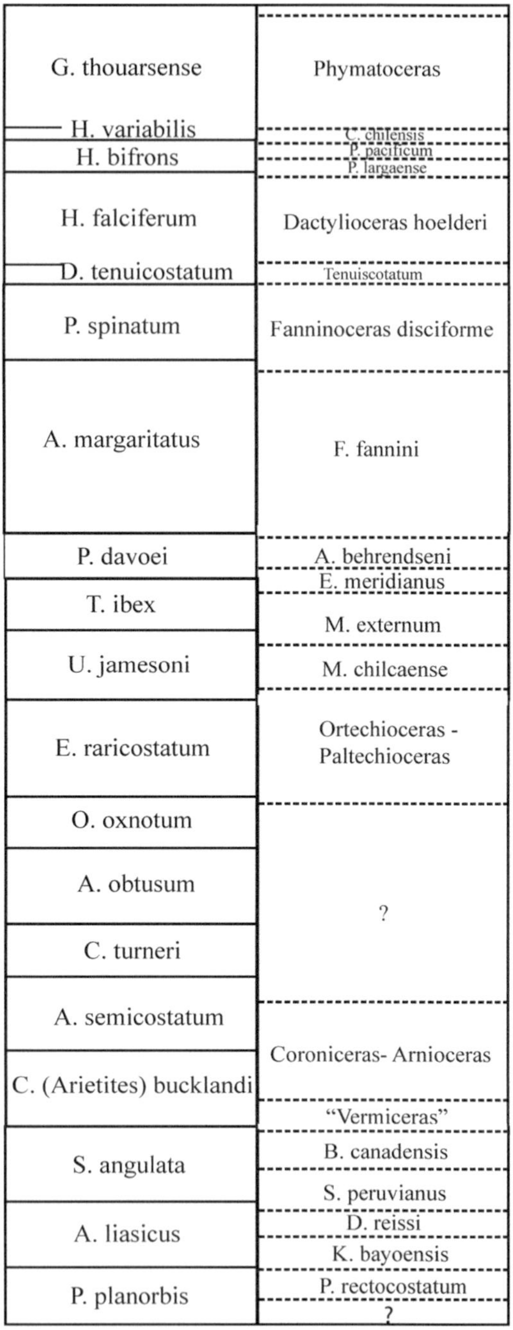

Fig. 5 Proposed correlation between the Andean (right) and standard ammonite zones for the Lower Jurassic according to Riccardi (2008)

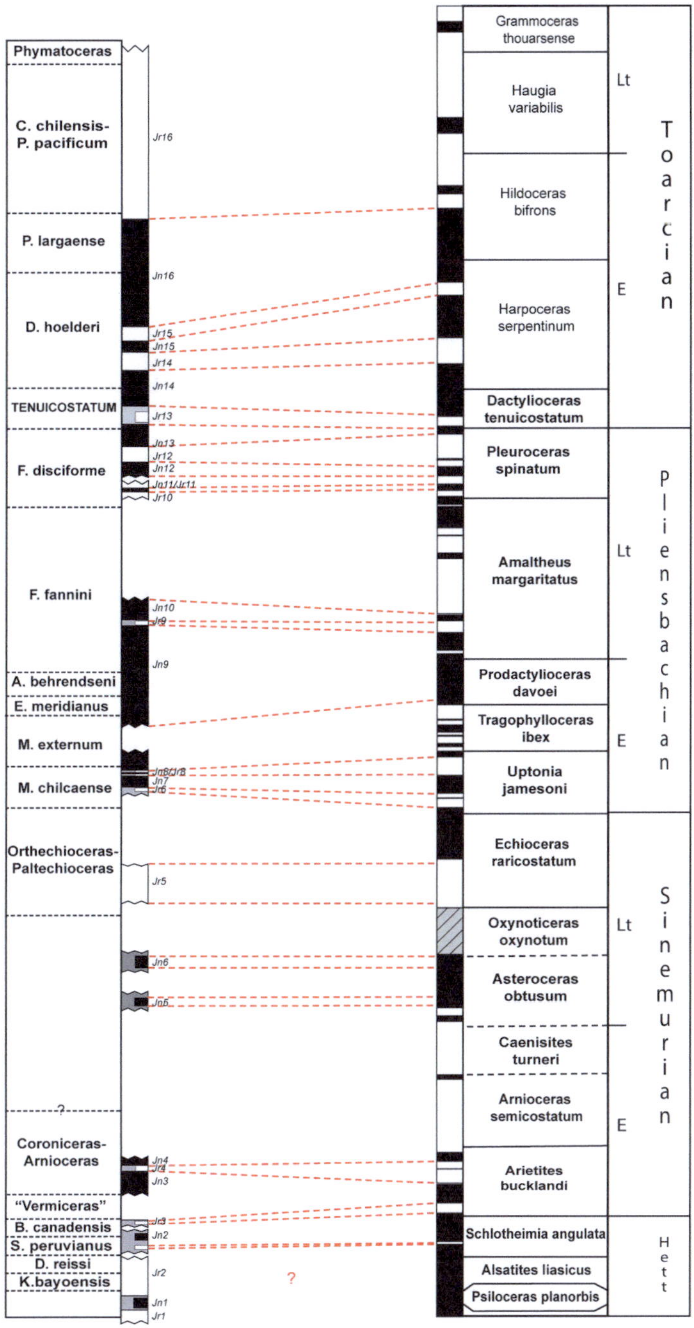

Fig. 6 Bio-magnetostratigraphic correlation between the local scale (left) and the GTS2016 (right). *Jn*: normal polarity zone, *Jr*: reverse polarity zone

the same, magnetostratigraphy in the Lower Jurassic enabled a more precise dating of sampled levels, such as the identification of the *Tenuicostatum* Zone (Fig. 3) in the Toarcian that was not recorded by fossils.

2.5 Lower Jurassic: Discussion

2.5.1 Hettangian–Sinemurian

The most remarkable difference in the correlation between both magnetostratigraphic scales occurs at the base of the Hettangian (Fig. 6). In the studied sections, this interval comprises the Andean *P. rectocostatum* Zone (*Jr1* to *Jr2*) equivalent to the *P. planorbis* Standard Zone, *K. bayoensis* (*Jr2*) correlatable with the lower *A. liasicus* Standard Zone, *D. reissi* (*Jr2*) correlatable with the middle *A. liasicus* Standard Zone, *S. peruvianus* Zone (*Jr2*, *Jn2*) equivalent to upper *A. liasicus* and *S. angulata* Standard Zones and finally, *B. canadensis* Zone (*Jn2*, *Jr3*) equivalent to upper *S. angulata* Standard Zone. A dominantly reverse polarity (*Jr1* to *Jr3*) was isolated within an interval where the GTS2016 shows normal polarity only. The Hettangian in the GTS has been controversial and in fact, has changed drastically over the years. Previously to the GTS2012 (*e.g.*, 2004, 2008), the Hettangian had been assigned normal polarity only at its base (*P. planorbis* Zone) followed by dominantly reverse in *A. liasicus* Zone through *S. angulate* Zone, and passing on to the Sinemurian (*C. bucklandi* Zone). Nonetheless, it is worth noting that after *P. planorbis* Zone, none of these polarities were tied to ammonites (Ogg 2004). Polarities of the GTS previous to 2012 had been compiled from sections in western Austria that characteristically present limited stratigraphic extent and contain sparse ammonites (Ogg and Hinnov 2012). Later, the GTS2012 showed the entire Hettangian and lower Sinemurian with normal polarity. The change was substantiated considering the magnetostratigraphies obtained at three localities: (1) Paris basin (Yang et al. 1995), (2) UK (Hounslow et al. 2004) and (3) eastern USA (Kent et al. 1995; Kent and Olsen 2008). In the Paris basin, polarities were tied to ammonite zones only in the Sinemurian. There, they found dominantly reverse in the Lower through Middle Sinemurian, followed by dominantly normal in the Upper Sinemurian (Yang et al. 1995). The Hettangian, on the other hand, bearing normal polarity, was dated with palynology and sequence stratigraphy. In UK, studies carried out at the St. Audrie's Bay ammonites were recorded only at the base of the Hettangian (*P. planorbis* Zone) bearing normal polarity (Hounslow et al. 2004). In USA, one of the studies was carried out in the Newark rift basin, and the isolated normal polarity was assigned to the lower Hettangian based on the occurrence of the palynofloral *Corollina meyeriana* Zone, radiometric ages in basalts and cyclostratigraphy (Kent et al 1995). The other magnetostratigraphy is from or the Hartford rift basin, and the recorded dominantly normal polarity with three minor reversals was assigned to the Hettangian based on palynology and vertebrate biostratigraphy (Kent and Olsen 2008). Accordingly, it is most likely that the lower Hettangian bears normal polarity (*P. planorbis* Zone). However, the normal polarity assigned in the GTS to the rest of the Hettangian (*A. liasicus* though *S. angulata* Zones) was never tied to ammonites. This, we consider, prevents a straightforward correlation between polarity zones in the Neuquén Basin tied to ammonites and the GTS2016.

Further up in the Sinemurian, the Andean "*Vermiceras*" Zone correlatable with the lower *A. bucklandi* Zone is shown with no polarity due to the intrusion of sills. The sampled levels within the *Coroniceras–Arnioceras* Zone assigned to polarities *Jn3*, *Jr4* and *Jn4* in Arroyo Malo, are interpreted to correlate with the *A. bucklandi* Zone, thus missing the upper *Coroniceras–Arnioceras* Zone. Upwards, *Jn5–Jn6* were tentatively correlated to the only extended normal polarity interval in the GPTS assigned to the *A. obtusum* Zone of Late Sinemurian age.

2.5.2 Pliensbachian–Toarcian

The best biostratigraphic and magnetostratigraphic correlations are achieved in this time interval due to the better definition of ammonite zones on the one hand, and a greater number of paleomagnetic sites, on the other (Fig. 6). The first sites are located at the base of the Puesto Araya section, below the *M. externum* Zone bearing reverse and normal polarities (*Jr6–Jr8*, *Jn7–Jn9*). The magnetostratigraphy allowed to place these sites within *M. chilcaense* Zone that correlates with the *U. jamesoni* Standard Zone. The following *M. externum* Zone was sampled only partially, and the magnetostratigraphy suggests that the corresponding normal polarity (*Jn9*) correlates with that of the *U. jamesoni–T. ibex* Zones boundary. Upwards, *E. meridianus* and *A. behrendseni* Zones bear also normal polarity (*Jn9*) that can correlate with that of the *P. davoei* Standard Zone. The following *F. fannini* Zone was sampled only partially according to the magnetostratigraphy, and isolated normal and reverse polarities (*Jn9–Jn10*, *Jr9*) from such interval are correlated with those of the lower *A. margaritatus* Zone, meaning that the upper half of *F. fannini* Zone is missing. *F. disciforme* Zone is represented in the Rajapalo section, and sampled sites bear reverse and normal polarities (*Jr10–Jr12*, *Jn11–Jn13*) that are correlated to those of *P. spinatum* Standard Zone. Upwards, reverse and normal polarities (*Jr13*, *Jn14*) allowed to identify the *Tenuicostatum* Zone which in the section, shows no fossils record (Fig. 3). The following *D. hoelderi* Zone is represented in Rajapalo–Chacay Melehue and mostly, in Arroyo Blanco. Sites bear reverse and normal polarities that correspond to *Jn14* represented mostly in Rajapalo, *Jr14–Jr16* (Arroyo Blanco), and *Jn15–Jn16* (Arroyo Blanco and Chacay Melehue), and correlate with those of *H. serpentinum* Standard Zone. Upwards, *P. largaense* Zone yields normal and reverse polarities (*Jn16*, *Jr16*). The magnetostratigraphy allowed to recognize that *P. largaense* Zone was sampled entirely only in Arroyo Blanco, and that is equivalent to the lower half of the *H. bifrons* Zone. Finally, the *P. pacificum–C. chilensis* and the lowermost upper Toarcian *Phymatoceras* Zones, bear reverse polarity (*Jr16*) correlated with the dominant reverse of the upper *H. bifrons* to *G. thouarsense* Standard Zones.

3 Middle Jurassic: Geology and Paleomagnetic Sampling

The paleomagnetic study was carried out in the Lajas Formation cropping out along the southern margin of Covunco creek (Portada Covunco section, Fig. 1), which is part of the Agrio fold and thrust belt. The Lajas Formation represents the regressive facies of the Cuyo Group. In southern Neuquén, the Cuyo Group starts in the Early Pliensbachian with a regional marine transgression, which is evidenced by the overlay of the shallow-water limestones of the Chachil Formation, and the deep marine deposits of the Los Molles Formation on the continental facies of the Precuyo Group, that straddles the Intraliassic (Gulisano et al. 1984) unconformity. This succession is followed by the deltaic, shallow-marine Lajas Formation (Early Bajocian–Early Callovian) and fluvial Challacó Formation (Bathonian–Early Callovian). The Lajas Formation is a well-known deltaic succession (Gulisano 1981; Gulisano et al. 1984), interpreted as tide-dominated deltas (Brandsaeter et al. 2005), wave dominated deltas (Spalletti 1995; Gugliotta et al. 2015) and more recently fluvial-dominated deltas (Zavala and González 2001; Canale et al. 2015; Kurcinka et al. 2017; Steel et al. 2018). Facies at Portada Covunco consist of fine to coarse sandstones and fine conglomerates, massive or with cross-stratification and current ripple lamination, interpreted to have been deposited in mouth bars of fluvial-dominated deltas (Canale et al. 2015).

The age of the Lajas Formation is diachronic from south to north. In the central and southern areas of the basin is assigned to the Upper Aalenian–Middle Bathonian from sequence stratigraphy and palynology and to the Callovian to the north (Westermann and Riccardi 1972; Riccardi and Westermann 1991; Gulisano et al. 1984; Riccardi and Gulisano 1992; Zavala 1993, 1996a, 1996b; García et al. 1994; Martínez et al. 2002; García et al. 2006; Zavala and González 2001; among others). The Portada Covunco section is interpreted as Upper Bajocian to Middle Bathonian based on sequence stratigraphy data (Zavala and González 2001). However, palynological studies indicate an Early Callovian age for the uppermost part of this unit (Martinez et al. 2002). Therefore, magnetostratigraphy can be an efficient tool in refining the age of the Lajas Formation in the Covunco section. Furthermore, Iglesia Llanos et al. (2019) found in the middle to upper part of Los Molles Formation which lies underneath, in a locality 20 km away from Arroyo Covunco, ammonites that represent the first record for Los Molles in that locality. These in situ ammonites indicate an Upper Bajocian age for the middle part of this unit. Hence, stratigraphic and paleontological information would aim to Upper Bajocian to Middle Bathonian ages for the Lajas Formation in this part of the basin.

The paleomagnetic study in the Lajas Formation comprised 29 sites distributed along the c. 290 m-thick section (Fig. 7). At each site, four oriented cores were drilled with a portable gasoline-powered drill.

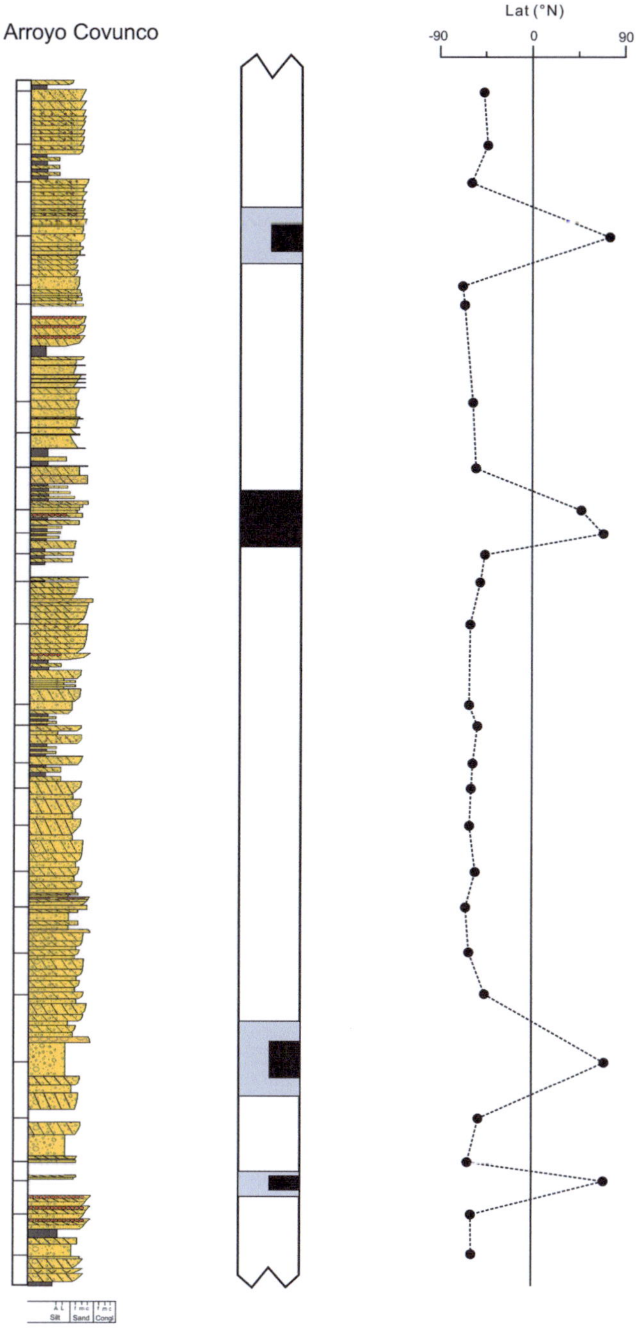

Fig. 7 Litholog, paleomagnetic sites, magnetostratigraphy and VGP in the Lajas Formation at Portada Covunco. Black: normal polarity, white: reverse polarity. Half-column within gray intervals correspond to polarities isolated from single sites (taken from Merino 2017)

3.1 Paleomagnetic Analysis and Field Tests

Cores were sliced into one to two standard specimens, yielding a total of 280 that were processed and analyzed.

Demagnetization procedures were carried out at the Laboratório de Paleomagnetismo do Departamento de Geofísica–Instituto de Astronomia, Geofísica e Ciências Atmosféricas, Universidad de Sao Paulo, Brasil, whereas rock magnetic studies were performed in these specimens also at the Laboratorio de Paleomagnetismo D. Valencio. Magnetic components were analyzed with the paleomagnetic software created in the Utrecht University.

The paleomagnetic analysis showed two systematic components. The softer component was erased at either <250 °C or <15 mT, bearing mostly northern declinations with negative inclinations (Fig. 8a). The harder component was isolated between 250 and 550 °C or 15 and 50 mT and when corrected to the paleohorizontal, directions yielded NE (SW) declinations with negative (positive) inclinations. The latter is interpreted to correspond to ChRM that also showed straight trajectories to the origin and curved trajectories defining great circles. Components with straight trajectories were calculated with PCA using the Paldir software (Utrecht University). Fmd were

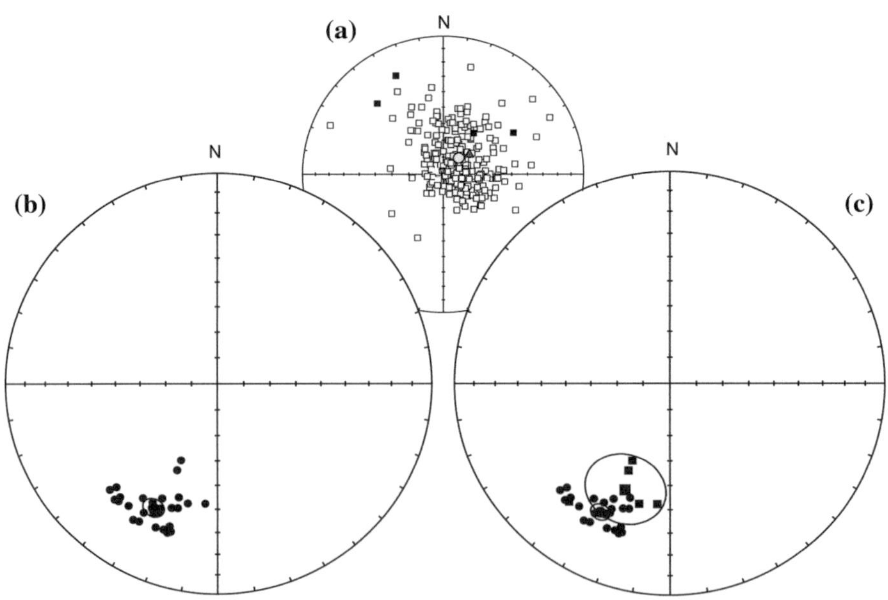

Fig. 8 Stereoplots with final mean directions of the Lajas Formation. **a** Secondary with tectonic directions that coincides (Decl. = 350.5°, Incl. = −59.4°, α_{95} = 2.5°) with the present-day local field, **b** ChRM with tectonic correction (reverse polarity) and **c** reversal test that proves that the normal and reverse polarity mean directions are statistically undistinguishable. Symbols: full (open) symbols = lower (upper) hemisphere; yellow circle = direction of present-day local magnetic field; red triangle = mean direction of the secondary component

calculated using Fisherian statistics with the Palfish software (Utrecht University). Specimens with curved trajectories were analyzed using remagnetization circles and fmd were calculated with the Palfit software (Utrecht University) that combines remagnetization circles with reference directions. Only fmd with $N \geq 3$, $\alpha_{95} = \leq 15°$ and $K \geq 10$ were considered. The calculated mean direction of the ChRM, when corrected to the paleohorizontal, is Decl = 207.2°, Incl = 34.0°, α_{95} = 3.8°, K = 52.5, N = 28—reverse polarity—(Fig. 8b). These mean normal and reverse polarity directions are statistically undistinguishable when calculated in the same hemisphere (Fig. 8c). They pass the statistical reversal test of McFadden and McElhinny (1990) as Rc. Therefore, normal and reverse directions are antiparallel and belong to the same population, and the fact that they come from the same stratigraphic section strongly suggests that the isolated magnetization is primary (Fig. 8c).

The soft component yields a mean direction that when corrected to the paleohorizontal, is coincident with the dipolar local field (Fig. 8a). This component is interpreted as a remagnetization likely acquired during formation of the fold and thrust belt produced by the circulation of fluids (Van der Voo and French 1977; Zegers et al. 2003).

Complementary rock magnetic studies were carried out to identify the origin of the magnetic carriers. Acquisition of isothermal remanence magnetization (IRM) curves performed in nine specimens from nine sites, show that they reach saturation at about 300 mT, which supports that the main carrier of the magnetic remanence is titanomagnetite (e.g., Heslop et al. 2002). Thermomagnetic curves, on the other hand, performed in six specimens of representative lithologies, indicate an increase in susceptibility near 540 °C in the high-temperature curves, which can be indicative of titanium-poor magnetite grains as the main carrier of the magnetic remanence. Thus, based on the demagnetization diagrams and rock magnetic studies, it is interpreted that the isolated primary magnetization is carried by titanomagnetite.

3.2 *Anisotropy of Magnetic Susceptibility*

The anisotropy of magnetic susceptibility (AMS) method records the preferred orientation of anisotropic magnetic minerals, i.e., paramagnetic, diamagnetic and ferromagnetic minerals that make up the rock. The susceptibility of anisotropic samples is a second-order tensor which is represented by three mean susceptibility axes (k_1 maximum, k_2 intermediate, k_3 minimum). AMS is used to determine petrofabric of rocks and in the case of sedimentary rocks, eventually, directions of paleocurrents. Sedimentary rocks generally show three main petrofabrics (e.g., Hrouda 1982; Rochette et al. 1992; Borradaile and Henry 1997): (i) primary or sedimentary, (ii) secondary or tectonic and (iii) mixed, that result from the superposition of (i) and (ii). In the case of non-deformed primary sedimentary (i), the susceptibility ellipsoid is strongly oblate, with the minimum k_3 axes orthogonal to the bedding plane, a distinct magnetic foliation and a poor or absent magnetic lineation. Directions of paleocurrents are given by the clustering of the k_1 axis (e.g., Borradaile and Henry

1997; Lanza and Meloni 2006). At Arroyo Covunco, clustering of the k_1 axis and imbrication of the k_3 axis indicate that paleocurrents flow from the SW to the NE. This is fully consistent with paleocurrent directions measured in coarser-grained lithologies in the same section.

3.3 Magnetostratigraphy

In the GTS, the Bajocian through Bathonian features a time interval in which the Earth's magnetic field changed polarity at a very fast rate (Chrons M45 to M26), so fast that a gap is shown within Chron M39 with no defined ("mixed") polarities (Ogg and Hinnov 2012; Ogg et al. 2016). In addition, there is no easy availability of oceanic crust of this age. For this reason, there are two magnetostratigraphic scales for this time lapse, one that is derived from deep-tow magnetic data on the relative fast-spreading Japanese lineations, and another composite scale made up of polarities obtained on-land and calibrated with ammonites, nannofossils in the Sub-Mediterranean and Sub-Boreal realms. The two scales are slightly different, in a manner that the "oceanic" is made up of dominantly reverse with many thin normal polarities distributed in a relatively regular fashion. The "on-land" scale, on the other hand, even though reveals a high rate of reversals, shows a more distinct pattern of dominantly normal and dominantly reverse polarities. For instance, the base of the Lower Bajocian yields reverse, dominantly normal (upper M44–lower M43), and dominantly reverse polarities, whereas the "oceanic" scale in the same interval shows dominantly reverse polarity. Ogg and Hinnov (2012) emphasize that the results of direct modeling from deep-tow surveys for pre-M25r marine magnetic anomalies are dissimilar, depending on whether anomalies are projected onto near-surface or mid-depths. Moreover, the authors remark that on-land magnetostratigraphies for the Oxfordian through Bajocian are indeed consistent with an intermediate geophysical model. Hence, we ponder the "on-land" magnetostratigraphic scale from Spain as the reference in the GTS.

At Portada Covunco, VGP were calculated from mean directions (Fig. 7). Accordingly, four normal and five reverse polarity zones were identified in 29 sampling sites distributed along the c. 290 m-thick section. The section starts with a reverse polarity zone (sites 1 and 2), followed by a normal polarity zone (site 3), reverse (sites 4 and 5), normal (site 6), reverse (sites 7–18, c. 120 m.), normal (sites 19 and 20). Above, there are reverse (sites 21–25), normal (site 26) and reverse polarity zones (sites 27–29).

3.4 Middle Jurassic: Discussion

The magnetostratigraphy at Portada Covunco shows dominantly reverse polarity with thin normal intervals. These thin normal polarities are placed in the lower part (single sites), in the middle (two sites) and upper part (single site) of the section. When correlated with the GTS2016 (Fig. 9), the best fit achieved is with the Lower–Middle Bathonian (Chrons M41 through lower M39). This lapse bears dominantly reverse polarity that comprises two thin normal zones at the base with two other thicker normal zones at the top. Should we move down the GPTS into the Upper Bajocian or up in the Upper Bathonian, the pattern of polarities changes to dominantly normal (Chrons M42 and M39, respectively).

Accordingly, results point to an Early Bajocian age (base of Chron M41) for the first site in the Lajas Formation (Fig. 9). Upwards, normal polarities at sites 3 and 6 would correlate with the two normal within M41. This is followed by a large reverse interval that is correlated with the upper part of M41—lower half of Chron M40 that comprises the Early–Middle Bathonian boundary. The thicker normal zone at sites 19–20 is correlated with the lower M40 and thus, the uppermost normal polarity zone would be equivalent to the top of M40-base of M39. At the top, the reverse zone (sites 27–29) would be equivalent with the lower M39 assigned to the uppermost Middle Bathonian. This age agrees with that interpreted by Zavala and González (2001), as well as is consistent with the Upper Bajocian age assigned from ammonites in the underlying Los Molles Formation (Iglesia Llanos et al. 2019).

From the ages determined from magnetostratigraphy and stratigraphic thicknesses, minimum sedimentation rates were calculated. Results show two discernible rates, one from the base up to c. 60 m of 211 m/Ma and the other from 60 to 258 m of 415 m/Ma.

4 Upper Jurassic–Lower Cretaceous: Geology and Paleomagnetic Sampling

The Vaca Muerta Formation is a thick rhythmic alternation of dark bituminous mudstones, marlstones and limestones deposited as result of a rapid and widespread paleo-Pacific Early Tithonian marine transgression in the Neuquén Basin (Legarreta and Uliana 1991). The Vaca Muerta Formation is a first-class unconventional gas and oil reservoir. The studied section crops out along the Arroyo Loncoche in southern Mendoza (Fig. 1) and is made up of c. 280 m-thick ammonite-bearing marine rocks. In a decimeter scale, the unit shows rhythmic alternations of marlstones, and limestones that were intruded in the lower part, by a c. 20 m thick andesitic sill.

A thorough analysis of facies in the studied section is published in Kietzmann et al. (2008, 2011a, b, 2014). In general terms, four facies associations were characterized in this area (FA-1 to FA-4), that represent basinal to middle carbonate ramp deposits.

The Vaca Muerta Formation comprises the Jurassic/Cretaceous boundary (Leanza 1945, 1981), which so far, had not been precisely placed. Jurassic–Cretaceous Andean biostratigraphy is well defined by ammonites (Fig. 10) and secondly, by microfossils such as calcareous nannofossils (Bown and Concheyro 2004; Ballent

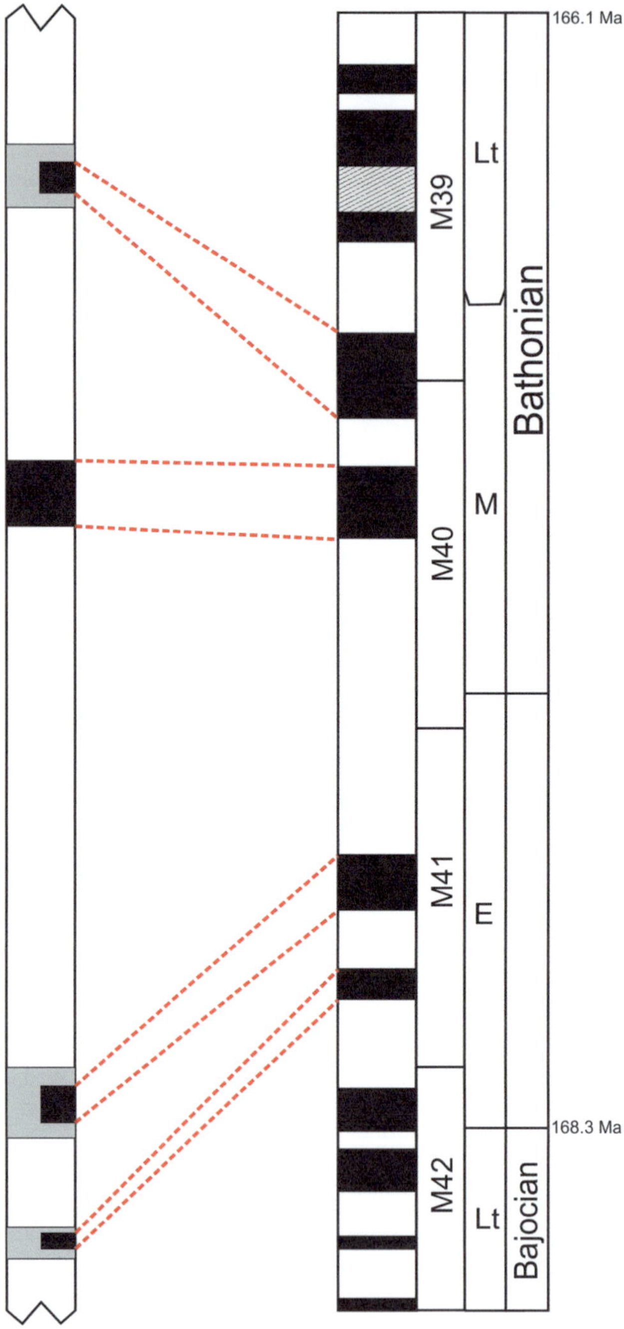

Fig. 9 Proposed magnetostratigraphic correlation between the local scale (left) and the GPTS16 (right)

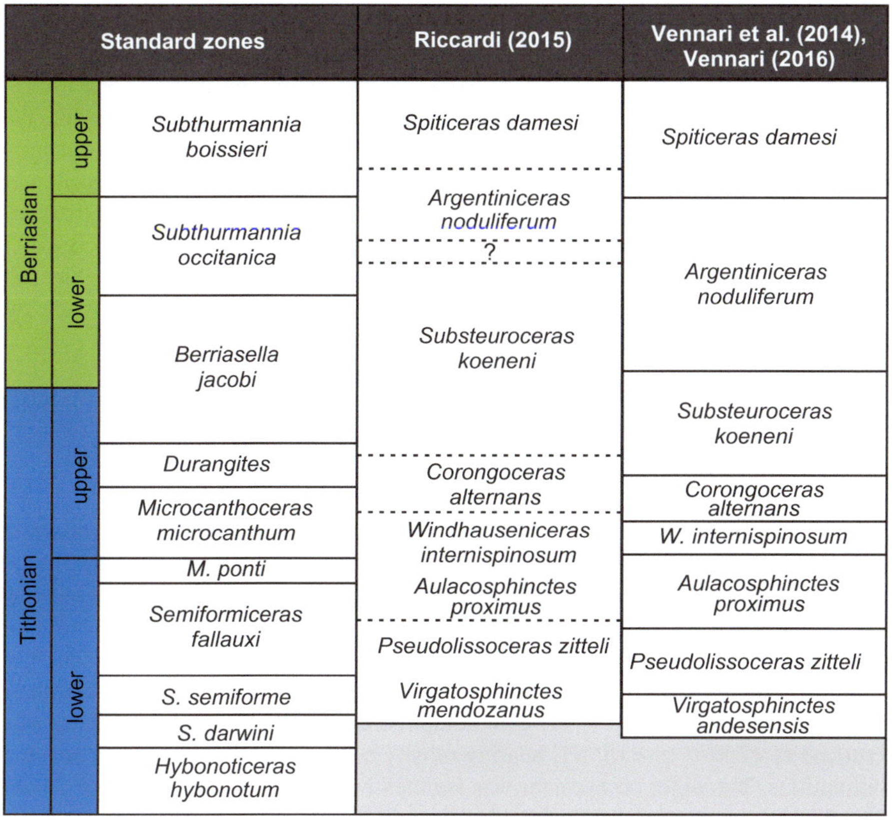

Fig. 10 Biostratigraphic subdivision of the Tithonian–Valanginian interval in the Neuquén Basin, showing correlation between Andean and standard Ammonite Zones according to Riccardi (2015) and Vennari et al. (2016)

et al. 2011). Ammonites from the studied section indicate that the Vaca Muerta Formation deposited from the Early Tithonian (*V. andesensis* Zone, ex *V. mendozanus*) through the *S. damesi* Zone assigned to the Late Berriasian (Kietzmann et al. 2011a, b, 2014, 2015).

Nannofossil events at Arroyo Loncoche were published by Lescano and Kietzmann (2010) and Kietzmann et al. (2011b). Calpionellids are generally poorly preserved due to recrystallization and are still poorly studied in the Neuquén Basin (Fernández Carmona and Riccardi 1999; Kietzmann 2017; López Martinez et al. 2017). Calpionellids from the Arroyo Loncoche section were described by Kietzmann (2017) who recognized the *Chitinoidella* and *Crassicollaria* Zones in the lower and middle part of the section. Additionally, eight calcareous dinoflagellate cyst zones were established by Ivanova and Kietzmann (2017) from the same section. A detailed cyclostratigraphic study was performed by Kietzmann et al. (2015) at Arroyo Loncoche that was recently revised in Kietzmann et al. (2018) (see Chap.

"Orbital Controls and High-Resolution Cyclostratigraphy of Late Jurassic–Early Cretaceous in the Neuquén Basin"). Vertical arrangement of cycles shows a well-ordered hierarchy of cycles within the Milankovitch band. The section contains 487 elementary cycles (~20.5 ky), grouped in 88 bundles (~100 ky) and 22 superbundles (~405 ky). Through the correlation of the resultant paleomagnetic and cyclostratigraphic data, it was possible to date the Vaca Muerta Formation with unprecedented precision (Iglesia Llanos et al. 2017) (see Chap. "Orbital Controls and High-Resolution Cyclostratigraphy of Late Jurassic–Early Cretaceous in the Neuquén Basin"). The paleomagnetic study involved 55 sampling horizons distributed along the 280 m-thick section (Fig. 11). On average, four oriented cores were drilled at each site with a portable gasoline-powered drill.

4.1 Paleomagnetic Analysis and Field Tests

From each core, at least two standard specimens were sliced, making up a total of c. 450 specimens that were processed and analyzed. Demagnetization procedures were performed at the Laboratorio de Paleomagnetismo D. Valencio, IGEBA, Universidad de Buenos Aires. Magnetic components were analyzed with the paleomagnetic software created in the Utrecht University. Microprobe and thin sections analysis were also employed in the identification of magnetic mineralogies. The paleomagnetic analysis showed two systematic components. The softer component was also removed at <250°C or <15 mT, bearing mostly northern declinations with negative inclinations. The other component was isolated between 250 and 550 °C or 15 and 50 mT. When corrected to the paleohorizontal, directions yielded NE (SW) declinations with negative (positive) inclinations (Fig. 12a). The latter is interpreted as the ChRM also shows straight trajectories to the origin and curved trajectories defining great circles. Components with straight trajectories were calculated with PCA using the Paldir software (Utrecht University). Fmd were calculated using fisherian statistics with the Palfish software (Utrecht University). Specimens with curved trajectories were analyzed using remagnetization circles and fmd were calculated with the Palfit software (Utrecht University) that combines remagnetization circles with reference directions. We considered sites with $N \geq 3$, $\alpha95 = \leq 15°$ and $k \geq 10$.

A thorough paleomagnetic analysis of the Vaca Muerta Formation is presented in Iglesia Llanos et al. (2017). The calculated mean direction of the ChRM, when corrected to the paleohorizontal in reverse polarity, is: Decl $= 196.4°$, Incl. $= 51.2°$, $\alpha_{95} = 2.8°$, $K = 48.4$, $N = 53$. The ChRM passed the reversal test and baked contact tests. In the first case, the overlap of the mean normal and reverse directions (Fig. 12a) shows that they are statistically undistinguishable, and the statistical test of McFadden and McElhinny (1990) classifies it as Rb. The result indicates that the normal and reverse directions are part of the same population and, taking into account that they belong to the same stratigraphic section, is a robust indicator that the isolated magnetization is primary. The baked contact test, on the other hand, was carried out where the sill intruded the host rock in the lower part of the section, both

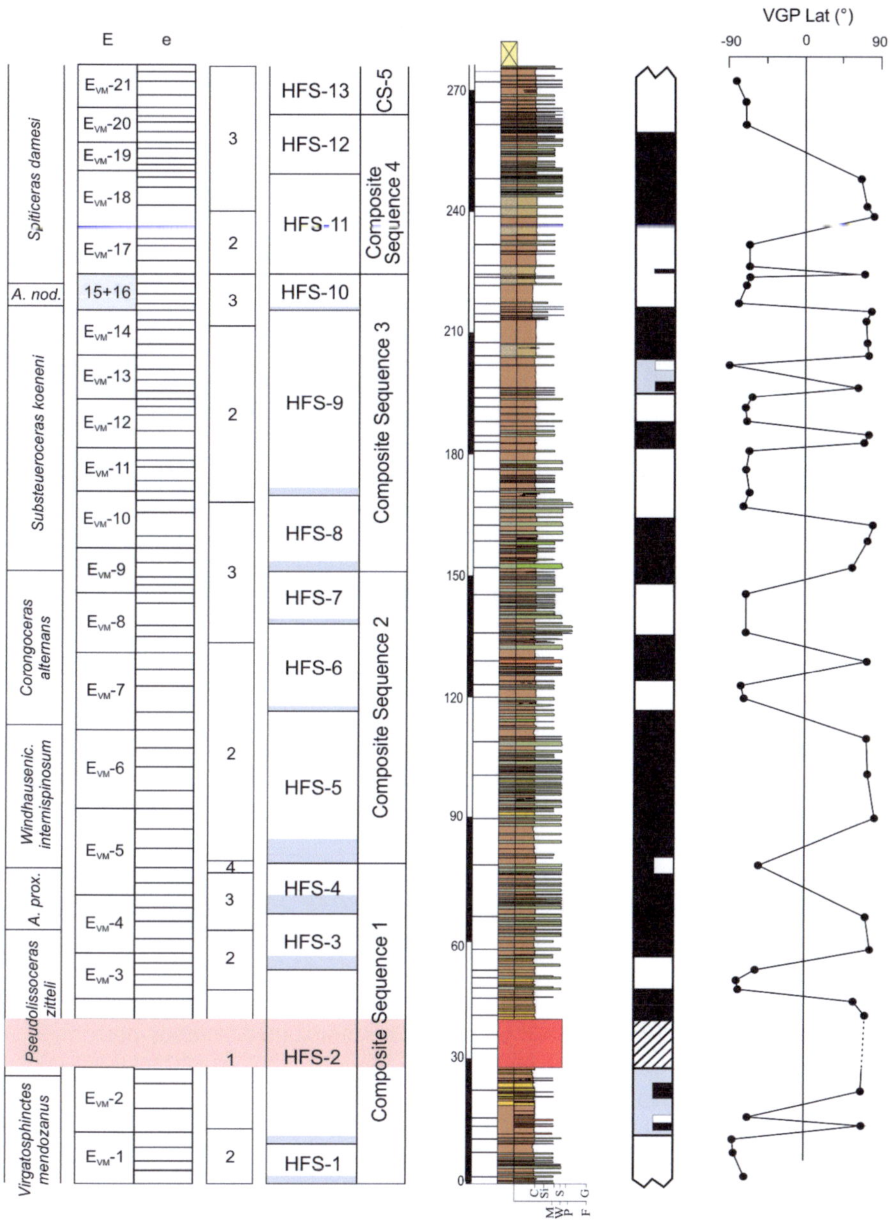

Fig. 11 Stratigraphic column of the Vaca Muerta Formation at Arroyo Loncoche. From left to right: Andean ammonite zones, low (E) and high (e) frequency eccentricity cycles (modified from Kietzmann et al. 2015), facies associations, sequence stratigraphy (Kietzmann et al. 2014), paleomagnetic horizons, polarities sequence and VGP latitudes. In facies associations 1: distal outer ramp to basin, 2: bioclastic outer ramp, 3: bioclastic middle ramp to proximal outer ramp, 4: Oyster autoparabiostrome dominated middle ramp. HFS ¼ high frequency sequence: gray-shaded bars are possible stratigraphic intervals where there may be condensation and omission of orbital cycles. Red in the stratigraphic column: sill

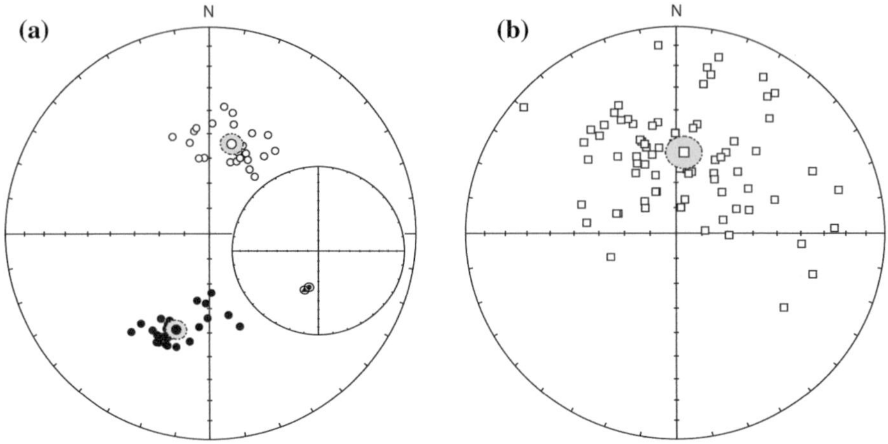

Fig. 12 Stereoplots final mean directions from the Vaca Muerta Formation **a** bedding-corrected mean site directions of the ChRM showing opposite polarities and antiparallel directions. Inset: bedding-corrected means of reverse and normal polarity populations (here shown in reverse polarity) with corresponding semi-angles of confidence. Note that both means are overlapped, indicating that they are statistically undistinguishable. These magnetizations are interpreted to be primary. **b** in situ directions from the soft component bearing northern declinations and negative inclinations yield. The mean coincides with the local dipolar field (= −55°) and is thus interpreted as a secondary magnetization produced by the local dipolar field. Symbols: full (open) symbols = lower (upper) hemisphere, gray circle = semi-angle of confidence

in the baked contact and far away from the intrusion. Since directions of ChRM pass the baked contact test, it indicates that the intrusions could not fully remagnetize the Jurassic–Cretaceous section.

The soft component yields a mean direction when considered in situ of: Decl = 4.5°, Incl. = −51.2°, $\alpha_{95} = 6.5°$, $K = 7.0$, $N = 78$ (Fig. 12b). This direction is fully coincident with that of the present local dipolar field (Dec = 0°, Inc = −55°) and thus is interpreted as a remagnetization produced most likely by the dipolar field.

Rock magnetic and petrographic studies were carried out in order to allow a better knowledge of the magnetic carriers. Thermomagnetic curves performed in cores from El Trapial show Curie temperatures between 570 and 580 °C that indicate that titanomagnetite is the main remanence carrier. In addition, these curves show the occurrence of pyrrhotite always associated with the sills. On the other hand, a thorough petrographical study was carried out in the type section of the Vaca Muerta Formation at Puerta Curacó (Neuquén). Accordingly, titanomagnetite was identified bearing anhedral, multi-domain-sized crystals that support single-domain crystals and secondly, pyrrhotite.

4.2 Anisotropy of Magnetic Susceptibility (AMS)

All specimens were subjected to AMS studies (Iglesia Llanos et al. 2017). The average magnetic susceptibility from sites (K_m) ranges from 8.4 × 10^{-5} to 3.1 × 10^{-6} SI, except in the sill where it amounts up to 4.3 × 10^{-3} SI (in Iglesia Llanos et al. 2017). The degree of anisotropy, P_j, is low and ranges from 1.156 (2 sites) to 1.004, F (foliation) from 1.102 to 1.002 and L (lineation) from 1.038 to 1.001. Shapes of the AMS ellipsoids are primarily oblate (AL20 to AL27 in the lower part of the section) to prolate-oblate with minor prolate or oblate shapes for the rest of the section. Shapes of the ellipsoids are in general, oblate with primary fabrics that vary from dominantly normal—minimum K_3 axes perpendicular to bedding plane—in the lower part of the section, to inverse—maximum K_1 axes perpendicular to bedding plane—in a few sites. The normal fabric is foliated and shows, when corrected to the horizontal, the K_3 axes very well grouped around the vertical, while the K_1 and K_2 axes are more or less dispersed along a girdle within the horizontal foliation plane. Inverse fabric, on the other hand, yields when corrected to the horizontal, well-grouped vertical K_1 axes with a strong lineation. Such inverse AMS fabrics are rather common in limestones and are suggestive of the occurrence of single-domain magnetite (Rochette 1988; Rochette et al. 1992).

4.3 Magnetostratigraphy

Virtual geomagnetic poles (VGP) calculated from fmd yielded 11 reverse and 10 normal polarity zones (Fig. 11). At c. 30 m from the base, the interval with no polarity (in red in the figure) indicates the position of the tertiary sill.

The equivalence between the Andean and Standard ammonite zones (Fig. 10) provided the reference to tie the magnetozones obtained at Arroyo Loncoche to the GTS2016 (Fig. 13). Results show an overall consistent correlation between both magnetostratigraphic scales, as well as it reveals changes in the sedimentation rates.

In order to tie the paleomagnetic floating scale to the GTS2016, Iglesia Llanos et al. (2017) followed the criteria in Kietzmann et al. (2015) using the base of the *W. internispinosum* Zone as the primary datum. Accordingly, the following paleomagnetic correlation was obtained (Fig. 13): from base to top, *V. andesensis* Zone comprises a set of reverse, normal, reverse and normal polarities. According to the biostratigraphic proposals, this zone is equivalent to the *S. semiforme* and *S. darwini* Standard Zones (Fig. 10). Would that be so, normal polarity only should be found (Fig. 13). The pattern of polarities from this interval correlates better to the upper part of *H. hybonotum* Standard Zone. The correlation between *V. andesensis* Zone and *H. hybonotum* Zone has been in fact, already suggested on the one hand, by Zeiss and Leanza (2010) based on ammonites and on the other, from calcareous dinoflagellate biostratigraphy (Ivanova and Kietzmann 2007). In this regard, cyclostratigraphic/paleomagnetic data from Kietzmann et al. (2015) questioned if *V.*

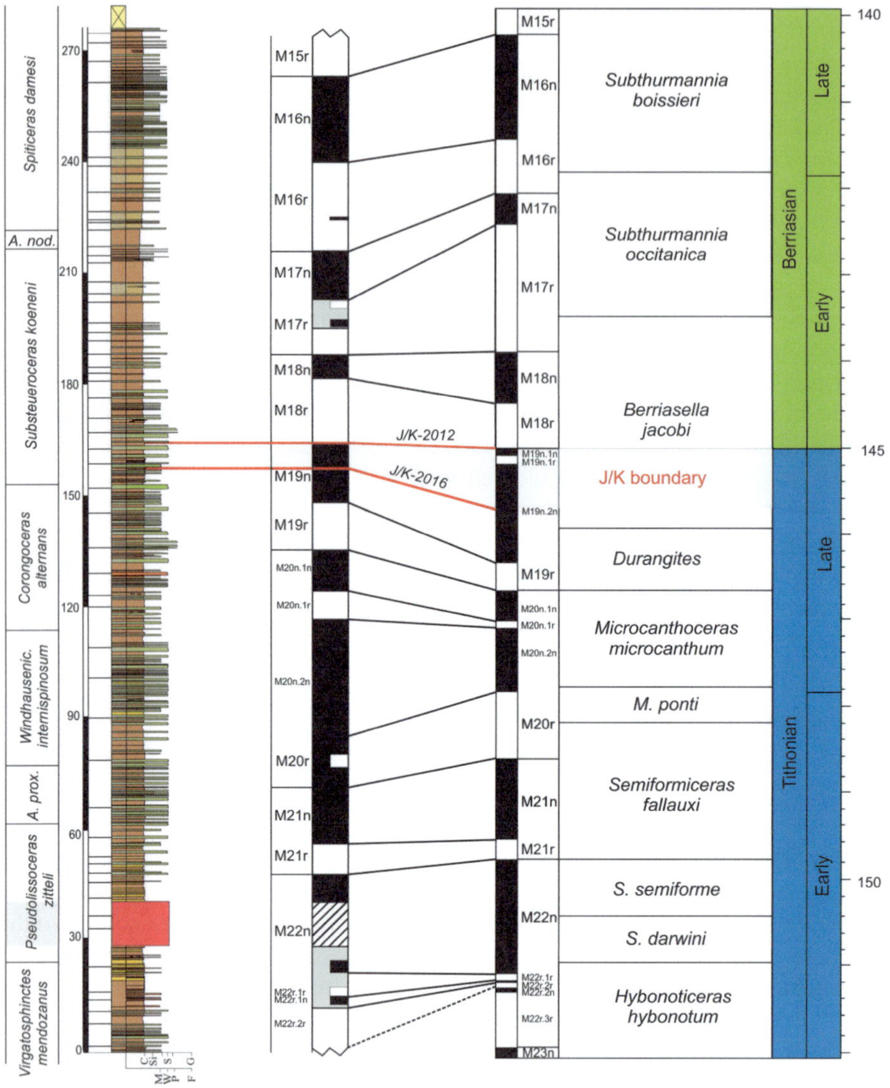

Fig. 13 Magnetostratigraphic correlation with the GTS2016 (Iglesia Llanos et al. 2017) tied to Andean ammonites and calibrated to standard zones. From left to right: Andean zones, sampling horizons, stratigraphic column, polarities sequence, interpreted Subchrons. The two accepted proposals for the position of the J-K boundary in the literature are shown: J/K-2012 places it between Subchrons M19n.1n and M18r Ogg and Hinnov (2012) and J/K-2016 in the middle of Subchron M19n.2n (Ogg et al. 2016)

andesensis Zone was equivalent to the *S. fallauxi* Zone (following the GTS2012) or even older than the top of *S. darwini* Zone in the Tethys (following the GTS2004). In accordance, *V. andesensis* spans Subchrons M22r (base) through M22n (top). And it is worth noting that the reverse polarity at the base of the section has been found in other three sections of Vaca Muerta: Puerto Curaco (Amigo 2016), Los Catutos (Kohan Martínez et al. 2018) and Las Loicas (Iglesia Llanos 2015, unpublished).

The following *P. zitteli* Zone bears normal, reverse and normal polarities that would correlate with M22n through M21n. *A. proximus* Zone comprises a set of normal and reverse polarities that are correlated with M21n and M20r, *W. internispinosum* bears normal polarity that is correlated with M20n.2n. Upwards, *C. alternans* Zone comprised normal, reverse, normal, reverse, normal and reverse polarities that would represent magnetosubzones M20n.1r through M19n. *S. koeneni* Zone comprises normal, reverse, normal, reverse, normal, reverse, normal and reverse polarities that are correlated with M19n to M16r, where the J-K boundary should be placed. The precise position of this boundary based on ammonites varies depending on which biostratigraphic proposal is considered. Both proposals agree that the boundary is located within the *S. koeneni* Zone, but one proposal places it in the lower part of the zone (Leanza 1996; Riccardi 2008, 2015) and the other at the top (e.g., Aguirre-Urreta et al. 2014; Vennari et al. 2014; López Martínez et al. 2017; Lena et al. 2018). Such controversy in the biostratigraphy makes paleomagnetism a relevant tool that would help in the placement of the boundary from the correlation of polarities. Indeed, in the GTS from Ogg and Hinnov (2012), the J-K boundary is located in a change of polarities, precisely between M19n.1n and M18r. Such timeline at Arroyo Loncoche is traced to the narrow interval between paleomagnetic sites AL16 and AL40, at *c.* 164 m from the base, in the lower third of *S. koeneni* Zone. More recently, the Berriasian Working Group proposed the placement of J-K boundary in the middle of the M19n.2n magnetosubzone (in Ogg et al. 2016). This way, the boundary in Arroyo Loncoche is constrained between sites AL38 and AL39, at c. 157.5 m from the base, in a lower position respect to the 2012 proposal but always in lower *S. koeneni* Zone (Fig. 13). The position of the J-K boundary is in very good agreement with the biostratigraphic proposal of Riccardi (2015). Furthermore, it matches the location assigned by cyclostratigraphy (Kietzmann et al. 2015). Further above, *A. noduliferum* Zone includes a dominant reverse with a minor opposite polarity, which is correlated with M16r. At the top, *S. damesi* Zone comprises reverse, normal and reverse polarities that are interpreted to represent M16r to M15r. Accordingly, we conclude that deposition of the Vaca Muerta Formation at Arroyo Loncoche spanned Subchrons M22r.2r through M15r.

4.4 Late Jurassic–Early Cretaceous: Discussion

The combination of magnetostratigraphy and cyclostratigraphy tied to the biostratigraphy at Arroyo Loncoche enabled a better knowledge of key issues regarding the Vaca Muerta Formation.

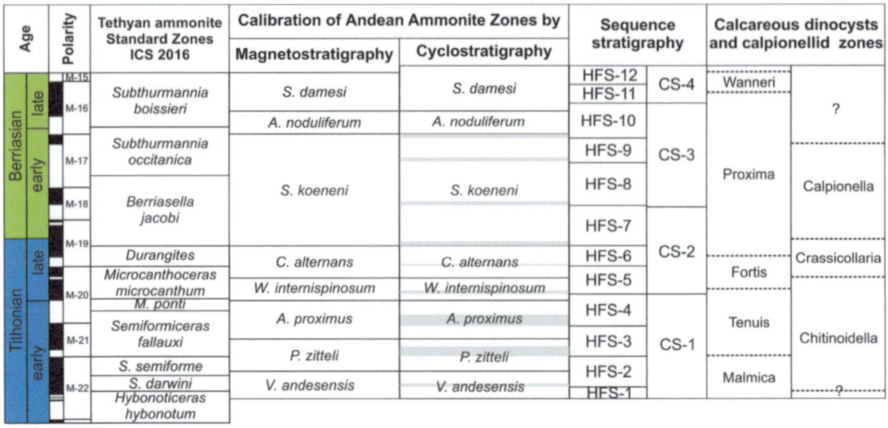

Fig. 14 Calibration of Andean Ammonite Zones using magnetostratigraphy (Iglesia Llanos et al. 2017), cyclostratigraphy based on low eccentricity cycles (Kietzmann et al. 2015), high frequency (HFS) and composite (CS) depositional sequences (Kietzmann et al. 2014), calcareous dinocyst zones (Ivanova and Kietzmann 2016) and calpionellid zones (González Tomassini et al. 2015). Gray-shaded bars indicate missing time with respect to the magnetostratigraphic calibration

One relevant issue is the age of the Tithonian transgression that encompassed the *V. andesensis* Zone. This Zone is traditionally correlated with the *S. darwini* and *S. semiforme* Standard Zones (Riccardi 2008, 2015; Vennari et al. 2014). However, the magnetic pattern recorded during *V. andesensis* Zone in at least three sections (e.g., Arroyo Loncoche, Puerto Curacao and El Ministerio) is not suitable with the standard zones (Fig. 14) bearing normal polarity (Iglesia Llanos et al. 2017). Accordingly, and bearing in mind former biostratigraphic data (Zeiss and Leanza 2010; Ivanova and Kietzmann 2017), we interpret that *V. andesensis* should be extended down to the top of *H. hybonotum* Zone (Figs. 13 and 14). Zeiss and Leanza (2010) divided the *V. mendozanus* Zone in two subzones, which were referred to the *H. hybonotum* and *S. darwini* Zones, respectively. On the other hand, cyclostratigraphy (Kietzmann et al. 2015) indicates that *V. andesensis* can be correlated (in time) with the upper part of *S. semiforme* Zone (following GTS2004) or upwards to the base of *S. fallauxi* Zone (following GTS2012). The reverse polarity at the base of the Vaca Muerta Formation restricts these two options to the *S. fallauxi* Zone, but still, the magnetic pattern does not fit the corresponding polarities from the GTS2016. In fact, Kietzmann et al. (2015) emphasize that basal biozones at the study section are found in distal facies of a carbonate ramp system and may contain minor discontinuities that are not detectable with biostratigraphic methods or field evidences, which eventually imply longer durations. Calibration of the Andean ammonite zones based on magnetostratigraphy actually enables the assessment of certain time lapses that are not represented in the cyclostratigraphy which could be related a priori with the stages of low-stand of sea level. Recently, Kietzmann et al. (2018) (see Chap. "Orbital Controls and High-Resolution Cyclostratigraphy of Late Jurassic–Early Cretaceous in the Neuquén Basin") reinterpreted their cyclostratigraphy based on new data published by Kohan Martínez et al. (2018), in which this difference is significantly reduced. So far, the most conclusive tool for this time interval in the basin is magnetostratigraphy.

The other key issue is the controversial placement of the Jurassic–Cretaceous boundary. The position of the J-K boundary is matter of debate worldwide, since it portrays no relevant faunal turnover or remarkable increase of faunal provinciality, especially in ammonites (e.g., Remane 1991; Wimbledon 2008; Wimbledon et al. 2011, 2013; Michalík and Reháková 2011; Ogg and Hinnov 2012 and references cited there). Complications in the regional, but particularly inter-regional correlation, constitute a major problem across the world. In the Neuquén Basin, ammonites convey a remarkable biostratigraphic resolution, although their autochthony prevents a straightforward correlation with the Standard zones, and consequently, of a unique (Fig. 10) biostratigraphic correlation scheme (e.g., Leanza 1981; Riccardi 2008, 2015; Vennari et al. 2014).

The Upper Jurassic through Cretaceous interval in the GTS2016 is derived from the marine M-sequence magnetic anomalies. Marine anomalies calibrated to biostratigraphy define the template against which magnetostratigraphic studies on-land are referenced. Because of this, a non-ambiguous correlation between a section on-land and the GPTS is likely only when a thorough biostratigraphic definition is attained (e.g., Ogg and Hinnov 2012). Although still controversial, there are three relevant definitions for the base of the Cretaceous System (Remane 1991; Wimbledon 2008; Grabowski 2011; Ogg and Hinnov 2012): 1) the base of the ammonite *Grandis* Subzone defined in the Colloque sur la Crétacé inferieur (1963) in the lower part of the *Calpionella* Zone, that is practically placed at the base of Subchron M18r, 2) the base of the *Jacobi* ammonite Zone defined in the Colloque sur la limite Jurassique-Crétacé (1973), approximately equivalent to the base of the *Calpionella* Zone and correlated with the upper part of the M19n.2n and 3) the base of the *Occitanica* ammonite Zone, correlated with the middle part of the *Calpionella* Zone and the lower part of Subchron M17r (Hoedemaeker 1990). Finally, in the recent years, the Berriasian Working Group has determined the Jurassic/Cretaceous boundary at the base of the *Calpionella* Zone within the middle part of magnetosubzone M19n.2n (in Ogg et al. 2016).

In the Neuquén Basin, it is agreed that the position of the boundary is located (Fig. 10) within the *S. koeneni* Zone (Riccardi 2008, 2015; Vennari et al. 2014). Salazar Soto (2012) based on a study on material mostly from the Lo Valdes Formation, Chile, has referred the *S. koeneni* Zone to the Berriasian. All these considerations led Riccardi (2015) to establish the correlation of *S. koeneni* Zone with *B. jacobi* through early *S. occitanica* Zones (Upper Tithonian – Lower Berriasian).

The cyclostratigraphic scale provided by Kietzmann et al. (2015, 2018) placed (in time) the Jurassic/Cretaceous boundary within the lower third of the *S. koeneni* Zone (Fig. 14). These data fully coincide with our magnetostratigraphic correlation, since both the base of magnetosubzone M18r and the middle part of magnetosubzone M19n.2n are located within this same interval (Iglesia Llanos et al. 2017).

5 Conclusions

5.1 Early Jurassic

A regional composite magnetostratigraphic scale was constructed from five sections located in southern Mendoza and northern Neuquén spanning the Hettangian–Toarcian. On the basis of diverse field tests for the paleomagnetic stability, the isolated magnetizations are interpreted to correspond to the time of deposition (extrusion) of the sedimentary (volcanic) rocks. Polarities were tied to Andean ammonite zones and from their correlation with the standard zones, calibrated to the GTS2016. The magnetostratigraphic scale yields 16 reverse (*Jr1–Jr16*) and 16 normal (*Jn1–Jn16*) polarity zones that encompass at least 19 ammonite zones. A consistent correlation was achieved with the GPTS16 that helped to unveil important issues.

The only remarkable difference in the magnetostratigraphic correlation takes place in the Hettangian. In Arroyo Malo, this stage bears ammonites of the *K. bayoensis* through *B. canadensis* Zones, and dominant reverse polarity (*Jr1–Jr3*) is recorded, whereas the GTS2016 shows normal polarity for the same interval. It is worth noting that the Hettangian in the GTS has been controversial due to the fact that reference sections of this age on-land bear ammonites only at the base (*P. planorbis* Zone, normal polarity), whereas the rest of the dominantly normal polarity has been dated with palynology and radiometric dating. In fact, this dominantly normal polarity assigned to the middle and upper Hettangian has changed drastically over the years from dominantly reverse (until 2012) to dominantly normal (post 2012). Hence, a straightforward correlation with polarities isolated at Arroyo Malo that have been tied to ammonite zones can certainly results a risky task. The best biostratigraphic and magnetostratigraphic correlations are achieved in the Pliensbachian–Toarcian due to the better definition of ammonite zones and a greater number of paleomagnetic sites. Magnetostratigraphy allowed to include the first barren sites in Puesto Araya (*Jr6–Jn8*) within *M. chilcaense* Zone (equivalent to *U. jamesoni* Zone). Upwards, it enabled to determine that *F. fannini* Zone was sampled only partially (i.e., the upper half is missing), that the barren *Tenuicostatum* Zone (Rajapalo) was fully represented (*Jn13–Jn14*) and that *P. largaense* Zone was sampled entirely only in Arroyo Blanco (*Jn16–Jr16*).

5.2 Middle Jurassic

The age of the Lajas Formation has always been estimated by its stratigraphic position. The application of the magnetostratigraphy is thus the best tool to refine the estimated age of the unit.

The resultant magnetostratigraphy comprises five reverse and four normal polarity zones within a pattern of dominantly reverse polarity. The correlation with the GTS2016 shows that the best fit of polarities is achieved in the Lower-uppermost

Middle Bathonian. According to the magnetostratigraphy, deposition of the Lajas Formation at Portada Covunco would have encompassed Chron M41 (Lower Bathonian) and lower Chron M39 (uppermost Middle Bathonian). Such time lapse is in agreement with the ages interpreted by other authors for the area. Minimum sedimentation rates calculated for the studied section show one of 211 m/Ma for the lower 60 m thick interval, and the other that almost doubles it (= 415 m/Ma) between the 60 and 258 m-thick interval.

5.3 Late Jurassic–Early Cretaceous

The magnetostratigraphic scale obtained in the Vaca Muerta Formation at Arroyo Loncoche comprises 11 reverse and 10 normal polarity zones, spanning the *V. andesensis* (Lower Tithonian)–*S. damesi* Zones (Upper Berriasian). The correlation of these polarity zones with the GPTS2016 was achieved from the calibration between the Andean and Standard ammonite zones. The combination of three chronostratigraphic tools such as biostratigraphy, magnetostratigraphy and cyclostratigraphy enabled dating the section with unprecedented precision. This has turned particularly relevant in the case of the lower *V. andesensis* Zone whose equivalence in the Tethyan Realm had not been well established so far, and in the position of the Jurassic–Cretaceous boundary, whose position in the Neuquén Basin had been hot matter of debate. Magnetostratigraphic data indicate that the J-K boundary should be placed at the lower part of *S. koeneni*, either at the M19n.1n-M18r boundary at c. 164 m from the base following Ogg and Hinnov (2012) or in the middle of magnetosubzone M19n.2n at 157.5 m from the base according to the most recent proposal from Berriasian Working Group (in Ogg et al. 2016). In any case, the position of the J-K boundary is thus in agreement with the biostratigraphic criteria from Riccardi (2008, 2015). The resultant magnetostratigraphy enabled to conclude that deposition of the Vaca Muerta Fm. at Arroyo Loncoche spanned Subchrons M22r.2r through M15r.

References

Aguirre-Urreta MB, Vennari VV, Lescano M et al (2014) Bioestratigrafía y geocronología de alta resolución de la Formación Vaca Muerta, Cuenca Neuquina. In: Abstracts of the 9 Congreso de Exploración y Desarrollo de Hidrocarburos, Mendoza, 3–7 Nov 2014

Amigo JD (2016) Magnetoestratigrafía de la Formación Vaca Muerta en la sección Puerta Curaco, norte de Neuquén. Degree Thesis, Universidad de Buenos Aires

Ballent S, Concheyro A, Náñez C et al (2011) Microfósiles mesozoicos y cenozoicos. In: Leanza HA, Arregui C, Carbone O et al (eds) Geología y Recursos Naturales de la Provincia del Neuquén. Asociación Geológica Argentina, Buenos Aires, pp 489–528

Borradaile GJ, Henry B (1997) Tectonic applications of magnetic susceptibility and its anisotropy. Earth Sci Rev 42(1):49–93

Bown P, Concheyro A (2004) Lower Cretaceous calcareous nannoplankton from the Neuquén Basin, Argentina. Mar Micropaleontol 52:51–84

Brandsaeter I, McIlroy D, Lia O et al (2005) Reservoir modelling and simulation of Lajas Formation outcrops (Argentina) to constrain tidal reservoirs of the Halten Terrace (Norway). Petrol Geosci 11:37–46

Canale N, Ponce JJ, Carmona NB, Drittanti DI, Olivera DE, Martínez MA, Bournod CN (2015) Sedimentología e Icnología de deltas fluvio-dominados afectados por descargas hiperpícnicas de la Formación Lajas (Jurásico Medio), Cuenca Neuquina, Argentina. Andean Geol 42(1)

Damborenea SE (1987) Early Jurassic Bivalvia of Argentina Pt.1, Stratigraphical Introduction and superfamilies Nuculanacae, Arcaceae, Mytilaceae and Pinnacea. Palaeontographica A 199:23–111

Fernández Carmona J, Riccardi AC (1999) Primer reporte de Calpionélidos calcáreos del Cretácico inferior–Berriasiano de la Provincia del Tethys en la República Argentina: Conexión Tethys-Pacífico. In: Abstracts of the 5 Simposio sobre o Cretáceo do Brasil, 29 Aug–2 Sept 1999

Fisher RA (1953) Dispersion on a sphere. Proc R Soc London A 217:295–305

García VM, Zavala CA, Quattrocchio ME (1994) Relación entre análisis palinológico y análisis de facies. Aplicación al Grupo Cuyo (Jurásico Medio) en la Cuenca Neuquina. Rev Asoc Geol Argent 49(1–2):184–195

García VM, Quattrocchio ME, Zavala CA et al (2006) Palinofacies, paleoambientes y paleoclima del Grupo Cuyo (Jurásico medio) en la Sierra de Chacaico, Cuenca Neuquina, Argentina. Rev Española Micropaleont 38(2–3):269–288

González Tomassini F, Kietzmann DA, Fantín MA et al (2015) Estratigrafía y análisis de facies de la Formación Vaca Muerta en el área de El Trapial, Cuenca Neuquina, Argentina. Petrotecnia 2015/2:78–89

Grabowski J (2011) Magnetostratigraphy of the Jurassic/Cretaceous boundary interval in the Western Tethys and its correlations with other regions: a review. Vol Jurassica 9:105–128

Gugliotta M, Flint SS, Hodgson DM, Veiga GD (2015) Stratigraphic record of river-dominated crevasse subdeltas with tidal influence (Lajas Formation, Argentina). J Sediment Res 85(3):265–284

Gulisano CA, Gutiérrez Pleimling A, Digregorio R (1984) Esquema estratigráfico de la secuencia Jurásica del oeste de la provincia de Neuquén. 9° Congreso Geológico Argentino. Actas 1:237–259

Gulisano CA, Gutiérrez Pleimling A (1995) The Jurassic of the Neuquén Basin. b) Mendoza Province, Asociación Geológica Argentina, Series E3, Buenos Aires, pp 103

Halls HC (1976) A least-squares method to find a remanence direction from converging remagnetization circles. Geophys J R Am Soc 45:291–304

Heslop D, Dekkers MJ, Kruiver PP et al (2002) Analysis of isothermal remanent magnetization acquisition curves using the expectation-maximization algorithm. Geophys J Int 148(1):58–64

Hoedemaeker PJ (1990) The Neocomian boundaries of the Tethyan Realm based on the distribution of ammonites. Cretaceous Res 11:331–342

Hounslow MW, Davydov VI, Klootwijk CT et al (2004) Magnetostratigraphy of the Carboniferous: a review and future prospects. Newsl Carboniferous Strat 22:35–41

Hrouda F (1982) Magnetic anisotropy of rocks and its application in geology and geophysics. Surv Geophys 5(1):37–82

Iglesia Llanos MP (1997) Magnetoestratigrafía y Paleomagnetismo del Jurásico Inferior marino de la Cuenca Neuquina, República Argentina. Ph.D. thesis, Universidad de Buenos Aires

Iglesia Llanos MP (2009) Estudio Paleomagnético del Jurásico Marino de la Cuenca Neuquina: Correlación Magnetoestratigráfica del Jurásico Inferior y una Nueva Curva de Deriva Polar Aparente para América del Sur. Rev Asoc Geol Argent 65(2):387–399

Iglesia Llanos MP (2012) Palaeomagnetic study of the Jurassic from Argentina: magnetostratigraphy and palaeogeography of South America. Rev Paléobiol, SP11:151–168

Iglesia Llanos MP, Kietzmann DA, Kohan Martínez M et al (2017) Magnetostratigraphy of the Upper Jurassic-Lower Cretaceous from Argentina: implications for the J-K boundary in the Neuquén Basin. Cretaceous Res 70:189–208

Iglesia Llanos MP, Kietzmann DA, Kohan Martínez M et al (2019) Magnetostratigraphy of a Middle Jurassic delta system (Lajas Formation), Portada Covunco section, southern Neuquén Basin, Argentina. J S Am Earth Sci 94:102235

Iglesia Llanos MP, Prezzi CB (2013) The role of true polar wander on the Jurassic palaeoclimate. Int J Earth Sci 102:745–759

Iglesia Llanos MP, Riccardi AC, Singer SE (2006) Palaeomagnetic study of Lower Jurassic marine strata from the Neuquén Basin, Argentina: a new Jurassic apparent polar wander path for South America. Earth Planet Sci Lett 252(3–4):379–397

Iglesia Llanos MP, Riccardi AC (2000) The Neuquén composite section: magnetostratigraphy and biostratigraphy of the marine lower Jurassic from the Neuquén basin (Argentina). Earth Plan Sc Lett 181:443–457

Iglesia Llanos MP, Vizán H (1995) Magnetoestratigrafía y Paleomagnetismo del Pliensbaquiano y Toarciano marino de la Argentina. Geol Colombiana 20:143–146

Ivanova DK, Kietzmann DA (2016) Calcareous dinoflagellate cysts from the Tithonian - Valanginian Vaca Muerta Formation in the southern Mendoza area of the Neuquén Basin, Argentina. In: Abstracts of the VI Simposio Argentino del Jurásico, Malargüe, 18–19 April 2016

Ivanova DK, Kietzmann DA (2017) Calcareous dinoflagellate cysts from the Tithonian - Valanginian Vaca Muerta Formation in the southern Mendoza area of the Neuquén Basin, Argentina. J S Am Earth Sci 77:150–169

King RF (1955) Remanent magnetism of artificially deposited sediments. Mon Not R Astron Soc Geophys Suppl 7:115–134

Kent DV, Olsen PE (2008) Early Jurassic magnetostratigraphy and paleolatitudes from the Hartford continental rift basin (eastern North America): Testing for polarity bias and abrupt polar wander in association with the central Atlantic Magmatic province. J Geophys Res 113:B06105

Kent DV, Olsen PE, Witte WK (1995) Late Triassic-earliest Jurassic geomagnetic polarity sequence and paleolatitudes from drill cores in the Newark rift basin, eastern North America. J Geophys Res 100(B8):14965–14998

Kietzmann DA (2017) Chitinoidellids from the early Tithonian–early Valanginian Vaca Muerta formation in the Northern Neuquén Basin, Argentina. J S Am Earth Sci 76:152–164

Kietzmann DA, Palma RM, Bressan GS (2008) Facies y microfacies de la rampa tithoniana-berriasiana de la Cuenca Neuquina (Formación Vaca Muerta) en la sección del arroyo Loncoche – Malargüe, provincia de Mendoza. Rev Asoc Geol Argent 63(4):696–713

Kietzmann DA, Martín-Chivelet J, Palma RM et al (2011a) Evidence of precessional and eccentricity orbital cycles in a Tithonian source rock: the mid-outer carbonate ramp of the Vaca Muerta Formation, Northern Neuquén Basin, Argentina. AAPG Bull 95(9):1459–1474

Kietzmann DA, Blau J, Riccardi AC et al (2011b) An interesting finding of chitinoidellids (Clapionellidea Bonet) in the Jurassic-Cretaceous boundary of the Neuquén Basin. In: Abstracts of the 18 Congrreso Geologico Argentino, Neuquén, 4–6 May 2011

Kietzmann DA, Palma RM, Riccardi AC et al (2014) Sedimentology and sequence stratigraphy of a Tithonian - Valanginian carbonate ramp (Vaca Muerta Formation): A misunderstood exceptional source rock in the Southern Mendoza area of the Neuquén Basin, Argentina. Sediment Geol 302:64–86

Kietzmann DA, Palma RM, Iglesia Llanos MP (2015) Cyclostratigraphy of an orbitally-driven Tithonian-Valanginian carbonate ramp succession, Southern Mendoza, Argentina: Implications for the Jurassic-Cretaceous boundary in the Neuquén Basin. Sedediment Geol 315:29–46

Kirschvink JL (1980) The least-squares line and plane and the analysis of palaeomagnetic data. Geophys J Roy Astron Soc 62:699–718

Kohan Martínez M, Kietzmann DA, Iglesia Llanos MP (2018) Magnetostratigraphy and cyclostratigraphy of the Tithonian interval from the Vaca Muerta Formation, southern Neuquén Basin, Argentina. J S Am Earth Sci 85:209–228

Kurcinka C, Dalrymple RW, Gugliotta M (2017) Facies and architecture of river dominated to tide-influenced mouth bars in the lower Lajas Formation (Jurassic), Argentina. AAPG Bull 102:885–912

Lanza R, Meloni A (2006) The Earth's magnetism. Springer, Berlin-Heidelberg
Leanza HA (1981) The Jurassic-Cretaceous boundary beds in West Central Argentina and their ammonite zones. Neues Jahrb Geol P-A 161:62–92
Leanza HA (1996) Advances in the ammonite zonation around the Jurassic/Cretaceous boundary in the andean realm and correlation with tethys. In: Jost Wiedmann Symposium, Abstracts, T€ubingen, pp 215e219
Legarreta L, Uliana MA (1991) Jurassic–Cretaceous Marine Oscillations and Geometry of Back Arc Basin, Central Argentina Andes. In: McDonald, DIM (ed) Sea level changes at active plate margins: Process and product. International Association of Sedimentologists, SP 12:429–450
Lena L, López-Martínez R, Lescano M et al (2018) Cross-continental age calibration of the Jurassic/Cretaceous boundary. Solid Earth Discuss. https://doi.org/10.5194/se-2018-57
Kietzmann DA, Lescano M (2010) Nanofósiles Calcáreos de la Formación Vaca Muerta (Tithoniano inferior- Valanginiano inferior) en la región sudoccidental de la Provincia de Mendoza. In: Abstracts of the 10 Congreso Argentino de Paleontologia y Bioestratigrafía y 7 Congreso Latinoamericano de Paleontología, La Plata, 20–24 Sept 2010
López-Martínez R, Aguirre-Urreta B, Lescano M et al (2017) Tethyan calpionellids in the Neuquén Basin (Argentine Andes), their significance in defining the Jurassic/Cretaceous boundary and pathways for Tethyan-Eastern Pacific connections. J S Am Earth Sci 78:116–125
Martínez MA (2002) Palynological zonation of the Lajas Formation (Middle Jurassic) of the Neuquén Basin, Argentina. Ameghiniana 39(2):221–240
Martínez MA, Quattrocchio ME, Zavala CA (2002) Análisis palinofacial de la Formación Lajas (Jurásico Medio), Cuenca Neuquina, Argentina. Significado paleoambiental y paleoclimático. Rev Española Micropaleont 34(1):81–104
McFadden PL (1990) A new fold test for palaeomagnetic studies. Geophys J Int 103:163–169
McFadden PL, McElhinny MV (1988) The combined analysis of remagnetization circles and direct observations in palaeomagnetism. Earth Planet Sc Lett 87:163–189
McFadden P, McElhinny MW (1990) Classification of the reversal test in palaeomagnetism. Geophys J Int 103:725–729
McIlroy D (2007) Palaeoenvironmental controls on the ichnology of tide-influenced facies with an example from a macrotidal tide-dominated deltaic depositional system, Lajas Formation, Neuquén Province, Argentina. Sediment–Organism Interactions: A Multifaceted Ichnology. SEPM (Society for Sedimentary Geology). Spec Publ 88:195–213
Mena M (1994) Cirdi, Software para el cálculo de direcciones mediante el análisis de círculos máximos. Laboratorio de Paleomagnetismo D. Valencio, Universidad de Buenos Aires
Merino G (2017) Magnetoestratigrafía y Sedimentología de la Formación Lajas (Jurásico Medio) en la sección Arroyo Covunco, provincia de Neuquén. Degree thesis, Universidad de Buenos Aires
Michalik J, Reháková D (2011) Possible markers of the Jurassic/Cretaceous boundary in the Mediterranean Tethys: a review and state of art. Geosci Front 2(4):475–490
Ogg JG (2004) The Jurassic period. In: Gradstein F, Ogg JG, Smith A (eds) A geologic time scale. Cambridge University Press, Cambridge, pp 307–343
Ogg JG, Hinnov LA (2012) The Jurassic Period. In: Gradstein FM, Ogg J, Schmitz M, Ogg G (eds) The geologic time scale 2012. Elsevier, Amsterdam, pp 731–792
Ogg JG, Ogg GM, Gradstein FM (2016) A concise geologic time scale. Elsevier, Amsterdam
Remane J (1991) The Jurassic-Cretaceous boundary: problems of definition and procedure. Cretaceous Res 12(5):447–453
Riccardi AC (1983) The Jurassic of Argentina and Chile. In: Moullade M, Naim AEM (eds) The Phanerozoic geology of the world II, The Mesozoic B:201–203
Riccardi AC (2008) The marine Jurassic of Argentina: a biostratigraphic framework. Episodes 31:326–335
Riccardi A (2015) Remarks on the Tithonian–Berriasian ammonite biostratigraphy of west central Argentina. Vol Juras 13:23–52

Riccardi AC, Gulisano CA (1992) Unidades limitadas por discontinuidades. Su aplicación al Jurásico Andino. Rev Asoc Geol Argent 45(3–4): 346–364

Riccardi AC, Iglesia Llanos MP (1999) Primer hallazgo de amonites en el Triásico de la Argentina. Rev Asoc Geol Argent 54(3):298–300

Riccardi AC, Damborenea SE, Manceñido MO et al (1991) Hettangian and Sinemurian (Lower Jurassic) biostratigraphy of Argentina. J S Am Earth Sci 4(3):159–170

Riccardi AC, Damborenea SE, Manceñido MO et al (2004) The Triassic/Jurassic boundary in the Andes of Argentina. Riv Ital Paleontol S 110:69–76

Riccardi AC, Westermann GEG (1991) Middle Jurassic Ammonoid Fauna and Biochronology of the Argentine-Chilean Andes. Part III: Bajocian-Callovian Eurycephalitinae, Stephanocerataceae. Palaeontogr A 216:1–110

Rochette P (1988) Inverse magnetic fabric carbonate bearing rocks. Earth Planet Sci Lett 90:229–237

Rochette P, Jackson M, Aubourg C (1992) Rock magnetism and the interpretation of anisotropy of magnetic susceptibility. Rev Geophys 30(3):209–226

Salazar Soto C (2012) The Jurassic-Cretaceous boundary (Tithonian–Hauterivian) in the Andean Basin of Central Chile: ammonites, bio- and sequence stratigraphy and palaeobiogeography. Ph.D. thesis, Rupecht-Karls-Universität Heildeberg

Spalletti L (1995) Depósitos de tormenta en un frente deltaico. Jurásico medio de la Cuenca Neuquina, República Argentina. Rev Soc Geol España 8:261–272

Steel E, Simms AR, Steel R, Olariu C, Marzo M (2018) Hyperpycnal delivery of sand to the continental shelf: Insights from the Jurassic Lajas Formation, Neuquén Basin, Argentina. Sedimentology 65(6):2149–2170

Torsvik TH (1992) Iteractive analysis of palaeomagnetic data. NGU, Trondheim

Torsvik TH, Briden JC, Smethurst MA (2000) Super-IAPD Interactive analysis of palaeomagnetic data. www.geodynamics.no/software.htm

Van der Voo R, French RB (1977) Paleomagnetism of the Late Ordovician Juniata Formation and the remagnetization hypothesis. J Geophys Res 82(36):5796–5802

Vennari VV (2016) Tithonian ammonoids (Cephalopoda, Ammonoidea) from the Vaca Muerta Formation, Neuquén Basin, West-Central Argentina. Palaeontogr A 306(1–6):85–165

Vennari V, Lescano M, Naipauer M et al (2014) New constraints on the Jurassic-Cretaceous boundary in the High Andes using high-precision U-Pb data. Gondwana Res 26:374–385

Volkheimer W (1978) Descripción geológica de la Hoja 27b, Cerro Sosneado, Provincia de Mendoza. Bol Serv Geol Nac 151:1–85

Westermann GEG (1988) Duration of Jurassic stages based on average and scaled subzones. In: Agterberg FP, Rao CN (eds) Recent advances in quantitative stratigraphic correlation. Hindustan Publishing, Delhi, pp 90–100

Westermann GEG, Riccardi AC (1972) Middle Jurassic ammonoid fauna and biochronology of the Argentine-Chilean Andes. Part I: Hildocerataceae. Palaeontographica A 140:1–116

Wimbledon WAP (2008) The Jurassic-Cretaceous boundary: an age-old correlative enigma. Episodes 31(4):423–428

Wimbledon WAP, Casellato CE, Reháková D et al (2011) Fixing a basal Berriasian and Jurassic/Cretaceous (J/K) boundary—is there perhaps some light at the end of the tunnel? Riv Ital Paleontol S 117:295–307

Wimbledon WAP, Reháková D, Pszczółkowski A et al (2013) An account of the bio- and magnetostratigraphy of the Upper Tithonian—Lower Berriasian interval at Le Chouet, Drôme (SE France). Geol Carpathica 64(6):437–460

Yang Z, Moreau G, Bucher H et al (1995) Hettangian and Sinemurian magnetostratigraphy from Paris Basin. J Geophys Res 101(B4):8025–8042

Zavala CA (1993) Estratigrafía y análisis de facies de la Formación Lajas (Jurásico medio) en el sector suroccidental de la Cuenca Neuquina. Provincia del Neuquén. República Argentina. Ph.D. Thesis, Universidad del Sur

Zavala CA (1996a) Sequence stratigraphy in continental to marine transitions. An example from the Middle Jurassic Cuyo Group, south Neuquén Basin, Argentina. GeoRes Forum 1(2):285–293

Zavala CA (1996b) High-resolution sequence stratigraphy in the Middle Jurassic Cuyo Group, South Neuquén Basin, Argentina. GeoRes Forum 1(2):295–303
Zavala C, González R (2001) Estratigrafía del Grupo Cuyo (Jurásico inferior-medio) en la Sierra de la Vaca Muerta, Cuenca Neuquina. Bol Inf Petrol 65:40–54
Zegers TE, Dekkers MJ, Bailly S (2003) Late Carboniferous to Permian remagnetization of Devonian limestones in the Ardennes: role of temperature, fluids, and deformation. J Geophys Res Solid Earth 108(B7):2357
Zeiss A, Leanza HA (2010) Upper Jurassic (Tithonian) ammonites from the lithographic limestones of the Zapala region, Neuquén Basin, Argentina. Beringeria 41:23–74

Orbital Controls and High-Resolution Cyclostratigraphy of Late Jurassic–Early Cretaceous in the Neuquén Basin

Diego A. Kietzmann, María Paula Iglesia Llanos and Melisa Kohan Martínez

Abstract Detailed cyclostratigraphic analyses have been made from seven Tithonian–Hauterivian sections of the Vaca Muerta and Agrio Formations, exposed in southern Mendoza area of the Neuquén Basin. Both lithostratigraphic units are characterized by decimeter-scale rhythmic alternations of marlstones and limestones, showing a well-ordered hierarchy of cycles, including elementary cycles, bundles, and superbundles. According to biostratigraphic data, elementary cycles have a periodicity of ~18–21 ky, which correlates with the precessional cycle of the Earth's axis. Spectral analysis based on time series of elementary cycle thicknesses allowed us to identify frequencies of ~400 ky, and ~90–120 ky, which we interpret as the modulation of the precessional cycle by the Earth's orbital eccentricity. A third band frequency of ~40 ky was also identified that can be assigned to the obliquity cycle. Cyclostratigraphy enabled the construction of almost continuous floating astronomical time scale for the Tithonian–Hauterivian, for which a minimum duration of 5.67 myr for the Tithonian, 5.27 myr for the Berriasian, >3.45 myr for the Valanginian, and 5.96 myr for the Hauterivian have been assessed. Additionally, the likely transference mechanisms of the orbital signal to the sedimentary record are analyzed, proposing the coexistence of carbonate exportation and dilution as the dominant mechanisms.

Keywords Cyclostratigraphy · Astronomical time scale · Jurassic–Cretaceous · Neuquén basin

D. A. Kietzmann (✉) · M. P. Iglesia Llanos · M. K. Martínez
Facultad de Ciencias Exactas y Naturales, Departamento de Ciencias Geológicas, Universidad de Buenos Aires, Ciudad Universitaria, Pabellón II, Intendente Güiraldes, 2160, C1428EHA Ciudad Autónoma de Buenos Aires, Argentina
e-mail: diegokietzmann@gl.fcen.uba.ar

CONICET-Universidad de Buenos Aires, Instituto de Geociencias Básicas, Ambientales y Aplicadas de Buenos Aires (IGeBA), Buenos Aires, Argentina

D. Kietzmann and A. Folguera (eds.), *Opening and Closure of the Neuquén Basin in the Southern Andes*, Springer Earth System Sciences,
https://doi.org/10.1007/978-3-030-29680-3_9

1 Introduction

Cyclostratigraphy has become an important tool for determining Jurassic–Cretaceous geologic time and establishing floating astronomical time scales (ATS). Almost the entire Jurassic and Cretaceous have been calibrated using the long-term eccentricity cycle, which is the most stable cycle during the Phanerozoic. Many important works involving the Upper Jurassic–Lower Cretaceous were dedicated to demonstrating the presence of Milankovitch cycles in the sedimentary record. Nevertheless, only a few works have carried out detailed cyclostratigraphic studies for stratigraphic time calibration (Hinnov and Hilgen 2012; Ogg and Hinnov 2012a, b). Particularly, marlstone–limestone rhythmic successions have been cited as key examples of climate-forced sedimentation resulting from different palaeoenvironmental factors, including fluctuations in the supply of calcareous biogenic sediments produced by plankton (productivity cycles: Einsele and Ricken 1991), and fluctuations in the input of fine terrigenous sediments (dilution cycles: Arthur et al. 1984; Einsele and Ricken 1991). Additional potential mechanisms include diagenetic redistribution of calcium carbonate (dissolution cycles, Einsele and Ricken 1991; Westphal et al. 2010), fluctuations in carbonate-mud exportation from shallow areas (carbonate dilution cycles, Pittet and Strasser 1998), and fluctuations in superficial fertility (fertility cycles, Premoli Silva et al. 1989).

The Late Jurassic–Early Cretaceous in the Neuquén Basin represented a time of tectonic quietude that prompted the development of thick carbonate and mixed successions (Legarreta and Uliana 1991, 1996), included in the Mendoza Group (Figs. 1

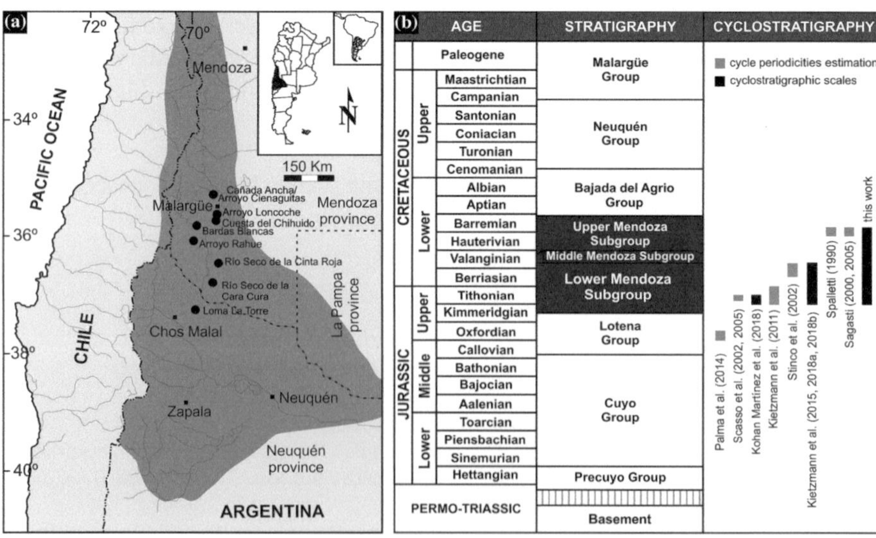

Fig. 1 **a** Location map of the Neuquén Basin showing studied localities. **b** Stratigraphic chart for the Neuquén Basin, showing stratigraphic coverage of cyclostratigraphic studies for the Upper Jurassic and Lower Cretaceous

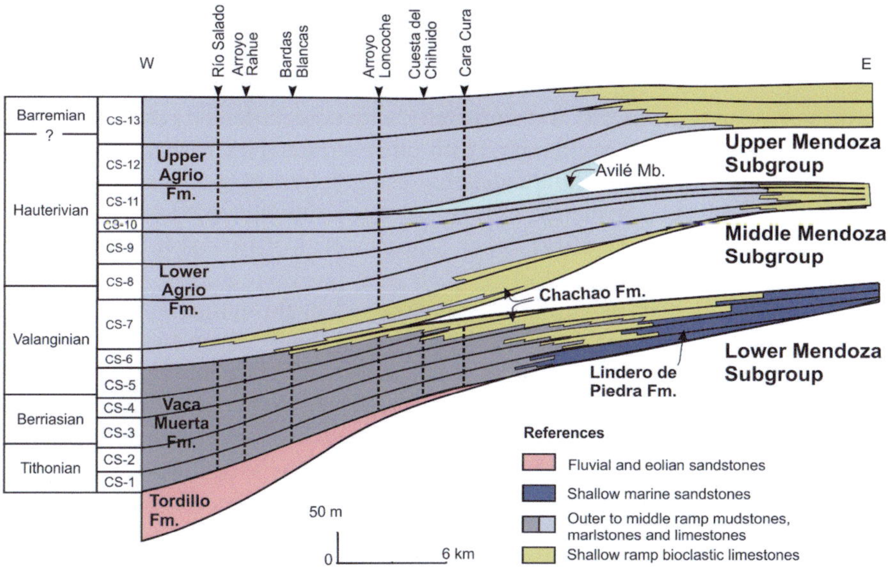

Fig. 2 Litostratigraphic subdivision of the Mendoza Group in the Southern Mendoza area, showing the location of the studied sections

and 2). Despite the notable limestone–marlstone rhythmicity of these successions, only a few works have dealt with this subject. The first study of cyclicity in Argentina was carried out by Spalletti et al. (1990) for the Upper Hauterivian–Lower Barremian interval using well logs data from the Upper Agrio Formation in the Filo Morado block. These authors recognized periodicities of 17–23, 114, 422, and 1143 ky, consistent with the Milankovitch frequency band. Later, Sagasti (2000, 2005) replicated these types of studies in outcrops of the same age, using lithological limestone–marlstone couples, and found similar periodicities of 10–30, 45–47, 103–127, and ~400 ky. At the same time, Scasso et al. (2002, 2005) estimated the periodicity for the rhythmicity of late Early–early Late Tithonian Los Catutos Member of the Vaca Muerta Formation, obtaining a periodicity of 18.2 ky for limestone–marlstone elementary cycles. Kietzmann et al. (2011) studied the Tithonian interval at the Arroyo Loncoche section, obtaining a periodicity of 21.1 ky for the elementary cycle, and 90–120 ky for bundles of elementary cycles, which is consistent with the modulation of the precessional cycle caused by the Earth's orbital eccentricity. A more detailed study was published later by Kietzmann et al. (2015), who analyzed four Tithonian–Early Valanginian sections of the Vaca Muerta Formation, and obtained an elementary cycle of ~20 ky, bundle periodicities of 90–120, and ~400 ky for superbundles. These authors proposed for the first time a floating orbital scale for the Tithonian–Lower Valanginian in the Neuquén Basin and established minimum durations of Andean ammonite zones. Under the light of new data provided by Kohan Martínez et al. (2018) from Los Catutos Member, Kietzmann et al. (2018a) revised this floating orbital scale introducing three new low-frequency eccentricity cycles and tied to the magnetostratigraphy obtained by Iglesia Llanos et al. (2017) (see Chap. "Magnetostratigraphy of the Jurassic Through Lower Cretaceous in the Neuquén Basin"). This enabled the calibration of the Tithonian–Berriasian in the Neuquén Basin, chronostratigraphically and temporally, with unprecedented precision. In fact, the combination of cyclostratigraphy with other stratigraphic disciplines, such as magnetostratigraphy, biostratigraphy, chemostratigraphy, and geochronology, is contributing to establishing an increasingly robust chronostratigraphic framework in the Andes (Iglesia Llanos et al. 2017; Kohan Martínez et al. 2018; and Kietzmann et al. 2018a, b).

The aim of this study is to provide an update on cyclostratigraphic studies performed in the Upper Jurassic–Lower Cretaceous of the Neuquén Basin, as well as discussing their applicability in the temporal calibration of the stratigraphic record. In this study, two new complete Early Tithonian–Early Valanginian stratigraphic sections are added to the high-precision correlation made by Kietzmann et al. (2018a). The same analysis is presented in the Late Valanginian–Early Barremian interval where two new sections are described in detail by the authors and compared with two sections published by Sagasti (2000). We also analyze the different proxies and mechanisms that allowed the accumulation and preservation of the orbital controls that acted in the Neuquén Basin during this time interval.

2 Upper Jurassic–Lower Cretaceous Stratigraphy

The Late Jurassic–Early Cretaceous in the Neuquén Basin is characterized by a generalized thermal subsidence regime interrupted by localized tectonic events (Legarreta and Uliana 1991; Vergani et al. 1995; Ramos and Folguera 2005; Ramos 2010). This tectonic regime allowed the depocenters to be filled by continental deposits during lowstands of sea-level, and marine mixed siliciclastic-carbonate, carbonate and evaporitic systems during transgressive and highstand sea-level stages (Legarreta and Gulisano 1989; Legarreta and Uliana 1991; Legarreta et al. 1993).

After a period of low sea-level stand, during the Late Oxfordian–Kimmeridgian, a rapid Early Tithonian transgression led to the development of two thick mixed-carbonate ramp systems. The first established during the Early Tithonian–Early Valanginian and is represented by the Lower Mendoza Subgroup, whose distal facies are included in the Vaca Muerta Formation (Legarreta and Uliana 1991; Kietzmann et al. 2008, 2014, 2016). The second system took place during the Early Valanginian–Early Barremian and is represented by the Middle and Upper Mendoza Subgroup that includes the other thick rhythmic alternation of the Agrio Formation (Legarreta and Uliana 1991; Spalletti et al. 2001; Sagasti 2000, 2005) (Fig. 2).

2.1 Tithonian–Early Valanginian

In the Mendoza area, the Lower Mendoza Subgroup starts with the continental deposits of the Tordillo Formation (Kimmeridgian–Lower Tithonian), on top of which the distal marine deposits of the Vaca Muerta Formation (Early Tithonian–Early Valanginian) were abruptly deposited. The latter includes basinal to middle carbonate ramp deposits, which laterally pass to shallow-water clastic marine facies of the Lindero de Piedra Formation (Tithonian) and the middle to inner ramp oyster-deposits of the Chachao Formation (Early Valanginian), making up an homoclinal carbonate ramp system (e.g., Carozzi et al. 1981; Mitchum and Uliana 1985). The top of the Lower Mendoza Subgroup is marked in the southernmost part of the basin by the so-called Intravalanginian unconformity, a tectonically diacronich unconformity, which is not recorded in the Mendoza sector, and the Vaca Muerta/Chachao system is followed by the Agrio Formation.

Biostratigraphy is well defined based on ammonites (Leanza and Hugo 1977; Riccardi 2008, 2015; Riccardi et al. 2011). Calcareous nannofossils, radiolarians, organic and calcareous dinoflagellates, and calpionellids (Quattrocchio et al. 2003; Bown and Concheyro 2004; Ballent et al. 2004, 2011; Kietzmann and Palma 2009; Ivanova and Kietzmann 2017; Kietzmann, 2017) are also important markers for the calibration of ammonite zones, but still need a thorough review to accurately establish the stratigraphic position of the most important bioevents.

In the Vaca Muerta Formation, a magnetostratigraphic study has been carried out by Iglesia Llanos et al. (2017). These authors recognized 11 reverse and 10 normal polarity zones, which were calibrated based on the correlation between Andean and Tethyan ammonite zones, and the Geomagnetic Polarity Time Scale compiled by Ogg and Hinnov (2012a, b). Accordingly, the authors conclude that the deposition of the Vaca Muerta Formation took place during the M22r.2r to M15r Subchrons (see Chap. "Magnetostratigraphy of the Jurassic Through Lower Cretaceous in the Neuquén Basin").

A sequence stratigraphic framework was established by Kietzmann et al. (2014, 2016) for the southern Mendoza and northern Neuquén sectors. These authors recognized two hierarchies of depositional sequences: (1) composite depositional sequences (CS) for large-scale sequences and (2) high-frequency depositional sequences (HFS) for those of small scale. Because of their average duration, composite sequences are considered equivalent to third-order sequences, whereas high-frequency sequences are interpreted as fourth-order sequences. In this chapter, we consider only composite sequences (Fig. 2), due to the importance of removing sea-level trends for spectral analysis. High-frequency sequences are too close to the superbundle hierarchy, and therefore, their removal could generate fictitious frequencies.

2.2 Upper Valanginian–Lower Barremian

In the Mendoza area, the Middle Mendoza Subgroup involves the upper part of the Chachao Formation and the Lower Agrio Formation (Pilmatué Member). The Upper Mendoza Subgroup includes the Avilé and Upper (Agua de la Mula) Members of the Agrio Formation, extending up to the top of the Chorreado Member. The Chorreado Member should be considered from a lithostratigraphic point of view, as part of the Agrio Formation, although it should be included within the Huitrín Meseosequence based on genetic relationships from sequence stratigraphy (Gutiérrez Pleimling et al. 2011).

Biostratigraphy of this interval is well defined based on ammonites (Leanza and Hugo 1977; Aguirre-Urreta et al. 1999, 2005, 2011). Microfossils, especially calcareous nannofossils, organic-walled dinoflagellates, and foraminifera are well tied to ammonite zones (Bown and Concheyro 2004; Aguirre-Urreta et al. 2005; Ballent et al. 2011; Lescano and Concheyro 2014).

The Middle Mendoza Subgroup was divided into five depositional sequences by Legarreta and Gulisano (1989), while the Upper Mendoza Subgroup was divided into other five sequences (Fig. 2). However, with the purpose of removing eustatic trends for spectral analysis, we only considered the tendencies generated by the two second-order sequences that define both subgroups.

3 Methodology

Detailed sedimentological sections from the Vaca Muerta and Agrio formations were described bed-by-bed and studied for cyclicity. Cyclostratigraphic analysis is based on the differentiation of decimetric-scale carbonate/siliciclastic lithofacies couplets or elementary cycles. Bundles and superbundles of cycles were also recognized in the field but searched applying time-series analysis with the software PAST3.21 (Hammer et al. 2001; Hammer and Harper 2006) using the REDFIT procedure of Schulz and Mudelsee (2002). The thickness of the elementary cycles was used as a proxy. Time was introduced as constant time intervals once the duration of the elementary cycles was estimated: The time represented in the stratigraphic section divided by the total number of elementary cycles. The time series is introduced in the software in the form of two columns with time and thickness values, so that the periodicity of the elementary cycle is used as scale in both procedures (thickness/time conversion).

The following primary premises are assumed for spectral analyses: (1) Sections consist of a succession of elementary cycles with similar thickness within the range of 20–40 cm, so they are considered of similar time duration, (2) obtained spectrograms are statistically reliable, since the number of cycles is large enough to obtain statistically significant results. According to Weedon (2003), the sampling density must be at least 12 times the lowest frequency. In our sections, the time series is at least 25 times longer than the low-frequency eccentricity cycle, (3) vertical changes in facies and bed thicknesses throughout the sections are related to periodic climate factors and/or to other non-periodic factors. Each data set was corrected prior to spectral analysis, subtracting the mean value and trends generated by changes in sea-level, allowing data centering and variance stabilization.

4 Cyclostratigraphic Analysis

4.1 New Stratigraphic Sections for the Construction of the Upper Jurassic–Lower Cretaceous ATS in the Neuquén Basin

In this work, we added two new complete Early Tithonian–Early Valanginian stratigraphic sections from the Vaca Muerta Formation. The first one is located at Cañada Ancha on the margin of the Salado River (Figs. 1 and 2). This section extends from the Early Tithonian *Virgatosphinctes andesensis* Zone to the Early Valanginian *Olcostephanus (O.) atherstoni* Zone and is represented mainly by outer ramp facies (Kietzmann et al. 2014). The second section locates at the southern part of the Cara Cura range, in the dry bed of stream known as Río Seco de la Cara Cura. This section extends from the Early Tithonian *Virgatosphinctes andesensis* Zone to the Early Valanginian *Lissonia riveroi* Zone and is represented mainly by outer and middle ramp facies (Kietzmann et al. 2014).

For the cyclostratigraphic calibration of the Late Valanginian–Hauterivian, we used two detailed sections published by Sagasti (2000) for the Upper Hauterivian–Lower Barremian at the Salado River (Arroyo Cienaguitas) and the northern Cara Cura range (Río Seco de la Cinta Roja). In addition, we included a complete Upper Valanginian–Lower Barremian section from the Loncoche creek, and the lower part of the Pilmatue Member at Loma La Torre of Late Valanginian–Early Hauterivian age. Successions are dominated by similar outer ramp facies to those of the Vaca Muerta Formation (Sagasti and Poire 1998; Sagasti 2000, 2005; Kietzmann and Paulin 2019; Palazzolo 2019).

4.2 Cycles Staking Pattern

Both in the Vaca Muerta Formation and in the Agrio Formation, we recognized a well-ordered hierarchy of cycles, including elementary cycles, bundles of four or five elementary cycles, and superbundles (four or five bundles). Elementary cycles have a regular thickness in the order of 20–40 cm, and therefore, they can be considered as units of similar temporal duration. Each elementary cycle consists of two hemicycles of similar thickness, which can be represented by limestone/marlstone,

marlstone/marlstone, or marlstone/mudstone combinations depending to the facies in which occurs.

Elementary cycles are grouped into sets of 4–5 elementary cycles (bundles), which are grouped into sets of 4–5 bundles (superbundles). The 5:1 ratio (5 elementary cycles per bundles) is commonly attributed to high-frequency eccentricity (95 and 125 ky), while the 4:1 ratio (4 bundles per superbundles) as low-frequency eccentricity (405 ky). Bundles and superbundles start with a thick elementary cycle and show a higher proportion of marlstones toward the top. This characteristic stacking pattern is well exhibited in stratigraphic sections with low sedimentation rates, such as in Cuesta del Chihuido and Arroyo Loncoche (Fig. 3). Bundles and superbundles can be dominated by marlstone/marlstone or limestone/marlstone elementary cycles. This descriptive division correlates with facies and system tracts, so that variations in the proportion of marlstones or limestones have no real genetic connotation respect to the transference mechanisms of the orbital signal to the sedimentary record. Yet, it relates with the proximal-distal trend within the carbonate ramp system: Basinal to outer

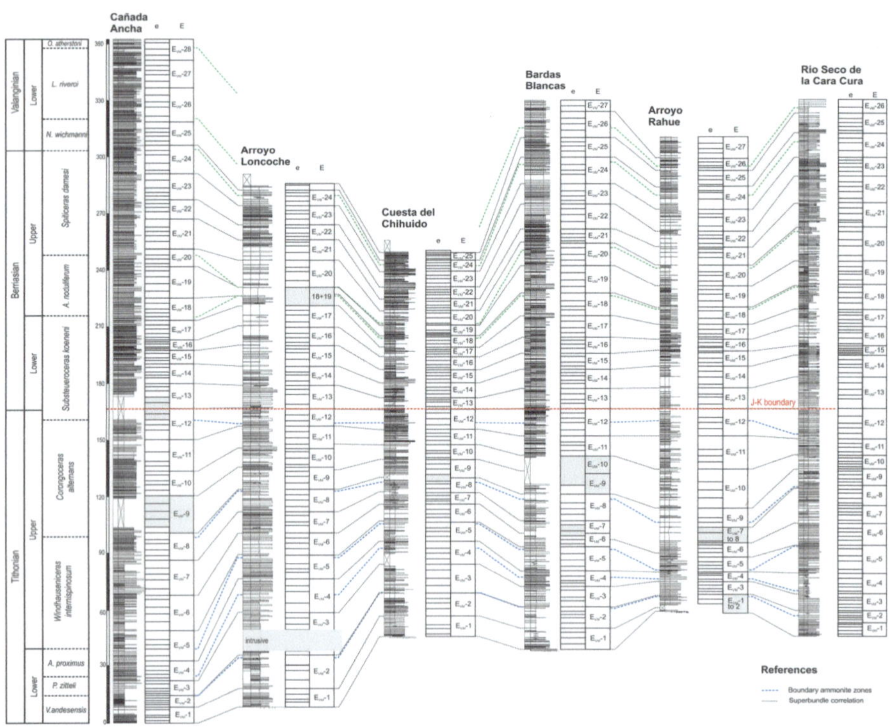

Fig. 3 High-resolution cyclostratigraphic correlation of studied stratigraphic sections of the Vaca Muerta Formation, using low-frequency eccentricity cycle (E) and high-frequency eccentricity cycle (e) (modified from Kietzmann et al. 2018a). Jurassic–Cretaceous boundary according to magnetostratigraphy (Iglesia Llanos et al. 2017; see Chap. "Magnetostratigraphy of the Jurassic Through Lower Cretaceous in the Neuquén Basin"), ammonite zones after Kietzmann et al. (2014, 2018a)

ramp deposits are dominated by marlstone/marlstone elementary cycles. Bundles or superbundles can occasionally start with a limestone/marlstone elementary cycle, and conversely, outer to middle ramp deposits are dominated by limestone/marlstone elementary cycles (Figs. 3, 4 and 5).

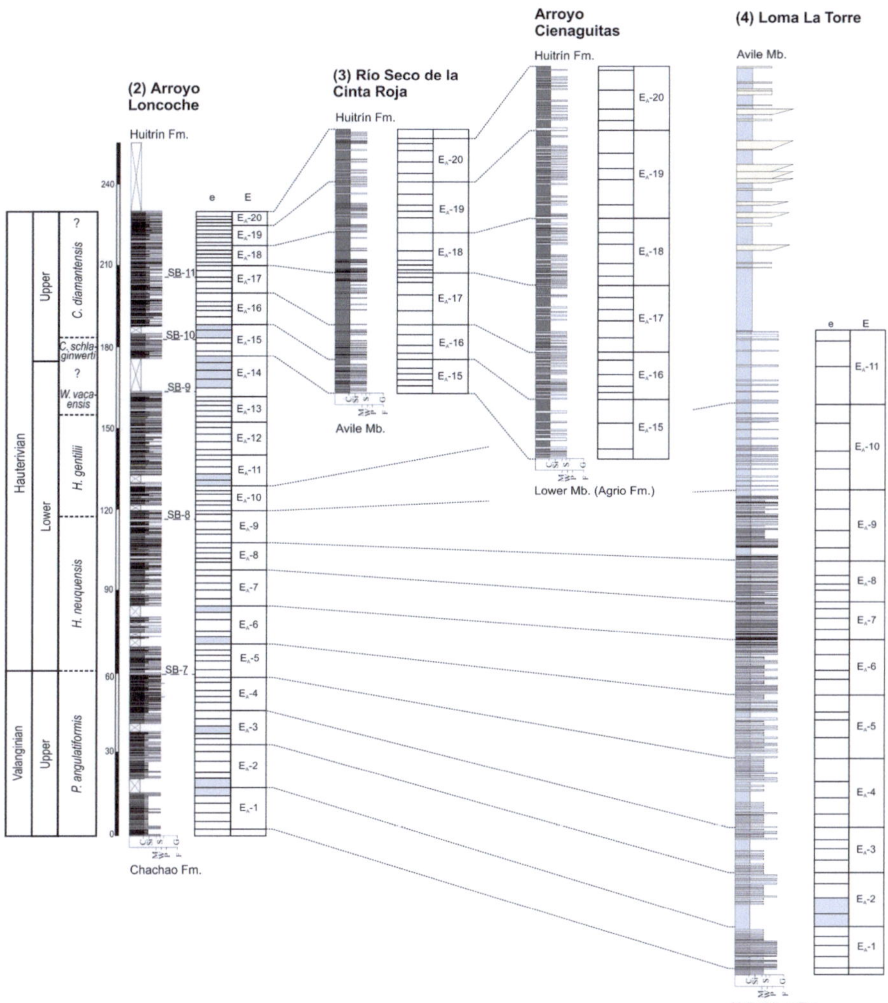

Fig. 4 High-resolution cyclostratigraphic correlation of studied stratigraphic sections of the Agrio Formation, using low-frequency eccentricity cycle (E) and high-frequency eccentricity cycle (e). Ammonite zones after Sagasti and Poire (1998), and Sagasti (2002, 2005)

4.3 *Elementary Cycle Periodicity and Spectral Analysis in the Vaca Muerta Formation*

According to ammonite biostratigraphic data from the studied sections, the Vaca Muerta Formation spans the Early Tithonian–Early Valanginian (Leanza and Hugo 1977). Detailed magnetostratigraphic correlations by Iglesia Llanos et al. (2017) in the Tithonian–Berriasian Arroyo Loncoche section indicate that the Vaca Muerta Formation spans the M22r.2r–M15r Subchrons (uppermost *Hybonoticeras hybonotum* to *Subthurmannia boissieri* Standard Zones). These data point to a time interval of *c.* 13 Ma (Ogg et al. 2016) for the Vaca Muerta Formation in Cañada Ancha and Río Seco de la Cara Cura sections. The Cañada Ancha section contains 654 elementary cycles, which are grouped in 130 bundles and 28 superbundles, whereas the Río Seco de la Cara Cura section includes 635 elementary cycles, 127 bundles, and 26 superbundles. Dividing the time-length of these sections by the number of elementary cycles, the periodicity of limestone/marlstone cycles has duration of ~21 ky, which can be attributed to the precessional cycle of the Earth. Data from the previous studied sections are consistent with this average value (Table 1).

Spectra obtained from time series of elementary cycle thickness are very consistent with each other (Fig. 4). At Cañada Ancha, REDFIT spectrum shows a peak above

Table 1 Cycle periodicities from the Vaca Muerta Formation (Lower Tithonian–Lower Valanginian)

Stratigraphic section	Estimated time (My)	Number of elementary cycles	Elementary cycle periodicity (ky)	Bundle periodicity (ky)	Superbundle periodicity (ky)
Previous studies					
Arroyo Loncoche	~10	487	20.53	118 91	410
Cuesta del Chihuido	~10	495	20.20	95	403
Bardas Blancas	~11	501	21.9	111	390
Arroyo Rahue	~11	510	21.56	89	395
El Ministerio	~3	168	18	120	400
This study					
Cañada Ancha	~13	654	19.9	102 84	480
Río Seco de la Cara Cura	~13	635	20.47	97	406

Data from previous studies include works of Kietzmann et al. (2011, 2015, 2018a) and Kohan Martínez et al. (2018)

Fig. 5 Limestone–marlstone rhythmic successions in the Neuquén Basin: **a** Upper Tithonian at Cuesta del Chihuido section (Vaca Muerta Formation). **b** Lower Hauterivian at Arroyo Loncoche section (Agrio Formation). References: (E) superbundles or low-frequency eccentricity cycles (e) bundles or high-frequency eccentricity cycles

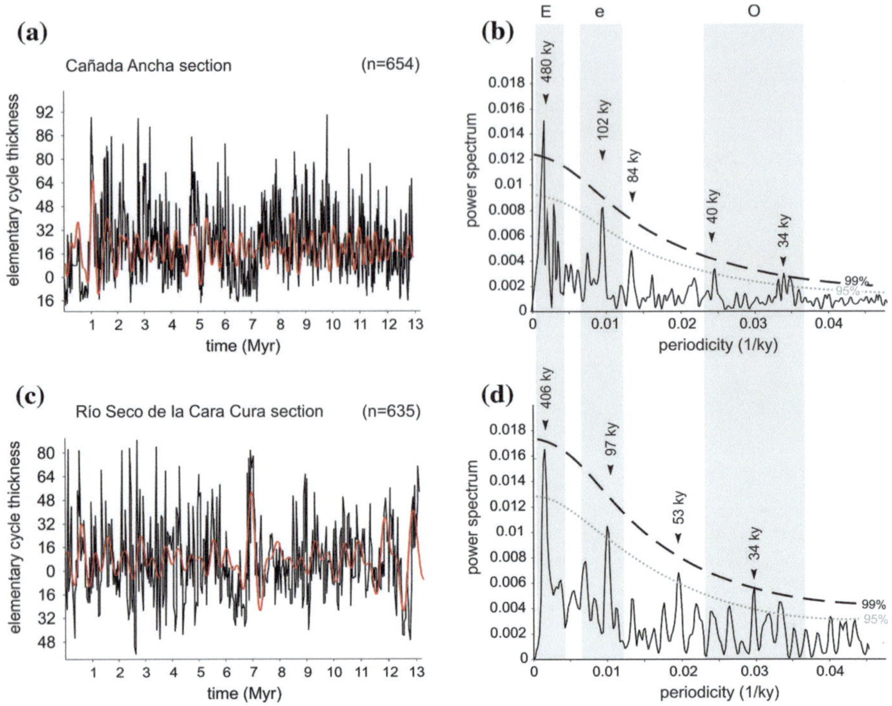

Fig. 6 Time series of elementary cycle thicknesses and REDFIT spectrum of new studied localities from the Vaca Muerta Formation, showing: low-frequency (E) and high-frequency (e) eccentricity cycle, obliquity (O) cycle and precession (P) cycle. **a**, **b** Cañada Ancha Section; **c**, **d** Río de la Cara Cura section. In **a** and **c,** red line indicates the modeled low-frequency eccentricity

the 95% confidence level and two peaks above the 99% confidence level (Fig. 6a, b). The first one has a periodicity of 403 ky which is consistent with the low-frequency eccentricity cycle. The other two peaks reveal periodicities of 103 cycles ky and 86 ky, respectively. These periodicities can also be attributed to the high-frequency eccentricity cycle. At Río Seco de la Cara Cura, a first peak of 410 ky and two peaks of 111 and 91 ky were observed, but an additional peak of 39 ky is observed above the 95% confidence level that can be assigned to the obliquity signal (Fig. 4). Similar results were observed in the other five sections of the Vaca Muerta Formation by Kietzmann et al. (2015, 2018a, b) for the Arroyo Loncoche, Cuesta del Chihuido, Bardas Blancas and Arroyo Rahue sections, as well as in El Ministerio section (Kohan Martínez et al. 2018) (Table 1). The calculated periodicities fit well with the so-called precession-eccentricity syndrome (PES) defined by Fischer et al. (2004), which is characteristic of mid- and low latitudes (Berger and Loutre 1994).

4.4 Elementary Cycle Periodicity and Spectral Analysis in the Agrio Formation

The Agrio Formation in the studied sections spans the Late Valanginian–Late Hauterivian (Leanza and Hugo 1977; Sagasti and Poire 1998; Sagasti 2000, 2005). Ammonite biostratigraphy suggests the correlation with the *Busnardoites campylotoxus* to the *Pseudothurmannia ohmi* Standard Zones (Aguirre-Urreta et al. 2005). However, magnetostratigraphic studies still need to be carried out in the Agrio Formation to tie the Andean ammonite zones in the international chronostratigraphic scale. These data point to a time interval of *c.* 8 Ma for the Agrio Formation at Arroyo Loncoche. The other partial sections from the Agrio Formation comprise approximately 3 Ma at Río Seco de la Cinta Roja, Arroyo Cianaguitas, and Loma La Torre.

The Arroyo Loncoche section, which spans the Late Valanginian–Late Hauterivian, contains 435 elementary cycles, which are grouped in 87 bundles and 20 superbundles. The Late Hauterivian–Early Barremian is represented also in the Arroyo Cienaguitas and Río Seco de la Cinta Roja sections. The Arroyo Cienaguitas contains 155 elementary cycles, 31 bundles, and 6 superbundles, Río Seco de la Cinta Roja section 135 elementary cycles, 27 bundles, and 6 superbundles, and Loma La Torre section (Upper Valanginian–Lower Hauterivian) 164 elementary cycles, 41 bundles, and 11 superbundles. After dividing the time-length of these sections for the number of elementary cycles, the periodicity of limestone/marlstone cycles has durations of ~18 to 19 ky that can be also attributed to the precessional cycle of the Earth (Table 2).

Table 2 Cycle periodicities from the Agrio Formation (Upper Valanginian–Lower Barremian)

Stratigraphic section	Estimated time (My)	Number of elementary cycles	Elementary cycle periodicity (ky)	Bundle periodicity (ky)	Superbundle periodicity (ky)
Previous studies					
Arroyo Cienaguitas	~3	103	~25	126 103 69	378
Río Seco de la Cinta Roja	~3	112	~18	125 117 77.5 65	875 319
This study					
Arroyo Loncoche	~8	435	18.4	97	390
Loma La Torre	~3	164	18.3	90	400

Previous studies include works of Sagasti (2000, 2005)

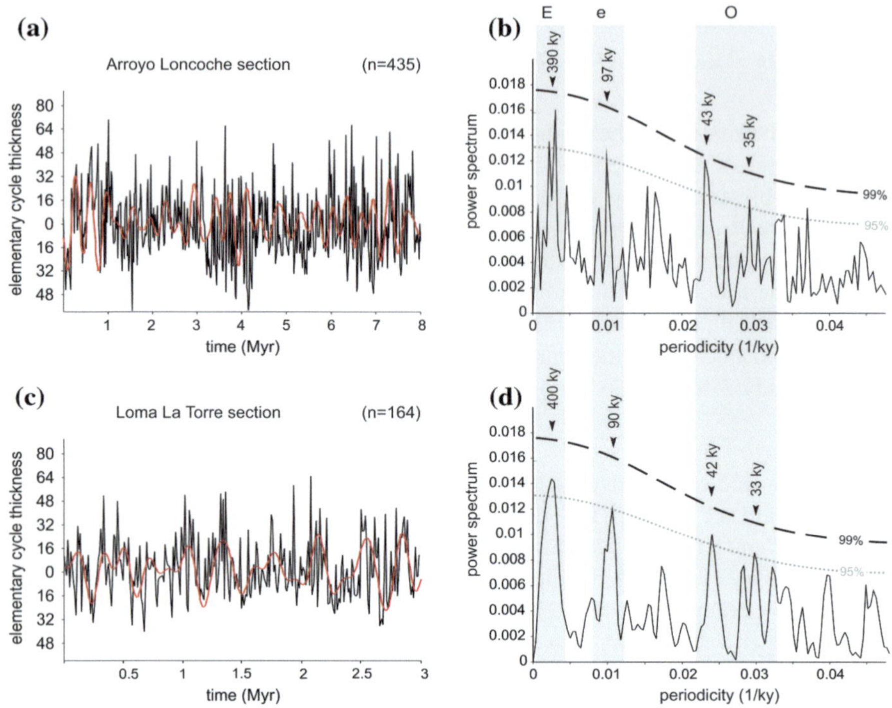

Fig. 7 Time series of elementary cycle thicknesses and REDFIT spectra of new studied localities from the Agrio Formation, showing low-frequency (E) and high-frequency (e) eccentricity cycles, obliquity (O) cycles and precession (P) cycles. **a**, **b** Arroyo Loncoche section; **c**, **d** Loma La Torre section. In **a** and **c,** red line indicates the modeled low-frequency eccentricity

Spectra obtained from time series of elementary cycle thicknesses are very consistent among each other (Fig. 7). At Arroyo Loncoche section, redfir spectrum shows four peaks above the 95% confidence level (Fig. 7a, b). The first one presents a periodicity of 390 ky which is consistent with the low-frequency eccentricity cycle. The other three peaks have periodicities of 97, 43 and 35 ky that can be attributed to the high-frequency eccentricity cycle and the obliquity cycle. At Loma La Torre section, a similar pattern is recognized (Fig. 7c, d), with a first peak of 400 ky and a second peak of 90 ky that are consistent with the low- and high-frequency eccentricity cycles, and two peaks of 42 and 33 ky that are assigned to the obliquity signal (Kietzmann and Paulin 2019).

4.5 High-Resolution Calibration of the Early Tithonian–Late Hauterivian in the Neuquén Basin

The cyclicity analysis of the ten studied stratigraphic sections enabled the construction of a floating astronomical scale for the Early Tithonian–Late Hauterivian in the Neuquén Basin (Fig. 8). Accordingly, minimum durations of ammonite zones within the Mendoza Group are the following:

(1) The Lower Tithonian ammonite zones are distributed in 5 low-frequency eccentricity cycles (E-cycles) so that considering the 405 ky periodicity, this interval represents 2.025 myr. The lower part of the Lower Tithonian is not represented, and 2 E-cycles would thus be missing. The *Virgatosphinctes andesensis* Zone contains two E-cycles, with a minimum duration of 0.81 myr. The *Pseudolissoceras zitteli* Zone shows 1.5 E-cycles with a minimum duration of 0.61 myr, whereas the *Aulacosphinctes proximus* Zone shows 1.5 E-cycles with a minimum duration of 0.61 myr.

2) The Upper Tithonian comprises 7 E-cycles, representing 2.835 myr. The *Windhauseniceras internispinosum* Zone comprises 3 E-cycles with a minimum duration of 1.21 myr, whereas the *Corongoceras alternans* Zone contains also 3 E-cycles with a minimum duration of 1.21 myr. According to magnetostratigraphic data from the same section (Iglesia Llanos et al. 2017) the J-K boundary, in the middle part of Magnetosubzone M19n.2n, one E-cycle from the following *Substeueroceras koeneni* Zone is part of the Upper Tithonian.

(3) The Lower Berriasian comprises 6 E-cycles, representing 2.43 myr and corresponds to the *Substeueroceras koeneni* Zone. The Upper Berriasian includes 7 E-cycles, representing other 2.84 myr. The *Argentiniceras noduliferum* Zone shows 2 E-cycles with a minimum duration of ~0.81 myr, and the *Spiticeras damesi* Zone contains 4 E-cycles, with a minimum duration of ~1.62 myr. According to magnetostratigraphy (Iglesia Llanos et al. 2017), the lowermost part of the *Neocomites wichmanni* Zone should be included in the Upper Berriasian extending it in 1 E-cycle (0.405 myr).

(4) The Lower Valanginian represented in the Vaca Muerta Formation includes 3 E-cycles distributed in 3 ammonite zones. The upper part of the *Neocomites wichmanni* Zone contains 2 high-frequency eccentricity cycles (E-cycles) with a minimum duration of 0.203 myr. The *Lissonia riveroi* Zone contains 2 E-cycles with a minimum duration of 0.81 myr. The lowermost part of the *Olcostephanus (O.) atherstoni* Zone is also represented in the Vaca Muerta Formation with 2 E-cycles and a minimum duration of 0.203 myr. The rest of this zone is represented in the Chachao Formation and in the basal part of the Agrio Formation, but not in the studied sections of the latter formation, its minimum duration has not been assessed in this study. The Upper Valanginian is represented by the *Pseudofavrella angulatiformis* Zone that contains 4 E-cyles and at least 1 e-cycle, with a minimum duration of 1.72 myr.

(5) The Lower Hauterivian ammonite zones are distributed in 8 E-cycles and 4 E-cycles, so that this interval represents 3.6 myr. The *Holcoptichites neuquensis*

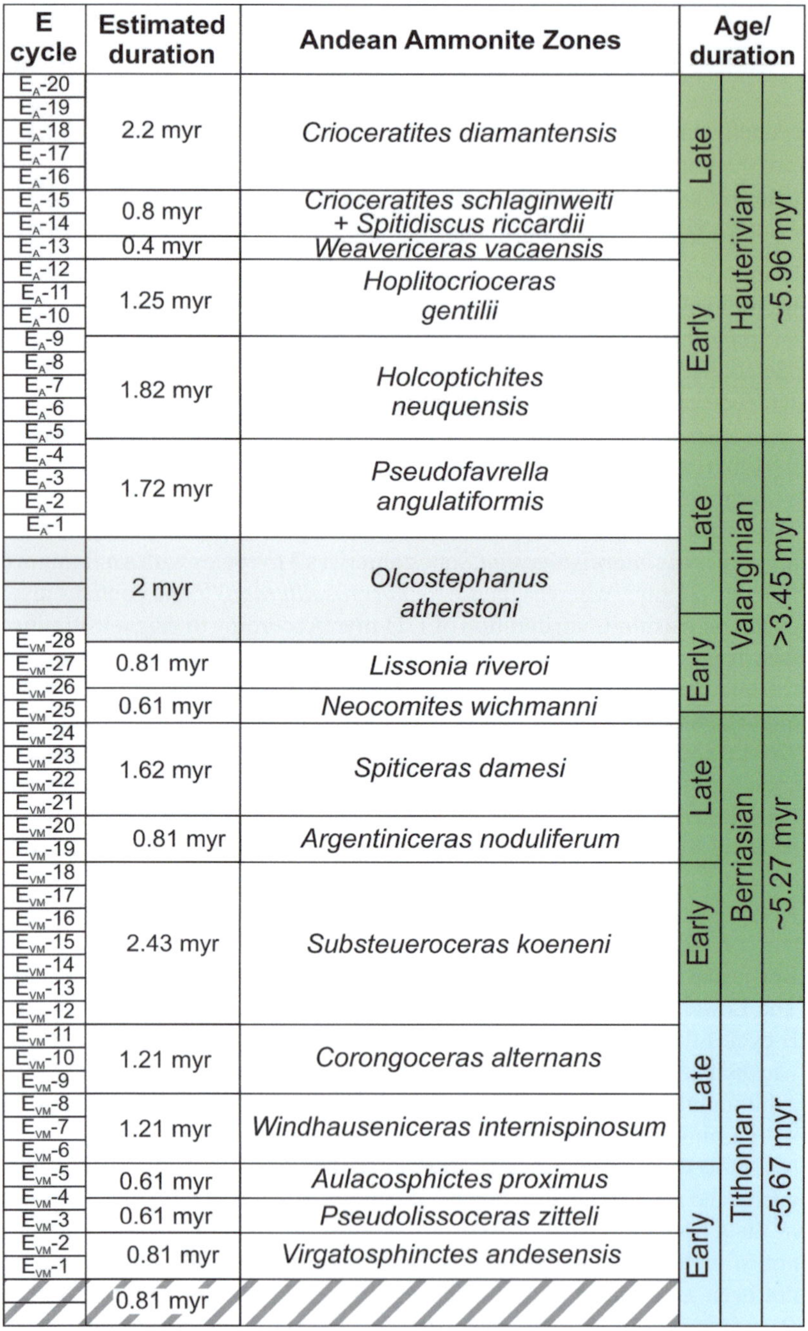

Fig. 8 Astronomical calibration of the Tithonian–Hauterivian in the Neuquén Basin, using 405 ky low-frequency eccentricity cycles. The lowermost Tithonian interval (oblique stripes) is not represented in the Vaca Muerta Formation and the illustrated cycles are taken from Huang et al. (2010). Gray-shaded interval in the *Olcostephanus (O.) atherstoni* Zone does not present cyclostratigraphic data

Zone contains 4 E-cycles and 2 E-cycles with a minimum duration of 1.82 myr. The *Hoplitocrioceras gentilii* Zone includes 3 E-cycles and 1 e-cycle showing a minimum duration of 1.25 myr. The *Weavericeras vacaensis* Zone is poorly defined in the studied sections, but it contains at least one E-cycle, having therefore a minimum duration of 0.4 myr.

(6) The Upper Hauterivian ammonite zones ares distributed in 4.5 E-cycles with a minimum duration of 1.82 myr. The *Spitidiscus riccardi* and the *Crioceratites schlaginwerti* Zones are poorly defined in the studied sections, but they contain at least 2 E-cycle, having therefore a minimum duration of 0.8 myr, and the *Crioceratites diamantensis* Zone contains 5 E-cycles with a minimum duration of 2.2 myr. Nevertheless, more detailed studies are needed to determine whether the studied sections reach or not the Early Barremian.

Based on the number of low-frequency eccentricity cycles within the Vaca Muerta Formation, we estimated a minimum duration of each stage in the Neuquén Basin (Fig. 8):

1) The Tithonian would have a minimum duration of 4.86 myr; however, the basal part of the Tithonian is not represented in the Vaca Muerta Formation. Considering ammonite zones durations reported by Huang et al. (2010) for the Kimmeridge Clay Formation, Kietzmann et al. (2018a) interpreted that at least two low-frequency eccentricity cycles are missing, and therefore, the estimated minimum duration for the Tithonian is 5.67 myr.

(2) Considering the position of the Jurassic–Cretaceous boundary obtained by magnetostratigraphy (Iglesia Llanos et al. 2017; see Chap. "Magnetostratigraphy of the Jurassic Through Lower Cretaceous in the Neuquén Basin") the Berriasian presents a minimum duration of 5.27 myr (Kietzmann et al. 2018a).

3) The Valanginian should be studied in more detail, since the basal part (*Olcostepahus (O.) atherstoni* Zone) has not been analyzed yet. However, the results of this work indicate that it would have at least a minimum duration of 3.45 myr. Cyclostratigraphic studies in the Vocotian Basin (France) and the Subbetic Domain (Spain) indicate a minimum duration of 5.9 myr (Huang et al. 1993) and 5.08 myr (Martínez et al. 2013) so that this zone would comprise at least 4 E-cycles with a duration of c.a. 2 myr.

4) The minimum duration of the Hauterivian is estimated in 5.96 myr; however, it is necessary to carefully analyze the upper part of the studied sections to determine whether it can reach in fact the Early Barremian. Previous cyclostratigraphic studies in the Vocotian Basin (France) and the Subbetic Domain (Spain) indicate a minimum duration of 5.3 myr (Huang et al. 1993) and 5.9 myr (Martínez et al. 2015), very consistent with our results, but considerably longer than the durations proposed in the GTS2012 and 2016 (Ogg et al. 2016).

5 Orbital Controls on the Sedimentation in the Neuquén Basin

Marlstone–limestone rhythmic deposits are important examples of climate-forced sedimentation resulting from different palaeoenvironmental responses to orbital forcing (Arthur et al. 1984; Einsele and Ricken 1991; Berger and Loutre 1994). However, some authors have claimed a diagenetic origin for these cycles (Ricken 1986, 1987; Munnecke and Samtleben 1996; Westphal et al. 2000, 2004, 2010). Criteria for differentiating primary from diagenetic origin deposits were described by Einsele (1982) and Einsele and Ricken (1991), based on simple geological criteria. Recently, Westphal et al. (2010) stated that the occurrence of palynomorphs, calcite fossils in both calcareous and terrigenous hemicycles, sedimentary structures and textures, and variations in clay mineralogy constitute unequivocal criteria for interpreting a primary origin of the deposits. The Vaca Muerta Formation includes several groups of planktonic and benthic invertebrates of aragonitic, calcitic, and organic composition, which are present in both calcareous and clastic hemicycles. Additionally, although there is evidence of diagenetic overprint (Catalano et al. 2018), taphonomic, sedimentary structures, and petrographic evidences are irrefutable signals that support the primary origin of rhythmicity in the Vaca Muerta Formation (Kietzmann et al. 2015).

Four main transference mechanisms were proposed for primary rhythmic successions (Einsele and Ricken 1991), including fluctuations in the production of calcareous biogenic particles by plankton or "productivity cycles," in the input of fine-grained terrigenous sediments or "dilution cycles," of the lysocline or "dissolution cycles," and in the minimum oxygen zone or "redox cycles." The first two mechanisms were proposed in the Neuquén Basin: Productivity and dilution mechanisms were suggested in different intervals for the Vaca Muerta Formation by Concheyro et al. (2006) and Palma et al. (2008) based on the limestone/marlstone ratio, whereas the dilution mechanism was interpreted for the Agrio Formation by Sagasti (2000, 2005) based on sedimentological and geochemical data. However, Kietzmann et al. (2011, 2015) proposed that elementary cycles are driven by "carbonate exportation" (Pittet and Strasser 1998; Bádenas et al. 2003). These cycles are originated by fluctuation in shallow-water carbonate production, which in turn, involves changes in carbonate basinward exportation.

The estimated water depth for the Vaca Muerta and Agrio formations ranges between 50 and 200 m (Sagasti 2005; Kietzmann et al. 2008, 2014), so dissolution mechanism did not produce the rhythmicity, since the carbonate compensation depth (CCD) during the Upper Jurassic–Lower Cretaceous was at 2–3 km (Arthur et al. 1985). Similarly, redox cycles cannot be considered as a potential mechanism, since any changes in the icnological and faunal associations are observed at elementary cycles scale (Doyle et al. 2005; Kietzmann and Palma 2009; Kietzmann et al. 2014).

Productivity (sensu Einsele 1982) and fertility (sensu Premoli Silva et al. 1989) mechanisms would not explain the cyclicity in the Vaca Muerta Formation, since sedimentation was controlled by hydrodynamic factors instead of a strong pelagic input;

the abundance of transported skeletal particles and particles derived from shallow-water areas, such as dasycladacean algae, agglutinated foraminifera, and abundant crustacean microcoprolites in both carbonate and marly facies, are strong evidences of shallow-water input (Kietzmann et al. 2014, 2015). Similar observations were made by Sagasti (2005) for the Agrio Formation, where pelagic calcareous material is represented by a poorly diversified association of calcareous nannofossils that is considered by this author a subsidiary component. In fact, the most abundant components in mudstones and limestones are micritic pellet (Kietzmann and Palma 2011, 2014). Recent SEM studies show that these pellets are made up of coccoliths that can produce lamination and were produced by shrimps (callianassids) (Kietzmann et al. 2010; Kietzmann and Palma 2014). Callianassids dwell from shallow to deep marine areas and can be detritivorous or suspension feeders. In fact, the middle and outer ramp facies of the Vaca Muerta and Agrio formations are rich in pellets, but also show a high level of bioturbation produced by crustaceans, such as *Thalassinoides* or *Rhizocorallium* (Sagasti and Poire 1998; Doyle et al. 2005; Kietzmann and Palma 2009). As mentioned by Kietzmann and Palma (2011), pellets are mostly transported, appearing most frequently in medium-sand size, and hydrodynamically behaving as sand. Therefore, cyclicity can be interpreted in terms of "productivity" related to variations in carbonate exportation. The four productivity and dilution intervals identified in the Vaca Muerta Formation by Concheyro et al. (2006) and Palma et al. (2008) are related to changes in sea-level. The two marlstone-rich intervals attributed to dilution cycles coincide with transgressive system tracts, whereas carbonate-rich intervals attributed to productivity cycles coincide with regressive system tracts, where the increase in the proportion of carbonates results from a decrease in accommodation that induces an increase in carbonate basinward exportation from shallower areas (Kietzmann et al. 2015).

The dilution mechanism was proposed by Sagasti (2005) for the Agrio Formation, based on the idea of high-frequency paleoclimatic fluctuation in the North Patagonian Massif, which was the main active source area during the Agrio carbonate ramp deposition (Legarreta and Uliana 1991; Eppinger and Rosenfeld 1996; Sagasti 2005). In that model, fluctuations in clastic dilution occurred as from recurring latitudinal migration of climate zones forced by the precessional cycle. The alternation of humid and arid climatic conditions in the North Patagonian Massif caused changes in the amount of terrigenous sediment supply. During one phase of the precessional cycle, the humid conditions promoted an increase in rainfall, and consequently in runoff and sediment supply to the marine basin. During the arid phase, terrigenous input was inhibited, and carbonate material exported from shallow areas, and could be accumulated in the distal areas.

An important question for the Agrio dilution model is that even if orbital forcing would have provided the trigger, it is not clear if it was effective during the Late Jurassic–Early Cretaceous. Climate response during Neogene through Holocene was much more sensitive due to the presence of large polar ice masses and glaciers that increase sensitivity of the system to disturbance (Berger 2013). In contrast, carbonate systems were highly sensitive to orbitally induced temperature changes during greenhouse conditions (Fischer 1982; Goldhammer et al. 1990, Strasser et al.

1999), particularly in shallow water, where these fluctuations controlled the carbonate production rate and sea-level amplitude (Algeo and Wilkinson 1988; Drummond and Wilkinson 1993; Burgess et al. 2001; Kemp et al. 2016). In this scenario, the carbonate exportation mechanism would be more likely. However, since mudstone volume in units of the Mendoza Group is striking, it is difficult to asses which of the mechanisms remained constant and which one fluctuated, so that the sedimentary pattern observed in the Vaca Muerta and Agrio Formations would result from the coexistence of both mechanisms.

6 Conclusions

The Vaca Muerta and Agrio formations in southern Mendoza are characterized by decimeter-scale rhythmic alternations of marlstones and limestones, showing a well-ordered hierarchy of cycles, including elementary cycles, bundles, and superbundles. Based on spectral analysis, elementary cycles have a periodicity of ~21 ky, which correlates with the precessional cycle of Earth's axis; bundles and superbundles have periodicities of ~90 to 120 ky and ~400 ky, which we interpret as the modulation of the precessional cycle by the Earth's orbital eccentricity. Obliquity periodicities have also been recognized, but not in all the studied sections.

Cyclostratigraphic data allowed us to build a floating astronomical time scale for the Tithonian–Hauterivian in the Neuquén Basin: (1) The Tithonian would have a minimum duration of 5.67 myr. (2) Considering the position of the Jurassic–Cretaceous boundary obtained by magnetostratigraphy, the Berriasian presents a minimum duration of 5.27 myr. (3) The Valanginian should be studied in more detail, since the *Olcostepahus (O.) atherstoni* Zone has not been analyzed yet, however, it would have at least a minimum duration of 3.45 myr. (4) The minimum duration of the Hauterivian is estimated in 5.96 myr, which is very consistent with previous cyclostratigraphic studies in the Tethys but differs with the 3.1 myr in the GTS2012 and 3.9 y the GTS2016.

Acknowledgements We are especially grateful to A. C. Riccardi (Universidad Nacional de La Plata y Museo, Argentina) and H. A. Leanza (Museo de Ciencias Naturales Bernandino Rivadavía, Argentina) for the helpful discussions regarding the biostratigraphy of the Neuquén Basin. We thank Sebastián Paulin and Franco Palazzolo for the help in the fieldwork. This research has been done under the framework of the PICT-2015-0206 and PICT-2016-3762 projects supported by Agencia Nacional de Promoción Científica y Tecnológica.

References

Aguirre-Urreta MB, Concheyro A, Lorenzo M et al (1999) Advances in the biostratigraphy of the Agrio Formation (Lower Cretaceous) of the Neuquén Basin, Argentina: ammonites, palynomorphs, and calcareous nannofossils. Palaeogeogr Palaeocl 150:33–47

Aguirre-Urreta MB, Rawson PF, Concheyro GA et al (2005) Lower Cretaceous biostratigraphy of the Neuquén Basin. In: Veiga GD, Spalletti LA, Howell JA, Schwarz E (eds) The Neuquén Basin, Argentina: a case study in sequence stratigraphy and basin dynamics. The Geological Society, London, SP 252, pp 57–81

Aguirre-Urreta B, Lazo DG, Griffin M et al (2011) Megainvertebrados del Cretácico y su importancia bioestratigráfica. In: Leanza HA, Arregui C, Carbone O, Danieli JC, Vallés JM (eds) Geología y Recursos Naturales de la Provincia del Neuquén. Asociación Geológica Argentina, Buenos Aires, pp 465–488

Algeo TJ, Wilkinson BH (1988) Periodicity ofmesoscale Phanerozoic sedimentary cycles and the roleof Milankovitch orbital modulation. J Geol 96:313–322

Arthur MA, Dean WE, Stow DAV (1984) Models for the deposition of Mesozoic fine-grained organic-carbon-rich sediment in the deep sea. In: Stow DAV, Piper DJW (eds) Fine-grained sediments: deep-water processes and facies. The Geological Society, London, SP 15, pp 527–560

Arthur MA, Dean WE, Pollastro R et al (1985) A comparative geochemical study of two transgressive pelagic limestone units, Cretaceous western interior basin, U.S. In: Pratt LM, Kauffman EG, Zelt FB (eds) Fine-grained deposits and biofacies of the Cretaceous western interior seaway: evidence of cyclic sedimentary processes. Soc Econ Paleontol Mineral, Field Trip Guidebook 4, pp 16–27

Bádenas B, Aurell M, Rodríguez-Tovar FJ, Pardo-Igúzquiza E (2003) Sequences stratigraphy and bedding rhythms of an outer ramp limestones succession (Late Kimmeridgian, Northeast Spain). Sediment Geol 161:153–174

Ballent SC, Ronchi DI, Angelozzi GN (2004) Microfósiles calcáreos tithonianos (Jurásico superior) en el sector oriental de la cuenca Neuquina, Argentina. Ameghiniana 41:13–24

Ballent S, Concheyro A, Náñez C et al (2011) Microfósiles mesozoicos y cenozoicos. In: Leanza HA, Arregui C, Carbone O, Danieli JC, Vallés JM (eds) Geología y Recursos Naturales de la Provincia del Neuquén. Asociación Geológica Argentina, Buenos Aires, pp 489–528

Berger WH (2013) On the Milankovitch sensitivity of the Quaternary deep-sea record. Clim Past 9:2003–2011

Berger A, Loutre MF (1994) Astronomical forcing through geological time. In: de Boer PL, Smith DG (eds) Orbital Forcing and cyclic sequences. IAS, SP 19, pp 15–24

Bown P, Concheyro A (2004) Lower Cretaceous calcareous nannoplankton from the Neuquén Basin, Argentina. Mar Micropal 52:51–84

Burgess PM, Wright VP, Emery D (2001) Numerical forward modelling of peritidal carbonate parasequence development: implications for outcrop interpretation. Basin Res 13:1–16

Carozzi AV, Bercowski F, Rodriguez M et al (1981) Estudio de microfacies de la Formación Chachao (Valanginiano), Provincia de Mendoza. In: Abstracts of the 8 Congreso Geológico Argentino, San Luis, 20–26 Sept 1981

Catalano JP, Scasso RS, Kietzmann DA et al (2018) Carbonate sedimentology and diagenesis of Vaca Muerta Formation in Puerta Curaco, Neuquén Basin. In: Abstracts of the 10° Congreso de Exploración y Desarrollo de Hidrocarburos, Mendoza, 7–9 Nov 2018

Concheyro A, Palma RM, Lescano M et al (2006) Nanofósiles calcáreos en los episodios de productividad y dilución de la Formación Vaca Muerta. In: Abstracts of the 9 Congreso Argentino de Paleontología y Bioestratigrafía, Córdoba, 18–22 Sept 2006

Doyle P, Poire DG, Spalletti LA et al (2005) Relative oxygenation of the Tithonian–Valanginian Vaca Muerta–Chachao formations of the Mendoza Shelf, Neuquén Basin, Argentina. In: Veiga GD, Spalletti LA, Howell JA et al (eds) The Neuquén Basin, Argentina: a case study in sequence stratigraphy and basin dynamics. The Geological Society, London, SP 252, pp 185–206

Drummond CN, Wilkinson BH (1993) Carbonatecycle stacking patterns and hierarchies of orbitallyforced eustatic sea-level change. J Sed Petrol 63:369–377

Einsele G (1982) Limestone-Marl cycles (periodicities): diagnosis, significance, causes—a review. In: Einsele G, Seilacher A (eds) Cyclic and event stratification. Springer, Berlin, pp 8–53

Einsele G, Ricken W (1991) Limestone-Marl alternation—an overview. In: Einsele G, Ricken W, Seilacher A (eds) Cycles and events in stratigraphy. Springer, Berlin, Heidelberg, New York, pp 23–47

Eppinger KJ, Rosenfeld U (1996) Western margin and provenance of sediments of the Neuquén Basin (Argentina) in the late jurassic and early cretaceous. Tectonophysics 259:229–244

Fischer AG (1982) Long-term climatic oscillations recorded in stratigraphy. Geophysics Study Committee, National Research Council, Climate in Earth History. National Academy Press, Washington DC, pp 97–104

Fischer AG, D'Argenio B, Premoli Silva I, Weissert H, Ferreri V (2004) Cyclostratigraphic approach to earth's history: an introduction. In: D'Argenio B, Fischer AG, Premoli Silva I, Weissert H, Ferreri V (eds) Cyclostratigraphy: Approaches and case histories. Soc Econ Paleontol Mineral, SP 81, pp 5–16

Goldhammer RK, Dunn PA, Hardie LA (1990) Depositional cycles, composite sea-level changes, cycle stacking patterns, and the hierarchy of stratigraphic forcing: examples from Alpine Triassic platform carbonates. Geol Soc Am Bull 102:535–562

Gutiérrez Pleimling A, Olea G, Suárez M et al (2011) El Miembro Chorreado de la Formación Huitrín (Cretácico Temprano). In: Leanza HA, Arregui C, Carbone O et al (eds) Geología y Recursos Naturales de la Provincia del Neuquén. Asociación Geológica Argentina, Buenos Aires, pp 175–180

Hammer Ø, Harper DAT (2006) Paleontological data analysis. Blackwell, Oxford

Hammer Ø, Harper DAT, Ryan PD (2001) PAST: paleontological statistics software package for education and data analysis. Palaeont Electr 4(1):1–9

Hinnov L, Hilgen F (2012) Cyclostratigraphy and astrochronology. In: Gradstein FM, Ogg JG, Schmitz MD, Ogg GM (eds) The geologic time scale. Elsevier, Oxford, pp 63–84

Huang C, Ogg JG, Gradstein FM (1993) A quantitative study of Lower Cretaceous cyclic sequences from the Atlantic Ocean and the Vocotian Basin (SE France). Paleoceanography 8(2):275–291

Huang C, Hesselbo SP, Hinnov L (2010) Astrochronology of the late Jurassic Kimmeridge Clay (Dorset, England) and implications for Earth system processes. Earth Planet Sci Lett 289:242–255

Iglesia Llanos MP, Kietzmann DA, Kohan Martínez M et al (2017) Magnetostratigraphy of the Upper JurassiceLower Cretaceous from Argentina: implications for the J-K boundary in the Neuquén Basin. Cretac Res 70:189–538

Ivanova DK, Kietzmann DA (2017) Calcareous dinoflagellate cysts from the Tithonian - Valanginian Vaca Muerta Formation in the southern Mendoza area of the Neuquén Basin, Argentina. J S Am Earth Sci 77:150–169

Kemp DB, Van Manen SM, Pollitt DA et al (2016) Investigating the preservation of orbital forcing in peritidal carbonates. Sedimentology 63:1701–1718

Kietzmann DA (2017) Chitinoidellids from the early Tithonian - early Valanginian Vaca Muerta Formation in the Northern Neuquén Basin, Argentina. J S Am Earth Sci 76:152–164

Kietzmann DA, Blau J, Fernández DE, Palma RM (2010) Crustacean microcoprolites from the upper jurassic—lower cretaceous of the Neuquén Basin, Argentina: systematics and biostratigraphic implications. Acta Palaeontol Pol 55(2):277–284

Kietzmann DA, Palma RM (2009) Microcrinoideos saccocómidos en el Tithoniano de la Cuenca Neuquina. ¿Una presencia inesperada fuera de la región del Tethys? Ameghiniana 46:695–700

Kietzmann DA, Palma RM (2011) Las tempestitas peloidales de la Formación Vaca Muerta (Tithoniano-Valanginiano) en el sector surmendocino de la Cuenca Neuquina. La Am J Sedimentology Basin Anal 18:121–149

Kietzmann DA, Palma RM (2014) Early Cretaceous crustacean microcoprolites from Sierra de la Cara Cura, Neuquén Basin, Argentina: Taphonomy, environmental distribution, and stratigraphic correlation. Cret Res 49:214–228

Kietzmann DA, Paulin SM (2019) Cyclostratigraphy of an upper Valanginian – lower Hauterivian mixed siliciclastic-carbonate ramp succession (Pilmatué Member of the Agrio Formation), Loma La Torre section, northern Neuquén Basin, Argentina. Cret Res 98:26–46

Kietzmann DA, Palma RM, Bressan GS (2008) Facies y microfacies de la rampa tithoniana-berriasiana de la Cuenca Neuquina (Formación Vaca Muerta) en la sección del arroyo Loncoche – Malargüe, provincia de Mendoza. Rev Asoc Geol Argent 63:696–713

Kietzmann DA, Martín-Chivelet J, Palma RM et al (2011) Evidence of precessional and eccentricity orbital cycles in a Tithonian source rock: the mid-outer carbonate ramp of the Vaca Muerta Formation, Northern Neuquén Basin, Argentina. AAPG Bull 95:1459–1474

Kietzmann DA, Palma RM, Riccardi AC et al (2014) Sedimentology and sequence stratigraphy of a Tithonian-Valanginian carbonate ramp (Vaca Muerta Formation): a misunderstood exceptional source rock in the Southern Mendoza area of the Neuquén Basin, Argentina. Sediment Geol 302:64–86

Kietzmann DA, Palma RM, Iglesia Llanos MP (2015) Cyclostratigraphy of an orbitally-driven Tithonian-Valanginian carbonate ramp succession, Southern Mendoza, Argentina: implications for the Jurassic-Cretaceous boundary in the Neuquén Basin. Sediment Geol 315:29–46

Kietzmann DA, Ambrosio A, Suriano J et al (2016) The Vaca Muerta-Quintuco system (Tithonian–Valanginian) in the Neuquén Basin, Argentina: a view from the outcrops in the Chos Malal fold and thrust belt. AAPG Bull 100:743–771

Kietzmann DA, Iglesia Llanos MP, Kohan Martinez M (2018a) Astronomical calibration of the Tithonian-Berriasian in the Neuquén Basin, Argentina: a contribution from the Southern Hemisphere to the Geologic Time Scale. In: Montenari M (ed) Cyclostratigraphy and astrochronology. Elsevier, Stratigraphy & Timescales 3, pp 327–355

Kietzmann DA, Iglesia Llanos MP, Ivanova DKMA et al (2018b) Toward a multidisciplinary chronostratigraphic calibration of the Jurassic-Cretaceous transition in the Neuquén Basin. Rev Asoc Geol Argent 75(2):175–187

Kohan Martínez M, Kietzmann DA, Iglesia Llanos MP et al (2018) Magnetostratigraphy and cyclostratigraphy of the Tithonian interval from the Vaca Muerta Formation, southern Neuquén Basin, Argentina. J S Am Earth Sci 85:209–228

Leanza HA, Hugo CA (1977) Sucesión de amonites y edad de la Formación Vaca Muerta y sincrónicas entre los Paralelos 35º y 40º l.s. Cuenca Neuquina-Mendocina: Rev Asoc Geol Argent 32:248–264

Legarreta L, Gulisano CA (1989) Análisis estratigráfico de la Cuenca Neuquina (Triásico Superior-Terciario Inferior). In: Chebli GA, Spalletti LA (eds) Cuencas Sedimentarias Argentinas. Universidad de Tucumán, Serie Correlación Geológica, vol 6, pp 221–243

Legarreta L, Uliana MA (1991) Jurassic–Cretaceous Marine Oscillations and Geometry of Back Arc Basin, Central Argentina Andes. In: McDonald DIM (ed) Sea level changes at active plate margins: process and product. IAS, SP 12, pp 429–450

Legarreta L, Uliana MA (1996) The Jurassic succession in west central Argentina: stratal patterns, sequences, and paleogeographic evolution. Palaeogeogr Palaeocl 120:303–330

Legarreta L, Gulisano C, Uliana M (1993) Las secuencias sedimentarias jurásico-cretácicas. In: Ramos VA (ed) Geología y Recursos Naturales de la Provincia de Mendoza. Asociación Geológica Argentina, Buenos Aires, pp 87–114

Lescano M, Concheyro A (2014) Nanocónidos del Grupo Mendoza (Cretácico Inferior) en la Provincia del Neuquén, República Argentina: Taxonomía, Cronoestratigrafía e Implicancias Paleogeográficas. Ameghiniana 51:466–499

Martínez M, Deconick J-F, Pellenard P et al (2013) Astrochronology of the Valanginian stage from reference sections (Vocotian Basin, France) and palaeoenvironmental implications for the Weissert event. Palaeogeogr Palaeocl 376:91–102

Martínez M, Deconick J-F, Pellenard P et al (2015) Astrochronology of the Valanginian-Hauterivian stages (Early Cretaceous): chronological relationships between the Paraná-Etendeka large igneous province and the Weissert and Faraoni events. Global Planet Change 131:158–173

Mitchum RM, Uliana M (1985) Seismic stratigraphy of carbonate depositional sequences, Upper Jurassic-Lower Cretaceous, Neuquén Basin, Argentina. In: Berg BR, Woolverton DG (eds) Seismic Stratigraphy 2. An integrated approach to hydrocarbon analysis. Tulsa, AAPG Mem 39:255–83

Munnecke A, Samtleben C (1996) The formation of micritic limestones and the development of limestone-marl alternations in the Silurian of Gotland, Sweden. Facies 34:159–176
Ogg JG, Hinnov L (2012a) The Jurassic Period. In: Gradstein FM, Ogg JG, Schmitz M, Ogg G (eds) The Geologic time scale 2012. Elsevier, Amsterdam, pp 731–792
Ogg JG, Hinnov L (2012b) The Cretaceous Period. In: Gradstein FM, Ogg JG, Schmitz M, Ogg G (eds) The geologic time scale 2012. Elsevier, Amsterdam, pp 793–854
Ogg JG, Ogg GM, Gradstein FM (2016) A concise geologic time scale. Elsevier, Amsterdam
Palazzolo F (2019) Cicloestratigrafía de la Formación Agrio (Valanginaiano-Hauteriviano) en la sección del arroyo Loncoche, Cuenca Neuquina surmendocina. Degree thesis, Universidad de Buenos Aires
Palma RM, Martín-Chivelet J, López Gómez J et al (2008) High-resolution cyclostratigraphy analysis from a Tithonian alternating marl-limestones succession Vaca Muerta Fomation, Neuquén Basin, Mendoza, Argentina. In: Abstracts of the IGCP 506 Fifth Symposiums, Hammamet, 28–31 Mar 2008
Pittet B, Strasser A (1998) Depositional sequences in deep-shelf environments formed through carbonate mud import from the shallow platform (late Oxfordian, German Swabian Alb and eastern Swiss Jura). Eclogae Geol Helve 91:149–169
Premoli Silva I, Erba E, Tornaghi I (1989) Paleoenvironmental signals and changes in surface fertility in mid-Cretaceous C org-rich facies of the fucoid marls (central Italy). Geobios 11:225–236
Quattrocchio ME, Martínez MA, García VM et al (2003) Palinoestratigrafía del tithoniano-Hauteriviano del centro-oeste de la Cuenca Neuquina, Argentina. Rev Esp Micropaleont 354:51–74
Ramos VA (2010) The tectonic regime along the Andes: present-day and Mesozoic regimes. Geol J 45:2–25
Ramos VA, Folguera A (2005) Tectonic evolution of the Andes of Neuquén: constraints derived from the magmatic arc and Foreland deformation. In: Veiga GD, Spalletti LA, Howell JA, Schwarz E (eds) The Neuquén Basin, Argentina: a case study in Sequence stratigraphy and basin dynamics. The Geological Society, London, SP 252, pp 15–35
Riccardi AC (2008) The marine Jurassic of Argentina: a biostratigraphic framework. Episodes 31:326–335
Riccardi AC (2015) Remarks on the Tithonian-Berriasian ammonite biostratigraphy of west central Argentina. Volumina Jurassica 13:23–52
Riccardi AC, Damborenea SE, Manceñido MO et al (2011) Megainvertebrados jurásicos y su importancia geobiológica. In: Leanza HA, Arregui C, Carbone O, Danieli JC, Vallés JM (eds) Geología y Recursos Naturales de la Provincia del Neuquén. Asociación Geológica Argentina, Buenos Aires, pp 441–464
Ricken W (1986) Diagenetic bedding: a model for Limestone–Marl alternations. Springer, Berlin
Ricken W (1987) The carbonate compaction law: a new tool. Sedimentology 34:1–14
Sagasti G, Poire D (1998) Asociaciones icnológicas de la porción basal de la Formación Agrio, arroyo Loncoche, provincia de Mendoza. Rev Asoc Argentina Sediment 5:105–118
Sagasti G (2000) Ciclos de Milankovitch en el Cretácico Inferior de la Cuenca Neuquina Surmendocina, Argentina. In: Abstracts of the 2 Congreso Latinoamericano de Sedimentología y 8 Reunión Argentina de Sedimentología, Mar del Plata, 14–17 Mar 2000
Sagasti G (2005) Hemipelagic record of orbitally-induced dilution cycles in Lower Cretaceous sediments of the Neuquén Basin. In: Veiga GD, Spalletti LA, Howell JA, Schwarz E (eds) The Neuquén Basin, Argentina: A case study in sequence stratigraphy and basin dynamics. The Geological Society, London, SP 252, pp 231–250
Scasso RA, Alonso SM, Lanés S et al (2002) Petrología y geoquímica de una ritmita marga-caliza del Hemisferio Austral: El Miembro Los Catutos (Formación Vaca Muerta), Tithoniano medio de la Cuenca Neuquina. Rev Asoc Geol Argent 57:143–159
Scasso RA, Alonso SM, Lanés S et al (2005) Geochemistry and petrology of a Middle Tithonian limestone-marl rhythmite in the Neuquén Basin, Argentina: depositional and burial history. In:

Veiga GD, Spalletti LA, Howell JA, Schwarz E (eds) The Neuquén Basin, Argentina: A case study in sequence stratigraphy and basin dynamics. The Geological Society, London, SP 252:207–229
Schulz M, Mudelsee M (2002) REDFIT: estimating red-noise spectra directly from unevenly spaced paleoclimatic time series. Comput Geosci 28:421–426
Spalletti LA, Del Valle A, Kielbowicz A (1990) Análisis cicloestratigráfico del intervalo Hauteriviano superior-Barremiano en el área de Filo Morado, Cuenca Neuquina. In: Abstracts of the III Reunión Argentina de Sedimentología, San Juan, 20–24 May 1990
Spalletti LA, Poiré DG, Schwarz E et al (2001) Sedimentologic and sequence stratigraphic model of a Neocomian marine carbonate siliciclastic ramp: Neuquén Basin, Argentina. J S Am Earth Sci 14:609–624
Strasser A, Pittet B, Hillgärtner H et al (1999) Depositional sequences in shallow carbonatedominated sedimentary systems: concepts for a high-resolution analysis. Sediment Geol 128:201–221
Vergani GD, Tankard AJ, Belotti HJ et al (1995) Tectonic evolution and paleogeography of the Neuquén basin, Argentina. In: Tankard AJ, Suárez SR, Welsink HJ (eds) Petroleum basins of South America. AAPG Memoir 62:383–402
Weedon G (2003) Time-series analysis and cyclostratigraphy. Examining stratigraphic record of environmental cycles. Cambridge University Press, New York, p 259
Westphal H, Head MJ, Munnecke A (2000) Differential diagenesis of rhythmic limestone alternations supported by palynomorph evidence. J Sed Res 70:715–725
Westphal H, Böhm F, Bornholdt S (2004) Orbital frequencies in the sedimentary record: distorted by diagenesis? Facies 50:3–11
Westphal H, Hilgen F, Munnecke A (2010) An assessment of the suitability of individual rhythmic carbonate successions for astrochronological application. Earth-Sci Rev 99:19–30

Sedimentology and Sequence Stratigraphy of the Agrio Formation (Late Valanginian–Earliest Barremian) and the Closure of the Mendoza Group to the North of the Huincul High

Pablo José Pazos, Marcos Comerio, Diana Elizabeth Fernández, Carolina Gutiérrez, María Candela González Estebenet and Arturo Miguel Heredia

Abstract The Agrio Formation (late Early Valanginian–earliest Barremian) is an environmentally complex marine and continental succession that involves unconformities and flooding surfaces that respond to tectonic thermal subsidence. Internally, the Pilmatué Member contains five third-order sequences eustatically controlled. The Avilé Member starts over a regional unconformity that is the result of erosion and bypass by tectonic quiescence and only punctuated subsidence permits to explain the abnormal thickness in some areas. The Agua de la Mula Member starts with an isochronous and geologically instantaneous inundation which is better explained by tectonic subsidence rather than global eustatism. It contains four sequences but of fourth order. It also shows a sedimentary input from the east in some areas, largely neglected in the literature. The Chorreado Member, from a sequence stratigraphy point of view, is part of the Mendoza Group as it does not represent a basin expansion after a minor unconformity during the Barremian. Contrarily, the unconformity that marks the base of the Troncoso Member of the Huitrín Formation is an evidence of intense regional basin reorganization. The depocentres in the discussed intervals and units shift to the northwest while the thickness in proximal areas in the two marine members of the Agrio Formation point out to accommodation space created by tectonism. This is the first sequence stratigraphic model for the interval in two decades after the first absolute ages and latest biozone calibration provided for the Agrio Formation.

Keywords Agrio Formation · Sequence stratigraphic model · Tectonic controls

---

P. J. Pazos (✉) · D. E. Fernández · C. Gutiérrez · M. C. G. Estebenet · A. M. Heredia
Universidad de Buenos Aires, Facultad de Ciencias Exactas Y Naturales, Departamento de Ciencias Geológicas, CONICET—Universidad de Buenos Aires, Instituto de Estudios Andinos Don Pablo Groeber (IDEAN), Buenos Aires, Argentina
e-mail: pazos@gl.fcen.uba.ar

M. Comerio
Centro de Tecnología de Recursos Minerales y Cerámica (CETMIC), CONICET. La Plata, Buenos Aires, Argentina

D. Kietzmann and A. Folguera (eds.), *Opening and Closure of the Neuquén Basin in the Southern Andes*, Springer Earth System Sciences,
https://doi.org/10.1007/978-3-030-29680-3_10

1 Introduction

The Agrio Formation is a unit composed of marine and continental intervals. In the former, bituminous shales, marls, muddy siltstones, sandstones and limestones were deposited as a result of marine transgressions and regressions from the palaeo-Pacific Ocean between the Late Valanginian and earliest Barremian in the Neuquén Basin of Patagonia (Weaver 1931; Groeber 1946; Legarreta and Uliana 1991). Outcrops of this unit have also been documented in the Aconcagua region (Aguirre-Urreta and Lo Forte 1996) and reach Chile. The unit is widely exposed in the thrust and folds belts of the Cordillera Principal over more than 600 km in a N–S trend, and it is also present in the subsurface of the Neuquén Embayment totalizing an area of more than 100,000 km^2 (Marchese 1971; Legarreta and Uliana 1991). The knowledge of the Agrio Formation is important for biostratigraphic and palaeontological correlations between the Gondwana and Tethyan faunas (Riccardi 1984; Bown and Concheyro 2004; Aguirre-Urreta et al. 2007, Aguirre-Urreta and Rawson 2012; Paolillo et al. 2018), as well as for palaeocological and taphonomical studies (Aguirre-Urreta et al. 2005, 2007; Lazo et al. 2005, 2009; Lescano and Concheyro 2014; Cataldo and Lazo 2016), besides sedimentological-ichnological and stratigraphical approaches (Legarreta and Gulisano 1989; Sagasti 2005; Spalletti et al. 2001a, Spalletti et al. 2001b; Archuby et al. 2011; Fernández and Pazos 2012, 2013; Pazos et al. 2012; Guler et al. 2013; Comerio 2016; Schwarz et al. 2018). In addition, the relevance of the Agrio Formation is related to hydrocarbon exploration, including both conventional and unconventional reservoirs (Uliana and Legarreta 1993; Boll et al. 2014).

The pioneer work by Weaver (1931) divided the Agrio Formation in a lower and an upper part entirely of marine origin and a middle part of continental sandstones known as Avilé Member, which constitutes the most persistent stratigraphic marker in Neuquén (Weaver 1931). In these days, the lower and upper marine members are known as Pilmatué and Agua de la Mula members (Leanza and Hugo 2001). In the Agrio Formation, some stratigraphic uncertainties are not completely resolved yet, like the lower and upper contacts of the unit, the order (frequency) and number of stratigraphic sequences and the role of major allocyclic controls (eustatic sea-level fluctuations and/or tectonic activity). For instance, ammonoid sub-biozones have been used to denote the passage from the underlying Mulichinco Formation to the Agrio Formation but the base of the unit is not completely coeval with a synchronous transgression containing a specific ammonoid sub-biozone (cf. Aguirre-Urreta and Rawson 1997, 1999; Aguirre-Urreta et al. 2005; Schwarz et al. 2006). Similarly, the top of the upper member is subject of controversy. In many areas of the basin, overlying "classical" Agrio Formation deposits, a discrete several metre thick, or even thicker in subsurface, succession of siliciclastic, carbonatic and evaporitic deposits named Chorreado Member are considered either as part of the Huitrín Formation (Groeber 1946; Legarreta et al. 1981; Legarreta 2002; Gutiérrez Pleimling et al. 2011) or included in the upper part of the Agrio Formation (Leanza and Hugo 2001; Leanza 2003). These stratigraphic discrepancies are inconvenient when a sequence stratigraphic model is applied, and both possibilities will be discussed in this work in detail. Internally, the knowledge about sequence stratigraphy and facies arrangement of the marine members of the Agrio Formation is unbalanced. For the upper Agua de la Mula Member, several sequence stratigraphic models have been suggested (e.g. Brinkmann 1994; Spalletti et al. 2001a, b; Archuby et al. 2011; Guler et al. 2013), while the lower Pilmatué Member is less understood in terms of internal architecture and stratigraphic framework. However, some works discussed different and contrasting stratigraphic scenarios like several sequences (Legarreta and Gulisano 1989; Lazo 2007) or a succession of parasequences (Schwarz et al. 2018). The integration of information from different approaches suggests that the original storm wave-dominated model for this mixed siliciclastic-carbonate ramp is more complex due to tidal and fluvial controls (e.g. Pazos et al. 2012; Fernández and Pazos 2013). Additionally, for the middle continental Avilé Member there are thickness changes not fully understood yet (e.g. Veiga et al. 2009).

The aim of this work is: (i) to discuss the lower and upper contact of the unit taking into account observations in key outcrops; (ii) to propose an integrated sequence stratigraphic framework for the Pilmatué Member and discuss previous proposals for the Agua de la Mula Member; and (iii) to analyse the allocyclical controls active during the deposition of the Agrio Formation.

2 Geological Setting

2.1 *The Neuquén Basin*

The Neuquén Basin, located in west-central Argentina and Chile between 32° and 41° S, recorded the Meso-Cenozoic geological evolution of the eastern Southern Central Andes. The basin was bounded by relatively tectonically stable areas: the North Patagonian Massif to the southeast and the San Rafael-Las Matras blocks to the north-east. To the west, the boundary was related to the evolution of the Andean arc comprehended between these latitudes (Fig. 1) (see Chap. "Early Andean Magmatism in Southern Central Chile, 33º–40ºS"). It covers an area of more than 100,000 km^2 and it reaches 7,000 m in thickness. The sedimentary infill was deposited during three different tectonic stages: (1) a Late Triassic–Early Jurassic rift phase; (2) a Middle Jurassic–Early Cretaceous thermal sag; and (3) a Late Cretaceous–early Cenozoic foreland basin (Vergani et al. 1995; Legarreta and Uliana 1991; Mpodozis and Ramos 2008; Naipauer and Ramos 2016; Horton 2018). The synrift deposits, represented by several isolated depocentres, are included in the Precuyano Cycle and are composed of continental and pyroclastic sediments with interbedded volcaniclastic rocks (Gulisano 1981; Carbone et al. 2011) (see Chap. "The Syn-Rift of the Neuquén Basin (Precuyano and Lower Cuyano Cycle): Review of Structure, Volcanism, Tectono-Stratigraphy and Depositional Scenarios"). From the Middle Jurassic until the Early Cretaceous, the regional thermal subsidence regime controlled the retroarc stage of the Neuquén Basin with the development of transgressive-regressive cycles

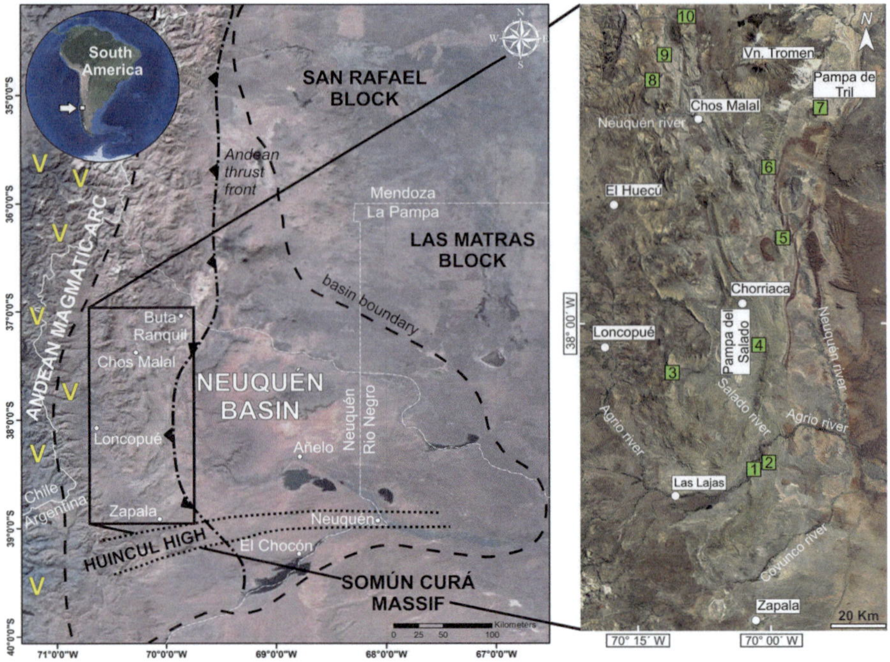

Fig. 1 Neuquén Basin in central-western Argentina: TM satellite image showing the basin boundaries and main morphostructural elements (modified from Naipauer et al. 2015). Study area with main stratigraphic sections discussed in this work: (1) Bajada Vieja; (2) Bajada del Agrio; (3) Portada Quintuco; (4) Agua de la Mula; (5) Cerro Rayoso; (6) Mina San Eduardo; (7) Loma La Torre; (8) Cerro La Parva; (9) Caepe Malal; and (10) Tricao Malal

included in the Cuyo, Lotena, Mendoza and Bajada del Agrio Groups (see Chaps. "Jurassic Uplift along the Huincul Arch and its Consequences in the Stratigraphy of the Cuyo and Lotena groups. Neuquén Basin, Argentina" and "Environmental Controls and Facies Architecture of a Jurassic Carbonate Episode (La Manga Formation), Mendoza province, Neuquén Basin"). In the Neuquén province, the Mendoza Group (Kimmeridgian–lowest Barremian) includes the Tordillo, Vaca Muerta, Mulichinco and Agrio Formations (Fig. 2), reaching about 2000 m thick of marine and continental deposits (Groeber 1946; Leanza and Hugo 2001). The Bajada del Agrio Group (Barremian–Albian) includes the Huitrín and Rayoso Formations, composed of continental, evaporitic and marginal-marine deposits, which predated the final disconnection of the palaeo-Pacific Ocean with the Neuquén Basin (Leanza 2003; Lazo et al. 2017). During the Late Cretaceous, a regional compressive tectonic regime led to the basin supply inversion (Ramos and Folguera 2005; Tunik et al. 2010), and after that sedimentation continued with continental successions and a shallow-water marine transgression (Fig. 2) from the Atlantic Ocean (Legarreta and Gulisano 1989; Aguirre-Urreta et al. 2011) with overlapping phases of extensional and contractional deformation (Ramos and Folguera 2005).

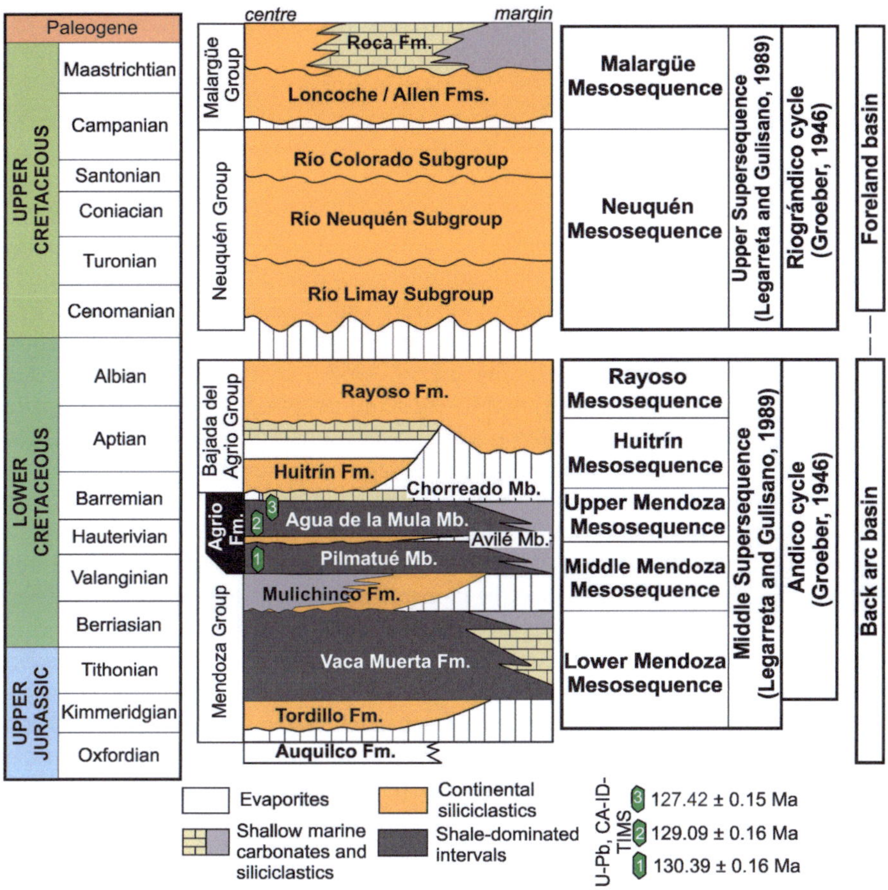

Fig. 2 Upper Jurassic–Palaeogene stratigraphic schematic chart for the Neuquén Basin, at the Neuquén Province with main outcropping lithostratigraphic units (modified from Howell et al. 2005). The Agrio Formation from base to top includes the Pilmatué, Avilé, Agua de la Mula and Chorreado Members (Leanza and Hugo 2001; Leanza 2003). Note U–Pb, CA-ID-TIMS zircon ages obtained from tuff levels in the Pilmatué (Aguirre-Urreta et al. 2017) and the Agua de la Mula Members (Aguirre-Urreta et al. 2015)

Due to the Andean deformation, the Neuquén Basin is subdivided into two main areas: the Cordillera Principal to the west and the Neuquén Embayment to the east (Yrigoyen 1972; Ramos 1999). The latter shows the original morphology of the eastern border of the basin south of 37°S (Legarreta and Uliana 1991). In the embayment, the Mesozoic successions are recorded in the subsurface, being slightly deformed by the Andean deformation. The Cordillera Principal of Argentina and Chile is characterized by the development of a series of N–S thin- and thick-skinned fold and thrust belts (e.g. Agrio, Chos Malal, Malargüe) with excellent exposures of the Mesozoic successions (e.g. Giambiagi et al. 2003; Turienzo et al. 2014; Rojas Vera et al. 2015).

2.2 Stratigraphic Overview

The marine and continental successions of the Neuquén Basin have been subdivided in different sedimentary cycles with multiple time-scale hierarchies (e.g. Groeber 1946; Legarreta and Gulisano 1989; Legarreta and Uliana 1991). The Agrio Formation integrates the Andean Cycle of Groeber (1946), comprising the Mendoza and Bajada del Agrio Groups (Stipanicic et al. 1968; Leanza 2003), that according to the present knowledge correspond to the Kimmeridgian–Albian (e.g. Naipauer and Ramos 2016). Based on outcrop analysis and subsurface-seismic data, Legarreta and Gulisano (1989) presented a regional sequence stratigraphic scheme for the Neuquén Basin where the Andean Cycle, as was defined by Groeber (1946), partially coincides with the Middle Supersequence defined by these authors (Fig. 2). According to this scheme, the record of the Agrio Formation lies within the Middle Mendoza Mesosequence (Mm) and the Upper Mendoza Mesosequence (Ms). These mesosequences show a low-angle to sigmoidal shape with thickness ranging from 250 to 1000 m and, internally, they are characterized by five broad shallowing-upward sequences: Mm1–Mm5 and Ms1–Ms5, respectively (Legarreta et al. 1981; Legarreta and Gulisano 1989). The Mm begins with a lowstand systems tract consisting of continental and shallow-marine facies (Mulichinco Formation) that grades northward of the basin into shallow-marine carbonates (Chachao Formation). Both components are replaced upwards by open to shallow-marine facies (Pilmatué Member) that represent the transgressive and highstand systems tracts. A similar trend is exhibited in the Ms. It starts with lowstand continental deposits (Avilé Member) abruptly covered by a rapid inundation with open or occasionally marginal-marine facies depending on the position in the basin (Agua de la Mula Member). These facies also represent the transgressive and highstand systems tracts throughout the unit. Towards the north and northwest of the basin, the uppermost part of Ms includes shallow-marine carbonates, clastic sediments and evaporate deposits (Chorreado Member).

Due to the predominance of basinal conditions, in particular in areas towards the north-western region of the basin, the lowstand deposits of the Mm and Ms can be absent (e.g. Legarreta et al. 1981). In the eastern and southern part of the basin, particularly in the subsurface of the Neuquén Embayment, there is a unit contemporaneous with the Agrio Formation named Centenario Formation (Digregorio 1972) that mainly consists of continental (south) and shallow-marine (east) facies that constitute very important oil reservoirs (Cabaleiro et al. 2002), but their internal evolution is not completely understood.

Based on biostratigraphic schemes, the base of the Agrio Formation is considered a diachronous surface over the Mulichinco Formation (Aguirre-Urreta and Rawson 1997; Aguirre-Urreta et al. 2005). Ammonoid biozones, calcareous nannofossil events and well-constrained volcanic zircon U-Pb ages indicate that the Agrio Formation was deposited during the late Early Valanginian to Early Barremian (Aguirre-Urreta et al. 2005, 2015, 2017). According to these works, the entire unit ranges from the uppermost part of the Early Valanginian (*Olcostephanus* (*O.*) *atherstoni* sub-biozone; recorded only at one locality) to the Early Barremian (*Sabaudiella riverorum* Biozone). However, at most localities of the Neuquén Basin, the base of the Agrio Formation is Late Valanginian (*Pseudofavrella angulatiformis* Biozone), thus indicating the diachronic nature of the base of this unit (Aguirre-Urreta and Rawson 1997, 1999).

3 Materials and Methods

This study comprises the analysis of the Agrio Formation in the Neuquén province to the north of the Huincul High (Agrio and Chos Malal fold and thrust belts), following a S–N and SE–NW trend, coinciding broadly with proximal–distal depositional settings (Fig. 1). The Pilmatué Member was logged in detail (1:100) at Bajada Vieja (38°25′44.62″S–70°03′49.62″W) and Mina San Eduardo (37°31′30.84″S–70°01′14.28″W) localities. The transition from the Mulichinco Formation to the overlaying Pilmatué Member was also analysed at Portada Quintuco (38°08′49.60″S–70°22′08.63″W) and Cerro La Parva (37°15′47.35″S–70°26′34.43″W). Six localities were also logged in detail (1:100) for the study of the Agua de La Mula Member: Bajada del Agrio (38°25′35.16″S–69°59′39.18″W); Agua de la Mula (38°03′20.39″S–70°01′57.97″W); Cerro Rayoso (37°44′43.03”S–69°56’56.35″W); Mina San Eduardo (37°31′49.93″S–70°00′12.80″W); Loma La Torre (37° 19′ 43.05″–69° 49′43.04″); and Caepe Malal (37° 11′ 49.15″–70° 23′25.10″).

In this work, some classical sequence stratigraphic models are used for correlation of sequences bounded by unconformities within well-constrained chronostratigraphic frameworks (e.g. Vail et al. 1977; Van Wagoner et al. 1990; Embry and Johannenssen 1992). Diagnostic surfaces are defined following Van Wagoner et al. (1990) and ramp divisions follow Burchette and Wright (1992). The marine parts of the Agrio Formation are well known as a mixed siliciclastic-carbonatic homoclinal ramp (Spalletti et al. 2001a, 2001b; Lazo et al. 2005; Cichowolski et al. 2012; Pazos et al. 2012; Fernández and Pazos 2012; Comerio et al. 2018; Schwarz et al. 2018). Accordingly, in this work the unit is divided into: (1) distal and proximal outer ramp settings; (2) intermediate position with distal and proximal middle ramp facies; and (3) inner ramp to marginal-marine settings. Timing and correlations of the depositional sequences were established based on ammonoid biozones according to the schemes proposed by Aguirre-Urreta et al. (2005, 2015, 2017) which represent for outcrop studies the age-diagnostic fossil intervals for easy stratigraphic location in the unit. The stratigraphic divisions used for the Valanginian–Barremian are following the proposal of Martinez et al. (2015).

4 Palaeogeography and Sedimentological Framework

The sedimentary record of the Agrio Formation in terms of palaeoenvironments and facies distribution needs to be approached considering palaeogeographic divisions related to tectonic features of the basin that precede the deposition of the unit (e.g. Huincul High) or, that are the result of the much modern Andean tectonism (e.g. fold and thrust belts). The Neuquén Basin is particularly known for presenting an internal sedimentary evolution that differs to the south and north of the Huincul High in contemporaneous deposits. This element is a morphostructural unit with an east–west trend (Fig. 1) that was particularly active during the Lower Jurassic (Freije et al. 2002; Naipauer and Ramos 2016). However, at the time of the Agrio Formation deposition, the differentiation into dominantly carbonatic (south from the Huincul High) and siliciclastic and mixed siliciclastic-carbonatic deposits (north from the Huincul High) is clearly distinguishable and suggestive of restricted and more open marine circulation, respectively. In order to compare the sedimentary record, where main controls like tectonic subsidence and palaeocirculation are similar, the sedimentology and sequence stratigraphy will be focused on outcrops and subsurface data to the north of this morphostructural unit. Legarreta (2002) provided very complete palaeogeographic maps for the basin evolution at different times that coincide with major sea-level changes. The facies maps referred to the Agrio Formation show facies belts deepening to the northwest following the axis of the basin with a SE-NW orientation and a starving basin area to the west, while the eastern border was inferred in subsurface. The facies maps were depicted for the Upper Valanginian (Pilmatué Member), Lower Hauterivian (Avilé Member) and Upper Hauterivian (Agua de la Mula Member). Interestingly, the geographic distribution of the continental wedge (Avilé Member) is more restricted, with thickness changes mainly attributed to eustatic controls (see Veiga et al. 2009). The Chorreado Member was also mapped by Gutiérrez Pleimling et al. (2011) and exhibits a restricted distribution and a complex internal facies concatenation. The distribution of the Avilé and Chorreado Members is based on a combination of outcrop and subsurface data, as both units contain very good quality hydrocarbon reservoirs (e.g. Gulisano and Gutiérrez Pleimling 1988; Gulisano et al. 2001; Veiga et al. 2002, 2011; Rossi 2001; Rossi and Masarik 2002; Barrionuevo 2002; Valenzuela 2002; Gutiérrez Pleimling et al. 2011).

At first glance, the facies belt presented by Legarreta (2002) does not show any particular evidence of a secondary axis for the sedimentary input or an explanation about the notorious differences in thickness between proximal and distal areas. The Pilmatué Member (90–520 m thick) is a marine succession represented by organic carbon-bearing mudstones towards its base which grades upwards to laminated claystones-siltstones and heterolithic deposits interbedded with bioclastic limestones, calcarenites, sandstones and coral biostromes (e.g. Sagasti 2005; Lazo 2007; Cichowolski et al. 2012; Garberoglio et al. 2013). In this unit, shoreface deposits with prograding bars and open marine settings dominate (Zavala et al. 2011; Schwarz et al. 2018), but also dinosaur trampling surfaces are found indicating very shallow to subaerial conditions (Heredia et al. 2018). The Avilé Member (10–60 m, or exceptionally

180 m thick) has a continental origin with fluvial, aeolian and lacustrine/playa-lake deposits (e.g. Rossi 2001; Rossi and Masarik 2002; Veiga et al. 2002, 2009, 2011). The Agua de la Mula Member (70–495 m thick) is better understood that the Pilmatué Member in terms of sedimentary processes and facies architecture. Traditionally, it contains a fair-weather and storm wave deposits (Spalletti et al. 2001a, 2001b; Lazo et al. 2005; Sagasti 2005; Comerio et al. 2018) but also marginal-marine facies affected by tidal processes (Tunik et al. 2009, Fernández and Pazos 2012, Pazos et al. 2012) and fluvial discharges (Aguirre-Urreta et al. 1993) documented to the top of the unit. Dinosaur trackways and other indicators of exposure were also documented in the Agua de la Mula Member in a palaeogeographic position where fully marine facies were expected (see Pazos et al. 2012, Fernández and Pazos 2013, 2015).

5 Sequence Stratigraphy and Bounding Surfaces

Sequence Stratigraphy as a facies correlation technique has evolved enormously in the last decades (see Catuneau et al. 2009). It is still based on the recognition of bounding surfaces with different hierarchy and origin. In particular, for the Agrio Formation and also the Huitrín Formation, this approach was pioneered by Legarreta and Gulisano (1989), Legarreta and Uliana (1991), Gutiérrez Pleimling (1991), Rossi (2001), Spalletti et al. (2001a, 2001b) and Legarreta (2002). In these works, the Exxon School concepts were applied and the eustatic sea-level changes were considered the main allocyclic controls in the sequence architecture. The lower boundary of the unit is interpreted as diachronic taking into account that depending on the locality the *Pseudofavrella angulatiformis* Biozone is above or within the typical basal organic-rich black shales of the Pilmatué Member (Aguirre-Urreta and Rawson 1997, 1999). An alternative interpretation was suggested by Schwarz et al. (2006, Fig. 14) that considered the preceding interval to the black shales as the last highstand system tract of the Mulichinco Formation. Both schemes are irreconcilable: one implies diachronism and the other an isochronous transgression. Internally, the Pilmatué Member was subdivided into five sequences (Legarreta and Gulisano 1989; Lazo 2007) but sedimentological explanations cast doubts about the stratigraphic framework (Spalletti et al. 2011). Sea-level fluctuations probably were the main allocyclic control in the origin of an exceptional nautilid concentration as part of a carbonate transgressive systems tract in a storm-dominated depositional setting. This carbonate-dominated tract is followed by a highstand systems tract mainly composed of siliciclastic facies with a coarsening-upward trend (Cichowolski et al. 2012).

The Avilé Member was interpreted as a result of a forced regression of low-frequency order (Rossi 2001; Legarreta 2002; Veiga et al. 2009). It is not a simple forced regression as was originally envisaged because it contains high order sequences inside, probably climatically controlled or as a response to sea-level fluctuations (Veiga et al. 2009). Ammonoid biozones remain constant all over the basin both underlying and overlying the continental member. Veiga et al. (2002) related the preservation of the aeolian deposits due to the upward movement of the phreatic level following the model by Kocurek et al. (2001). It is important to point out that the fluvial and aeolian facies of Avilé Member in Neuquén Province are replaced by lacustrine/playa-lake facies in Mendoza Province (Rossi and Masarik 2002; Veiga et al. 2011). In the north of Neuquén province, the Avilé Member is abnormally thick (180 m) and mostly fluvial compared with the 10–60 m fluvio-aeolian thick section in most of the basin (Veiga et al. 2009, 2011), something not clearly explained yet (see discussion in Sect. 7).

The marine transgression of the Agua de la Mula Member that uniformly covers the Avilé Member deposited black shales over an aeolian sandstone landscape suggesting a very rapid transgression without erosion, but showing an abrupt facies dislocation. Internally, it was divided into three or four stratigraphic sequences (Spalletti et al. 2001b; Archuby et al. 2011). Recently, in an integrated sedimentological-palaeontological approach four sequences were defined and exhibit a progradational stacking pattern (Guler et al. 2013). These sequences are dominated by storm deposits in a ramp setting (Comerio et al. 2018) at least for the three lower ones but the upper one shows well-defined tidal controls (Fernández and Pazos 2012; Pazos et al. 2012). The uppermost levels of the Agrio Formation end with marginal-marine deposits in several localities that suggest the complete infill of the basin close to the south of Buta Ranquil city (Fig. 1). To the north of this region and in most Mendoza Province, the architecture of the Agrio Formation is completely different showing a rhythmic motif with a dominance of fine-grained siliciclastic and carbonate outer ramp deposits (Sagasti 2005). In general terms, the sequence stratigraphic framework is better understood in the Agua de la Mula Member than in the Pilmatué Member.

The upper boundary of the Agrio Formation is still matter of debate. As was pointed out earlier in this work, lithostratigraphically the Chorreado Member is part of the Huitrín Formation after the redefinition of Legarreta and Boll (1982). From a sequence stratigraphy point of view, this unit is a part of the Upper Mendoza Mesosequence which also includes the Avilé and the Agua de la Mula Member (Legarreta and Gulisano 1989). In a detailed analysis of the Chorreado Member, it is concluded that the unit is internally complex in facies and that it is controlled by the pre-existent facies belt left by the Agrio Formation and is almost absent to the south of the Pichi Neuquén river (Gutiérrez Pleimling 1991, Fig. 1). In the Neuquén province, the isopach map depicted by Gutiérrez Pleimling (1991, plate 18) and Gutiérrez Pleimling et al. (2011) clearly shows that the depositional axis is very contrasting with the regional trend for the entire Agrio Formation. Several authors considered the base of the Troncoso Member of the Huitrín Formation like the "master" sequence boundary (Legarreta 2002; Gutiérrez Pleimling 1991; Gutiérrez Pleimling et al. 2011; Veiga et al. 2005), with a regional reorganization of the depositional system. This indirectly supports that the Chorreado Member is well situated as part of the Upper Mendoza Mesosequence, resulting largely disputable the criteria to include it as part of the Huitrín Formation.

6 The Pilmatué Member Analysis

The Pilmatué succession is divided into five (I to V) sequences, four of which are internally organized in transgressive and highstand systems tracts. The transgressive surfaces that denote the beginning of a regional inundation are usually dominated by grey or black shales. The stratigraphic expression of the sequences is usually asymmetric with thick (highstand) regressive packages. These sequences were recognized in proximal localities like Bajada Vieja (BV) and distal ones like Mina San Eduardo (MSE). They are illustrated in Fig. 3 following a regional N–S cross-section where the entire unit thins from 520 m in the southernmost locality (BV) to 420 m in the northernmost locality (MSE). Sequence I was also analysed at Portada Quintuco (PQ) and Cerro La Parva (CP) localities, to study the regional flooding at the time of the onset of the Pilmatué transgression (Fig. 4). Due to the presence of the *Olcostephanus* (*O.*) *atherstoni,* sub-biozone sequence I begins at CP in the uppermost part of the Early Valanginian (Aguirre-Urreta and Rawson 1997, 1999). Nevertheless, in the majority of other localities, sequence I starts in the Late Valanginian (Fig. 5) because it encases in the *Pseudofavrella angulatiformis* Biozone (Aguirre-Urreta and Rawson 1997, 1999; Aguirre-Urreta et al. 2005, 2017). Sequence II encompasses the uppermost part of *Pseudofavrella angulatiformis* Biozone of the latest upper Valanginian, while sequences III, IV and V correspond to the lower Hauterivian due to the presence of *Holcoptychites neuquensis*, *Hoplitocrioceras gentilii* and *Weavericeras vacaensis* ammonoid biozones (Aguirre-Urreta et al. 2005), respectively, in all localities (Fig. 5). This biostratigraphic–sequence stratigraphic scheme was previously analysed by Lazo (2007) but in the present work a more distal section (MSE) is included.

6.1 *Sequence I*

Sequence I (S-I) is 100–200 m thick, with the maximum thickness at CP (Fig. 4). There, the relative depositional time span is larger: in this locality S-I encompasses the *O.* (*O.*) *atherstoni* Biozone and part of the *P. angulatiformis* Biozone (Aguirre-Urreta and Rawson 1997, 1999), while in BV, PQ and MSE the S-I is restricted to a part of *P. angulatiformis* Biozone (e.g. Lazo 2007). S-I occurs above marginal-marine (PQ and MSE) or upper shoreface (CP) deposits of the Mulichinco Formation (Fig. 6a) that represent the latest components (highstand) of a long-term (second order) lowstand systems tract (Schwarz et al. 2006).

The stacking pattern of the transgressive systems tract (TST-I) is undoubtedly thinning-upward (retrogradational) in CP, PQ and BV, while in MSE is aggradational. In CP, TST-I begins with illite-rich, 10–15 m thick, mudstones that intercalate with m thick oyster-dominated beds. In BV, similar deposits were interpreted as proximal outer ramp to middle ramp facies (Toscano et al. 2018). In CP, this siliciclastic-dominated interval grades into marls and micritic limestones (23 m thick, Figs. 4

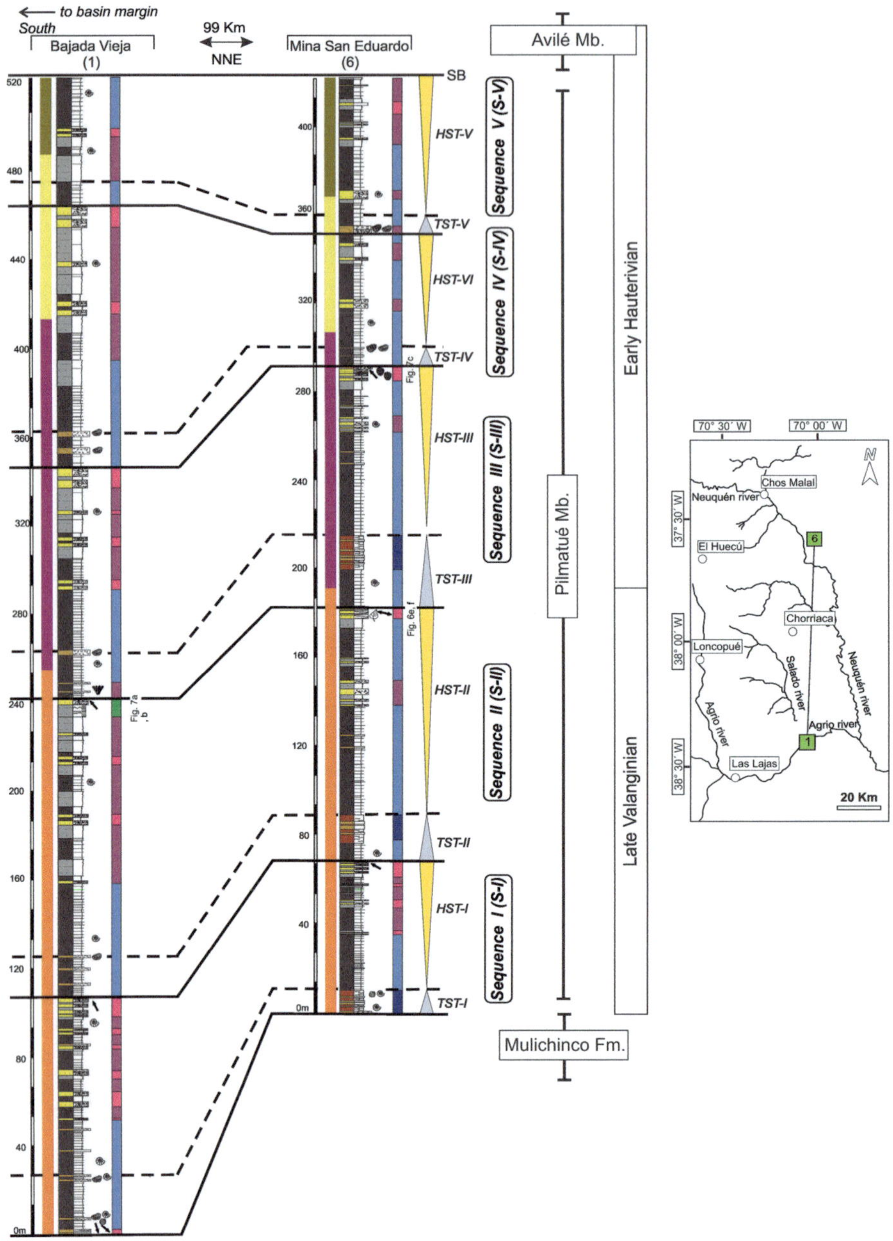

Fig. 3 Logged section for the Pilmatué Member including sequences I to V. See legend in Fig. 4

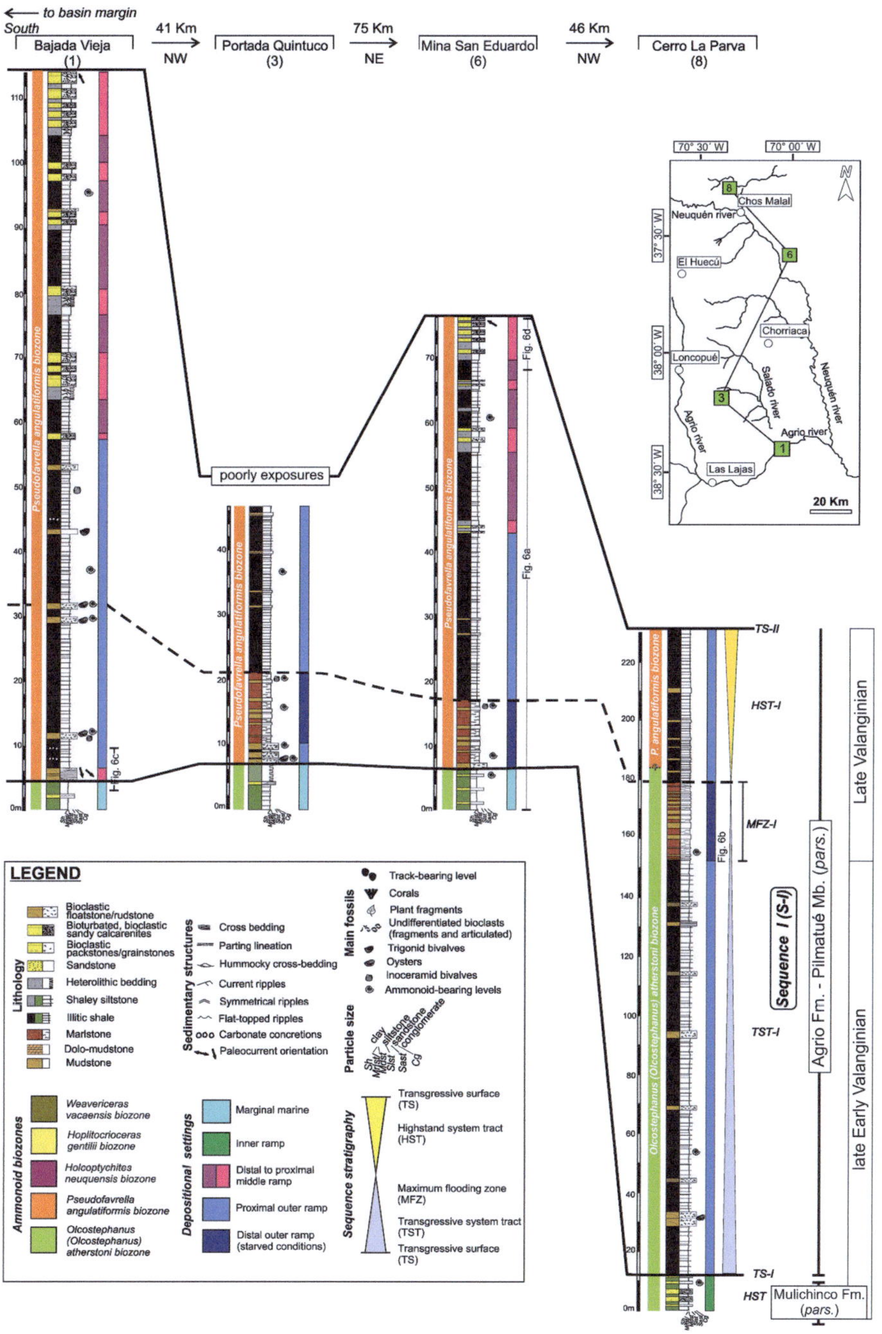

Fig. 4 Detail of the sequence I of the Pilmatué Member

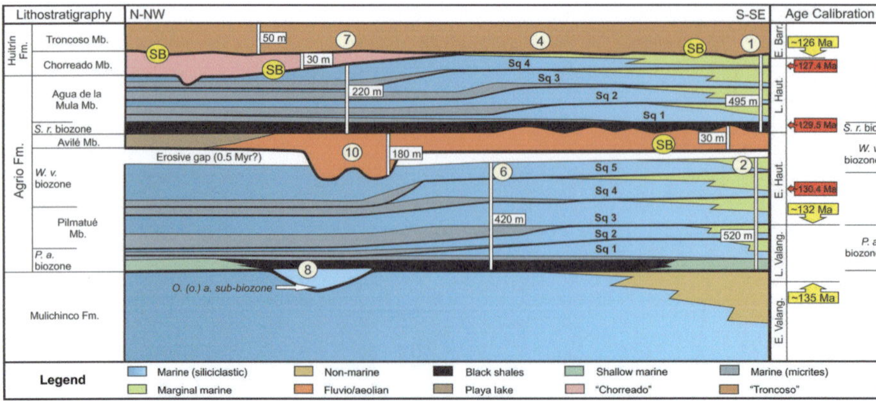

Fig. 5 Sequence stratigraphic framework of the Agrio Formation. The numbers of the localities follow those of Fig. 1: (1) Bajada del Agrio; (2) Bajada Vieja; (3) Agua de la Mula; (6) Mina San Eduardo; (7) Loma La Torre; (8) Cerro La Parva; and (10) Tricao Malal. Vertical bars = thickness (not to scale). SB = sequence boundary. Sq = sequences. Suggested ages in yellow from Martinez et al. (2015, Fig. 8). Biozones of interest and absolute ages in red from Aguirre-Urreta and Rawson (1997) and Aguirre-Urreta et al. (2005, 2015, 2017). Thickness of Troncoso Member near Loma La Torre after Veiga and Vergani (2011). Thickness of Chorreado Member after Gutierrez Pleimling et al. (2011). Thickness of Avilé Member in locality 10 after Veiga et al. (2009). Thicknesses in localities 1, 2, 6 and 7 from this work

and 6b) that represent distal outer ramp deposits (starved basinal facies) accumulated under poorly oxygenated bottom waters (e.g. Comerio et al. 2018). Accordingly, these organic-rich mudstones (starved basinal facies) characterize a maximum flooding zone (MFZ-I). In MSE, the TST-I starts with the starved facies (10 m thick, Figs. 4 and 6a), while in the proximal-most section (BV), this facies is absent. At BV, amalgamated bioclastic deposits showing cross-stratification with sets indicating a southward palaeoflow mark the beginning of the Pilmatué Member (Figs. 4 and 6c).

The highstand systems tract (HST-I) is 40–100 m thick. In proximal (BV) and distal (MSE) sections, HST-I starts with siliciclastic muddy-dominated intervals depleted in organic components and contains a great diversity of marine macro-invertebrates. In this locality, these deposits were interpreted as proximal outer-to middle ramp (Lazo 2006). In BV and MSE, this interval is replaced upwards by coarsening-upward heterolithic packages capped in some cases by fine- to medium-grained sandstones pervasively burrowed or mottled (massive in appearance). These shallowing-upward deposits were interpreted as part of middle to inner ramp settings by Cichowolski et al. (2012) in the Salado section. They represent high-frequency relative sea-level oscillations or stacked parasequences that range from 6 to 18 m thick (Figs. 4 and 6d). Accordingly, in BV and MSE the stacking pattern of HST-I has a clear progradational tendency.

In distal localities (PQ and CP), each parasequence of HST-I starts with a lower micritic interval and ends with an upper clastic mudstone interval. This expression

Fig. 6 Sedimentology and key surfaces of sequences I and II of the Pilmatué Member. A. Passage from Mulichinco Formation to the Pilmatué Member (Agrio Formation) at Mina San Eduardo. S-I = Sequence I. S-II = Sequence 2. B. Starved basinal facies of S-I in Cerro la Parva. Person for scale. C. Amalgamated bioclastic deposits with cross-stratification at the base of the Pilmatué Member in Bajada Vieja. Person for scale. D. Top of S-I in Mina San Eduardo showing a clear progradational tendency at the top of HST-I. Person for scale. E. Channelized sandstone bodies at the top of the HST-II in Mina San Eduardo. Graphic scale with arrows = approximately 6 m F. Note plant remains recorded in this sandstone level. Graphic scale = 3 cm

of parasequences is typical of mixed, carbonate–siliciclastic systems in distal areas (Bohacs and Lazar 2010).

6.2 *Sequences II to V*

In sequences II to V in both studied localities (BV and MSE, Fig. 3), the sequence boundaries are synchronic surfaces. For this reason, they are analysed together and some differences from the proximal to the distal localities are highlighted. In sequences II to V, transgressive systems tracts (TST-II to TST-V) are 10–45 m thick with the maximum thickness in BV. The most distinctive characteristic of TSTs is that they are represented by carbonate deposits with notorious facies changes between proximal (BV) and distal (MSE) depositional settings. The TSTs (TST-II to TST-IV) in MSE show a retrogradational stacking pattern containing starved basinal facies that mark the upper boundary for the TSTs and represent the MFZs. In BV, the starved facies are not present and the stacking pattern is mostly aggradational in all TSTs. In this locality, transgressions include basal intervals with amalgamated shell beds and in some cases with coral colonies and gregarious oysters (Lazo 2006; Toscano et al. 2018) that suggest low siliciclastic input. These basal intervals are similar to those documented and interpreted as transgressive bioclastic condensed system tracts in the Salado section by Cichowolski et al. (2012).

In contrast with the TSTs, the highstand systems tracts (HSTs) are relatively thick (80–40 m) and show an aggradational to progradational stacking pattern. They are characterized by siliciclastic and mixed deposits (siliciclastic mudstones, heterolithic intervals, sandstones, bioclastic sandstones and coquinas) associated with a great diversity of benthic organisms that indicate middle to inner ramp settings (e.g. Lazo 2006). In particular, the top of HST-II at MSE shows channelized sandstone bodies with a basal erosion surface, containing plant fragments (Figs. 3 and 6e, f) and mud intraclasts that suggest a connection with fluvial discharges derived from the eastern margin of the basin. This interval correlates in BV with heterolithic (mainly wavy and flaser bedding) deposits (Figs. 3 and 7a) with oscillatory or combined-flows sedimentary structures showing flat-topped ripples (Fig. 7b) and diverse trace fossils. These evidences suggest an important regression within the top of *P. angulatiformis* Biozone during the latest Late Valanginian. Also, the top of the HST-III shows dinosaur tracks (Figs. 3 and 7c) and dolomitization features that indicate salinity fluctuations related to mixture of marine-meteoric waters and subaerial exposure similar as was interpreted for the Agua de la Mula Member (Tunik et al. 2009; Pazos et al. 2012). Finally, HST-V exhibits siliciclastic muddy-dominated intervals that intercalate in MSE with thin storm events indicating deposition above the storm wave base within a distal middle ramp. In BV and MSE, the passage from marine facies of the Pilmatué Member to the continental deposits of the Avilé Member is abrupt. The surface that separates marine from continental units is interpreted as a sequence boundary (Legarreta and Gulisano 1989; Veiga et al. 2002) occurring in the upper part of the lower Hauterivian within the *Weavericeras vacaensis* Biozone (Rawson and Aguirre-Urreta 2012).

7 Discussion

7.1 The Pilmatué Member

The base of sequence I of the Pilmatué Member is diachronous. At CP locality, it begins with transgressive deposits of the late Early Valanginian and is covered by organic- and micritic-rich mudstones (starved basinal facies) of the Late Valanginian (Fig. 5) that correspond to the maximum flooding zone (MFZ-I). In the rest of the

Fig. 7 Sedimentology and key surfaces of the Pilmatué Member (sequences II and III), of the Agua de la Mula Member and Huitrín Formation. **a**. Heterolithic deposits at the top of the HST-II in Bajada Vieja (Pilmatué Member). Graphic scale = 1.5 m. **b**. The heterolithic deposits show flat-topped ripples (Pilmatué Member in Bajada Vieja). Graphic scale = 10 cm. **c**. Trampling surface at the top of HST-III (Pilmatué Member in Mina San Eduardo). Graphic scale = 30 cm. Arrows point to tracks. **d**. Tidal flat deposits at the top of the Agua de la Mula Member in Loma La Torre. Rock pick = 33 cm. e. The Chorreado Member and the Troncoso Member in Loma La Torre. Person for scale

studied localities, the MFZ-I overlies the Mulichinco Formation and contains the *Pseudofavrella angulatiformis* sub-biozone of the Late Valanginian (Fig. 5). In general, the MFZ-I is composed of organic- and micritic-rich mudstones (Fig. 6b) but in BV locality it exhibits more siliciclastic input due to a more proximal location in the basin. In MSE, Zavala et al. (2011) suggested that, overlying the "lower black shales" of the base of Pilmatué Member, a *ca.* 120 m thick package of Hauterivian amalgamated sandstones lies that they related to the Cortaderas Alignment. The observations in the MSE area permitted to conclude that under any circumstance the basal shales are Hauterivian because as was mentioned before they correspond to the *P. angulatiformis* Biozone that is Late Valanginian (Aguirre-Urreta and Rawson 1999; Aguirre-Urreta et al. 2005). The interval coincides with part of our sequence I in the area and we did not find a 120 m thick sandstone package approximately

4 km to the north of the section logged by Zavala et al. (2011), neither did Schwarz et al. (2018). The overlying sequences (S-II to S-IV) are arranged asymmetrically: short transgressive retrogradational system tracts and thicker highstand system tracts that exhibit a progradational stacking pattern (Fig. 4). In MSE, two features resulted crucial to understand the depositional architecture. Firstly, amalgamated sandstones with a well-developed erosive base and cross bedding containing plant remain at the top of the second sequence record the lowest fluvial input documented for the Pilmatué Member (Fig. 6e, f). Secondly, the subaerial dinosaur trampling surface (Heredia et al. 2018) at the top of the highstand of this sequence (Fig. 7c) clearly indicates that the succession can be subdivided into stacked sequences rather than only parasequences as was suggested by Schwarz et al. (2018). The five sequences proposed by Lazo (2007) in BV locality were also recognized in the present work. Nevertheless, instead of a thinning-upward stacking pattern for each sequence (Fig. 3 of Lazo 2007), we document a retrogradational thinning-upward transgressive system tract followed by prograding (thickening upward) siliciclastic deposits of the highstand system tract (Fig. 4). As it was also shown in the study of Cichowolski et al. (2012) carbonate-bearing facies dominate in the transgressive systems tract and siliciclastic deposits containing hummocky cross-stratification and other inshore sedimentary structures dominate in the highstand interval. The carbonate facies resulted from primary production or re-deposition in absence of siliciclastic material due to that during transgressions siliciclastic particles were retained close to the coast and later started to advance during the beginning of the highstand.

7.2 The Avilé Member

The Avilé Member developed on top of a regional erosional surface (Legarreta 2002) that represents a master sequence boundary across the basin (Veiga et al. 2009). The overall deposition of this continental succession was controlled by eustatic sea-level oscillations, based on the partial connection of the Neuquén Basin with the palaeo-Pacific Ocean during the accumulation of the unit in the late Early Hautervian (Veiga et al. 2009). However, the thickness differences of approximately 100 m between some localities are not explainable by merely eustatic sea-level drops: tectonic control related to localized subsidence is necessarily involved. When compared to the position of the depocentre of the Pilmatué Member, the main depocentre for the Avilé Member shifted north (Fig. 8) and a very abnormal thickness (180 m) was documented in Tricao Malal (Veiga et al. 2009, 2011). This constitutes another evidence of tectonic subsidence incision, especially compared to what was documented further east in subsurface (Veiga and Vergani 1993).

Towards the top of the Avilé Member previously to the marine inundation represented by distal outer ramp deposits of the Agua de la Mula Member (Comerio et al. 2018), the rising of the phreatic zone allowed the preservation of aeolian and fluvial deposits (south) or playa-lake deposits (north) which indicate the earliest stages of the transgressive systems tract (Veiga et al. 2009).

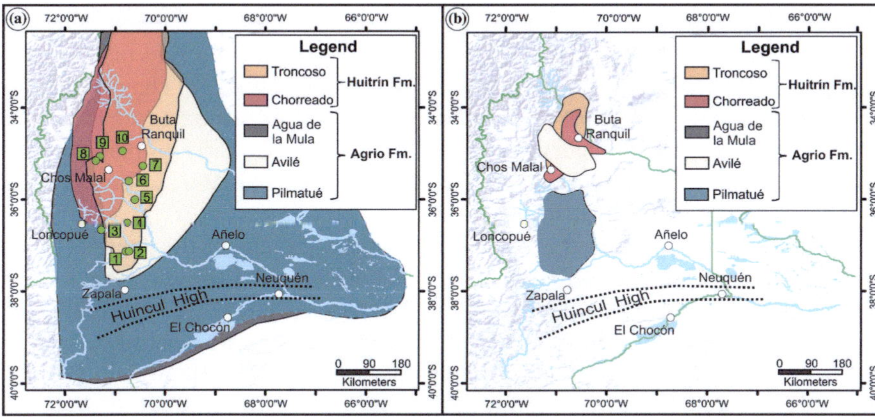

Fig. 8 Areal distribution of Agrio Formation and part of the Huitrín Formation and their maximum thickness areas. Overimposed areas are not in stratigraphic order. **a**. Regional distribution of preserved thicknesses compiled from Legarreta (2002) and localities discussed in this work (see Fig. 1). **b**. Maximum thickness distribution. Information for Avilé, Chorreado and Troncoso Members after Rossi (2001), Gutierrez Pleimling et al. (2011, Fig. 4) and Veiga and Vergani (2011, Fig. 2), respectively. Information for Pilmatué and Agua de la Mula Members from this work

7.3 The Agua de La Mula Member

The marine Agua de la Mula Member is better understood than the Pilmatué Member, regarding sequence stratigraphy. The logged sections included in the present work support the same number of sequences (four) as was proposed in a complete approach by Guler et al. (2013). The base of the unit, unlike the diachronic base of the Pilmatué Member, is an isochronous transgression present in all the studied localities (Fig. 5). This is based on the observation that for the distal outer ramp deposits of the Agua de la Mula Member (Comerio et al. 2018) the ammonoids of the *Spitidiscus ricardii* Biozone (Rawson and Aguirre-Urreta, 2012) are recorded regionally evidencing an isochronous transgression during the Late Hauterivian.

The thickness of the Agua de la Mula Member as occurs in the Pilmatué Member, paradoxically, is maximum to the south or proximal area in the basin and thinning to the north (Figs. 5 and 8b). In the distal outer ramp area, a maximum water depth of 200 m could be estimated (Comerio et al. 2018). Therefore, the 495 and 420 m thicknesses recorded at relatively proximal areas (Bajada del Agrio and Agua de la Mula localities, respectively) cannot be explained only taking into account the available accommodation space by sea-level changes. It certainly indicates that subsidence prevented the rapid progradation of clastics that were retained in proximal areas. While the rest of the basin is almost starved of sand-sized siliciclastic particles, the Cerro Rayoso locality is an exception: there, m thick amalgamated sandstones occur and the palaeocurrent directions are E–W instead of N–S or SE–NW that indicates a sediment input from the east (Pazos et al. 2012). This exception was also suggested from subsurface near Cerro Rayoso for the Pilmatué Member (Schwarz et al. 2018). Clearly, the eastern margin of the basin is a supply area neglected in previous models. The depocentres of both marine members close to the proximal settings indicate active subsidence rather than sea-level (eustatic) change alone as was suggested for the Lower Cretaceous in the basin due to post-extensional thermal subsidence (Manceda and Figueroa 1995; Horton 2018). The early proposed models based only on global sea-level changes (Legarreta and Uliana 1991; Legarreta 2002; Howell et al. 2005) were inadequate.

Stratigraphically upward, the sequence boundary and age of the deposits of the Chorreado Member are not well constrained. For instance, in Loma La Torre locality tidal flat deposits observed in the top of the Agrio Formation (Fig. 7d), similarly to what is found in other localities (see Sect. 4), are covered by transgressive shallow-marine carbonates, no more than 2 m thick assignable to the Chorreado Member (Fig. 7e). This unit includes a complex distribution of facies both continental, marine and evaporates and constitutes a high-frequency sequence deposited in a short time controlled by the geometry of the NW prograding clinoforms of the Agrio Formation (Gutiérrez Pleimling 1991). This explains the maximum thickness with a complex trend (Gutiérrez Pleimling et al. 2011) suggesting the infilling of the remnant accommodation space to the north and a much localized onlap onto the Agua de la Mula Member.

7.4 Huitrín Formation Interval

The next sequence boundary is a regional unconformity and includes an important erosion event with bypass that shows the basin reorganization (Fig. 8): the deposition of coarse-grained fluvial sandstones and aeolian deposits (Troncoso Member "inferior"), that sometimes show climbing features (Fig. 7e), preceding a marine inundation with evaporates (Troncoso Member "superior") and carbonate deposits of the La Tosca Member (Legarreta 2002; Strömback et al. 2005; Veiga et al. 2005). La Tosca Member was originally considered as Aptian in age and it was later constrained to the Barremian through macro-invertebrates (Lazo and Damborenea 2011) or nannoplankton (Lazo et al. 2017). This implies that at least two intrabarremian unconformities took place, reinforcing tectonism as a neglected control in the basin evolution. From a sequence stratigraphy point of view, the Chorreado Member is still a part of the Mendoza Group as was suggested by Legarreta and Gulisano (1989) and Leanza and Hugo (2001). While during the Late Hauterivian subsidence was dominant, it was reduced in the Barremian with two events of basin erosion that produced the reorganization of depocentres at different positions (Fig. 8b).

The geologically instantaneous transgression at the base of the Agua de la Mula Member is common in glacioeustatic controlled sequences (Catuneanu and Zecchin 2013). However, during a greenhouse time as occurred in the Cretaceous (e.g. McArthur et al. 2007) it is very difficult to attribute such abrupt inundation to glacioeustatism. Therefore, tectonism is the dominant control in the basin evolution for the closure of the Mendoza Group and to the lower part of the Bajada del Agrio Group which constitutes the last palaeo-Pacific sea ingression in the basin.

7.5 *Hierarchy and Frequency*

The age-interval and the age constrained stratigraphic surfaces of the Agrio Formation (upper Mendoza Group) and Huitrín Formation (Bajada del Agrio Group) have changed considerably from the earliest versions of Legarreta and Gulisano (1989) and Legarreta and Uliana (1991, where sequence boundaries were based on the global sea-level change curve of Haq et al. (1987). According to the proposal of Legarreta and Uliana (1991), the time spanned in the stratigraphic interval mentioned before is approximately 18 Myr. The time spanned suggested for this interval was modified substantially in later works (e.g. Veiga et al. 2002). Based on cyclostratigraphic studies (Martinez et al. 2015) and new ideas considering ammonite biozones, calcareous nannofossils and high precision U-Pb ages (Aguirre-Urreta et al. 2015, 2017) the time spanned in this interval is approximately 10 Myr (Fig. 5).

The base of the Agrio Formation represents a transgressive systems tract of a long-term second-order sequence that includes the Mulichinco Formation as the lowstand systems tract (Legarreta and Gulisano 1989; Legarreta and Uliana 1991; Schwarz et al. 2006). Within the Pilmatué Member, a high precision absolute age was obtained from one tuff level (CA-ID TIMS U-Pb age of 130.39 $\pm$ 0.16 Ma) interbedded with mudstones containing ammonites of the *Holcoptychites agrioensis* sub-biozone, *Holcoptychites Neuquénsis* Biozone, of the Early Hauterivian (Aguirre-Urreta et al. 2017). Considering that the member starts in the late Early Valanginian/Late Valanginian, the five sequences documented in the Pilmatué Member have a total duration of approximately 5 Myr (ages obtained from Martinez et al. 2015). If this is a correct assumption, each of the five sequences has a duration of approximately 1 Myr, which is in agreement with a third- or fourth-order hierarchy (Vail et al. 1991). However, according to Martinez et al. (2015) the duration of the Valanginian is different to the one proposed by Cohen et al. (2013) (see Chap. "Orbital Controls and High-Resolution Cyclostratigraphy of Late Jurassic–Early Cretaceous in the Neuquén Basin"). Therefore, the proposed duration is highly tentative.

The deposition of the Avilé Member involved about 0.5 Myr during the late Early Hauterivian (Aguirre-Urreta and Rawson 1997; Veiga et al. 2009). The sequence boundary (SB) that denotes the base of Avilé Member was interpreted as a low-order (third-order hierarchy) lowstand wedge (Legarreta and Uliana 1991; Veiga et al. 2009) that contains higher frequency (fourth–fifthorder) sequences inside (Veiga et al. 2009). The authors studied the relationship between bed load and suspension facies and concluded that internal surfaces indicate higher frequency controls probably due to climatic seasonality.

In the Agua de la Mula Member, the pioneering U-Pb SHRIMP age obtained by Aguirre-Urreta et al. (2008) for a tuff level situated immediately overlying the

Avilé Member within *Spitidiscus riccardii* Biozone was re-dated with a more precise method (U-Pb, CA-ID-TIMS) obtaining a different age of 129.09 ± 0.16 Ma (Aguirre-Urreta et al. 2015). In addition, the authors analysed by this method another tuff level close to the top of the unit (*Paraspiticeras groeberi* Biozone) providing an age of 127.42 ± 0.15 Ma. Taking into account that the Agua de la Mula Member is an Upper Hauterivian–Lower Barremian succession (Aguirre-Urreta and Rawson 2012; Aguirre-Urreta et al. 2015), using the boundaries between the Hauterivian and Barremian proposed by Martinez et al. (2015) the unit has a total duration of approximately 3.5 Myr. Accordingly, and as was also shown in the analysis of Guler et al. (2013) the Avilé and the Agua de la Mula members are related to a third-order asymmetrical sequence, while the four defined sequences within the Agua de la Mula Member represent high-frequency (fourth-order hierarchy) fluctuations in the sea level (Fig. 5).

8 Conclusions

The salient conclusions of the study of the closure of the Mendoza Group involved necessarily a discussion of the Huitrín, due to that lithostratigraphic and sequence stratigraphic approaches are not completely coincident in nature, and the origin of the unconformity that separates the Agua de la Mula Member of the Agrio Formation from the Chorreado Member of the Huitrín Formation is not understood.

The time involved in the deposition of the Agrio Formation is approximately 10 Myr following the proposal of Martinez et al. (2015) which is considerably shorter than the 18 Myr suggested in previous schemes. Then, it implies that the five prograding sequences documented in the Pilmatué Member were deposited in 5 Myr and are interpreted as third-order sequences. The lower one starts with a diachronic inundation. It begins in the early Late Valanginian; the onset of the transgression from the west is only documented at Cerro La Parva (CP) locality. Conversely, evidence of subaerial exposition with dinosaur trackways is only documented in Mina San Eduardo (MSE) where the sedimentary input from the east is confirmed, as was previously suggested for the Agua de la Mula Member and for the Pilmatué Member in subsurface. This persistent area of sedimentary supply was traditionally neglected in regional models for the unit. It does not have an equivalent record in the Avilé Member, reinforcing that the basin reorganization with localized subsidence permits to explain the deposition of 180 m thick fluvial deposits documented in Tricao Malal (TM) and also disregards eustatic sea-level changes as the main control in the member deposition as it was traditionally suggested. The envisaged time gap between the Pilmatué and Avilé Members is probably around 0.5 Myr as indicated by previous studies.

The Agua de la Mula Member contains four sequences, organized in asymmetrical (thin) transgressive-(thick) highstand system tracts. The first transgressive system tract of the member documents an isochronous transgression that in a greenhouse period resembles a postglacial transgression and can only be explained by tectonism.

The marginal-marine facies documented all over the eastern part of the basin in the uppermost sequence, explain that the Chorreado Member follows a depositional trend controlled by the remnant accommodation space after Agua de la Mula Member deposition. This trend does not evidence a basin expansion. The unconformity that separates the Chorreado and Troncoso Members is a master sequence boundary that documents the true reorganization of the basin. From a sequence stratigraphy point of view, the Chorreado Member is still part of the Mendoza Group. Taking into account that the time spanned in the deposition of the "Chorreado" up to the La Tosca Member of the Huitrín Formation involves only the Barremian, and also new areas of sedimentary supply, the master sequence boundary at the base of the Troncoso Member is interpreted as tectonically rather than eustatically controlled.

This chapter shows how depocenters evolved in different ways between the Late Valanginian and the Barremian and confirms the previously proposed thermal subsidence, punctuated by rapid episodes that permitted almost geologically instantaneous transgressions. Contrarily, tectonic calm produced erosion and bypass and it is documented in the unconformity that precedes the Avilé Member. The proposed sequence stratigraphic scheme indicates that eustatic sea-level changes only controlled the internal architecture of the Agrio Formation and the marine transgression of the Chorreado Member, but tectonism was the main control and it has been largely neglected as an allocyclic control in this interval of the basin infill.

Acknowledgements The authors acknowledge to the University of Buenos Aires and CONICET for the financial support with the grants to P.J. Pazos. We are particularly thankful to Beatriz Aguirre-Urreta for the clarifying information about ammonoid biozones. Sergio E. Cocca contributed with some excellent field pictures and in his name we acknowledge to the Servicio Geológico Minero de Neuquén for the field support and hospitality. Finally, we acknowledge to the editors for the possibility to contribute in this special publication about the Neuquén Basin. It is the contribution R-265 of the Instituto de Estudios Andinos: Don Pablo Groeber.

References

Aguirre-Urreta MB, Gutiérrez Pleimling A, Leanza HA (1993) La ubicación estratigráfica de Spitidiscus (Ammonoidea) en el Cretácico inferior de la Cuenca Neuquina. In: Abstract of the 12 Congreso Geológico Argentino y 2° Congreso de Exploración de Hidrocarburos, Mendoza, 10–15 Oct 1993

Aguirre-Urreta MB, Lescano M, Schmitz MD et al (2015) Filling the gap: new precise early cretaceous radioisotopic ages from the andes. Geol Mag 152:557–564

Aguirre-Urreta MB, Lo Forte GL (1996) Los depósitos tithoneocomianos. In: Ramos VA (ed) Geología de la región del Aconcagua, provincias de San Juan y Mendoza, República Argentina. Dirección Nacional del Servicio Geológico, Anales 24, pp 179–230

Aguirre-Urreta MB, Mourgues FA, Rawson PF et al (2007) The Lower Cretaceous Chañarcillo and Neuquén Andean basins: ammonoid biostratigraphy and correlations. Geol J 42:143–173

Aguirre-Urreta MB, Pazos PJ, Lazo DG et al (2008) First U-Pb SHRIMP age of the Hauterivian stage, Neuquén Basin, Argentina. J S Am Earth Sci 26:91–99

Aguirre-Urreta MB, Rawson PF (1997) The ammonite sequence in the Agrio Formation (Lower Cretaceous), Neuquén Basin, Argentina. Geol Mag 134:449–458

Aguirre-Urreta MB, Rawson PF (1999) Stratigraphic position of Valanginites, Lissonia and Acantholissonia in the Lower Valanginian (Lower Cretaceous) ammonite sequence of the Neuquén basin, Argentina. In: Olorrz F, Rodriguez-Tovar R (eds) advancing research on living and fossil-cephalopods. Kluwer Academic, New York, pp 521–529

Aguirre-Urreta MB, Rawson PF (2012) Lower Cretaceous ammonites from the Neuquén Basin, Argentina: a new heteromorph fauna from the uppermost Agrio Formation. Cretaceous Res 35:208–216

Aguirre-Urreta MB, Rawson PF, Concheyro GA et al (2005) Lower cretaceous biostratigraphy of the Neuquén Basin. In: Veiga GD, Spalletti LA, Howell JA et al (eds) The Neuquén Basin: a case study in sequence stratigraphy and basin dynamics. The Geological Society, London, SP 252, pp 57–81

Aguirre-Urreta MB, Schmitz M, Lescano M et al (2017) A high precision U-Pb radioisotopic age for the Agrio Formation, Neuquén Basin, Argentina: implications for the chronology of the Hauterivian stage. Cretaceous Res 75:193–204

Aguirre-Urreta MB, Tunik M, Naipauer M et al (2011) Malargüe Group (Maastrichtian–Danian) deposits in the Neuquén Andes, Argentina: implications for the onset of the first Atlantic transgression related to Western Gondwana break-up. Gondwana Res 19:482–494

Archuby FM, Wilmsen M, Leanza HA (2011) Integrated stratigraphy of the upper Hauterivian to lower Barremian Agua de la Mula member of the Agrio Formation, Neuquén Basin, Argentina. Acta Geol Pol 61:1–26

Barrionuevo M (2002) Reservorios del Miembro Agrio Superior de la Formación Agrio. In: Schiuma M, Hinterwimmer G, Vergani G (eds) Rocas Reservorio de las Cuencas Productivas de la Argentina. Instituto Argentino del Petroleo y el Gas, Mar del Plata, pp 447–456

Bohacs KM, Lazar OR (2010) Sequence stratigraphy in fine-grained rocks. In: Schieber J, Lazar OR, Bohacs KM (eds) Sedimentology and stratigraphy of shales: expressions and correlation of depositional sequences in the devonian of tennessee, Kentucky, and Indiana. AAPG 2010 Annual Convention, Field Guide for SEPM Field Trip 10

Boll A, Alonso J, Fuentes F, Vergara M et al (2014) Factores controlantes de las acumulaciones de hidrocarburos en el Sector Norte de la Cuenca Neuquina, entre los ríos Diamante y Salado, provincia de Mendoza, Argentina. In: Abstracts of the 9 Congreso de Exploración y Desarrollo de Hidrocarburos, Mendoza, 3–7 Nov 2014

Bown PR, Concheyro A (2004) Lower Cretaceous calcareous nannoplankton from the Neuquén Basin, Argentina. Mar Micropal 52:51–84

Brinkmann HD (1994) Facies and sequences of the Agrio Formation (Lower Cretaceous) in the central and southern Neuquén Basin, Argentina. Zbl Geo Pal I:309–317

Burchette TP, Wright VP (1992) Carbonate ramp depositional systems. Sediment Geol 79:3–57

Cabaleiro A, Cazau L, Lasalle D et al (2002) Los reservorios de la Formación Centenario. In: Schiuma M, Hinterwimmer G, Vergani G (eds) Rocas reservorio de las cuencas productivas de la Argentina. Instituto Argentino del Petroleo y el Gas, Buenos Aires, pp 407–426

Carbone O, Franzese J, Limeres M et al (2011) El ciclo Precuyano (Triásico Tardío-Jurásico Temprano) en la Cuenca Neuquina. In: Leanza H, Arregui C, Carbone O et al (eds) Geología y Recursos Naturales de la provincia del Neuquén. Asocaicón Geológica Argentina, Buenos Aires, pp 63–76

Cataldo CS, Lazo DG (2016) Taxonomy and paleoecology of a new gastropod fauna from dysoxic outer ramp facies of the Lower Cretaceous Agrio Formation, Neuquén Basin, west central Argentina. Cretaceous Res 57:165–189

Catuneanu O, Abreu V, Bhattacharya JP et al (2009) Towards the standardization of sequence stratigraphy. Earth-Sci Rev 92:1–33

Catuneanu O, Zecchin M (2013) High-resolution sequence stratigraphy of clastic shelves II: controls on sequence development. Mar Petrol Geol 39:26–38

Cichowolski M, Pazos PJ, Tunik M et al (2012) An exceptional storm accumulation of nautilids in the lower Cretaceous of the Neuquén Basin, Argentina. Lethaia 45:121–138

Cohen KM, Finney SC, Gibbard PL et al (2013) The ICS international stratigraphic chart. Episodes 36:199–204
Comerio M (2016) Estudio mineralógico de las arcillas del Miembro Agua de la Mula—Formación Agrio, en un marco estratigráfico secuencial, en el Engolfamiento Neuquino. Universidad de Buenos Aires, PhD tesis
Comerio M, Fernández DE, Pazos PJ (2018) Sedimentological and ichnological characterization of muddy storm related deposits: the upper Hauterivian ramp of the Agrio Formation in the Neuquén Basin, Argentina. Cretaceous Res 85:78–94
Digregorio JH (1972) Neuquén. In: Leanza AF (ed) Geología Regional Argentina. Centenario de la Academia Nacional de Ciencias, Cordoba, pp 439–506
Embry AF, Johannessen EP (1992) T-R sequence stratigraphy, facies analysis and reservoir distribution in theuppermost Triassic-Lower Jurassic succession, western Sverdrup Basin, Arctic Canada. In: Vorren TO, Bergsager E, Dahl-Stamnes OA et al (eds) Arctic Geology and Petroleum Potential 2, Norw Petrol Soc, pp 121–146
Fernández DE, Pazos PJ (2012) Ichnology of marginal marine facies of the Agrio Formation (Lower Cretaceous, Neuquén Basin, Argentina) in its type locality. Ameghiniana 49:505–524
Fernández DE, Pazos PJ (2013) Xiphosurid trackways in a Lower Cretaceous tidal flat in Patagonia: palaeoecological implications and the involvement of microbial mats in trace-fossil preservation. Palaeogeogr Palaeocl 375:16–29
Fernández DE, Pazos PJ (2015) Ichnological research in Lower Cretaceous marginal-marine facies from Patagonia: outcrop studies, SEM examinations and paleontological/sedimentological integration. Neues Jahrb Geol P-A 277:177–188
Freije H, Azúa G, González R, Ponce JJ et al (2002) Actividad tectónica sinsedimentaria en el Jurásico del sur de la Cuenca Neuquina. In: Abstracts of the 5 Congreso de Exploración y Desarrollo de Hidrocarburos, Mar del Plata, 29 Oct–2 Nov 2002
Garberoglio RM, Lazo DG, Palma RM (2013) An integrate analysis of an Hauterivian coral biostrome from the Agrio Formation, Neuquén Basin, west-central Argentina. Cretaceous Res 43:97–115
Giambiagi LB, Alvarez PP, Godoy E et al (2003) The control of pre-existing extensional structures in the evolution of the southern sector of the Aconcagua fold and thrust belt. Tectonophysics 369:1–19
Groeber P (1946) Observaciones geológicas a lo largo del meridiano 70° 1. Hoja Chos Malal. Rev Asoc Geol Argent 1:177–208
Guler MV, Lazo DG, Pazos PJ et al (2013) Palynofacies analysis and palynology of the Agua de la Mula Member (Agrio Formation) in a sequence stratigraphy framework, lower Cretaceous, Neuquén Basin, Argentina. Cretaceous Res 41:65–81
Gulisano CA (1981) El ciclo cuyano en el norte de Neuquén y sur de Mendoza. Congreso Geológico Argentino. In: Abstarcts of the 8 Congreso Geológico Argentino, vol 8. San Luis, 20–26 Sep 1981
Gulisano CA, Gutiérrez Pleimling A (1988) Depósitos eólicos del Miembro Avilé (Formación Agrio, Cretácico inferior) en el norte del Neuquén, Argentina. In: Abstracts of the 2 Reunión Argentina de Sedimentología, Buenos Aires, 1–5 Aug 1988
Gulisano CA, Minniti SM, Rossi GC et al (2001) The agrio petroleum system: hydrocarbon contribution and key elements. In: Neuquén Basin, Argentina. New technologies and new play concepts in Latin America. AAPG 2001, Hedberg Research Conference, Mendoza, 5–9 Nov 2001
Gutiérrez Pleimling A (1991) Estratigrafía de la Formación Huitrín: un estudio puntual sobre la ruta nacional nº 40, provincia del Neuquén. Bol Inf Petrol 9:85–100
Gutiérrez Pleimling A, Olea G, Suárez M et al (2011) El Miembro Chorreado de la Formación Huitrín. In: Leanza H, Arregui C, Carbone O et al (eds) Geología y Recursos Naturales de la Provincia del Neuquén. Asociación Geológica Argentina, Neuquén, pp 175–180
Haq BU, Hardenbol J, Vail PR (1987) Chronology of fluctuating sea levels since the triassic. Science 235:1156–1166
Heredia AM, Pazos PJ, Fernández DE et al (2018) Las huellas de dinosaurios como evidencia de exposición subaérea del Miembro Pilmatué de la Formación Agrio (Cretácico Inferior) de la

Cuenca Neuquina. In: Abstracts of the 32 Jornadas Argentinas de Paleontología de Vertebrados, Corrientes, 14–17 May 20018
Horton BK (2018) Sedimentary record of Andean mountain building. Earth-Sci Rev 178:279–309
Howell JA, Schwarz E, Spalletti LA et al (2005) The Neuquén Basin: an overview. In: Veiga GD, Spalletti LA, Howell JA et al (eds) The Neuquén Basin: a case study in sequence stratigraphy and basin dynamics. The Geological Society, London, SP 252, pp 1–14
Kocurek G, Robinson NI, Sharp JM (2001) The response of the water table in coastal aeolian systems to changes in sea level. Sediment Geol 139:1–13
Lazo DG (2006) Análisis tafonómico e inferencia del grado de mezcla temporal y espacial de la macrofauna del Miembro Pilmatué de la Formación Agrio, Cretácico Inferior de cuenca Neuquina. Ameghiniana 43:311–326
Lazo DG (2007) Análisis de biofacies y cambios relativos del nivel del mar en el Miembro Pilmatué de la Formación Agrio, Cretácico Inferior de cuenca Neuquina, Argentina. Ameghiniana 44:73–89
Lazo DG, Cataldo CS, Luci L, Aguirre-Urreta MB (2017) Groeber y los invertebrados fósiles del Miembro La Tosca, Cretácico Inferior de la Cuenca Neuquina: una historia de controversias paleontológicas. Rev Asoc Geol Argent 74:9–39
Lazo DG, Cichowolski M, Rodríguez DL et al (2005) Lithofacies, palaeoecology and palaeoenvironments of the Agrio Formation, Lower Cretaceous of the Neuquén Basin, Argentina. In: Veiga GD, Spalletti LA, Howell JA, Schwarz E (eds) The Neuquén Basin: A case study in sequence stratigraphy and basin dynamics. The Geological Society, London, AP 252, pp 295–315
Lazo DG, Concheyro GA, Ottone EG et al (2009) Bioestratigrafía integrada de la Formación Agrio en su localidad tipo, Cretácico temprano de cuenca Neuquina. Rev Asoc Geol Argent 65:322–341
Lazo DG, Damborenea SE (2011) Barremian bivalves from the Huitrín Formation, west-central Argentina: taxonomy and paleoecology of a restricted marine association. J Paleont 85:721–749
Leanza HA (2003) Las sedimentitas huitrinianas y rayosianas (Cretácico Inferior) en el ámbito central y meridional de la cuenca Neuquina, Argentina. Servicio Geológico Minero Argentino, Serie Contribuciones Técnicas-Geología, Buenos Aires, pp 1–31
Leanza HA, Hugo CA (2001) Hoja Geológica Zapala, Hoja 3969-I. Servicio Geológico Minero Argentino, Boletín 275, Buenos Aires, pp 1–128
Legarreta L (2002) Eventos de desecación en la Cuenca Neuquina: depósitos continentales y distribución de hidrocarburos. In: Abstracts of the 5 Congreso de Exploración y Desarrollo de Hidrocarburos, Mar del Plata, 29 Oct–2 Nov 2002
Legarreta L, Boll A (1982) Formación Huitrín. Análisis estratigráfico y esquema prospectivo. Yacimientos Petrolíferos Fiscales, Buenos Aires
Legarreta L, Gulisano CA (1989) Análisis estratigráfico secuencial de la Cuenca Neuquina (Triásico Superior-Terciario Inferior, Argentina). In: Chebli GA, Spalletti LA (eds) Cuencas Sedimentarias Argentinas, Simposio de Cuencas Sedimentarias Argentinas. Universidad de Tucumán, vol 6, Serie Correlación Geológica, pp 221–243
Legarreta L, Kozlowski E, Boll A (1981) Esquema estratigráfico y distribución de facies del Grupo Mendoza en el ámbito sur mendocino de la Cuenca Neuquina. In: Abstarcts of the 8 Congreso Geológico Argentino, San Luis, pp 20–26
Legarreta L, Uliana MA (1991) Jurassic–Cretaceous marine oscillations and geometry of back-arc basin, Central Argentina Andes. In: McDonald DIM (ed) Sea level changes at active plate margins: process and product. Int Ass Sediment SP 12:429–450
Lescano M, Concheyro A (2014) Nanocónidos del Grupo Mendoza (Cretácico Inferior) en la provincia del Neuquén, República Argentina: taxonomía, cronoestratigrafía e implicancias paleogeográficas. Ameghiniana 51:466–499
Manceda R, Figueroa D (1995) Inversion of the Mesozoic Neuquén rift in the Malargüe fold and thrust belt, Mendoza, Argentina. In: Tankard AJ, Suárez R, Welsink HJ (eds) Petroleum Basins of South America. AAPG Memoir 62:369–382

Marchese HG (1971) Litoestratigrafía y variaciones faciales de las sedimentitas mesozoicas de la Cuenca Neuquina, provincia de Neuquén, República Argentina. Rev Asoc Geol Argent 26:343–410

Martinez M, Deconinck JF, Pellenard PM et al (2015) Astrochronology of the Valanginian-Hauterivian stages (Early Cretaceous): chronological relationships between the Paraná-Etendeka large igneous province and the Weissert and the Faraoni events. Global Planet Change 131:158–173

McArthur JM, Janssen NMM, Reboulet S et al (2007) Palaeotemperatures, polar ice-volume, and isotope stratigraphy (Mg/Ca, δ18O, δ13C, 87Sr/86Sr): the early Cretaceous (Berriasian, Valanginian, Hauterivian). Palaeogeogr Palaeocl 248:391–430

Mpodozis C, Ramos VA (2008) Tectónica Jurásica en Argentina y Chile: extensión, subducción oblicua, rifting, deriva y colisiones? Rev Asoc Geol Argent 63:481–497

Naipauer M, Ramos VA (2016) Changes in source areas at Neuquén Basin: Mesozoic evolution and tectonic setting based on U-Pb ages on zircons. In: Folguera A, Naipauer M, Sagripanti L, Ghiglione M, Orts D, Giambiagi L (eds) Growth of the Southern Andes. Springer Earth System Sciences, Amsterdam, pp 33–61

Naipauer M, Tunik M, Marques JC et al (2015) U-Pb detrital zircon ages of Upper Jurassic continental successions: implications for the provenance and absolute age of the Jurassic-Cretaceous boundary in the Neuquén Basin. The Geological Society, London, SP 399:131–154

Paolillo MA, Guler MV, Lazo DG et al (2018) Early Cretaceous dinoflagellate cysts from the Agrio Formation at its type locality (Neuquén Basin, Argentina) and their biostratigraphic implications. Ameghiniana. https://doi.org/10.5710/amgh.26.04.2018.3176

Pazos PJ, Lazo DG, Tunik MA et al (2012) Paleoenvironmental framework of dinosaur track-sites and other ichnofossils in Early Cretaceous mixed siliciclastic–carbonate deposits in the Neuquén Basin, northern Patagonia (Argentina). Gondwana Res 22:1125–1140

Ramos VA (1999) Las provincias geológicas del territorio argentino. In: Caminos R (ed) Geología Argentina. Instituto de Geología y recursos minerales, Anales 29(3), Buenos Aires, pp 41–96

Ramos VA, Folguera A (2005) Tectonic evolution of the Andes of Neuquén: constraints derived from the magmatic arc and foreland deformation. In: Veiga GD, Spalletti LA, Howell JA, Schwarz E (eds) The Neuquén Basin: a case study in sequence stratigraphy and Basin dynamics. The Geological Society, London, SP 252, pp 15–35

Rawson PF, Aguirre-Urreta MB (2012) Lower Cretaceous ammonites from the Neuquén Basin, Argentina: The Hauterivian genus *Spitidiscus*. Cretaceous Res 33:97–105

Riccardi AC (1984) Las asociaciones de amonitas del Jurásico, Cretácico de la Argentina. In: Abstracts of the 9 Congreso Geológico Argentino, San Carlos de Bariloche, pp 5–9 Nov 1984

Rojas Vera EA, Mescua J, Folguera A et al (2015) Evolution of the Chos Malal and Agrio fold and thrust belts, Andes of Neuquén: insights from structural analysis and apatite fission track dating. J S Am Earth Sci 64:418–433

Rossi GR (2001) Arenisca Avilé: Facies, ambientes sedimentarios y estratigrafía de una regresión forzada del Hauteriviano inferior de la Cuenca Neuquina. Universidad Nacional de La Plata, PhD tesis

Rossi GR, Masarik C (2002) Acumulaciones de hidrocarburos en areniscas eólicas. Un ejemplo del hauteriviano de la Cuenca Neuquina. In: Abstracts of the 5 Congreso de Exploración y Desarrollo de Hidrocarburos, Mar del Plata, 29 Oct–2 Nov 2002

Sagasti G (2005) Hemipelagic record of orbitally-induced dilution cycles in Lower Cretaceous sediments of the Neuquén Basin. In: Veiga GD, Spalletti LA, Howell JA, Schwarz E (eds) The Neuquén Basin, Argentina: a case study in sequence stratigraphy and Basin Dynamics. The Geological Society, London, SP 252:231–250

Schwarz E, Spalletti LA, Howell JA (2006) Sedimentary response to a tectonically induced sea-level fall in a shallow back-arc basin: the Mulichinco Formation (Lower Cretaceous), Neuquén Basin, Argentina. Sediment Geol 53:55–81

Schwarz E, Veiga GD, Alvarez Trentini G, Isla MF, Spalletti LA (2018) Expanding the spectrum of shallow-marine, mixed carbonate–siliciclastic systems: processes, facies distribution and depositional controls of a siliciclastic-dominated example. Sedimentology 65:1558–1589

Spalletti LA, Poiré DG, Pirrie D et al (2001a) Respuesta sedimentológica a cambios en el nivel de base en una secuencia mixta clástica–carbonática del Cretácico de la Cuenca Neuquina, Argentina. Rev Soc Geol España 14:57–74

Spalletti LA, Poiré DG, Schwarz E et al (2001b) Sedimentologic and sequence stratigraphic model of a Neocomian marine carbonate–siliciclastic ramp: Neuquén Basin, Argentina. J S Am Earth Sci 14:609–624

Spalletti LA, Veiga GD, Schwarz E (2011) La Formación Agrio (Cretácico Temprano) en la Cuenca Neuquina. In: Leanza HA, Arregui C, Carbone O et al (eds) Geología y recursos naturales del Neuquén. Asocaicón Geológica Argentina, Buenos Aires, pp 145–160

Stipanicic PN, Rodrigo F, Baulíes OL et al (1968) Las formaciones presenonianas en el denominado Macizo Nordpatagónico. Rev Asoc Geol Argent 23:67–98

Strömback A, Howell JA, Veiga GD (2005) The transgression of an erg–sedimentation and reworking/ soft-sediment deformation of aeolian facies: Cretaceous Troncoso Member, Neuquén Basin, Argentina. In: Veiga GD, Spalletti LA, Howell JA et al (eds) The Neuquén Basin, Argentina: a case study in sequence stratigraphy and Basin Dynamics. The Geological Society, London, SP 252:163–184

Toscano AG, Lazo DG, Luci L (2018) Taphonomy and paleoecology of lower cretaceous oyster mass occurrences from west-central Argentina and evolutionary paleoecology of gregariousness in oysters. Palaios 33:237–255

Tunik MA, Folguera A, Naipauer M et al (2010) Early uplift and orogenic deformation in the Neuquén Basin: Constraints on the Andean uplift from U-Pb and Hf isotopic data of detrital zircons. Tectonophysics 489:258–273

Tunik MA, Pazos PJ, Impicini A et al (2009) Dolomitized tidal cycles in the Agua de la Mula Member of the Agrio Formation (Early Cretaceous), Neuquén Basin, Argentina. Latin Am J Sediment Bas An 16:29–43

Turienzo M, Sánchez N, Dimieri L et al (2014) Tectonic repetitions of the Early Cretaceous Agrio Formation in the Chos Malal fold-and-thrust belt, Neuquén basin, Argentina: geometry, kinematics and structural implications for Andean building. J S Am Earth Sci 53:1–19

Uliana MA, Legarreta L (1993) Hydrocarbons habitat in a Triassic-to-Cretaceous sub-Andean setting: Neuquén Basin, Argentina. J Petrol Geol 16:397–420

Vail PR, Audemard F, Bowman SA et al (1991) The stratigraphic signatures of tectonics, eustasy and sedimentology–an overview. In: Einsele G, Ricken W, Seilacher A (eds) Cycles and Events in Stratigraphy. Springer, Berlin, pp 617–659

Vail PR, Mitchum RM Jr, Thompson S (1977) Seismic stratigraphy and global changes of sea level, part four: global cycles of relative changes of sea level. AAPG Memoir 26:83–98

Valenzuela M (2002) Los reservorios del miembro Avilé de la Formación Agrio. In: Schiuma M, Hinterwimmer G, Vergani G (eds) Rocas Reservorio de las Cuencas Productivas de la Argentina. Instituto Argentino del Petróleo y el Gas, Mar del Plata, pp 433–446

Van Wagoner JC, Mitchum RM, Campion KM et al (1990) Siliciclastic sequence stratigraphy in well logs, cores and outcrops. AAPG, Methods in Exploration, nro 7

Veiga GD, Howell JA, Strömbäck A (2005) Anatomy of a mixed marine-nonmarine lowstand wedge in a ramp setting. The record of the Barremian-Aptian complex relative sea-level fall in the central Neuquén basin, Argentina. In: Veiga GD, Spalletti LA, Howell JA et al (eds) The Neuquén Basin, Argentina: a case study in sequence stratigraphy and basin dynamics. The Geological Society, London, SP 252, pp 163–184

Veiga GD, Spalletti LA, Flint S (2002) Aeolian/fluvial interactions and high-resolution sequence stratigraphy of a non-marine lowstand wedge: the Avilé Member of the Agrio Formation (Lower Cretaceous), central Neuquén Basin, Argentina. Sedimentology 49:1001–1019

Veiga GD, Spalletti LA, Flint S (2009) Anatomy of a fluvial lowstand wedge: the Avilé Member of the Agrio Formation (Hauterivian) in central Neuquén Basin (NW Neuquén province), Argentina.

In: Nichols G, Williams E, Paola C (eds) Sedimentary Environments, Processes and Basins. A tribute to Peter Friend. Int Ass Sediment SP 38:341–365

Veiga GD, Spalletti LA, Schwarz W (2011) El Miembro Avilé de la Formación Agrio (Cretácico Temprano). In: Leanza HA, Arregui C, Carbone O, Danieli JC, Vallés JM (eds) Geología y recursos naturales del Neuquén. Asociación Geológica Argentina, Buenos Aires, pp 161–174

Veiga GD, Vergani G (1993) Depósitos de nivel bajo: nuevo enfoque sedimentológico y estratigráfico del Miembro Avilé en el Norte del Neuquén. Argentina. In: Abstracts of the 12 Congreso Geológico Argentino y 2 Congreso de Exploración de Hidrocarburos. Actas, Mendoza, 10–15 Oct 1993

Veiga GD, Vergani G (2011) El miembro Troncoso inferior de la Formación Huitrín (Cretácico temprano) In: Leanza H, Arregui C, Carbone O, Danieli JC, Vallés JM (eds) Geología y Recursos Naturales de la provincia del Neuquén. Asociación Geológica Argentina, Buenos Aires, Buens Aires, pp 181–188

Vergani G, Tankard AJ, Belotti J et al (1995) Tectonic evolution and paleogeography of the Neuquén Basin, Argentina. In: Tankard AJ, Suárez R, Welsink HJ (eds) Petroleum basins of South America. AAPG Memoir 62:383–402

Weaver CE (1931) Paleontology of the Jurassic and Cretaceous of West Central Argentina. University of Washington, Memoir 1, Washington, pp 1–469

Yrigoyen MR (1972) Cordillera Principal. In: Leanza AF (ed) Geología Regional Argentina. Academia Nacional de Ciencias de Córdoba, Córdoba, pp 345–364

Zavala C, Arcuri M, Di Meglio M, Zorzano A (2011) Las capas de San Eduardo: 130 metros de arenas en el Miembro Inferior de la Fm. Agrio, Cretácico Inferior, Cuenca Neuquina. In: Abstracts of the 8 Congreso de Exploración y Desarrollo de Hidrocarburos, Mar del Plata, 8–12 Nov 2011

Closure of the Neuquén Basin in the Southern Andes

Middle Jurassic-Late Cretaceous Paleogeography of the Western Margin of the Neuquén Basin (34° 30′–36° S)

Felipe Tapia, Marcia Muñoz, Marcelo Farías, Reynaldo Charrier and Daniela Astaburuaga

Abstract U–Pb dating of detrital and igneous zircons from the retroarc deposits of the Neuquén Basin has shed light over the Mesozoic evolution of the western border of South America, yet the coeval arc and forearc regions remain mostly indirectly characterized. Furthermore, recent paleogeographic reconstructions consider the arc and forearc regions as a tectonically stable and static region at least until Late Cretaceous. In this chapter, we aim to contribute to the Middle Jurassic-Late Cretaceous paleogeographic reconstructions of the western margin of South America from a western point of view integrating the coeval arc and forearc evolution, between 34° 30′ and 36° S. We focus here in the deposits exposed along the Chilean slope of the Principal Cordillera and use four new detrital zircon age data to determine their ages and main source areas. These ages are compared with 38 published U–Pb detrital zircon ages and integrated into a series of paleogeographic cross sections which illustrate the Mesozoic evolution along the Southern Central Andes encompassing the forearc, arc, and retroarc regions. Our data show that the arc and forearc regions were active at least since the Middle Jurassic. Evidence for this tectonic activity corresponds to the development of forearc basins in the Middle Jurassic and Early Cretaceous times. New ages along the Chilean slope of the Andes allow suggesting an early beginning for the compressive period during the latest Early Cretaceous. The formation of a geographic barrier, as a consequence of the compressive regime, would explain the differences in the sediments provenance between western and eastern deposits during the latest Late Cretaceous. Finally, the almost complete record

F. Tapia (✉)
Departamento de Ciencias Geológicas, Facultad de Ciencias Exactas y Naturales, Universidad de Buenos Aires. Consejo Nacional de Investigaciones Científicas y Técnicas, Instituto de Estudios Andinos "Don Pablo Groeber" (IDEAN), Int. Güiraldes 2160 (C1428EGA), Buenos Aires, Argentina
e-mail: ftapiasilva@gmail.com

M. Muñoz · R. Charrier
Escuela de Ciencias de la Tierra, Facultad de Ingeniería, Universidad Andres Bello, Campus República, Salvador Sanfuentes, 2357 Santiago, Chile

M. Farías · D. Astaburuaga
Departamento de Geología, Universidad de Chile, Casilla 13518, Correo 21, Santiago, Chile

D. Kietzmann and A. Folguera (eds.), *Opening and Closure of the Neuquén Basin in the Southern Andes*, Springer Earth System Sciences,
https://doi.org/10.1007/978-3-030-29680-3_11

of Mesozoic ages in the detrital and volcanic deposits of the western slope of the Southern Central Andes constitutes a counter-argument about the null or waning activity proposed for the Middle Jurassic or Late Cretaceous from U–Pb detrital zircon analysis of the eastern Mesozoic deposits. Conversely, our data indicate a continued activity of the arc-related volcanism and magmatism throughout all the Mesozoic time.

Keywords Middle Jurassic-Late cretaceous · Chilean slope · Western Neuquén Basin · U–Pb dating

1 Introduction

In the last years, U–Pb dating of detrital and igneous zircons from marine and continental sedimentary deposits of the retroarc Neuquén Basin (Fig. 1a) has shed light over the Mesozoic evolution of the western border of South America. For such sedimentary deposits, the identification of the source areas has pointed out major paleogeographic changes during the Gondwana break-up as the cessation of the syn-rift volcanism and the subsequent onset of the Andean subduction (Naipauer et al. 2015), intraplate Late Jurassic deformation (Naipauer et al. 2012), or the passage from a back-arc to a foreland stage of the Neuquén basin (Balgord and Carrapa 2016; Di Giulio et al. 2017, 2012; Fennell et al. 2017; Horton et al. 2016).

The new U–Pb ages obtained along the retroarc area have allowed unraveling the Mesozoic evolution of this Andean segment, yet the coeval arc and forearc regions (Fig. 1a) remain indirectly characterized. In fact, the occurrence of magmatic lull periods in the Mesozoic evolution of the Southern Central Andes has been largely inferred from the absence of the corresponding age components in the zircon record (Balgord 2017; Fennell et al. 2017). Moreover, recent paleogeographic reconstructions for central Chile and west-central Argentina consider the arc and forearc regions as a tectonically stable and relatively static region at least until the Late Cretaceous (cf. Horton et al. 2016). Conversely, the arc and forearc regions evidence subsidence and uplift episodes triggering the development of arc and forearc basins during their Mesozoic evolution (e.g., Lo Prado forearc basin; Charrier et al. 2007).

In this chapter, we aim to contribute to the Middle Jurassic-Late Cretaceous paleogeographic reconstructions of the western margin of South America from a western Andean point of view, integrating the coeval arc and forearc evolution. The latter are widely represented in numerous sedimentary, volcanic, and volcanoclastic deposits of this age exposed along the Principal Cordillera of Chile–Argentina and the areas further west until the Coastal Cordillera (Fig. 1b). Although all such deposits are relevant for a comprehensive paleogeographic analysis, only the Principal Cordillera area has been more intensively studied. We focus here in the marine and continental sedimentary deposits along with volcanic and volcanoclastic rocks of the Principal Cordillera, between 34° 30′ and 36° S, and use four new detrital zircon age data to determine their ages and main source areas. These data are compared with 38

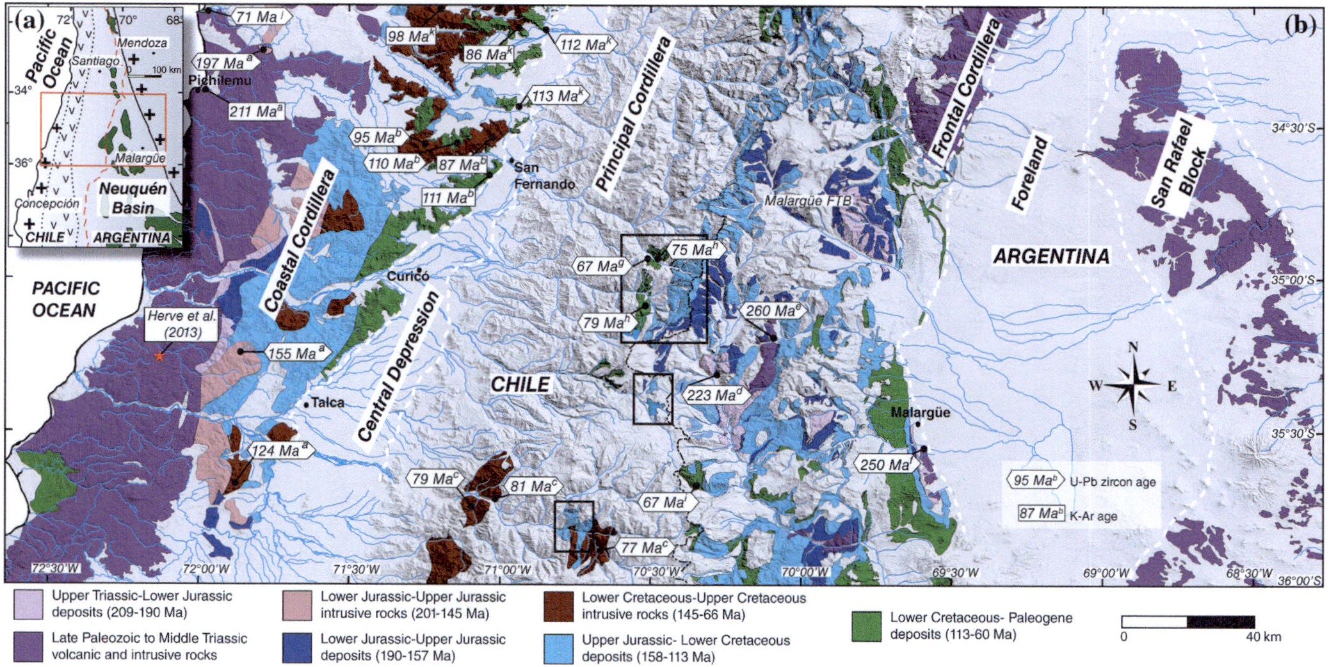

Fig. 1 **a** Paleogeographic map of the western margin of South America (32°–38° S) showing the distribution of the Mesozoic volcanic arc (black vees), depocenters (green polygons), and pre-Jurassic basement (black crosses) of the Neuquén Basin. Red polygon indicates the location of the Fig. 1b. **b** Geological map of the study area showing the main geographic, tectonic, and morphostructural features of the Southern Central Andes, along with the outcrops of the Upper Paleozoic and Mesozoic rocks (modified from Charrier et al. 2015). Black rectangles show the location of the maps displayed in Fig. 3. Red star indicates the location of the Mesozoic samples of Hervé et al. (2013). U–Pb and K–Ar ages of Upper Paleozoic-Upper Cretaceous igneous rocks are shown, and their references correspond to: a, Vásquez et al. (2011); b, Contreras et al. (in press); c, Tapia et al. (2015a, b); d, Naipauer et al. (2015); e, Mescua et al. (2015); f, Llambías et al. (2005); g, Mosolf et al. (2018); h, Muñoz et al. (2018); i, Iannelli et al. (2018)

published U–Pb detrital zircon ages and integrated into a series of paleogeographic cross sections which illustrate the Mesozoic evolution along the Southern Central Andes encompassing the forearc, arc, and retroarc regions.

2 Geological Background

The Andean orogen constitutes the main geological feature developed along the current South American continental margin. Overall, this corresponds to a doubly vergent (Armijo et al. 2010) deformed and uplifted belt developed in response to the eastward subduction of the Farallón-Nazca Plate underneath the continent. The study area of this work (34° 30′–36° S) is hosted in the Southern Central Andes (Gansser 1973). There, the Andean margin can be subdivided, between the coastline and the foreland, in four major continental morphostructural units (Fig. 1b). From west to east, these correspond to (i) Coastal Cordillera, (ii) Central Depression, (iii) Principal Cordillera, and (iv) the Frontal Cordillera (which disappears south of 34° 40′ S) and the foreland. These morphostructural units are composed of igneous, metamorphic, and sedimentary rocks formed since Paleozoic times (Charrier et al. 2007, 2015).

The tectonic and geological evolution of the Southern Central Andes has been related to an almost continuous subduction regime. The Neoproterozoic to Late Paleozoic evolution is characterized by terrane accretion and westward arc migration alternating with periods of rifting and extensional basin formation (Charrier et al. 2007, 2015, and references therein). Lasting until the Late Devonian (Ramos et al. 1986), this period was followed by a subduction setting associated with the development of a magmatic arc in the western margin of Gondwana and eastern retroarc basins during Carboniferous time (Limarino and Spalletti 2006). The arc expanded and migrated toward the foreland during the Early Permian, in a process related to the San Rafael tectonic phase (Charrier et al. 2007, 2015). Contemporaneously, the extensive intermediate to acid magmatic episode of the Choiyoi province was emplaced at the SW margin of the Gondwana during the Early Permian to Late Permian-Early Triassic (Sato et al. 2015).

A major paleogeographic reorganization took place during the Late Permian-Early Triassic to Late Triassic associated with extensional conditions affecting the previous Gondwanan orogen (Charrier et al. 2007, 2015). On the Chilean slope, this period is known as the Pre-Andean cycle (cf. Charrier et al. 2015). Generalized extension resulted in the regional scale development of NW-trending rift basins (Charrier 1979; Coloma et al. 2017; González et al. 2018; Uliana et al. 1989) and widely distributed magmatism (Oliveros et al. 2017). Particularly, along west-central Argentina and central Chile, a series of rifts were characterized by narrow asymmetric half-grabens bounded by predominantly normal faults parallel to previous Paleozoic sutures of the accreted terranes (Ramos and Kay 1991). Among such domains, the most important are the Cuyo and Ischigualasto basins in Argentina, and the San Felix and El Quereo-Los Molles basins in Chile.

During the latest Late Triassic and Early Jurassic (205–186 Ma), a new rifting period affected the western margin of Gondwana which lead to the early development of the Neuquén Basin. This rift stage was characterized by isolated depocenters controlling the eruption of acidic and mafic volcanic rocks grouped in the Precuyano cycle (Fig. 2) in Argentina, a process that marks the onset of the sedimentation in the Neuquén Basin (Franzese and Spalletti 2001; Vergani et al. 1995) (see Chap. "The Syn-Rift of the Neuquén Basin (Precuyano and Lower Cuyano Cycle): Review of Structure, Volcanism, Tectono-Stratigraphy and Depositional Scenarios"). On the Chilean slope of the Southern Central Andes, this period has been correlated to the latest second stage of the Pre-Andean cycle (Fig. 2) (Charrier et al. 2015). In the study area, the Hettangian–Sinemurian marine deposits of the Laguna de Tilicura

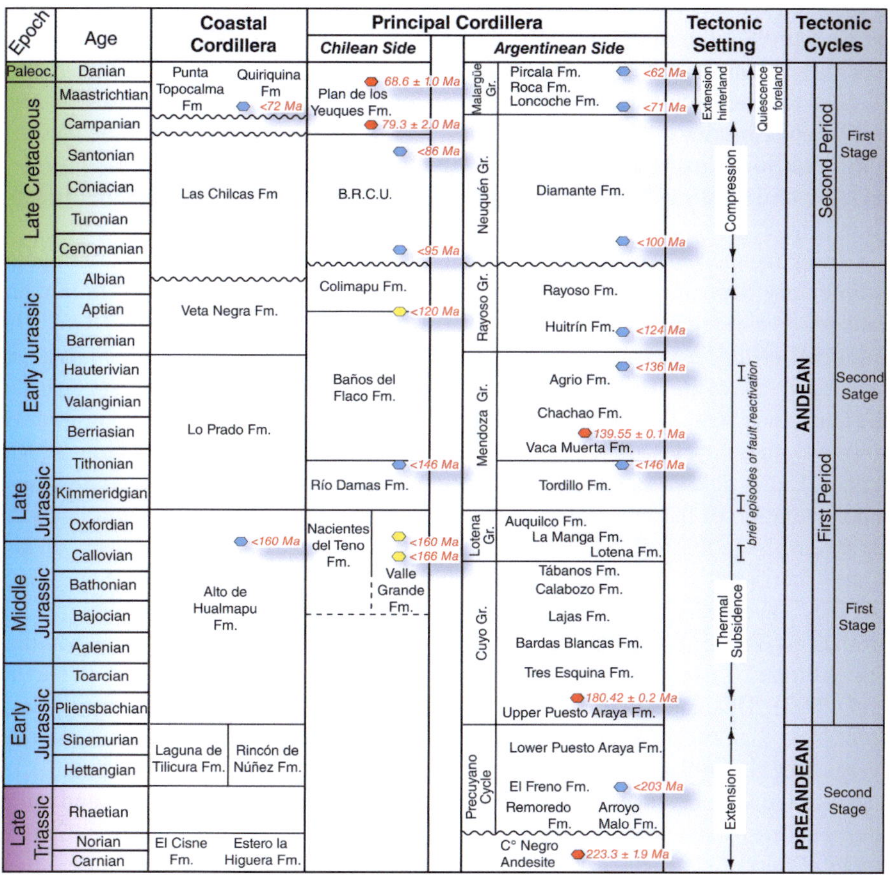

Fig. 2 Chronostratigraphic chart for Jurassic to Cretaceous deposits of the Coastal and Principal Cordilleras at ~35° S. Colored diamonds represent U–Pb zircon ages in igneous (red) and sedimentary (blue) rocks reported in the literature and from this work (yellow). Tectonic cycles are taken from Charrier et al. (2015)

and Rincón de Núñez formations, that crop out in the Coastal Cordillera (Fig. 1b), have been assigned to the upper part of the Pre-Andean cycle (Fig. 2) (Charrier et al. 2015). The fossil content (Corvalán 1976) would allow correlating these units with the marine deposits of the Arroyo Malo Formation and the lower section of the Puesto Araya Formation of the Precuyano cycle (Fig. 2) (cf. Lanés 2005).

From Early Jurassic to Late Cretaceous, the previous rifting stage was followed by a period characterized by the development of a magmatic arc, in a narrow position along the current Coastal Cordillera (Fig. 1b), and a back-arc basin to the east (Charrier et al. 2007; Mpodozis and Ramos 1989; Ramos 2000). The latter was dominated by thermal subsidence and a Sinemurian–Pleinsbachian marine transgression from the Panthalassic Ocean (Franzese and Spalletti 2001; Legarreta and Uliana 1996; Mazzini et al. 2010). These tectonic–paleogeographic changes would mark the onset of the first period of the Andean cycle (cf. Charrier et al. 2015). However, recent studies have proposed to redefine the Andean cycle placing the beginning of this cycle during the Triassic or even before that (Oliveros et al. 2017). Although we agree in the need to redefine the cycle boundaries, in this chapter, we will follow the previous division of Charrier et al (2007) that considers the placing of the arc in the western margin of South America and the development of a thermal subsidence-controlled back-arc basin during the Early Jurassic as indicators of the beginning of the Andean cycle (Fig. 2).

During the thermal subsidence stage, more than 4000 m of marine and continental sedimentary deposits accumulated in the Neuquén Basin, representing three main transgressive–regressive cycles (Howell et al. 2005; Legarreta and Uliana 1996). The sediments of the basin were separated into four different lithostratigraphic groups in the eastern side of the Andes (Fig. 2): Cuyo, Lotena, Mendoza, and Rayoso (Legarreta and Gulisano 1989). On the Chilean slope of the Principal Cordillera, at the latitude of this study, these cycles are represented by the Nacientes del Teno, Valle Grande, Río Damas, Baños del Flaco, and Colimapu formations (Fig. 2). The characteristics, age, and correlation of the latter Mesozoic units and their paleogeographic implications are described in the following section. The forearc and arc-related deposits of this stage (Charrier et al. 2007) crop out along the Coastal Cordillera in Chile (Fig. 1b), and they have been grouped in the Alto de Hualmapu, Lo Prado, and Veta Negra formations (Fig. 2). The Alto del Hualmapu Formation is correlated with the Middle-Upper Jurassic La Horqueta Formation cropping out to the north of the 34° S (Charrier et al. 2007). This volcanic unit consists of a continental andesitic and volcanoclastic succession that unconformably underlies the Early Cretaceous Lo Prado Formation (Vergara et al. 1995 and references therein).

Since the Late Cretaceous until present, different pulses of contraction and uplift alternated with extensional episodes controlled the evolution of the Southern Central Andes (Charrier et al. 2007, 2015). The occurrence of compressive conditions since the Late Cretaceous marks the beginning of the second period of the Andean Cycle (Fig. 2). During this time, the Late Paleozoic and Mesozoic rocks of the Coastal Cordillera underwent deformation (Arancibia 2004; Boyce et al. 2014) and accelerated exhumation (Gana and Zentilli 2000; Parada et al. 2005; Willner et al. 2005). Synorogenic deposits associated with this crustal shortening (Fig. 2) correspond to the Lower-Upper Cretaceous Las Chilcas Formation (Boyce et al. 2014) which crops out along the eastern and western side of the Coastal Cordillera and Central Depression between 33° and 35° S (Fig. 1b). Afterward, shortening migrated to the east during the Late Cretaceous, accompanied by the migration of the magmatic arc (Folguera et al. 2011; Folguera and Ramos 2011). During this time, deformation was associated with the development of the first phase of fold-and-thrust belts in the eastern slope of the Southern Central Andes (Fennell et al. 2017; Folguera et al. 2011; Mescua et al. 2013; Ramos and Folguera 2005; Zamora Valcarce et al. 2006; Zapata and Folguera 2005) (see Chap. "The Late Cretaceous Orogenic System: Early Inversion of the Neuquén Basin and Associated Synorogenic Deposits, 35°–38° S"). Coevally, synorogenic deposits were accumulated in the Cretaceous foreland basin, and they were subsequently involved in the deformation of the fold-and-thrust belt during the eastward migration of the orogenic front (Fennell et al. 2017). These synorogenic deposits correspond to the Brownish Red Clastic Unit (BRCU) in Chile and Neuquén Group in Argentina (Fig. 2). The development of Campanian–Maastrichtian extensional intra-arc basins in the hinterland of the orogen marks the end of the Late Cretaceous shortening (Iannelli et al. 2018; Muñoz et al. 2018). Throughout this period, the volcanic and volcanic clastic deposits of the Plan de los Yeuques Formation and Los Angeles Unit (Fig. 2) accumulated in the hinterland while in the eastern area the marine Malargüe Group (Fig. 2) was deposited along the foreland as result of the Maastrichtian–Danian Atlantic transgression (Aguirre-Urreta et al. 2011).

Shortening locus and the magmatic arc remained in the foreland area until the Eocene, when the arc retreated to the west and a period of extension affected the western flank of the Andes (Charrier et al. 2007, 2015). Contractional tectonics recommenced along the western slope of the Southern Central Andes during the early Miocene (Charrier et al. 2002). Then, the deformation migrated to the east producing the development of eastern fold-and-thrust belts that characterize this Andean segment (Folguera et al. 2011). The arc magmatism migrated to the east up to 600 km from the current Pacific trench simultaneously with the deformation, since the Miocene reaching the foreland during the late Pliocene (Folguera et al. 2011) (see Chap. "The Miocene Magmatism in the Malargüe and Chos Malal Fold and Thrust Belts"). After that, the arc returns to western Principal Cordillera reaching during the Quaternary its current location.

3 Analytical Methods

To obtain maximum depositional ages and to carry out a provenance analysis, we collected four samples from the Mesozoic sedimentary units cropping out along the Chilean slope of the Principal Cordillera between 34° and 36° S (Fig. 3b, c). Zircon grains were separated from total rock samples using standard crushing, washing, heavy liquid, and paramagnetic procedures at facilities of the Departamento de Geología of the Universidad de Chile. U–Pb determinations in single zircon crystals

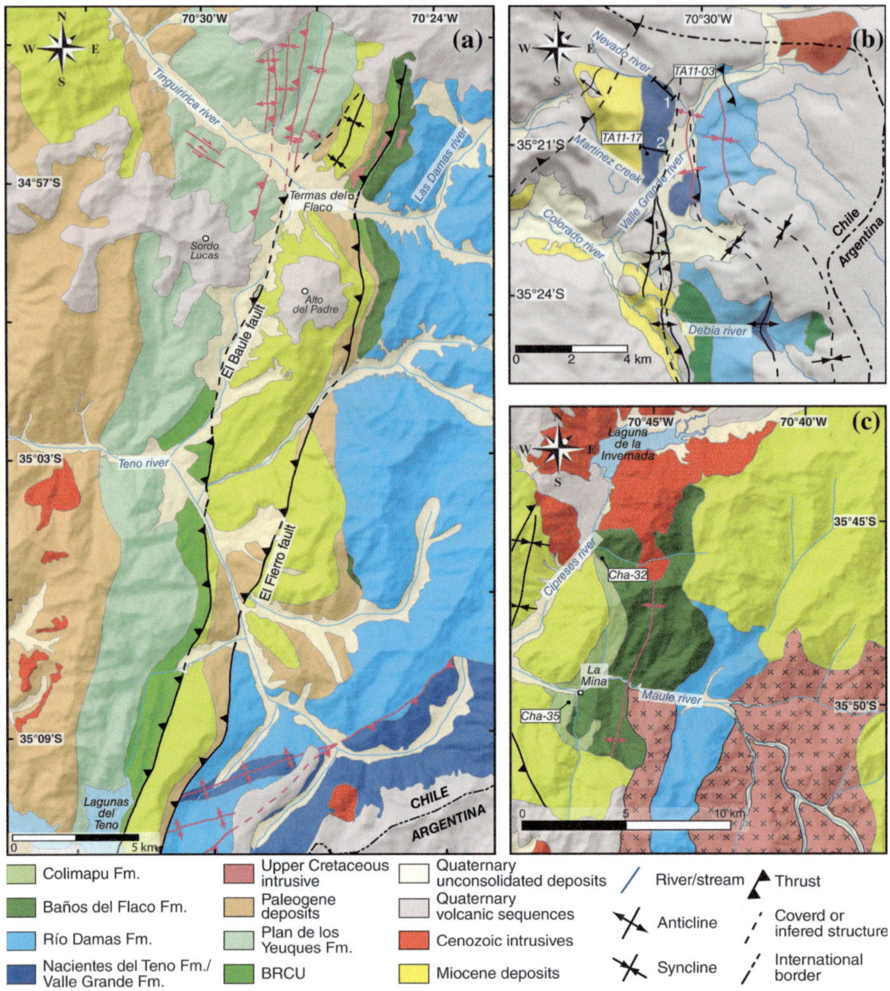

Fig. 3 Geological maps of the three study areas along the Principal Cordillera of Chile between 35° and 36° S. **a** Tinguiririca valley (modified from Muñoz et al. 2018), **b** Valle Grande valley (modified from Tapia et al. 2015a), and **c** Maule valley (modified from Astaburuaga 2014). Red structures correspond to faults and folds with evidence of pre-Neogene activity. Thick black lines in (**b**), also signaled with numbers (1, 2), indicate the section measured in this study (1) and by Gonzalez and Vergara (1962) (2). The locations of the samples dated in this study are indicated in (**b**) and (**c**)

were performed at the Centro de Geociencias - Universidad Autónoma de México by laser ablation inductively coupled plasma mass spectrometry (LA-ICP-MS). The presented probability density plots with stacked histograms were calculated with the Density Plotter software (Vermeesch 2012), while weighted mean $^{206}Pb/^{238}U$ age calculations were carried out using ISOPLOT/EX 3.7 (Ludwig 1999). For all age calculations, U–Pb data with a discordance over 15% were rejected. For detrital zircon samples, the maximum depositional age was determined through the weighted mean age of the youngest cluster of three or more grain overlapping at 2σ level following Dickinson and Gehrels (2009). Detailed analytical methods and the full data set are reported in the Data Repository files (Tables 1, 2, 3, and 4).

4 Mesozoic Stratigraphy of the Chilean Principal Cordillera

In Central Chile, between 34° 30′ and 36° S, the Mesozoic stratigraphic record exposed along the Principal Cordillera is discontinuous, which prevents its study in a single area. For this reason, we focus on the geology of the three different areas in this region where outcrops of Mesozoic deposits cover continuously the period from Middle Jurassic to Late Cretaceous (Figs. 1b and 3). From north to south, these correspond to (1) the Tinguiririca valley (35° S, Fig. 3a), (2) Valle Grande valley (35° 20′ S, Fig. 3b), and (3) Maule valley (35° 50′ S, Fig. 3c). The pioneer studies developed in this whole region were performed by Burckhard (1900), Davidson and Vicente (1973), González and Vergara (1962) and Klohn (1960), who carried out a detailed description of the Mesozoic rocks and proposed the first hypotheses about their deposition and later deformation.

The oldest record of the Neuquén Basin infill in the study area corresponds to Middle-Upper Jurassic Nacientes del Teno Formation (Klohn 1960), which was deposited during the thermal subsidence stage (Fig. 2). This unit crops out along the headwaters of the Teno river (Fig. 3a) and has been correlated to the south with the Valle Grande Formation (González and Vergara 1962) (Fig. 2), a unit which has been studied in detail in this work in the homonymous river (Fig. 3b). At this location, the Valle Grande Formation does not have an exposed base and conformably underlies the Upper Jurassic Río Damas Formation (Fig. 2). It consists of <500 m of marine sedimentary deposits, of lower Callovian-upper Oxfordian age, divided in two members, as shown in the generalized stratigraphic column performed in this work along the Nevado river (Fig. 4). The basal portion of the lower member consists of *ca.* 30 m of fossiliferous limestones and shales with sandstone intercalations (Fig. 4). An erosive contact separates the basal portion from an overlying 75-m-thick succession of red sandstones and sandy breccias. Limestone and shale clasts in a very coarse sandstone above of the contact (Fig. 4) evidence the erosion and recycling of the underlying levels. Moreover, this sandstone dominated interval shows current ripples with different directions, reflecting variable flow direction likely related to a shallow and proximal marine environment. Greenish and reddish coarse sandstones and conglomeratic sandstones dominate the middle part of the lower member of the Valle Grande Formation. The above characteristics along with plant remains and desiccation cracks would indicate nonmarine or nearshore environment for the middle part of the lower member of the formation. The upper part of the Nevado river

Table 1 U–Pb zircon data for sample TA1103 from the Valle Grande Formation

TA1117DZ_31	546	1.16	33.597	3.13%	0.0527	2.47%	189.1	5.8	316.9	55.3	189.1	5.8
TA1117DZ_30	294	1.4	29.414	2.50%	0.0564	2.36%	215.5	5.3	469.5	51.3	215.5	5.3
TA1117DZ_29	744	0.3	6.5484	2.37%	0.0714	1.14%	916.1	20.2	969.5	23.1	916.1	20.2
TA1117DZ_28	400	1.34	28.0958	2.46%	0.0503	1.65%	225.5	5.4	209.7	37.8	225.5	5.4
TA1117DZ_27	1,646	0.75	23.4283	2.64%	0.0529	1.18%	269.4	7	325.6	26.5	269.4	7
TA1117DZ_26	815	0.69	16.7581	2.29%	0.0546	1.20%	373.6	8.3	394	26.7	373.6	8.3
TA1117DZ_25	676	0.39	22.8034	2.31%	0.0521	1.32%	276.7	6.3	290.9	30	276.7	6.3
TA1117DZ_24	671	0.88	16.5353	2.41%	0.056	1.38%	378.5	8.9	451.8	30.3	378.5	8.9
TA1117DZ_23	369	1.52	29.1279	2.41%	0.0513	1.89%	217.6	5.2	253.9	42.9	217.6	5.2
TA1117DZ_22	231	0.93	6.0009	2.12%	0.0817	1.39%	993.6	19.5	1238.9	27	1238.9	27
TA1117DZ_21	530	0.61	21.1402	1.85%	0.0525	1.11%	297.9	5.4	309.1	25.2	297.9	5.4
TA1117DZ_20	552	1.13	32.2016	1.95%	0.0529	1.39%	197.1	3.8	322.7	31.3	197.1	3.8
TA1117DZ_19	141	0.53	6.8704	2.44%	0.0799	1.09%	876	20	1193.4	21.4	1193.4	21.4
TA1117DZ_18	1,535	0.33	32.1635	1.89%	0.0499	1.14%	197.4	3.7	190.8	26.2	197.4	3.7
TA1117DZ_17	126	1.24	28.3765	1.83%	0.0511	1.72%	223.3	4	247.3	39.1	223.3	4
TA1117DZ_16	126	1.24	28.3765	1.83%	0.0511	1.72%	223.3	4	247.3	39.1	223.3	4
TA1117DZ_15	89	3.27	29.5501	2.46%	0.061	2.10%	214.5	5.2	640.4	44.6	214.5	5.2
TA1117DZ_14	430	1.34	30.9279	2.82%	0.0551	1.24%	205.1	5.7	415.9	27.5	205.1	5.7
TA1117DZ_13	923	1.22	33.0295	2.06%	0.0538	1.10%	192.3	3.9	364.3	24.5	192.3	3.9
TA1117DZ_12	794	0.63	23.213	2.03%	0.0524	1.25%	271.9	5.4	302.2	28.3	271.9	5.4
TA1117DZ_11	458	1.69	35.2644	1.97%	0.0544	1.54%	180.3	3.5	387.8	34.2	180.3	3.5
TA1117DZ_10	113	0.77	40.1902	1.93%	0.0547	2.02%	158.4	3	400.6	44.6	158.4	3
TA1117DZ_9	616	0.95	23.8188	2.01%	0.0531	1.05%	265.1	5.2	333.8	23.7	265.1	5.2
TA1117DZ_8	1,219	0.83	21.919	1.68%	0.0533	1.03%	287.6	4.7	343	23.2	287.6	4.7
TA1117DZ_7	781	1.26	35.5159	2.34%	0.0503	1.60%	179	4.1	211	36.7	179	4.1
TA1117DZ_6	116	1.09	34.5436	2.15%	0.054	1.93%	184	3.9	371.8	42.8	184	3.9
TA1117DZ_5	830	1.56	39.618	2.15%	0.0502	1.29%	160.7	3.4	206.1	29.7	160.7	3.4
TA1117DZ_4	671	1.07	39.6056	1.95%	0.0518	1.47%	160.7	3.1	275.6	33.4	160.7	3.1
TA1117DZ_3	933	0.37	20.6106	1.92%	0.0523	1.09%	305.4	5.7	299.5	24.6	305.4	5.7
TA1117DZ_2	216	1.34	35.1719	1.69%	0.0501	1.57%	180.7	3	198.7	36	180.7	3
TA1117DZ_1	657	0.81	19.285	1.76%	0.0528	1.04%	325.9	5.6	318.6	23.6	325.9	5.6
	U (ppm)	Th/U	238U/206Pb	±1s	207Pb/206Pb	±1s	206Pb/238U	±1s	207Pb/206Pb	±1s	Best age (Ma)	±1s
TA1103_15	549	0.54	39.3691	1.99%	0.0488	1.14%	161.7	3.2	135.9	26.6	161.7	3.2
TA1103_6	245	0.88	38.2299	1.99%	0.0483	1.43%	166.5	3.3	115.9	33.5	166.5	3.3
TA1103_8	175	0.75	38.2026	2.24%	0.0487	1.87%	166.6	3.7	132.4	43.4	166.6	3.7
TA1103_16	459	0.63	37.8897	1.99%	0.0489	1.27%	167.9	3.3	140.7	29.6	167.9	3.3
TA1103_27	310	0.64	35.5162	1.82%	0.0513	1.48%	179	3.2	255.1	33.8	179	3.2
TA1103_9	218	0.7	35.4773	2.01%	0.0499	1.41%	179.2	3.6	192.2	32.4	179.2	3.6
TA1103_53	391	0.55	35.1883	1.80%	0.0512	1.17%	180.6	3.2	248.6	26.8	180.6	3.2
TA1103_25	464	0.96	34.9597	1.76%	0.0501	1.17%	181.8	3.2	201.6	26.9	181.8	3.2
TA1103_14	653	0.68	34.7248	1.99%	0.0537	1.22%	183	3.6	357.7	27.3	183	3.6
TA1103_34	785	0.22	33.6664	1.77%	0.0499	0.96%	188.7	3.3	188.5	22.1	188.7	3.3
TA1103_46	687	0.86	33.435	1.93%	0.0493	1.31%	190	3.6	164.3	30.4	190	3.6
TA1103_7	184	1.65	33.3819	2.14%	0.0495	1.45%	190.3	4	173.7	33.6	190.3	4
TA1103_24	724	0.47	33.3535	1.70%	0.0497	0.95%	190.4	3.2	182.9	22.1	190.4	3.2
TA1103_35	1,700	0.17	33.2487	1.70%	0.0493	0.79%	191	3.2	163.5	18.3	191	3.2
TA1103_47	540	0.84	33.1138	2.24%	0.0499	1.76%	191.8	4.2	191.8	40.5	191.8	4.2
TA1103_57	374	0.49	33.1037	2.97%	0.0497	1.27%	191.9	5.6	178.7	29.3	191.9	5.6
TA1103_28	455	0.72	32.8871	1.66%	0.0501	1.09%	193.1	3.2	198.5	25.1	193.1	3.2
TA1103_59	528	0.45	32.8254	2.95%	0.051	1.07%	193.5	5.6	239	24.6	193.5	5.6
TA1103_40	739	0.53	32.5384	1.81%	0.0513	1.03%	195.1	3.5	256.5	23.4	195.1	3.5
TA1103_45	339	0.59	31.5852	2.22%	0.0933	3.09%	200.9	4.4	1493.3	57.3	200.9	4.4
TA1103_23	917	1.35	31.3595	1.78%	0.0616	0.89%	202.4	3.5	659.4	18.9	202.4	3.5
TA1103_18	180	0.93	29.126	2.02%	0.0515	1.53%	217.6	4.3	261.9	34.7	217.6	4.3
TA1103_12	774	1.02	28.9394	2.04%	0.0501	0.88%	219	4.4	198.1	20.3	219	4.4
TA1103_17	253	1.16	28.8705	2.09%	0.051	1.39%	219.5	4.5	242.2	31.6	219.5	4.5
TA1103_39	162	1.4	26.9368	1.99%	0.0536	1.46%	235	4.6	356	32.6	235	4.6
TA1103_55	569	0.88	26.3381	3.24%	0.0571	1.21%	240.2	7.6	494.9	26.5	240.2	7.6
TA1103_29	406	1.46	24.7907	1.71%	0.0509	1.04%	254.9	4.3	237	23.9	254.9	4.3
TA1103_52	69	1.05	24.6282	2.37%	0.0528	2.88%	256.6	6	318.7	64.2	256.6	6
TA1103_37	641	0.49	23.241	1.76%	0.0522	0.90%	271.6	4.7	296.2	20.4	271.6	4.7
TA1103_26	904	0.17	23.1152	1.65%	0.0518	0.79%	273	4.4	278.4	18.1	273	4.4
TA1103_60	576	0.27	23.0585	2.90%	0.0518	0.80%	273.7	7.8	274.9	18.2	273.7	7.8
TA1103_3	400	0.93	22.9986	1.98%	0.0518	1.08%	274.4	5.3	275.7	24.6	274.4	5.3
TA1103_48	568	0.5	22.5774	1.87%	0.0521	0.88%	279.4	5.1	291.4	20	279.4	5.1

(continued)

Table 1 (continued)

TA1103_58	301	0.94	22.5271	2.88%	0.0529	1.16%	280	7.9	324.1	26.1	280	7.9
TA1103_36	581	0.47	22.2581	1.75%	0.0517	0.93%	283.3	4.9	271.7	21.3	283.3	4.9
TA1103_31	166	1.02	22.1044	1.86%	0.0515	1.29%	285.2	5.2	263	29.3	285.2	5.2
TA1103_1	563	0.47	21.9742	1.95%	0.0522	0.93%	286.9	5.5	294.6	21.2	286.9	5.5
TA1103_21	465	0.5	21.9447	2.12%	0.0531	0.96%	287.3	6	334.1	21.5	287.3	6
TA1103_4	516	0.74	21.7121	1.92%	0.0545	0.98%	290.3	5.5	390.3	21.9	290.3	5.5
TA1103_30	380	1.08	21.4241	1.70%	0.0528	1.01%	294.1	4.9	318.7	22.7	294.1	4.9
TA1103_56	862	0.82	20.7397	2.89%	0.0533	0.78%	303.6	8.6	342.9	17.5	303.6	8.6
TA1103_33	304	0.58	20.2361	1.74%	0.052	1.08%	310.9	5.3	285.8	24.6	310.9	5.3
TA1103_22	412	1.03	19.6362	2.03%	0.0525	0.95%	320.2	6.3	306.1	21.4	320.2	6.3
TA1103_5	193	0.53	19.2126	2.07%	0.0525	1.26%	327.1	6.6	308.5	28.5	327.1	6.6
TA1103_50	549	0.55	19.1212	1.82%	0.0539	0.81%	328.6	5.8	365	18.2	328.6	5.8
TA1103_54	168	0.77	18.9598	1.75%	0.0512	1.33%	331.3	5.6	249.1	30.4	331.3	5.6
TA1103_41	394	0.64	18.4692	1.75%	0.0546	0.91%	339.9	5.8	394.6	20.3	339.9	5.8
TA1103_43	286	0.78	18.1072	1.78%	0.0525	1.15%	346.5	6	307	25.9	346.5	6
TA1103_32	351	0.54	18.0576	1.86%	0.0848	3.12%	347.5	6.3	1310.7	59.3	347.5	6.3
TA1103_11	288	0.86	16.7	1.93%	0.0555	1.09%	374.9	7	432.1	24.1	374.9	7
TA1103_10	1,378	1.64	13.8608	2.02%	0.0595	0.71%	449.1	8.7	584.8	15.2	449.1	8.7
TA1103_44	821	0.42	6.2099	2.17%	0.0829	0.74%	962.5	19.4	1268	14.4	1268	14.4

section is dominated by an alternation of fossiliferous and calcareous sandstones and limestones (Fig. 4), which are in turn overlaid by 200 m of evaporitic deposits along with minor sandstone and mudstone intercalations (Fig. 4a). According to the correlation with the eastern unit, the latter member corresponds to the upper Oxfordian evaporitic Auquilco Formation (Fig. 2) accumulated in the Neuquén Basin after a regional regression episode (Legarreta and Gulisano 1989).

Two U–Pb detrital zircon age determinations were performed in this work in the Valle Grande Formation. The first determination was performed in the middle portion of this unit, above the described erosive contact (Figs. 3b and 4a). The obtained results indicate a maximum depositional age of 165.5 ± 3.2 Ma (sample TA11-03; Fig. 5; Table 1), thus signaling that the change in sedimentary conditions occurred during the middle Callovian or before. This is also supported by the fossiliferous content of the basal portion of this formation, reported by González and Vergara (1962) in a section nearby the area of the Martínez Creek (Figs. 3b and 4b). There, the presence of *Sphaeroceras rotundus* Torquist (Fig. 4) marks the early Callovian *Bodenbenderi* biozone (Riccardi 2008), thus indicating such age for the basal portion of the Valle Grande Formation.

A second U–Pb detrital zircon age determination was performed toward the top of the lower member (Fig. 4) of the Valle Grande Formation in the section studied by González and Vergara (1962) (Fig. 4). The sample analyzed corresponds to a sandstone level and the maximum depositional age obtained corresponds to 160.6 ± 2.8 Ma (sample TA11-17; Fig. 5; Table 2). Immediately below the dated sample (Fig. 4), the reported fossiliferous fauna corresponds to *Parawedekindia* sp., and *Euaspidoceras* sp. (Fig. 4) (González and Vergara 1962). The former marks the late Callovian-early Oxfordian *P. Pressulus* biozone while the latter fossil indicates the late early Oxfordian-middle Oxfordian *Subvinalesphinctes pseudokranaus* biozone (Riccardi 2008). The fossil content and the new U–Pb age reported here indicate an early-middle Oxfordian age for the top of the lower member of this unit (Riccardi 2008). Likewise, this evidence confirms the late Oxfordian age for the evaporitic upper member of the Valle Grande Formation.

Table 2 U–Pb zircon data for sample TA1117 from the Valle Grande Formation

	U (ppm)	Th/U	238U/206Pb	±1s	207Pb/206Pb	±1s	206Pb/238U	±1s	207Pb/206Pb	±1s	Best age (Ma)	±1s
TA1117DZ_98	693	0.89	25.9102	2.42%	0.0519	1.45%	244.1	5.8	281.9	32.8	244.1	5.8
TA1117DZ_97	750	0.64	25.3404	3.10%	0.0546	1.81%	249.5	7.6	397.5	40	249.5	7.6
TA1117DZ_96	1,309	1.33	27.7767	3.00%	0.053	2.33%	228	6.7	327.4	52	228	6.7
TA1117DZ_95	376	2.07	25.8017	2.96%	0.0525	2.15%	245.1	7.1	309.3	48.2	245.1	7.1
TA1117DZ_94	894	0.31	20.4492	2.55%	0.0526	1.51%	307.8	7.7	312.6	34	307.8	7.7
TA1117DZ_93	3,652	2.22	59.283	3.34%	0.0563	1.56%	107.8	3.6	465.6	34.1	107.8	3.6
TA1117DZ_92	805	0.92	20.1264	2.70%	0.0541	1.48%	312.6	8.2	375.6	33.1	312.6	8.2
TA1117DZ_91	268	1.23	24.2086	6.94%	0.0651	4.08%	260.9	17.7	778.8	83.4	260.9	17.7
TA1117DZ_90	356	0.85	38.3552	3.41%	0.0543	3.70%	165.9	5.6	382.1	81.2	165.9	5.6
TA1117DZ_89	962	0.39	30.3338	2.31%	0.0532	1.64%	209.1	4.7	337.9	36.7	209.1	4.7
TA1117DZ_88	508	2.24	26.3762	2.39%	0.0511	1.53%	239.9	5.6	244.1	34.8	239.9	5.6
TA1117DZ_87	1,267	1.04	25.4567	2.49%	0.0509	1.48%	248.4	6.1	235.5	33.7	248.4	6.1
TA1117DZ_86	382	1.19	27.7222	2.74%	0.0524	2.21%	228.4	6.1	303.4	49.5	228.4	6.1
TA1117DZ_85	1,061	1.32	29.7223	2.82%	0.0521	1.65%	213.3	5.9	288.5	37.4	213.3	5.9
TA1117DZ_84	2,233	0.93	28.408	2.43%	0.0523	1.15%	223	5.3	300.4	26	223	5.3
TA1117DZ_83	1,000	0.72	33.9771	2.37%	0.0509	1.24%	187	4.4	234.7	28.4	187	4.4
TA1117DZ_82	439	1.6	25.5169	2.45%	0.0525	1.63%	247.8	6	307.5	36.7	247.8	6
TA1117DZ_81	1,085	0.93	29.9206	2.48%	0.0521	1.18%	211.9	5.2	288.2	26.8	211.9	5.2
TA1117DZ_80	295	1.6	30.319	2.46%	0.0566	1.87%	209.2	5.1	474.2	40.9	209.2	5.1
TA1117DZ_79	596	0.68	16.3363	2.25%	0.0549	1.26%	383	8.4	406.8	27.9	383	8.4
TA1117DZ_78	979	0.93	979.0225	5.38%	0.0641	4.46%	6.6	0.4	745.5	91.5	6.6	0.4
TA1117DZ_77	695	0.58	17.9197	2.39%	0.0528	1.28%	350.1	8.1	321.3	28.8	350.1	8.1
TA1117DZ_76	868	0.93	23.8251	2.34%	0.0525	1.28%	265	6.1	309.3	29	265	6.1
TA1117DZ_75	106	1.06	25.6945	4.19%	0.0534	4.49%	246.1	10.1	345.7	98.5	246.1	10.1
TA1117DZ_74	1,827	0.28	15.5279	2.94%	0.0697	1.22%	402.3	11.4	921	24.9	402.3	11.4
TA1117DZ_73	320	1.27	20.3517	2.76%	0.0548	2.10%	309.2	8.3	403.4	46.3	309.2	8.3
TA1117DZ_72	806	0.8	35.4042	2.84%	0.0567	2.36%	179.6	5	478.4	51.3	179.6	5
TA1117DZ_71	712	0.66	23.779	2.31%	0.0525	1.45%	265.6	6	308.9	32.7	265.6	6
TA1117DZ_70	769	0.77	21.2682	2.85%	0.0542	1.73%	296.2	8.2	378.5	38.5	296.2	8.2
TA1117DZ_69	791	1.08	22.6612	2.98%	0.0518	2.10%	278.4	8.1	278.7	47.3	278.4	8.1
TA1117DZ_68	101	2.63	18.4772	3.51%	0.0571	3.80%	339.8	11.6	493.8	81.6	339.8	11.6
TA1117DZ_67	603	0.97	34.6341	2.69%	0.0542	1.77%	183.5	4.9	378.4	39.3	183.5	4.9
TA1117DZ_66	1,542	0.43	34.1575	2.20%	0.05	1.19%	186	4	195.9	27.3	186	4
TA1117DZ_65	901	0.79	38.9815	2.45%	0.0508	2.00%	163.3	4	229.9	45.6	163.3	4
TA1117DZ_64	698	0.93	21.6309	2.25%	0.0514	1.30%	291.3	6.4	258.7	29.5	291.3	6.4
TA1117DZ_63	302	1.1	32.6145	2.48%	0.0511	2.12%	194.7	4.8	246.4	48.2	194.7	4.8
TA1117DZ_62	434	1.1	36.4135	3.01%	0.0511	2.34%	174.6	5.2	247.4	53.1	174.6	5.2
TA1117DZ_61	364	1.36	41.8602	2.76%	0.0712	2.33%	152.2	4.1	963.8	46.9	152.2	4.1
TA1117DZ_60	1,454	0.88	22.7752	2.18%	0.0534	1.09%	277	5.9	346.3	24.4	277	5.9
TA1117DZ_59	654	1.04	34.9919	2.32%	0.0497	1.80%	181.6	4.2	178.7	41.5	181.6	4.2
TA1117DZ_58	798	0.63	17.2067	2.67%	0.0532	1.74%	364.2	9.5	335.4	39.1	364.2	9.5
TA1117DZ_57	101	1.53	26.0017	2.97%	0.0764	3.20%	243.3	7.1	1104.8	62.7	243.3	7.1
TA1117DZ_56	300	1.08	32.1013	2.43%	0.0521	2.12%	197.8	4.7	290.1	47.6	197.8	4.7
TA1117DZ_55	186	1.07	38.0557	2.78%	0.0547	2.81%	167.2	4.6	401.8	61.8	167.2	4.6
TA1117DZ_54	1,504	0.43	36.3437	3.35%	0.0501	1.90%	175	5.8	198	43.6	175	5.8
TA1117DZ_53	444	0.95	2.5994	2.32%	0.1278	1.05%	2098.3	41.5	2068.6	18.4	2068.6	18.4
TA1117DZ_52	959	0.5	21.0011	2.69%	0.0541	1.47%	299.9	7.9	373.7	32.7	299.9	7.9
TA1117DZ_51	335	1.99	39.7409	4.10%	0.0616	2.90%	160.2	6.5	659.5	61	160.2	6.5
TA1117DZ_50	630	0.64	37.4056	2.34%	0.0501	1.85%	170.1	3.9	198.8	42.3	170.1	3.9
TA1117DZ_49	262	1.02	30.6796	2.80%	0.0671	2.68%	206.8	5.7	840.8	54.8	206.8	5.7
TA1117DZ_48	7,493	1.97	39.3417	2.54%	0.0921	1.34%	161.8	4.1	1468.8	25.3	161.8	4.1
TA1117DZ_47	1,003	0.59	7.7313	2.36%	0.0653	1.16%	784.1	17.4	785.4	24.2	784.1	17.4
TA1117DZ_46	669	0.63	39.0197	3.14%	0.0531	2.20%	163.1	5	334.8	49	163.1	5
TA1117DZ_45	325	0.92	39.849	2.56%	0.0509	2.19%	159.8	4	236.9	49.8	159.8	4
TA1117DZ_44	788	0.55	27.8556	3.02%	0.0516	1.38%	227.4	6.8	268.2	31.4	227.4	6.8
TA1117DZ_43	1,854	0.91	23.7706	2.78%	0.0527	1.34%	265.6	7.2	317.5	30.2	265.6	7.2
TA1117DZ_42	268	0.93	22.3863	2.40%	0.0544	1.77%	281.7	6.6	388.5	39.2	281.7	6.6
TA1117DZ_41	277	1.19	24.7727	2.48%	0.0524	2.02%	255.1	6.2	303.6	45.5	255.1	6.2
TA1117DZ_40	1,037	0.56	23.421	2.33%	0.0521	1.36%	269.5	6.2	288.5	30.8	269.5	6.2
TA1117DZ_39	980	1.41	31.4248	2.94%	0.0674	2.04%	201.9	5.8	849.1	41.9	201.9	5.8
TA1117DZ_38	467	1.83	27.5837	3.08%	0.064	2.55%	229.6	6.9	741	53.1	229.6	6.9
TA1117DZ_37	718	1.16	24.8474	2.40%	0.0516	1.40%	254.4	6	267.9	31.7	254.4	6
TA1117DZ_36	473	0.7	33.2712	2.30%	0.0508	1.84%	190.9	4.3	229.5	42	190.9	4.3
TA1117DZ_35	684	0.94	34.2487	2.34%	0.0499	1.49%	185.5	4.3	188.1	34.2	185.5	4.3
TA1117DZ_34	1,019	1.06	17.6681	2.38%	0.0549	1.18%	354.9	8.2	407.2	26.2	354.9	8.2
TA1117DZ_33	852	1.08	27.9651	2.25%	0.0507	1.38%	226.5	5	225.2	31.5	226.5	5
TA1117DZ_32	166	1.66	31.3336	3.22%	0.0509	3.41%	202.5	6.4	238.5	76.9	202.5	6.4

Table 3 U–Pb zircon data for sample CHA32 from the Baños del Flaco Formation

	U (ppm)	Th/U	206Pb/238U	±1s	207Pb/206Pb	±1s	206Pb/238U	±1s	207Pb/206Pb	±1s	Best age (Ma)	±1s
Zircon_01_008	360	0.51	0.02183	0.00028	0.05105	0.00153	139	2	243	68	139	2
Zircon_02_009	322	0.65	0.02154	0.00021	0.05047	0.00187	137	1	217	84	137	1
Zircon_03_010	711	0.82	0.02061	0.00015	0.05197	0.00130	131.5	0.9	284	56	132	0.9
Zircon_04_011	227	0.64	0.02235	0.00029	0.05583	0.00240	142	2	446	94	142	2
Zircon_05_012	287	0.53	0.02028	0.00028	0.05455	0.00191	129	2	394	77	129	2
Zircon_06_014	474	0.6	0.02058	0.00016	0.05201	0.00140	131	1	286	60	131	1
Zircon_08_016	846	0.92	0.02078	0.00016	0.05053	0.00106	133	1	219	48	133	1
Zircon_09_017	1083	1.01	0.01957	0.00014	0.04772	0.00095	124.9	0.9	85	45	125	0.9
Zircon_10_018	671	0.84	0.02016	0.00020	0.04716	0.00113	129	1	57	53	129	1
Zircon_100_126	240	0.48	0.02040	0.00019	0.05075	0.00218	130	1	229	96	130	1
Zircon_11_020	853	0.86	0.02036	0.00012	0.04954	0.00094	129.9	0.8	173	43	130	0.8
Zircon_12_021	1166	0.86	0.02058	0.00017	0.04923	0.00103	131	1	159	48	131	1
Zircon_13_022	304	0.5	0.02042	0.00019	0.05287	0.00190	130	1	323	80	130	1
Zircon_14_023	540	0.55	0.01952	0.00017	0.05400	0.00228	125	1	371	94	125	1
Zircon_15_024	1856	0.68	0.02095	0.00017	0.04910	0.00128	134	1	153	59	134	1
Zircon_17_027	858	0.72	0.01998	0.00014	0.04974	0.00099	127.5	0.9	183	45	128	0.9
Zircon_18_028	397	0.68	0.02051	0.00019	0.05306	0.00245	131	1	331	102	131	1
Zircon_19_029	381	0.77	0.02137	0.00016	0.04979	0.00144	136	1	185	65	136	1
Zircon_20_030	1107	1.05	0.02135	0.00019	0.04877	0.00102	136	1	137	48	136	1
Zircon_21_032	960	1.01	0.01803	0.00016	0.04802	0.00110	115	1	100	52	115	1
Zircon_22_033	759	0.98	0.01963	0.00022	0.05282	0.00164	125	1	321	69	125	1
Zircon_23_034	637	0.69	0.01976	0.00016	0.04864	0.00204	126	1	131	92	126	1
Zircon_24_035	797	0.87	0.02024	0.00014	0.05074	0.00122	129.2	0.9	229	54	129	0.9
Zircon_26_038	307	0.53	0.02117	0.00018	0.05143	0.00185	135	1	260	81	135	1
Zircon_27_039	490	0.57	0.02109	0.00018	0.05003	0.00140	135	1	196	64	135	1
Zircon_28_040	492	0.42	0.02029	0.00017	0.05089	0.00142	129	1	236	63	129	1
Zircon_29_041	698	0.59	0.02069	0.00015	0.05002	0.00100	132	0.9	196	45	132	0.9
Zircon_30_042	357	0.58	0.02140	0.00020	0.05368	0.00156	136	1	358	64	136	1
Zircon_31_044	767	0.72	0.02117	0.00017	0.04918	0.00113	135	1	156	53	135	1
Zircon_32_045	399	0.8	0.02041	0.00020	0.05269	0.00174	130	1	315	74	130	1
Zircon_33_046	1424	1.3	0.02131	0.00018	0.04955	0.00094	136	1	174	43	136	1
Zircon_34_047	768	0.88	0.02056	0.00014	0.04915	0.00108	131.2	0.9	155	50	131	0.9
Zircon_35_048	270	0.54	0.02063	0.00021	0.06045	0.00470	132	1	620	168	132	1
Zircon_36_050	445	0.72	0.02058	0.00024	0.04945	0.00301	131	2	169	140	131	2
Zircon_37_051	294	0.59	0.02085	0.00017	0.05189	0.00187	133	1	281	84	133	1
Zircon_38_052	403	0.69	0.01823	0.00021	0.06546	0.00352	116	1	789	116	116	1
Zircon_39_053	594	0.86	0.02281	0.00041	0.04959	0.00124	145	3	176	60	145	3
Zircon_40_054	581	0.94	0.02087	0.00021	0.05186	0.00150	133	1	279	68	133	1
Zircon_41_056	222	0.55	0.02163	0.00031	0.04804	0.00279	138	2	101	130	138	2
Zircon_42_057	286	0.49	0.01998	0.00022	0.05565	0.00195	128	1	438	80	128	1
Zircon_43_058	1188	1.15	0.01965	0.00015	0.04963	0.00094	125.4	0.9	178	45	125	0.9
Zircon_44_059	819	0.63	0.01930	0.00014	0.05014	0.00115	123.2	0.9	201	55	123	0.9
Zircon_45_060	830	1.12	0.02046	0.00012	0.05016	0.00115	130.6	0.8	202	55	131	0.8
Zircon_46_062	990	0.96	0.01992	0.00014	0.04998	0.00100	127.2	0.9	194	48	127	0.9
Zircon_47_063	624	0.83	0.02031	0.00016	0.04979	0.00139	130	1	185	67	130	1
Zircon_48_064	416	0.5	0.02112	0.00017	0.05087	0.00132	135	1	235	61	135	1
Zircon_49_065	630	0.69	0.02022	0.00022	0.04860	0.00151	129	1	129	74	129	1
Zircon_50_066	907	0.74	0.02055	0.00016	0.04969	0.00119	131	1	181	57	131	1
Zircon_51_068	942	0.86	0.02045	0.00016	0.05088	0.00122	130	1	235	57	130	1
Zircon_52_069	680	0.72	0.02005	0.00014	0.04990	0.00105	128	0.9	190	50	128	0.9
Zircon_53_070	452	0.51	0.02031	0.00022	0.05394	0.00291	130	1	369	125	130	1
Zircon_54_071	801	0.99	0.02087	0.00014	0.04856	0.00107	133.2	0.9	127	53	133	0.9
Zircon_55_072	563	0.77	0.02152	0.00026	0.05438	0.00207	137	2	387	88	137	2
Zircon_56_074	439	0.7	0.02136	0.00016	0.05266	0.00158	136	1	314	70	136	1
Zircon_57_075	478	0.63	0.02115	0.00019	0.04897	0.00157	135	1	146	77	135	1
Zircon_58_076	712	0.89	0.02045	0.00015	0.04955	0.00119	130.5	0.9	174	57	131	0.9
Zircon_60_078	1364	1.04	0.02010	0.00013	0.04868	0.00097	128.3	0.8	132	45	128	0.8
Zircon_61_080	461	0.62	0.01939	0.00018	0.05306	0.00289	124	1	332	122	124	1
Zircon_62_081	593	0.53	0.02105	0.00015	0.04845	0.00126	134.3	0.9	121	59	134	0.9
Zircon_63_082	469	0.66	0.01998	0.00022	0.05307	0.00149	128	1	332	62	128	1
Zircon_64_083	675	0.64	0.01889	0.00015	0.05256	0.00126	120.6	0.9	310	53	121	0.9
Zircon_65_084	844	0.93	0.01980	0.00019	0.05354	0.00123	126	1	352	51	126	1
Zircon_66_086	237	0.58	0.02161	0.00024	0.05814	0.00244	138	2	535	90	138	2
Zircon_68_088	314	0.51	0.02197	0.00020	0.04972	0.00169	140	1	182	77	140	1
Zircon_69_089	914	0.75	0.02056	0.00017	0.04842	0.00097	131	1	120	45	131	1
Zircon_70_090	841	0.71	0.02095	0.00016	0.05069	0.00122	134	1	227	54	134	1

Table 4 U–Pb zircon data for sample CHA35 from the Colimapu Formation

Zircon_71_092	872	1.22	0.02089	0.00023	0.05038	0.00166	133	1	213	74	133	1
Zircon_73_094	345	0.78	0.02186	0.00020	0.04608	0.00196	139	1	2	83	139	1
Zircon_74_095	540	0.57	0.01989	0.00014	0.05081	0.00132	127	0.9	232	58	127	0.9
Zircon_75_096	1670	1.29	0.01941	0.00023	0.04875	0.00088	124	1	136	41	124	1
Zircon_76_098	503	0.67	0.02058	0.00016	0.05070	0.00122	131	1	227	54	131	1
Zircon_77_099	726	0.87	0.01981	0.00027	0.04732	0.00194	126	2	65	85	126	2
Zircon_78_100	890	0.8	0.02108	0.00023	0.04855	0.00121	134	1	126	57	134	1
Zircon_79_101	615	0.57	0.01991	0.00014	0.04974	0.00124	127.1	0.9	183	56	127	0.9
Zircon_80_102	466	0.5	0.02016	0.00017	0.04949	0.00153	129	1	171	70	129	1
Zircon_81_104	308	0.52	0.02111	0.00017	0.05612	0.00202	135	1	457	78	135	1
Zircon_82_105	711	0.76	0.02068	0.00015	0.04993	0.00110	132	0.9	192	50	132	0.9
Zircon_83_106	431	0.74	0.02093	0.00015	0.04879	0.00166	133.5	0.9	138	76	134	0.9
Zircon_84_107	589	0.67	0.02027	0.00015	0.05155	0.00129	129.4	0.9	266	55	129	0.9
Zircon_85_108	932	1.13	0.02049	0.00014	0.05754	0.00155	130.8	0.9	512	57	131	0.9
Zircon_86_110	465	0.52	0.02118	0.00015	0.04918	0.00133	135.1	0.9	156	60	135	0.9
Zircon_87_111	450	0.61	0.02029	0.00015	0.04840	0.00131	129.5	0.9	119	60	130	0.9
Zircon_89_113	334	0.51	0.02121	0.00023	0.05394	0.00167	135	1	369	67	135	1
Zircon_90_114	1607	1.19	0.01970	0.00018	0.04808	0.00087	126	1	103	41	126	1
Zircon_91_116	656	0.64	0.01892	0.00013	0.05067	0.00147	120.8	0.8	226	64	121	0.8
Zircon_92_117	731	0.67	0.02000	0.00013	0.05164	0.00098	127.7	0.8	270	42	128	0.8
Zircon_93_118	488	0.65	0.01953	0.00014	0.05464	0.00175	124.7	0.9	398	69	125	0.9
Zircon_94_119	897	0.89	0.02013	0.00016	0.05069	0.00106	128	1	227	46	128	1
Zircon_96_122	271	0.55	0.02090	0.00022	0.05512	0.00289	133	1	417	112	133	1
Zircon_97_123	744	0.72	0.02288	0.00019	0.04957	0.00124	146	1	175	56	146	1
Zircon_98_124	871	0.73	0.02023	0.00014	0.05036	0.00131	129.1	0.9	212	58	129	0.9
Zircon_99_125	737	0.72	0.01989	0.00014	0.04930	0.00108	127	0.9	162	49	127	0.9
Zircon_07_015	313	0.46	0.02056	0.00020	0.05628	0.00180	131	1	463	68	131	1
Zircon_16_026	381	0.65	0.02320	0.00023	0.04647	0.00197	148	1	22	85	148	1
Zircon_25_036	455	0.93	0.01972	0.00026	0.05267	0.00354	126	2	315	146	126	2
Zircon_59_077	1235	0.67	0.02229	0.00042	0.05177	0.00155	142	3	275	66	142	3
Zircon_67_087	643	0.65	0.01966	0.00015	0.04911	0.00142	125.5	0.9	153	64	126	0.9
Zircon_72_093	1383	1.04	0.01936	0.00017	0.05184	0.00161	124	1	278	68	124	1
Zircon_88_112	350	0.64	0.02133	0.00026	0.04803	0.00223	136	2	101	97	136	2
Zircon_95_120	1043	1.33	0.02169	0.00024	0.04978	0.00115	138	2	185	52	138	2
	U (ppm)	Th/U	206Pb/238U	±1s	207Pb/206Pb	±1s	206Pb/238U	±1s	207Pb/206Pb	±1s	Best age (Ma)	±1s
Zircon_01_008	592	1	0.0207	0.0003	0.0551	0.0035	132	2	415	131	132	2
Zircon_02_009	422	0.95	0.0195	0.0002	0.0521	0.0027	125	1	292	108	125	1
Zircon_03_010	160	0.42	0.0218	0.0003	0.0568	0.0040	139	2	483	144	139	2
Zircon_04_011	231	0.51	0.0212	0.0003	0.0558	0.0035	135	2	445	131	135	2
Zircon_05_012	296	0.62	0.0202	0.0003	0.0661	0.0044	129	2	810	130	129	2
Zircon_06_014	685	0.72	0.0215	0.0003	0.0642	0.0039	137	2	749	120	137	2
Zircon_07_015	421	0.51	0.0204	0.0002	0.0557	0.0032	130	1	440	125	130	1
Zircon_08_016	175	0.6	0.0212	0.0004	0.0695	0.0043	135	2	914	124	135	2
Zircon_09_017	2020	0.27	0.0207	0.0004	0.0734	0.0068	132	2	1024	184	132	2
Zircon_10_018	463	0.84	0.0203	0.0002	0.0647	0.0038	130	1	764	119	130	1
Zircon_11_020	745	0.78	0.0212	0.0003	0.0604	0.0047	135	2	617	164	135	2
Zircon_12_021	108	0.6	0.0202	0.0005	0.0962	0.0152	129	3	1551	307	129	3
Zircon_13_022	92	0.47	0.0214	0.0003	0.0695	0.0054	136	2	915	155	136	2
Zircon_14_023	1191	0.93	0.0213	0.0002	0.0492	0.0012	136	1	159	54	136	1
Zircon_15_024	490	0.89	0.0201	0.0003	0.0527	0.0032	128	2	316	131	128	2
Zircon_16_026	2393	0.66	0.0199	0.0002	0.0480	0.0023	127	1	98	99	127	1
Zircon_17_027	269	0.55	0.0207	0.0002	0.0559	0.0022	132	1	447	84	132	1
Zircon_18_028	382	1.33	0.0202	0.0002	0.0506	0.0027	129	1	224	116	129	1
Zircon_19_029	133	0.46	0.0205	0.0003	0.0645	0.0059	131	2	758	189	131	2
Zircon_20_030	228	0.69	0.0241	0.0007	0.1792	0.0204	153	4	2645	188	2645	188
Zircon_21_032	428	0.49	0.0220	0.0006	0.1110	0.0145	140	3	1815	239	140	3
Zircon_22_033	646	0.95	0.0205	0.0002	0.0597	0.0023	131	1	593	80	131	1
Zircon_24_035	1752	1.57	0.0198	0.0001	0.0509	0.0010	126.5	0.8	237	45	127	0.8
Zircon_25_036	436	0.94	0.0212	0.0002	0.0584	0.0026	135	1	546	93	135	1
Zircon_26_038	738	0.91	0.0198	0.0002	0.0490	0.0020	127	1	148	88	127	1
Zircon_27_039	194	0.59	0.0204	0.0003	0.0589	0.0039	130	2	563	141	130	2
Zircon_28_040	539	0.7	0.0212	0.0002	0.0561	0.0019	135	1	458	73	135	1
Zircon_29_041	261	0.74	0.0202	0.0002	0.0699	0.0051	129	2	924	147	129	2
Zircon_30_042	269	0.68	0.0214	0.0002	0.0558	0.0027	137	2	445	106	137	2
Zircon_31_044	630	0.78	0.0196	0.0002	0.0699	0.0035	125	1	924	100	125	1
Zircon_32_045	773	1	0.0195	0.0002	0.0508	0.0017	125	1	231	74	125	1
Zircon_33_046	222	0.53	0.0220	0.0010	0.0840	0.0240	140	6	1293	602	140	6
Zircon_34_047	107	0.6	0.0199	0.0004	0.0593	0.0046	127	2	579	166	127	2

(continued)

Table 4 (continued)

Zircon_35_048	354	0.76		0.0206	0.0002	0.0510	0.0025		131	1	242	108	131	1
Zircon_36_050	446	0.71		0.0198	0.0003	0.0577	0.0045		126	2	520	166	126	2
Zircon_37_051	366	0.76		0.0212	0.0002	0.0596	0.0024		135	1	588	86	135	1
Zircon_39_053	822	1.29		0.0210	0.0002	0.0599	0.0017		134	1	601	61	134	1
Zircon_40_054	1326	0.66		0.0208	0.0002	0.0523	0.0014		132	1	299	57	132	1
Zircon_41_056	1719	0.85		0.0214	0.0002	0.0514	0.0015		136	1	261	66	136	1
Zircon_42_057	177	0.48		0.0207	0.0003	0.0594	0.0030		132	2	583	108	132	2
Zircon_43_058	1004	0.86		0.0199	0.0002	0.0500	0.0014		127.2	0.9	193	63	127	0.9
Zircon_44_059	123	0.65		0.0225	0.0004	0.0676	0.0065		143	2	856	197	143	2
Zircon_45_060	2614	0.1		0.0191	0.0002	0.0507	0.0022		122	1	227	95	122	1
Zircon_46_062	5622	0.96		0.0190	0.0001	0.0540	0.0011		121.6	0.8	371	45	122	0.8
Zircon_47_063	198	0.95		0.0215	0.0003	0.0502	0.0034		137	2	205	146	137	2
Zircon_48_064	176	0.51		0.0217	0.0002	0.0556	0.0030		138	1	436	117	138	1
Zircon_49_065	222	0.59		0.0196	0.0003	0.0635	0.0058		125	2	726	192	125	2
Zircon_50_066	1712	0.72		0.0192	0.0001	0.0509	0.0011		122.3	0.8	238	49	122	0.8
Zircon_51_068	1363	0.31		0.0455	0.0004	0.0622	0.0021		287	3	680	71	287	3
Zircon_52_069	177	0.48		0.0204	0.0004	0.0819	0.0093		130	3	1244	222	130	3
Zircon_53_070	161	0.61		0.0215	0.0005	0.0940	0.0183		137	3	1508	373	137	3
Zircon_54_071	220	0.55		0.0203	0.0003	0.0706	0.0065		129	2	945	180	129	2
Zircon_55_072	434	0.7		0.0197	0.0002	0.0719	0.0050		126	2	983	133	126	2
Zircon_56_074	251	0.53		0.0203	0.0002	0.0731	0.0055		129	2	1016	142	129	2
Zircon_57_075	180	0.64		0.0204	0.0004	0.0600	0.0074		130	3	604	256	130	3
Zircon_58_076	317	0.84		0.0193	0.0002	0.0613	0.0049		123	1	650	162	123	1
Zircon_59_077	96	0.49		0.0207	0.0003	0.0686	0.0063		132	2	888	178	132	2
Zircon_60_078	972	1.17		0.0202	0.0002	0.0467	0.0015		129	0.9	34	63	129	0.9
Zircon_61_080	751	0.75		0.0198	0.0002	0.0516	0.0028		126	1	270	115	126	1
Zircon_62_081	370	0.62		0.0213	0.0003	0.0637	0.0053		136	2	733	164	136	2
Zircon_63_082	229	0.35		0.0218	0.0004	0.0527	0.0074		139	2	317	279	139	2
Zircon_64_083	166	0.58		0.0212	0.0003	0.0511	0.0030		135	2	245	125	135	2
Zircon_65_084	622	1.31		0.0205	0.0002	0.0681	0.0055		131	1	872	156	131	1
Zircon_66_086	278	0.67		0.0211	0.0003	0.0709	0.0094		134	2	954	262	134	2
Zircon_67_087	1304	1.08		0.0198	0.0001	0.0537	0.0017		126.3	0.9	357	65	126	0.9
Zircon_68_088	179	0.66		0.0202	0.0003	0.0542	0.0048		129	2	380	187	129	2
Zircon_70_090	263	0.62		0.0201	0.0002	0.0540	0.0034		128	1	372	132	128	1
Zircon_71_092	523	0.61		0.0206	0.0002	0.0572	0.0030		131	1	501	107	131	1
Zircon_72_093	235	0.47		0.0201	0.0005	0.0711	0.0142		128	3	960	399	128	3
Zircon_73_094	275	1.02		0.0190	0.0003	0.0573	0.0037		121	2	505	132	121	2
Zircon_74_095	3869	0.45		0.0190	0.0002	0.0540	0.0020		121.1	1	370	76	121	1
Zircon_76_098	431	0.49		0.0213	0.0003	0.0553	0.0039		136	2	423	146	136	2
Zircon_77_099	263	0.64		0.0192	0.0003	0.0600	0.0050		123	2	602	171	123	2
Zircon_78_100	1532	0.3		0.0228	0.0002	0.0519	0.0013		145	1	282	51	145	1
Zircon_79_101	378	0.62		0.0197	0.0002	0.0603	0.0036		126	1	615	120	126	1
Zircon_81_104	3762	0.61		0.0194	0.0002	0.0712	0.0020		124	1	962	53	124	1
Zircon_82_105	662	0.99		0.0196	0.0002	0.0583	0.0043		125	1	540	151	125	1
Zircon_84_107	249	0.64		0.0202	0.0002	0.0568	0.0041		129	2	483	149	129	2
Zircon_85_108	268	0.71		0.0209	0.0003	0.0529	0.0036		133	2	323	141	133	2
Zircon_23_034	413	0.33		0.0754	0.0005	0.0589	0.0017		469	3	563	56	469	3
Zircon_38_052	130	0.56		0.0285	0.0020	0.2872	0.0540		181	12	3402	290	3402	290
Zircon_75_096	158	0.57		0.0363	0.0028	0.3870	0.0640		230	18	3859	243	3859	243
Zircon_83_106	107	0.51		0.0324	0.0025	0.3366	0.0651		205	16	3647	294	3647	294

Based on fossil content and U–Pb determinations, the Valle Grande Formation can be assigned to the early Callovian-late Oxfordian according to the original proposal of González and Vergara (1962). This age range supports the chronological correlation of such unit with the uppermost section of the Cuyo Group and Lotena Group, in the Neuquén Basin (Fig. 2).

A new sedimentary cycle in the Neuquén Basin is recorded in the Chilean stratigraphy in the Río Damas Formation with a total thickness of 3700 m (Charrier et al. 2007). The lower section (>2000 m) is made of fine-grained facies which overlay the Upper Oxfordian evaporitic deposits of the Valle Grande Formation (Fig. 2). After that, 1000 m of coarse breccias with volcanic intercalations dominate the upper section of the unit (Klohn 1960). The Rio Damas Formation is the western equivalent of the continental deposits grouped in the Tordillo Formation in the eastern

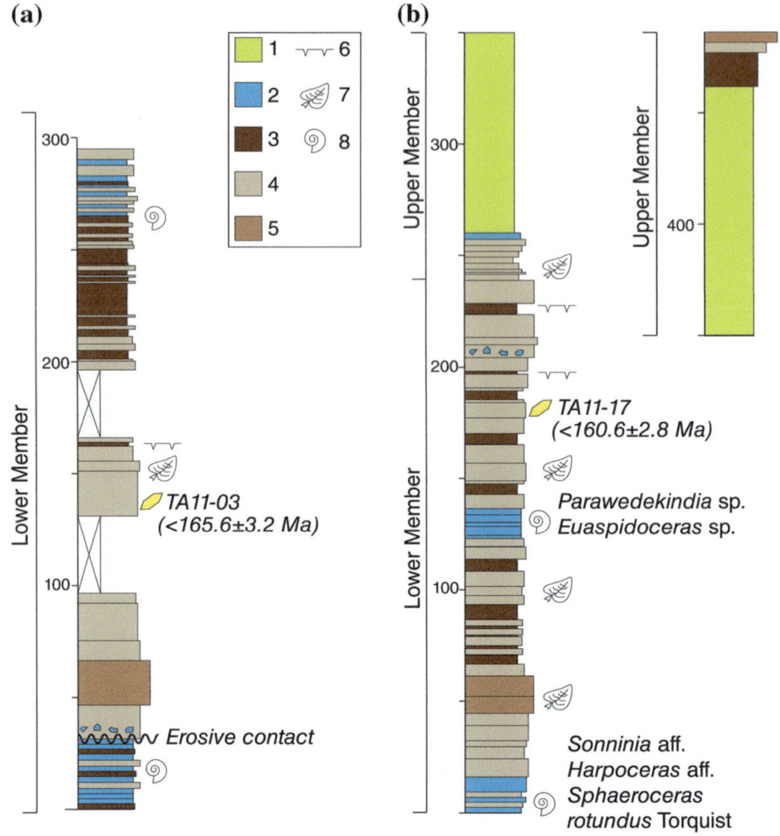

Fig. 4 Stratigraphic column of the Valle Grande Formation. The bold-cursive letters below each column indicate the section measured in this study along the Nevado river (**a**) and the section measured by González and Vergara (1962) near of the Martinez creek (**b**). See locations in Fig. 3b. 1. Gypsum; 2. Limestone; 3. Mudstone; 4. Sandstone; 5. Conglomerate; 6. Drying cracks; 7. Plants; 8. Ammonite

side of the Neuquén Basin (Charrier et al. 1996; Davidson and Vicente 1973). For the middle portion of this unit, a maximum depositional age of 146.6 ± 2.1 Ma (Fig. 5) (U–Pb detrital age; Rossel et al. 2014) agrees with the recent Tithonian maximum depositional age assigned to the Tordillo Formation in Argentina (Naipauer et al. 2014, 2015; Vennari et al. 2014). The volcanic rocks in the uppermost portion of the Río Damas Formation have been interpreted to represent magmatism in the back-arc domain, with subduction-related affinity, derived from a depleted sub-arc asthenospheric mantle (Rossel et al. 2014).

The marine, fossiliferous and mainly calcareous series of the Baños del Flaco Formation (Fig. 2) mark the sea level rise that occurred during the Late Jurassic. This unit conformably and transitionally overlies the Río Damas Formation at its type locality, situated in the Termas del Flaco area (Fig. 3a). Here, it reaches a thickness

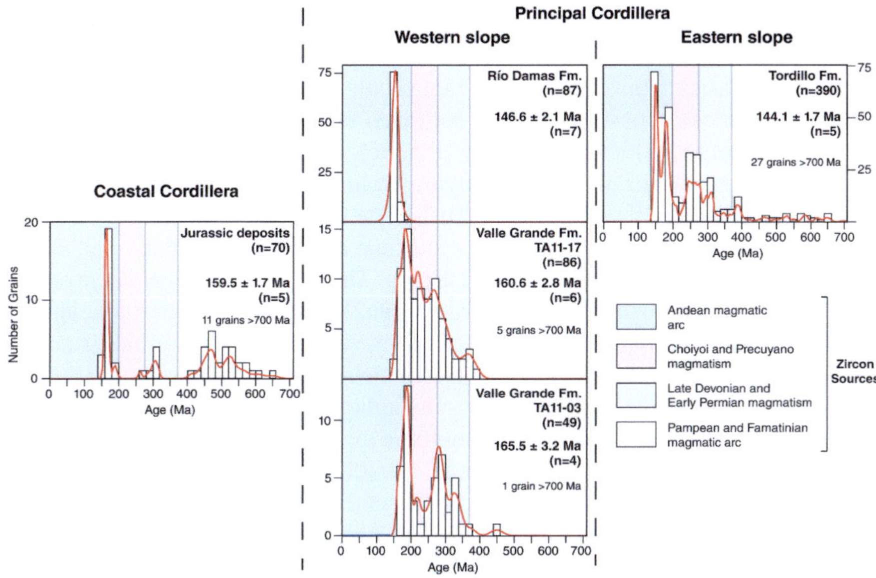

Fig. 5 Comparative Frequency histograms and relative probability density plots of the U–Pb detrital zircon ages performed in Jurassic sedimentary rocks of the Southern Central Andes between 34° and 35° S. Displayed age data are taken from: (i) Valle Grande Formation: this work, (ii) Coastal Cordillera deposits: compiled from (Hervé et al. 2013), (iii) Río Damas Formation: Rossel et al. (2014), (iv) Tordillo Formation: compiled from Horton et al. (2016) and Naipauer et al. (2015). Weighted mean of the youngest U–Pb ages with 2σ error bars for individual samples is indicated. n: number of samples

of <400 m and is subdivided into two members (Salazar 2012). The lower member consists of 191-m-thick conglomerate, limestone, and calcareous sandstone. The upper member reaches a thickness of 178 m and is composed of siltstone, calcareous sandstone, and sandy limestone (Salazar 2012). Based on the fossil content, the Baños del Flaco Formation has been assigned to the middle Tithonian to lower Berriasian (Covacevich et al. 1976; Salazar 2012) in the Termas del Flaco type locality (Fig. 3a), an assignation that supports the correlation with the lower Tithonian-lower Valanginian Vaca Muerta Formation of the Mendoza Group in Argentina (Fig. 2).

It is noteworthy that the restricted platform facies at the top of the Neocomian marine deposits in other areas of Chile and Argentina (Charrier 1981; Kietzmann et al. 2014) are not observed in the Termas del Flaco region. Moreover, the contact of the Baños del Flaco Formation with the overlying red continental Upper Cretaceous strata (BRCU series; see below) is abrupt, which suggests that an episode of emergence and erosion, or an interruption of sedimentation, occurred before of the accumulation of the latter deposits in the Late Cretaceous (Charrier et al. 1996; Mescua et al. 2013).

Upper Jurassic-lower Cretaceous marine deposits of the Baños del Flaco Formation crop out also in the Valle Grande area (Fig. 3b). Here, a 300 m section of this unit of calcareous sandstones and black shales overlies the Upper Jurassic Río

Damas Formation; in turn, the top of the sequence is unknown (Fig. 3b). Based on the fossil content, these Lower Cretaceous marine deposits are assigned to the middle Tithonian-Berriasian (Burckhard 1900; Tapia et al. 2011).

The southern outcrops of the Baños del Flaco Formation are exposed along the Maule river nearby the La Mina locality (Fig. 3c). Here, the base of the unit is not exposed, and the contact with the overlying Colimapu Formation corresponds to an erosive surface (Astaburuaga et al. 2012). The series consists of a 400-m-thick succession of limestone and calcareous sandstone at the lower sections and coarse sandstone and fine conglomerates toward the top. The presence of the *Holcoptychites neuquensis* Douvillé (González and Vergara 1962) supports a Hauterivian age for these marine deposits along the Maule river. Moreover, a new U–Pb detrital zircon age determined in this work for a sample collected toward the top of this series (Fig. 3c) further constrains the age of this unit. Results indicate a maximum depositional age of 120.1 ± 1.1 Ma (Fig. 6; Table 3) constraining the deposition of this portion of the Baños del Flaco Formation to the Hauterivian-middle Aptian.

Unconformably overlying the Baños del Flaco Formation, the red detritic continental deposits of the Colimapu Formation are exposed nearby the La Mina locality (Fig. 3c). This unit is composed of conglomerates, coarse sandstones, and siltstones intercalations which are in turn subordinated to tuffs toward the top of the series (Astaburuaga et al. 2012). A maximum depositional age of 121.8 ± 0.8 Ma (Fig. 6; Table 4) was obtained in this work by detrital zircon U–Pb determinations in sandstone of this series near its contact with the underlying Baños del Flaco Formation (Fig. 3c). This determination confirms a Middle Aptian age for the base of the red continental deposits of the Colimapu Formation.

Based on the new ages, the outcrops of the Baños del Flaco Formation along the Maule river correlate with the upper deposits of the Mendoza Group in Argentina (Agrio Formation, Fig. 2) and the lower units of the overlying Rayoso Group (Huitrín Formation, Fig. 2). In addition, the continental deposits of the Colimapu Formation correlate to the east with the continental Rayoso Formation (Fig. 2). Thereby, the Mesozoic sedimentary successions exposed along the Maule river (Fig. 3c) would represent the final stage of the last transgressive–regressive cycle record in the Neuquén Basin (Legarreta and Gulisano 1989) representing the red detrital deposits of the Colimapu Formation the disconnection with the paleo-Pacific ocean.

Younger continental Late Cretaceous deposits crop out along the western slope of the Principal Cordillera. These are well exposed in an N–S trending belt from the Termas del Flaco locality toward the south into the Teno river valley (Fig. 3a) and have been grouped in the BRCU series (*Brownish Red Clastic Unit*; Charrier et al. 1996). This unit corresponds to a succession of clastic continental deposits, of marked brownish red coloration, which was originally reported by Charrier et al. (1996) from studies in the upper Tinguiririca valley (Fig. 3a). Here, the unit shows a wedge shape with a westward growth reaching a maximum thickness of ~230 m (Muñoz et al. 2018). It is composed of a basal sedimentary breccia followed by a series of conglomerates and conglomeratic sandstones with intercalations of finer levels of mudstones and sandstones (Charrier et al. 1996; Muñoz et al. 2018). The BRCU series overlays in apparent conformity the Baños del Flaco Formation and

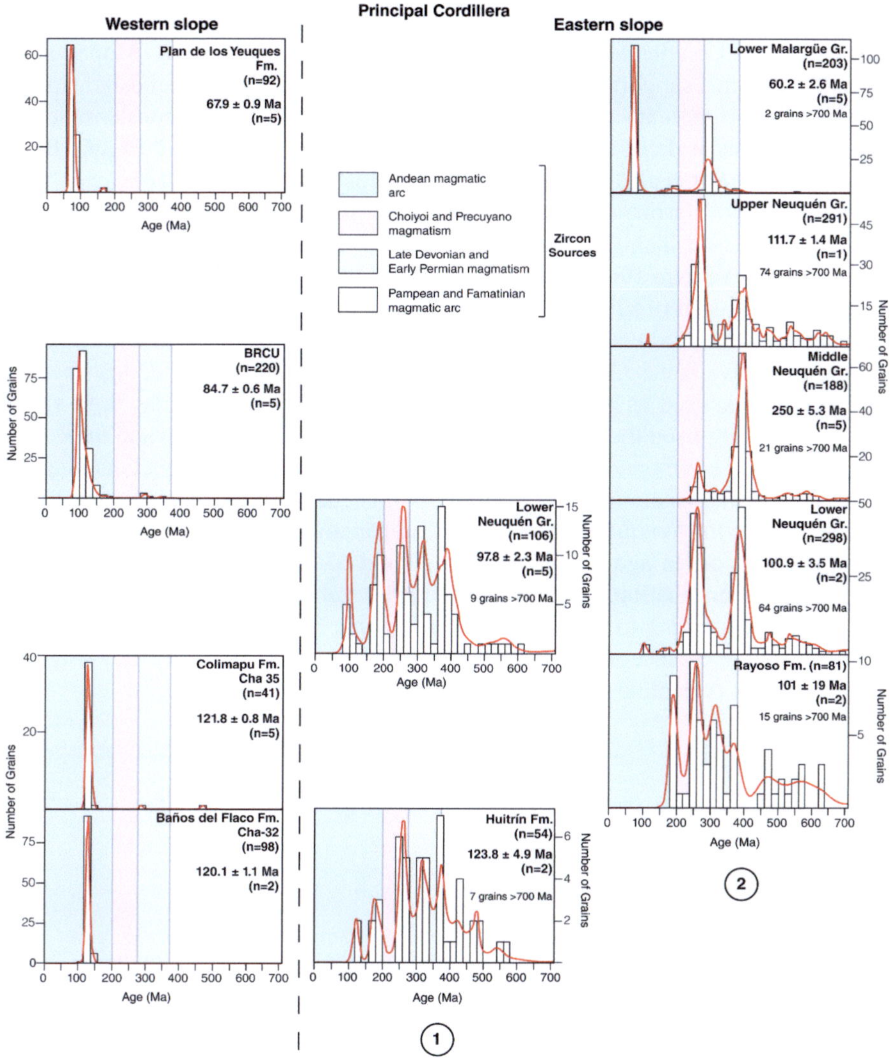

Fig. 6 Comparative Frequency histograms and relative probability density plots for the U–Pb detrital ages of the Cretaceous deposits in the western and eastern slope of the Principal Cordillera. Circled numbers indicate the western (1) and eastern (2) location of the samples along the Argentinean Principal Cordillera. Ages for the Chilean western slope of the Andes are from this work and Muñoz et al. (2018), while those from the eastern Argentinean slope are compiled from Balgord and Carrapa (2016), Di Giulio et al. (2017, 2012), Fennell et al. (2017), Horton et al. (2016), Tunik et al. (2010)

is overlain discordantly by Paleogene rocks (Fig. 3a). U–Pb detrital zircon ages assign the BRCU series to the latest Cenomanian-Early Campanian (Muñoz et al. 2018, Fig. 6), thus supporting a chronological correlation of these deposits with the Neuquén Group present in Argentina (Fig. 2). This has led to the consideration of the BRCU series as deposited during the foreland basin stage of the Neuquén Basin evolution (Charrier et al. 1996; Mescua et al. 2013; Muñoz et al. 2018).

The youngest Mesozoic unit on the Chilean Principal Cordillera corresponds to the continental and mainly volcanic series of the Plan de los Yeuques Formation (González and Vergara 1962). This unit crops out in the Tinguiririca and Teno rivers (Fig. 3a) and consists of 2200 m of lava flows, tuffs, and volcanic breccias, with minor sedimentary intercalations, defining an apparent bimodal suite between the effusive and the explosive facies (Muñoz et al. 2018). The base of the Plan de los Yeuques Formation is exposed as a pseudo-concordant marked contact in the Teno valley (Fig. 3a), passing from the fine stratified red beds of the BRCU series to the coarse stratified massive tuffs and lavas of the former unit. Its upper contact is an angular unconformity that separates it from Eocene-Oligocene deposits (Muñoz et al. 2018). For the Plan de los Yeuques Formation, a Campanian–Danian age was determined, which is based on an age range between 79 and 63 Ma obtained from numerous radiometric determinations in tuff and volcanic levels (Mosolf et al. 2018; Muñoz et al. 2018). According to geochemical and isotopic analysis, this unit corresponds to the latest Late Cretaceous magmatic arc and can be correlated with the Infiernillo unit exposed in Argentina (Iannelli et al. 2018).

5 Zircon Provenance and Paleogeographic Evolution

5.1 Potential Sediment Sources

The Mesoproterozoic-lower Mesozoic basement exposed along the Southern Central Andes constituted an important sediment source for the Mesozoic deposits of the Neuquén Basin (Naipauer and Ramos 2016). Conversely, such deposits largely lack the age component >1700 Ma associated with the eastern cratonic region of South America (2900–2500 and 2200–1700 Ma) (cf. Naipauer and Ramos, 2016).

The southern Frontal Cordillera, San Rafael, and Las Matras blocks located along the northern and eastern margin of the Neuquén Basin (Fig. 1b) correspond to the most important source areas of Mesoproterozoic-lower Neoproterozoic zircons (1300–900 Ma) (Naipauer and Ramos 2016). These Proterozoic rocks correspond to the "Grenvillian" basement of accreted terranes during the Neoproterozoic-Paleozoic (Ramos 2010). The southern Frontal Cordillera and San Rafael block are also potential sources of Neoproterozoic-Early Devonian zircons because rocks related to magmatism of the Pampean (700–500 Ma) and Famatinian (500–360 Ma) cycles are exposed along such domains.

To the west, the main source of Neoproterozoic-Paleozoic zircons corresponds to the Coastal Cordillera (Fig. 1b). This is composed of the upper Carboniferous Coastal batholith (300–320 Ma) (Deckart et al. 2014), which is associated with the Gondwanian magmatic arc, and the Paleozoic metamorphic series (Hervé 1988). The former unit is a potential source of the Carboniferous zircons. In turn, the latter metamorphic series show a spectrum of U–Pb detrital zircon ages which signals them as a potential western source of both late Proterozoic-Early Devonian (650–400 Ma) and Mississippian zircons (330–360 Ma) (Hervé et al. 2013). It is noteworthy that the upper Paleozoic metasedimentary units exposed along the San Rafael block and Frontal Cordillera, which were deposited into the late Carboniferous retroarc basins (e.g., San Rafael Basin; Limarino and Spalletti 2006), could also be a source of Late Carboniferous detrital zircons. These units are more suitable than the Coastal batholith to be source areas for the eastern deposits of the Neuquén Basin considering that the detritus would need to cross over the coeval Late Jurassic volcanic arc to supply sediments to the eastern area of the basin.

Another important component of the pre-Mesozoic basement is the Choiyoi magmatic province (280–250 Ma) developed after the Carboniferous magmatism and the San Rafael orogenic phase (Sato et al. 2015). Plutonic and intrusive rocks of this magmatic province form the most extensive basement of the basin (Fig. 1b) cropping out in the Frontal Cordillera, San Rafael block, and into basement blocks inside the basin such as the Tordillo High and Bardas Blancas anticline (Naipauer and Ramos 2016).

Finally, the Andean magmatic arc (<200 Ma) is the main Mesozoic sediment source of the Neuquén Basin. This group contains Jurassic-Cretaceous intrusive and volcanic rocks exposed along the western and eastern Coastal Cordillera (Fig. 1b).

5.2 *Zircon Provenance*

The zircon provenance analysis is presented in the following according to three timeframes corresponding to major changes in the Mesozoic paleogeography. These time ranges correspond to (1) Middle-Late Jurassic; (2) Late Jurassic-Early Cretaceous; and (3) latest Early-Late Cretaceous.

5.2.1 Middle-Late Jurassic

Detrital zircon U–Pb results for the two samples of the Middle-Upper Jurassic Valle Grande Formation show dominant peak of Jurassic zircons which indicate provenance from the western magmatic arc in the Coastal Cordillera (Fig. 5). A series of secondary peaks, from the Middle Permian to Late Triassic (Fig. 5), point out to a sediment supply from the Choiyoi and Precuyano magmatism. Source areas of these zircons correspond either to the San Rafael block and Frontal Cordillera in the east or structural basement highs into the basin. Similarly, the peaks of late Carboniferous age also suggest provenance from the latter morphostructures.

U–Pb detrital zircon ages of metasedimentary Jurassic deposits exposed in the Coastal Cordillera (Fig. 5) also show a spectrum with a dominant Jurassic peak which marks the provenance from the coeval magmatic arc (Fig. 5). In addition, the presence of Carboniferous age peaks signals the Paleozoic Coastal batholith as a source of Gondwanian zircons, an inference also supported by the proximity of the Mesozoic deposits to the Paleozoic outcrops (Hervé et al. 2013, Fig. 1b). Moreover, these Jurassic deposits show ages >400 Ma (Fig. 5) which could represent evidence for the recycling of Paleozoic metamorphic basement sources in the Coastal Cordillera. Unlike the Valle Grande Formation samples, Middle Permian-Late Triassic sources are not represented in the pattern of the detrital ages of Jurassic coastal unit (Fig. 5), thus indicating that the eastern basement related to the Choiyoi and Precuyano magmatism was not a sediment source for the western Mesozoic deposits.

In summary, the results indicate that the Middle Jurassic deposits currently exposed along the Coastal Cordillera were supplied by local sources as the Paleozoic units and the coeval magmatic arc, while the sediments of the Valle Grande Formation were derived from the coeval magmatic arc and the eastern basement sources. These differences mark the presence of a geographic barrier which prevented the sediment supply from both the Paleozoic units to the east and the eastern basement sources to the west. According to the zircon provenance analysis, this barrier would correspond to the coeval magmatic arc which separated the retroarc and forearc domains (Fig. 7a). This paleogeographic model implies the development of a forearc basin to the west of the magmatic arc and the east of the Paleozoic-composed high (Vergara et al. 1995, Fig. 7a). This hypothesis is also supported by the angular unconformity separating the Lower and Middle Jurassic deposits in the Coastal Cordillera at the latitude of this study (Thiele and Morel 1981). On the previous model, the occurrence of the forearc basin had been recognized only for the deposition of the Lo Prado Formation during the Early Cretaceous (Charrier et al. 2007; Jara and Charrier 2014). Accordingly, the Middle-Late Jurassic forearc basin identified in this study would have been established previously to the Lo Prado forearc basin.

5.2.2 Late Jurassic-Early Cretaceous

For the Late Jurassic-Early Cretaceous deposits exposed along the study area, the detrital zircon analysis shows a change in the source areas when compared with both the previous period and the contemporaneous eastern deposits. The detrital zircon spectrum of Upper Jurassic Rio Damas Formation, in the western slope of the Principal Cordillera, is characterized by a unimodal distribution of Jurassic ages (Fig. 5), thus evidencing a sediment provenance from a single source area related to the western magmatic arc (Rossel et al. 2014). Regarding the previous period, this fact contrasts with the western Middle-Upper Jurassic Valle Grande Formation where both western and eastern sources areas can be recognized. Concerning the

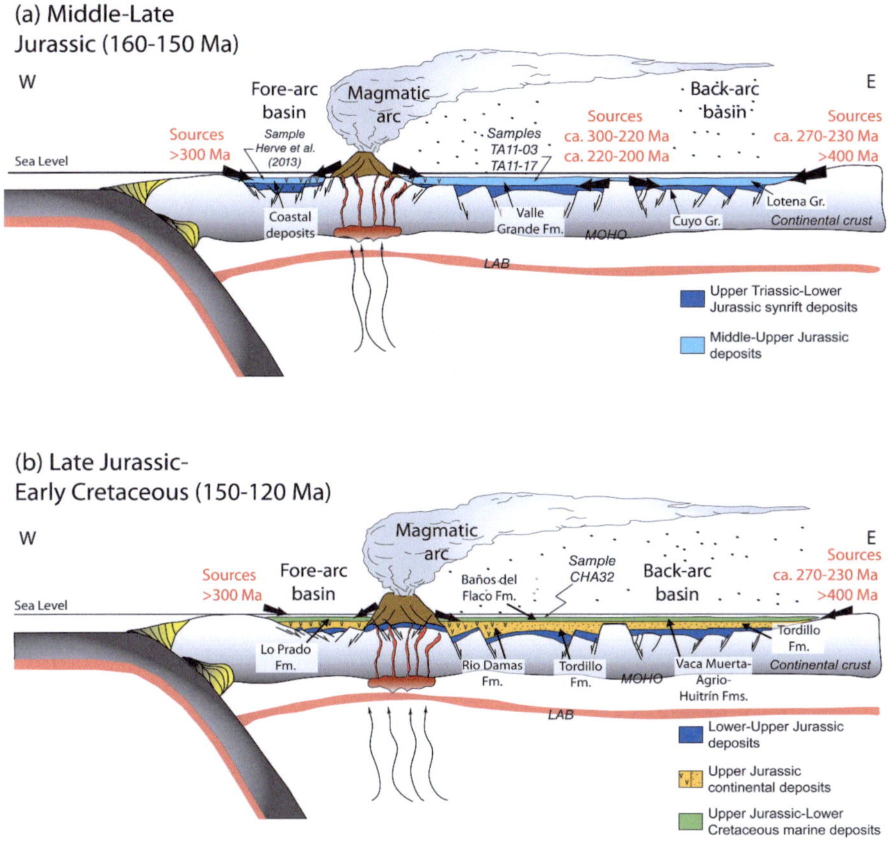

Fig. 7 Schematic cross section of central Chile and western Argentina illustrating the main paleogeographic features during Middle-Late Jurassic (**a**) and Late Jurassic-Early Cretaceous (**b**)

contemporaneous eastern deposits, the zircon spectra of the Tordillo Formation indicate that the primary source areas were the Jurassic Andean magmatic arc along with the eastern basement region, therefore contrasting with the unimodal distribution of the Río Damas Formation (Fig. 5).

The differences exposed about the zircon provenance between the Rio Damas and Valle Grande formations could be related to the significant paleogeographic reorganization which occurred during the Late Jurassic-Early Cretaceous (cf. Charrier et al. 2015). During this time, the Late Jurassic magmatic arc migrated to the east (Charrier et al. 2007) and underwent a progressive emersion due to the establishment of a transpressional regime along the Coastal Cordillera of central Chile (Creixell et al. 2011). In this new setting, the study area would have been placed proximal to the magmatic arc which would also explain the presence of a >1000-m-thick series of arc-signature lavas in the upper portion of the Río Damas Formation (Rossel et al. 2014). In turn, the Tordillo Formation corresponds to the distal deposits accumulated in the back-arc basin during the Late Jurassic, an area which was supplied by sediments from both the western volcanic arc and the eastern basement regions (Fig. 7b). The contemporaneous paleogeography is completed by the development of a new forearc basin (Lo Prado forearc basin, Charrier et al. 2007; Jara and Charrier 2014) located to the west of the volcanic arc (Fig. 7b). This change in the configuration is supported by an uplift period during the Late Jurassic (Vergara et al. 1995) as evidenced by the unconformity between the Upper Jurassic and Lower Cretaceous deposits in the Coastal Cordillera (Thomas 1958; Piracés and Maksaev 1977). In this sense, the development of the Lo Prado forearc basin could be the trigger for the uplift of the volcanic and volcanoclastic deposits of the Coastal Cordillera during the Early Cretaceous.

According to the detrital zircon analysis, the described proximal position to the magmatic arc of the study area is a characteristic that started during the Late Jurassic and continued until the Late Cretaceous (Fig. 6). Contrarily to the Chilean Mesozoic units, the Early Cretaceous Vaca Muerta, Agrio, and Huitrín Formations (Fig. 2), exposed along the Argentina slope of the Andes, indicate a sediment provenance mainly from the basement of the basin and the eastern foreland cratonic region (Naipauer and Ramos 2016). This is evidenced in the scarce of even null presence of zircons with Late Jurassic and Cretaceous ages derived from the magmatic arc (Fig. 6) (Naipauer and Ramos, 2016). Evidence of sediment supply from the coeval volcanic arc has been only recorded by ash-fall tuffs interbedded in the Vaca Muerta (Vennari et al. 2014) and upper Agrio Formations (Aguirre-Urreta et al. 2008) dated respectively in ~139 and 132 Ma.

For the Agrio and Huitrín formations, the decrease in the relative abundance of Late Jurassic and Cretaceous zircon ages and the increase in the abundance of older ages imply that the sediments were derived from erosion of the foreland basement located to the east and southeast of the basin (Tunik et al. 2010). This age pattern can be explained by the base-level fall and the creation of incised valleys in the eastern cratonic margin of the Neuquén Basin (Naipauer and Ramos 2016). An alternative hypothesis would be that the exhumation of the eastern basement region is related to the regional uplift that occurred during the Paraná plume impact and the opening of the South Atlantic (Naipauer and Ramos 2016). From our data, it is not possible to elucidate the differences in the zircon provenance pattern between western and eastern deposits. Instead, the maximum depositional age of the Baños del Flaco Formation along with the presence of the overlying continental Colimapu Formation in the Maule river constrains the final disconnection of the basin with the paleo-Pacific Ocean in the western region, in agreement with the proposals for the eastern area (Veiga et al. 2005).

5.2.3 Latest Early-Late Cretaceous

The western deposits of the Neuquén Basin represented by the continental sediments of the Colimapu Formation mark the beginning of the regional regression at the end of the Early Cretaceous. As for the previous stage, the western deposits show a sediment provenance only from the magmatic arc, while the contemporaneous Rayoso Formation, in Argentina, lacks Cretaceous zircons and shows dominant Precambrian to Paleozoic basement sources (Fig. 6). This evidence suggests that the paleogeographic conditions developed during the previous stage continued during the latest Late Cretaceous. However, there is also evidence along the arc and forearc regions at the latitude of this study that arguments for significant paleogeographic changes after 120 Ma.

Volcanoclastic, volcanic, and sedimentary deposits of the Las Chilcas Formation define the eastern boundary of the Coastal Cordillera between 33° and 35° 30′ S. Recent studies in the Las Chilcas type locality (~33° S) have shown the synorogenic nature of these deposits (Boyce et al. 2014), which were associated with the latest Early Cretaceous contractional episode (Charrier et al. 2007). Near the San Fernando locality (Fig. 1), this unit is unconformably underlain by the Early Cretaceous Lo Prado Formation (Contreras and Schilling, In press), a stratigraphic relation that is observed along the Coastal Cordillera of Central Chile (cf. Charrier et al. 2007). Moreover, the Las Chilcas Formation is intruded by 95 Ma arc-related rocks (Fig. 1). U–Pb zircon ages carried out in volcanic levels of this formation indicate an Albian depositional age (Fig. 1) (Contreras and Schilling, in Press; Godoy et al. 2009). Accordingly, the compressive deformation along the Coastal Cordillera at ~35° S would have begun at 113 Ma with the deposition of Las Chilcas Formation along the eastern slope of the latest Early Cretaceous subduction orogen (Fig. 8a). Moreover, the deposition age of the Las Chilcas Formation and the beginning of crustal shortening in the study area are in agreement with an accelerated exhumation period, registered between 113 and 80 Ma, along the Paleozoic metamorphic basement of the western Coastal Cordillera (Willner et al. 2005).

Along the eastern slope of the Andes, at the studied latitudes, a change in the paleogeography has also been proposed during the latest Early Cretaceous. This has been attributed to the Andean uplift event which would be responsible for the unconformity observed between the Rayoso Group and the Neuquén Group (Veiga et al. 2005). In this scenario, the difference in provenance areas between western and eastern uppermost Lower Cretaceous deposits of the Neuquén Basin would reflect the presence of a geographic barrier separating both domains (Fig. 8a). The development of this feature could be the result of a transient large-scale continental folding (Di Giulio et al. 2017) or the development of the forebulge during the beginning of the compressive period (Balgord and Carrapa 2016). In either case, the development of a geographic barrier by crustal uplift during the latest Early Cretaceous could explain the erosion of the upper portion of the Baños del Flaco Formation in the Termas del Flaco locality (Fig. 8a). Noteworthy, the sedimentation in this place was resumed with the accumulation of the BRCU during the Late Cretaceous, most likely when the forebulge migrated eastward and the wedge-top or foredeep was placed in the Termas del Flaco locality.

Recently, Muñoz et al. (2018) have shown the differences in provenance areas between the western and eastern Upper Cretaceous deposits of the Neuquén Basin along the study area. On one hand, Upper Cretaceous BRCU series has a sediment

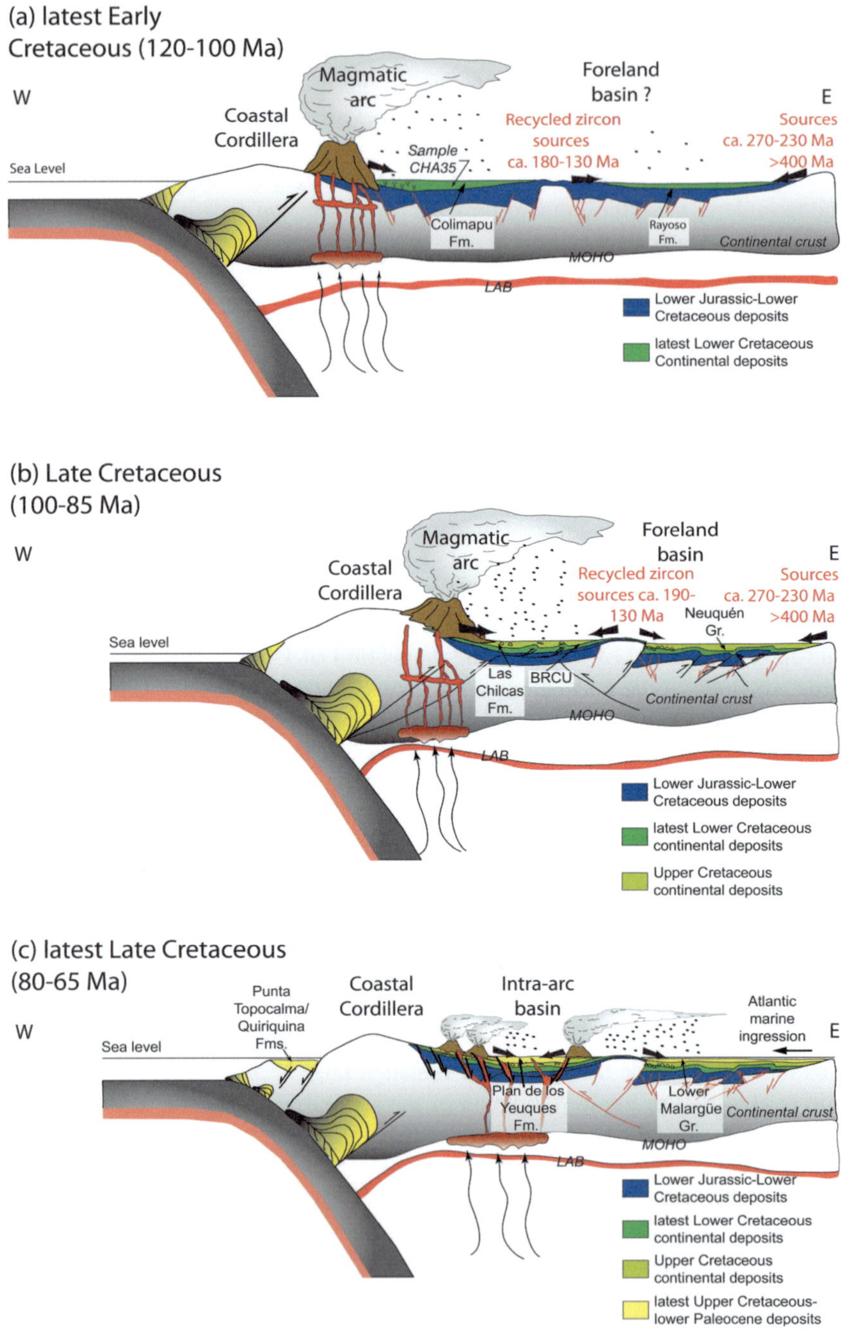

Fig. 8 Schematic cross section of central Chile and western Argentina illustrating the main paleogeographic features of the latest Early Cretaceous (**a**), Late Cretaceous (**b**), and latest Late Cretaceous (**c**). Red faults correspond to active structures during previous stages

provenance from Cretaceous sources (Fig. 6), which can be related to both the coeval Cretaceous volcanic arc, located west of the study area along the Coastal Cordillera, and the Lower Cretaceous sedimentary sequences as a recycled component. On the other hand, detrital zircon ages of the lower Neuquén Group are characterized by a Jurassic-Early Cretaceous arc component (<200 Ma) and a minor Early Triassic-late Devonian component (200–400 Ma; Fig. 5b). In turn, the middle and upper Neuquén Group (Fig. 5b) exhibit only components of Early Triassic-late Devonian (200–400 Ma) and cratonic (>400 Ma) ages (Fennell et al. 2017).

The difference between the BRCU and Neuquén Group provenance was interpreted as evidence of a different paleogeographic setting for deposition of these partly coeval units (Muñoz et al. 2018). Following this, the authors argued for the presence of a positive relief separating the BRCU and Neuquén Group depocenters which would be associated with the growth of the Late Cretaceous fold-and-thrust belt (Fig. 8b). The BRCU, cropping out to the west of the Neuquén Group (Fig. 1b), would have deposited under this configuration in the wedge-top depozone of the Late Cretaceous foreland basin system, along with the lower portion of the eastern Upper Cretaceous deposits (Muñoz et al. 2018). After that, the western depocenter was disconnected from the rest of the foreland basin system (Fig. 8) evolving probably as a disconnected piggyback basin from the connected wedge-top and foredeep depocenters that developed to the east of the study area (Muñoz et al. 2018). The latter domains, in turn, would have corresponded to the domains where the middle and upper Neuquén Group were accumulated (Fig. 8b) (Fennell et al. 2017).

Finally, the volcanic rocks of the Plan de los Yeuques Formation record an eastward migration of the magmatic arc at least since ~80 Ma (Muñoz et al. 2018). Based on the nature, extensional structures, and geochemical characteristics of the latter unit, Muñoz et al. (2018) invoked a change from compressional to extensional conditions in the arc area during the Late Cretaceous. Their model suggests that deposition of the Plan de los Yeuques occurred in an intra-arc basin setting where extension and crustal thinning would be responsible for the subsidence in the arc area. Coevally, the sedimentation along the eastern region was dominated by marine and fluvial deposits of the lower Malargüe Group (Horton et al. 2016; Balgord and Carrapa 2016). The detrital zircon ages of this unit indicate a provenance exclusively from the latest Upper Cretaceous volcanic arc contrasting with the underlying Neuquén Group (Fig. 6). Accordingly, the deposition of the Malargüe Group would have occurred into an overfilled foreland basin, when the Upper Cretaceous synorogenic deposits covered the basement of the Neuquén Basin and the eastern basement region. This scenario has been invoked to propose neutral to extensional tectonic conditions dominating the western margin of South America during the end of the Cretaceous. Such conditions could be related to a slab roll-back period (Horton 2018) which is in accordance with the development of the extensional intra-arc basin of the Plan de los Yeuques Formation. During this period, Upper Cretaceous marine deposits were also accumulated on the western flank of the Coastal Cordillera (Encinas et al. 2014). The accumulation of such deposits was controlled by normal faulting (Becerra et al. 2013), most likely associated with slab roll-back that would have experimented the western margin of Gondwana during the latest Cretaceous.

6 Conclusions

Our data show that the arc and forearc regions were active at least since the Middle Jurassic. During such time, the western margin of the Neuquén Basin was characterized by a Paleozoic basement high and a volcanic arc which bounded the forearc basin. Moreover, these features acted as source areas to the sediments of the basin. This configuration continued during the Early Cretaceous, but both the volcanic arc and the forearc basin underwent an eastward migration during this time. New ages along the Chilean slope of the Andes allow suggesting that a compressive period began during the latest Early Cretaceous which is also consistent with the period of the maximum exhumation rates estimated for the western Paleozoic basement of the Coastal Cordillera. The beginning of such period along the western margin of South America triggered the development of a geographic barrier inhibiting the sediments supply from the contemporaneous volcanic arc to the eastern retroarc basin, a hypothesis that can explain the differences in the sediment provenance between western and eastern deposits during the latest Late Cretaceous. Such scenario could also disregard the null or waning activity inferred from U to Pb detrital zircon analysis of the eastern Mesozoic deposits during the Middle Jurassic or Late Cretaceous (cf. Balgord 2017). This is indeed the case when considering the almost complete record of Mesozoic ages present in the coeval western detrital and volcanic deposits of the Andean segment that records the continued activity of the arc-related magmatism throughout all the Mesozoic time.

Acknowledgements We are grateful to María Pía Rodríguez for their constructive criticism of early versions of the manuscript, and many useful suggestions which improved it considerably. This work was funded by the Chilean government through the project FONDECYT1161806 (CONICYT, Chilean national research agency). This is a contribution of the Instituto de Estudios Andinos Don Pablo Groeber.

Supplementary Data

U–Pb dating was performed by the laser ablation inductively coupled plasma mass spectrometry (LA-ICP-MS) employing a Thermo Xseries QICPMS (FD samples) and a Thermo ICapQcquadrupole ICPMS (remaining samples), in both cases coupled with a Resolution M050 excimer laser workstation. The isotopic ratios, ages, and errors were calculated according to Petrus and Kamber (2012). In the data tables, analytical determinations are reported at 2σ uncertainties, which were propagated according to Paton et al. (2010). Analyzed spots were 23 μm, using an analytical protocol modified from Solari et al. (2010), and these were selected after the study of transmitted light and cathodoluminescence images. For each spot, a set of trace elements were also measured in order to monitor possible inclusions in zircon crystals, using the NIST-SRM 610 glass as a reference standard and 29Si as an internal standard (15.323% assumed stoichiometric value).

References

Aguirre-Urreta MB, Pazos PJ, Lazo DG, Fanning CM, Litvak VD (2008) First U–Pb SHRIMP age of the Hauterivian stage, Neuquén Basin, Argentina. J S Am Earth Sci 26(1):91–99

Aguirre-Urreta MB, Tunik MA, Naipauer M et al (2011) Malargue Group (Maastrichtian-Danian) deposits in the Neuquén Andes, Argentina: implications for the onset of the first Atlantic transgression related to Western Gondwana break-up. Gondwana Res 19:482–494

Arancibia G (2004) Mid-cretaceous crustal shortening: evidence from a regional-scale ductile shear zone in the coastal range of central Chile (32° S). J S Am Earth Sci 17:209–226

Armijo R, Rauld R, Thiele R et al (2010) The West Andean thrust, the San Ramon fault, and the seismic hazard for Santiago, Chile. Tectonics 29(2). https://doi.org/10.1029/2008tc002427

Astaburuaga D (2014) Evolución estructural del límite Mesozoico-Cenozoico de la Cordillera Principal entre 35° 30′ y 36° S, región del Maule. Thesis, Universidad de Chile, Chile

Astaburuaga D, Farías M, Charrier R et al (2012) Geología y estructuras del límite Mesozoico-Cenozoico de la Cordillera Principal entre 35° 30′ y 36° S, región el Maule, Chile. In: Abstracts of the 13 Congreso Geológico Chileno, Antofagasta, 5–9 Aug 2012

Balgord EA (2017) Triassic to Neogene evolution of the south-central Andean arc determined by detrital zircon U–Pb and Hf analysis of Neuquén Basin strata, central Argentina (34° S–40° S). Lithosphere 9(3):453–462

Balgord EA, Carrapa B (2016) Basin evolution of Upper Cretaceous-Lower Cenozoic strata in the Malargüe fold-and-thrust belt: Northern Neuquén Basin, Argentina. Basin Res 28:183–206

Becerra J, Contreras-Reyes E, Arriagada C (2013) Seismic structure and tectonics of the southern Arauco Basin, south-central Chile (~38° S). Tectonophysics 592:53–66

Boyce D, Charrier R, Tapia F et al (2014) Mid-Cretaceous compressional deformation in Central Chile: the beginning of the Andean building. In: Abstracts of the AGU Fall Meeting. San Francisco, 15–19 Dec 2014

Burckhard C (1900) Profils géologiques transversaux de la Cordillère Argentino-Chilienne. Stratigraphie et Tectonique. A M La Plata, Sec Geol Min:1–189

Charrier R (1979) El Triásico de Chile y regiones adyacentes en Argentina: Una reconstrucción paleogeográfica y paleoclimática. Comunicaciones 26:1–47

Charrier R (1981) Geologie der chilenischen Hauptkordillere zwischen 34 °und 343°0” südlicher Breite und ihre tektonische, magmatische und paläogeographische Entwicklung. Berliner Geowissenschaftliche Abhandlungen. PhD thesis, Freie Universität Berlin

Charrier R, Baeza O, Elgueta S et al (2002) Evidence for Cenozoic extensional basin development and tectonic inversion south of the flat-slab segment, southern Central Andes, Chile (33°–36° S.L.). J S Am Earth Sci 15:117–139

Charrier R, Pinto L, Rodríguez MP (2007) Tectonostratigraphic evolution of the Andean Orogen in Chile. In: Moreno T, Gibbons W (eds) The Geology of Chile. The Geological Society, London, pp 21–114

Charrier R, Ramos VA, Tapia F et al (2015) Tectono-stratigraphic evolution of the Andean Orogen between 31 and 37 S (Chile and Western Argentina). In: Sepúlveda S, Giambiagi LB, Moreiras SM et al (eds) Geodynamic Processes in The Andes of Central Chile and Argentina. The Geological Society, London, SP 399, pp 13–61

Charrier R, Wyss AR, Flynn JJ et al (1996) New evidence for late Mesozoic early Cenozoic evolution of the Chilean Andes in the Upper Tinguiririca Valley (35 degrees S), Central Chile. J S Am Earth Sci 9:393–422

Coloma F, Valin X, Oliveros V et al (2017) Geochemistry of Permian to Triassic igneous rocks from northern Chile (28°–30° 15′ S): implications on the dynamics of the proto-Andean margin. Andean Geol 44:147–178

Contreras JP, Schilling ME Geología del área San Fernando - Curicó, regiones del Libertador General Bernardo O´Higgins y del Maule, Escala 1:100.000, in: Carta Geológica de Chile, Serie Geológica Básica. SERNAGEOMIN, p 50 (In Press)

Corvalán DJ (1976) El Triásico y Jurásico de Vichuquén-Tilicura y de Hualañe, Provincia de Curicó. Implicancias Paleogeográficas. In: Abstracts of the 1 Congreso Geológico Chileno. Santiango, 2–7 Aug 1976

Covacevich V, Varela J, Vergara M (1976) Estratigrafía y sedimentación de la Formación Baños del Flaco al sur del río Tinguiririca, Cordillera de los Andes, Provincia de Curicó, Chile. In: Abstracts of the 1 Congreso Geológico Chileno. Santiango, 2–7 Aug 1976

Creixell C, Parada MÁ, Morata D, Vásquez P, Pérez de Arce C (2011) Middle-late jurassic to early cretaceous transtension and transpression during arc building in central Chile: evidence from mafic dike swarms. Andean Geol 38:37–63

Davidson J, Vicente JC (1973) Características paleogeográficas y estructurales del área fronteriza de las nacientes del Teno (Chile) y Santa Elena (Argentina) (Cordillera Principal, 35° a 35° 15′ de latitud sur), V Congreso Geológico Argentino, Córdoba, 22–28 Oct 1976

Deckart K, Hervé F, Fanning CM et al (2014) Geocronología U–Pb e isótopos de Hf-O en circones del batolito de la Costa Pensilvaniana, Chile. Andean Geo 41:49–82

Di Giulio A, Ronchi A, Sanfilippo A et al (2017) Cretaceous evolution of the Andean margin between 36° S and 40° S latitude through a multi-proxy provenance analysis of Neuquén Basin strata (Argentina). Basin Res 29:284–304

Di Giulio A, Ronchi A, Sanfilippo A et al (2012) Detrital zircon provenance from the Neuquén Basin (south-central Andes): cretaceous geodynamic evolution and sedimentary response in a retroarc-foreland basin. Geology 40:559–562

Dickinson WR, Gehrels GE (2009) Use of U–Pb ages of detrital zircons to infer maximum depositional ages of strata: a test against a Colorado Plateau Mesozoic database. Earth Planet Sci Lett 288:115–125

Encinas A, Stinnesbeck W, Valencia VA (2014) Primera datación radiométrica (U–Pb, LA-ICP-MS, en zircones detríticos) de la Formación Punta Topocalma: Observaciones sobre la sedimentación marina durante el cretácico tardío en Chile central. Andean Geol 41:436–445

Fennell LM, Folguera A, Naipauer M et al (2017) Cretaceous deformation of the southern Central Andes: synorogenic growth strata in the Neuquén Group (35° 30′–37° S). Basin Res 29:51–72

Folguera A, Orts DL, Spagnuolo MG et al (2011) A review of late cretaceous to quaternary palaeogeography of the Southern Andes. Biol J Linn Soc 103:250–268

Folguera A, Ramos VA (2011) Repeated eastward shifts of arc magmatism in the Southern Andes: a revision to the long-term pattern of Andean uplift and magmatism. J S Am Earth Sci 32:531–546

Franzese JR, Spalletti LA (2001) Late triassic-Early jurassic continental extension in SouthWestern Gondwana: tectonic segmentation and pre-break-up rifting. J South Am Earth Sci 14:257–270

Gana P, Zentilli M (2000) Historia termal y exhumación de intrusivos de la Cordillera de la Costa de Chile central. In: Abstratcs of the IX Congreso Geológico Chileno. Puerto Varas, 31 July–4 April 2000

Gansser A (1973) Facts and theories on the Andes: twenty-sixth William Smith Lecture. J Geol Soc 129(2):93–131

Godoy E, Schilling M, Solari M, Fock A (2009) Geología del área Rancagua-San Vicente de Tagua-Tagua, Región del Libertador Bernardo O'Higgins. Carta Geológica de Chile, Serie Geológica Básica No. 118. Servicio Nacional de Geología y Minería, Santiago

González Ferrán O, Vergara M (1962) Reconocimiento geologico de la cordillera de los andes entre los paralelos 35° y 38°. Sur Bol Inst Inv Geol 1:116

González J, Oliveros V, Creixell C et al (2018) The triassic magmatism and its relation with the pre-andean tectonic evolution: geochemical and petrographic constrains from the high andes of north central Chile (29° 30′–30° S). J S Am Earth Sci 87:95–112

Hervé F (1988) Late Paleozoic subduction and accretion in southern Chile. Episodes 11:183–188

Hervé F, Calderón M, Fanning CMM et al (2013) Provenance variations in the late Paleozoic accretionary complex of central Chile as indicated by detrital zircons. Gondwana Res 23:1122–1135

Horton BK, Fuentes F, Boll A et al (2016) Andean stratigraphic record of the transition from backarc extension to orogenic shortening: a case study from the northern Neuquén Basin, Argentina. J S Am Earth Sci 71:17–40
Horton BK (2018) Sedimentary record of Andean mountain building. Earth-Sci Rev 178:279–309
Howell JA, Schwarz E, Spalletti LA et al (2005) The Neuquén Basin: an overwiew. In: Howell J, SchwarzE, Spalletti L, Veiga GD (eds) The Neuquén Basin, Argentina: A case study in sequence stratigraphy and basin dynamics. The Geological Society, London, SP 252, pp 1–14
Iannelli SB, Fennell LM, Litvak VD et al (2018) Geochemical and tectonic evolution of late cretaceous to early paleocene magmatism along the Southern Central Andes (35–36° S). J S Am Earth Sci 87:137–156
Jara P, Charrier R (2014) Nuevos antecedentes estratigráficos y geocronológicos para el Meso-Cenozoico de la Cordillera Principal de Chile entre 32° y 32° 30′ S: Implicancias estructurales y paleogeográficas. Andean Geol 41:174–209
Kietzmann DA, Palma RM, Riccardi AC et al (2014) Sedimentology and sequence stratigraphy of a Tithonian-Valanginian carbonate ramp (Vaca Muerta formation): a misunderstood exceptional source rock in the Southern Mendoza area of the Neuquén Basin, Argentina. Sediment Geol 302:64–86
Klohn C (1960) Geología de la Cordillera de los Andes de Chile Central. Provincias de Santiago, O'higgins, Colchagua y Curicó. Bol Inst Inv Geol 8
Lanés S (2005) Late triassic to early jurassic sedimentation in northern Neuquén Basin, Argentina: tectonosedimentary evolution of the first transgression. Geol Acta 3:81–106
Legarreta L, Gulisano CA (1989) Análisis estratigráfico secuencial de la Cuenca Neuquina (Triásico superior-Terciario inferior). In: Chebli G, Spalletti LA (eds) Cuencas Sedimentarias Argentinas. Universidad de Tucumán, vol 6, Serie Correlación Geológica, pp 221–243
Legarreta L, Uliana MA (1996) The Jurassic succession in west-central Argentina: stratal patterns, sequences and paleogeographic evolution. Palaeogeogr Palaeoclimatol 120:303–330
Limarino CO, Spalletti LA (2006) Paleogeography of the upper Paleozoic basins of southern South America: an overview. J S Am Earth Sci 22:134–155
Llambías EJ, Sato AM, Basei MAS (2005) El basamento prejurásico medio en el anticlinal Chihuido, Malargüe: Evolución magmática y tectónica. Rev Asoc Geol Argentina 60:567–578
Ludwig KR (1999) Isoplot/EX version 2.10: a geochronological toolkit for Microsoft Excel, vol 1. Berkeley Geochronol Cent SP, pp 1–49
Mazzini A, Svensen H, Leanza HA et al (2010) Early jurassic shale chemostratigraphy and U–Pb ages from the Neuquén Basin (Argentina): implications for the toarcian oceanic anoxic event. Earth Planet Sci Lett 297:633–645
Mescua JF, Giambiagi LB, Ramos VA (2013) Late cretaceous uplift in the Malargüe fold-and-thrust belt (35° S), southern Central Andes of Argentina and Chile. Andean Geol 40:102–116
Mescua JF, Tapia F, Farias M et al (2015) Edades U–Pb y correlaciones del Paleozoico de las Nacientes del río Salado, y la ocurrencia de la Fase San Rafaélica en la Cordillera Principal de Menndoza. In: Abstracts of the 16 Reunión de Tectónica, General Roca, 15–22 Oct 2015
Mosolf JG, Gans PB, Wyss AR et al (2018) Late Cretaceous to Miocene volcanism, sedimentation, and upper-crustal faulting and folding in the Principal Cordillera, central Chile: field and geochronological evidence for protracted arc volcanism and transpressive deformation. Geol Soc Am Bull 131:252–273
Mpodozis C, Ramos VA (1989) The Andes of Chile and Argentina in: Ericksen GE, Cañas Pinochet MT, Reinemund JA (Eds.), Geology of the Andes and Its Relation to Hydrocarbon and Mineral Resources. Min Res Earth Sci Series, pp 59–90
Muñoz M, Tapia F, Persico M et al (2018) Extensional tectonics during late Cretaceous evolution of the Southern Central Andes: evidence from the Chilean main range at ~35° S. Tectonophysics 744:93–117
Naipauer M, García Morabito E, Marques JC et al (2012) Intraplate late jurassic deformation and exhumation in western central Argentina: constraints from surface data and U–Pb detrital zircon ages. Tectonophysics 524–525:59–75

Naipauer M, Ramos VA (2016) Changes in Source Areas at Neuquén Basin: Mesozoic evolution and tectonic setting based on U–Pb ages on zircons. In: Folguera A, Naipauer M, Sagripanti L et al (eds) Growth of the Southern Andes. Springer Earth System Sciences, Switzerland, pp 1–269

Naipauer M, Tapia F, Mescua JF et al (2015) Detrital and volcanic zircon U–Pb ages from southern Mendoza (Argentina): an insight on the source regions in the northern part of the Neuquén Basin. J S Am Earth Sci 64:434–451

Naipauer M, Tunik MA, Marques JC et al (2014) U–Pb detrital zircon ages of Upper Jurassic continental successions: implications for the provenance and absolute age of the Jurassic-Cretaceous boundary in the Neuquén Basin. In: Sepúlveda S, Giambiagi LB, Moreiras SM et al (eds) Geodynamic Processes in The Andes of Central Chile and Argentina. The Geological Society, London, SP 399, pp 131–154

Oliveros V, González J, Espinoza M et al (2017) The early stages of the volcanic arc in the Southern Central Andes. In: Folguera A, Contreras-Reyes E et al (eds) The Evolution of the Chilean-Argentinean Andes. Springer, pp 185–212

Parada MA, Féraud G, Fuentes F et al (2005) Ages and cooling history of the Early Cretaceous Caleu pluton: testimony of a switch from a rifted to a compressional continental margin in central Chile. J Geol Soc London 162:273–287

Paton C, Woodhead JD, Hellstrom JC et al (2010) Improved laser ablation U–Pb zircon geochronology through robust downhole fractionation correction. Geoch Geophy Geosy 11(3). https://doi.org/10.1029/2009gc002618

Petrus JA, Kamber BS (2012) VizualAge: a novel approach to laser ablation ICP-MS U–Pb geochronology data Reduction. Geostand Geoanal Res 36:247–270

Piracés R, Maksaev V (1977) Geología de la Hoja Quillota [unpublished report with map at 1:250 000 scale]: Santiago, Ins Inves Geolo: 140 p

Ramos VA (2010) The Grenville-age basement of the Andes. J S Am Earth Sci 29:77–91

Ramos VA (2000) The Southern Central Andes. In: Abstracts of the 31 international geological congress. Rio de Janeiro, 6–17 Aug 2000

Ramos VA, Folguera A (2005) Tectonic evolution of the Andes of Neuquén: constraints derived from the magmatic arc and foreland deformation. In: Veiga GD, Spalletti LA, Howell JA, Schwarz E (eds) The Neuquén Basin, Argentina: a case study in sequence stratigraphy and basin dynamics. The Geological Society, London, Special Publications 252, pp 15–35

Ramos VA, Jordan TE, Allmendinger RW et al (1986) Paleozoic Terranes of the Central Argentine-Chilean Andes. Tectonics 5:855–880

Ramos VA, Kay SM (1991) Triassic rifting and associated basalts in the Cuyo basin, central Argentina. In: Harmon RS, Rapela CW (eds) Andean magmatism and its tectonic setting. The Geological Society, America, SP 265, pp 79–92

Riccardi AC (2008) El Jurasico de la Argentina y sus amonites. Rev Asoc Geol Argentina 63:625–643

Rossel P, Oliveros V, Mescua JF et al (2014) El volcanismo jurásico superior de la Formación Río Damas-Tordillo (33°–35.5° S): antecedentes su sobre petrogénesis, cronología, proveniencia e implicancias tectónicas. Andean Geol 41:529–557

Salazar C (2012) The Jurassic-Cretaceous Boundary (Tithonian-Hauterivian) in the Andean Basin, Central Chile: Ammonite fauna, Bio-and Sequence Stratigraphy and Palaeobiogeography. Ph.D. Thesis, Univertsität Heidelberg

Sato AM, Llambías EJ, Basei MAS et al (2015) Three stages in the Late Paleozoic to Triassic magmatism of southwestern Gondwana, and the relationships with the volcanogenic events in coeval basins. J S Am Earth Sci 63:48–69

Solari LA, Gímez-Tuena A, Bernal JP et al (2010) U–Pb zircon geochronology with an integrated la-icp-ms microanalytical workstation: achievements in precision and accuracy. Geostand Geoanal Res 34:5–18

Tapia F, Farías M, Naipauer M et al (2015a) Late Cenozoic contractional evolution of the current arc-volcanic region along the southern Central Andes (35° 20' S). J Geodyn 88:36–51

Tapia F, Muñoz M, Farías M et al (2015b). Hallazgo de estratos de edad Cretácico Tardío en el curso alto del río Tinguiririca (~34° 55′ S). Parte II : implicancias Tectónicas. In: Abstratcs of the XIV Congreso Geológico Chileno. Concepción, 4–8 Oct 2015

Tapia F, Farías M, Rubilar A (2011) Depósitos Marinos Del Jurásico Superior en el curso superior del río Colorado de Lontué (35° 23′ S), VII Región, Chile. In: Abstratcs of the 18 Congreso Geológico Argentino. Neuquén, 2–6 May 2011

Thiele R, Morel R (1981) Tectónica Triásico-Jurásica en la Cordillera de la Costa, Al norte y sur del Río Mataquito (34° 45′–35° 15′ Lat. S); Chile. Rev Geol Chile 13/14:49–61

Thomas H (1958) Geología de la Cordillera de la Costa entre el Valle de La Ligua y la Cuesta de Barriga: Santiago. Inst Inv Geol 2: 86 p

Tunik MA, Folguera A, Naipauer M et al (2010) Early uplift and orogenic deformation in the Neuquén Basin: constraints on the Andean uplift from U–Pb and Hf isotopic data of detrital zircons. Tectonophysics 489:258–273

Uliana MA, Biddle J, Cerdan JJ (1989) Mesozoic Extension and the Formation of Argentine Sedimentary Basins. In: Tankard AJ, Balkwill HR (eds) Extensional tectonics and stratigraphy of the North Atlantic Margins. AAPG Memoir 46, pp 599–614

Vásquez P, Glodny J, Franz G et al (2011) Early Mesozoic Plutonism of the Cordillera de la Costa (34°–37° S), Chile: Constraints on the Onset of the Andean Orogeny. J Geol 119:159–184

Veiga GD, Howell JA, Strömback A (2005) Anatomy of a mixed marine-non- marine lowstand wedge in a ramnp setting. The record of a Barremian-Aptian complex relative sea-level fall in the central Neuquén Basin, Argentina. In: Veiga GD, Spalletti LA, Howell JA et al (eds) The Neuquén Basin, Argentina: a case study in sequence stratigraphy and basin dynamics. The Geological Society, London, SP 252, pp 139–162

Vennari VV, Lescano M, Naipauer M et al (2014) New constraints on the Jurassic—cretaceous boundary in the high Andes using high-precision U–Pb data. Gondwana Res 263:74–385

Vergani GD, Tankard AJ, Belotti HJ et al (1995) Tectonic evolution and paleogeography of the Neuquén Basin, Argentina. AAPG Special Volumen, pp 383–402

Vergara M, Levi BL, Nyström JO et al (1995) Jurassic and early cretaceous island arc volcanism, extension, and subsidence in the coast range of central Chile. Geol Soc Am Bull 706:1427–1440

Vermeesch P (2012) On the visualisation of detrital age distributions. Chem Geol 312(313):190–194

Willner AP, Thomson SN, Kröner A et al (2005) Time markers for the evolution and exhumation history of a late Palaeozoic paired metamorphic belt in North-Central Chile (34°–35° 30′ S). J Petrol 46:1835–1858

Zamora Valcarce G, Zapata T, del Pino D et al (2006) Structural evolution and magmatic characteristics of the Agrio fold-and-thrust belt. In: Kay SM, Ramos VA (eds) Evolution of an Andean Margin: a tectonic and magmatic view from the Andes to the Neuquén Basin. The Geological Society, America, SP 407, pp 125–145

Zapata T, Folguera A (2005) Tectonic evolution of the Andean Fold and Thrust Belt of the southern Neuquén Basin, Argentina. In: Veiga GD, Spalletti LA, Howell J, Schwarz E (eds) The Neuquén Basin: Argentina: A Case Study in Sequence Stratigraphy and Basin Dynamics. The Geological Society, London, SP 252, pp 37–56

The Late Cretaceous Orogenic System: Early Inversion of the Neuquén Basin and Associated Synorogenic Deposits (35°–38° S)

Lucas Fennell, Pablo Borghi, Federico Martos, Eduardo Agustín Rosselot, Maximiliano Naipauer and Andrés Folguera

Abstract The first regional tectonic uplift was registered in the Neuquén Basin at ca. 100 Ma, resulting in the inception of the first foreland basin of the Andes at these latitudes. The infill of this foreland basin is represented by the nonmarine deposits of the Neuquén Group, characterized by the presence of growth strata and fold-thrust belt detrital derivation, typical of wedge-top depozones. The presence of a long-lasting hiatus and angular unconformities in structures west of the wedge-top area indicate that these constituted the main detrital source areas of the Late Cretaceous foreland basin. These structures were uplifted by tectonic inversion of pre-existing normal faults, a mechanism that was also responsible for isolated basement block uplifts detected in the foredeep depozone. The regional unconformity between the synorogenic deposits of the Neuquén Group and the back-bulge deposits of the Bajada del Agrio Group represents the migration of the forebulge, which constituted another important detrital source of the foreland basin during its mature phase. An active fold-thrust belt, disconnected basement block uplifts and the identification of the typical depozones of present foreland basins provide a complete picture of the Late Cretaceous orogenic system responsible for the early inversion of the Neuquén Basin.

Keywords Neuquén group basin · Foreland basin · Early andean uplift · Flexural subsidence · Cretaceous

1 Introduction and Geological Setting

The Late Cretaceous represents a period of important palaeogeographic reorganization in the Neuquén Basin, whose stratigraphy records the transition between a

L. Fennell (✉) · F. Martos · E. A. Rosselot · M. Naipauer · A. Folguera
CONICET—Universidad de Buenos Aires, Instituto de Estudios Andinos Don Pablo Groeber (IDEAN), Buenos Aires, Argentina
e-mail: lucasfennell90@gmail.com

P. Borghi
Wintershall Energía S. A, Buenos Aires, Argentina

D. Kietzmann and A. Folguera (eds.), *Opening and Closure of the Neuquén Basin in the Southern Andes*, Springer Earth System Sciences,
https://doi.org/10.1007/978-3-030-29680-3_12

retroarc extensional settings towards a contractional system at ca. 100 Ma (Vergani et al. 1995; Manceda and Figueroa 1995; Tunik et al. 2010; Balgord and Carrapa 2016; Fennell et al. 2017a). This transition is marked in the Neuquén Basin by a regional unconformity over the deposits of the Lower Cretaceous Mendoza and Bajada del Agrio Groups (Groeber 1946), and the inception of a Late Cretaceous flexural foreland basin (Ramos and Folguera 2005). The infill of this foreland basin is represented by the nonmarine red beds of the Neuquén Group and the equivalent Diamante Formation, whose different naming responds to the difficulty of making regional correlations at the moment of their definition (Groeber 1947), although these deposits nowadays can be perfectly correlated (Bettini et al. 1978; Fennell et al. 2017b). Therefore, in order to avoid unnecessary naming, these deposits will be addressed as the Neuquén Group throughout the rest of the chapter, whose study is possible due to the complex deformational scenario that affected the Neuquén Basin since the Southern Central Andes' initial uplift (Horton 2018a). Of all the morphostructural units in which the Southern Central Andes are segmented (see Folguera et al. 2016), the best exposures of these units are found in the Principal Cordillera, where the orogenic front has cannibalized and incorporated the Late Cretaceous foreland basin into the Andean wedge before reaching its current position (Fig. 1).

Together with the Huincul ridge (Silvestro and Zubiri 2008), the area comprised between 35° S and 38° S represents one of the best-known areas in terms of structural kinematic evolution along the entire Neuquén Basin (Fig. 1), in which an important Late Cretaceous component has been recognized (Orts et al. 2012; Mescua et al. 2014; Rojas Vera et al. 2015; Folguera et al. 2015; Sagripanti et al. 2016; Branellec et al. 2016; Sánchez et al. 2018). This area is characterized by the occurrence of a series of long wavelength double-plunging anticlines (Fig. 2), whose deepest level of exposure reaches the upper Palaeozoic to Lower Triassic structural basement (Llambías et al. 2007). Nonmarine deposits of the Neuquén Group are found surrounding these thick-skinned structures, although their real regional extent remains obscured by subsequent exhumation processes and the thick Cenozoic volcanic and nonmarine synorogenic cover (Silvestro et al. 2005; Silvestro and Atencio 2009; Kay et al. 2006; Horton et al. 2016). Nevertheless, by integrating field observations and isopach maps based on subsurface information (Bettini et al. 1978; Manacorda et al. 2002), a wedge-shaped geometry for the Late Cretaceous sequences becomes evident, with maximum thicknesses recorded to the west and onlapping relationships over the San Rafael Block and the North Patagonian massif to the east (Figs. 1 and 2). In order to make a complete description of the Late Cretaceous foreland basin system, the main structures in the Argentine territory between 35°–38° S were divided in three sectors: western, central and eastern sectors (Fig. 2).

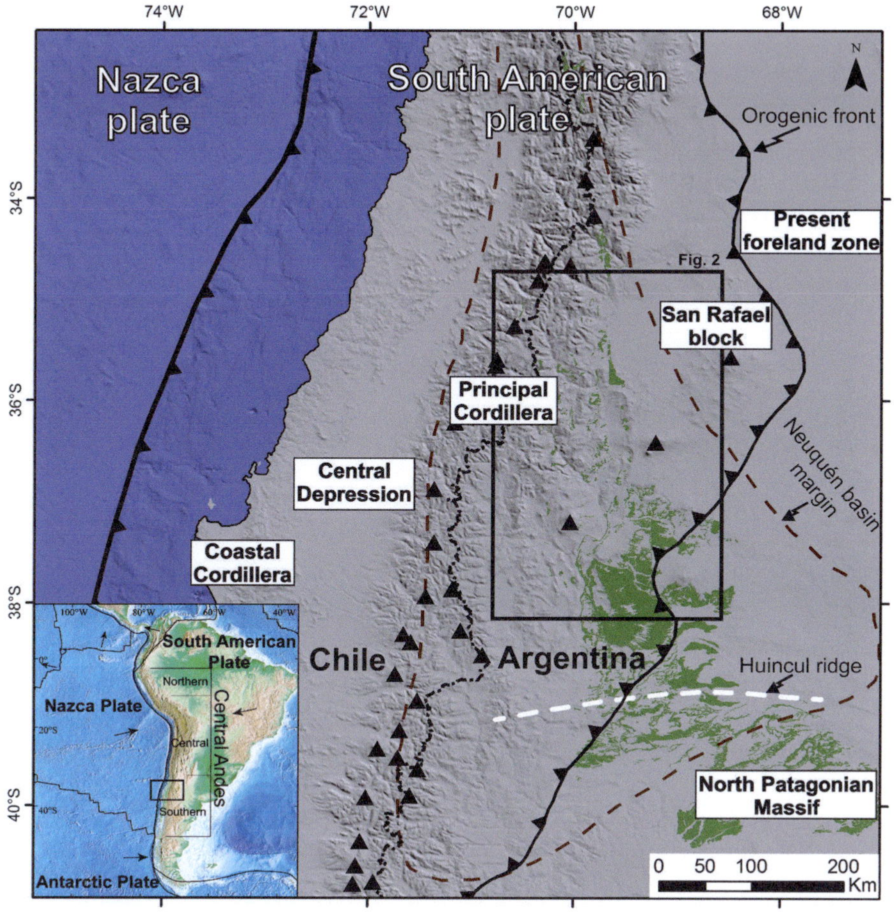

Fig. 1 Morphostructural units in which the Southern Central Andes are segmented and location of the study area within the Neuquén Basin. Outcrops of the Late Cretaceous Neuquén Group nonmarine foreland basin deposits are highlighted in green, and the location of the Neuquén Basin margin, the Huincul ridge and the current orogenic front are taken from Sagripanti et al. (2011)

2 Western Sector

Within the western sector, the northernmost analysed structure corresponds to the Río del Cobre anticline, whose backlimb exposes a ca. 100 Ma hiatus between Cretaceous units and late Eocene volcaniclastic rocks, indicating as much as 5 km of erosion (Mescua et al. 2013). Southeast of this structure, the Dedos Silla anticline presents partly eroded and condensed sections of the Late Cretaceous red beds (Legarreta and Kozlowski 1984; Mescua et al. 2013), evidencing active uplift during or after sedimentation (Branellec et al. 2016). Due to its NE trend, the southern plunge of this structure reaches the international boundary at approximately 35° 40′ S, where Lower

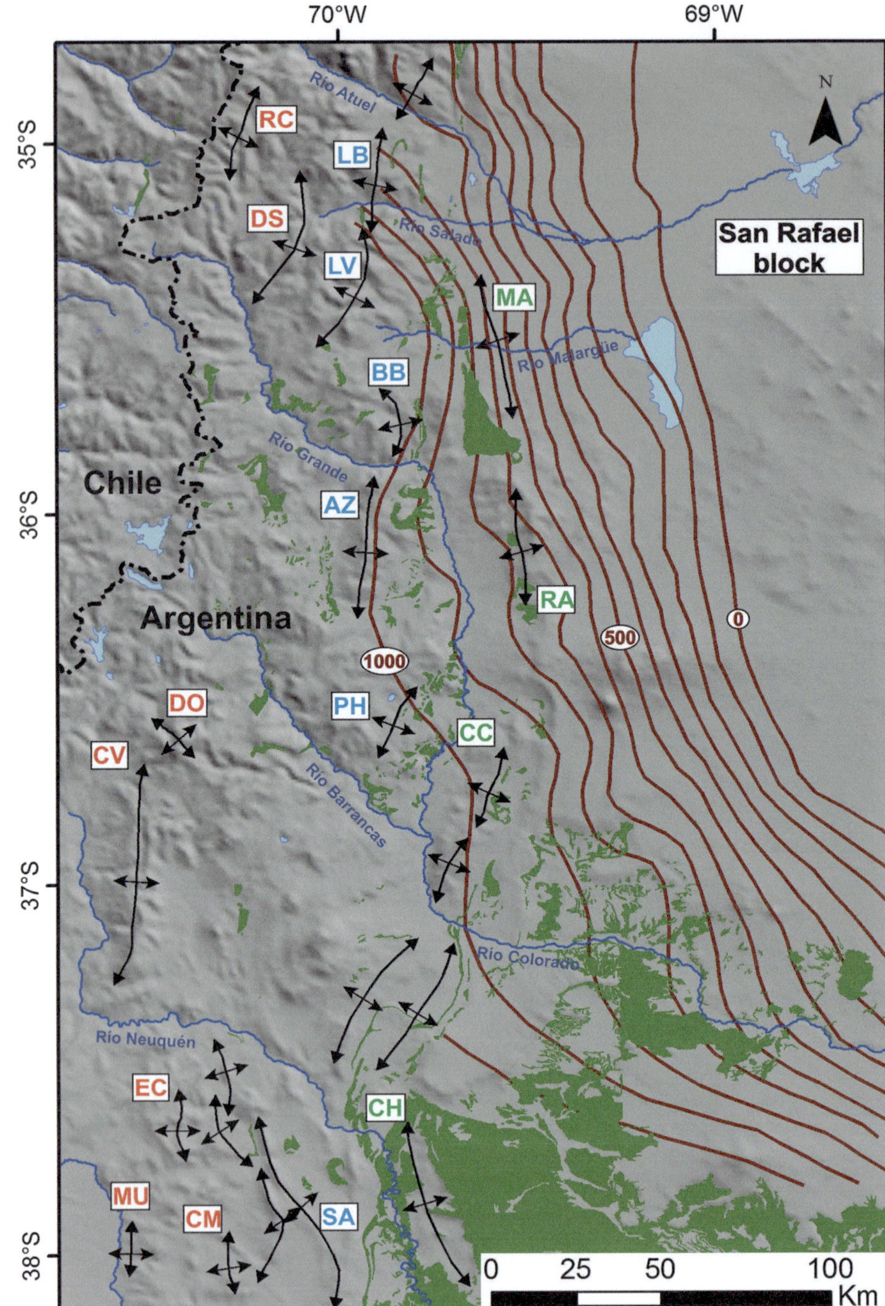
70°W
69°W
35°S
36°S
37°S
38°S
N
Río Atuel
Río Salado
Río Malargüe
Río Grande
Río Barrancas
Río Colorado
Río Neuquén
San Rafael block
Chile
Argentina
RC
DS
LB
LV
MA
BB
AZ
RA
DO
CV
PH
CC
EC
CH
MU
CM
SA
1000
500
0
0 25 50 100
Km

◂**Fig. 2** Map showing the main structures and the wedge-shaped regional geometry of the Late Cretaceous foreland basin, as evidenced by the isopachs obtained from outcrops (in green) and subsurface information of the Neuquén Group (taken from Manacorda et al. 2002). Abbreviations in red correspond to western anticlines: RC, Río del Cobre; DS, Dedos Silla; DO, Domuyo, CV, Cordillera del Viento; EC, El Cholar; MU, Mulichinco; CM, Cerro Mocho; in blue correspond to the central anticlines: LB, Los Blancos; LV, La Valenciana; BB, Bardas Blancas; AZ, Sierra Azul; PH, Puntilla de Huincán; SA, Salado; and in green correspond to the eastern anticlines: MA, Malargüe; RA, Ranquil Co; CC, Cara Cura; CH, Chihuidos

Cretaceous marine deposits of the Mendoza Group are unconformably covered by the latest Cretaceous volcaniclastic deposits of the Los Ángeles unit (Fennell et al. 2017c; Iannelli et al. 2018), suggesting an important non-depositional or erosive event during Late Cretaceous times (Fig. 3a).

South of the Río del Cobre and Dedos Silla anticlines, an important Late Cretaceous depocentre, located between the Ríos Grande and Barrancas withholds ~1.6 km of preserved thicknesses of the Neuquén Group (Orts et al. 2012), which separates these structures from the Domuyo and Cordillera del Viento double-plunging anticlines (Fig. 2). The Cordillera del Viento anticline represents one of the best examples of Late Cretaceous structures along the entire Neuquén Basin, due to the fact that the Mendoza Group deposits are unconformably overlain by the Neuquén Group deposits in its eastern limb (Groeber 1946), while preserving an even longer hiatus in its western limb (Fig. 3b). This hiatus can be observed at 37° S along the Neuquén River, where the structural basement of the Neuquén Basin represented by the Upper Permian to Lower Triassic Choiyoi Group (Llambías et al. 2007) is unconformably overlain by the Eocene volcaniclastic rocks of the Cayanta Formation (Fig. 3b) (Jordan et al. 2001).

This structural trend is completed by the El Cholar, Mulichinco and Cerro Mocho structures (Fig. 2), whose crosscutting and unconformable relationships with latest Cretaceous to Palaeocene magmatic rocks allow their characterization as Late Cretaceous uplifted structures (Ramos 1981; Zamora Valcarce et al. 2009; Rojas Vera et al. 2015).

3 Central Sector

In the central sector, structures such as the Los Blancos and La Valenciana anticlines (Fig. 2) present numerous remnants of the Neuquén Group deposits, which due to their position respect to the western Río del Cobre and Dedos Silla structures, suggest that these could have been associated with synorogenic sedimentation during Late Cretaceous times (Mescua et al. 2014; Branellec et al. 2016). Although no growth strata have been reported in the vicinities of these structures yet, the fact that latest Cretaceous marine and lacustrine deposits of the Malargüe Group (Barrio 1990;

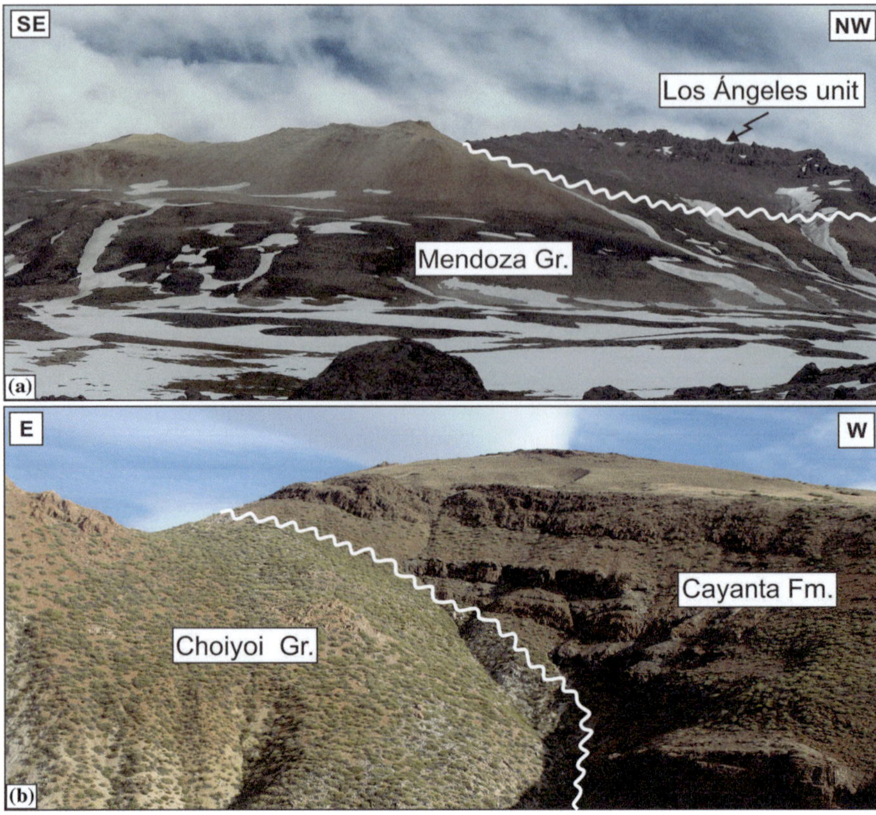

Fig. 3 **a** Unconformable relation between the Lower Cretaceous marine deposits of the Mendoza Group and the latest Cretaceous volcaniclastic deposits of the Los Ángeles unit near the international border between Chile and Argentina at 35° 40′ S. **b** A ca. 200 Ma hiatus is evidenced by the onlap of the Eocene volcaniclastic rocks of the Cayanta Formation over the Upper Permian to Lower Triassic crystalline rocks of the Choiyoi Group along the Río Neuquén at 37° S (Heredia, pers. comm.)

Parras and Griffin 2013) onlap against these structures indicate that these could have been positive features during Late Cretaceous times (Bettini et al. 1978).

A similar setting is observed in the Bardas Blancas anticline to the south, although the presence of growth strata associated with this structural trend documented in the Sierra Azul anticline suggests its Late Cretaceous growth (Fig. 2). In the eastern limb of the Sierra Azul anticline, Fennell et al. (2017a) identified progressive unconformities and thickness variations in the upper terms of the Neuquén Group (Fig. 4a), which are associated with seismetes preserved in associated fine-grained facies (Fig. 4b). The coarser facies of the Neuquén Group in this area are composed of limestones at the base, which are gradually replaced by acid volcanic clasts (Borghi et al. 2017) and intraclasts of the Neuquén Group towards the top (Fig. 4c). The combination of these observations, as well as the documentation of onlap relations in its western limb by Orts et al. (2012), converts the Sierra Azul anticline in one of the type examples of

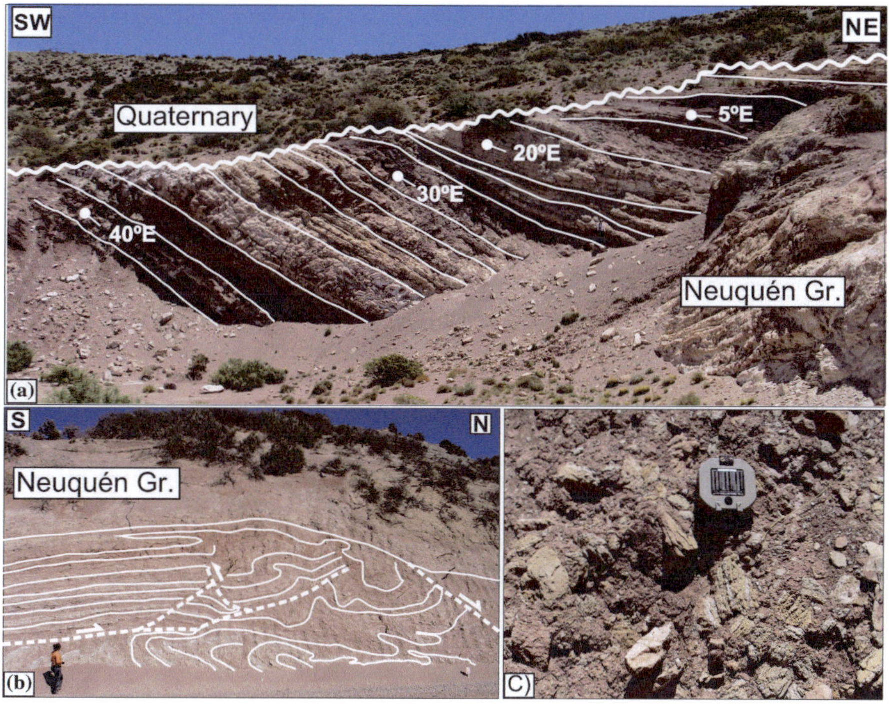

Fig. 4 **a** Growth strata in the upper terms of the Neuquén Group in the eastern limb of the Sierra Azul anticline. **b** Seismetes preserved in fine-grained facies of the Neuquén Group associated with the growth of the Sierra Azul anticline. **c** The coarser facies in the upper terms of the Neuquén Group are mainly composed of intraclasts, evidencing the progressive cannibalization of its deposits during the uplift of the Sierra Azul anticline

growing structures below the level of sedimentation during Late Cretaceous times. Although less documented given the extensive Neogene volcanic cover, the Puntilla de Huincán anticline to the south (Fig. 2) represents another example of this type of structures, where seismic profiles show thinning of the Late Cretaceous sequences towards its hinge (Galarza et al. 2009).

South of the Río Barrancas, Late Cretaceous depositional geometries and structures are covered by the voluminous Quaternary volcanic cover of the Tromen volcano (Sagripanti et al. 2016), although structures such as the Salado anticline share the behaviour with the rest of the structures within the central sector (Fig. 2). This structure in particular has been analysed by Cobbold and Rossello (2003) through a seismic profile and well data, where they show active thrusting throughout almost all of the Late Cretaceous, evidenced by the presence of a Late Cretaceous footwall basin east of the Salado anticline whose thickness exceeds by 1.8 km the thickness recorded in the hanging wall.

4 Eastern Sector

The eastern sector is represented by isolated structures whose uplift has been attributed mainly to the Cenozoic, and in particular to the last 20 Myrs (Silvestro et al. 2005; Zamora Valcarce et al. 2009; Silvestro and Atencio 2009; Sagripanti et al. 2011; Álvarez Cerimedo et al. 2013). However, evidences of uplift have been reported in many of these structures, whose spatial development coincides with major displacements in the isopachs of the Neuquén Group (Fig. 2).

The Ranquil Co anticline is the northernmost representative of these structures, where Groeber (1947) had already hinted the increasing thickness of the Neuquén Group towards both limbs of the structure (see Fennell et al. 2017b for a view of the scheme). This structure was recently inspected by Fennell et al. (2017a), confirming the thickness variations in the Neuquén Group (Fig. 5a) and detecting the presence of progressive unconformities in its core (Fig. 5b). These evidences suggest that the lower Neuquén Group deposits in this area present growth geometries, whose coarse facies are mainly composed of acid volcanic clasts (Fig. 5c), supported by sandstone petrographic descriptions (Fennell et al. 2017b).

The Cara Cura anticline to the south (Fig. 2) also presents evidences of active uplift during Late Cretaceous times, where syndepositional deformation has been reported in the coarse facies of the basal terms of the Neuquén Group (Fennell et al. 2017a). Upon closer inspection, the basal contact of the Neuquén Group in the western limb

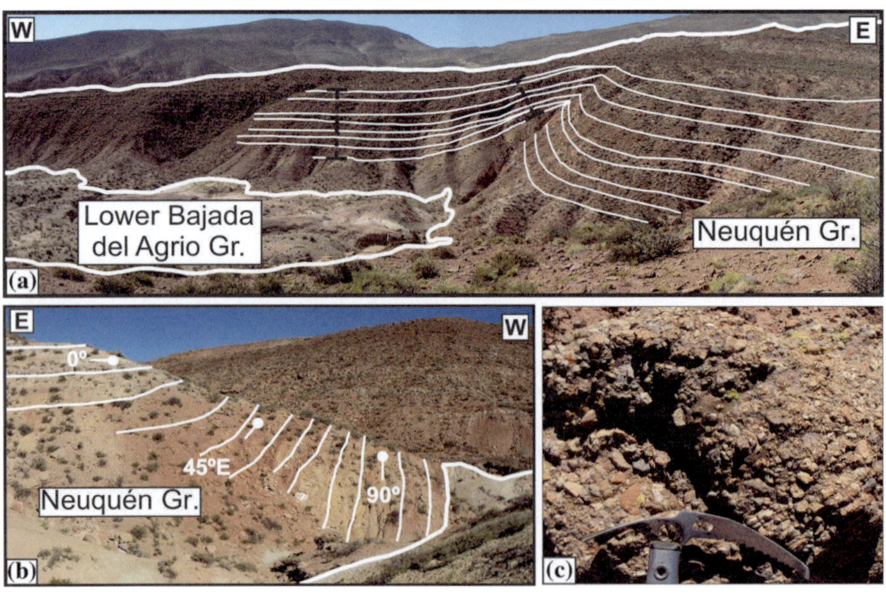

Fig. 5 **a** Thinning of the Neuquén Group deposits towards the hinge of the Ranquil Co anticline. **b** Progressive unconformities in the core of the Ranquil Co anticline. **c** Acid volcanic clasts dominate the coarse facies of the lower terms of the Neuquén Group in the Ranquil Co anticline

of the Cara Cura anticline is erosive and angular (Fig. 6a), highlighting the absence of the upper terms of the Bajada del Agrio Group and becoming the first description in this area of this highly documented angular unconformity in other sectors of the Neuquén Basin (Groeber 1946; Ramos and Folguera 2005). Internal angular and erosive unconformities are common in the lower Neuquén Group deposits west of the Cara Cura anticline (Fig. 6b), whose clastic composition is dominated by limestone

Fig. 6 **a** View of the angular unconformity between the lower terms of the Bajada del Agrio Group and the red beds of the Neuquén Group west of the Cara Cura anticline. **b** Internal erosive and angular unconformities in the lower terms of the Neuquén Group adjacent to the Cara Cura anticline. **c** The coarse facies of the lower Neuquén Group in the vicinities of the Cara Cura anticline are mainly composed by limestone clasts

pebbles (Fig. 6c), in accordance with sandstone petrographic analyses (Fennell et al. 2017b).

Moving southwards, the Chihuido anticline (Fig. 2) appears as another important Late Cretaceous structure in the eastern sector, where the angular unconformity between the deposits of the Neuquén and Bajada del Agrio Groups was first described (Groeber 1946; Ramos and Folguera 2005). In this area, deformational and liquefaction structures are interpreted as seismetes, whose nature and distribution suggest they were triggered by earthquakes of >6 magnitude in the Richter scale (Sánchez et al. 2013).

5 Late Cretaceous Orogenic Erosion and Structural Domains

The structures grouped into the western, central and eastern sectors present defined structural characteristics during the deposition of the Neuquén Group, which allow their separation into different structural domains. Structures in the western sector present evidences of non-deposition or exhumation due to active erosion during Late Cretaceous times, as highlighted by the presence of a long-lasting hiatus and important angular unconformities, which are among the most direct consequences of tectonic uplift. Orogenic erosion can be analysed through thermochronology, which is based on the production of an isotope or radiation damage in minerals such as apatites and zircons, reflecting the path taken by a particular rock in its way to the surface (Reiners and Brandon 2006). One of the most popular thermochronologic methods are the apatite fission track analyses (AFTA), which have been performed in some of the structures located in the western sector, revealing cooling ages between ~70 and 40 Ma (Fig. 7) (Folguera et al. 2015; Rojas Vera et al. 2015). Moreover, an Ar/Ar cooling age obtained from a pluton intruding the Cordillera del Viento anticline has been interpreted as the product of a 6 km tectonic uplift (Fig. 7) (Kay et al. 2006), in accordance with AFTA recently performed in this structure (Sánchez et al. 2018) (see Chapter "Structural and thermochronological constraints on the exhumation of the Chos Malal fold and thrust belt ~37° S"). Since erosion usually follows rock and surface uplift in space and time, the cooling ages obtained from the AFTA and Ar/Ar dating in the western structures could be reflecting their erosion after their initial growth (Reiners and Brandon 2006).

The central sector groups structures where thickness variations and progressive unconformities have been described during the deposition of the Neuquén Group. These evidences suggest that structural uplift occurred coeval to sedimentation, which might have resulted in positive topographic features affecting the local drainage patterns during brief periods of time. In this case, the only structure targeted for AFTA was the Bardas Blancas anticline (Fig. 7), with cooling ages spanning a wide temporal range between 120 and 52 Ma (Folguera et al. 2015), which might indicate active erosion during and/or after deposition of the Neuquén Group.

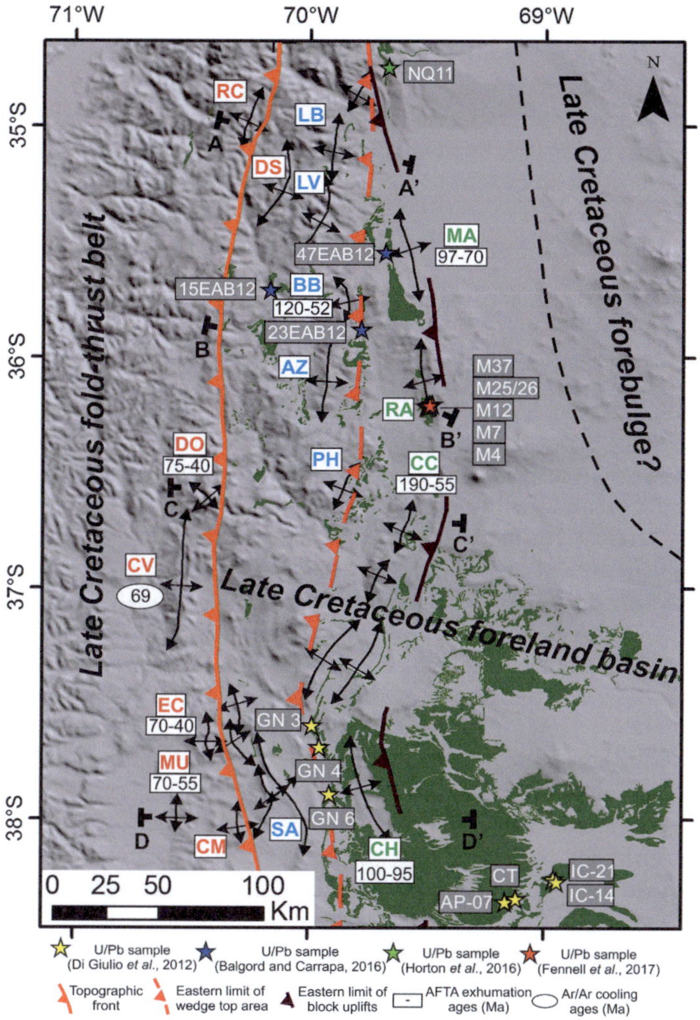

Fig. 7 The Late Cretaceous foreland basin system was characterized by a western sector of exhumed and actively eroded structures, which sourced sediments to the foreland basin, where structures were growing below the level of sedimentation after the eastern migration of the forebulge up to a position that would have coincided with the current location of the San Rafael Block. Abbreviations of the main structures are the same as in Fig. 2

On the other hand, structures located in the eastern sector show punctuated uplift events and limited amount of erosion, highlighted by mild angular unconformities between the Bajada del Agrio and Neuquén Group deposits. These structures are isolated, and seem to have been only active during the deposition of the lower terms of the Neuquén Group. These observations are supported by AFTA performed in some of these structures such as the Malargüe, Cara Cura and Chihuido anticlines

(Fig. 7), where cooling ages overlap with evidences of Late Cretaceous contractional deformation (Zamora Valcarce et al. 2009; Folguera et al. 2015). However, it should be noted that although the Malargüe anticline does not show any direct evidence of Late Cretaceous uplift, the fault responsible for its growth shows minor displacements both to the north in the Río Atuel area (Boll et al. 2014) and to the south in the La Batra area (Fennell et al. 2017a).

6 Late Cretaceous Detrital Zircon Provenance and Source Areas

Changes in detrital source areas are considered as an indirect evidence of change in the regional slope affecting the basins paleodrainage, which can be linked to rock exhumation in response to tectonic uplift. In order to study these provenance polarity reversals, the U–Pb dating of detrital zircons gained relevance during the last few years in the Neuquén Basin, since the pioneering study of Tunik et al. (2010). This study compared the U–Pb age patterns of detrital zircons of the Lower and Upper Cretaceous units, showing that eastern provenance characterized the zircons of the Bajada del Agrio Group, while western provenance dominated the zircons of the Neuquén Group. This study was later complemented by Di Giulio et al. (2012) at 38° S (Fig. 7), who show a second polarity reversal towards the top of the Neuquén Group, associated with the growing importance of the eastern source areas once the Late Cretaceous fold-thrust belt entered a mature phase. However, the replication of this type of analyses by Balgord and Carrapa (2016) at 36° S (Fig. 7) arrived to a different result, since a constant western provenance is recorded throughout the deposits of the Bajada del Agrio and Neuquén Groups. These apparent contradictory results were supported by AFTA performed by Di Giulio et al. (2017) in detrital apatites from the Neuquén Group, which reported Cretaceous cooling ages independent on their U–Pb crystallization ages (Fig. 8a). These observations led Fennell et al. (2017a) to propose that the detrital zircon provenance in the Neuquén Group also responds to the position of the sample within the basin, which they supported with a complete set of U–Pb age patterns of detrital zircons from the Ranquil Co anticline showing no time-dependant source variations (Fig. 7).

In order to understand these spatio-temporal variations, it is important to note that source areas of zircons with U–Pb ages younger than ~200 Ma can only be found west of the basin or be directly sourced from the coeval magmatic arc, sources of ~200–400 Ma zircons are ubiquitous along the basin, and zircons older than ~400 Ma are only sourced from the eastern cratonic area (Naipauer and Ramos 2016). Nevertheless, since the deposits of the Neuquén Group are also a consequence of the erosion of the Neuquén basin's sedimentary infill, zircon recycling is another important process that should be taken into account. In consequence, the Late Cretaceous foreland basin can be divided into different areas showing a distinct U–Pb age pattern of detrital zircons (Fig. 8).

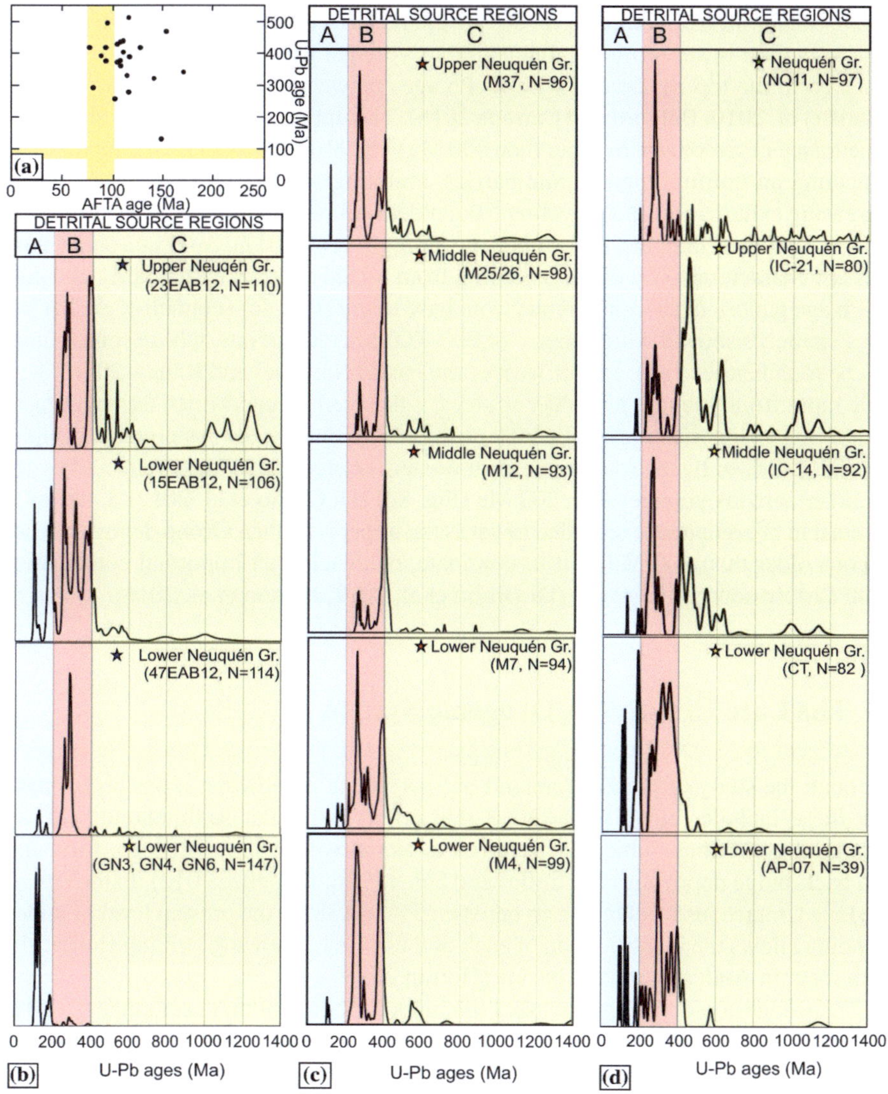

Fig. 8 **a** AFTA and U–Pb dating of detrital apatites show that the Neuquén Group detritus (depositional age highlighted in yellow) are the consequence of erosion during Cretaceous times, independently on their crystalization age (error bars not included, taken and modified from Di Giulio et al. 2017). **b** Samples in the westernmost sector of the basin were sourced from the west, and show an increasing level of exhumation throughout the section (taken and modified from Di Giulio et al. 2012 and Balgord and Carrapa 2016). **c** Samples located in a central position show a predominance of 200–400 Ma zircons, whose sources are ubiquitous along the basin (taken and modified from Fennell et al. 2017a). **d** U–Pb age patterns of detrital zircons from samples located east of growing structures show an initial western detrital derivation, which gradually changes towards an eastern derivation, indicating an increasing importance of the cratonic sources once the Late Cretaceous fold-thrust belt enters its mature phase (taken and modified from Di Giulio et al. 2012 and Horton et al. 2016). Location of all the U–Pb samples is given in Fig. 7

The westernmost samples (Fig. 8b) are characterized by a western detrital derivation, with their lower terms receiving zircons younger than 200 Ma, being gradually replaced to the top by zircons with U–Pb ages between 200 and 400 Ma (Fig. 8b) (Tunik et al. 2010; Balgord and Carrapa 2016). The upper Neuquén Group shows the appearance of zircon grains older than 400 Ma (Fig. 8b) (Balgord and Carrapa 2016), indicating an uplifted topographic barrier inhibiting western detrital derivation, or increasing exhumation and recycling from older sedimentary units, which agree with the description of their clastic components (Borghi et al. 2017). A second set of samples are characterized by detritus coming from mixed sources with a fairly constant U–Pb age pattern of detrital zircons from base to top (Fig. 8c) (Fennell et al. 2017a). These patterns show the dominance of 200–400 Ma old zircons, whose source areas can be found either to the east or west of the basin (Naipauer and Ramos 2016), with little input from younger and older sources, which can be interpreted as recycling of underlying sedimentary units (Fennell et al. 2017a). Finally, the easternmost samples are characterized by a strong input from western sources in its base, reflected by the abundant zircons younger than 200 Ma (Fig. 8d) (Di Giulio et al. 2012). Later on, a reversal in provenance is recorded towards the upper Neuquén Group deposits, when zircons older than 400 Ma start to dominate, evidencing an important contribution from eastern sources (Fig. 8d) (Di Giulio et al. 2012; Horton et al. 2016).

7 The Late Cretaceous Orogenic System

Through the study of present foreland basin systems worldwide, it has been shown that these include four discrete areas that reflect particular sedimentological and structural conditions, which are referred to as the wedge-top, foredeep, forebulge and back-bulge depozones (DeCelles and Giles 1996; DeCelles 2012). In the case of the Late Cretaceous foreland basin between 35° and 38° S, the present level of understanding allows the identification of each of these areas, showing striking similarities with modern Andean foreland basins (Horton 2018b).

The wedge-top depozone includes the mass of sediment that accumulates on top of the frontal part of the orogenic wedge, which is characterized by the presence of growth strata, tectonic unconformities and derivation of immature sediments from the fold-thrust belt (DeCelles and Giles 1996; DeCelles 2012; Capaldi et al. 2017). In the Late Cretaceous foreland basin, the structures described in the western sector correspond to the frontal part of the orogenic wedge, marking the topographic front during deposition of the Neuquén Group (Fig. 7). Structures in the central sector correspond to structures growing below the level of sedimentation in the wedge-top depozone accumulating textural and compositional immature detritus (Fig. 7), resulting in important thickness variations, progressive unconformities and evidences of syndepositional deformation.

The foredeep depozone consists of the cratonicward tapering wedge of sediment derived predominantly from the fold-thrust belt, with minor contributions from the stable craton (DeCelles and Giles 1996; DeCelles 2012; Capaldi et al. 2017).

Although this depozone is usually characterized as tectonically inactive, in areas with pre-existing heterogeneities inherited from former tectonic events, fault reactivation leading to basement block uplifts appears as an additional feature (e.g., Jordan and Allmendinger 1986; Mitra and Mount 1998; Ramos et al. 2002). Such structures should have been common in the Late Cretaceous foreland basin given the wide variety of pre-existing structures (Manceda and Figueroa 1995; Bechis et al. 2014), although their presence in the Late Cretaceous geological record has been underestimated. Structures present in the eastern sector could have behaved as disconnected basement block uplifts during the Late Cretaceous (Fig. 7), since they correspond to inverted structures with a punctuated uplift event during the initial stages of synorogenic sedimentation. The synchronicity in the uplift of these structures could be explained by a first stage of early fault inversion throughout the entire Neuquén Basin. This early basin inversion was later deactivated and followed by the development of the fold-thrust belt to the west, in agreement with results obtained from analogue modelling of tectonic inversion of former extensional sedimentary basins (e.g., Amilibia et al. 2005; Bonini et al. 2012).

The forebulge depozone corresponds to the region of potential flexural uplift along the cratonic side of the foredeep, usually characterized by condensed stratigraphic sections or low-angle unconformities that are used to track its position through time (Decelles and Giles 1996; DeCelles 2012). Balgord and Carrapa (2016) interpret the passage of the forebulge based on a 25–30 Myr unconformity between the lower terms of the Bajada del Agrio Group and the Neuquén Group, while foredeep and forebulge geometries have been interpreted in an area located east of the basement block uplift area towards the middle terms of the Neuquén Group (Sánchez and Asurmendi 2014). Sánchez and Asurmenti (2014) also detect periods of substantial detrital influx coming from the fold-thrust belt to the west, which is a common feature of distal foredeep and forebulge areas (DeCelles 2012). Although the final position of the forebulge is unknown, the isopach map of the Neuquén Group suggests that the position of the forebulge between 35° and 38° S would have coincided with the current position of the San Rafael Block (Figs. 2 and 7). In this context, the back-bulge deposits are represented by the Bajada del Agrio Group, which are found partially or completely eroded due to the passage of the forebulge before deposition of the Neuquén Group.

Since all the components of the Late Cretaceous foreland basin system have been identified, the Late Cretaceous orogenic system between 35° and 38° S can be schematized based on a series of previously published structural sections (Fig. 9) (Orts et al. 2012; Mescua et al. 2014; Folguera et al. 2015; Rojas Vera et al. 2015).

In the northernmost section, located at ~35° S (Fig. 7), the frontal part of the fold-thrust belt is represented by the Río del Cobre anticline, and the wedge-top area by the Dedos Silla anticline, where the clastic components indicate exhumation of Late Jurassic and Early Cretaceous units to the west (Fig. 9) (Mescua et al. 2013). The transition towards the foredeep in this area is not clearly defined, although clastic provenance in the eastern Neuquén Group outcrops indicates erosion of a peripheral bulge located further east (Mescua et al. 2013), where evidence of active thrusting involving the basement has been hinted through seismic sections by Boll et al. (2014).

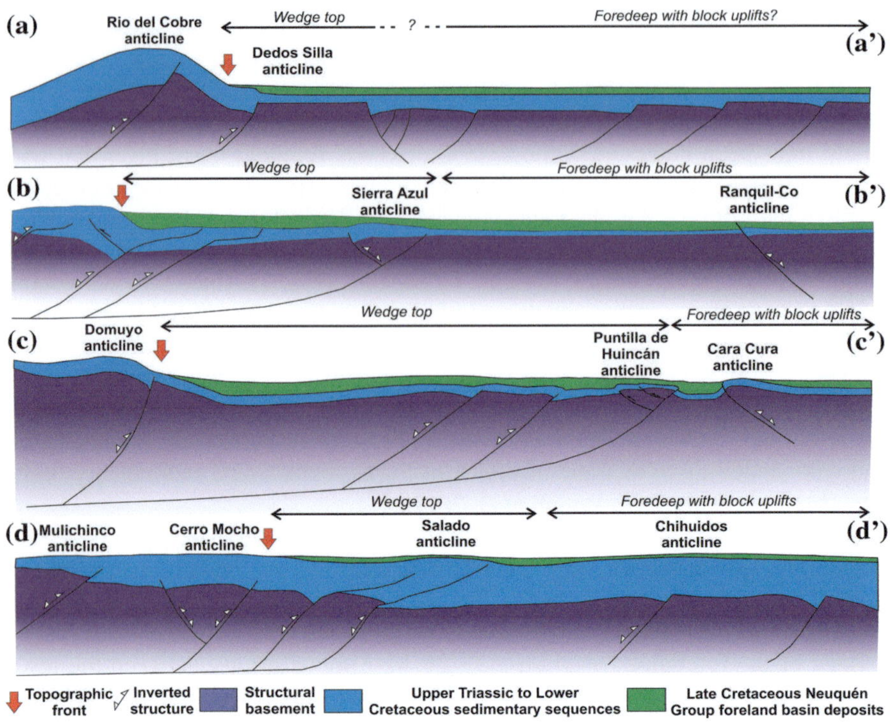

Fig. 9 Schematic representation of the Late Cretaceous orogenic system between 35° and 38° S through a series of published structural sections (A–A′ from Mescua et al. 2014; B–B′ from Orts et al. 2012; C–C′ from Folguera et al. 2015; D–D′ from Rojas Vera et al. 2015). Location of the structural cross sections is given in Fig. 7

The structural section performed around 36° S (Fig. 7) shows the Sierra Azul anticline as part of the wedge-top area (Fig. 9), receiving sediments from the fold-thrust belt to the west, as evidenced by the complete unroofing sequence detected in the outcrops of the Neuquén Group in this area (Borghi et al. 2017). The Ranquil Co anticline in the eastern sector of the section would represent a basement block uplift within the foredeep area, deforming strata from the Neuquén Group sourced from both the eastern cratonic area and the active fold-thrust belt to the west (Fig. 9).

At 37° S (Fig. 7), the exhumed and actively eroded frontal part of the Cretaceous orogenic wedge is represented by the Domuyo anticline, defining a wide wedge-top area including the Puntilla del Huincán anticline (Fig. 9). To the east, the Cara Cura anticline would have been disconnected from the main decollement and would have been active only during the initial stages of contractional deformation, being partially eroded and sourcing the limestone clasts to the western depocentre (Fig. 9).

Finally, the southern section at 38° S (Fig. 9) shows the exhumation of the Mulichinco and Cerro Mocho anticlines in the inner sector, the growth of the Salado anticline in the wedge-top area, and the Chihuidos anticline as a basement block uplift in the eastern foredeep area (Fig. 9).

A general analysis of the Late Cretaceous orogenic system indicates that inversion mechanisms were predominant along the uplifted structures, which can be separated into a group of connected structures growing in the fold-thrust belt and wedge-top areas, and a group of isolated structures acting as basement block uplifts in the foredeep depozone (Fig. 9). The effect of this contractional stage outlined the present structural framework between 35° and 38° S, being its impact in the geological record more intense towards the southern structures, in agreement with a regional trend of increasing deformation during Late Cretaceous times towards higher latitudes along the Andes (see Gianni et al. 2018).

Acknowledgements We acknowledge Nemesio Heredia and Bruno Colavitto for some of the photographs presented in this manuscript, which also greatly benefited from lively discussions with Victor Ramos, Facundo Fuentes, René Manceda, Ernesto Cristallini, Guido Gianni, Andrés Echaurren, Felipe Tapia, Emilio Rojas Vera and Mark Brandon through the years. This is the R-316 contribution of the Instituto de Estudios Andinos "Don Pablo Groeber".

References

Álvarez Cerimedo J, Orts DL et al (2013) Mecanismos y fases de construcción orogénicos del frente oriental andino (36° S, Argentina). Andean Geol 40(3):504–520

Amilibia A, McClay KR, Sábat F et al (2005) Analogue modelling of inverted oblique rift systems. Geol Acta 3(3):251–271

Balgord EA, Carrapa B (2016) Basin evolution of upper Cretaceous-lower cenozoic strata in the Malargüe fold-and-thrust belt: northern Neuquén Basin, Argentina. Basin Res 28(2):183–206

Barrio CA (1990) Paleogeographic control of the upper Cretaceous tidal deposits, Neuquén Basin, Argentina. J S Am Earth Sci 3(1):31–49

Bechis F, Cristallini EO, Giambiagi LB et al (2014) Transtensional tectonics induced by oblique reactivations of previous lithospheric anisotropies during the Late Triassic to Early Jurassic rifting in the Neuquén basin: insights from analog models. J Geodynamics 79:1–17

Bettini FH, Pombo RA, Mombru CA et al (1978) Consideraciones sobre el diastrofismo andino en la vertiente oriental de la cordillera principal entre los 34°30′ y los 37° de latitud sur. In: Abstracts of the 7 Congreso Geológico Argentino, Neuquén, 5–9 Nov 1978

Boll A, Alonso J, Fuentes F et al (2014) Factores controlantes de las acumulaciones de hidrocarburos en el sector norte de la cuenca neuquina, entre los ríos Diamante y Salado, provincia de Mendoza, Argentina. In: Abstracts of the 9 Congreso de Exploración y Desarrollo de Hidrocarburos, Mendoza, 3–4 Nov 2014

Bonini M, Sani F, Antonielli B (2012) Basin inversión and contractional reactivation of inherited normal faults: a review based on previous and new experimental models. Tectonophysics 522–523:55–88

Borghi P, Gómez Omil R, Fennell L et al (2017) Nuevas evidencias del levantamiento del sur de los Andes Centrales (36° S) durante la depositación del Grupo Neuquén. In Abstracts of the 20 Congreso Geológico Argentino, San Miguel de Tucumán, 7–11 Aug 2017

Branellec M, Niviere B, Callot JP et al (2016) Mechanisms of basin contraction and reactivation in the basement-involved Malargüe fold-and-thrust belt, Central Andes (34–36° S). Geol Mag 153(5/6):926–944

Capaldi TN, Horton BK, McKenzie NR et al (2017) Sediment provenance in contractional orogens: the detrital zircon record from modern rivers in the Andean fold-thrust belt and foreland basin of western Argentina. Earth Planet Sci Lett 479:83–97

Cobbold PR, Rosello EA (2003) Aptian to recent compressional deformation, foothills of the Neuquén Basin, Argentina. Marine Petrol Geol 20:429–443

DeCelles PG (2012) Foreland basin systems revisited: variations in response to tectonic settings. In: Busby C, Azor A (eds) Tectonics of sedimentary basins: recent advances. Wiley-Blackwell, Oxford, pp 405–426

DeCelles PG, Giles KA (1996) Foreland basin systems. Basin Res 8(2):105–123

Di Giulio A, Ronchi A, Sanfilippo A et al (2012) Detrital zircon provenance from the Neuquén Basin (south-central Andes): Cretaceous geodynamic evolution and sedimentary response in a retroarc-foreland Basin. Geology 40:559–562

Di Giulio A, Ronchi A, Sanfilippo A et al (2017) Cretaceous evolution of the Andean margin between 36° S and 40° S latitude through a multi-proxy provenance analysis of Neuquén Basin strata (Argentina). Basin Res 29(3):284–304

Fennell LM, Folguera A, Naipauer M et al (2017a) Cretaceous deformation of the southern Central Andes: synorogenic growth strata in the Neuquén Group (35° 30′–37° S). Basin Res 29(S1):51–72

Fennell LM, Naipauer M, Folguera A (2017b) El movimiento Intersenoniano de Pablo Groeber en el norte de Neuquén y sur de Mendoza: bases de la primera orogenia andina. Rev Asoc Geol Argentina 74(1):59–73

Fennell LM, Iannelli SB, Folguera A et al (2017) Interrupciones extensionales en el desarrollo de la faja plegada y corrida de Malargüe (36° S). In: Abstarcts of the 20 Congreso Geológico Argentino, San Miguel de Tucumán, 7–11 Aug 2017

Folguera A, Bottesi G, Duddy I et al (2015) Exhumation of the Neuquén Basin in the southern Central Andes (Malargüe fold and thrust belt) from field data and low-temperature thermochronology. J S Am Earth Sci 64(2):381–398

Folguera A, Naipauer M, Sagripanti L et al (2016) An introduction to the Southern Andes (33–50° S): Book structure. In: Folguera A, Naipauer M, Sagripanti L, Ghiglione M, Orts DL, Giambiagi L (eds) Growth of the Southern Andes. Springer, pp 1–7

Galarza BJ, Zamora Valcarce G, Folguera A et al (2009) Geología y Evolución tectónica del Frente Cordillerano a los 36° 30′ S: bloques de Yihuin-Huaca y Puntilla de Huincán. Mendoza. Rev Asoc Geol Argent 65(1):170–191

Gianni GM, Dávila FM, Echaurren A et al (2018) A geodynamic model linking Cretaceous orogeny, arc migration, foreland dynamic subsidence and marine ingression in southern South America. Earth-Sci Reviews 185:437–462

Groeber P (1946) Observaciones Geológicas a lo largo del meridiano 70: 1. Hoja Chos Malal. Rev Asoc Geol Argent 1(3):117–208

Groeber P (1947) Observaciones geológicas a lo largo del meridiano 70: 3, Hojas Domuyo, Mari Mahuida, Huarhuar-co y parte de Epu Lauken, 4, Hojas Bardas Blancas y Los Molles. Rev Asoc Geol Argent 2(4):347–433

Horton BK (2018a) Tectonic regimes of the central and southern Andes: responses to variations in plate coupling during subduction. Tectonics 37:402. https://doi.org/10.1002/2017TC004624

Horton BK (2018b) Sedimentary record of Andean mountain building. Earth-Sci Rev 178:279–309

Horton BK, Fuentes F, Boll A et al (2016) Andean stratigraphic record of the transition from backarc extension to orogenic shortening: a case study from the northern Neuquén Basin, Argentina. J S Am Earth Sci 71:17–40

Iannelli SB, Fennell LM, Litvak VD et al (2018) Geochemical and tectonic evolution of Late Cretaceous to early Paleocene magmatism along the Southern Central Andes (35–36° S). J S Am Earth Sci. https://doi.org/10.1016/j.jsames.2017.12.008

Jordan TE, Allmendinger RW (1986) The Sierras Pampeanas of Argentina: a modern analogue of rocky mountain foreland deformation. Am J Sci 286:737–764

Jordan TE, Burns WM, Veiga R et al (2001) Extension and basin formation in the southern Andes caused by increased convergence rate: a mid-Cenozoic trigger for the Andes. Tectonics 20(3):308–324

Kay SM, Burns M, Copeland P (2006) Upper Cretaceous to Holocene magmatism and evidence for transient Miocene shallowing of the Andean subduction zone under the northern Neuquén Basin. In: Kay SM, Ramos VA (eds) Evolution of an Andean margin: a tectonic and magmatic view from the Andes to the Neuquén Basin (35–39° S). Geol Soc Am, SP 407, pp 19–60

Legarreta L, Kozlowski E (1984) Secciones condensadas del Jurásico-Cretácico de los Andes del sur de Mendoza: estratigrafía y significado tectosedimentario. In: abstracts of the 9 Congreso Geológico Argentino, San Carlos de Bariloche, 5–89 Nov 1984

Llambías EJ, Leanza HA, Carbone O (2007) Evolución tectono-magmática durante el Pérmico al Jurásico Temprano en la Cordillera del Viento (37° 05′ S–37° 15′ S): nuevas evidencias geológicas y geoquímicas del inicio de la cuenca Neuquina. Rev Asoc Geol Argentina 62(2):217–235

Manacorda L, Reinante SME, Cazau L, Penna E (2002) Los reservorios del Grupo Neuquén. Abstracts of the V Congreso de Exploración y Desarrollo de Hidrocarburos, Mar del Plata, 29 Oct–2 Nov 2002

Manceda R, Figueroa D (1995) Inversion of the Mesozoic Neuquén rift in the Malargüe fold and thrust belt, Mendoza, Argentina. In: Tankard AJ, Suárez R, Welsink HJ (eds) Petroleum basins of South America. AAPG Memoir, vol 62, pp 369–382

Mescua JF, Giambiagi LB, Ramos VA (2013) Late Cretaceous uplift in the Malargüe fold-and-thrust belt (35° S), southern Central Andes of Argentina and Chile. Andean Geol 40(1):102–116

Mescua JF, Giambiagi LB, Tassara A et al (2014) Influence of pre-Andean history over Cenozoic foreland deformation: structural styles in the Malargüe fold-and-thrust belt at 35° S, Andes of Argentina. Geosphere 10(3):585–609

Mitra S, Mount VS (1998) Foreland basement-involved structures. AAPG Bulletin 82(1):70–109

Naipauer M, Ramos VA (2016) Changes in source areas at Neuquén basin: mesozoic evolution and tectonic setting based on U–Pb ages on zircons. In: Folguera A, Naipauer M, Sagripanti L et al (eds) Growth of the Southern Andes. Springer, New York, pp 33–61

Orts DL, Folguera A, Giménez M et al (2012) Variable structural controls through time in the Southern Central Andes (~36° S). Andean Geol 39(2):220–241

Parras A, Griffin M (2013) Late Cretaceous (Campanian/Maastrichtian) freshwater to restricted marine mollusk fauna from the Loncoche Formation, Neuquén Basin, west-central Argentina. Cretaceous Res 40:190–206

Ramos VA (1981) Descripción geológica de la Hoja 33 c Los Chihuidos Norte, provincia del Neuquén. Bol Serv Geol Nac 182:1–103

Ramos VA, Folguera A (2005) Tectonic evolution of the Andes of Neuquén: constraints derived from the magmatic arc and foreland deformation. In: Veiga GD, Spalletti LA, Howell JA et al (eds) The Neuquén Basin, Argentina: a case study in sequence stratigraphy and basin dynamics. Geol Soc, London, SP, vol 252, pp 15–35

Ramos VA, Cristallini EO, Pérez DJ (2002) The Pampean flat-slab of the Central Andes. J S Am Earth Sci 15:59–78

Reiners PW, Brandon MT (2006) Using thermochronology to understand orogenic erosion. Annu Rev Earth Planet Sci 34:419–466

Rojas Vera EA, Mescua J, Folguera A et al (2015) Evolution of the Chos Malal and Agrio fold and thrust belts, Andes of Neuquén: insights from structural analysis and apatite fission track dating. J S Am Earth Sci 64:418–433

Sagripanti L, Bottesi G, Naipauer M et al (2011) U/Pb ages on detrital zircons in the southern central Andes Neogene foreland (36°–37° S): constraints on Andean exhumation. J S Am Earth Sci 32(4):555–566

Sagripanti L, Folguera A, Fennell L et al (2016) Progression of the deformation in the Southern Central Andes (37° S). In: Folguera A, Naipauer M, Sagripanti L, Ghiglione M, Orts DL, Giambiagi L (eds) Growth of the Southern Andes. Springer, New York, pp 115–132

Sánchez ML, Asurmendi E (2014) Modelo de depósito de la Formación Cerro Lisandro: lóbulos de desembocadura y deltas de tipo Gilbert. Cretácico superior, región central de cuenca Neuquina, Argentina. Rev Mex Cs Geol 31(2):141–162

Sánchez ML, Asurmendi E, Armas P (2013) Subgrupo Río Colorado (Grupo Neuquén): registros de paleosismicidad en la cuenca de antepaís andina, Cuenca Neuquina, Provincias de Neuquén y Río Negro. Rev Asoc Geol Argent 70(1):96–114

Sánchez NP, Coutand I, Turienzo M et al (2018) Tectonic evolution of the Chos Malal fodl.and.thrust belt (Neuquén Basin, Argentina) from (U-Th)/He and fission track thermochronometry. Tectonics 37:1907–1929. https://doi.org/10.1029/2018tc004981

Silvestro J, Zubiri M (2008) Convergencia Oblicua: Modelo estructural alternativo para la dorsal neuquina (39° S)—Neuquén. Rev Asoc Geol Argent 63(1):49–64

Silvestro J, Atencio M (2009) La cuenca Cenozoica del Río Grande y Palauco: Edad, evolución y control estructural, faja plegada de Malargüe (36° S). Rev Asoc Geol Argent 65(1):154–169

Silvestro J, Kraemer P, Achilli F et al (2005) Evolución de las cuencas sinorogéncias de la Cordillera Principal entre 35°–36° S, Malargüe. Rev Asoc Geol Argent 60(4):627–643

Tunik M, Folguera A, Naipauer M et al (2010) Early uplift and orogenic deformation in the Neuquén Basin: Constraints on the Andean uplift from U–Pb and Hf isotopic data of detrital zircons. Tectonophysics 489:258–273

Vergani GD, Tankard J, Belotti J et al (1995) Tectonic evolution and paleogeography of the Neuquén basin, Argentina. In: Tankard AJ, Suárez R, Welsink HJ (eds) Petroleum Basins of South America. AAPG Memoir, vol 62, pp 383–402

Zamora Valcarce G, Zapata T, Ramos VA et al (2009) Evolución tectónica del frente andino en Neuquén. Rev Asoc Geol Argent 65(1):192–203

Structural and Thermochronological Constraints on the Exhumation of the Chos Malal Fold and Thrust Belt (~37° S)

Natalia Sánchez, Martín Turienzo, Isabelle Coutand, Fernando Lebinson, Vanesa Araujo and Luis Dimieri

Abstract This chapter reviews the evolution of the Chos Malal fold and thrust belt based on structural and thermochronological data. This fold and thrust belt can be divided into an inner-western zone, characterized by exposed basement-involved structures that are inserted in the sedimentary cover and generated a wide region with thin-skinned deformation, and an outer-eastern zone, where blind thrusts, involving basement rocks, produce deformation in the cover restricted to the deformation front where the main hydrocarbon deposits of the region are located. Zircon (U–Th)/He and apatite fission-track cooling ages at the Cordillera del Viento, ranging from 72 to 51 Ma, evidence a period of uplift and exhumation in the inner zone during the Late Cretaceous–Paleocene. These data are in agreement with the notable unconformity and hiatus observed between Paleogene volcanics and Paleozoic basement rocks. The contraction in this zone continued during the Miocene giving rise to most of the thick and thin-skinned structures of the Chos Malal fold and thrust belt, as revealed by apatite fission-track ages between ~15 and 10 Ma. Furthermore, folded volcanic sequences with $^{40}Ar/^{39}Ar$ ages of ~15 Ma and intrusive rocks with U–Pb ages of 11.5 Ma evidence this compressive episode. The deformation advanced toward the foreland during the Late Miocene, where samples from the basement-involved Las Yeseras–Pampa Tril anticlines show apatite fission-tracks ages of ~9–7 Ma, which led to the uplift and exhumation of the outer zone of the Chos Malal fold and thrust belt.

Keywords Thermochronology · Andean exhumation · Chos Malal fold and thrust belt

N. Sánchez (✉) · M. Turienzo · F. Lebinson · V. Araujo · L. Dimieri
Departamento de Geología, Universidad Nacional del Sur. INGEOSUR-CONICET, Av. Alem 1253 - Cuerpo B´ 2º Piso, 8000 Bahía Blanca, Argentina
e-mail: natalia.sanchez@uns.edu.ar

I. Coutand
Dalhousie University, 1459 Oxford Street, PO BOX 15000, Halifax, Canada

D. Kietzmann and A. Folguera (eds.), *Opening and Closure of the Neuquén Basin in the Southern Andes*, Springer Earth System Sciences,
https://doi.org/10.1007/978-3-030-29680-3_13

1 Introduction

Constraining the age of the individual structures is one of the main tasks in order to understand the tectonic evolution of fold and thrust belts (FTB) and their associated hydrocarbon deposits. The Chos Malal FTB, located in the northwest of the Neuquén province (Fig. 1), is formed by tectonic structures of varied scale that involve both basement and cover rocks of the Neuquén Basin. The excellent outcrops and the abundant subsurface information have made possible to study in detail the structural style of this belt, characterized both in the inner and outer zones by a close link between the thick and thin-skinned structures (Kozlowski et al. 1996, 1998; Nocioni 1996; Zapata et al. 1999, Sánchez et al. 2014, 2015, 2018; Turienzo et al. 2014, 2018).

The tectonic evolution of the Neuquén Andes has been addressed from different geological studies, and in general, it has been recognized that the orogenic building took place through contractional stages during the Late Cretaceous, Paleogene and Miocene (Cobbold and Rossello 2003; Ramos and Folguera 2005; Folguera et al. 2007; Ramos et al. 2011; Rojas Vera et al. 2015; Sagripanti et al. 2016, Sánchez et al. 2018) (see Chap. The Late Cretaceous orogenic system: early inversion of the Neuquén Basin and associated synorogenic deposits, 35°–38° S).

The increasing availability of geo- and thermo-chronological data has allowed reaching important advances regarding the exhumation age of the main structures that form the Neuquén fold and thrust belt. The purpose of this chapter is to integrate the existing information about the age of igneous rocks (^{40}Ar/^{39}Ar, U–Pb; Fig. 1), cooling ages from detrital and igneous apatites and zircons (U–Th/He, fission tracks, Fig. 2), and structural data from balanced cross-sections, in order to analyze the evolution of the Chos Malal FTB. A kinematic reconstruction is proposed constraining the age of the structures that formed this FTB, which show a normal sequence of deformation toward the foreland from the Late Cretaceous to the Late Miocene.

2 Geological Framework

The Neuquén Basin has a long and heterogeneous geological history linked to the complex tectonic framework that affected the western edge of Gondwana from the Late Paleozoic to the present controlling its stratigraphy and structural evolution. In the north of the Neuquén province, the basement of the basin is composed of Carboniferous volcano-sedimentary successions (Fig. 3), which were deformed during the San Rafael orogeny (Zöllner and Amos 1973; Llambías et al. 2007; Danieli et al. 2011; Giacosa et al. 2014). Subsequently, these rocks were intruded by Permian granitoids and unconformably covered by Permian–Early Triassic volcanic rocks in a post-orogenic extensional setting (Llambías et al. 2007; Llambías and Sato 2011).

The opening of the Neuquén Basin began with depocenters of dominantly NW-SE orientation controlled by normal faults, during the Late Triassic–Early Jurassic

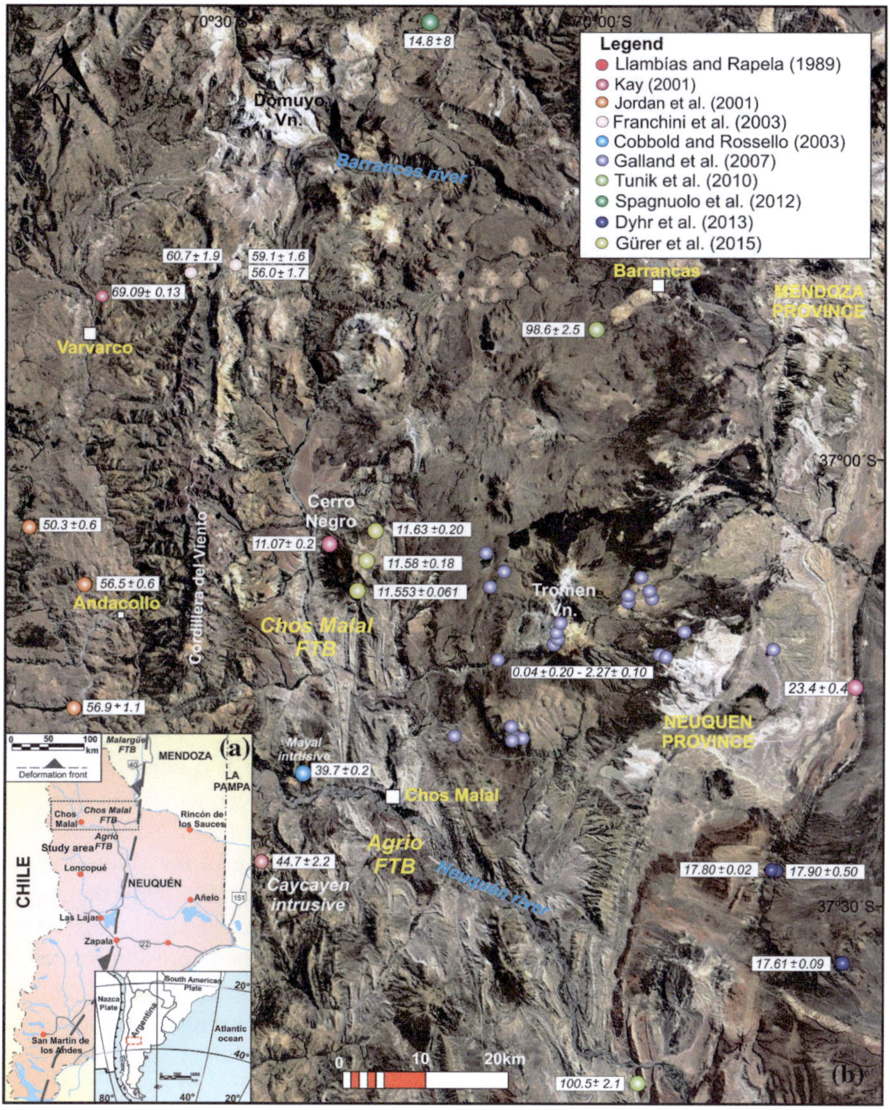

Fig. 1 **a** Location of the study area in the Chos Malal FTB where thermochronological data were obtained. **b** Published $^{40}Ar/^{39}Ar$ and U–Pb ages in the region

stage of rifting (Vergani et al. 1995; Franzese and Spalleti 2001; Carbone et al. 2011). Both basement rocks and syn-rift sequences are well exposed in the study area along the Cordillera del Viento (Fig. 4). From Early Jurassic to Early Cretaceous, the development of a steeply dipping, active subduction zone and the associated magmatic arc along the western margin of Gondwana led to back-arc subsidence (Vergani et al. 1995; Howell et al. 2005; Arregui et al. 2011). This period of thermal

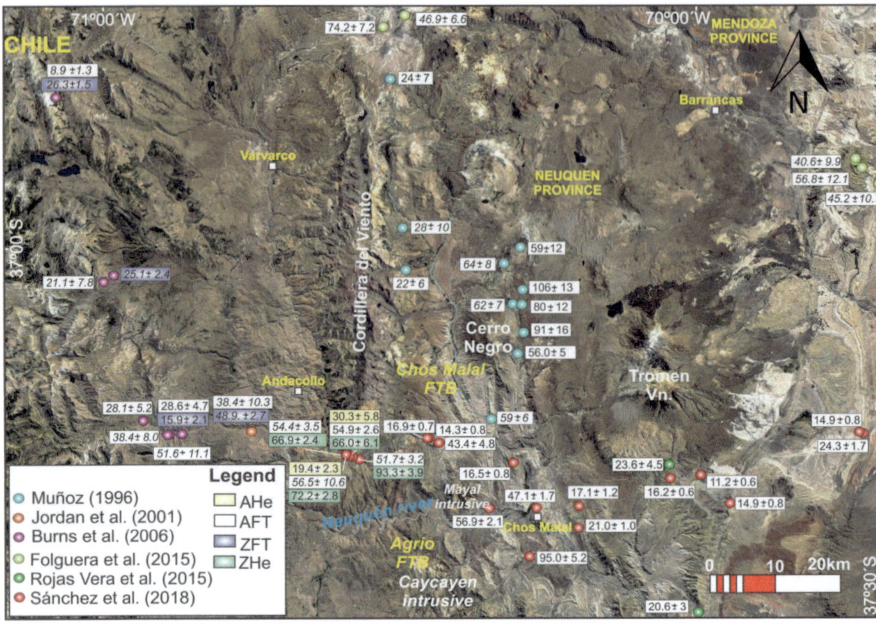

Fig. 2 Thermochronological data in the Chos Malal FTB. AHe apatite (U–Th)/He; AFT apatite fission tracks; ZFT zircon fission tracks and ZHe zircon (U–Th)/He

subsidence is accounted for more than 4,000 m of marine and continental rocks which constitute most of the petroleum systems of the Neuquén Basin (Vergani et al. 1995, 2011; Legarreta et al. 2005). Figure 3 summarizes the names and ages of the different geological units deposited during this stage in the studied sector of the basin. In the Aptian, the Neuquén Basin lost its connection with the Pacific Ocean and a continental depositional environment prevailed during the rest of the Cretaceous. The thick sequences of Late Cretaceous red beds are interpreted as accumulated in a foreland basin, in which the flexural subsidence would be related to the beginning of the Andean compression (Cobbold and Rosello 2003; Ramos and Folguera 2005; Tunik et al. 2010; Di Giulio et al. 2015; Fennell et al. 2016). Meso-Cenozoic magmatism in the Neuquén Basin was strongly linked to the evolution of the subducting plates along the Andean margin (Ramos and Folguera 2005; Kay et al. 2006; Folguera and Ramos 2011). In the north of the Neuquén province, this magmatic activity is represented by numerous volcanic and intrusive rocks (Fig. 1), whose age and relationship with the tectonic evolution of the Chos Malal FTB will be analyzed in Sect. 4 of this chapter.

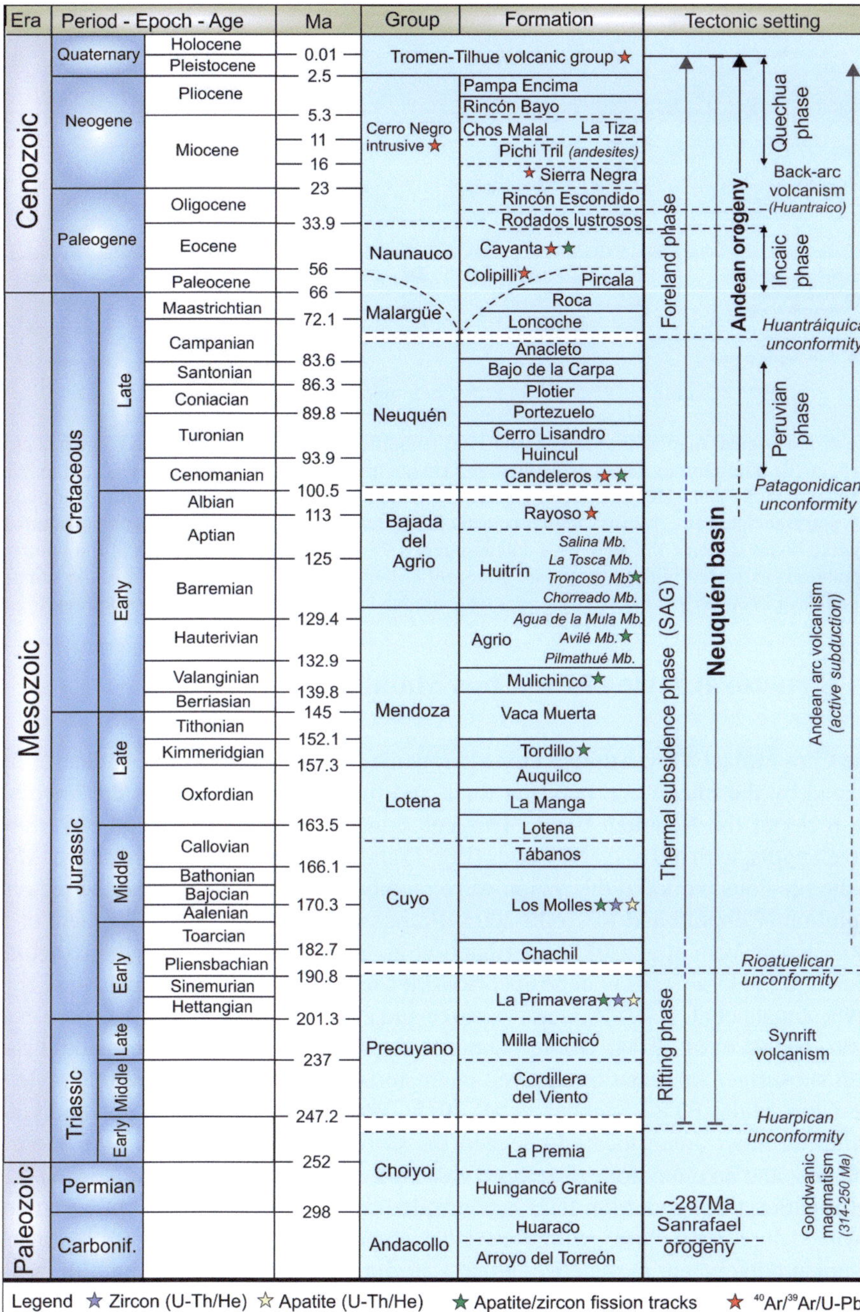

Fig. 3 Stratigraphic column summarizing the nomenclature, age and tectonic setting of the units exposed in the study area (based on Gulisano and Gutierrez Pleimling 1995; Leanza 2003, 2009; Leanza et al. 2005, 2013; Llambías et al. 2007; Tunik et al. 2010). The stars indicate the geological units with their geochronological and/or termochronological ages

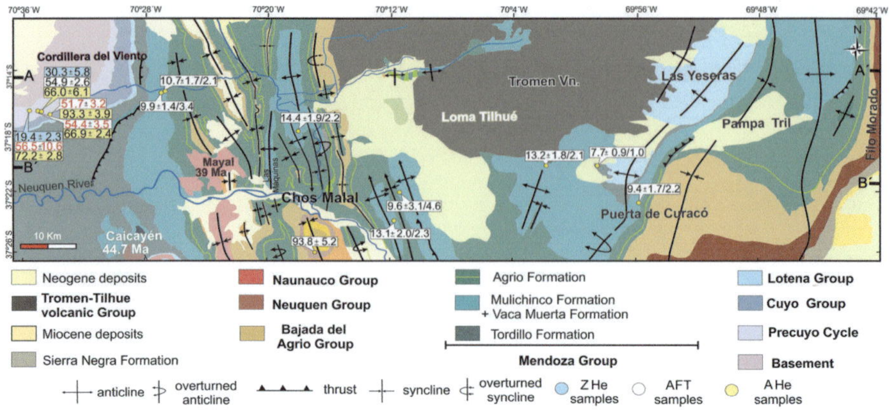

Fig. 4 Geological map of the Chos Malal FTB (modified from Sánchez 2015 and Sánchez et al. 2015, 2018). The samples dated using apatite AHe (apatite (U–Th)/He) ages are in light blue, AFT ages are in white and ZHe (zircon (U–Th)/He) ages are in yellow. Numbers in red are AFT pooled ages and associated 1σ error for samples passing the χ^2 test ($P(\chi^2) > 5\%$). The other black numbers in white boxes indicate the AFT ages and associated 95% confidence level errors of the youngest populations extracted with the software Binomfit (Brandon 1992, 1996, 2002) from the samples failing the $\chi 2$ test ($P(\chi^2) < 5\%$. These ages were used to reconstruct the kinematic evolution (Fig. 6)

3 Structural Style of the Chos Malal FTB

The Chos Malal FTB is a thick-skinned orogenic wedge product of the Andean uplift, formed by the interaction between thick and thin-skinned structures that involves the rocks of the Neuquén Basin. This belt contains fault-related folds of variable wavelengths, with NS to NNW trends (Fig. 4) and dominant eastward vergence. Most of the previous works in the region were mainly oriented to determine the tectonic evolution (Cobbold and Rossello 2003; Ramos and Folguera 2005; Folguera et al. 2007; Rojas Vera et al. 2015; Sagripanti et al. 2016) and to analyze the structures related to the hydrocarbon deposits (Ploszkiewicz and Viñes 1987; Kozlowski et al. 1998; Zapata et al. 2001; Zamora Valcarce and Zapata 2005; Zamora Valcarce et al. 2006; Fantín et al. 2014; Gómez Omil et al. 2014). The integration of field data with subsurface information allowed characterizing in detail the structural style of the Chos Malal FTB (Sánchez et al. 2015; Turienzo et al. 2018) and subdivides it into an inner zone, located between the Cordillera del Viento and the Tromen Volcano, and an outer zone developed eastward of this volcanic center (Fig. 5). The deformation throughout the FTB is controlled by large thrusts that form basement-involved duplex structures, with a lower detachment located at a depth of ~14 km and an upper detachment in the cover along Late Jurassic evaporites. Restoration of the structural cross-sections A and B indicates values of tectonic shortening of 24.7 km (18%) and 22.2 km (16.5%), respectively (Turienzo et al. 2018).

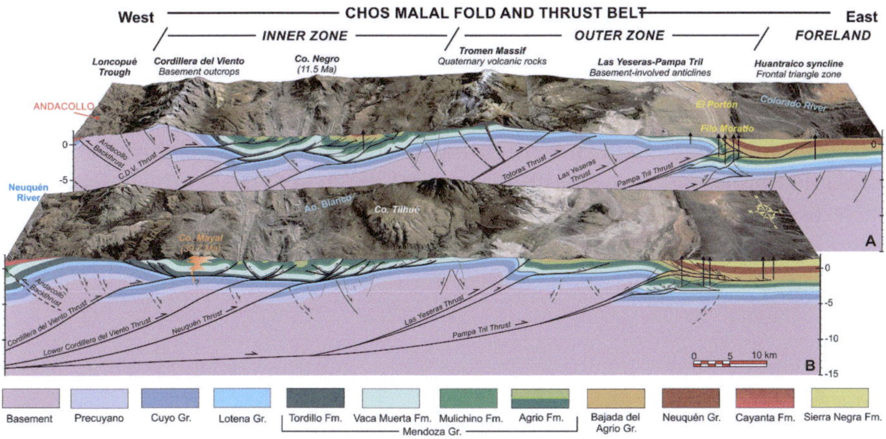

Fig. 5 Structural cross-sections across the Chos Malal FTB (modified from Turienzo et al. 2018). Reconstruction of the inner zone is based on Sánchez et al. (2015). See location on Fig. 4

3.1 *Inner Zone*

The main morphostructural unit is the Cordillera del Viento, a NS-trending, basement-involved anticline, which is approximately 15 km wide and 65 km long (Figs. 1 and 5). Paleozoic basement rocks crop out in the western slope of this range, forming the core of the anticline that was strongly exhumed and eroded. Paleogene volcanic rocks cover unconformably the folded Paleozoic and Mesozoic sequences, whose uplift would be enhanced through back thrusts that some authors interpreted as inverted normal faults (Zapata et al. 1999; Cobbold and Rossello 2003, Folguera et al. 2007). Syn-rift deposits exposed in the Cordillera del Viento are bounded by regional NW and minor NE normal faults (Sagripanti et al. 2014), most of them only partially inverted since they preserve their extensional character. A complete succession of Middle Triassic–Early Cretaceous strata, with dips of ~30°–40° E forms the eastern slope of the Cordillera del Viento (Figs. 4 and 5). This homoclinal sequence was interpreted as the forelimb of a first-order, basement-involved, fault-bend anticline, which moved onto the cover generating second-order folds (Kozlowski et al. 1998; Sánchez et al. 2015, 2018; Turienzo et al. 2018). Repetition of the Late Jurassic layers evidences a deep detachment within the stratigraphic sequence, which probably locates along the Oxfordian evaporites (Auquilco Formation) on top of the Lotena Group (Sánchez et al. 2015). To the east of the Cordillera del Viento, two other basement-involved wedges in depth were interpreted in seismic lines. These structures have the same detachment level and the displacement transmitted contributed to create additional thin-skinned structures in the inner zone of the Chos Malal FTB (Fig. 5). The deformation in the sedimentary cover is represented by numerous second-order folds, involving the Late Jurassic-Cretaceous sequences, which have N to NNW trends and wavelengths of ~3–5 km (Sánchez et al. 2015). Along the northern cross-section, all the second-order anticlines have vergence to the east, while along the southern cross-section there are folds with both east and west vergence (Fig. 5). All these thin-skinned anticlines were reconstructed as fold-propagation and fault-bend folds, or as a combination of both mechanisms. In the fault-bend anticlines, the thrusts propagate to an upper detachment on Early Cretaceous shales (Pilmathué Mb. of the Agrio Formation), thus transferring the deformation to the overlying sequences. Such displacement allows the creation of third-order structures, like the complex repetitions are seen on the Agrio Formation in the Chacay Melehue area (Turienzo et al. 2014, 2018).

3.2 Outer Zone

The structural style in the outer zone of the Chos Malal FTB is dominated by large east-vergent, basement-involved anticlines, with complex thin-skinned deformation restricted to the mountain front (Fig. 5). The geometry of the major structures is relatively simple, formed by two large doubly plunging anticlines (Las Yeseras and Pampa Tril) that have gently dipping back limbs and steeply dipping forelimbs. A wide frontal syncline, cored by Miocene volcanic rocks, connects the FTB with the undeformed foreland. Well, data throughout the mountain front show that basement rocks in the hangingwall of the Pampa Tril thrust are ~3–4 km higher than in the footwall (Fig. 5). Different structural models have been proposed in order to explain such basement uplift, including: (i) duplex structures connected with a shallow back thrust forming a triangle zone (Ploszkiewicz and Viñes 1987; Ramos and Barbieri 1989; Nocioni 1996); (ii) a basement-involved fault-propagation fold transported into the cover (Viñes 1989; Kozlowski et al. 1998; Allmendinger et al. 2004) and (iii) positive inversion of previous normal faults (Zamora Valcarce and Zapata 2005; Zamora Valcarce et al. 2006). Taking into account the previous studies, examining well and seismic data, and evaluating the structural behavior at the inner zone, Turienzo et al. (2018) analyzed the complex geometry at the mountain front of the Chos Malal FTB. The basement-involved Pampa Tril thrust was initially connected with the cover within the Jurassic evaporites to create thin-skinned structures, and then out-of-sequence faulting in the hangingwall led to the uplift of basement rocks to its present structural relief (Fig. 5). Displacement of this fault generated a passive roof back thrust within the Cretaceous evaporites of the Bajada del Agrio Group, creating a triangle zone as it was proposed in some of the previous studies in the area. The geometry of the thin-skinned structures below this triangle zone is not observed in the seismic lines but can be accurately reconstructed using well data. The Filo Morado anticline is a very tight fold, with steeply dipping limbs, which involves the Late Jurassic–Early Cretaceous successions (Zapata et al. 2001; Zamora Valcarce and Zapata 2005; Zamora Valcarce et al. 2006). To the north, along the same structural trend, a comparable anticline constitutes the El Portón oil field while to the south, a second-order anticline below the Pampa Tril basement thrust sheet (Fig. 5) could constitute an unexplored structural trap (Turienzo et al. 2018).

4 Kinematic Evolution

The structural development of the Chos Malal FTB is generally interpreted to be mostly occurred through contractional stages during the Late Cretaceous, Paleogene and Miocene (Kozlowski et al. 1998; Cobbold et al. 1999; Cobbold and Rossello 2003; Folguera et al. 2006, 2007, 2015; Ramos and Folguera 2005; Rojas Vera et al. 2015; Sagripanti et al. 2016, Sánchez et al. 2018). The spatial and temporal evolutions are of the structures during the Andean orogeny illustrated in Fig. 6, integrates the structural and thermochronological data studied by Sánchez et al. (2018) and the geochronological information existing throughout the region. Tectono-stratigraphic studies of the Late Cretaceous red beds in the Neuquén Basin suggest that they were deposited in response to the onset of Andean contraction (Cobbold and Rosello 2003; Ramos and Folguera 2005; Fennell et al. 2016), between ~100 and 90 Ma (Tunik et al. 2010; Di Giulio et al. 2015). It should be noted that these data come from sectors of the basin outside the study area and that at the moment in the Chos Malal FTB there is no evidence of structures of that age.

Kay et al. (2006) obtained an $^{40}Ar/^{39}Ar$ age in biotites of 69.09 ± 0.13 Ma for the Varvarco pluton, northwest of the Cordillera del Viento (Fig. 1b), which they interpret as a cooling age associated with the uplift of this basement-involved structure. In samples from Late Triassic–Early Jurassic rocks south of the Cordillera del Viento, Sánchez et al. (2018) obtained zircon (U–Th)/He cooling ages, whose effective closure temperature is 180 ± 20 °C (Reiners et al. 2004), ranging from 72.2 ± 2.8 to 66.0 ± 6.1 Ma. These data evidence an early stage of deformation in the inner zone of the Chos Malal FTB that could be associated with the uplift of the western flank of the Cordillera del Viento, through the inversion of preexisting eastward dipping normal faults and then by the growth of the Andacollo back thrusts (ABT, Fig. 6b).

In the same samples from the south of the Cordillera del Viento, Sánchez et al. (2018) found apatite fission-track (AFT) ages between 51.7 ± 3.2 and 56.5 ± 10.7 Ma (Figs. 4 and 7). In apatite crystals, tracks are annealed at temperatures above 150–120 °C, depending on the cooling rate and the composition of the crystal (e.g., Green et al. 1985; Ketcham et al. 1999; Reiners and Brandon 2006). Paleogene volcanic rocks, with an $^{40}Ar/^{39}Ar$ age on hornblende and fission tracks in zircons of 56 ± 1.1 to 48.9 ± 2.7 Ma (Jordan et al. 2001), lie in angular unconformity on the Paleozoic basement rocks which indicates a significant Pre-Eocene uplift. This episode of deformation is interpreted in association with the growth of the basement-involved Cordillera del Viento thrust (CdVT), which propagated eastwards on the Late Jurassic evaporites creating second-order structures in the cover (Fig. 6c). Subvolcanic rocks that form the Sierra del Mayal, dated with $^{40}Ar/^{39}Ar$ on whole rock and yielding an age of 39.7 ± 0.2 Ma (Cobbold and Rossello 2003), displace the axis of the syncline located to the east of the Cordillera del Viento, which evidences its post-folding intrusion (Fig. 4). Apatite fission-track analyses conducted in neighboring structures, in the Malargüe and Agrio FTBs north and south, respectively, show significant exhumation between 75 and 55 Ma (Folguera et al. 2015), and 70–40 Ma

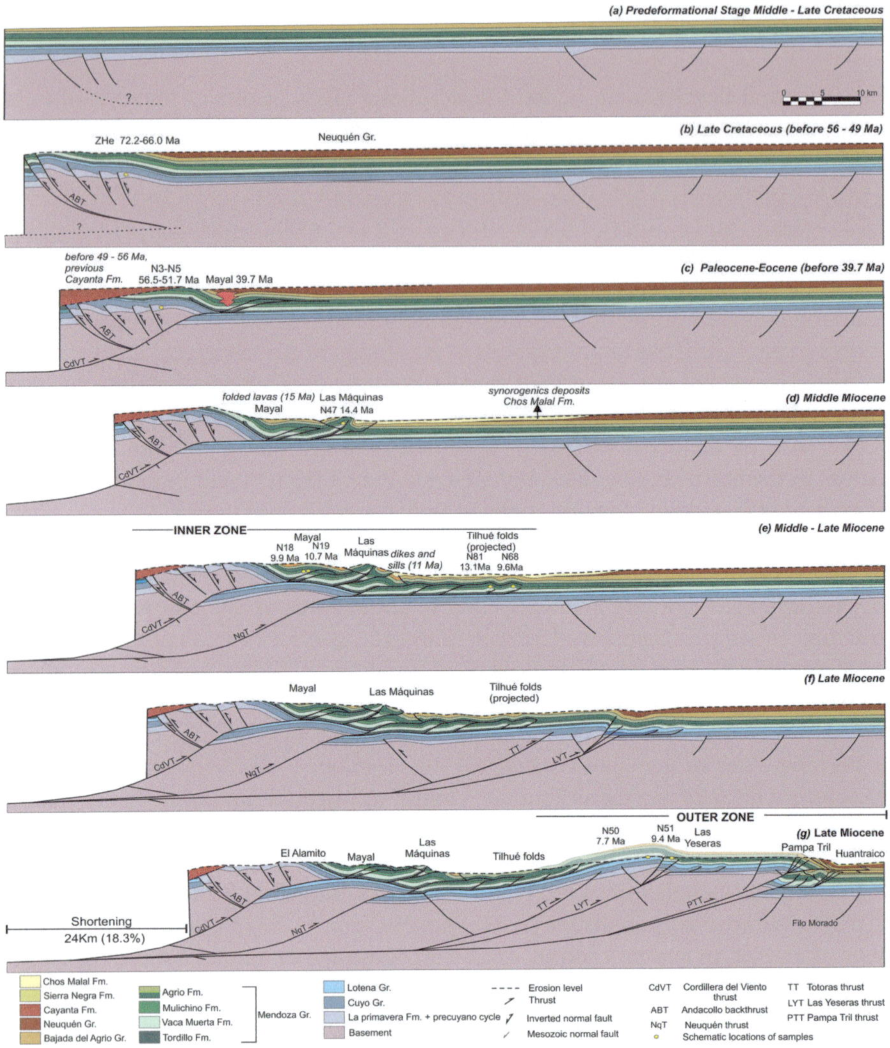

Fig. 6 Structural-thermal evolution of the Chos Malal FTB (based on Sánchez et al. 2018). **a** Pre-contractional stage; **b**, **c** onset of Andean compression during the Late Cretaceous–Paleogene, through faulting and uplift of the Cordillera del Viento; **d**, **e** related thick- and thin-skinned structures completed the deformation in the inner zone; **f**, **g** contraction progressed toward the foreland through the basement-involved Las Yeseras thrust (LYT) and Pampa Tril thrust (PTT), and the thin-skinned Filo Morado anticline. Out-of-sequence branch of the PTT created a large basement uplift, whose eastward displacement was compensated by back thrusting in a triangle zone

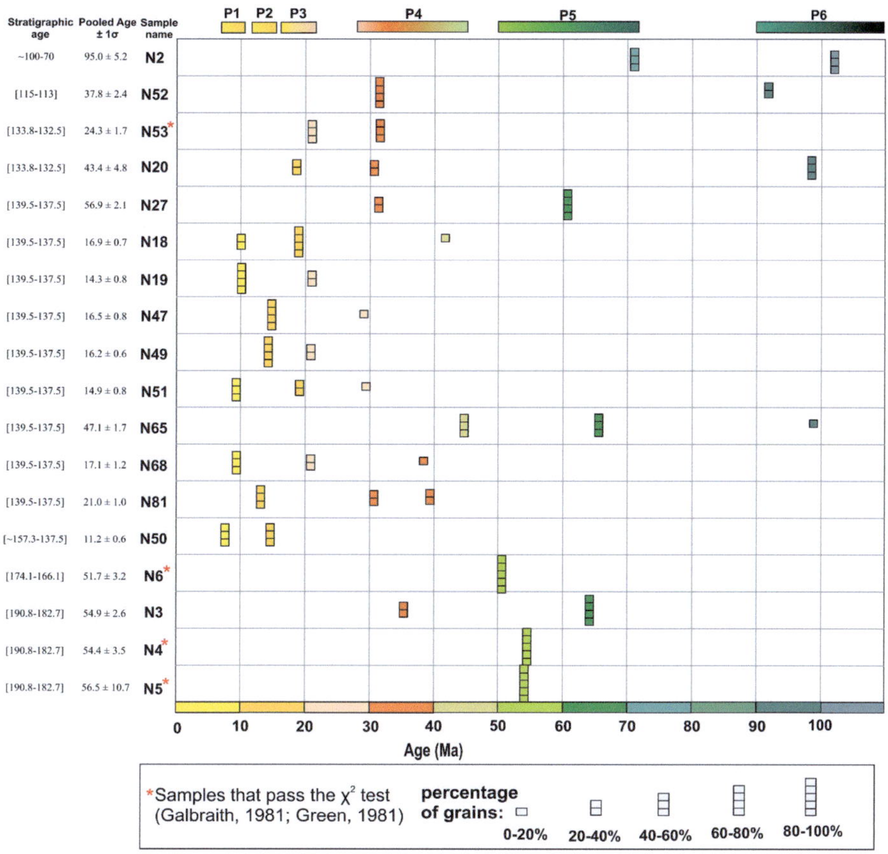

Fig. 7 Binomial peak-fit ages (*P*1–*P*6) were determined using Binomfit (Brandon 1996, 2002) based on Sánchez et al. (2018). *P*1 and *P*2 were used to reconstruct the kinematic evolution (Fig. 6)

(Rojas Vera et al. 2015). These data agree with the structural history in the inner zone of the Chos Malal FTB (Fig. 6 b and c), and overall, they support a Late Cretaceous–Paleogene contraction on these Andean belts.

Growth of thick-skinned structures in the inner zone of the Chos Malal FTB and related thin-skinned folds continued during the Neogene (Fig. 6d, e). Analysis of apatite (U–Th)/He in two samples south of Cordillera del Viento (Figs. 2 and 4), whose effective closing temperature is 70 ± 20° C (e.g., Shuster et al. 2006; Wolf et al. 1998), showed ages of 30.3 ± 5.8 and 19.4 ± 2.3 Ma (Sánchez et al. 2018). Although these data suggest a Late Oligocene–Early Miocene exhumation, more structural and thermochronological data are required in order to interpret their tectonic meaning. Folded volcanic sequences in the northern sector of the Chos Malal FTB (Fig. 1b), dated by $^{40}Ar/^{39}Ar$ on amphibole and plagioclase at 14.51 ± 0.12 Ma and 14.8 ± 8.0 Ma, respectively (Spagnuolo et al. 2012), support a Neogene contraction (Folguera et al. 2007). Sánchez et al. (2018) obtained minimum cooling apatite fission-track (AFT) ages for samples located on second-order folds throughout the inner zone of the Chos Malal FTB (Figs. 2 and 4). Most of the samples (except N4, N5, N6 and N53) fail the $\chi 2$ test (Galbraith 1981; Green 1981) with P ($\chi 2$) < 5% (Fig. 7), indicating that fission-track grain ages distributions are non-Poissonian, and represent a mixture of components or peak ages (Brandon 1992, 1996, 2002). With the exception of samples N5, N4 and N6, the samples have a wide range of cooling ages such that it is likely that the youngest grains have been selectively reset to form the youngest age components (Garver et al. 1999). According to published vitrinite reflectance data, the grains should have been subjected to maximum burial temperatures from 119.5 ± 9.9 to 215.6 ± 9.0 °C between stratigraphic depths of 1225 and 3715 m (Muñoz 1996). Therefore, Sánchez et al. (2018) considered that the youngest populations P1 and P2, in the range of 7.7 ± 0.9/1.0 to 15.0 ± 1.4/3.2 Ma, which corresponds to partially reset samples, are the result of Middle–Late Miocene cooling/exhumation (Figs. 4 and 7). Sample N47, over the Las Máquinas anticline that would be associated with the eastward push of the Cordillera del Viento anticline (Fig. 6d), has a youngest AFT peak yielding an age of 14.4 ± 1.9/2.2 Ma (Fig. 7). In sequence, thrusting produced a new basement-involved wedge and consequently more fault-related folds in the cover (Fig. 6e). The youngest population of AFT ages on this second thick-skinned structure are 9.9 ± 1.4/3.4 and 10.7 ± 1.7/2.1 Ma, while ages on the thin-skinned folds are 13.1 ± 2.0/2.3 and 9.6 ± 3.1/4.6 Ma (Sánchez et al. 2018). These comparable cooling ages suggest a genetic linkage between thick and thin-skinned structures. Las Máquinas anticline was intruded by dykes and sills that form the Cerro Negro (Fig. 1b), with U–Pb ages ranging from 11.55 ± 0.06 Ma to 11.63 ± 0.2 Ma, whose emplacement in the anticline hinge was coeval with the compressional tectonics (Gürer et al. 2016). All these geochronological and structural data indicate that the inner zone of the Chos Malal FTB underwent a significant contraction and exhumation between the Middle and Late Miocene (Figs. 6 d and e).

The basement-cored Las Yeseras and Pampa Tril anticlines dominate the structural configuration in the outer zone of the Chos Malal FTB (Figs. 6 f and g). East-dipping back-arc lavas in the region of Huantraico (Fig. 1b), whose $^{40}Ar/^{39}Ar$ ages vary from 23.4 ± 0.4 Ma to 17.61 ± 0.09 Ma (Kay and Copeland 2006, Dyhr et al. 2013), indicate a Post-Early Miocene folding; although the presence of overlying clastic sequences suggest a Late Miocene deformation (Garrido et al. 2012). The AFT ages, in samples N50 and N51 collected in the flanks of the Las Yeseras anticline (Figs. 4 and 7), have youngest peaks of 7.7 ± 0.9/1 and 9.4 ± 1.7/2.2 Ma (Sánchez et al. 2018). These AFT ages reflect a Late Miocene contraction in the outer zone of the Chos Malal FTB (Fig. 6g), where cooling could be related to the growth of the Las Yeseras anticline itself or more likely to displacements above the Pampa Tril thrust.

5 Implications for the Petroleum Systems in the Neuquén Basin

The structural evolution of fold and thrust belts and creation of structural traps determine the degree of maturity of potential source rocks attained at different times and controls the hydrocarbon preservation and the migration paths. Understanding the geometry and age of the main tectonic structures and the associated uplift and erosion is essential to evaluate hydrocarbon generation, properties of the reservoirs and seal rocks, and migration paths. Most of the oilfields in the northwest of the Neuquén province are located to the east of the deformation front, being only the Filo Morado and El Portón fields situated within the fold and thrust belt (Fig. 5). The hydrocarbon accumulations in the Filo Morado and El Portón oilfields occur in complex structural traps associated with thin-skinned structures that were developed in front of the major east-vergent basement-cored Pampa Tril anticline (Ploszkiewicz and Viñes 1987; Zapata et al. 2001; Allmendinger et al. 2004; Zamora Valcarce et al. 2006; Turienzo et al. 2018). The source rocks in these oilfields are the organic-rich shales of the Vaca Muerta Formation and subordinately the Agrio Formation (Fig. 3). The most important reservoir units are the sandstones with high primary porosity of the Lower Troncoso Member of the Huitrín Formation and the Avilé Member of the Agrio Formation. The regional generation and expulsion of hydrocarbons in the Neuquén Basin started after the accumulation of the Late Cretaceous Neuquén Group (Kozlowski et al. 1998; Gómez Omil et al. 2014). In the area of Chos Malal, the maturity of the source rock evolved in direct association with the structures of the fold and thrust belt. Uplift of the Cordillera del Viento and the associated thin-skinned structures produced hydrocarbon migration into structural traps located in the inner zone, while hydrocarbons generated in response to the uplift of the Pampa Tril basement thrust sheet are preserved in the Filo Morado oilfield and below the frontal syncline (Kozlowski et al. 1998). Most previous studies agreed in that the Vaca Muerta and Agrio formations entered in the oil generation window since the ~80 Ma, and that the complete maturation and the main structural traps were developed during the Cenozoic tectonics approximately between 50 and 10 Ma (Kozlowski et al. 1998; Vergani et al. 2011; Legarreta and Villar 2012; Gómez Omil et al. 2014). The structural-thermal evolution for the Chos Malal FTB (Sánchez et al. 2018) constrains the sequential development of the thick and thin-skinned structures throughout the belt (Fig. 6). AFT ages on the Las Yeseras–Pampa Tril support that the outer zone of this FTB was deformed in the Late Miocene, around ~7–9 Ma (Fig. 6g). Detailed reinterpretation of subsurface information at the deformation front allowed to identify and to characterize a complex deformation pattern, linking thick- and thin-skinned structures to the Pampa Tril and Filo Morado anticlines, and hence identifying new exploration leads (Turienzo et al. 2018). The deformation and exhumation of the structures in the FTB strongly controlled the evolution of the oil system (maturity, trapping, preservation, migration, pressure, etc.) not only on the mountain front but also on the adjacent foreland (Kozlowski et al. 1998; Gómez Omil et al. 2014, Ceballos and Rivero 2014).

6 Concluding Remarks

The integration of structural, geochronological and thermochronological data allowed to constraint the kinematics of the Chos Malal FTB, which show a normal sequence of deformation toward the foreland zone, from the Late Cretaceous to the Late Miocene. The deformation in this belt began at the inner zone during the Late Cretaceous–Paleogene, supported by zircon (U–Th)/He and apatite fission-track cooling ages between 72 and 51 Ma, by means of back thrusting and east-vergent thrusting in the Cordillera del Viento. The structural development of the inner zone was completed in the Middle Miocene, where closely related thick and thin-skinned structures were created between ~15–10 Ma according to fission-track analysis in detrital apatite and published $^{40}Ar/^{39}Ar$ and U–Pb ages of volcanic and subvolcanic rocks. Finally, uplift and exhumation of the outer zone of the Chos Malal FTB, during the Late Miocene, is documented by apatite fission-tracks ages of ~9–7 Ma on the basement-cored Las Yeseras–Pampa Tril anticlines. This detailed review about the structural evolution of the Chos Malal FTB contributes to understand the Andean tectonics and to evaluate the hydrocarbon systems in the Neuquén Basin.

Acknowledgements This work was supported by the CONICET PUE0047CO, CONICET PIP 0583, FONCYT PICT 2015-0419 y SECYTUNS PGI 24/h136 and by an Emerging Leaders of the America Program scholarship from the Department of Foreign Affairs, Trade and Development (Global Affairs Canada) awarded to I. Coutand for N. Sanchez. We are grateful to the Departmento de Geología of the Universidad Nacional del Sur and INGEOSUR-CONICET for logistical and financial support. We also acknowledge the Subsecretaría de Energía, Minería e Hidrocarburos of the Neuquén Province for providing the seismic lines and well information, and the Municipalidad of Chos Malal and Andacollo who facilitated our stay during the field works.

References

Allmendinger R, Zapata T, Manceda R et al (2004) Trishear kinematic modeling of structures, with examples from the Neuquén Basin, Argentina. In: McClay K (ed) Thrust tectonics and hydrocarbon systems. AAPG Memoir 82:356–371

Arregui C, Carbone O, Leanza HA (2011) Contexto tectosedimentario. In: Leanza HA, Arregui C, Carbone O et al (eds) Geología y Recursos Naturales de la Provincia del Neuquén. Asociación Geológica Argentina, Buenos Aires, pp 29–36

Brandon MT (1992) Decomposition of fission-track grain-age distributions. Am J Sci 292:535–564

Brandon MT (1996) Probability density plots for fission-track age distributions. Rad Meas 26:663–676

Brandon MT (2002) Decomposition of mixed grain age distributions using Binomfit. OnTrack 24:13–18

Carbone O, Franzese J, Limeres M et al (2011) El Ciclo Precuyano (Triásico Tardío - Jurásico Temprano) en la Cuenca Neuquina. In: Leanza HA, Arregui C, Carbone O et al (eds) Geología y Recursos Naturales de la Provincia del Neuquén. Asociación Geológica Argentina, Buenos Aires, pp 63–76

Ceballos M, Rivero M (2014) El proceso de exhumación Neógeno y sus consecuencias en las acumulaciones de hidrocarburos del borde noreste de la Cuenca Neuquina, Argentina. In: Abstracts of the 9 Congreso de Exploración y Desarrollo de Hidrocarburos, IAPG, Mendoza, 5–9 Nov 2014
Cobbold P, Rossello E (2003) Aptian to recent compressional deformation, foothills of the Neuquén Basin Argentina. Mar Petrol Geol 20:429–443
Cobbold P, Diraison M, Rossello E (1999) Bitumen veins and Eocene Transpression, Neuquén Basin, Argentina. Tectonophysics 314:423–442
Danieli JC, Coppolecchia M, Elissondo M (2011) El Grupo Andacollo (Paleozoico Tardío). In: Leanza HA, Arregui C, Carbone O (eds) Geología y Recursos Naturales de la Provincia del Neuquén. Asociación Geológica Argentina, Buenos Aires, pp 49–52
Di Giulio A, Ronchi A, Sanfilippo A et al (2015) Cretaceous evolution of the Andean margin between 36° S and 40° S latitude through a multi–proxy provenance analysis of Neuquén Basin strata (Argentina). Basin Res 29:284–304
Dyhr CT, Holm PM, Llambias EJ et al (2013) Subduction controls on Miocene back–arc lavas from Sierra de Huantraico and La Matancilla and new 40Ar/39Ar dating from the Mendoza region, Argentina. Lithos 179:67–83
Fantín J, Manceda R, Palacio B et al (2014) Caracterización de las fracturas naturales de las Formaciones Vaca Muerta y Mulichinco en la estructura de Filo Morado, Cuenca Neuquina, Argentina. In: Abstracts of the 9 Congreso de Exploración y Desarrollo de Hidrocarburos, IAPG, Mendoza, 5–9 Nov 2014
Fenell L, Folguera A, Naipauer M et al (2016) Cretaceous deformation of the Southern Central Andes: synorogenic growth strata in the Neuquén Group (35° 30′–37° S). Basin Res 291:51–72
Folguera A, Ramos VA (2011) Repeated eastward shifts of arc magmatism in the Southern Andes: a revision to the long-term, pattern of Andean uplift and magmatism. J South Am Earth Sci 32(4):531–546
Folguera A, Ramos VA, González Díaz E et al (2006) Miocene to Quaternary deformation of the Guañacos fold and thrust belt in the Neuquén Andes between 37° and 37° 30′ S. In: Kay, S. M, Ramos VA (eds) Evolution of an Andean margin: a tectonic and magmatic view from the Andes to the Neuquén Basin (35°–39° S latitude). Geological Society of America, SP 407, pp 247–266
Folguera A, Ramos VA, Zapata TR et al (2007) Andean evolution at the Guañacos and Chos Malal fold and thrust belts (36° 30′–37° S). J Geodyn 44:129–148
Folguera A, Bottesi G, Duddy I et al (2015) Exhumation of the Neuquén Basin in the southern Central Andes (Malargüe fold and thrust belt) from field data and low-temperature thermochronology. J S Am Earth Sci 64:381–398
Franchini M, López-Escobar L, Schalamuk IBA, Meinert L (2003) Magmatic characteristics of the Paleocene Cerro Nevazón region and other Late Cretaceous to Early Tertiary calc-alkaline subvolcanic to plutonic units in the Neuquén Andes, Argentina. J S Am Earth Sci 16(5):399–421
Franzese J, Spalletti L (2001) Late Triassic-Early Jurassic continental extension in southwestern Gondwana: tectonic segmentation and pre–break–up rifting. J S Am Earth Sci 14:257–270
Galbraith RF (1981) On statistical models for fission track counts. Math Geol 13:438–471
Galland O, Hallot E, Cobbold PR, Ruffet G, de Bremond D'Ars J (2007) Volcanism in a compressional Andean setting: a structural and geochronological study of Tromen volcano (Neuquén province, Argentina). Tectonics 26(4)
Garrido A, Kramarz A, Forasiepi A et al (2012) Estratigrafía, mamíferos fósiles y edad de las secuencias volcanosedimentarias eoceno–miocenas de la sierra de Huantraico–sierra Negra y cerro Villegas (provincia del Neuquén, Argentina). Andean Geol 39(3):482–510
Garver JI, Brandon MT, Roden–Tice M et al (1999) Exhumation history of orogenic highlands determined by detrital fission–track thermochronology. In: Ring U, Brandon MT, Lister GS et al (eds) Exhumation processes: normal faulting, ductile flow and erosion. The Geological Society, London, 154, pp 283–304
Giacosa R, Allard J, Foix N et al (2014) Stratigraphy, structure and geodynamic evolution of the Paleozoic rocks in the Cordillera del Viento (37° S latitude, Andes of Neuquén, Argentina). J Iberian Geol 40:331–348

Gómez Omil R, Caniggia J, Borghi P (2014) La Formación Vaca Muerta en la faja plegada de Neuquén y Mendoza. Procesos que controlaron su madurez. In: Abstracts of the 9 Congreso de Exploración y Desarrollo de Hidrocarburos, IAPG, Mendoza, 5–9 Nov 2014

Green PF (1981) A newlook at statistics in fission track dating. Meas Nucl Tracks Radiat 5:77–86

Green PF, Duddy IR, Gleadow AJW et al (1985) Fission track annealing in apatite: track length measurements and the form of the Arrhenius plot. MeasNuclTracks Radiat 10:323–328

Gulisano C, Gutiérrez Pleimling A (1995) Field guide the Jurassic of the Neuquén Basin, a) Neuquén province. Asociación Geológica Argentina, Buenos Aires, SP 158, p 111

Gürer D, Galland O, Corfu F et al (2016) Structure and evolution of volcanic plumbing systems in fold–and–thrust belts: a case study of the Cerro Negro de Tricao Malal, Neuquén Province, Argentina. Geol Soc Am Bull 128(1–2):315–331

Howell JA, Schwarz E, Spalletti LA et al (2005) The Neuquén Basin: an overview. In: Veiga GD, Spalletti LA, Howell JA et al (eds) The Neuquén Basin, Argentina: a case study in sequence stratigraphy and basin dynamics. The Geological Society, London, SP 352, pp 1–14

Jordan T, Burns W, Veiga R (2001) Extension and basin formation in the southern Andes caused by increased convergence rate: a mid–Cenozoic trigger for the Andes. Tectonics 20:308–324

Kay S (2001) Tertiary to Recent Magmatism and Tectonics of the Neuquén Basin between 36 °05′ and 38°S Latitude, 80 pp. Repsol-YPF (unpublished report), Buenos Aires

Kay SM, Copeland P (2006) Early to middle Miocene back–arc magmas of the Neuquén Basin: geochemical consequences of slab shallowing and the westward drift of South America. In: Kay SM, Ramos VA (eds) Evolution of an Andean margin: a tectonic and magmatic view from the Andes to the Neuquén Basin (35°–39° S lat). Geol Soc Am, SP 407, pp 185–213

Kay S, Burns W, Copeland PC et al (2006). Upper Cretaceous to Holocene magmatism and evidence for transient Miocene shallowing of the Andean subduction zone under the northern Neuquén Basin. In: Kay SM, Ramos VA (eds) Evolution of an Andean margin: a tectonic and magmatic view from the Andes to the Neuquén Basin (35°–39° S lat). Geological Society of America, SP 407, pp 19–60

Ketcham RA, Donelick RA, Carlson WD (1999) Variability of apatite fission-track annealing kinetics; III, Extrapolation to geological time scales. Am Mineral 84(9):1235–1255. https://doi.org/10.2138/am-1999-0903

Kozlowski EE, Cruz CE, Sylwan CA (1996) Geología estructural de la zona de Chos Malal, Cuenca Neuquina, Argentina. In: Abstracts of the 13 Congreso Geológico Argentino y 3 Congreso de Exploración de Hidrocarburos, AGA-IAPG, Buenos Aires, 13–18 Oct 1996

Kozlowski EE, Cruz CE, Sylwan CA (1998) Modelo exploratorio en la faja corrida de la Cuenca Neuquina, Argentina. Bol Inf Petrol 55:4–23

Leanza HA (2003) Las sedimentitas Huitrinianas y Rayosianas (Cretácico inferior) en el ámbito central y meridional de la Cuenca Neuquina, Argentina. SEGEMAR, Contribuciones Técnicas, Geología 2:1–31

Leanza HA (2009) Las principales discordancias del Mesozoico de la Cuenca Neuquina según observaciones de superficie. Rev Mus Argent Cs Nat 11:145–184

Leanza HA, Llambías EJ, Carbone O (2005) Unidades estratigráficas limitadas por discordancias en los depocentros de la cordillera del Viento y la sierra de Chacaicó durante los inicios de la Cuenca Neuquina. In: Abstracts of the 5 Congreso de Exploración y Desarrollo de Hidrocarburos, IAPG, Mar del Plata, 29 Oct–2 Nov 2002

Leanza H, Mazzini A, Corfu F et al (2013) The Chachil Limestone (Pliensbachian–earliest Toarcian) Neuquén Basin, Argentina: U–Pb age calibration and its significance on the Early Jurassic evolution of southwestern Gondwana. J S Am Earth Sci, 42171–185

Legarreta L, Villar HJ (2012) Las facies generadoras de hidrocarburos de la Cuenca Neuquina. Petrotecnia, 14–39

Legarreta L, Villar HJ, Laffitte C et al (2005) Cuenca Neuquina. In: Abstracts of the VI Congreso de Exploración y Desarrollo de Hidrocarburos, IAPG, Mar del Plata, 15–19 Nov 2005

Llambías EJ, Leanza HA, Carbone O (2007) Evolución tectono–magmática durante el Pérmico al Jurásico Temprano en la Cordillera del Viento (37° 05′ S–37° 15′ S): nuevas evidencias geológicas y geoquímicas del inicio de la cuenca Neuquina. Rev Asoc Geol Argent 62:217–235
Llambías EJ, Sato AM (2011) Ciclo Gondwánico: la provincia magmática Choiyoi en Neuquén. In: Leanza HA, Arregui C, Carbone O et al (eds) Geología y Recursos Naturales de la Provincia del Neuquén. Asociación Geológica Argentina, Buenos Aires, pp 53–62
Muñoz N (1996) The thermal evolution of Jurassic and Cretaceous source rocks in the Malargüe thrust belt, Argentina: implications for hydrocarbon exploration. Royal Holloway University of London, p 98
Nocioni A (1996) Estudio estructural de la Faja Plegada y Corrida de la Cuenca Neuquina– Surmendocina. In: Abstractcs of the 13 Congreso Geológico Argentino y 3 Congreso de Exploración de Hidrocarburos, 13–18 Oct 1996
Ploszkiewicz V, Viñes R (1987) Filo Morado: Un descubrimiento exploratorio en cinturón plegado. Bol Inf Petrol, Tercera Época IV(10):97–102
Ramos V, Barbieri M (1989) El volcanismo Cenozoico de Huantraico: Edad y relaciones isotópicas iniciales, provincia del Neuquén. Rev Asoc Geol Argent 43:210–223
Ramos VA, Folguera A (2005) Tectonic evolution of the Andes of Neuquén: Constraints derived from the magmatic arc and foreland deformation. In: Veiga GD, Spalletti LA, Howell JA, Schwarz E (eds) The Neuquén Basin: a case study in sequence stratigraphy and basin dynamics. The Geological Society, London, SP 252, pp 15–35
Ramos VA, Mosquera A, Folguera A (2011) Evolución tectónica de los Andes y del Engolfamiento Neuquino adyacente. In: Leanza HA, Arregui C, Carbone O et al (eds) Geología y Recursos Naturales de la Provincia de Neuquén. Asociación Geológica Argentina, Buenos Aires, pp 335–348
Reiners PW, Spell TL, Nicolescu S, Zanetti KA (2004) Zircon (U-Th)/He thermochronometry: He diffusion and comparisons with 40Ar/39Ar dating. Geochimica et Cosmochimica Acta 68(8):1857–1887
Reiners PW, Brandon MT (2006) Using thermochronology to understand orogenic erosion. An Review Earth Planet Sci 34:419–466
Rojas Vera E, Mescua J, Folguera A et al (2015) Evolution of the Chos Malal and Agrio fold and thrust belts, Andes of Neuquén: Insights from structural analysis and apatite fission track dating. J S Am Earth Sci 64(2):418–433
Sánchez N (2015) Evolución tectónica de las estructuras andinas en la región del río Neuquén (~37° 20′ L.S) faja corrida y plegada de Chos Malal, provincia de Neuquén. Ph.D. thesis, Universidad Nacional del Sur, Bahía Blanca
Sánchez N, Turienzo M, Dimieri L et al (2014) Reconstrucción estructural a los 37° 18′ S, faja corrida y plegada de Chos Malal, Provincia de Neuquén. Rev Asoc Geol Argent 71(2):233–246
Sánchez N, Turienzo M, Lebinson F et al (2015) Structural style of the Chos Malal fold–and–thrust belt, Neuquén Basin, Argentina: relationship between thick and thin–skinned tectonics. J S Am Earth Sci 64(2):399–417
Sánchez N, Coutand I, Turienzo M et al (2018) Tectonic evolution of the Chos Malal fold–and–thrust belt (Neuquén Basin, Argentina) from (U–Th)/He and fission–track thermochronometry. Tectonics 37:1907–1929
Sagripanti L, Folguera A, Giménez M et al (2014) Geometry of middle to late Triassic extensional deformation pattern in the Cordillera del Viento (Southern Central Andes): a combined field and geophysical study. J Iber Geol 40:349–366
Sagripanti L, Folguera A, Fennell L et al (2016) Progression of the Deformation in the Southern Central Andes (37°S). In: Folguera A, Naipauer M, Sagripanti L et al (eds) Growth of the Southern Andes. Springer, New York, pp 115–132
Spagnuolo M, Litvak V, Folguera A et al (2012) Neogene magmatic expansion and mountain building processes at the southern Central Andes, 36°–37° S, Argentina. J Geodyn 53:81–95
Shuster DL, Flowers RM, Farley KA (2006) The influence of natural radiation damage on helium diffusion kinetics in apatite. Earth Planet Sci Lett 249:148–161

Tunik M, Folguera A, Naipauer M, Pimentel MM et al (2010) Early uplift and orogenic deformation in the Neuquén Basin: constraints on the Andean uplift from U-Pb and Hf isotopic data of detrital zircons. Tectonophysics 489(1–4):258–273

Turienzo M, Sánchez N, Dimieri L et al (2014) Tectonic repetitions of the Early Cretaceous Agrio Formation in the Chos Malal fold–and–thrust belt, Neuquén basin, Argentina: geometry, kinematics and structural implications for Andean building. J S Am Earth Sci 53:1–19

Turienzo M, Sánchez N, Lebinson F et al (2018) The structure of the Southern Central Andes (Chos Malal fold and thrust belt). In: Folguera et al (eds) The evolution of the Chilean–Argentinean Andes, Springer, New York, pp 419–451

Vergani G, Tankard AJ, Belotti HJ et al (1995) Tectonic Evolution and Paleogeography of the Neuquén Basin, Argentina. In: Tankard J, Suarez SR, Welsink J (eds) Petroleoum Basin of South America. AAPG Memoir 62:383–402

Vergani GD, Arregui C, Carbone O (2011) Sistemas petroleros y tipos de entrapamientos de la Cuenca Neuquina. Asociación Geologica Argentina, Neuquén, pp 645–656

Viñes R (1989) Interpretación de la estructura de Filo Morado. I Congreso Nacional de Exploración de Hidrocarburos, IAPG, Mar del Plata, 15–21 Apr 1989

Wolf RA, Farley KA, Kass DM (1998) Modeling of the temperature sensitivity of theapatite (U-Th)/He thermochronometer. Chem Geol 148:105–114

Zamora Valcarce G, Zapata T (2005) Estilo estructural del frente de la faja plegada Neuquina a los 37° S. VI Congreso de Exploración y Desarrollo de Hidrocarburos, IAPG, Mar del Plata, 15–19 Nov 2005

Zamora Valcarce G, Zapata T, Ansa A et al (2006) Three–dimensional structural modeling and its application for development of the El Portón field, Argentina. AAPG Bull 90(3):307–319

Zapata T, Brissón I, Dzelalija F (1999) La Estructura de la faja plegada y corrida andina en relación con el control del basamento de la Cuenca Neuquina. Bol Inf Petrol 64:112–121

Zapata T, Dzelalija F, Olivieri G (2001) Desarrollo de reservorios fracturados de la formación Mulichinco usando modelado estructural 3D: yacimiento Filo Morado, Cuenca Neuquina, Argentina. Bol Inf Petrol 66:38–47

Zöllner W, Amos AJ (1973) Descripción geológica de la Hoja 32b, Chos Malal (Prov. Neuquén), Bull. 143, SEGEMAR, Buenos Aires, p 91

The Andean Foreland Evolution of the Neuquén Basin: A Discussion

Facundo Fuentes and Brian K. Horton

Abstract The Neuquén Basin of west-central Argentina preserves a complex structural and stratigraphic record of the Andean orogeny marked by intermittent periods of foreland sediment accumulation, bypass, and regional exhumation and erosion. Nonmarine deposits of the Cenomanian–Campanian Neuquén Group are the earliest of foreland basin affinity, with clear input from western sediment sources and relatively rapid subsidence rates. However, available provenance and structural data fail to demonstrate a Late Cretaceous regional fold-thrust belt capable of creating rapid accommodation and voluminous sediment supply, suggesting that the magmatic arc may have played an important role as a flexural load and sediment source. The Upper Cretaceous basin architecture defined by the Neuquén Group seems to mimic that of underlying Jurassic–Lower Cretaceous units, indicating either a continued thermal contribution to basin subsidence and/or flexure concentrated along previously thinned lithosphere. Fine-grained marine and nonmarine facies of the overlying Campanian–Eocene Malargüe Group were deposited under slowing rates of subsidence and unclear tectonic conditions. These deposits are capped by a condensed section or unconformity of late Eocene to earliest Miocene age. This foreland hiatus is contemporaneous with the development of extensional basins in the orogen interior, possibly resulting from slab rollback and decoupling along the subduction plate margin, with sediment bypass and minor erosion in the foreland basin aided by a global sea-level lowstand. At ca. 20 Ma, proximal areas adjacent to the advancing fold-thrust belt started accumulating coarse clastic fluvial and alluvial fan deposits in an unambiguous flexural foreland basin. However, Neogene deposition farther east remained irregular, with only relatively thin accumulations in localized zones. In the distal foreland, erosional exhumation rather than sediment accumulation was the dominant process since late Eocene to Oligocene time.

F. Fuentes (✉)
YPF S.A., Buenos Aires, Argentina
e-mail: facundo.fuentes@ypf.com

B. K. Horton
Department of Geological Sciences and Institute for Geophysics, Jackson School of Geosciences, University of Texas at Austin, Austin, USA

D. Kietzmann and A. Folguera (eds.), *Opening and Closure of the Neuquén Basin in the Southern Andes*, Springer Earth System Sciences,
https://doi.org/10.1007/978-3-030-29680-3_14

Keywords Thermal versus flexural subsidence · Late Cretaceous · Paleogene · Neogene · Coupling versus decoupling subduction · Sedimentation rates · Neutral regime

1 Introduction

The Neuquén Basin marks a pivotal change from a central Andean zone of high retroarc shortening and thick foreland basin accumulation to a northern Patagonian zone with modest orogenic shortening and very thin foreland basin fill. In the central Andes of Perú, Bolivia, and northern Argentina, shortening in excess of 100–300 km (e.g., Kley et al. 1999; McQuarrie 2002; Ramos et al. 2004) is accompanied by foreland basin fill up to 5–7 km thick (e.g., Zapata and Allmendinger 1996; Hernández et al. 1999; DeCelles et al. 2011). However, at the 33–40° S latitudes of the Neuquén Basin, there is a major transition to the Patagonian Andes, with less than 30–60 km of shortening (e.g., Giambiagi et al. 2012) and a foreland basin succession less than 1–2 km thick (e.g., Horton et al. 2016).

Within the Neuquén Basin, the most complete retroarc record of synorogenic sedimentation is preserved in the Malargüe region (Fig. 1), with strata of foreland basin affinity spanning the last 100 Myr (Fig. 2). This stratigraphic package defines Late Cretaceous and Neogene periods of rapid foreland accumulation separated by an enigmatic phase between ca. 55 and 20 Ma. This Eocene–earliest Miocene phase is represented by a tabular interval of fine-grained deposits characterized by slow sedimentation rates and a capping disconformity that marks a regional hiatus. Thereafter, Neogene synorogenic sedimentation is recorded by an upward coarsening stratigraphic succession typical of a proximal foreland basin adjacent to a fold-thrust belt.

The mid-Cenozoic lull in sedimentation and resulting foreland unconformity have important tectonic connotations, because such long-lasting hiatuses and/or extreme cases of stratigraphic condensation can reflect contradictory conditions in foreland basins. For high-shortening orogens, 10–20 Myr phases of little to no net sedimentation are commonly associated with cratonward forebulge migration during pronounced shortening and thrust belt advance (e.g., DeCelles and Giles 1996; DeCelles 2012). These non-angular unconformities or condensed sections are well-documented in foreland basins worldwide. On the other hand, in foreland basins associated with thrust belts of modest or unsteady shortening, hiatuses of similar duration (10–20 Myr) may reflect basin abandonment or sediment bypass, possibly related to changes in mechanical coupling between the subducting and overriding plates (Horton and Fuentes 2016; Horton 2018a, b).

This contribution provides a discussion of the Andean foreland basin evolution in the Neuquén Basin region, which exemplifies retroarc foreland basins in low-shortening, low-propagation thrust belt settings resulting from irregular plate coupling, and provides a framework for comparison with less understood regions farther south in Patagonia.

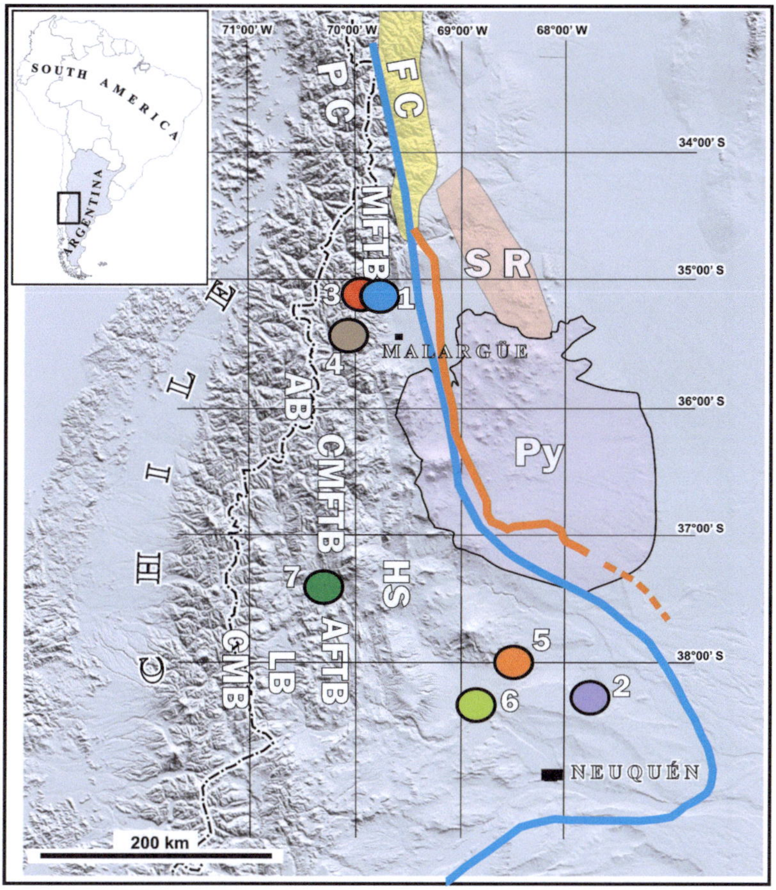

Fig. 1 Regional map and digital elevation model (DEM) of the Neuquén Basin region, including eastern basin limits for the Vaca Muerta Formation (blue line) and Neuquén Group (orange line). Colored circles indicate well/outcrop locations for sediment accumulation diagrams (Fig. 4). AB: Abanico Basin, AFTB: Agrio fold-thrust belt, CMB: Cura Mallín Basin, CMFTB: Chos Malal fold-thrust belt, FC: Frontal Cordillera, HS: Huantraico syncline, LB: Loncopué Basin, PC: Principal Cordillera, Py: Payenia volcanic field, SR: San Rafael block. From the DEM, note that regional exhumation dominates the Neuquén Basin region, with low-relief areas of sediment accumulation limited to northern areas near Malargüe

2 Regional Setting

The Neuquén Basin and associated fold-thrust belt (Fig. 1) are situated above a normal-dipping segment of the subducting Nazca slab. Since the Late Triassic–Early Jurassic, the basin system evolved first as backarc extensional depocenters, then as a large thermal sag, and finally as a flexural foreland basin with a triangular map pattern (including the prominent east-trending Neuquén Embayment) that accumulated up to

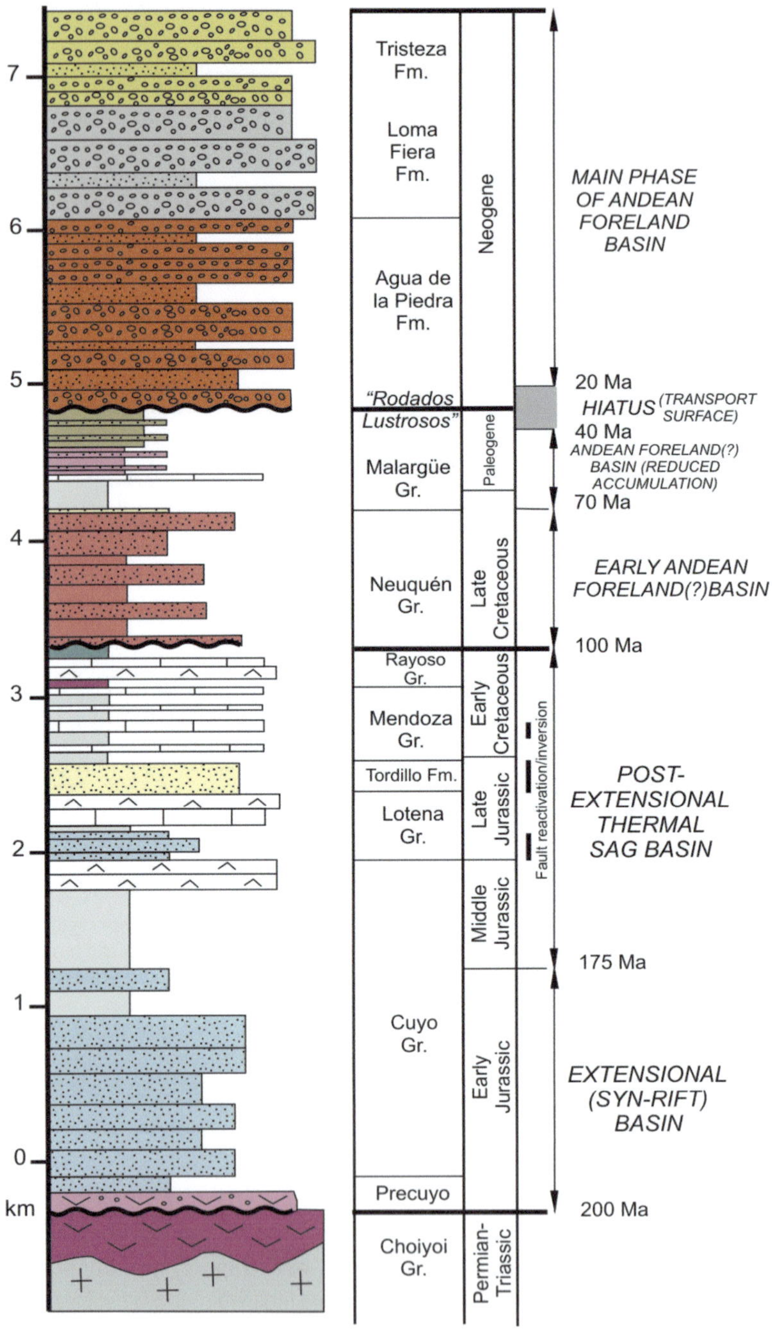

Fig. 2 Generalized Mesozoic–Cenozoic stratigraphic column for the Neuquén Basin and Malargüe foreland basin system, including stratigraphic names and interpreted tectonic phases (modified from Horton et al. 2016)

~7 km of marine and nonmarine basin fill (e.g., Legarreta and Gulisano 1989; Vergani et al. 1995). Sediment accumulation was locally interrupted by brief inversion events (e.g., Vergani et al. 1995) (Fig. 2). By mid-Cretaceous time, opening of the South Atlantic and rapid westward advance of the South American plate over subducting ocean lithosphere promoted a contractional regime with the development of a retroarc fold-thrust belt and associated foreland basin (Coney and Evenchick 1994). Along the Neuquén Basin, the retroarc orogen is divided into the Malargüe, Chos Malal, and Agrio segments of the Andean fold-thrust belt (Fig. 1). Although the structure of these segments is relatively well known (e.g., Kozlowski et al. 1997; Zapata et al. 2002; Zamora Valcarce 2007; Giambiagi et al. 2009; Kraemer et al. 2011; Rojas Vera et al. 2015; Fuentes et al. 2016), less work has been done to link thrust belt deformation with sedimentation in the adjacent foreland basin. The evolution of the foreland basin has been analyzed at a local level (e.g., Zapata et al. 2002; Kraemer et al. 2011; Fuentes et al. 2016; Horton et al. 2016) or within temporally restricted windows (e.g., Tunik et al. 2010; Balgord and Carrapa 2016; Fennel et al. 2017).

Today the Neuquén Basin undergoes regional erosion along most of its length. Only the northern zone adjacent to the Malargüe fold-thrust belt preserves a Neogene record of sedimentation (Fig. 1). This preservation is partially controlled by topographic damming due to the San Rafael basement block and the extensive magmatic topography of the Payenia volcanic field. The relatively complete record of the Andean foreland basin in the Malargüe region is known from stratigraphic sections preserved in the hanging walls of frontal thrust belt structures (e.g., Kraemer et al. 2011; Fuentes et al. 2016; Horton et al. 2016) and from hydrocarbon exploration wells (e.g., Boll et al. 2014). In other zones of the Neuquén Basin, Andean foreland basin deposits are exhumed to variable stratigraphic levels and thus preserve a less complete record of foreland sedimentation during growth of the Andean fold-thrust belt.

3 Stratigraphy

The switch from principally postextensional thermal subsidence to regional contraction and flexural subsidence in the Neuquén Basin likely occurred at ca. 100 Ma and is marked by clastic red beds of the Neuquén Group (e.g., Ramos and Folguera 2005; Ramos 2010; Tunik et al. 2010; Horton and Fuentes 2016) (Fig. 2). This Cenomanian–Campanian unit up to >1.5 km thick was deposited principally by fluvial and alluvial fan processes (Cazau and Uliana 1973; Legarreta and Gulisano 1989). Where constrained by regional seismic reflection data, the regional thickness distribution of the Neuquén Group mimics that of underlying units of the Neuquén Basin, with a depositional area that resembles that of the Jurassic–Lower Cretaceous thermal sag-related units (Fig. 1). However, seismic reflection data indicate that the depositional limit of the Neuquén Group expanded farther to the east, onlapping basement along distal margins. Farther north, the Neuquén Group projects to the west of the Frontal Cordillera, similar to underlying units. Locally, the Neuquén Group shows a low angle unconformity with preceding units of the Rayoso Group and Centenario formations, as in the western part of the basin, the Chiuhidos High, the Entre Lomas system, and the Northeastern platform (Vergani et al. 1995; Mosquera 2008). In the northern part of the Neuquén Basin, lithofacies distributions and clast-size trends within the Neuquén Group indicate considerable westward transport of sediment away from eastern sources of sediment (Legarreta and Uliana 1998). Modern provenance studies based on U–Pb isotopic age data for regional sample suites, however, indicate an important additional contribution from western magmatic arc sources (Tunik et al. 2010; Di Giulio et al. 2012; Balgord and Carrapa 2016; Horton et al. 2016) (Fig. 3). In addition, hinterland outcrops show locally derived sediments from western sources (Broens and Pereira 2005; Mescua et al. 2013; Rosselot et al. 2017; Borghi et al. 2017; Mackaman-Lofland et al. 2019).

The overlying Malargüe Group is represented by a mixed marine and nonmarine succession of clastic, carbonate, and evaporite rocks that can reach up to 800 m in thickness. The Malargüe Group is composed of four distinctive units in western regions: the Loncoche, Roca, Pircala, and Coihueco formations, with correlative units in the eastern and southeastern basin regions (Legarreta et al. 1989). The Loncoche Formation is dominated by gray siltstone and claystone of nonmarine to fully marine origin. Upsection, the Roca Formation is a distinctive shallow-marine carbonate followed by the Pircala and Coihueco Formations. The Pircala is dominated by red siltstone and minor sandstone of distal fluvial systems and includes shallow lacustrine carbonates at its top. The capping unit, the Coihueco Formation, is a distinctive brown, tan, and olive unit composed of poorly consolidated mudstone and muddy sandstone deposited in poorly organized, relatively distal fluvial floodplain settings. Sediment was mostly supplied from the western magmatic arc, as shown by regional correlations (e.g., Legarreta et al. 1989) and U–Pb provenance signatures (Horton et al. 2016).

Whereas fossil assemblages from the base of the Malargüe Group indicate a Campanian age (Legarreta et al. 1989, and references therein), the upper age limit is less clear but has been assigned to the Paleocene or early Eocene (e.g., Uliana and Dellapé 1981; Legarreta et al. 1989; Parras et al. 1998; Aguirre Urreta et al. 2011). Recently published detrital zircon U-Pb ages for the upper Malargüe Group (Pircala and Coihueco formations) in the northern Neuquén Basin include ages as young as middle Eocene (48–38 Ma) (Horton and Fuentes 2016; Horton et al. 2016) (Fig. 3). New U–Pb results for two samples from the upper levels of the Coihueco Formation (~20–30 m below the upper contact) indicate maximum depositional ages of 40.7 ± 0.4 Ma (sample COI12, $n = 6$, with a youngest single grain of 39.4 ± 0.5 Ma) and 40.0 ± 0.4 Ma (sample COI13, $n = 7$, with a youngest single grain of 38.0 ± 0.4 Ma) (Fig. 3). Over large segments of the Neuquén Basin, mostly along the east-trending Neuquén Embayment, the Malargüe Group is the youngest known stratigraphic unit.

The subsequent sedimentary history is poorly understood and varies from place to place. The most complete stratigraphic record is found in the northern Neuquén Basin adjacent to the Malargüe fold-thrust belt. Here, the Coihueco Formation is overlain disconformably by a thin (<20 m) but distinctive stratigraphic unit known as the "Rodados Lustrosos" (Groeber 1951; Yrigoyen 1993) composed of poorly consolidated

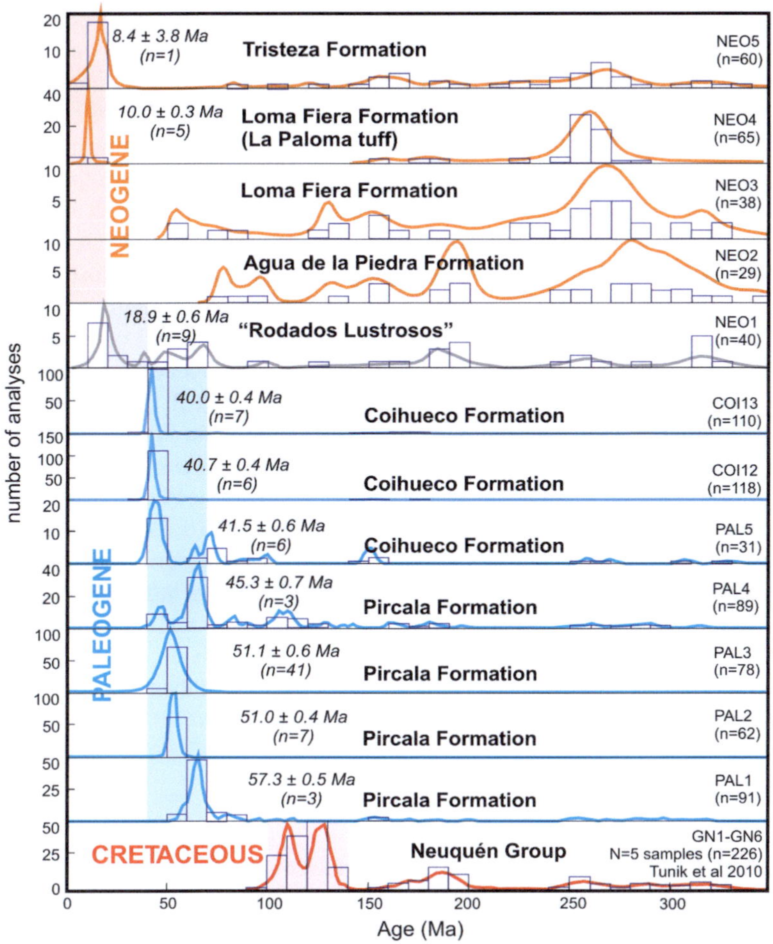

Fig. 3 Detrital zircon U–Pb age distributions depicted as age probability functions (curves) and age histograms (open rectangles); shaded vertical bars denote the young age populations for Upper Cretaceous, Paleogene, and Neogene samples of the Malargüe fold-thrust belt region (modified from Horton and Fuentes 2016, with additional samples COI12 and COI13 from the uppermost Coihueco Formation – see Appendix 1). Notice the prevalence of magmatic arc sources for the Neuquén and Malargüe groups, and the cosmopolitan character of zircon ages in the Neogene deposits, typical of fold-thrust belt sources. The "Rodados Lustrosos" represents a hiatus or condensed section from ca. 40–20 Ma

pebble to cobble conglomerate. Clasts within this unit are well-rounded, highly polished and exhibit oxidized and desert varnished surfaces. Although this unit has not been directly dated, recent studies suggest that it represents a highly condensed stratigraphic interval or a transport ("bypass") surface representing a roughly 40–20 Ma period of variable weathering, erosion, and severely limited sediment accumulation (Horton et al. 2016; Horton 2018b). To the south, in the Huantraico syncline area, this thin unit locally sits unconformably on top of Neuquén Group beds (Garrido et al. 2012), suggesting complete erosional removal of the Malargüe Group.

In the northern part of the basin, early Neogene deposition is documented by the Agua de la Piedra Formation, which denotes the abrupt appearance of coarse clastic deposits consisting of red and orange sandstone and conglomerate deposited by large-scale braided fluvial and fluvial megafan systems with local eolian fields. Trough cross strata show dominantly west to east paleocurrents (Horton et al. 2016), in agreement with sediment sources in the deforming fold-thrust belt. Detrital zircon U–Pb ages indicate that this unit spans from roughly 20 Ma to at least 15 Ma (Horton et al. 2016). The overlying Loma Fiera and Tristeza formations contain more-proximal facies, with cobble to boulder conglomerate indicative of alluvial fan sedimentation adjacent to the Andean thrust front, and growth strata evidencing syndepositional deformation in the adjacent thrust belt structures (Fuentes et al. 2016). The youngest U–Pb age populations for detrital zircons of the Loma Fiera Formation are ca. 12–10 Ma, whereas they are <8 Ma for the overlying Tristeza Formation (Horton et al. 2016). At the top, the Rio Diamante Formation of probable Pliocene age is capped by Quaternary alluvial and volcanic deposits (Turienzo 2010; Boll et al. 2014).

Similar clastic synorogenic deposits of Neogene age are described in intermontane basins west of Malargüe (Silvestro and Kraemer 2005), including the Agua de la Piedra Formation and a series of coarse-grained sedimentary and volcanic rocks of the Molle, Butaló, Pincheira, and Coyoco formations. Similarly, in the Rio Grande and Palauco sub-basins, Silvestro and Atencio (2009) separated the Neogene succession into multiple units, including widespread volcanic deposits.

4 Basin Evolution

Unlike better understood regions of the Andean retroarc foreland basin to the north, the Neuquén Basin segment has experienced a complex evolution that remains relatively poorly understood. Several major questions include:

(1) When was the onset of foreland basin evolution? Does the Cenomanian–Campanian Neuquén Group represent true flexural foreland basin conditions?
(2) What was the tectonic setting during deposition of thin, distal facies of the Campanian–lower Eocene Malargüe Group?
(3) What is the significance of the mid-Cenozoic disconformity and the "Rodados Lustrosos" capping the Malargüe Group?
(4) What was the relation of this hiatus to extension in hinterland and arc regions?
(5) How pervasive was this hiatus beyond the Neuquén Basin?
(6) Why was the development of the Neogene foreland basin locally restricted?
(7) Was the bulk of the Neuquén Basin originally covered by Neogene foreland basin strata, and if so, when was it exhumed to present levels?

Deposition of the Neuquén Group has been attributed by many authors as the first signal of retroarc foreland basin conditions starting at ca. 100 Ma (Ramos and Folguera 2005; Ramos 2010; Tunik et al. 2010). This view is generally based on the following criteria: (a) The Neuquén Group is composed of thick nonmarine red beds that, across large portions of the basin, sit unconformably on older deposits (mostly over the Rayoso Group); (b) it registers provenance from western, mostly volcanic sources; (c) contractional growth strata have been recognized in different parts of the basin (Cobbold and Rosello 2003; Galarza et al. 2009; Orts et al. 2012; Fennell et al. 2017); and (d) local outcrops show proximal facies with derivation from underlying units (e.g., Mescua et al. 2013). However, it must be noted that: (a) there is a general pattern of grain-size reduction from east to west (e.g., Legarreta et al. 1989); (b) provenance data show considerable sediment contributions from eastern sources (e.g., Di Giulio et al. 2012); (c) growth strata have been interpreted from ambiguous field relationships or poor-quality seismic data that contrast with regional seismic data lacking in clear growth geometries; and (d) brief phases of local contraction have been described for Middle Jurassic to Early Cretaceous stages of basin evolution without the establishment of an integrated foreland basin system (Vergani et al. 1995; Cobbold and Rosello 2003).

As Miall (2009) pointed out, a key indicator to define a foreland basin should be a subsidence pattern linked to contemporaneous loading by a fold-thrust belt (Beaumont 1981; Jordan 1981; DeCelles 2012). A set of burial diagrams distributed along the basin indicate increased subsidence in the Late Cretaceous at ca. 100 Ma (Fig. 4), coincident with initial Neuquén Group deposition. However, some back-stripping attempts have led other authors to suggest thermal sag conditions associated with lithospheric cooling (e.g., Legarreta and Uliana 1991).

The areal distribution of the Neuquén Group largely mimics that of the underlying units of the Neuquén Basin with further onlap onto the basin margins. With lithospheric flexure as the main driver of Late Cretaceous subsidence, some additional mechanism is necessary to preserve the former basin shape: either a continuation of thermal subsidence or inheritance of a previously thinned crust capable of concentrating differential subsidence. If the Neuquén Group accumulated in a foreland basin governed exclusively by flexural subsidence, the accommodation configuration would have departed from that of the former thermal sag basin.

The available lines of evidence—including changes in the subsidence rate, local basal unconformities, delivery of volcanic detritus from the west, coarse clastic input from the east, indications of contractional structures, and basin architecture—are insufficient to define a purely flexural foreland basin. However, provenance data indicate positive western topography likely related to an increased arc magmatism and embryonic thrust belt structures. A purely classic foreland basin context for the Neuquén Group is difficult to reconcile with available data. It is possible that remaining thermal subsidence from the former postextensional basin stage was important in the generation of accommodation and that the magmatic arc rather than a highly shortened thrust belt provided an important contribution to crustal loading and basin flexure.

The tectonic conditions during deposition of the Malargüe Group remain unclear. Legarreta et al. (1989) suggested an asymmetric basin with coarser facies to the west. In their view, the Malargüe Group records the main reversal in sediment polarity in the Neuquén Basin. A key feature of this unit, however, is the substantially slower

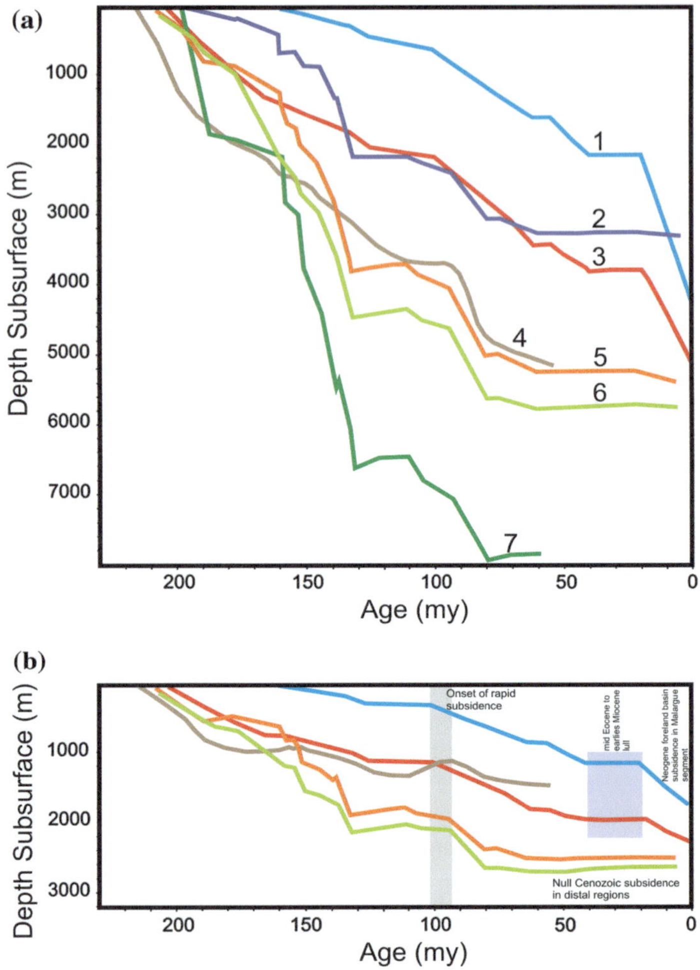

Fig. 4 Plots of **a** sediment accumulation and **b** tectonic subsidence for 7 localities distributed across the Neuquén Basin (see Fig. 1 for locations). Curve 4 is from Manceda and Figueroa (1995). Note the acceleration of subsidence at ca. 100 Ma, the basin-wide reduction in accommodation from 40 to 20 Ma, and the renewed rapid subsidence starting at ca. 20 Ma in localized areas (in particular, in curves 1 and 3)

accumulation rates relative to the underlying Neuquén Group (Fig. 4). At a regional scale, this period reflects subsidence dominated by sediment loading and lithospheric cooling, with minimal flexure (Uliana and Biddle 1988), although Gianni et al. (2018) suggest Maastrichtian–Danian dynamic subsidence along large parts of Patagonia. In any case, the slow subsidence of the Malargüe Group has been interpreted to reflect a neutral margin with no major thrust loading and limited foreland basin subsidence (Horton and Fuentes 2016). Provenance data show direct derivation from magmatic arc sources (Fig. 3) with limited contributions from thrust belt or basement sources

(Horton and Fuentes 2016). Detrital zircon age constraints for the youngest preserved sections of the Malargüe Group indicate that this relative tectonic quiescence of the early Cenozoic persisted until ca. 40 Ma.

The middle Eocene to earliest Miocene marks a major basin-wide change as relatively continuous retroarc sedimentation was interrupted by a long period of minimal to no net sedimentation with localized zones of modest erosion. In the northern Neuquén Basin, this protracted ~40–20 Ma period is registered only by the presence of thin gravels of the "Rodados Lustrosos," which can be interpreted to reflect basin abandonment or sediment bypass. Erosional exhumation likely related to isostatic rebound may be registered in thermochronological data that indicate slow cooling at ca. 40 Ma in the Malargüe area and ca. 30 Ma in the Neuquén Embayment (Rocha et al. 2018; See Chap. Thermochronological constraints on the exhumation of the Malargüe fold and thrust belt). As noted by Horton and Fuentes (2016), this period of foreland stratigraphic condensation, bypass, and erosion is coincident with the development of well-known late Eocene to earliest Miocene extensional basins in hinterland regions, the magmatic arc, and the forearc, including the Abanico and Cura-Mallin basins, and possible early evolution of the Loncopue Trough (Jordan et al. 2001; Burns et al. 2006; Folguera et al. 2010; Rojas Vera et al. 2010; Winocur et al. 2015; Mackaman-Lofland et al. 2019). In foreland regions along the northeastern sector of the Neuquén Basin, mafic volcanic rocks of the Palauco basin have been dated between 24 and 18.5 Ma (Silvestro and Atencio 2009; Dyhr et al. 2013) and related to extensional conditions (Dyhr et al. 2013; Fennel et al. 2017). Rather than locally restricted processes, extensional conditions in hinterland and arc regions persisted for hundreds of kilometers from the San Juan province (~28° S) to northern Patagonia (~46° S) and have been related to a nearly stable position of South America with minimum absolute plate motion during the Oligocene (Ramos 2010; Horton 2018a). This record of extension in the orogen interior and the absence of foreland basin sedimentation led Horton and Fuentes (2016) to postulate a slab rollback configuration with limited mechanical coupling along the convergent margin. In foreland regions, the absence of thrust-load flexure was accompanied by little to no accommodation, sediment bypass, and unconformity development. It must be noted, however, that the Oligocene global sea-level lowstand would have further inhibited foreland accommodation, particularly for the narrow continental width of southern South America.

In the northern Neuquén Basin, renewed sedimentation in the foreland basin commenced at ca. 20 Ma with deposition of the Agua de la Piedra Formation, the first Cenozoic unit of unequivocal flexural foreland basin origin. Sedimentary facies and provenance data indicate sedimentation in fluvial and alluvial fan systems sourced largely from developing thrust sheets to the west (Horton et al. 2016), and sediment accumulation histories show a sharp acceleration typical of an advancing foredeep (Fig. 4). Similar conditions of foreland basin sedimentation were continuous southward into the Rio Grande and Palauco sub-basins (Silvestro and Atencio 2009). Farther south, in the Huantraico syncline, lower Miocene deposits capping the "Rodados Lustrosos" (Garrido et al. 2012) suggest that establishment of this foredeep was largely synchronous during early Miocene shortening across the Malargüe, Chos Malal, and Agrio segments of the retroarc fold-thrust belt.

Early Miocene foreland basin sedimentation was of regional extent and not restricted to the Neuquén Basin. Increasing subsidence rates, unambiguous thrust belt provenance, and eastward advance of a foreland flexural wave is recognized across large segments of the central and southern Andes, including northwestern Argentina (Lomas de Olmedo and Metán-Alemanía sub-basins; Starck 2011), San Juan province (Precordillera basins; Levina et al. 2014), Mendoza province (Mariño Formation; Buelow et al. 2018), and Patagonia (Ramos et al. 2015). However, unlike areas to the north, Neogene foreland basin sedimentation in the Neuquén Basin was confined to a relatively narrow strip adjacent to the frontal fold-thrust belt. Although current exhumation levels along most of the Neuquén Basin (defined principally by Cretaceous–Paleogene exposures of the Neuquén and Malargüe Groups) preclude identification of possible post-20 Ma deposits, burial models based on calibrated vitrinite maturity values indicate that no major Neogene sedimentary overburden existed along the bulk of the basin. However, seismic reflection profiles in the Malargüe region show that the Neogene foreland basin persisted farther east of the former basin limit, indicating that, for the first time, the basin depositional limit was independent of the precursor Mesozoic basin configuration.

In the Neuquén Embayment, basin abandonment and modest erosional exhumation seem to have been the dominant process since middle to late Eocene time. Along the northeastern margin of the Neuquén Embayment, Ceballos and Rivero (2014) estimated ~1 km of exhumation starting at ca. 30 Ma (Fig. 5). The youngest accumulated strata here consist of 50–100 m of Malargüe Group deposits, in agreement with cooling ages reported by Rocha et al. (2018).

It is expected that sediment bypass and regional exhumational processes along the Neuquén Basin starting in middle to late Eocene time would have generated depositional counterparts in subsiding regions located farther east. The main candidates include the onshore Pampean Basin (Folguera et al. 2015) and deepwater sediments in various offshore basins of the Atlantic Ocean.

5 Conclusions

The Andean retroarc foreland within the Neuquén Basin has undergone a complex evolution that departs from classic models of orogenic systems with high-magnitude shortening, large-scale thrust belt advance, and persisting mechanical coupling along the subduction plate boundary. Such classic foreland basin systems show a vertical stacking of depozones and the possible effects of cyclical buildup and removal of dense orogenic roots (DeCelles and Giles 1996; DeCelles 2012; DeCelles et al. 2009, 2011, 2015). In the Neuquén Basin region, however, modest amounts of orogenic shortening and apparently irregular and unsteady plate coupling led to a foreland basin evolution that included two separate episodes of rapid flexural subsidence with an intervening phase of basin abandonment and sediment bypass (e.g., Horton and Fuentes 2016; Horton 2018a, 2018b). The hiatus may have involved modest isostatic rebound, as predicted in some models of foreland basin evolution during tectonic quiescence (e.g., Beaumont 1981; Heller et al. 1988; Jordan 1995).

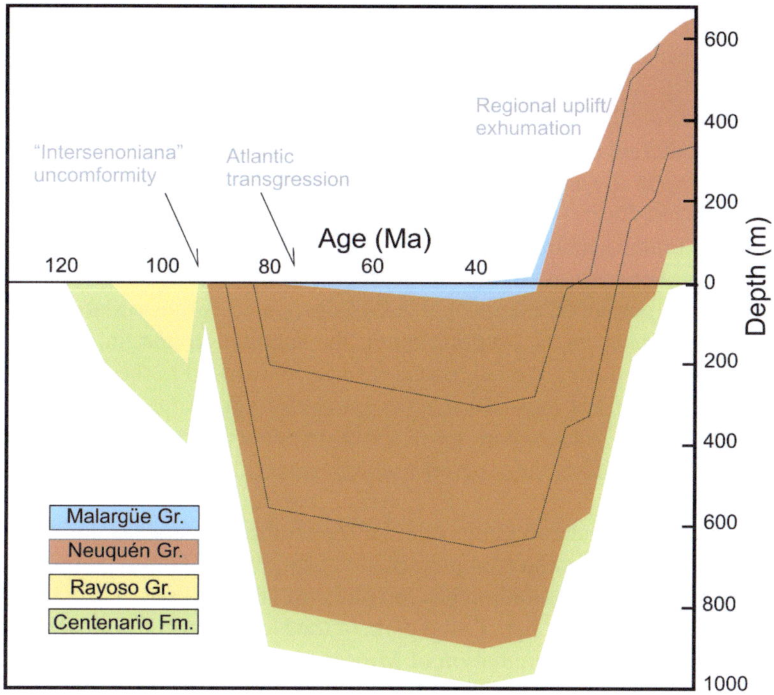

Fig. 5 Geohistory diagram showing the burial and exhumation history for the Neuquén Basin, from well data along the northeastern margin of the Neuquén Embayment ("northeastern platform"; based on Cevallos and Rivero 2014). Note the phases of rapid burial (Late Cretaceous), slow burial (latest Cretaceous–Paleogene), and basin exhumation (Oligocene–present)

There are still major uncertainties in our understanding of the evolution of the Andean foreland basin and its linkages to the developing orogenic belt, with much work to be done. Such work will require a holistic approach, given that most work on paired thrust belt-foreland basin systems are of local scale or lack detailed temporal constraints.

The overall evolution of the Andean foreland within the Neuquén Basin system can be summarized into four stages.

(1) Earliest foreland basin conditions commenced at ca. 100 Ma and are registered in nonmarine clastic deposits of the Cenomanian–Campanian Neuquén Group (>1.5 km thick). However, the existing data are somewhat contradictory. The lack of a clear record of an integrated regional-scale thrust belt suggests that the flexural load may have been largely generated by the Andean magmatic arc. Although the Neuquén Group recorded a "reversal in sediment polarity" (Coney and Evenchick 1994; Horton 2018b), it may not be considered a classic foreland basin in the sense that subsidence may have been driven largely by mechanisms unrelated to crustal shortening.

(2) The overlying Upper Cretaceous–Paleogene Malargüe Group is characterized by diminished subsidence with provenance dominated by magmatic arc sources to the west. A neutral margin with limited shortening and associated flexural subsidence can be interpreted for this unit, but the precise tectonic setting of the Andean belt and depositional basin remains somewhat obscure.
(3) A late Eocene to earliest Miocene stratigraphic hiatus or condensed interval in the Neuquén Basin (locally represented by the "Rodados Lustrosos") records severely limited accommodation characterized by basin abandonment, sediment bypass, and/or erosion. Significant segments of the Andean hinterland and magmatic arc experienced extensional conditions, possibly due to slab rollback and diminished mechanical coupling along the subduction margin. A global sea-level lowstand during the Oligocene further contributed to the lack of foreland sedimentation. This mid-Cenozoic event marks a fundamental shift along the Neuquén Basin, as large portions of the basin never again accumulated substantial sediments and instead experienced regional abandonment and slow erosional exhumation.
(4) Accumulation of unambiguous flexural foreland basin deposits during regional shortening started in the Neuquén Basin at ca. 20 Ma. However, this Neogene record of rapid accumulation was restricted to proximal sectors adjacent to the developing fold-thrust belt. Large parts of the Neuquén Embayment underwent slow erosion with accumulation of Miocene and younger sediments limited to isolated zones.

The overall Cenozoic foreland basin evolution of the Neuquén basin was controlled by regional-scale factors and influenced by second-order local conditions. Following rapid accumulation and a reversal in sedimentary polarity (Neuquén Group), slow subsidence along a neutral margin dominated the Andean retroarc of the southern central Andes during latest Cretaceous to middle Eocene time. This episode was followed by regional extensional conditions in large part of hinterland regions of Argentina and Chile from the middle Eocene to earliest Miocene. The Oligocene tectonic lull and sea-level lowstand inhibit sediment accumulation in foreland regions. The onset of rapid accumulation in the foreland basin at ca. 20 Ma was synchronous from Bolivia to north Patagonia. However, the latitude of the Neuquén basin is pivotal, as the thick Neogene sedimentary record typical of the northern segments of the Central Andes is not expressed in the foreland of southern Mendoza southward into Patagonia.

Acknowledgements We thank the editors of this volume, Andrés Folguera and Diego Kietzmann, for the invitation to present this work. We gratefully acknowledge informative discussions with Ignacio Brisson, Victor Ramos, Guillermo Fratti, Andrés Boll, and Daniel Starck. We thank Lucas Fennell for a careful review that helped improve this contribution.

Appendix

Appendix 1. U-Pb geochronologic results

Grain #	[U]	U/Th	207Pb*	±	206Pb*	±	error	206Pb*	±	207Pb*	±	206Pb*	±	Best age	±
	(ppm)		235U*	(%)	238U	(%)	corr.	238U*	(Ma)	235U	(Ma)	207Pb*	(Ma)	(Ma)	(Ma)
								Age (Ma)		Age (Ma)		Age (Ma)			
COI12: Coihueco Formation; n=118 (35.13761°S, 69.60687°W)															
COI12_13	152	6.8	0.03481	2.4	0.00591	1.1	0.45	38.0	0.4	34.7	0.8	NA	NA	38.0	0.4
COI12_87	509	2.0	0.03976	1.9	0.00595	1.2	0.64	38.2	0.5	39.6	0.7	122.1	34.2	38.2	0.5
COI12_42	55	7.0	0.03748	2.9	0.00625	1.1	0.38	40.2	0.4	37.4	1.1	NA	NA	40.2	0.4
COI12_101	136	5.4	0.03761	2.1	0.00632	1.2	0.58	40.6	0.5	37.5	0.8	NA	NA	40.6	0.5
COI12_55	172	4.1	0.04030	1.9	0.00634	1.1	0.57	40.7	0.4	40.1	0.7	2.4	36.7	40.7	0.4
COI12_66	71	6.2	0.03272	13.6	0.00635	1.1	0.08	40.8	0.4	32.7	4.4	NA	NA	40.8	0.4
COI12_41	1339	2.0	0.04193	1.4	0.00636	1.2	0.81	40.9	0.5	41.7	0.6	90.8	19.7	40.9	0.5
COI12_94	302	6.8	0.04182	2.0	0.00637	1.4	0.71	40.9	0.6	41.6	0.8	81.1	33.6	40.9	0.6
COI12_56	40	8.7	0.03176	13.0	0.00638	1.7	0.13	41.0	0.7	31.8	4.1	NA	NA	41.0	0.7
COI12_78	140	7.7	0.03911	3.2	0.00638	0.9	0.28	41.0	0.4	39.0	1.2	NA	NA	41.0	0.4
COI12_75	391	4.4	0.04047	2.4	0.00639	1.9	0.81	41.0	0.8	40.3	0.9	NA	NA	41.0	0.8
COI12_73	474	5.8	0.04050	1.8	0.00639	1.3	0.72	41.1	0.5	40.3	0.7	NA	NA	41.1	0.5
COI12_95	238	3.2	0.04042	2.1	0.00640	1.3	0.62	41.2	0.5	40.2	0.8	NA	NA	41.2	0.5
COI12_118	429	4.8	0.04238	1.2	0.00641	0.9	0.72	41.2	0.4	42.1	0.5	98.1	19.8	41.2	0.4
COI12_103	679	6.1	0.04190	1.4	0.00641	1.1	0.75	41.2	0.4	41.7	0.6	71.2	22.8	41.2	0.4
COI12_23	177	4.5	0.04182	2.1	0.00641	1.3	0.60	41.2	0.5	41.6	0.9	65.0	40.5	41.2	0.5

(continued)

(continued)

Appendix 1. U-Pb geochronologic results

COI12_96	211	4.8	0.04077	3.4	0.00642	1.3	0.39	41.2	0.5	40.6	1.4	2.5	76.1	41.2	0.5
COI12_2	90	6.9	0.03581	2.6	0.00642	1.1	0.43	41.2	0.5	35.7	0.9	NA	NA	41.2	0.5
COI12_105	86	5.2	0.04222	2.5	0.00642	1.0	0.38	41.3	0.4	42.0	1.0	84.3	55.6	41.3	0.4
COI12_64	353	4.0	0.04202	1.9	0.00642	1.3	0.69	41.3	0.5	41.8	0.8	72.5	32.4	41.3	0.5
COI12_68	219	5.8	0.03917	1.9	0.00642	1.1	0.58	41.3	0.4	39.0	0.7	NA	NA	41.3	0.4
COI12_47	366	5.8	0.04182	1.7	0.00643	1.0	0.61	41.3	0.4	41.6	0.7	59.6	32.3	41.3	0.4
COI12_34	218	4.9	0.04205	2.1	0.00643	1.3	0.65	41.3	0.6	41.8	0.8	70.5	37.3	41.3	0.6
COI12_83	388	4.3	0.04429	2.1	0.00643	1.4	0.65	41.3	0.6	44.0	0.9	192.4	36.8	41.3	0.6
COI12_62	329	4.7	0.03860	2.0	0.00643	1.1	0.57	41.3	0.5	38.5	0.7	NA	NA	41.3	0.5
COI12_48	264	5.6	0.03982	1.7	0.00644	1.2	0.70	41.4	0.5	39.6	0.7	NA	NA	41.4	0.5
COI12_100	73	6.7	0.03550	5.5	0.00644	1.2	0.22	41.4	0.5	35.4	1.9	NA	NA	41.4	0.5
COI12_31	97	6.8	0.04102	2.4	0.00644	1.2	0.50	41.4	0.5	40.8	1.0	8.2	49.7	41.4	0.5
COI12_90	121	4.1	0.04148	2.1	0.00644	1.1	0.55	41.4	0.5	41.3	0.9	33.2	42.2	41.4	0.5
COI12_14	212	4.3	0.04140	2.1	0.00645	1.3	0.61	41.4	0.5	41.2	0.8	26.5	39.3	41.4	0.5
COI12_65	105	5.0	0.03891	3.9	0.00645	1.0	0.26	41.5	0.4	38.8	1.5	NA	NA	41.5	0.4
COI12_59	176	4.7	0.04347	2.4	0.00645	1.4	0.58	41.5	0.6	43.2	1.0	140.9	45.0	41.5	0.6
COI12_21	239	6.4	0.04331	2.2	0.00646	1.2	0.57	41.5	0.5	43.1	0.9	131.1	41.9	41.5	0.5
COI12_43	912	1.8	0.04227	1.5	0.00646	1.2	0.81	41.5	0.5	42.0	0.6	72.9	21.2	41.5	0.5
COI12_27	104	6.4	0.04241	7.0	0.00646	1.3	0.18	41.5	0.5	42.2	2.9	80.3	162.8	41.5	0.5
COI12_39	55	7.2	0.04361	3.3	0.00647	1.2	0.36	41.6	0.5	43.3	1.4	142.0	72.7	41.6	0.5
COI12_17	160	5.5	0.04058	2.2	0.00647	1.0	0.44	41.6	0.4	40.4	0.9	NA	NA	41.6	0.4
COI12_106	75	6.7	0.04458	3.2	0.00648	1.3	0.40	41.6	0.5	44.3	1.4	191.8	69.0	41.6	0.5

(continued)

(continued)

Appendix 1. U-Pb geochronologic results

COI12_115	186	4.5	0.04257	2.0	0.00648	1.1	0.53	41.6	0.4	42.3	0.8	82.9	40.7	41.6	0.4
COI12_74	179	9.3	0.04387	2.1	0.00648	1.3	0.63	41.6	0.5	43.6	0.9	152.7	38.0	41.6	0.5
COI12_50	1033	2.8	0.04310	1.5	0.00648	1.0	0.67	41.7	0.4	42.8	0.6	109.6	25.6	41.7	0.4
COI12_52	313	4.4	0.04890	2.9	0.00649	1.2	0.43	41.7	0.5	48.5	1.4	399.2	59.1	41.7	0.5
COI12_110	58	6.4	0.04276	2.9	0.00649	1.2	0.40	41.7	0.5	42.5	1.2	88.3	63.3	41.7	0.5
COI12_92	240	8.9	0.04057	2.3	0.00649	1.2	0.54	41.7	0.5	40.4	0.9	NA	NA	41.7	0.5
COI12_60	158	4.9	0.03820	2.4	0.00649	1.4	0.59	41.7	0.6	38.1	0.9	NA	NA	41.7	0.6
COI12_112	48	7.0	0.04018	3.8	0.00649	1.6	0.42	41.7	0.7	40.0	1.5	NA	NA	41.7	0.7
COI12_82	60	6.9	0.05089	6.8	0.00649	2.0	0.29	41.7	0.8	50.4	3.3	485.3	142.9	41.7	0.8
COI12_70	214	6.0	0.04574	2.9	0.00650	1.3	0.45	41.7	0.6	45.4	1.3	244.1	59.7	41.7	0.6
COI12_104	556	3.8	0.04154	1.7	0.00650	1.2	0.74	41.8	0.5	41.3	0.7	17.0	27.0	41.8	0.5
COI12_19	285	5.9	0.04198	2.1	0.00650	1.4	0.70	41.8	0.6	41.8	0.8	39.9	35.7	41.8	0.6
COI12_114	56	7.4	0.03944	3.5	0.00650	1.7	0.47	41.8	0.7	39.3	1.4	NA	NA	41.8	0.7
COI12_16	123	4.9	0.03780	3.1	0.00651	1.3	0.44	41.8	0.6	37.7	1.1	NA	NA	41.8	0.6
COI12_117	49	7.2	0.02785	8.3	0.00651	1.3	0.16	41.8	0.6	27.9	2.3	NA	NA	41.8	0.6
COI12_85	112	6.3	0.04193	2.6	0.00652	1.1	0.41	41.9	0.4	41.7	1.1	31.5	56.7	41.9	0.4
COI12_99	384	2.8	0.04288	1.7	0.00652	1.2	0.74	41.9	0.5	42.6	0.7	84.4	27.0	41.9	0.5
COI12_91	527	3.9	0.04228	1.6	0.00652	1.1	0.66	41.9	0.4	42.0	0.7	49.3	29.2	41.9	0.4
COI12_84	299	6.7	0.04779	2.5	0.00652	1.7	0.70	41.9	0.7	47.4	1.1	334.3	40.4	41.9	0.7
COI12_4	177	2.9	0.04333	2.3	0.00653	1.3	0.56	41.9	0.5	43.1	1.0	106.3	44.7	41.9	0.5
COI12_12	813	4.1	0.04268	1.7	0.00653	1.2	0.75	42.0	0.5	42.4	0.7	70.3	25.9	42.0	0.5
COI12_89	313	4.2	0.04976	3.7	0.00653	1.0	0.26	42.0	0.4	49.3	1.8	423.0	80.0	42.0	0.4

(continued)

(continued)

Appendix 1. U-Pb geochronologic results

COI12_40	92	5.9	0.03651	1.9	0.00654	1.0	0.50	42.0	0.4	36.4	0.7	NA	NA	42.0	0.4
COI12_109	830	2.9	0.05119	2.3	0.00654	1.3	0.58	42.0	0.6	50.7	1.1	483.4	41.7	42.0	0.6
COI12_5	22	12.5	0.03047	22.5	0.00654	1.9	0.09	42.0	0.8	30.5	6.8	NA	NA	42.0	0.8
COI12_107	238	5.5	0.04333	2.2	0.00654	1.3	0.60	42.0	0.6	43.1	0.9	102.0	42.4	42.0	0.6
COI12_81	1110	4.1	0.04255	1.6	0.00654	1.2	0.75	42.0	0.5	42.3	0.7	57.9	25.4	42.0	0.5
COI12_15	674	3.9	0.04350	1.6	0.00655	1.1	0.71	42.1	0.5	43.2	0.7	108.1	26.4	42.1	0.5
COI12_6	48	8.5	0.02633	3.9	0.00655	1.5	0.37	42.1	0.6	26.4	1.0	NA	NA	42.1	0.6
COI12_7	74	6.0	0.03944	3.2	0.00657	1.2	0.37	42.2	0.5	39.3	1.3	NA	NA	42.2	0.5
COI12_51	1026	5.2	0.04290	1.6	0.00657	1.3	0.82	42.2	0.6	42.7	0.7	66.0	22.3	42.2	0.6
COI12_58	139	4.9	0.04369	2.5	0.00658	1.2	0.47	42.3	0.5	43.4	1.1	108.2	52.8	42.3	0.5
COI12_24	81	6.0	0.03707	3.5	0.00658	1.3	0.38	42.3	0.6	37.0	1.3	NA	NA	42.3	0.6
COI12_79	147	7.5	0.04823	4.2	0.00658	1.3	0.31	42.3	0.6	47.8	1.9	334.5	89.7	42.3	0.6
COI12_28	120	6.8	0.04335	2.4	0.00658	1.1	0.46	42.3	0.5	43.1	1.0	87.2	51.4	42.3	0.5
COI12_45	181	5.9	0.04257	1.8	0.00659	0.9	0.49	42.3	0.4	42.3	0.7	42.7	37.8	42.3	0.4
COI12_113	88	4.2	0.04108	3.0	0.00659	1.1	0.38	42.4	0.5	40.9	1.2	NA	NA	42.4	0.5
COI12_71	502	4.2	0.04524	1.9	0.00660	1.2	0.65	42.4	0.5	44.9	0.8	183.2	33.7	42.4	0.5
COI12_97	418	4.3	0.04537	2.0	0.00660	1.0	0.52	42.4	0.4	45.1	0.9	189.4	38.9	42.4	0.4
COI12_11	349	8.2	0.04351	1.8	0.00660	1.1	0.64	42.4	0.5	43.2	0.7	89.3	32.2	42.4	0.5
COI12_116	67	4.3	0.04614	3.2	0.00660	1.1	0.33	42.4	0.4	45.8	1.4	225.8	70.1	42.4	0.4
COI12_33	223	5.8	0.04542	2.0	0.00661	1.3	0.66	42.4	0.6	45.1	0.9	188.9	35.4	42.4	0.6
COI12_36	84	6.5	0.05342	5.0	0.00661	1.2	0.24	42.5	0.5	52.8	2.6	553.2	107.1	42.5	0.5
COI12_102	224	6.3	0.04186	2.8	0.00662	1.2	0.42	42.5	0.5	41.6	1.1	NA	NA	42.5	0.5

(continued)

(continued)

Appendix 1. U-Pb geochronologic results

COI12_26	380	3.6	0.04334	2.0	0.00662	1.2	0.60	42.5	0.5	43.1	0.8	73.2	37.8	42.5	0.5
COI12_119	570	4.1	0.04392	1.5	0.00663	0.9	0.61	42.6	0.4	43.6	0.6	101.1	28.4	42.6	0.4
COI12_25	232	4.5	0.04382	1.9	0.00663	1.2	0.62	42.6	0.5	43.5	0.8	95.3	35.7	42.6	0.5
COI12_18	376	4.2	0.04225	1.8	0.00663	1.5	0.79	42.6	0.6	42.0	0.8	7.9	26.8	42.6	0.6
COI12_49	24	8.0	0.05745	7.0	0.00665	1.5	0.21	42.7	0.6	56.7	3.9	696.9	146.0	42.7	0.6
COI12_54	1000	3.3	0.04374	1.4	0.00666	1.0	0.71	42.8	0.4	43.5	0.6	81.8	24.0	42.8	0.4
COI12_111	261	3.3	0.04463	1.9	0.00666	1.2	0.64	42.8	0.5	44.3	0.8	127.0	34.9	42.8	0.5
COI12_67	173	4.4	0.04292	3.3	0.00667	1.4	0.42	42.8	0.6	42.7	1.4	33.0	71.8	42.8	0.6
COI12_10	98	5.3	0.03882	3.2	0.00667	1.2	0.37	42.9	0.5	38.7	1.2	NA	NA	42.9	0.5
COI12_22	521	4.5	0.04269	1.6	0.00669	1.3	0.80	43.0	0.6	42.4	0.7	10.9	23.5	43.0	0.6
COI12_44	229	5.4	0.04529	2.8	0.00670	1.2	0.42	43.0	0.5	45.0	1.2	149.9	59.8	43.0	0.5
COI12_29	110	9.1	0.03444	8.7	0.00671	1.2	0.13	43.1	0.5	34.4	2.9	NA	NA	43.1	0.5
COI12_20	114	4.6	0.04126	5.3	0.00671	1.2	0.23	43.1	0.5	41.0	2.1	NA	NA	43.1	0.5
COI12_32	399	6.4	0.04342	1.7	0.00672	1.3	0.79	43.2	0.6	43.2	0.7	41.6	25.3	43.2	0.6
COI12_76	105	7.4	0.05638	4.0	0.00673	1.4	0.35	43.3	0.6	55.7	2.2	630.0	80.7	43.3	0.6
COI12_30	267	5.6	0.04194	2.1	0.00680	1.3	0.62	43.7	0.6	41.7	0.9	NA	NA	43.7	0.6
COI12_61	274	5.0	0.07352	7.6	0.00680	1.3	0.17	43.7	0.6	72.0	5.3	1157.0	149.0	43.7	0.6
COI12_63	106	7.9	0.03988	3.5	0.00680	1.0	0.28	43.7	0.4	39.7	1.3	NA	NA	43.7	0.4
COI12_9	29	10.8	0.05980	9.8	0.00682	1.8	0.19	43.8	0.8	59.0	5.6	729.4	203.5	43.8	0.8
COI12_77	682	4.6	0.07505	7.2	0.00683	1.2	0.17	43.9	0.5	73.5	5.1	1187.9	140.6	43.9	0.5
COI12_37	209	6.1	0.04647	1.6	0.00684	1.1	0.69	43.9	0.5	46.1	0.7	162.6	26.8	43.9	0.5
COI12_86	251	6.2	0.04524	2.7	0.00684	1.7	0.62	43.9	0.7	44.9	1.2	99.3	50.6	43.9	0.7

(continued)

(continued)

Appendix 1. U-Pb geochronologic results

COI12_3	84	4.2	0.05408	5.2	0.00684	1.8	0.35	44.0	0.8	53.5	2.7	504.3	108.2	44.0	0.8
COI12_72	57	6.6	0.04027	3.1	0.00689	1.2	0.38	44.2	0.5	40.1	1.2	NA	NA	44.2	0.5
COI12_1	138	6.1	0.04521	2.1	0.00689	1.0	0.46	44.2	0.4	44.9	0.9	80.2	44.2	44.2	0.4
COI12_8	150	3.2	0.04504	2.3	0.00694	0.8	0.36	44.6	0.4	44.7	1.0	53.1	51.1	44.6	0.4
COI12_93	78	6.0	0.04272	3.7	0.00712	1.4	0.37	45.8	0.6	42.5	1.5	NA	NA	45.8	0.6
COI12_35	107	3.5	0.05843	7.1	0.00721	2.1	0.30	46.3	1.0	57.7	4.0	560.2	147.0	46.3	1.0
COI12_57	117	7.5	0.05805	6.3	0.00721	1.2	0.18	46.3	0.5	57.3	3.5	544.1	134.5	46.3	0.5
COI12_108	106	6.5	0.10718	13.3	0.00755	2.8	0.21	48.5	1.3	103.4	13.0	1678.8	240.5	48.5	1.3
COI12_38	201	5.3	0.05057	2.4	0.00760	1.2	0.49	48.8	0.6	50.1	1.2	111.2	50.0	48.8	0.6
COI12_53	199	5.6	0.22982	1.8	0.00831	1.3	0.72	53.4	0.7	210.1	3.4	2830.3	20.4	53.4	0.7
COI12_0	219	5.5	0.16667	1.4	0.02354	1.0	0.68	150.0	1.4	156.5	2.0	256.2	23.8	150.0	1.4
COI12_69	177	4.1	0.16300	1.6	0.02429	1.1	0.64	154.7	1.6	153.3	2.3	131.6	29.8	154.7	1.6
COI12_98	87	3.9	0.18554	1.9	0.02794	1.2	0.67	177.6	2.2	172.8	2.9	107.5	32.5	177.6	2.2
COI12_88	154	7.4	0.44061	1.4	0.05908	1.0	0.71	370.0	3.5	370.7	4.2	374.8	21.6	370.0	3.5

COI13: Coihueco Formation; n=111 (35.13755°S, 69.60700°W)															
COI13_70	266	1.9	0.04487	2.2	0.00613	1.3	0.59	39.4	0.5	44.6	1.0	331.7	40.4	39.4	0.5
COI13_81	77	3.8	0.03690	3.9	0.00622	1.3	0.32	40.0	0.5	36.8	1.4	NA	NA	40.0	0.5
COI13_12	82	3.0	0.03728	3.4	0.00623	1.2	0.35	40.1	0.5	37.2	1.3	NA	NA	40.1	0.5
COI13_108	308	1.7	0.03962	2.0	0.00624	1.4	0.70	40.1	0.6	39.5	0.8	1.2	34.0	40.1	0.6
COI13_44	405	2.0	0.04531	2.2	0.00624	1.2	0.53	40.1	0.5	45.0	1.0	313.5	42.9	40.1	0.5
COI13_10	87	2.7	0.03910	2.4	0.00627	1.2	0.50	40.3	0.5	38.9	0.9	NA	NA	40.3	0.5
COI13_72	207	1.7	0.03765	2.2	0.00627	1.6	0.71	40.3	0.6	37.5	0.8	NA	NA	40.3	0.6
COI13_39	589	1.4	0.04028	1.4	0.00631	0.9	0.64	40.5	0.4	40.1	0.6	13.9	26.4	40.5	0.4
COI13_117	134	2.6	0.03622	2.3	0.00631	1.1	0.50	40.6	0.5	36.1	0.8	NA	NA	40.6	0.5
COI13_4	113	2.7	0.03863	4.6	0.00631	1.1	0.24	40.6	0.5	38.5	1.8	NA	NA	40.6	0.5
COI13_105	169	2.2	0.03958	2.6	0.00632	1.5	0.59	40.6	0.6	39.4	1.0	NA	NA	40.6	0.6
COI13_88	122	2.4	0.04073	3.4	0.00632	1.0	0.30	40.6	0.4	40.5	1.4	36.5	78.0	40.6	0.4
COI13_28	232	2.0	0.04034	2.5	0.00633	1.4	0.57	40.7	0.6	40.2	1.0	10.5	49.3	40.7	0.6
COI13_99	473	1.1	0.04135	1.7	0.00633	1.2	0.74	40.7	0.5	41.1	0.7	66.5	26.4	40.7	0.5
COI13_73	116	2.3	0.04020	3.3	0.00634	1.1	0.33	40.7	0.4	40.0	1.3	NA	NA	40.7	0.4
COI13_110	104	2.8	0.04255	2.9	0.00634	1.0	0.36	40.8	0.4	42.3	1.2	131.1	63.1	40.8	0.4
COI13_98	272	1.4	0.04112	2.1	0.00634	1.2	0.58	40.8	0.5	40.9	0.8	49.7	40.8	40.8	0.5
COI13_29	566	0.9	0.04209	1.8	0.00635	1.0	0.59	40.8	0.4	41.9	0.7	101.7	33.6	40.8	0.4
COI13_53	427	1.4	0.04576	2.3	0.00635	1.4	0.62	40.8	0.6	45.4	1.0	295.7	40.6	40.8	0.6
COI13_67	164	2.0	0.04094	2.3	0.00635	0.9	0.41	40.8	0.4	40.7	0.9	35.4	50.4	40.8	0.4
COI13_34	95	2.7	0.04019	2.9	0.00635	1.1	0.37	40.8	0.4	40.0	1.1	NA	NA	40.8	0.4

(continued)

(continued)

COI13: Coihueco Formation; n=111 (35.13755°S, 69.60700°W)															
COI13_5	125	2.3	0.03740	2.5	0.00636	1.2	0.48	40.9	0.5	37.3	0.9	NA	NA	40.9	0.5
COI13_113	191	1.6	0.03836	2.4	0.00636	1.4	0.57	40.9	0.6	38.2	0.9	NA	NA	40.9	0.6
COI13_7	875	2.2	0.04153	1.5	0.00637	1.0	0.68	40.9	0.4	41.3	0.6	63.3	25.5	40.9	0.4
COI13_97	112	2.8	0.04067	2.7	0.00638	1.4	0.53	41.0	0.6	40.5	1.1	10.3	54.6	41.0	0.6
COI13_104	49	2.7	0.04808	4.4	0.00638	1.6	0.36	41.0	0.7	47.7	2.1	397.8	92.7	41.0	0.7
COI13_57	1089	1.3	0.04411	2.0	0.00639	1.3	0.67	41.1	0.5	43.8	0.8	197.7	33.6	41.1	0.5
COI13_87	63	3.7	0.03317	4.2	0.00639	1.5	0.35	41.1	0.6	33.1	1.4	NA	NA	41.1	0.6
COI13_9	140	2.6	0.04144	2.0	0.00640	1.0	0.49	41.2	0.4	41.2	0.8	45.6	41.3	41.2	0.4
COI13_92	136	1.9	0.04079	2.5	0.00641	1.4	0.55	41.2	0.6	40.6	1.0	6.4	49.2	41.2	0.6
COI13_78	644	1.4	0.04115	1.7	0.00641	1.3	0.76	41.2	0.5	40.9	0.7	27.4	26.0	41.2	0.5
COI13_51	122	2.4	0.04001	2.9	0.00641	1.4	0.50	41.2	0.6	39.8	1.1	NA	NA	41.2	0.6
COI13_69	90	2.2	0.03880	3.2	0.00641	1.2	0.39	41.2	0.5	38.7	1.2	NA	NA	41.2	0.5
COI13_26	236	1.5	0.04044	2.2	0.00641	1.3	0.62	41.2	0.6	40.3	0.9	NA	NA	41.2	0.6
COI13_6	130	2.2	0.03900	2.6	0.00641	1.2	0.44	41.2	0.5	38.8	1.0	NA	NA	41.2	0.5
COI13_82	181	2.3	0.04309	2.1	0.00641	1.4	0.67	41.2	0.6	42.8	0.9	134.5	37.0	41.2	0.6
COI13_23	166	1.9	0.03830	4.4	0.00642	1.4	0.31	41.2	0.6	38.2	1.6	NA	NA	41.2	0.6
COI13_15	269	2.8	0.04364	1.8	0.00642	1.2	0.66	41.2	0.5	43.4	0.8	163.0	31.6	41.2	0.5
COI13_30	249	2.9	0.04204	1.9	0.00643	1.4	0.71	41.3	0.6	41.8	0.8	71.2	32.2	41.3	0.6
COI13_25	364	1.8	0.04152	2.0	0.00644	1.3	0.68	41.4	0.6	41.3	0.8	36.0	34.4	41.4	0.6
COI13_89	188	2.0	0.04243	2.5	0.00644	1.3	0.51	41.4	0.5	42.2	1.0	86.9	51.8	41.4	0.5
COI13_3	57	2.7	0.04347	3.5	0.00645	1.0	0.30	41.4	0.4	43.2	1.5	142.5	78.0	41.4	0.4
COI13_71	105	2.8	0.04116	2.3	0.00645	1.1	0.47	41.4	0.4	41.0	0.9	12.6	48.0	41.4	0.4

(continued)

(continued)

COI13: Coihueco Formation; n=111 (35.13755°S, 69.60700°W)															
COI13_90	474	1.6	0.04089	1.6	0.00645	1.0	0.65	41.5	0.4	40.7	0.6	NA	NA	41.5	0.4
COI13_84	369	1.2	0.04038	2.3	0.00645	1.3	0.58	41.5	0.5	40.2	0.9	NA	NA	41.5	0.5
COI13_2	428	2.8	0.04229	1.6	0.00646	1.0	0.60	41.5	0.4	42.1	0.7	73.6	30.8	41.5	0.4
COI13_85	307	3.0	0.04223	2.1	0.00646	1.4	0.65	41.5	0.6	42.0	0.9	68.2	37.5	41.5	0.6
COI13_14	109	2.5	0.04055	3.8	0.00647	0.9	0.23	41.6	0.4	40.4	1.5	NA	NA	41.6	0.4
COI13_102	91	2.9	0.03877	5.2	0.00648	1.6	0.30	41.6	0.6	38.6	2.0	NA	NA	41.6	0.6
COI13_91	199	2.0	0.04111	2.3	0.00648	1.4	0.59	41.6	0.6	40.9	0.9	NA	NA	41.6	0.6
COI13_17	434	2.5	0.04147	1.6	0.00649	1.1	0.66	41.7	0.5	41.3	0.7	15.7	29.6	41.7	0.5
COI13_16	100	2.2	0.04091	4.1	0.00649	1.1	0.27	41.7	0.5	40.7	1.6	NA	NA	41.7	0.5
COI13_66	84	2.1	0.04048	3.6	0.00650	1.4	0.38	41.8	0.6	40.3	1.4	NA	NA	41.8	0.6
COI13_59	382	1.9	0.04253	1.8	0.00650	1.2	0.67	41.8	0.5	42.3	0.8	70.6	32.2	41.8	0.5
COI13_13	173	2.5	0.04091	2.7	0.00651	1.4	0.54	41.8	0.6	40.7	1.1	NA	NA	41.8	0.6
COI13_27	52	2.5	0.03829	4.2	0.00651	1.4	0.34	41.8	0.6	38.1	1.6	NA	NA	41.8	0.6
COI13_116	159	3.5	0.06127	5.2	0.00651	1.4	0.27	41.8	0.6	60.4	3.1	875.5	104.3	41.8	0.6
COI13_119	177	1.6	0.04317	2.1	0.00652	1.2	0.55	41.9	0.5	42.9	0.9	100.2	41.6	41.9	0.5
COI13_46	111	2.7	0.04006	2.5	0.00652	1.3	0.54	41.9	0.6	39.9	1.0	NA	NA	41.9	0.6
COI13_115	1078	3.9	0.05779	3.7	0.00653	1.2	0.32	41.9	0.5	57.0	2.1	748.7	74.8	41.9	0.5
COI13_40	52	3.7	0.04204	3.4	0.00653	1.3	0.37	42.0	0.5	41.8	1.4	32.7	76.6	42.0	0.5
COI13_42	107	2.0	0.04140	2.6	0.00653	1.2	0.47	42.0	0.5	41.2	1.0	NA	NA	42.0	0.5
COI13_111	356	2.9	0.04341	1.7	0.00654	1.1	0.63	42.0	0.5	43.2	0.7	106.9	31.8	42.0	0.5
COI13_56	69	2.4	0.04051	3.1	0.00654	1.3	0.40	42.0	0.5	40.3	1.2	NA	NA	42.0	0.5
COI13_95	71	1.9	0.03742	4.4	0.00654	1.2	0.28	42.0	0.5	37.3	1.6	NA	NA	42.0	0.5

(continued)

(continued)

COI13: Coihueco Formation; n=111 (35.13755°S, 69.60700°W)															
COI13_101	593	1.4	0.04294	1.8	0.00654	1.4	0.78	42.1	0.6	42.7	0.8	78.7	27.3	42.1	0.6
COI13_107	195	1.4	0.04272	1.8	0.00656	1.2	0.67	42.1	0.5	42.5	0.8	61.5	32.5	42.1	0.5
COI13_93	61	1.8	0.06560	2.5	0.00656	1.1	0.44	42.1	0.5	64.5	1.5	1001.4	44.9	42.1	0.5
COI13_19	1065	1.3	0.04166	1.4	0.00656	1.2	0.85	42.2	0.5	41.4	0.6	NA	NA	42.2	0.5
COI13_8	1221	2.4	0.04729	1.7	0.00656	1.1	0.64	42.2	0.5	46.9	0.8	296.5	30.5	42.2	0.5
COI13_50	170	2.5	0.04283	2.6	0.00657	1.2	0.47	42.2	0.5	42.6	1.1	64.9	54.3	42.2	0.5
COI13_48	406	1.6	0.04314	1.4	0.00657	0.9	0.64	42.2	0.4	42.9	0.6	80.5	24.8	42.2	0.4
COI13_74	140	2.2	0.04181	2.3	0.00657	0.9	0.38	42.2	0.4	41.6	0.9	5.9	51.4	42.2	0.4
COI13_64	79	2.7	0.04084	2.8	0.00657	0.9	0.32	42.2	0.4	40.6	1.1	NA	NA	42.2	0.4
COI13_106	239	2.8	0.04261	2.4	0.00658	1.6	0.66	42.3	0.7	42.4	1.0	48.1	42.4	42.3	0.7
COI13_0	313	1.6	0.04257	2.3	0.00658	1.6	0.69	42.3	0.7	42.3	0.9	44.4	39.4	42.3	0.7
COI13_112	114	2.1	0.04203	2.6	0.00661	1.3	0.50	42.5	0.5	41.8	1.1	4.2	53.4	42.5	0.5
COI13_61	74	2.7	0.04582	2.7	0.00661	1.3	0.46	42.5	0.5	45.5	1.2	206.8	56.1	42.5	0.5
COI13_33	502	1.0	0.04331	1.9	0.00662	1.0	0.52	42.5	0.4	43.1	0.8	72.2	38.7	42.5	0.4
COI13_24	123	2.4	0.04196	2.6	0.00662	0.8	0.31	42.5	0.3	41.7	1.0	NA	NA	42.5	0.3
COI13_118	175	1.3	0.04177	2.3	0.00663	1.7	0.71	42.6	0.7	41.6	1.0	NA	NA	42.6	0.7
COI13_62	297	2.0	0.04368	1.6	0.00664	1.1	0.70	42.6	0.5	43.4	0.7	86.5	27.0	42.6	0.5
COI13_103	248	1.5	0.06610	4.6	0.00665	1.2	0.27	42.7	0.5	65.0	2.9	989.7	90.6	42.7	0.5
COI13_58	122	2.4	0.04236	2.1	0.00665	1.1	0.51	42.8	0.5	42.1	0.9	6.7	44.5	42.8	0.5
COI13_109	163	1.5	0.04185	2.7	0.00666	1.4	0.54	42.8	0.6	41.6	1.1	NA	NA	42.8	0.6
COI13_21	138	1.8	0.04523	2.7	0.00666	1.3	0.48	42.8	0.6	44.9	1.2	160.4	55.1	42.8	0.6
COI13_38	115	2.4	0.04418	3.8	0.00667	1.1	0.29	42.9	0.5	43.9	1.6	100.1	85.5	42.9	0.5

(continued)

(continued)

COI13: Coihueco Formation; n=111 (35.13755°S, 69.60700°W)															
COI13_83	36	3.9	0.04264	5.0	0.00669	1.6	0.32	43.0	0.7	42.4	2.1	9.5	113.8	43.0	0.7
COI13_43	180	1.7	0.04348	2.8	0.00670	1.6	0.57	43.0	0.7	43.2	1.2	54.0	55.2	43.0	0.7
COI13_41	77	2.4	0.04427	2.8	0.00672	1.2	0.43	43.2	0.5	44.0	1.2	88.7	59.4	43.2	0.5
COI13_55	155	2.2	0.04275	3.0	0.00672	1.7	0.55	43.2	0.7	42.5	1.3	4.3	60.4	43.2	0.7
COI13_47	108	2.7	0.04492	1.8	0.00675	0.8	0.44	43.3	0.3	44.6	0.8	113.9	37.4	43.3	0.3
COI13_77	412	1.5	0.04369	2.0	0.00677	1.5	0.75	43.5	0.6	43.4	0.8	37.7	31.5	43.5	0.6
COI13_96	292	2.0	0.04454	1.8	0.00678	1.2	0.70	43.6	0.5	44.2	0.8	81.1	30.2	43.6	0.5
COI13_45	191	2.2	0.04173	3.2	0.00679	1.7	0.52	43.6	0.7	41.5	1.3	NA	NA	43.6	0.7
COI13_22	71	2.7	0.04136	3.5	0.00681	1.3	0.38	43.7	0.6	41.1	1.4	NA	NA	43.7	0.6
COI13_20	92	2.1	0.03461	3.0	0.00682	1.3	0.44	43.8	0.6	34.5	1.0	NA	NA	43.8	0.6
COI13_80	171	2.2	0.05689	3.7	0.00688	1.3	0.36	44.2	0.6	56.2	2.0	602.3	74.5	44.2	0.6
COI13_75	78	2.5	0.04473	3.4	0.00710	1.0	0.30	45.6	0.5	44.4	1.5	NA	NA	45.6	0.5
COI13_60	184	1.2	0.05022	1.9	0.00841	1.1	0.59	54.0	0.6	49.8	0.9	NA	NA	54.0	0.6
COI13_1	110	2.8	0.06237	2.4	0.00971	1.1	0.46	62.3	0.7	61.4	1.4	28.7	50.6	62.3	0.7
COI13_36	93	2.1	0.07343	2.6	0.01165	1.2	0.46	74.7	0.9	72.0	1.8	NA	NA	74.7	0.9
COI13_18	417	2.9	0.15812	1.6	0.02341	1.4	0.86	149.2	2.1	149.1	2.3	147.1	19.3	149.2	2.1
COI13_35	243	1.4	0.18758	2.3	0.02434	1.4	0.60	155.0	2.1	174.6	3.7	448.2	41.4	155.0	2.1
COI13_100	352	2.2	0.17191	1.3	0.02533	0.9	0.72	161.3	1.5	161.1	1.9	158.2	21.2	161.3	1.5
COI13_37	70	1.8	0.18269	1.6	0.02668	1.0	0.66	169.8	1.8	170.4	2.5	178.7	27.4	169.8	1.8
COI13_31	252	6.5	0.18765	1.5	0.02744	1.3	0.86	174.5	2.3	174.6	2.4	175.7	18.1	174.5	2.3
COI13_114	651	2.7	0.32867	1.4	0.04556	1.2	0.86	287.2	3.3	288.5	3.4	299.2	16.0	287.2	3.3
COI13_11	134	2.5	0.73395	1.2	0.08985	0.8	0.69	554.6	4.3	558.9	5.1	576.2	18.4	554.6	4.3

(continued)

(continued)

COI13: Coihueco Formation; n=111 (35.13755°S, 69.60700°W)															
COI13_76	308	2.6	0.84416	1.4	0.10004	1.1	0.81	614.6	6.5	621.4	6.4	646.3	17.5	614.6	6.5
COI13_94	85	2.0	3.64141	1.7	0.26794	1.3	0.75	1530.3	17.8	1558.6	13.8	1597.2	21.2	1597.2	21.2

References

Aguirre Urreta B, Tunik M, Naipauer M et al (2011) Malargüe Group (Maastrichtian–Danian) deposits in the Neuquén Andes, Argentina: implications for the onset of the first Atlantic transgression related to Western Gondwana break-up. Gondwana Res 19:482–494

Balgord EA, Carrapa B (2016) Basin evolution of Upper Cretaceous-Lower Cenozoic strata in the Malargüe fold-and-thrust belt: Northern Neuquén Basin, Argentina. Basin Res 28:20–183

Beaumont C (1981) Foreland basins. Geophys J Roy Astr S 65:291–329

Boll A, Alonso A, Fuentes F et al (2014) Factores controlantes de las acumulaciones de hidrocarburos en el sector norte de la cuenca Neuquina, entre los ríos Diamante y Salado, Provincia de Mendoza, Argentina. In: Abstracts of the 9 Congreso de Exploración y Desarrollo de Hidrocarburos, IAPG, Mendoza, 5–9 Nov 2014

Borghi P, Gómez Omil R, Fennell L et al (2017) Nuevas evidencias del levantamiento del sur de los Andes Centrales (36°S) durante la depositación del Grupo Neuquén. In: Abstracts o the 20 Congreso Geológico Argentino, San Miguel de Tucumán, 7–11 Aug 2017

Broens S, Pereira DM (2005) Evolución estructural de la zona de transición entre las fajas plegadas y corridas de Aconcagua y Malargüe Provincia de Mendoza. Rev Asoc Geol Argent 60(4):685–695

Buelow E, Suriano J, Mahoney B, Kimbroughm DL, Mescua J, Giambiagi L, Hoke G (2018) Sedimentologic and stratigraphic evolution of the Cacheuta basin: Constraints on the development of the Miocene retroarc foreland basin, South-central Andes. Lithosphere 10:366–391

Burns WM, Jordan TE, Copeland P, Kelley SA (2006) The case for extensional tectonics in the Oligocene-Miocene southern Andes as recorded in the Cura Mallín Basin (36°–38° S). Geol Soc Am, SP 407:163–184

Cazau L, Uliana MA (1973) El Cretacico superior continental de la Cuenca Neuquina. In: Abstracts of the 5 Congreso Geologico Argentino, Córdoba, 22–28 Oct 1973

Cevallos M, Rivero MT (2014) El proceso de exhumación neógeno y sus consecuencias en las acumulaciones de hidrocarburos del borde noreste de la Cuenca Neuquina, Argentina. In: Abstracts of the 9 Congreso de Exploración y Desarrollo de Hidrocarburos, IAPG, Mendoza, 5–9 Nov 2014

Cobbold P, Rossello EA (2003) Aptian to recent compressional deformation, foothills of the Neuquén Basin, Argentina. Mar Petrol Geol 20:429–443

Coney PJ, Evenchick CA (1994) Consolidation of the American Cordilleras. J S Am Earth Sci 7:241–262

DeCelles PG (2012) Foreland basin systems revisited: variations in response to tectonic settings. In: Busby C, Azor A (eds) Tectonics of sedimentary basins: recent advances. Wiley, Oxford, pp 405–426

DeCelles PG, Giles KA (1996) Foreland basin systems. Basin Res 8:105–123

DeCelles PG, Carrapa B, Horton BK et al (2011) Cenozoic foreland basin system in the central Andes of northwestern Argentina: implications for Andean geodynamics and modes of deformation. Tectonics 30:TC6013. https://doi.org/10.1029/2011tc002948

DeCelles PG, Ducea MH, Kapp P et al (2009) Cyclicity in Cordilleran orogenic systems. Nat Geoscie 2:251–257

DeCelles PG, Zandt G, Beck SL et al (2015) Cyclical orogenic processes in the Cenozoic central Andes. In: DeCelles PG, Ducea MN, Carrapa B, Kapp PA (eds) Geodynamics of a Cordilleran Orogenic System: The Central Andes of Argentina and Northern Chile. Geological Society of America Memoir 212, 459–490

Di Giulio A, Ronchi A, Sanfilippo A et al (2012) Detrital zircon provenance from the Neuquén Basin (south-central Andes): Cretaceous geodynamic evolution and sedimentary response in a retroarc-foreland basin. Geology 40:559–562

Dyhr CT, Holm PM, Llambías EJ (2013) Geochemical constraints on the relationship between the Miocene-Pliocene volcanism and tectonics in the Palaoco and Fortunoso volcanic fields, Mendoza Region, Argentina: new insights from 40Ar/39Ar dating, Sr–Nd–Pb isotopes and trace elements. J Volcanol Geoth Res 266:50–68

Fennell LM, Folguera A, Naipauer M et al (2017) Cretaceous deformation of the southern central Andes: synorogenic growth strata in the Neuquén Group (35° 30′–37° S). Basin Res 29:51–72
Folguera A, Rojas Vera E, Bottessi G et al (2010) The Loncopué trough: a Cenozoic basin produced by extension in the southern central Andes. J Geodyn 49:287–295
Folguera A, Zárate M, Tedesco A et al (2015) Evolution of the Neogene Andean foreland basins of the Southern Pampas and Northern Patagonia (34°–41° S), Argentina. J S Am Earth Sci 64:452–466
Fuentes F, Horton BK, Starck D et al (2016) Structure and tectonic evolution of hybrid thick- and thin-skinned systems in the Malargüe fold-thrust belt, Neuquén Basin, Argentina. Geol Mag 153:1066–1084
Galarza BJ, Zamora Valcarce G et al (2009) Geología y Evolución tectónica del Frente Cordillerano a los 36° 30′ S: bloques de Yihuin-Huaca y Puntilla de Huincán, Mendoza. Rev Asoc Geol Argent 65(1):170–191
Garrido A, Kramarz A, Forasiepi A et al (2012) Estratigrafía, mamíferos fósiles y edad de las secuencias volcanosedimentarias eoceno-miocenas de la sierra de Huantraico-sierra Negra y cerro Villegas (provincia del Neuquén, Argentina). Andean Geol 39, 482–510
Giambiagi L, Ghiglione M, Cristallini E et al (2009) Kinematic models of basement/cover interactions: insights from the Malargüe fold and thrust belt, Mendoza, Argentina. J Struct Geol 31:1443–1457
Giambiagi L, Mescua J, Bechis F et al (2012) Thrust belts of the Southern Central Andes: along-strike variations in shortening, topography, crustal geometry, and denudation. Geol Soc Am Bull 124:1339–1351
Gianni GM, Dávila FM, Echaurren A, Fennell L et al (2018) A geodynamic model linking Cretaceous orogeny, arc migration, foreland dynamic subsidence and marine ingression in southern South America. Earth-Sci Rev 185:437–462
Groeber P (1951) La Alta Cordillera entre las latitudes 34° y 29°30′. Rev Instit Nac Invest Cs Nat, Mus Argent Cs Nat Bernardino Rivadavia, Cs Geol 1(5):235–352
Heller PL, Angevine CL, Winslow NS, Paola C (1988) Two-phase stratigraphic model of foreland-basin sequences. Geology 16:501–504
Hernández RM, Galli CI, Reynolds J (1999) Estratigrafía del Terciario en el noroeste argentino. In: González Bonorino G, Omarini R, Viramonte J (eds) Geología del Noroeste Argentino: Salta. Asociación Geológica Argentina, Buenos Aires, pp 316–328
Horton BK (2018a) Tectonic regimes of the central and southern Andes: responses to variations in plate coupling during subduction. Tectonics 37:402–429
Horton BK (2018b) Sedimentary record of Andean mountain building. Earth-Sci Rev 178:279–309
Horton BK, Fuentes F (2016) Sedimentary record of plate coupling and decoupling during growth of the Andes. Geology 44:647–650
Horton BK, Fuentes F, Boll ADF et al (2016) Andean stratigraphic record of the transition from backarc extension to orogenic shortening: a case study from the northern Neuquén Basin, Argentina. J S Am Earth Sci 71:17–40
Jordan TE (1981) Thrust loads and foreland basin evolution, Cretaceous, western United States. AAPG Bull 65:2506–2520
Jordan TE (1995) Retroarc foreland and related basins. In: Busby CJ, Ingersoll RV (eds) tectonics of sedimentary basins. Blackwell Science, Cambridge, Massachusetts, pp 331–362
Jordan TE, Burns WM, Veiga R et al (2001) Extension and basin formation in the southern Andes caused by increased convergence rate: a mid-Cenozoic trigger for the Andes. Tectonics 20:308–324
Kley J, Monaldi C, Salfity J (1999) Along-strike segmentation of the Andean foreland: causes and consequences. Tectonophysics 301:75–94
Kozlowski E, Cruz C, Sylwan C (1997) Modelo exploratorio en la faja corrida de la Cuenca Neuquina, Argentina. In: Abstracts of the 6 Simposio Bolivariano Exploración Petrolera en las Cuencas Subandinas, Bogotá, 14–17 Sept 1997

Kraemer P, Silvestro J, Achilli F et al (2011) Kinematics of a hybrid thick-thin-skinned fold and thrust belt recorded in Neogene syntectonic wedge-top basins, southern central Andes between 35° and 36° S, Malargüe, Argentina. In: McClay K, Shaw J, Suppe J (eds) Thrust fault-related folding. AAPG Memoir 94, 245–270

Legarreta L, Gulisano CA (1989) Análisis estratigráfico de la Cuenca Neuquina (Triásico superior-Terciario Inferior). In: Chebli GA, Spalletti LA (eds) Cuencas Sedimentarias Argentinas. Universidad de Tucumán, Serie Correlación Geológica, vol 6, pp 221–244

Legarreta L, Uliana MA (1991) Jurassic-Cretaceous marine oscillations and geometry of back-arc basin fill, central Argentine Andes. Int Ass Sediment SP 12:429–450

Legarreta L, Uliana MA (1998) Anatomy of hinterland depositional sequences: Upper Cretaceous fluvial strata, Neuquén Basin, west-central Argentina. Relative Role of Eustasy Climate and Tectonism in Continental Rocks. SEPM SP 59:83–92

Legarreta L, Kokogián DA, Boggetti DA (1989) Depositional sequences of the Malargue Group (Upper Cretaceous-lower Tertiary), Neuquén Basin, Argentina. Cretaceous Res 10:337–356

Levina M, Horton BK, Fuentes F et al (2014) Cenozoic sedimentation and exhumation of the foreland basin system preserved in the Precordillera thrust belt (31–32° S), southern central Andes, Argentina: Tectonics 33:1659–1680

Mackaman-Lofland C, Horton BK, Fuentes F et al (2019) Mesozoic to Cenozoic retroarc basin evolution during changes in tectonic regime, southern Central Andes (31–33° S): Insights from zircon U-Pb geochronology. J S Am Earth Sci 89:299–318

Manceda R, Figueroa D (1995) Inversion of the Mesozoic Neuquén rift in the Malargüe fold and thrust belt, Mendoza, Argentina. In: Tankard AJ, Suárez Soruco R, Welsink HJ (eds) Petroleum Basins of South America. AAPG Memoir 62:369–382

McQuarrie N (2002) Building a high plateau: the kinematic history of the central Andean fold–thrust belt, Bolivia. Geol Soc Am Bull 114:950–963

Mescua JF, Giambiagi LB, Ramos VA (2013) Late Cretaceous uplift in the Malargüe fold-and-thrust belt (35° S), southern central Andes of Argentina and Chile. Andean Geol 40:102–116

Miall AD (2009) Initiation of the Western Interior foreland basin. Geology 37:383–384

Mosquera A (2008) Mecánica de deformación de la Cuenca Neuquina (Triásico-Terciario). Ph.D. thesis, Universidad de Buenos Aires

Orts DL, Folguera A, Giménez M, Ramos VA (2012) Variable structural controls through time in the Southern Central Andes (~36° S). Andean Geol 39(2):220–241

Parras AM, Casadío S, Pires M (1998) Secuencias Depositacionales del Grupo Malargüe y el Límite Cretácico-Paleógeno, en el sur de la Provincia de Mendoza. Argentina. Asoc Paleont Argentina, SP, pp 61–69

Ramos VA (2010) The tectonic regime along the Andes: present-day and Mesozoic regimes. Geol J 45:2–25

Ramos VA, Folguera A (2005) Tectonic evolution of the Andes of Neuquén: constraints derived from the magmatic arc and foreland deformation. In: Veiga G, Spalleti LA, Howell JA, Schwarz E (eds) The Neuquén Basin: a case study in sequence stratigraphy and basins dynamics. Geol Soc, London, SP 252:15–35

Ramos VA, Zapata T, Cristallini E et al (2004) The Andean thrust system— Latitudinal variations in structural styles and orogenic shortening. In: McClay KR (ed) Thrust tectonics and hydrocarbon systems. AAPG Memoir 82:30–50

Ramos ME, Tobal JE, Sagripanti L, Folguera A, Orts DL, Giménez M, Ramos VA (2015) The North Patagonian orogenic front and related foreland evolution during the Miocene, analyzed from synorogenic sedimentation and U/Pb dating (~42° S). J S Am Earth Sci 64:467–485

Rocha E, Brisson IE, Fasola ME (2018) Modelado de sistemas petroleros a lo largo de la faja plegada Neuquina. In: Abstracts of the X Congreso de Exploración y Desarrollo de Hidrocarburos, IAPG, Mendoza, 5–9 Nov 2018

Rojas Vera EA, Folguera A, Zamora Valcarce G et al (2010) Neogene to Quaternary extensional reactivation of a fold and thrust belt: the Agrio belt in the Southern Central Andes and its relation to the Loncopue trough (38°–39° S). Tectonophysics 492:279–294

Rojas Vera EA, Mescua J, Folguera A et al (2015) Evolution of the Chos Malal and Agrio fold and thrust belts, Andes of Neuquén: Insights from structural analysis and apatite fission track dating. J S Am Earth Sci 64:418–433

Rosselot E, Martos F, Cristiano E et al (2017) Origen sinorogénico de la Formación Diamante en la Faja plegada y corrida del Aconcagua. In: Abstracts of the 20 Congreso Geológico Argentino, San Miguel de Tucumán, 7–11 Aug 2017

Silvestro J, Kraemer P (2005) Evolución tecto-sedimentaria de la Cordillera Principal en el sector surmendocino a los 35° 30' S: Faja plegada de Malargüe, Republica Argentina. In: Abstracts of the Congreso de Exploración y Desarrollo de Hidrocarburos, IAPG, Mar del Plata, 15–19 Nov 2005

Silvestro J, Atencio M (2009) La Cuenca Cenozo-ica del Río Grande y Palauco: Edad, Evolución y Control Estructural, Faja Plegada de Malargüe (36° S). Rev Asoc Geol Argent 65(1):154–169

Starck D (2011) Cuenca Cretácica-Paleógena del noroeste Argentino. In: Abstracts of the 8 Congreso de Exploración y desarrollo de Hidrocarburos, IAPG, Mar del Plata, 8–12 Nov 2009

Tunik M, Folguera A, Naipauer M et al (2010) Early uplift and orogenic deformation in the Neuquén Basin: Constraints on the Andean uplift from U-Pb and Hf isotopic data of detrital zircons. Tectonophysics 489:258–273

Turienzo M (2010) Structural style of the Malargüe fold and-thrust belt at the Diamante river area (34° 30′–34° 50′ S) and its linkage with the Cordillera frontal Andes of central Argentina. J S Am Earth Sci 29:537–556

Uliana MA, Biddle KT (1988) Mesozoic-Cenozoic paleogeographic and geodynamic evolution of southern South America. Rev Brasil Geocien 18:172–190

Uliana MA, Dellapé DA (1981) Estratigrafía y evolución paleoambiental de la sucesión Maastrichtiana-eoterciaria del engolfamiento Neuquino (Patagonia Septentrional). In: Abstratcs of the 8 Congreso Geológico Argentino, San Luis, 20–26 Sept 1981

Vergani GD, Tankard AJ, Belotti HJ et al (1995) Tectonic evolution and paleogeography of the Neuquén Basin, Argentina. AAPG Memoir 62:383–402

Winocur DA, Litvak VD, Ramos VA (2015) Magmatic and tectonic evolution of the Oligocene Valle del Cura basin, Main Andes of Argentina and Chile: Evidence for generalized extension. In:. Sepúlveda SA, Giambiagi LB, Moreiras SM, Pinto L, Tunik M, Hoke GD, Farías M (eds), Geodynamic Processes in the Andes of Central Chile and Argentina. The Geological Society, London, SP 399, pp 131–154

Yrigoyen M (1993) Los depositos sinorogenicos terciarios. In: Ramos VA (ed) Geologia y Recursos Naturales de Mendoza. In: Abstract of the 7 Congreso Geologico Argentino y 2 Congreso de Exploración de Hidrocarburos, Mendoza, 10–15 Oct 1993

Zamora Valcarce G (2007) Estructura y Cinemática de la faja plegada y corrida del Agrio, Cuenca Neuquina. Ph.D. thesis, Universidad de Buenos Aires

Zapata TR, Allmendinger RW (1996) Thrust-front zone of the Precordillera, Argentina: a thick-skinned triangle zone. AAPG Bull 80:359–381

Zapata TR, Córsico S, Dzelalija F et al (2002) La faja plegada y corrida del Agrio: Análisis estructural y su relación con los estratos terciarios de la cuenca neuquina, Argentina. V Congreso de Exploración y Desarrollo de Hidrocarburos, Mar del Plata, 29 Oct–2 Nov

Thermochronological Constraints on the Exhumation of the Malargüe Fold-Thrust Belt, Southern Central Andes

Alejandro Bande, Andrés Boll, Facundo Fuentes, Brian K. Horton and Daniel F. Stockli

Abstract The structural evolution of the southern Central Andes has been widely studied with a general consensus on its geometry and regional architecture. However, the timing of uplift is poorly constrained, with estimates ranging from Late Cretaceous to Pliocene. New thermochronometric results from the Malargüe fold-thrust belt (MFTB) reveal diachronous Neogene cooling and exhumation associated with contractional deformation in this sector of the Central Andes. The Malargüe fold-thrust belt is an excellent example of inversion structures superimposed on a backarc extensional basin (the Late Triassic–Cretaceous Neuquén Basin). As in many other inverted structural systems, preexisting discontinuities play a substantial role in the distribution of shortening and erosional exhumation. Apatite fission track and (U–Th)/He data together with thermal modeling help to define early Miocene (~20 Ma) rapid cooling in hinterland sectors of the MFTB coupled with thin-skinned shortening and retroarc foreland basin development to the east. By 10–7 Ma, continued shortening reactivated the easternmost frontal structures (e.g., Río Atuel fault) and partially inverted the Neuquén Basin. Subsequent deformation exhumed hinterland structures (e.g., Dedos-Silla block) by 5–2 Ma. These temporal constraints reveal that shortening-induced exhumation operated in a disparate manner, rather than following a systematic progression from hinterland to foreland. Additionally, thermal models of the fission track and (U–Th)/He data tightly constrain a long period (~40 Myr) of very slow cooling during an early–middle Cenozoic phase of neutral stress conditions.

Keywords Apatite fission track and (U–Th)/He thermochronology · Neogene exhumation · Malargüe fold-thrust belt

A. Bande (✉)
Tecpetrol S.A., Buenos Aires, Argentina
e-mail: alejandro.bande@tecpetrol.com

A. Boll
Independent Consultant, Buenos Aires, Argentina

F. Fuentes
YPF S.A., Buenos Aires, Argentina

B. K. Horton · D. F. Stockli
Jackson School of Geosciences, University of Texas at Austin, Austin, TX, USA

D. Kietzmann and A. Folguera (eds.), *Opening and Closure of the Neuquén Basin in the Southern Andes*, Springer Earth System Sciences,
https://doi.org/10.1007/978-3-030-29680-3_15

1 Introduction

Identifying the temporal–spatial pattern of deformation is crucial to reconstruct topographic evolution, structural kinematics, basin evolution, and the distribution of hydrocarbon resources. However, there are few constraints on the absolute timing of deformation and synorogenic deposition within the Neuquén Basin of west-central Argentina and associated fold-thrust belts (e.g., Malargüe, Chos Malal, Agrio fold-thrust belts). Recent studies highlight the applicability of diverse geochronological and thermochronological techniques. Whereas $^{40}Ar/^{39}Ar$, and U–Pb geochronological results for detrital materials and interbedded tuffs provide new depositional age constraints (e.g., Kraemer et al. 2011; Horton et al. 2016), the timing of exhumational cooling can be assessed from various low-temperature thermochronometers, including apatite fission track (AFT), zircon fission track (ZFT), and apatite (U–Th)/He (AHe) techniques (e.g., Levina et al. 2014; Folguera et al. 2015; Rojas Vera et al. 2015).

The geometry and kinematics of fold-thrust deformation in the Malargüe, Agrio, and Chos Malal fold-thrust belts have received much attention, leading to a broad consensus on the main structural configuration and basin geometries (e.g., Ploskiewicz 1987; Kozlowski et al. 1993; Manceda and Figueroa 1995; Ramos and Folguera 2005, Giambiagi et al. 2008; Silvestro and Atencio 2009; Kraemer et al. 2011; Boll et al. 2014; Fuentes et al. 2016). However, there remains an open debate about the structural kinematics and timing of deformation. Estimates for the onset of Andean deformation range from Late Cretaceous to Miocene (Kozlowski et al. 1993; Manceda and Figueroa 1995; Cobbold and Rossello 2003; Giambiagi et al. 2008; Tunik et al. 2010; Fennell et al. 2017). Additionally, several brief deformational episodes of variable magnitude affected the pre-Andean evolution of the Neuquén Basin (Vergani et al. 1995; Mosquera and Ramos 2006; Naipauer et al. 2012), adding further complexity to accurately constraining the history of deformation and exhumation in this part of the Andes.

The motivation of this contribution is to provide new thermochronological results from a suite of samples collected from major structures within the Malargüe fold-thrust belt to evaluate the absolute timing of exhumation and associated deformation. The classic contributions that defined the Malargüe fold-thrust belt (MFTB) by Kozlowski et al. (1993) and Manceda and Figueroa (1995) considered most of the Andean growth to have occurred during the Miocene deformational event and onwards. In contrast, studies south of the MFTB suggested the presence of older deformational phases of principally Late Cretaceous–Eocene age (e.g., Ramos 1981; Kozlowski et al. 1997; Cobbold and Rossello 2003; Tunik et al. 2010; Di Giulio et al. 2012; Rojas Vera et al. 2015). Recently, some authors (e.g., Orts et al. 2012; Spagnoulo et al. 2012; Mescua et al. 2013; Folguera et al. 2015; Fennell et al. 2017) have expanded the Late Cretaceous contractional phase to include the MFTB on the basis of stratigraphic, magmatic, and thermochronological evidence. Additionally, periods of negligible deformation have been postulated to play a significant role in the geodynamic evolution of the Andean orogenic belt at this latitude (Horton and Fuentes

2016; Horton 2018a). Our results provide new insights into temporal and spatial variations in cooling, which may reflect systematic trends, either along strike or across strike that can be compared to complementary datasets. Further, these results help enhance our understanding of the timing of potential fault reactivation and reheating events, with implications for trap formation, hydrocarbon generation, and migration.

2 Geological Setting

Subduction along the western edge of South America represents a long-lived process with complex spatial and temporal variations since the early Mesozoic (e.g., Mpodozis and Ramos 1990, 2008). In western Argentina at 34–40° S, the Neuquén Basin contains a retroarc stratigraphic record of these geodynamic changes incurred during Mesozoic and Cenozoic time. The northern segment of the Neuquén Basin is structurally deformed, displaying a broad north-trending axis consisting of a hybrid thick- and thin-skinned thrust belt and associated foreland basin (Fig. 1). Shortening-related structures involve Permian-Triassic igneous basement and the overlying 5–7 km thick basin-fill succession of clastic and subordinate carbonate to evaporitic strata.

2.1 Structural Geology

The Malargüe fold-thrust belt (MFTB) is characterized by large basement structures coupled with thin-skinned thrust systems developed both on top of these basement structures and along the eastern thrust front (Kozlowski et al. 1993; Manceda and Figueroa 1995; Giambiagi et al. 2008, 2009; Turienzo 2010; Fuentes et al. 2016). The structure of the foreland platform east of the MFTB is known from 2D and 3D seismic reflection data. Most subsurface structures in this foreland region consist of low-displacement basement faults with minor inversion related to the Andean deformation (Boll et al. 2014). Frontal structures of the MFTB include the Meson and Sosneado thrusts (Fig. 2), which carry Upper Cretaceous and Cenozoic strata. To the west, these thrusts merge at depth into the Rio Atuel fault (e.g., Fuentes et al. 2016), interpreted as an extensional fault reactivated during Cenozoic shortening. The Rio Atuel anticlinorium is the dominant structure of the frontal sector (Figs. 2 and 3) and is formed as a hanging-wall anticline above the Río Atuel fault, a master normal fault active during Jurassic extension (Giambiagi et al. 2008; Lanés et al. 2008; Bechis et al. 2010; Fuentes et al. 2016). The southern continuation of the Rio Atuel anticlinorium exposes lower Jurassic units and corresponds to the frontal Malargüe anticline, which exposes Permo-Triassic Choiyoi Group along its core (Kraemer et al. 2011; Boll et al. 2014). Farther west Late Triassic–Early Jurassic Cuyo group strata are exposed in a set of broad anticlines, the La Valenciana and Los Blancos anticlines (Fig. 3). Finally, in a western hinterland position, the Dedos-Silla block

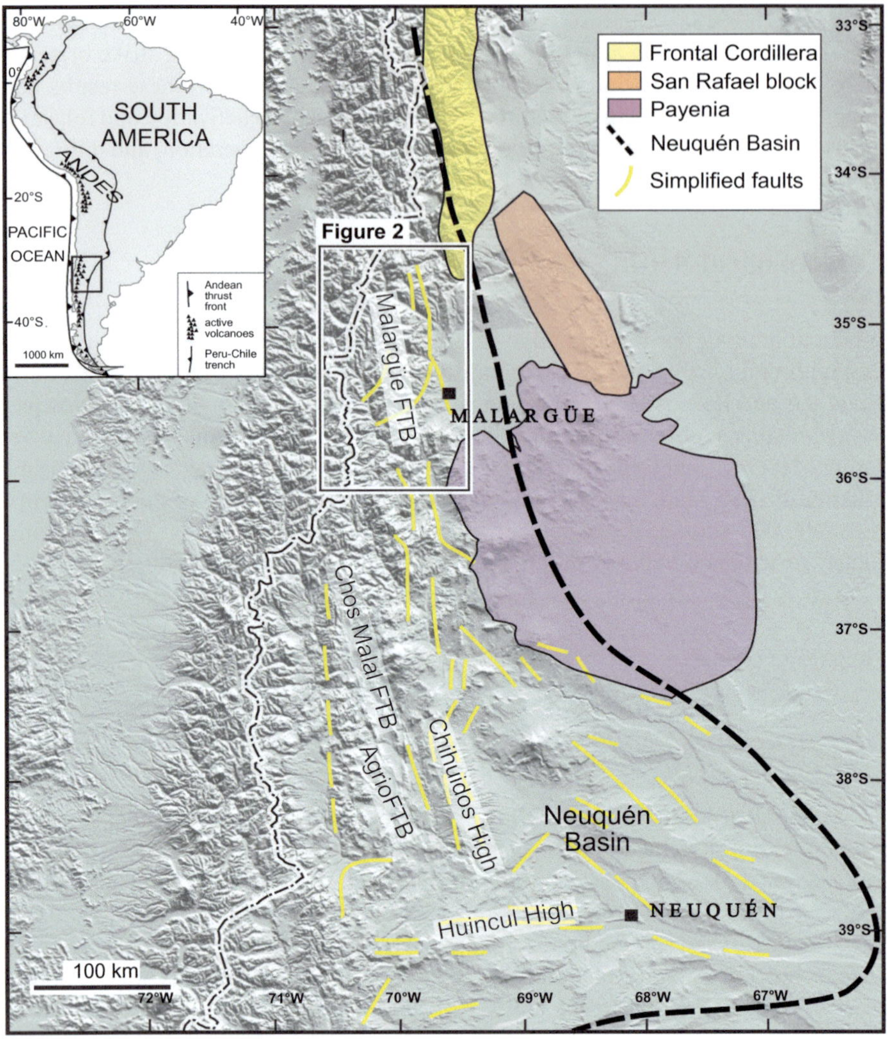

Fig. 1 Shaded relief map of the southern Central Andes at 33–40° S showing main structural elements (Malargüe fold-thrust belt, Chos Malal fold-thrust belt, Agrio fold-thrust belt, Chihuidos high, and Huincul high) relative to Cenozoic provinces (Frontal Cordillera, San Rafael basement block, and Payenia volcanic field) and the Neuquén Basin (dashed bold line representing eastern basin margin). Yellow line segments represent simplified traces of Late Triassic–Early Jurassic extensional faults (after Vergani et al. 1995; Manceda and Figueroa 1995; Bechis et al. 2010; Fuentes et al. 2016)

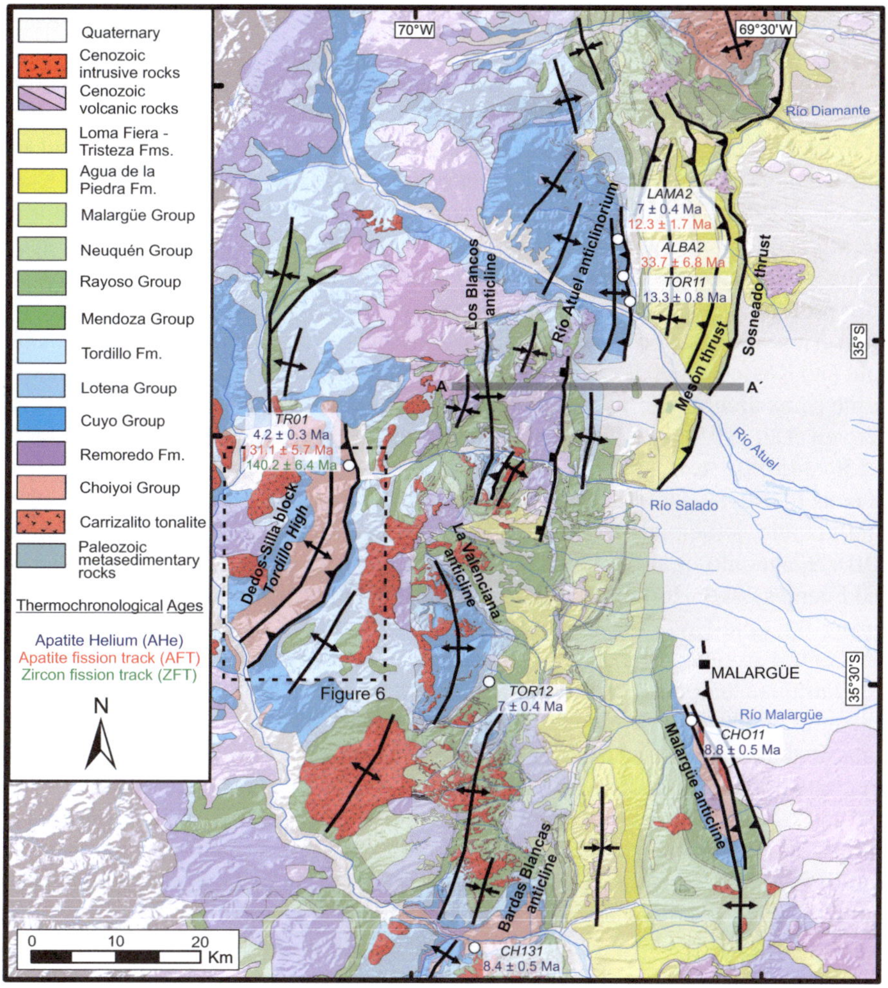

Fig. 2 Geologic map of the Malargüe fold-thrust belt (superimposed on a shaded relief image) showing major structures, trace of cross-section A–A' and location of thermochronological samples (after Kozlowski et al. 1989; Bechis 2009; Turienzo 2010)

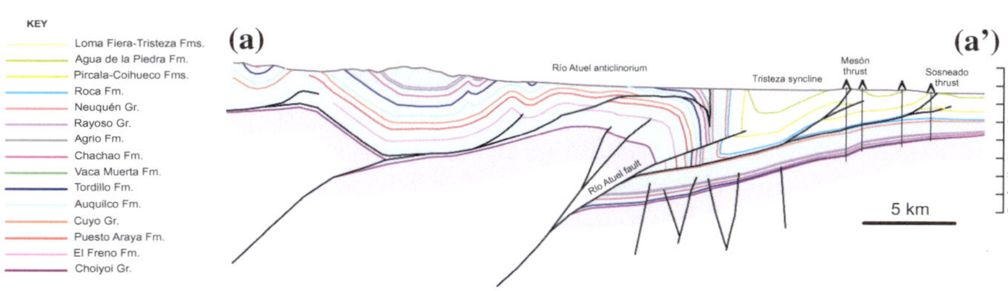

Fig. 3 East-west structural cross-section of the Malargüe fold-thrust belt and Malargüe foreland basin (for location see Fig. 2). Modified from Fuentes et al. (2016)

exposes Choiyoi Group basement and is interpreted as a basement high reactivated during the Cenozoic (Kozlowski et al. 1993; Manceda and Figueroa 1995).

There is general agreement that strong Miocene shortening drove Andean uplift in the MFTB (Kozlowski et al. 1993; Manceda and Figueroa 1995; Giambiagi et al. 2008; Ramos and Folguera 2009; Fuentes et al. 2016; Horton et al. 2016). Indeed, many authors (e.g., Silvestro and Kraemer 2005; Silvestro and Atencio 2009; Kraemer et al. 2011; Mescua et al. 2014) suggest at least two discrete events during the Miocene within the MFTB: a first one during the early–middle Miocene and a younger in the late Miocene–Pliocene. In addition, earlier Cenozoic deformation has been postulated in the northern Neuquén province: Kozlowski et al. (1997) inferred the Agrio fold-thrust belt was partially uplifted between 50 and 25 Ma, while Cobbold and Rossello (2003) suggested the presence of an Eocene transpressional event in the same area. Up to five discrete Mesozoic episodes have been proposed (e.g., Vergani et al. 1995), with the two most important events associated with Late Jurassic (~150–145 Ma) accumulation of the nonmarine Tordillo Formation and Late Cretaceous (~100–80 Ma) fluvial deposition of the Neuquén Group (Fig. 4). Based on stratigraphic, structural, and magmatic evidence, many authors (e.g., Tunik et al. 2010; Spagnoulo et al. 2012; Mescua et al. 2013; Rojas Vera et al. 2015; Horton and Fuentes 2016; Horton et al. 2016) suggest that deposition of the Neuquén Group during the Late Cretaceous was controlled by the onset of regional shortening in the Central Andes. Prior to Late Cretaceous shortening, brief episodes of punctuated intraplate deformation interrupted postrift thermal subsidence and modified the paleotopography of the Neuquén Basin (Vergani et al. 1995). In this context, the Late Jurassic deformational episode (~150–145 Ma) associated with Tordillo Formation accumulation is particular significant and variable across the Neuquén Basin. In the southern sector of the basin, a contractional event is interpreted (e.g., Mosquera and Ramos 2006; Naipauer et al. 2012), while in the northern sector deposition of the Tordillo Formation occurred during an extensive phase (e.g., Lo Forte 1996; Cegarra and Ramos 1996; Giambiagi et al. 2003; Mescua et al. 2008).

2.2 *Stratigraphy*

The Neuquén Basin initiated as a backarc basin during latest Triassic extension, producing fault-bounded depocenters with N-, NW-, and WSW-trending orientations (Digregorio et al. 1984, Vergani et al. 1995). The initial synrift phase (Late Triassic–Early Jurassic) is denoted by the Pre-Cuyo and Cuyo Groups, consisting of 1.5–2.5 km of volcanic, marine sandstone and shale deposits in localized depocenters with substantial lateral thickness variations dictated by syndepositional normal faults (Fig. 4). Backarc extension was followed by Middle Jurassic–Early Cretaceous thermal subsidence (sag phase), with punctuated inversion events (Vergani et al. 1995). Postrift deposition during this thermal sag phase involved deposition of 1–2 km of extensive marine shale and marginal marine evaporite and carbonate facies of the Lotena, Mendoza, and Rayoso Groups, with important nonmarine sandstone deposits

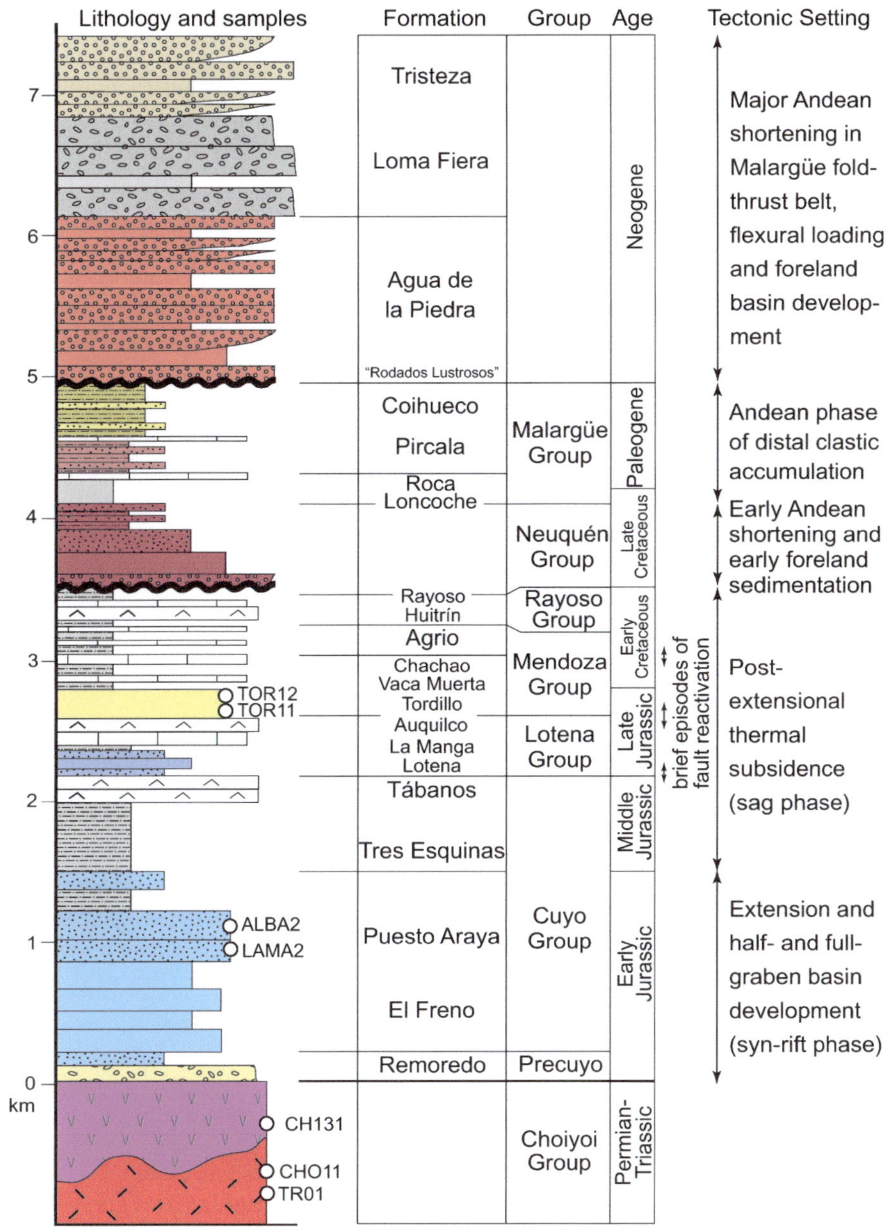

Fig. 4 Mesozoic-Cenozoic stratigraphic chart for the northern Neuquén Basin and Malargüe foreland basin showing lithology, stratigraphic nomenclature, and sample location. Modified from Horton et al. (2016)

of the Tordillo Formation signifying one of several brief (<5 Ma) fault-reactivation events (Legarreta and Uliana 1991).

Late Cretaceous–Cenozoic sedimentation was primarily related to shortening, but with several significant stratigraphic gaps and intra-Andean zones of extensional basin formation (Yrigoyen 1993; Folguera et al. 2010). Earliest Andean (Late Cretaceous) shortening is recorded by coarse clastic accumulation of the 0.5–1.5 km thick Neuquén Group, defined by a fluvial package which locally exhibits a basal angular unconformity (Legarreta and Uliana 1998). A subsequent Maastrichtian–Paleogene phase of distal clastic accumulation during unclear tectonic conditions involved nonmarine and localized marine conditions that generated the uniformly fine-grained intervals of shale, minor sandstone, and carbonate of the <0.5 km thick Malargüe Group (Loncoche, Roca, Pircala, and Coihueco formations) (Legarreta et al. 1989).

At 34–36° S, the fine-grained distal low-energy facies of the Paleogene Pircala and Coihueco Formations are capped by a sharp disconformity, marked by a distinctive conglomeratic unit referred to as the Rodados Lustrosos (Groeber 1947, 1951; Yrigoyen 1972, 1993). This 2–20 m thick unit at the base of the Agua de la Piedra Formation is composed of well-rounded, distinctively polished pebbles to small cobbles with extensively weathered, oxidized, and desert varnished surfaces and has been interpreted to represent conditions of no net accumulation in the foreland (Horton et al. 2016; Horton 2018a, b) (see Chap. "The Andean Foreland Evolution of the Neuquén Basin: A Discussion"). The overlying Neogene 1–3 km thick succession of coarse clastic deposits (Fig. 4) is described as large-scale fluvial system with possible fluvial megafans (Agua de la Piedra Formation) and alluvial fan sedimentation, with east-directed paleoflow away from a proximal and persistent source area (Loma Fiera and Tristeza formations) (Combina and Nullo 2011; Boll et al. 2014; Horton et al. 2016).

Andean magmatism in the Malargüe region records a complex history of eastward arc migration, with important late Cenozoic magmatic centers that sporadically invaded the foreland (Folguera and Ramos 2011). Some of these volcanic episodes have been related to potential flat-slab episodes during the Neogene (Kay et al. 2006; Folguera et al. 2009).

3 Methodology and Samples

A total of seven samples were collected from sedimentary and igneous units within the hanging walls of the principal structures forming the deformation front of the MFTB (Figs. 2 and 4). Thermochronological analyses of these samples help constrain the cooling and exhumation history of the northern Neuquén Basin.

3.1 Low-Temperature Thermochronology

This technique is based on temperature-dependent retention of radiogenic daughter products. Two commonly measured decay products are fission tracks, using the apatite and zircon fission track (AFT and ZFT) methods, and ^{4}He using the apatite (U–Th–Sm)/He method (AHe). Above certain temperatures, daughter products are not retained over geological timescales. This temperature is actually a zone, referred to as the partial annealing zone (PAZ) for AFT and ZFT (Gleadow and Duddy 1981) and the partial retention zone (PRZ) for ZHe (Wolf et al. 1998). Within the PAZ, fission tracks are progressively annealed (shortened) and eventually removed; in the PRZ, ^{4}He systematically escapes by diffusion. The precise temperatures for the PAZ and PRZ are dependent on the crystal kinetic characteristics and cooling rates (Brandon et al. 1998; Donelick et al. 1999; Reiners and Farley 2001; Ehlers and Farley 2003). Typically, the ZFT PAZ extends from ~180 to 350 °C (Tagami 2005), the AFT PAZ from ~60 to 120 °C (Ketcham et al. 1999), and the AHe PRZ from ~40 to 70 °C (Farley 2002). Slowly cooled samples generally have lower closure temperatures and older ages (Ehlers and Farley 2003). Since fission tracks form at a constant rate over time and initially have similar lengths (Green et al. 1989), the distribution of AFT lengths can be used to quantify the amount of annealing a sample has experienced (Ketcham et al. 1999). Thermal modeling can be used to identify possible cooling pathways (time-temperature histories; Ketcham et al. 2007).

3.2 Apatite and Zircon Fission Track (AFT and ZFT) Analyses

Apatite and zircon fission track analyses were carried out on three samples at Chronuscamp Research laboratory following laboratory procedures described in Soares et al. (2014). Sample TR01 (Table 1) was dated both by AFT and ZFT, while samples LAMA2 and ALBA2 were dated only by AFT. Ages were calculated following equations described by Soares et al. (2015). Available horizontal confined fission tracks (tracks in tracks) were measured on three AFT samples. However, in one sample (ALBA 2, Table 1) only three lengths were measured. Ages are reported as pooled ages.

3.3 Apatite (U–Th–Sm)/He (AHe) Analyses

Apatite (U–Th–Sm)/He analyses were carried out on six samples at the University of Texas at Austin using laboratory procedures described in Levina et al. (2014). At least three single-grain aliquots were measured per sample. Individual ages were calculated using standard alpha-ejection corrections and with FT corrections that

Table 1 Apatite and Zircon Fission Track (AFT, ZFT) data from the Malargüe fold-thrust belt

Sample	Mineral	Latitude (N°)	Longitude (E°)	Elev. (m)	n	Pooled Age (Ma)	±1σ	Rho(s) x10E+5	Rho(Zeta) xl0E+5	Zeta(mMS) x10E+3	P(i)-UNK μg/g	P(i)-STD μg/g	P(χ²)%	# lengths	Length (μm)	SD (μm)
TR01	Zircon	−35.18612	−70.0798	2210	20	**140.2**	**6.4**	3.78	8 96	1.3	346.8	415.8	80	0	ND	ND
TR01	Apatite	−35.18612	−70.0798	2210	24	**31.1**	**5.7**	1.44	1.39	1.53	8.1	6.8	30	25	9.1	2.21
LAMA2	Apatite	−34.80211	−69.71637	2410	25	**12.3**	**1.7**	2.49	1.39	1.53	35.5	6.8	80	48	9.45	2.48
ALBA2	Apatite	−34.S4781	−69.71334	2243	20	**33.7**	**6.8**	1.33	1 39	1.53	6.9	6.8	5	3	7.33	0.59

assume a homogeneous U and Th distribution (e.g., Farley et al. 1996; Farley 2002). Reported age uncertainties of 2σ are based on the reproducibility of replicate analysis of laboratory standards (Reiners et al. 2004). Estimated reproducibility uncertainties are ~8%. Both uncorrected and alpha-ejection corrected ages are reported (Table 2).

3.4 Inverse Thermal Modeling

To determine the cooling history recorded by individual samples, we used HeFTy software (v.1.8.3) (Ketcham 2005) to model both forward and inverse time-temperature paths for each sample. The principal aim of this modeling was to identify and estimate variations in cooling rate over time. Input data for the models includes AHe and AFT ages, track lengths, and the angle between each track and the crystallographic *c*-axis. Models were initially run with minimal constraints encompassing a broad time frame, from Jurassic to present. Subsequent models focused on the Cenozoic. In cases where apparent changes in cooling rate were depicted (defined as inflection points in the time-temperature plots), constraint boxes were defined and variably shifted (in time-temperature space) to test the goodness of fit and reliability of each model.

4 Results

The southernmost sample (CH131) was collected on the eastern flank of the Bardas Blancas anticline (Fig. 2) from Permo-Triassic Choiyoi Group granodiorites. The sample was analyzed using the AHe method for three aliquots (Table 2): Two aliquots show late Miocene ages (8.4 ± 0.5 and 9.9 ± 0.6 Ma). The third aliquot presents a middle Miocene age (15.8 ± 0.9 Ma) that may reflect internal U zoning or a different closure temperature produced by variations in composition or effective diffusion domain size (Gautheron et al. 2009).

A sandstone sample of the Jurassic Tordillo Formation (TOR12) was collected on the northern margin of the Malargüe River within the La Valenciana anticline (Fig. 2). The six AHe aliquots from this sample are tightly clustered between 7 and 8.9 Ma, showing a late Miocene age. South of the river, an igneous sample from the Choiyoi Group (CHO 11) was collected from the Malargüe anticline (Fig. 2). The youngest AHe aliquot indicates a late Miocene age (8.8 ± 0.5 Ma) that matches the ages obtained for Bardas Blancas and the La Valenciana anticline (Table 2). The remaining aliquots show older ages (20.8–33.8 Ma) that may reflect interval U zoning or different closure temperatures produced by compositional differences.

In the hanging wall of Río Atuel fault (Fig. 2), three sandstone samples were collected: two from the Jurassic Puesto Araya Formation (ALBA2 and LAMA2) and one from the Tordillo Formation (TOR11). AHe analyses of the LAMA2 sample (Table 2) show a younger aliquot of 7 ± 0.4 Ma and two older aliquots of 13.4 ± 0.8

Table 2 (U–Th)/He Ages of Apatite (AHe) in the Malargüe fold-thrust belt

Sample	Latitude (N°)	Logtitude (E°)	Elev. (m)	Corrected age Ma	Error (1σ) Ma	U (ppm)	Th (ppm)	147 Sm (ppm)	He (nmol/g)	[U]e	Ft
CH131-1	−35.8744	−69.8757	1455	**8.4**	**0.51**	1.8	10.6	289.1	0.2	5.6	0.58
CH131-2	−35.8744	−69.8757	1455	**9.9**	**0.59**	1.3	8.0	269.7	0.2	4.5	0.70
CH131-3	−35.8744	−69.8757	1455	**15.8**	**0.95**	3.2	16.8	251.2	0.5	8.3	0.61
LAMA2-2	−34.80211	−69.71637	2410	**7.0**	**0.42**	6.3	88.4	108.4	0.8	27.2	0.75
LAMA2-3	−34.80211	−69.71637	2410	**13.4**	**0.81**	6.5	125.0	76.9	1.6	35.7	0.59
LAMA2-4	−34.80211	−69.71637	2410	**17.9**	**1.07**	18.5	56.8	155.6	18	32.3	0.56
TR01-1	−35.1861	−70.0798	2210	**16.0**	**0.96**	7.9	13.9	432.7	0.9	13.2	0.70
TR01-2	−35.1861	−70.0798	2210	**5.4**	**0.32**	5.3	9.6	374.9	0.2	9.4	0.69
TR01-3	−35.1861	−70.0798	2210	**4.7**	**0.28**	9.1	20.5	482.4	0.3	16.2	0.67
TR01-4	−35.1861	−70.0798	2210	**4.2**	**0.25**	5.9	10.4	215.2	0.2	9.4	0.75
TR01-5	−35.1861	−70.0798	2210	**3.8**	**0.23**	6.8	14.6	380.7	0.2	12.0	0.57
TR01-6	−35.1861	−70.0798	2210	**2.9**	**0.17**	5.1	11.0	201.3	0.1	8.6	0.69
CHO11-1	−35.5507	−69.59414	1477	**25.2**	**1.51**	6.1	30.6	66.4	1.2	13.5	0.62
CHO11-2	−35.5507	−69.59414	1477	**33.8**	**2.03**	3.9	32.3	91.0	1.4	11.8	0.64
CHO11-3	−35.5507	−69.59414	1477	**20.8**	**1.25**	12.8	39.8	53.8	1.5	22.2	0.60
CHO11-4	−35.5507	−69.59414	1477	**8.3**	**0.53**	6.3	27.5	45.8	0.4	12.9	0.59
CHO11-5	−35.5507	−69.59414	1477	**23.3**	**1.70**	3.0	21 7	31.8	0 8	84	0.63
CHO11-6	−35.5507	−69.59414	1477	**17.3**	**1.04**	0.8	9.7	87.0	0.2	3.5	0.65
TOR11-1	−34.9616	−69.68468	1788	**23.8**	**1.73**	10.9	114	2B.9	17	13.7	0.79
TOR11-2	−34.9616	−69.68468	1788	**17.6**	**1.06**	6.0	12.9	26.4	0.7	9.1	0.78
TOR11-3	−34.9616	−69.68468	1788	**13.3**	**O.SO**	5 4	13.8	52.9	0.5	8.0	0.76

(continued)

Table 2 (continued)

Sample	Latitude (N°)	Logtitude (E°)	Elev. (m)	Corrected age Ma	Error (1σ) Ma	U (ppm)	Th (ppm)	147 Sm (ppm)	He (nmol/g)	[U]e	Ft
TOR11-4	−34.9616	−69.68468	1788	**13.9**	**0.83**	8.8	19.8	20.4	0.8	13.6	0.76
TOR11-5	−34.9616	−69.68468	1788	**71.9**	**4.31**	4.3	9.8	32.3	2.1	6.7	0.78
TOR11-6	−34.9616	−69.68468	1788	**17.9**	**1.03**	3.4	12.5	28.6	0.5	6.4	0.79
TOR12-1	−35.5148	−69.91351	2248	**7.6**	**0.45**	4.4	22.4	44.4	0.2	9.8	0.59
TOR12-2	−35.5148	−69.91351	2248	**7.0**	**0.42**	5.8	19.0	75.7	0.2	10.5	0.60
TOR12-3	−35.5148	−69.91351	2248	**8.9**	**0.53**	7.1	6.5	20.6	0.3	8.7	0.63
TOR12-4	−35.5148	−69.91351	2248	**8.7**	**0.52**	73 1	33.0	28.5	2.3	80.9	0.60
TOR12-5	−35.5148	−69.91351	2248	**7.9**	**0.47**	45.2	112.3	50.6	1.9	71.3	0.62
TOR12-6	−35.5148	−69.91351	2248	**8.3**	**0.53**	38.7	17.8	39.8	1.2	43.0	0.58

and 17.9 ± 1.1 Ma, respectively. AHe analyses of TOR11 show two aliquots that overlap within error (13.3 ± 0.8 and 13.9 ± 0.8 Ma) with the remaining aliquots showing older ages between 17.6 and 71 Ma. Both samples are detrital samples, increasing the probability of different compositions and closure temperatures among grains. Sample LAMA2 was analyzed by the AFT technique (Table 1) and also shows a late Miocene age (12.3 ± 1.7 Ma). The other Jurassic Puesto Araya sample, ALBA2, was only dated by AFT and exhibits a 33.7 ± 6.8 Ma age. Both AFT ages pass the chi-square test, consistent with a statistically homogeneous age population.

Finally, a Choiyoi Group sample (TR01) was collected from the Dedos-Silla block (Fig. 2) and analyzed by three thermochronometric techniques: AHe, AFT, and ZFT. The highest temperature thermochronometer (ZFT) show a 140.2 ± 6.4 Ma age (Table 1). AFT analyses indicate an Oligocene age (31.1 ± 5.7 Ma). For the AHe analyses, 5 of the 6 aliquots closely group with a Pliocene age (2.9–5.4 Ma), with a single aliquot showing a middle Miocene age (Table 2).

5 Interpretation of Ages and Discussion

Based on the new thermochronological results presented here, it is possible to infer patterns of cooling and associated shortening and exhumation within the Malargüe FTB (MFTB), along with comparisons to previous studies.

5.1 Late Miocene Rapid Cooling

The new results show that the frontal MFTB experienced rapid late Miocene cooling event at broadly 7–10 Ma (Fig. 7a). AFT and AHe ages together with modeling results collectively show that the Bardas Blancas structure, Valenciana anticline, and Malargüe anticline—all within the eastern frontal segment of the MFTB—were exhumed during late Miocene shortening. This timing is in agreement with evidence for coeval synorogenic sedimentation and the development of growth strata and internal unconformities within the Loma Fiera Formation (Boll et al. 2014; Fuentes et al. 2016; Horton et al. 2016). For the hanging wall of the Río Atuel fault, AFT and thermal modeling results for the Jurassic Puesto Araya sample LAMA2 reveal an Oligocene age (33 Ma) that reflects partial resetting. It is possible that this sample resided within the PAZ and never reached the 120 °C isotherm during post-depositional burial. The thermal model for this sample (Fig. 5) shows that rapid cooling was concentrated in the late Miocene (5–7 Ma), in agreement with the AHe data. However, slightly older cooling ages for Jurassic Puesto Araya sample ALBA2 (AFT age: 12.3 ± 1.7 Ma), Jurassic Tordillo sample TOR11 (AHe age: 13.3 ± 0.8 Ma), and other single aliquots may reflect early–middle Miocene exhumation. It is worth mentioning that all samples from this area (north of the Río Atuel) are detrital units, thus increasing the chance of variable grain composition and therefore

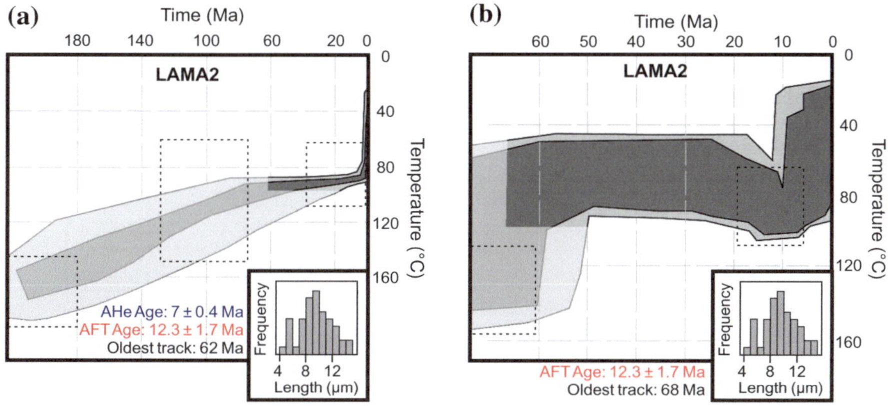

Fig. 5 Time-temperature histories from HeFTy thermal modeling of sample LAMA2. Unconstrained part of the models represented by transparent polygons. Dashed boxes represent model input constraints. Inset shows histogram of track-length distribution

different closure temperatures. Based on modeling results, a rate of ~10 °C/Myr can be estimated for these samples, suggesting relatively fast cooling.

Additional stratigraphic and structural datasets support our interpretation of strong inversion near the MFTB deformation front: (a) uplift of the Frontal Cordillera by ca. 5 Ma and related regional tilting (Boll et al. 2014); (b) subsurface growth–strata relationships for late Miocene–Pliocene Basin fill near the Malargüe anticline (Silvestro and Kraemer 2005; Silvestro and Atencio 2009); (c) growth–strata relationships reported along the eastern front of the Atuel anticlinorium, showing fault activity at 10.5–7.5 Ma (Fuentes et al. 2016; Horton et al. 2016); (d) initial synorogenic deposition in the frontal MFTB together with a regional unconformity at the base of the Loma Fiera Formation (Boll et al. 2014); (e) Loma Fiera and Tristeza Formations show a rapid shift to gravel-rich facies deposited in a proximal alluvial fan environment (Horton et al. 2016); and (f) complete Neogene unroofing of the foreland basin succession at 37° S (Sagripanti et al. 2012).

In the hinterland of the MFTB, within the Dedos-Silla block, sample TR01 shows clustered Pliocene AHe cooling ages. Thermal inverse models (Fig. 6a and b) incorporating AFT and AHe data define rapid cooling at 2–5 Ma, suggesting possible out-of-sequence reactivation of this block, as previously postulated (e.g., Kozlowski et al. 1993; Mescua et al. 2014). The associated thermal model (Fig. 6b) shows that samples resided at ~80 °C within the PAZ prior to Pliocene cooling. Based on these results, we suggest that the Oligocene AFT age (31.1 ± 5.7 Ma) corresponds to a partially reset age. The geological map relations of the Dedos-Silla block (Nullo et al. 2005) support the out-of-sequence interpretation: South of the Salado River, the Dedos-Silla basement fault truncates older folds involving Jurassic strata in its footwall (Fig. 6c). Giambiagi et al. (2008) suggested a foreland to hinterland sequence of basin inversion of the Atuel depocenter based on structural relations and interpretations, consistent with the thermochronological results presented here.

Fig. 6 **a** and **b** Time-temperature histories from HeFTy thermal modeling of sample TR01. Unconstrained part of the models represented by transparent polygons. Dashed boxes represent model input constraints. Inset shows histogram of track-length distribution. **C** Geologic map of the Dedos-Silla block showing location of sample TR01 in the hanging wall of the main reactivated thrust, which truncates folds in the footwall. See Fig. 2 for location

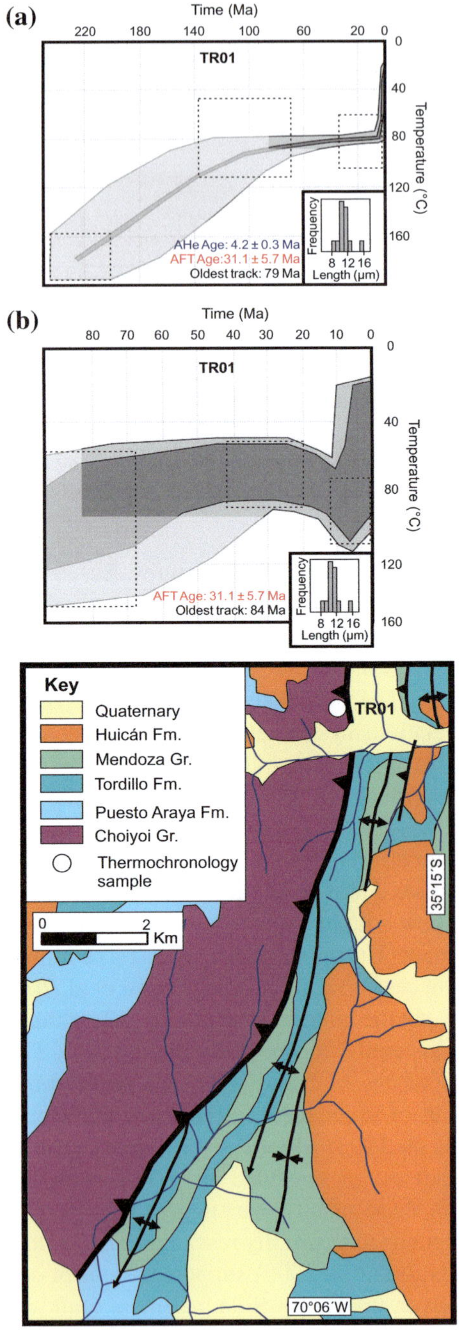

Overall, the new thermochronological data point to a late Miocene period of shortening and exhumation in the frontal sector of the MFTB (Fig. 7a), consistent with tectonic inversion of geometrically associated structures that involve crystalline basement at depth (Fuentes et al. 2016).

5.2 Early–Middle Miocene Shortening, Foreland Basin Development, and Hydrocarbon Implications

The fission track age, track-length distribution, and inverse modeling results for sample TR01 show a reheating event at 20–7 Ma of around ~20–30 °C (Fig. 6b). A similar reheating event was recorded by sample LAMA2 in the hanging wall of the Río Atuel Fault (Fig. 5b). Reheating may reflect flexural subsidence and burial during early–middle Miocene shortening of the western sector of the MFTB (Fig. 7b) (Silvestro and Kraemer 2005; Boll et al. 2014; Folguera et al. 2015). Neogene deposits on the eastern platform of the Malargüe foreland basin are up to 3 km thick and included the Agua de la Piedra, Loma Fiera, and Tristeza Formations (Boll et al. 2014). This upward thickening and coarsening succession is typical of an evolving foreland basin (DeCelles and Giles 1996). The lower-middle Miocene Agua de la Piedra Formation was deposited in a large-scale fluvial system involving possible fluvial megafans with localized aeolian redistribution, a depositional framework common to foredeep depozones (Horton et al. 2016). Together with a regional disconformity developed at its base, the Agua de la Piedra Formation reflects a change in sedimentation related to regional shortening. Similarly, the Río Grande Basin fill shows an early–middle Miocene shortening history for the MFTB prior to the final late Miocene phase of deformation and rock uplift (Silvestro and Atencio 2009).

Published thermochronological studies in the northern Neuquén and southern Mendoza provinces (Zamora Valcarce et al. 2009; Levina et al. 2014; Folguera et al. 2015, Rojas Vera et al. 2015; Sánchez et al. 2018) are consistent with our dataset suggesting Miocene (20–6 Ma) cooling and exhumation based on basement and detrital AFT ages and modeling results. However, Folguera et al. (2015) and Rojas Vera et al. (2015) presented results based only on AFT samples, disabling discrimination between early–middle Miocene (~20 Ma) and late Miocene–Pliocene (10–7 Ma) cooling episodes. Recently, Sánchez et al. (2018) used detrital AFT results to suggest that deformation occurred at 15–7 Ma in the Chos Malal fold-thrust belt with a stronger 9–7 Ma signal at the deformation front. On the other hand, Zamora Valcarce et al. (2009) interpret the existence of two events but do not present ages or raw thermochronological data.

The ages and models discussed here strongly suggest the presence of at least two discrete Neogene exhumational events (Fig. 7a and b). This interpretation is consistent with the final deformation phase postulated by Kozlowski et al. (1997), is in accordance with the second and third deformation episode suggested by Silvestro and Kraemer (2005) and Kraemer et al. (2011), and provides additional absolute ages

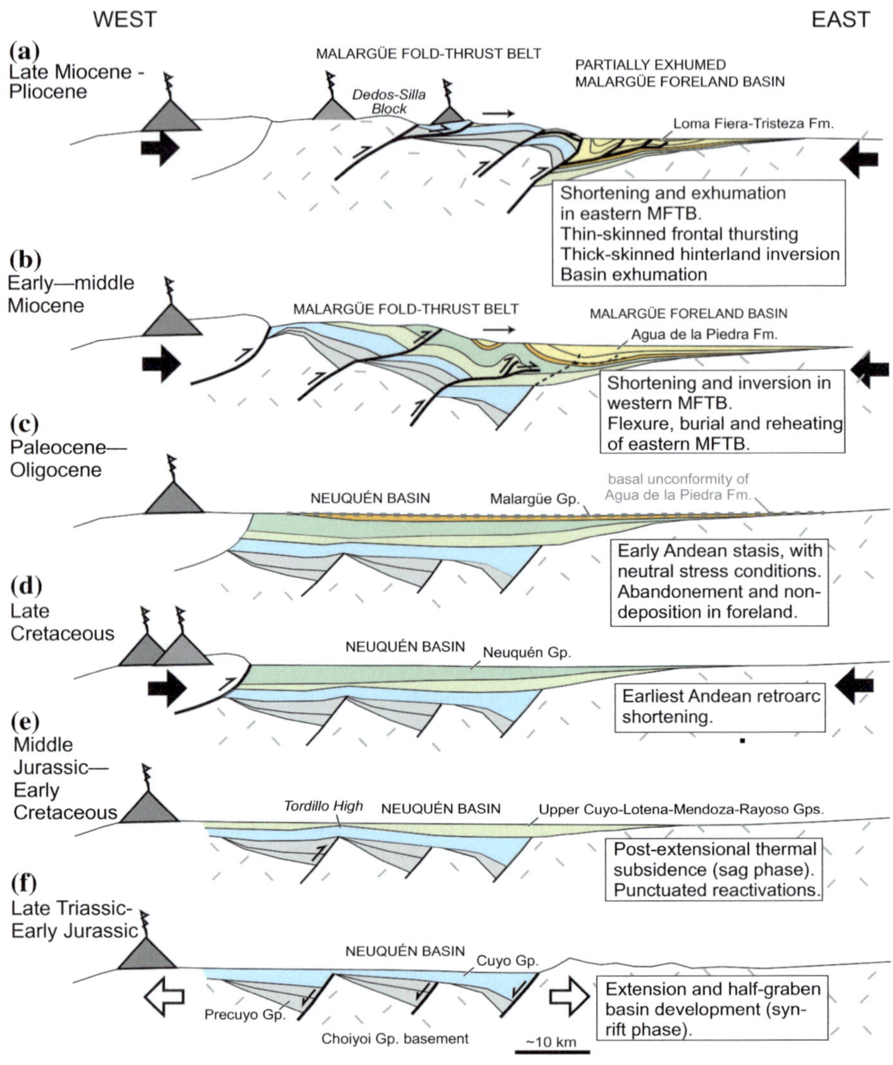

Fig. 7 Schematic cross-sections illustrating the evolution of the northern Neuquén Basin, Malargüe fold-thrust belt (MFTB) and foreland basin at ~35° S (Argentina) with interpreted fault activity and inferred regional stress regime (black arrows) (modified from Horton et al. 2016). **a** Eastward advance of shortening and exhumation in the MFTB, including basement reactivation of the Río Atuel fault, thin-skinned activation of frontal thrust faults, and out-of-sequence uplift of Choiyoi basement in the Dedos-Silla block. **b** Early–middle Miocene shortening and exhumation in the western MFTB together with formation of anticlinal structures in the eastern MFTB, development of the Malargüe foreland basin and reheating of Mesozoic strata. **c** Paleocene–Oligocene distal sedimentation in a low-accommodation setting (Malargüe Group) and late Eocene-Oligocene hiatus and unconformity development (basal Agua de la Piedra Formation boundary). **d** Late Cretaceous–early Andean shortening and foreland basin growth (Neuquén Group). **e** Middle Jurassic–Early Cretaceous postrift sedimentation (Lotena, Mendoza, and Rayoso groups) including punctuated uplift of the Tordillo High. **f** Early–Middle Jurassic extension and synrift accumulation (Pre-cuyo and Cuyo groups) in half-grabens, including basement-involved normal slip on the Río Atuel fault

and exhumational constraints to the final three deformational phases postulated by Boll et al. (2014). A detailed analysis of MFTB oil fields suggests that most of the traps were formed during the early–middle Miocene and the subsequent deformation phase (late Miocene–Pliocene) tilted and in some cases transported the structures (e.g., Puesto Rojas field), affecting the preservation of previously formed traps (Boll et al. 2014; Fuentes et al. 2016). Hydrocarbons trapped in the eastern platform and near the MFTB deformation front experienced long-distance migration from Vaca Muerta Formation generation pods or kitchens located to the west (Boll et al. 2014; Legarreta et al. 2008). In the area exists a lack of efficient carriers to conduct the hydrocarbon, being the basement faults the most plausible pathway used for migration (Boll et al. 2014). A second option is that hydrocarbons were generated and expulsed locally from the Jurassic Vaca Muerta Formation, within the hanging wall of the Rio Atuel anticlinorium. According to the thermal models, hanging wall rocks of the Río Atuel fault were at 90 °C reaching the oil generation window zone during the reheating event. However, the sample resided at this temperature for a short amount of time (<5 Myr) and likely expulsed a negligible volume of hydrocarbons. Therefore, most hydrocarbons generated and expulsed from the Vaca Muerta Formation likely migrated eastwards to foreland reservoir units using basement faults as carriers (Boll et al. 2014).

5.3 Early Cenozoic Tectonic Quiescence

Thermal modeling results allow the interpretation of a long-lived period (ca. 40 Myr) of very slow cooling rate (0.12 °C/Myr) (Figs. 5 and 6). This anomalous record reflects nondeposition, very slow sedimentation, or minor erosion (Fig. 7c). From late Campanian to middle Eocene time, a low subsidence regime accommodated limited deposition of low-energy, distal strata of the Malargüe Group (Yrigoyen 1972). Only ~450 m was deposited in ~35 Ma, implying low accumulation rate of ~13 m/Myr. Moreover, a major Paleogene hiatus (~40–20 Ma) is represented by a highly condensed section (<2–20 m thick) or disconformity demarcated by a distinctive regional conglomerate (named "Rodados Lustrosos") composed of resistant, well-rounded, polished clasts with highly weathered, oxidized, and desert varnished surfaces (Groeber 1951; Horton et al. 2016; Horton 2018b).

This calm tectonic period was recently described and dated from ~60 to ~20 Ma in the stratigraphic record and related to neutral stress conditions along the subduction margin (Horton et al. 2016; Horton and Fuentes 2016). Given the coeval late middle Eocene to earliest Miocene extension farther west and southwest (Burns et al. 2006; Charrier et al. 2007; Rojas Vera et al. 2014), we interpret this as a period of no thrust loading, limited flexural subsidence, and foreland sediment bypass, possibly amplified by minor isostatic rebound due to erosional unloading (Legarreta and Uliana 1991; Horton and Fuentes 2016; Horton 2018a, 2018b). Modeled time-temperature histories (Figs. 5b and 6b) are consistent with the interpretation.

5.4 Late Cretaceous Shortening and Early Andean Evolution

A broad tradition of relating deposition of the Neuquén Group to the onset of Andean shortening and uplift is supported by several datasets. Westward thickening of the Neuquén Group strata, provenance compositional changes, and conglomeratic facies and syndepositional thrust faults in western sectors indicate that Late Cretaceous deposition was synchronous with shortening and flexural subsidence (Ramos 1978; Tunik et al. 2010; Orts et al. 2012; Fennell et al. 2017). The 100–80 Ma time frame represents a critical transition from postextensional thermal subsidence to Andean retroarc shortening in the Neuquén Basin and its surroundings (e.g., Mpodozis and Ramos 1990; Cobbold and Rossello 2003; Zapata and Folguera 2005; Horton et al. 2016). The presented thermal models (Figs. 5a and 6a) are at the limit of confidence (oldest track modeled age between 62 and 84 Ma) to confidently identify a Late Cretaceous exhumational event but show slow monotonous cooling rate during the latest Cretaceous to Paleocene (70–50 Ma). In addition, we have not found any reset Late Cretaceous ages that might suggest early exhumation along the MFTB deformation front. Nevertheless, on the basis of AFT ages and modeling, Folguera et al. (2015) suggest the presence of a Late Cretaceous cooling event in the MFTB. Within the Bardas Blancas anticline, they present four AFT ages: Three of them show a late Miocene age and two of them (GC1026-5 and GC1026-6) overlap in error with our sample (CH131). The Late Cretaceous (120–52 Ma) cooling is modeled in one of the four samples. In the Malargüe anticline, Folguera et al. (2015) present five samples: Two of them (GC1026-1 and GC1026-17) show a late Miocene age. A third sample (GC1026-16) shows a Late Cretaceous age (82.4 $\pm$ 16.6 Ma) constrained by eight grains and one track length. The Late Cretaceous (97–70 Ma) cooling event was modeled in two samples (GC1026-16 and GC1026-18). Additionally, Rojas Vera et al. (2015) and Sánchez et al. (2018) based on AFT ages and modeling suggest the presence of a 75–55 Ma exhumation event related to Late Cretaceous–Paleocene shortening in the northern Neuquén and southern Mendoza provinces. We interpret that early deformation and exhumation was stronger in the western sector of the Andean belt, in the hinterland zones of the Malargüe, Chos Malal, and Agrio fold-thrust belts. This interpretation is consistent with other studies suggesting that Late Cretaceous deformation was focused in a narrow western belt adjacent to the magmatic arc (e.g., Mescua et al. 2013). However, recent provenance studies (e.g., Di Giulio et al. 2012; Balgord and Carrapa 2016; Horton et al. 2016) point to eastern-sourced detritus within the Neuquén Group, suggesting additional localized foreland uplift >50–100 km east of the modern thrust front.

Regionally, thermochronological ages and model results suggest that Late Cretaceous exhumation selectively affected structures within the Neuquén Basin. Hinterland sectors of the southern MFTB and Agrio fold-thrust belt experienced Cretaceous deformation, but this event did not reach the northern MFTB. The northernmost evidence of Late Cretaceous cooling might be located at Bardas Blancas anticline. This uneven distribution of exhumation points to a heterogeneous, spatially irregular distribution of Late Cretaceous shortening, consistent with a complex deformation scheme.

5.5 *Late Jurassic Localized Uplift*

ZFT results for sample TR01 show a Late Jurassic age (140.2 ± 6.4 Ma), which corresponds to a well-developed regional (Intramalmic) unconformity in the Neuquén Basin (Gulisano and Gutiérrez Pleimling 1995). This event is well-developed around the Huincul High, inverting normal faults, and controlled deposition of the nonmarine Tordillo Formation (Legarreta and Uliana 1991; Vergani et al. 1995). However, deposition of this unit in the northern sector of the Neuquén Basin occurred during extension reactivating and developing new normal faults (e.g., Lo Forte 1996; Cegarra and Ramos 1996; Giambiagi et al. 2003; Mescua et al. 2008). In the hinterland sector of the MFTB Davidson and Vicente (1972) and Legarreta and Kozlowski (1984) described important facies changes, dramatic stratal thinning, and the absence of selected stratigraphic units within the Mendoza Group. They relate these changes to the presence of a structural and topographic paleohigh (Tordillo High) during early Mesozoic time (Figs. 2 and 7e). Structural mapping and sections correlation in the Dedos-Silla block area suggest that the geometry of normal faults is dominated by west-dipping normal faults to the west and east-dipping normal faults to the east (Legarreta and Uliana 1999; Mescua et al. 2014). The Late Jurassic ZFT cooling age might suggest tectonic uplift of a horst structure in the hinterland sector of the MFTB. However, further ZFT or ZHe results are needed to confirm the presence and extent of the proposed Late Jurassic uplift.

5.6 *Conclusions*

We present new (U–Th)/He and fission track results that show rapid exhumation of the frontal sector of the Malargüe fold-thrust belt (MFTB) during the late Miocene, at ca. 10–7 Ma. The presence of preexisting basement-involved structures in this northern segment of the Neuquén Basin strongly influenced the distribution of exhumation, with maximum exhumation centered around the Rio Atuel fault. Interpretation of thermal modeling results places the onset of Andean hinterland deformation during the early Miocene (~20 Ma). Continued shortening affected the MFTB, enabling rapid cooling of frontal structures in the late Miocene and hinterland basement-involved structures (Dedos-Silla block) at 5–2 Ma. Overall, the MFTB experienced diachronous uplift with at least two events of rapid exhumation and related deformation.

The modeled early Cenozoic episode of slow cooling supports interpretations of neutral to extensional conditions, possibly due to diminished plate coupling along the subduction margin (Horton and Fuentes 2016; Horton 2018a). Finally, evidence for Late Cretaceous deformation is neither as widely distributed nor as sharply constrained as Miocene exhumation during major shortening within the MFTB. Maximum exhumation occurred in hinterland sectors of the southern MFTB and Agrio fold-thrust belt, with minimum exhumation focused in the northernmost sector of the MFTB and along the deformation front. The spatially irregular distribution of exhumation and deformation suggests a more complex history and mechanism of shortening in the Late Cretaceous relative to the robust Miocene record of deformation.

Acknowledgements This research was supported by funding from Tecpetrol S.A. and a U.S. National Science Foundation grant (EAR-1348031) to B. K. Horton. We acknowledge corrections and comments by José Mescua and editor Andrés Folguera that significantly improved the contribution. We thank Daniel Starck, Sebastián Ramirez, Kurt Constenius, and Mariya Levina for useful discussions that improved the manuscript.

References

Balgord EA, Carrapa B (2016) Basin evolution of Upper Cretaceous-Lower Cenozoic strata in the Malargüe fold-and-thrust belt: Northern Neuquén Basin, Argentina. Basin Res 28:183–206

Bechis F (2009) Deformación transtensiva de la cuenca Neuquina: Análisis a partir de ejemplos de campo y modelos análogos. PhD Thesis, Universidad de Buenos Aires

Bechis F, Giambiagi L, García V et al (2010) Kinematic analysis of a transtensional fault system: the Atuel depocentre of the Neuquén Basin, Central Andes, Argentina. J Struct Geol 32:886–899

Boll A, Alonso A, Fuentes F et al (2014) Factores controlantes de las acumulaciones de hidrocarburos en el sector norte de la cuenca Neuquina, entre los ríos Diamante y Salado, Provincia de Mendoza, Argentina. In: Abstracts of the 9 Congreso de Exploración y Desarrollo de Hidrocarburos, IAPG, Mendoza, 5–9 Nov 2014

Brandon MT, Roden-Tice MK, Garver JI (1998) Late Cenozoic exhumation of the Cascadia accretionary wedge in the Olympic Mountains, Northwest Washington State. Geol Soc Am Bull 110:985–1009

Burns WM, Jordan TE, Copeland P et al (2006) The case for extensional tectonics in the Oligocene-Miocene Southern Andes as recorded in the Cura Mallín basin (36°–38° S). In: Kay SM, Ramos VA (eds) Evolution of an Andean Margin: a tectonic and magmatic view from the Andes to the Neuquén Basin (35°–39° S lat). Geological Society America, SP 407, pp 163–184

Cegarra MI, Ramos VA (1996) La faja plegada y corrida del Aconcagua. In: Ramos VA (ed.) Geología de la región del Aconcagua, provincias de San Juan y Mendoza. SEGEMAR, Anales 24, p 387–422

Charrier R, Pinto L, Rodríguez MP (2007) Tectonostratigraphic evolution. In: Moreno T, Gibbons W (eds) The geology of Chile. Geological Society America, London, pp 21–114

Cobbold P, Rossello E (2003) Aptian to Recent compressional deformation in the foothills of the Neuquén basin Argentina. Mar Petrol Geol 20:429–443

Combina AM, Nullo F (2011) Cenozoic tectonic, volcanic and sedimentary cycles in southern Mendoza Province, Argentina (35º-37ºS and 69°30′ W). Andean Geol 38:198–218

Davidson MJ, Vicente JC (1972) Características paleogeográficas y estructurales del area fronteriza de las nacientes del Teno (Chile) y Santa Elena (Argentina), Cordillera Principal, 35° a 35° 15´de latitud Sur. In: Abstracts of the 5 Congreso Geológico Argentino, Córdoba, 23–26 Aug 2002

DeCelles PG, Giles KN (1996) Foreland basin systems. Basin Res 8:105–123

Di Giulio A, Ronchi A, Sanfilippo A, Tiepolo M, Pimentel M, Ramos VA (2012) Detrital zircon provenance from the Neuquén Basin (south-central Andes): Cretaceous geodynamic evolution and sedimentary response in a retroarcforeland basin. Geology 40:559–562

Digregorio RE, Gulisano CA, Gutiérrez Pleimling AR et al (1984) Esquema de la evolución geodinámica de la Cuenca Neuquina y sus implicancias paleogeográficas. 9 Congreso Geológico Argentino, San Carlos de Bariloche, 5–9 Nov 1984

Donelick RA, Ketcham RA, Carlson WD (1999) Variability of apatite fission-track annealing kinetics: II. Crystallographic orientation effects. Am Mineral 84(9):1224–1234

Ehlers TA, Farley KA (2003) Apatite (U–Th)/He thermochronometry: methods and applications to problems in tectonic and surface processes. Earth Planet Sci Lett 206:1–14

Farley KA (2002) (U–Th)/He dating; techniques, calibrations, and applications. Rev Mineral Geochem 47:819–843

Farley KA, Wolf RA, Silver LT (1996) The effects of long alpha-stopping distances on (U–Th)/He ages. Geochim Cosmochim Ac 60:4223–4229

Fennell LM, Folguera A, Naipauer M et al (2017) Cretaceous deformation of the southern Central Andes: synorogenic growth strata in the Neuquén Group (35° 30′–37° S). Basin Res 29:51–72

Folguera A, Ramos VA (2011) Repeated eastward shifts of arc magmatism in the Southern Andes: a revision to the long-term pattern of Andean uplift and magmatism. J S Am Earth Sci 32:531–546

Folguera A, Naranjo JA, Orihashi Y et al (2009) Retroarc volcanism in the northern San Rafael Block (34°–35° 30′ S), southern Central Andes: occurrence, age, and tectonic setting. J Volcanol Geoth Res 186:169–185

Folguera A, Rojas Vera E, Bottessi G et al (2010) The Loncopué trough: a Cenozoic basin produced by extension in the southern Central Andes. J Geodyn 49:287–295

Folguera A, Bottesi G, Duddy I et al (2015) Exhumation of the Neuquén Basin in the southern Central Andes (Malargüe fold and thrust belt) from field data and low temperature thermochronology. J S Am Earth Sci 64:381–398

Fuentes F, Horton BK, Starck D et al (2016) Structure and tectonic evolution of hybrid thick- and thin-skinned systems in the Malargüe fold-thrust belt, Neuquén basin, Argentina. Geol Magaz 153:1066–1084

Gautheron C, Tassan-Got L, Barbarand J et al (2009) Effect of alpha-damage annealing on apatite (U–Th)/He thermochronology. Chem Geol 266:157–170

Giambiagi LB, Álvarez P, Godoy E et al (2003) The control of pre-existing extensional structures on the evolution of the southern sector of the Aconcagua fold and thrust belt, southern Andes. Tectonophysics 369:1–19

Giambiagi LB, Bechis F, García V et al (2008) Temporal and spatial relationships of thick-and thick-and thin-skinned deformation: a case study from the Malargüe fold-and-thrust belt, southern Central Andes. Tectonophysics 459:123–139

Giambiagi L, Ghiglione M, Cristallini E et al (2009) Kinematic models of basement/cover interactions: insights from the Malargüe fold and thrust belt, Mendoza, Argentina. J Struct Geol 31:1443–1457

Gleadow AJW, Duddy IR (1981) A natural long-term annealing experiment for apatite. Nuclr Tracks 5:169–174

Green PF, Duddy IR, Laslett GM et al (1989) Thermal annealing of fission tracks in apatite, 4, Quantitative modeling techniques and extension to geological timescales. Chem Geol 79:155–182

Groeber P (1947) Observaciones geológicas a lo largo del meridiano 70°. 2: Hojas Sosneado y Maipo. Rev Asoc Geol Argent 2(2):141–176

Groeber P (1951) La Alta Cordillera entre las latitudes 34 & #xB0; y 29° 30′. Rev Inst Nac Inv Cs Nat Geología 1(5):235–352

Gulisano CA, Gutiérrez Pleimling AR (1995) El Jurásico de la Cuenca Neuquina: B Provincia de Mendoza, vol E3. Asociación Geológica Argentina, Guía de Campo, Buenos Aires

Horton BK (2018a) Tectonic regimes of the central and southern Andes: Responses to variations in plate coupling during subduction. Tectonics 37:402–429

Horton BK (2018b) Sedimentary record of Andean mountain building. Earth-Sci Rev 178:279–309

Horton BK, Fuentes F (2016) Sedimentary record of plate coupling and decoupling during growth of the Andes. Geology 44:647–650

Horton BK, Fuentes F, Boll A et al (2016) Andean stratigraphic record of the transition from backarc extension to orogenic shortening: a case study from the northern Neuquén Basin, Argentina. J S Am Earth Sci 71:17–40

Kay SM, Burns M, Copeland P et al (2006) Upper Cretaceous to Holocene magmatism and evidence for transient Miocene shallowing of the Andean subduction zone under the northern Neuquén Basin. In: Kay SM, Ramos VA (eds) Evolution of an Andean Margin: a tectonic and magmatic view from the Andes to the Neuquén Basin (35°–39° S lat). Geological Society America, SP 407, pp 19–60

Ketcham RA (2005) HeFTy: Forward and inverse modeling thermochronometer systems. Rev Mineral Geochem 58:596–597

Ketcham RA, Donelick RA, Carlson WD (1999) Variability of apatite fission-track annealing kinetics: III. Extrapolation to geological time scales. Am Mineral 84:1235–1255

Ketcham RA, Carter AC, Donelick RA et al (2007) Improved modeling of fission-track annealing in apatite. Am Mineral 92:799–810

Kraemer P, Silvestro J, Achilli F et al (2011) Kinematics of a hybrid thick-thin-skinned fold and thrust belt recorded in Neogene syntectonic wedge-top basins, southern central Andes between 35° and 36° S, Malargüe, Argentina. In: McClay K, Shaw J, Suppe J (eds) Thrust Fault-Related Folding. AAPG Memoir 94:245–270

Kozlowski E, Cruz C, Condat P et al (1989) Informe Geológico Malargüe Occidental, Cuenca Neuquina, provincia de Mendoza. YPF Internal Report and Map, Buenos Aires

Kozlowski E, Manceda R, Ramos VA (1993) Estructura. In: Ramos VA (ed) Geología y Recursos Naturales de Mendoza. Asociación geológica Argentina, Buenos Aires, pp 235–256

Kozlowski E, Cruz C, Sylwan C (1997) Modelo exploratorio en la faja corrida de la Cuenca Neuquina, Argentina. In: Abstracts of the 6 Simposio Bolivariano Exploración Petrolera en las Cuencas Subandinas, Bogotá, 14–17 Sept 1997

Lanés S, Giambiagi L, Bechis F, Tunik M (2008) Late Triassic-Early Jurassic successions from the Atuel depocenter: depositional systems, sequence stratigraphy and tectonic controls. Rev Asoc Geol Argent 63(4):534–548

Legarreta L, Kozlowski E (1984) Secciones condensadas del Jurásico-Cretácico de los Andes del sur de Mendoza: estratigrafía y significado tectosedimentario. In: Abstratcs of the 9 Congreso Geológico Argentino, San Carlos de Bariloche, 5–9 Nov 1984

Legarreta L, Uliana MA (1991) Jurassic-Cretaceous marine oscillations and geometry of back-arc basin fill, central Argentine Andes. Int As Sed SP 12:429–450

Legarreta L, Uliana MA (1998) Anatomy of hinterland depositional sequences: Upper Cretaceous fluvial strata, Neuquén Basin, west-central Argentina. In: Shanley KW, McCabe PJ (eds) Relative role of eustasy, climate, and tectonism in continental rocks. SEPM SP 59, pp 83–92

Legarreta L, Uliana MA (1999) El Jurásico y Cretácico de la Cordillera Principal y la Cuenca Neuquina. In: Caminos R (ed) Geología Argentina. SEGEMAR, Anales 29, BuenosAires, pp 399–416

Legarreta L, Kokogián DA, Boggetti DA (1989) Depositional sequences of the Malargue Group (Upper Cretaceous-lower Tertiary), Neuquén Basin, Argentina. Cretaceous Res 10:337–356

Legarreta L, Villar HJ, Cruz CE et al (2008) Revisión integrada de los sistemas generadores, estilos de migración-en-trampamiento y volumetría de hidrocarburos en los distritos productivos de la Cuenca Neuquina, Argentina. In: Abstratcs of the 7 Congreso de Exploración y Desarrollo de Hidrocarburos, IAPG, Mar del Plata, 5–8 Nov 2008

Levina M, Horton BK, Fuentes F et al (2014) Cenozoic sedimentation and exhumation of the foreland basin system preserved in the Precordillera thrust belt (31–32° S), southern central Andes, Argentina. Tectonics 33:1659–1680

Lo Forte GL (1996) Los depósitos jurásicos de la Alta Cordillera de Mendoza. In: Ramos VA (ed) Geología de la región del Aconcagua, provincias de San Juan y Mendoza. Dirección Nacional del Servicio Geológico, Subsecretaría de Minería de la Nación, Anales 24, pp 179–230

Manceda R, Figueroa D (1995) Inversion of the Mesozoic Neuquén rift in the Malargüe fold and thrust belt, Mendoza, Argentina. In: Tankard AJ, Suárez Soruco R, Welsink HJ (eds) Petroleum Basins of South America. AAPG Memoir 62:369–382

Mescua JF, Giambiagi LB, Bechis F (2008) Evidencias de téctonica extensional en el Jurásico Tardío (Kimeridgiano) del suroeste de la provincia de Mendoza. Rev Asoc Geol Argent 63(4):512–519

Mescua JF, Giambiagi LB, Ramos VA (2013) Late Cretaceous uplift in the Malargüe fold-and-thrust belt (35° S), southern central Andes of Argentina and Chile. Andean Geol 40:102–116

Mescua JF, Giambiagi LB, Tassara A, Gimenez M, Ramos VA (2014) Influence of pre-Andean history over Cenozoic foreland deformation: Structural styles in the Malargüe fold-and-thrust belt at 35° S, Andes of Argentina. Geosphere 10:585–609

Mosquera A, Ramos VA (2006) Intraplate deformation in the Neuquén Embayment. In: Kay SM, Ramos VA (eds) Evolution of an Andean Margin: a tectonic and magmatic view from the andes to the Neuquén Basin (35°–39° S lat). Geological Society America, SP 407, pp 97–123

Mpodozis C, Ramos VA (1990) The Andes of Chile and Argentina. In: Ericksen GE, Canas Pinochet MT, Reinemund JA (eds) Geology of the Andes and its relation to hydrocarbon and mineral resources: Circum-Pacific Council for Energy and Mineral Resources, vol 11, Earth Science Series, pp 59–90

Naipauer M, García Morabito E, Marques JC et al (2012) Intraplate Late Jurassic deformation and exhumation in western central Argentina: constraints from surface data and U–Pb detrital zircon ages. Tectonophysics 524–525:59–75

Nullo FE, Stephens G, Combina A et al (2005) Hoja Geológica 3569-III/3572-IV, Malargüe, provincial de Mendoza. Servicio Geológico Minero Argentino, Boletín 346, Buenos Aires, 85 p

Orts DL, Folguera A, Giménez M et al (2012) Variable structural controls through time in the Southern Central Andes (~36°S). Andean Geol 39(2):220–241

Ploskiewicz JV (1987) Las zonas triangulares de la faja fallada y plegada de la Cuenca Neuquina, Argentina. In: Abstracts of the 10 Congreso Geológico Argentino, San Miguel de Tucumán, 10–15 Oct 1987

Ramos VA (1978) Estructura. Geología y recursos naturales de la Provincia del Neuquén. Asociació Geológica Argentina, Buenos Aires, pp 99–118

Ramos VA (1981) Descripción geológica de la Hoja 33 c Los Chihuidos Norte, provincia del Neuquén. Servicio Geológico Nacional, Boletín 182, Buenos Aires, pp 1–103

Ramos VA, Folguera A (2005) Tectonic evolution of the Andes of Neuquén: constraints derived from the magmatic arc and foreland deformation. In: Veiga G, Spalleti LA, Howell JA, Schwarz E (eds) The Neuquén Basin: a case study in sequence stratigraphy and basins dynamics. Geological Society, London, SP 252, pp 15–35

Ramos VA, Folguera A (2009) Andean flat-slab subduction through time. In: Murphy JB, Keppie JD, Hynes AJ (eds) Ancient Orogens and modern analogues. Geological Society, London, SP 327, pp 31–54

Reiners PW, Farley KA (2001) Influence of crystal size on apatite (U–Th)/He thermochronology: an example from the Bighorn Mountains, Wyoming. Earth Planet Sci Lett 188:413–420

Reiners PW, Spell TL, Nicolescu S et al (2004) Zircon (U–Th)/He thermochronometry: He diffusion and comparisons with 40Ar/39Ar dating. Geochim Cosm Ac 68:1857–1887

Rojas Vera EA, Folguera A, Zamora Valcarce G et al (2014) Structure and development of the Andean system between 36°–39° S. J Geodyn 73:34–52

Rojas Vera EA, Mescua J, Folguera A, Becker TP, Sagripanti L, Fennell L, Orts D, Ramos VA (2015) Evolution of the Chos Malal and Agrio fold and thrust belts, Andes of Neuquén: Insights from structural analysis and apatite fission track dating. J S Am Earth Sci 64:418–433

Sánchez NP, Coutand I, Turienzo M et al (2018) Tectonic evolution of the Chos Malal fold-and-thrust belt (Neuquén Basin, Argentina) from (U–Th)/He and fission track thermochronometry. Tectonics. https://doi.org/10.1029/2018tc004981

Sagripanti L, Bottesi G, Kietzmann D et al (2012) Mountain building processes at the orogenic front. A study of the unroofing in Neogene foreland sequence (37° S). Andean Geol 39:201–219

Silvestro J, Atencio M (2009) La Cuenca Cenozoica del Río Grande y Palauco: Edad, Evolución y Control Estructural, Faja Plegada de Malargüe (36° S). Rev Asoc Geol Argent 65(1):154–169

Silvestro J, Kraemer P (2005) Evolución tecto-sedimentaria de la Cordillera Principal en el sector surmendocino a los 35° 30′ S: Faja plegada de Malargüe, Republica Argentina. In: Abstracts of the Congreso de Exploración y Desarrollo de Hidrocarburos, IAPG, Mar del Plata, 15–19 Nov 2005

Soares CJ, Guedes S, Hadler JC et al (2014) Novel calibration for LA-ICP-MS-based fission-track thermochronology. Phys Chem Miner 41:65–73

Soares CJ, Mertz-Kraus R, Guedes S et al (2015) Characterisation of apatites as potential Uranium reference materials for fission-track dating by LA-ICP-MS. Geostan Geoanal Res 39:305–313

Spagnoulo MG, Folguera A, Litvak V et al (2012) Late Cretaceous arc rocks in the Andean retroarc region at 36.5° S: Evidence supporting a Late Cretaceous slab shallowing. J S Am Earth Sci 38:44–56

Tagami T (2005) Zircon fission-track thermochronology and applications to fault studies. Rev Miner Geochem 58:95–122

Tunik M, Folguera A, Naipauer M et al (2010) Early uplift and orogenic deformation in the Neuquén Basin: Constraints on the Andean uplift from U–Pb and Hf isotopic data of detrital zircons. Tectonophysics 489:258–273

Turienzo M (2010) Structural style of the Malargüe fold and-thrust belt at the Diamante river area (34° 30′–34° 50′ S) and its linkage with the Cordillera frontal Andes of central Argentina. J S Am Earth Sci 29:537–556

Vergani GD, Tankard AJ, Belotti HJ et al (1995) Tectonic evolution and paleogeography of the Neuquén Basin, Argentina. AAPG Memoir 62:383–402

Wolf RA, Farley KA, Kass DM (1998) Modeling of the temperature sensitivity of the apatite (U–Th)/He thermochronometer. Chem Geol 148:105–114

Yrigoyen M (1972) Cordillera principal. In: Leanza AF (ed) Geología Regional Argentina. Academia Nacional de Ciencias, Córdoba, pp 345–364

Yrigoyen M (1993) Los depósitos sinorogénicos terciarios. In: Ramos VA (ed) Geología y Recursos Naturales de Mendoza. Asociación Geológica Argentina, Buenos Aires, pp 123–148

Zamora Valcarce G, Zapata T, Ramos VA et al (2009) Evolución tectónica del frente andino en Neuquén. Rev Asoc Geol Argent 65:192–203

Zapata T, Folguera A (2005) Tectonic evolution of the Andean fold and thrust belt of the southern Neuquén Basin, Argentina. In: Veiga GD, Spalletti LA, Howell JA, Schwarz E (eds) The Neuquén Basin, Argentina: a case study in sequence stratigraphy and basin dynamics. Geological Society, London, SP 252, pp 37–56

Late Cretaceous to Oligocene Magmatic Evolution of the Neuquén Basin

Sofía B. Iannelli, Lucas Fennell, Lucía Fernández Paz, Vanesa D. Litvak, Alfonso Encinas and Andrés Folguera

Abstract Geochemical variations in arc- and within-plate magmatic associations since Late Cretaceous times are analyzed and correlated with the main tectonic changes that influenced the Neuquén Basin evolution. The collision and southward migration of the Farallon-Aluk mid-ocean ridge along the Chilean trench since 80 Ma have played an important role in controlling the Late Cretaceous to Oligocene magmatic evolution of the arc and retroarc zones. The passage of this spreading center through the Chilean trench induced the development of geochemically distinct magmatic associations since Late Cretaceous to Eocene times associated with the extensional reactivation of the Cretaceous fold and thrust belt. Then, by Late Oligocene times, a major plate tectonic reorganization occurred when the Farallon plate broke apart and the resulting Nazca plate started an orthogonal subduction regime beneath the South American plate with higher convergence rates. Then extensional basins and associated magmatism developed at this time destabilizing the Paleogene fold and thrust belt and establishing a more homogeneous tholeiitic signature along the Andean axis.

Keywords Late Cretaceous magmatism · Oligocene magmatism · Extension · Farallon-Aluk mid-ocean ridge

1 Introduction

The main kinematic changes achieved during the Neuquén Basin evolution varied from (i) contractional deformation periods associated with stages of nearly orthogonal convergence and rapid westward displacement of the South American plate;

S. B. Iannelli (✉) · L. Fennell · L. Fernández Paz · V. D. Litvak · A. Folguera
Instituto de Estudios Andinos Don Pablo Groeber (IDEAN), Universidad de Buenos Aires-CONICET, Buenos Aires, Argentina
e-mail: sofia.iannelli@hotmail.com.ar

A. Encinas
Departamento de Ciencias de la Tierra, Universidad de Concepción, Concepción, Chile

D. Kietzmann and A. Folguera (eds.), *Opening and Closure of the Neuquén Basin in the Southern Andes*, Springer Earth System Sciences,
https://doi.org/10.1007/978-3-030-29680-3_16

(ii) stages of back-arc extension due to envigorated slab-roll back (e.g., Fennell et al. 2015; Folguera et al. 2011; Ramos and Folguera 2005a, b; Zamora Valcarce et al. 2006); and (iii) stages of dominant strike-slip regime along the arc zone during periods of oblique convergence (e.g., Cobbold and Rosello 2003; Mosolf et al. 2018).

The variations in tectonic style between Late Cretaceous to Oligocene times can be associated with the north–south passage of the Farallon-Aluk mid-ocean spreading ridge since ~80 Ma (e.g., Gianni et al. 2018; Somoza and Ghidella 2012; Müller et al. 2016), the oblique subduction of the Farallon plate beneath South America, and finally its break-up during the late Oligocene into the Cocos and Nazca plates (e.g., Cande and Kent 1992; Lonsdale 2005; Muñoz et al. 2000; Müller et al. 2016). These changes in plate dynamics (e.g., variable convergence rates, subduction angle, and direction of subduction) have been traditionally linked to variable geochemical signatures of magmatism emplaced in the western Neuquén Basin (e.g., Kay et al. 2006; Zamora Valcarce et al. 2006).

The aim of this chapter is to present a review of the different magmatic stages that occurred during the different deformational pulses associated with the closure of the Neuquén Basin between Late Cretaceous to Oligocene times (~34°–40° S), remarking the linkage between magmatism and the evolving tectonic regime.

2 Geological Setting of the Different Magmatic Stages in the Neuquén Basin

The geodynamic evolution of the Neuquén Basin since the Late Triassic has undergone periods of back-arc rifting and foreland basin development (e.g., Fennell et al. 2017; Folguera et al. 2011; Ramos and Folguera 2005a, b). The initial opening stage of the Neuquén Basin is represented by bimodal volcaniclastic sequences (Choiyoi Group) that mark the transition from an initial arc-like setting toward a within-plate magmatism associated with an extensional regime (e.g., Llambías et al. 2003; Heredia et al. 2012). Afterward, the Neuquén Basin infill is characterized by the alternation of transgressive and regressive cycles between Early Jurassic and Early Cretaceous times (see Martos et al., Palma et al., Zavala et al., Kietzmann et al., Pazos et al. chapters) (e.g., Gulisano and Gutiérrez Pleimling 1994; Legarreta and Uliana 1991). Then, the Late Cretaceous opening of the South Atlantic Ocean led to a contractional regime, and as a consequence of this, the Neuquén Basin back-arc evolved into a foreland basin (Somoza and Ghidella 2012; Ramos et al. 2011) that lately was affected by a latest Cretaceous–Paleocene Atlantic-derived transgression (Aguirre-Urreta et al. 2011).

The Late Cretaceous marine to continental foreland deposits that infilled the eastern Neuquén Basin was partially covered and intruded by the older Late Cretaceous to Paleogene magmatic association that we are analyzing in this chapter (e.g., Kay et al. 2006; Zamora Valcarce et al. 2006). The development of these eastern magmatic units was initially associated with an eastward arc migration into the foreland area, which led to the hypothesis of a shallow subduction setting for that time (e.g., Franchini et al. 2003; Spagnuolo et al. 2012; Zamora Valcarce et al. 2006). In this sense, Late Cretaceous–Paleocene magmatism, named as the Naunauco Group (~36° 30′–38° S), bears a calc-alkaline signature with a high slab influence (e.g., Franchini et al. 2003; Llambías and Aragón 2011; Zamora Valcarce et al. 2006). Northwards, Late Cretaceous–Paleocene magmatic rocks are widely described in the Chilean Andean side (e.g., Boyce 2015; Gana and Wall 1997; Mosolf 2018), being recently described through new U–Pb ages on the Argentinean side showing that retro-arc products of that age describe at least a 250 km long volcanic unit (~35° 30′–38° S) across the Neuquén Basin (Fennell et al. 2017; Iannelli et al. 2017b, 2018). This volcano-sedimentary sequence, named as Los Ángeles Unit (~67 Ma) on the Argentinean territory, shows mostly a basaltic to intermediate composition with a tholeiitic to enriched magma source (Iannelli et al. 2017b, 2018).

By Paleocene–Eocene times magmatism retracted to the southern (~37°–38° S) latitudes, being represented by the extrusive (Cayanta Formation) and intrusive (Collipili Formation) facies (Llambías and Rapela 1989). This magmatic sequence grouped as PVNM (*Provincia Volcánica Neuquino Mendocina*; Llambías and Rapela 1989) bears a calc-alkaline composition similar to the older (Late Cretaceous–Paleocene) Naunauco Group. However, in the southern extreme of the Neuquén Basin (~40° S), magmatism drastically changed showing a contrasting alkaline composition with minor slab influence that represents the northernmost outcrops of the within-plate Pilcaniyeu Belt (Aragón et al. 2011; Iannelli et al. 2016, 2017a; Rapela et al. 1988, 1983). Lately, latest Oligocene magmatism is controlled by extensional structures developed in the axial Andean zone in the Auca Pan (~29 Ma; Iannelli et al. 2017a; Ramos et al. 2014) and Cura Mallín (~27–16 Ma; Burns 2002; Jordan et al. 2001; Suárez and Emparán 1995) intra-arc basins. Extensional tectonics and tholeiitic Andean-type magmatism also developed further north (33°–36° S) in the Abanico Basin (~37–21 Ma) over the northernmost Neuquén Basin (Charrier et al. 2002; Godoy et al. 1999) (Fig. 1). This magmatism after ~24 Ma (Cura Mallín Formation) presents a tholeiitic composition and higher influx of slab fluids relative to older units (Auca Pan Formation; 29 Ma).

3 Lithological and Geochemical Features of the Late Cretaceous to Oligocene Magmatism in the Neuquén Basin

3.1 Late Cretaceous-Early Paleocene Magmatic Stage (70–56 Ma)

The Late Cretaceous–Early Paleocene magmatism in the Neuquén Basin is characterized by scattered sequences (Casé et al. 2008; Franchini et al. 2003; Iannelli et al. 2017b, 2018; Kay et al. 2006; Llambías and Aragón 2011; Mateo Fernández

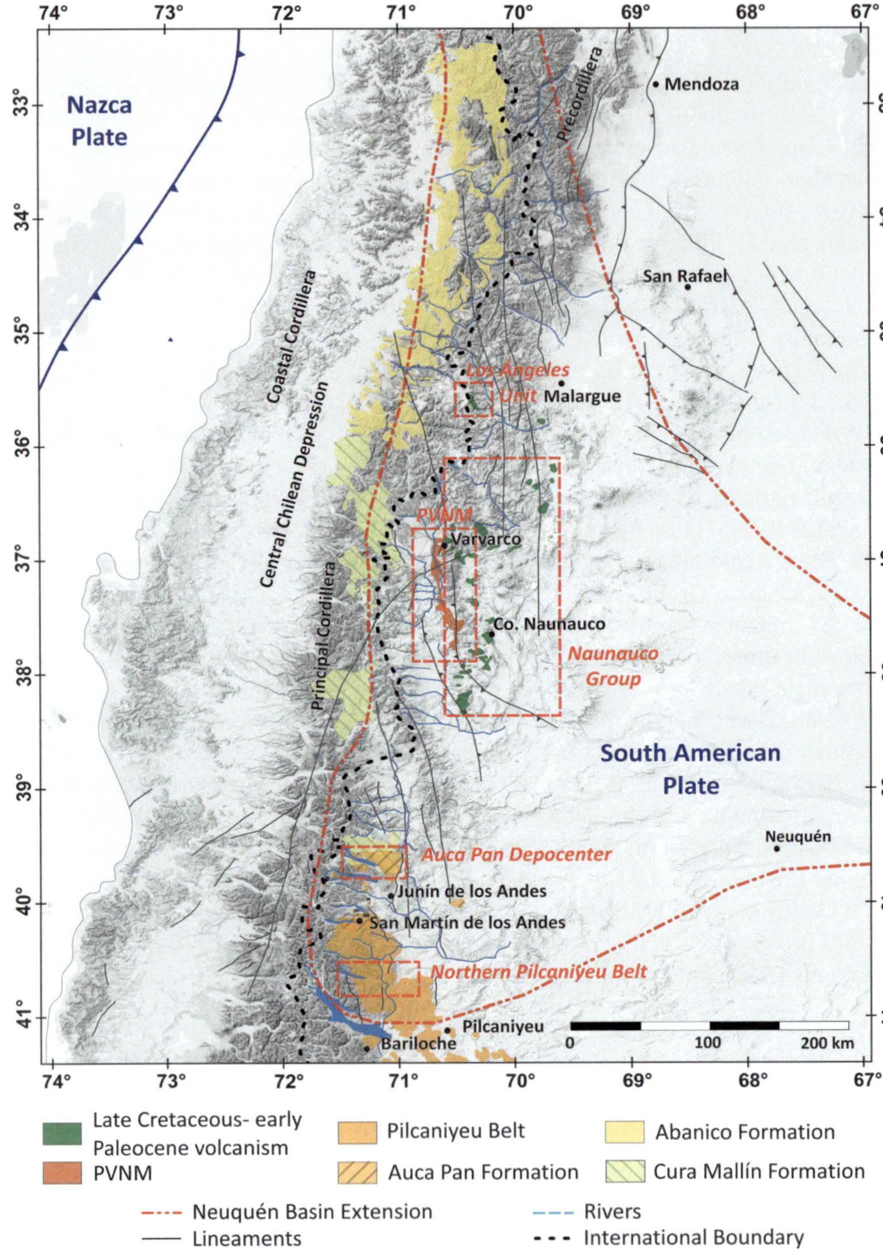

Fig. 1 Regional distribution of the main magmatic units along the Southern Central Andes since Late Cretaceous to Oligocene times (modified from Charrier et al. 2002, 2015; Giambiagi et al. 2012; Jordan et al. 2001; Kay et al. 2005, 2006; Litvak et al. 2015, 2018; Ramos et al. 2014; Rapela et al. 1983; Zamora Valcarce et al. 2006). The boundaries of the Neuquén Basin are indicated with a red dashed line

Caso et al. 2011; Zamora Valcarce et al. 2006). A migration of the magmatic arc locus toward the foreland zone is evidenced by the Late Cretaceous, coinciding with a magmatic lull over the Andean Chilean side between 36° and 39° S (Ramos and Folguera 2005a, b). This arc migration coincides with the subduction of the Farallon-Aluk spreading ridge southwards along the Chilean trench margin at the latitudes of the Neuquén Basin, since ~80 Ma (Somoza and Ghidella 2012; Müller et al. 2016).

The Late Cretaceous–Early Paleocene magmatic belt from 35° to 38° S presents a variable geochemical signature from north to south. The northern volcanic outcrops are represented by the Los Ángeles Unit, located at ~35° 30′ S (Figs. 1 and 2; Fennell et al. 2017; Iannelli et al. 2017b, 2018) that is described as a volcano-sedimentary sequence, conformed by basaltic to andesitic lava flows, interbedded with fine-grained lithic sandstones, locally intruded by basaltic dykes. An extensional

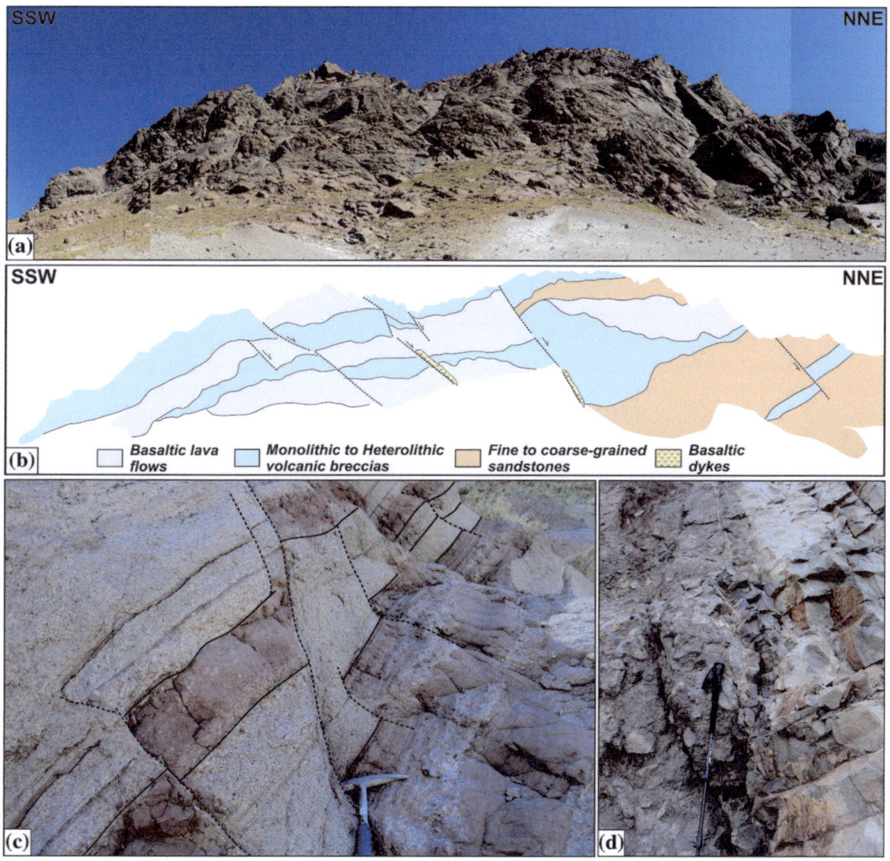

Fig. 2 **a** and **b** Latest Cretaceous lava flows, volcanic breccias and interbedded sandstones intruded by basaltic dykes along normal faults (modified from Iannelli et al. 2018); **c** minor scale normal faults controlling the deposition of the latest Cretaceous Los Ángeles Unit; and **d** contact between basaltic dykes intruding volcanic breccias of the same unit

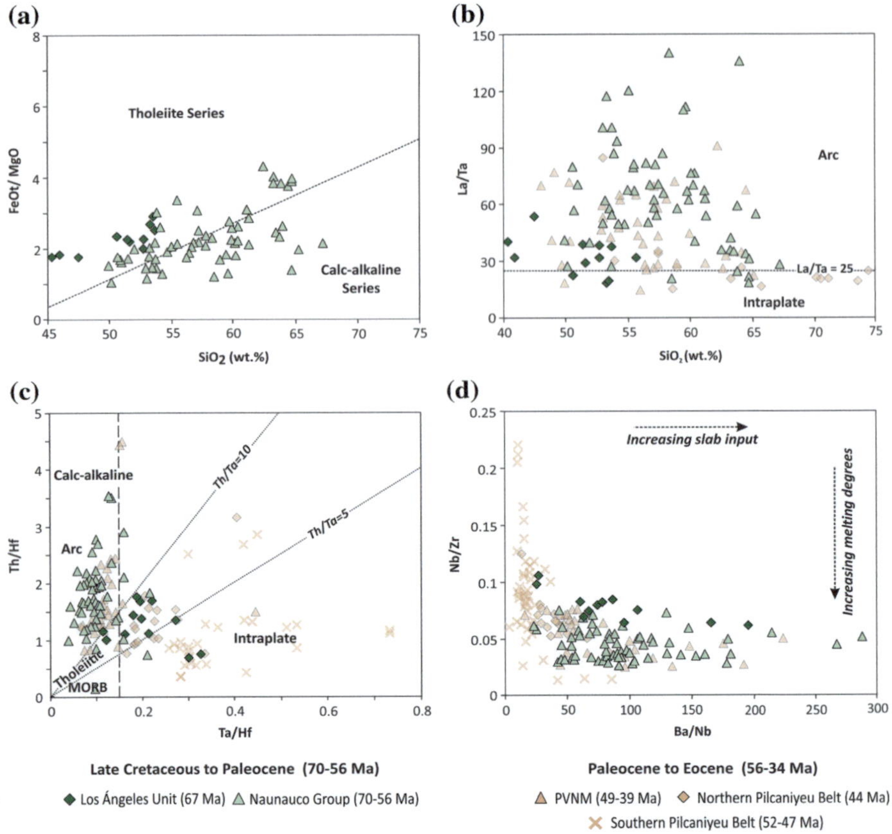

Fig. 3 **a** FeO_t/MgO versus SiO_2 (wt%) diagram showing the tholeiitic character of the Los Ángeles Unit in contrast to the calc-alkaline signature of Naunauco Group to the south; **b** La/Ta versus SiO_2 (wt%) showing the stronger arc-like character of Naunauco Group respect to Los Ángeles Unit; **c** the higher Ta/Hf ratios of Los Ángeles Unit indicates a more enriched composition than the Naunauco Group; and **d** Nb/Zr versus Ba/Nb indicates higher slab fluids and melting degrees for the Naunauco Group relative to the Los Ángeles Unit magmatic rocks. Geochemical data correspond to: Casé et al. (2008), Franchini et al. (2003), Iannelli et al. (2018), Kay et al. (2006), Mateo Fernández Caso et al. (2011), Zamora Valcarce et al. (2006)

setting is described during the emplacement of the Los Ángeles Unit magmatism (Figs. 2a and b; Fennell et al. 2017), where a series of E-W striking normal faults control the dip and thickness of this volcano-sedimentary sequence that presents growth structures (Fig. 2a) (Fennell et al. 2017). These normal faulting also control the intrusion of the basaltic dykes that cut the synextensional sequence (Fig. 2c and d; Fennell et al. 2017; Iannelli et al. 2018).

Geochemically, Los Ángeles Unit magmatic sequence grades from an arc-like magmatism with a tholeiitic composition in the base toward a more alkaline intraplate-like magmatism in the younger levels (Fig. 3; Iannelli et al. 2018). Slab-fluid signal and partial melting degrees decrease toward the last magmatic pulses as indicated by the decreasing La/Ta, Th/Ta (2.31–10.5) and Ba/Nb ratios together with the increasing Nb/Zr ratios (Fig. 3b, c and d; Iannelli et al. 2018). These magmas would have been emplaced in a normal thickness crust, as seen by their Sm/Yb values (2.19–2.74), typical of magmas equilibrating with low-pressure residual mineral assemblages (Iannelli et al. 2018).

Toward the southern latitudes (36° 30′–38° S), coeval Late Cretaceous–Early Paleocene volcanism is represented by the Naunauco Group (~74–56 Ma; Zamora Valcarce 2007) that has been described emplaced in syncline and anticline cores associated with a post-orogenic relaxational stage (e.g., Llambías and Aragón 2011; Zamora Valcarce et al. 2006). Contrastingly, the Naunauco magmatic rocks present a stronger calc-alkaline arc-like signature than the Los Ángeles Unit, displayed by their FeO_t/MgO (1.12–9.25), La/Ta (12.78–140) and Th/Hf (5.12–0.15) ratios (Fig. 3a, b and c) (Casé et al. 2008; Franchini et al. 2003; Iannelli et al. 2018; Mateo Fernández Caso et al. 2011; Zamora Valcarce et al. 2006). Higher slab-fluids influence and partial melting degrees characterize the Naunauco Group magmatic evolution when compared to the Los Ángeles Unit, as seen by their higher Ba/Nb and lower Nb/Zr ratios (Fig. 3d). Sm/Yb (1.37–2.94) ratios indicate initial emplacement under a similar crustal thickness for both Upper Cretaceous–Lower Paleocene magmatic sequences (Casé et al. 2008; Franchini et al. 2003; Iannelli et al. 2018; Kay et al. 2006; Mateo Fernández Caso et al. 2011; Zamora Valcarce et al. 2006).

3.2 Early Paleocene-Eocene Stage (56-34 Ma)

Arc-like magmatism appears by late Paleocene–Eocene times at 37°–38° S, represented by the *Provincia Volcánica Neuquino Mendocina* (Fig. 1) (PVNM; ~56–38 Ma; Llambías and Rapela 1989). At the end of this magmatic period, the Farallon-Aluk spreading ridge reached the Northern Patagonian latitudes (Breitsprecher and Thorkelson 2009), while northward the Farallon plate was subducting with a NE direction beneath the South American plate (Somoza and Ghidella 2012). These high degrees in obliquity of the Farallon plate subduction direction experience a slight decrease by 49 Ma (Pardo Casas and Molnar 1987; Somoza and Ghidella 2005; Müller et al. 2016). The emplacement of PVNM (37°–38° S) coincides with these changes, since it presents a time span between ~54.2 Ma and ~39.7 Ma (e.g., Llambías and Mailvicini 1978; Llambías and Rapela 1987, 1989; Rovere 1998).

During the Eocene, arc-like magmatism in the Neuquén Basin at 37°–38° S shows a higher slab-fluid influence (Ba/La: 16.01–101.1; Ce/Pb: 0.42–10.43) and partial melting degrees (Nb/Yb: 1.41–4.25) than older rocks from the Late Cretaceous Naunauco Group (~74–56 Ma) (Fig. 4) (Casé et al. 2008; Fernández Casó et al. 2011; Franchini et al. 2003; Kay et al. 2006; Zamora Valcarce et al. 2006). Kay et al. (2005, 2006) attributed this progressively higher arc-like signature to periods of forearc subduction erosion among other factors.

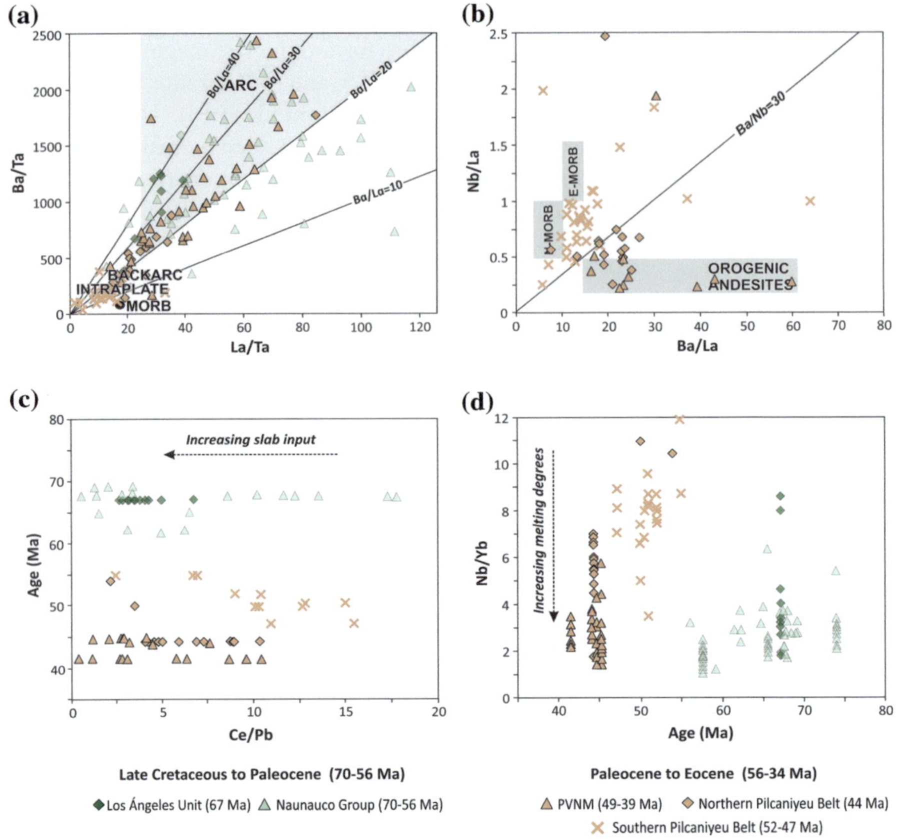

Fig. 4 **a** Ba/Ta versus La/Ta showing the contrasting behavior between the Pilcaniyeu Belt and northern PVNM magmatism; **b** increasing Nb/La indicates a tendency toward an E-MORB composition for the Pilcaniyeu Belt respect to the arc-like signature of the Eocene PVNM; and **c** and **d** Ce/Pb and Nb/Yb ratios indicate higher slab influence and melting degrees for PVNM magmatic rocks. Geochemical data correspond to: Aragón et al. (2011), Casé et al. (2008), Franchini et al. (2000), Iannelli et al. (2017), Kay et al. (2006), Llambías and Rapela (1989), Mateo Fernández Caso et al. (2011), Mazzoni et al. (1991), Rapela et al. (1988)

Toward northern latitudes (~35°–36° S), coeval magmatism is absent, while to the south (~40° S), Eocene magmatic rocks are described as part of the northern extreme of the Pilcaniyeu Belt (~44 Ma) (Aragón et al. 2011; Iannelli et al. 2016, 2017a; Rapela et al. 1983, 1988). The Pilcaniyeu Belt magmatism at 40° S is composed of a bimodal volcanic sequence of basal rhyolitic lavas and pyroclastic flows that grades toward basaltic to andesitic lava flow in the upper levels (Iannelli et al. 2016, 2017a). This sequence shows a contrasting geochemical signature relative to the northern coeval units (PVNM) (Fig. 4). The Pilcaniyeu Belt magmatic rocks at 40° S (~44 Ma) show an intraplate signature (La/Ta < 25) and a tendency toward an E-MORB composition as indicated by their higher Nb/La and lower Ba/La ratios than the PVNM (Fig. 4a and b; Iannelli et al. 2018). In contrast, northern PVNM magmatic rocks (~37°–38° S) present a typical arc-like signature, with higher slab-fluids influence and partial melting degrees than the Pilcaniyeu Belt as shown by their Ce/Pb and Nb/Yb values (Fig. 4c and d).

3.3 Latest Eocene–Early Miocene Stage (34–20 Ma)

During the latest Eocene–Oligocene, a major change in plate configuration occurred that led to the beginning of an extensional regime affecting whole sectors of the Southern Andean margin (e.g., Jordan et al. 2001; Muñoz et al. 2000). During this time period in the Neuquén Basin, a regional magmatic hiatus started by ~39 Ma and lasted at least 10 My (e.g., Kay et al. 2006; Muñoz et al. 2000), when the Farallon plate converged beneath the South American plate with high obliqueness (Somoza and Ghidella 2005). Then, magmatism resumed when the extensional regime achieved its maximum development by ~28–23 Ma, with a series of sedimentary–volcanic basins formed along the Southern Central Andes axis (e.g., Burns et al. 2006; Charrier et al. 1996; Jordan et al. 2001; Radic et al. 2002). This occurred by the latest Oligocene when the Farallon plate broke up into the Nazca and Cocos plates and a normal subduction regime with higher plate convergence rates started (Pardo Casas and Molnar 1987; Somoza and Ghidella 2005; Lonsdale 2005). These changes in tectonic configuration led to a widespread extensional regime that influenced the geochemical evolution of the arc-like Andean magmatism (e.g., Iannelli et al. 2017; Muñoz et al. 2000). Particularly, at the Neuquén Basin latitudes, the Cura Mallín intra-arc basin (~36°–38° 30′ S) and the Auca Pan depocenter to the south (~39° S) developed as part of this extensional system (Fig. 1) (Burns 2002; Burns et al. 2006; Kay et al. 2006; Ramos et al. 2014).

The Auca Pan depocenter developed immediately before the Farallon plate break-up (29.6 ± 1.2 Ma; Ramos et al. 2014), as an extensional depocenter with a volcanic infill composed of basaltic porphyritic lavas interbedded with volcanic breccias and minor silicic pyroclastic rocks (Fig. 5a and b; Iannelli et al. 2017a). Normal faulting and variations in strata thickness in relation to it have been described in the Oligocene Auca Pan depocenter deposits and in neighboring coetaneous sequences, as seen in the Rancahué Formation in the neighbor Aluminé depocenter (Franzese et al. 2011; García Morabito and Ramos 2012; Ramos et al. 2014). Geochemically, these early Oligocene sequences (~29 Ma) are characterized by arc-like signatures with a calc-alkaline magmatic source (FeO*/MgO: 1.18–5.56; Ba/La:13.83–20.76; Th/Hf: 0.37–3.53) but limited slab influence (e.g., Ba/Ce: 5.81–10.91; Figs. 6a, b and c) (Iannelli et al. 2017a).

Northward, the Cura Mallín Basin is developed with a younger life span between ~27 to 16 Ma (Jordan et al. 2001; Radic et al. 2002; Burns et al. 2006), being also associated with extensional faulting and strike-slip deformation at its basal sections (Burns et al. 2006; Radic et al. 2002). The Cura Mallín Basin infill is mostly composed of volcaniclastic rocks with minor lacustrine deposits and lava flows (Burns et al.

Fig. 5 **a** and **b** Variable lithological facies of the late Oligocene magmatism (Auca Pan Formation; ~29 Ma) at the southern Neuquén Basin latitudes (Iannelli et al. 2017a; Ramos et al. 2014)

2006; Suárez and Emparán 1995). This relatively younger volcanism bears a better developed arc-like signature and a tholeiitic magma source (FeO*/MgO: 1.39–7.42; La/Ta: 26.63–70.83; Ba/La: 17.05–34.74; Th/Hf: 0.11–1.75) relative to the older Auca Pan magmatic rocks (Fig. 6a and b) (Burns 2002; Kay et al. 2006; Utgé et al. 2009). Furthermore, after the Farallon plate break-up, magmatism showed higher slab-inputs and melting degrees, as stated by the higher Ba/Ce (7.65–15.68) and lower Nb/Y (0.29–0.26) ratios of Cura Mallín lavas, with respect to the Auca Pan Formation (Fig. 6c and d; Kay et al. 2006; Iannelli et al. 2017a; Utgé et al. 2009).

This extensional setting is also registered to the north with the development of the intra-arc Abanico Basin (30°–36° S) (e.g., Charrier et al. 1996, 2002; Kay et al. 2005; Muñoz et al. 2006; Piquer et al. 2017). The younger magmatic expressions of the northern Abanico Basin (~26–21 Ma) show a similar geochemical signature than the Cura Mallín lavas, presenting higher slab-fluids influence and melting degrees than the older Abanico magmatic sequences (~37–31 Ma) (Charrier et al. 1996; Kay et al. 2005; Muñoz et al. 2006; Piquer et al. 2017).

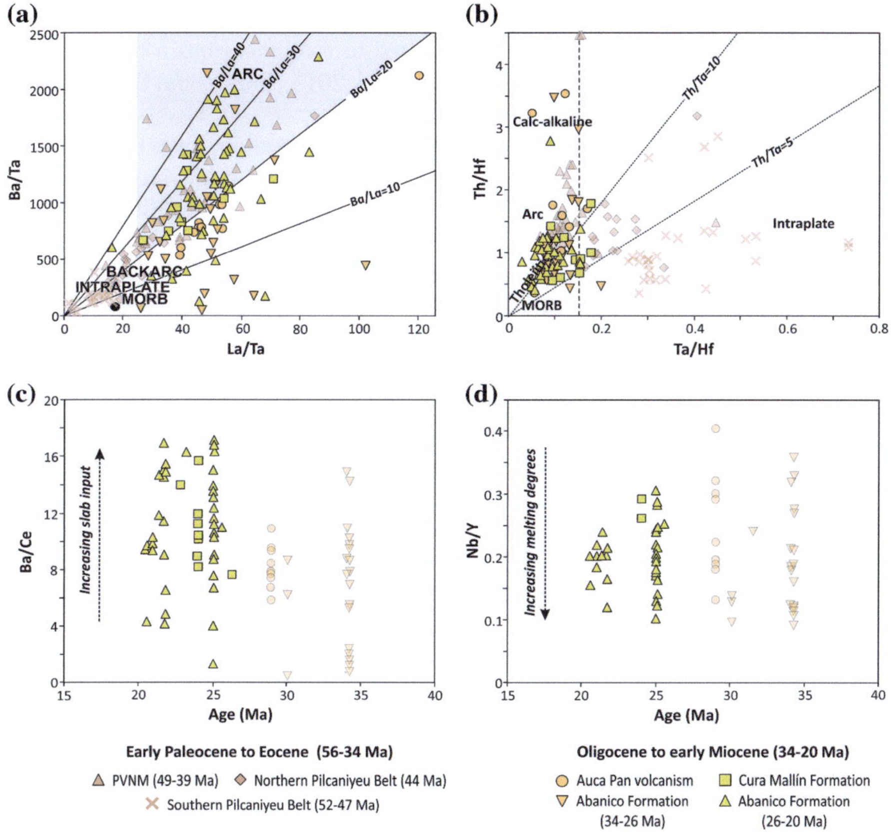

Fig. 6 **a** Ba/Ta versus La/Ta for Oligocene to early Miocene magmatism showing an arc-like behavior with higher Ba/Ta ratios for the Cura Mallín and younger Abanico sequences (26–20 Ma); **b** The restricted Th/Hf values for Cura Mallín and younger Abanico sequences indicate a more tholeiitic signature respect to the Auca Pan and older Abanico magmatic rocks; **c** and **d** Ba/Ce and Nb/Y show that younger than 25 Ma magmatism presents a higher slab influence and melting degrees. Geochemical data are taken from Aragón et al. (2011), Iannelli et al. (2017a), Kay et al. (2005, 2006), Llambías and Rapela (1989), Piquer et al. (2017) and Utgé et al. (2009)

4 Geodynamic Evolution of the Neuquén Basin and Its Magmatic Implications Between Late Cretaceous and Oligocene Times

4.1 Subduction of the Farallon-Aluk Spreading Ridge (70–56 Ma)

During the Late Cretaceous (~100–80 Ma), the beginning of a contractional regime provoked the initial contraction and uplift of the Andean margin (e.g., Folguera et al.

2011; Manceda and Figueroa 1995; Mosolf et al. 2018; Ramos 2010). Thus, the Neuquén Basin depositional dynamics changed to a foreland basin type with the deposition of synorogenic strata (e.g., Fennell et al. 2015; Horton and Fuentes 2016; Balgord and Carrapa 2016).

This tectonic setting was a consequence of the fragmentation of the Phoenix plate into the Chasca and Catequil plates (Seton et al. 2012), and the westward acceleration of the continent after the final break-up of Gondwana (Fig. 7a; e.g. Müller et al. 2016; Somoza and Ghidella 2012).

The collision and subsequent northward migration of Chasca-Catequil spreading ridge along the Andean margin caused the subduction of young lithosphere, which could have triggered a slab shallowing configuration and a waning in arc activity between 36° and 38° S latitudes (Fig. 7a) (Charrier et al. 2007; Fennell et al. 2015; Maloney et al. 2013; Ramos and Folguera 2005a, b).

By ~80 Ma, the Catequil Plate broke up into the Farallon and Aluk plates, and in consequence, the Farallon-Aluk spreading ridge started to subduct beneath the South American plate at 30° S (Fig. 7b; Müller et al. 2016; Seton et al. 2012). This spreading ridge migrated southward, reaching the Neuquén Basin latitudes by ~70 Ma (Fig. 7c) (Müller et al. 2016). The Latest Cretaceous to Paleocene period (75–60 Ma) has been recently described as an extensional period in the evolution of the Andean axis (Fennell et al. 2017; Tapia 2015; Muñoz et al. 2018) associated with the development of intra-arc basins. Volcano-sedimentary sequences with Late Cretaceous age controlled by extension have been described in the Chilean Andean slope at the latitudes of the Neuquén Basin (Fm. Plan de Los Yeuques; Muñoz et al. 2018) and also north of 33° S (e.g. Cornejo and Matthews 2001; Cornejo et al. 2003; Mpodozis and Allmendinger 1993; Charrier et al. 2007). At this stage, in the Neuquén Basin (36° 30′–38° S), a calc-alkaline arc-like magmatism of Late Cretaceous age (Naunauco Group) was installed in an eastward position relative to the location of the Early Cretaceous batholith (e.g., Franchini et al. 2003; Ramos and Folguera 2005a, b; Zamora Valcarce et al. 2006). To the north (35° 30′ S), coeval magmatism (Los Ángeles Unit) developed with a tholeiitic to alkaline composition with limited arc signature and a tendency toward an intraplate geochemical behavior (Fennell et al. 2017; Iannelli et al. 2017b, 2018). These contrasting geochemical compositions between two coetaneous magmatic sequences are interpreted as been related to the segmented nature of the Farallon-Aluk mid-ocean ridge (e.g., Müller et al. 2016; Thorkelson 1996). The alternated subduction of spreading centers and segments with more hydrated oceanic crust through transform zones that segment the former would have provoked the development of localyzed alkaline-type magmatism (Los Ángeles Unit; 35° 30′ S) intercalated with segments with calc-alkaline products (Naunauco Group; 36° 30′–38° S). Furthermore, the oblique subduction of this spreading ridge together with the contrasting subduction parameters between Farallon and Aluk plates could have provoked the opening of an incipient slab window at 35° 30′ S (Fennell et al. 2017; Iannelli et al. 2018; Somoza and Ghidella 2005). This incipient slab window would have allowed the influx of an enriched portion of mantle, which would explain the more alkaline and intraplate signature of the last magmatic pulses of northern Los Ángeles Unit. In contrast, the spreading ridge

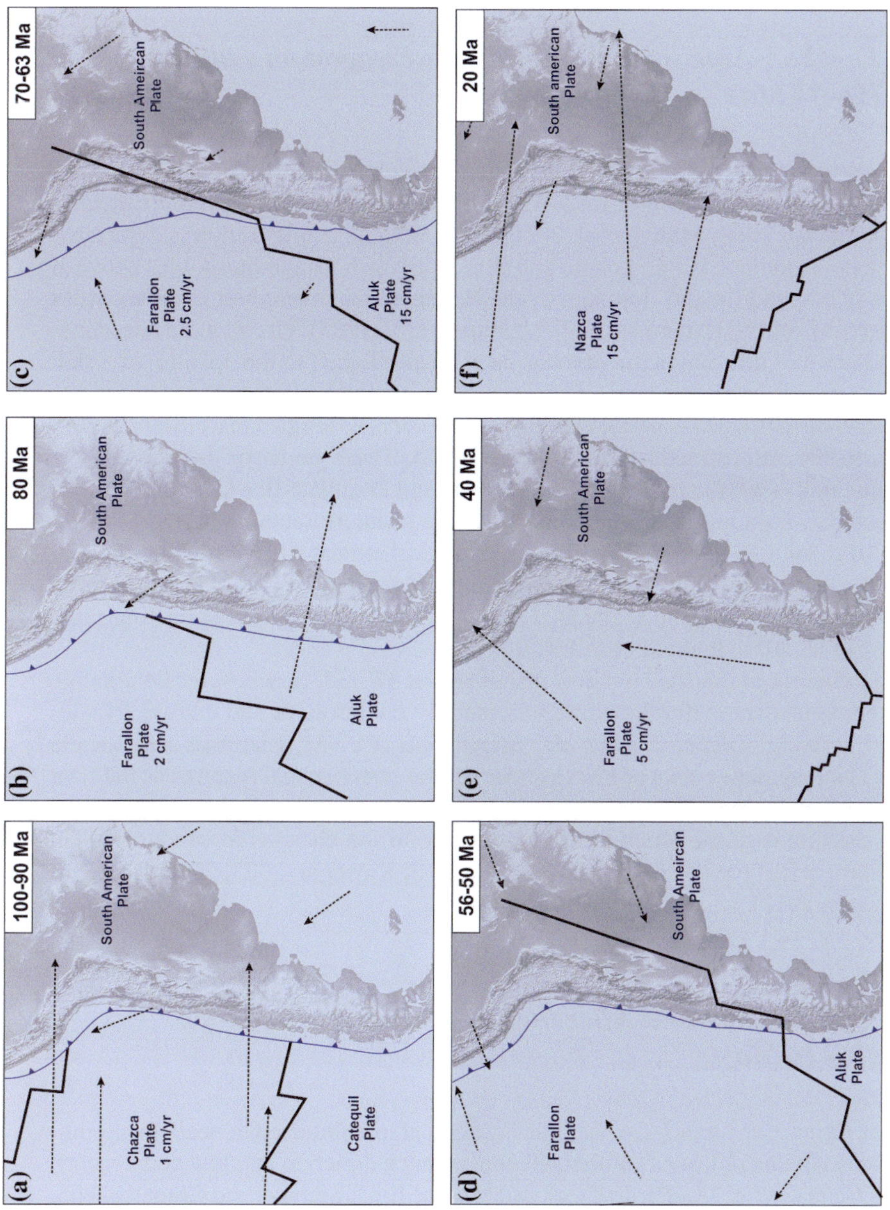

Fig. 7 Interaction between subducted Pacific mid-ocean ridges and South American Plate since Late Cretaceous to the early Miocene. Modified from Müller et al. (2016), Seton et al. (2012) and Somoza and Ghidella (2012)

orthogonal subduction at ~37° S could have triggered the eastward migration of the volcanic front and the development of a typical magmatic arc (Naunauco Group).

4.2 Farallon-Aluk Spreading Ridge at Patagonian Latitudes (56-34 Ma)

The Farallon-Aluk spreading ridge subducted through the Chilean trench toward the south reaching Patagonian latitudes (~42° S) by Eocene times (Fig. 7d) (Breitsprecher and Thorkelson 2009; Müller et al. 2016). This tectonic configuration is associated with the development of a magmatic gap along the North Patagonian Andes between 60 and 42 Ma and the development of the Pilaniyeu magmatic belt in an eastward position (~41–42° S) (Aragón et al. 2011; Rapela et al. 1983). The alkaline intraplate-like signature of this magmatic belt has been locally linked to the opening of a slab window (Aragón et al. 2011).

Toward the north (40°–39°S), the Pilcaniyeu Belt is developed in a closer position respect to the Andean axis (Rapela et al. 1983). These northern outcrops of the magmatic belt (~40°S) present a minor alkaline and intraplate-like signature with an incipient slab-fluid influx when compared to the southern sections (42°S) (Aragón et al. 2011; Iannelli et al. 2016, 2017a). This geochemical variation along the belt would also be an evidence of the decreasing influence of a slab window located southward at 42° S and the segmented nature of the Farallon-Aluk spreading ridge (Aragón et al. 2011; Iannelli et al. 2017a).

The influence of this slab window disappears at 37°–38° S, where a calc-alkaline arc-like magmatism is developed, represented by the intrusive and extrusive facies of the *Provincia Volcánica Neuquino Mendocina* (PVNM; Llambías and Rapela 1989). The magmatic rocks of PVNM bear similar geochemical features as the Late Cretaceous–Paleocene Naunauco Group but with a higher slab-fluids signal and partial melting degrees which have been related to the changes in subducted slab parameters at that time (e.g., Franchini et al. 2003; Kay et al. 2005, 2006; Llambías and Aragón 2011).

4.3 Times Before and After the Farallon Plate Break-Up (34–16 Ma)

By early Oligocene times (~34 Ma), the Farallon plate subducted beneath the Southern Andes with an oblique NE-directed convergence direction and low convergence rates of 6 cm/yr (Fig. 7e) (Pardo Casas and Molnar 1987; Somoza and Ghidella 2005, 2012). Afterward, a widespread extensional regime was installed along the Andean axis. The beginning of this extensional setting varies through latitude between 28 and 23 Ma (Fig. 7f) (e.g., Cande and Leslie 1986; Jordan et al. 2001; Lonsdale 2005; Müller et al. 2016).

Even though this extensional stage has been generally linked to the break-up of the Farallón Plate, extensional conditions started in specific parts along the Andean axis slightly earlier, such as in the intra-arc Auca Pan depocenter (~29 Ma; Ramos et al. 2014) and in the Abanico intra-arc basin (~36–31.7 Ma; Charrier et al. 1996; Muñoz et al. 2006). By ca. 28.3–25.8 Ma, after the Farallon plate break-up, the Nazca plate started subducting with an almost orthogonal configuration (N78° E), a relatively steep angle and a convergence rate of 15 cm/year (Fig. 7f; Pardo Casas and Molnar 1987; Somoza and Ghidella 2005). This new tectonic configuration led to a widespread development of the extensional regime and the development of the intra-arc basins along the Southern Central and Northern Patagonian Andes (e.g., Jordan et al. 2001; Muñoz et al. 2006). Under this new configuration, arc-related magmatic activity expanded toward the west, associated with the development of mid-Tertiary extensional basins (e.g., Encinas et al. 2015; Muñoz et al. 2000).

Particularly, in the Neuquén Basin latitudes (34°–40° S), some important differences have been noticed between the arc-related magmatic products before and after this event. For example, when considering the Auca Pan depocenter magmatism (~29 Ma; ~ 40° S), which developed immediately before the Farallon plate break-up, a calc-alkaline arc-like signature with slab-fluids input is seen in the geochemical composition of the volcanic lavas (Iannelli et al. 2016, 2017a; Ramos et al. 2014). However, immediately after, the volcanic expressions of the Cura Mallín Basin (Fig. 1) show a better developed arc-like signature with a tholeiitic source (Fig. 6) (Kay et al. 2006; Utgé et al. 2009). Increasing slab-fluids inputs and melting degrees are associated with more typical Andean-type magmatism after the break-up of the Farallon plate (Fig. 7f) (Kay et al. 2006; Iannelli et al. 2017a). Moreover, the similar geochemical behavior from the younger magmatic expressions of the northern Abanico intra-arc basin (~26–21 Ma; 30–34° S; Kay et al. 2005; Muñoz et al. 2006) shows that the new geodynamic setting also influenced northern magmatic evolution. These magmatic units present higher slab-fluid influence and melting degrees than the older magmatic sequences of the Abanico Basin at the same latitudes (~37–31 Ma; Muñoz et al. 2006; Charrier et al. 1996).

5 Conclusions

After considering the different magmatic stages that were emplaced in the Neuquén Basin (~34°–40° S) since the Late Cretaceous to the Oligocene, we can envisage a link between the different plate configurations through time and changes in geochemical evolution of coetaneous magmatism.

In particular, the collision and subduction of the Farallon-Aluk spreading ridge since ~80 Ma along the Andean trench and its southward migration have most likely controlled the emplacement of coeval magmatic units with contrasting geochemical signatures. The continuous passage of this segmented spreading ridge toward the south could have triggered the emplacement of coetaneous, but geochemically distinct, magmatic units until Eocene times.

Finally, when the Aluk Plate is completely subducted beneath the South American plate and the Farallon plate breaks-up into Nazca and Cocos plates, a widespread extensional regime installed affecting vast portions of the Southern Central and Patagonian Andes. After this late Oligocene major reconfiguration of the Andean margin, a more typical Andean arc-type magmatism developed, with a clearer and more homogeneous arc-related geochemical signature seen in the magmatic products associated with intra-arc extensional basins.

References

Aguirre-Urreta B, Tunik M, Naipauer M et al (2011) Malargüe Group (Maastrichtian-danian) Deposits in the Neuquén Andes: Implications for the onset of the first Atlantic transgression related to Western Gondwana break-up. Gondwana Res 19:482–494

Aragón E, D'Eramo F, Castro A et al (2011) Tectono-magmatic response to major convergence changes in the North Patagonian suprasubduction system; the Paleogene subduction–transcurrent plate margin transition. Tectonophysics 509:218–237

Balgord EA, Carrapa B (2016) Basin evolution of Upper Cretaceous-Lower Cenozoic strata in the Malargüe fold-and-thrust belt: northern Neuquén Basin, Argentina. Basin Res 28(2):183–206

Boyce D (2015) Modelo de Evolución Tectónica y Paleogeográfica del Margen Andino en Chile Central Durante el Cretácico Medio-Tardío: El Registro Estructural y Sedimentario en la Formación Las Chilcas. Ph.D. thesis, Universidad de Chile, Santiago

Breitsprecher K, Thorkelson DJ (2009) Neogene kinematic history of Nazca–Antarctic–Phoenix slab windows beneath Patagonia and the Antarctic Peninsula. Tectonophysics 464(1-4):10–20

Burns WM (2002) Tectonic and depositional evolution of the Tertiary Cura Mallín Basin in the southern Andes (36.5 to 38° S lat.). Ph.D. thesis, Cornell University

Burns WM, Jordan T, Copeland P et al (2006) Extensional tectonics in the Oligo-Miocene Southern Andes as recorded in the Cura-Mallín basin. In: Kay SM, Ramos VA (eds) Evolution of an Andean margin: a tectonic and magmatic view from the Andes to the Neuquén basin (35–39° S). Geological Society of America, SP 407, pp 163–184

Cande SC, Leslie RB (1986) Late Cenozoic tectonics of the southern Chile trench. J Geophys Res Solid Earth 91:471–496

Cande SC, Kent DV (1992) A new geomagnetic polarity time scale for the late Cretaceous and Cenozoic. J Geophys Res 97(B10):13917–13951

Casé AM, López-Escobar L, Danieli JC, Schalamuk A (2008) Butalón igneous rocks, Neuquén, Argentina: age, stratigraphic relationships and geochemical features. J S Am Earth Sci 26(2):188–203

Charrier R, Wyss AR, Flynn JJ et al (1996) New evidence for late Mesozoic-early Cenozoic evolution of the Chilean Andes in the upper Tinguiririca valley (35 S), central Chile. J S Am Earth Sci 9:393–422

Charrier R, Baeza O, Elgueta S et al (2002) Evidence for Cenozoic extensional basin development and tectonic inversion south of the flat-slab segment, southern Central Andes, Chile (33°–36° S.L.). J S Am Earth Sci 15:117–139

Charrier R, Pinto L, Rodríguez M (2007) Tectonostratigraphic evolution of the Andean orogen in Chile. In: Moreno T, Gibbons W (eds) The geology of Chile. The Geological Society, London, pp 21–114

Charrier R, Ramos VA, Tapia F et al (2015) Tectono-stratigraphic evolution of the Andean Orogen between 31 and 37° S (Chile and Western Argentina). In: Sepúlveda SA, Giambiagi LB, Moreiras

SM, Pinto L, Tunik M, Hoke GD, Farías M (eds) Geodynamic processes in the Andes of Central Chile and Argentina. The Geological Society, London, SP 399, pp 13–61

Cobbold PR, Rosello EA (2003) Aptian to recent contractional deformation, foothills of the Neuquén basin, Argentina. Mar Petrol Geol 20(5):429–443

Cornejo P, Matthews S (2001) Evolution of magmatism from the uppermost Cretaceous to Oligocene, and its relationship to changing tectonic regime. In: Abstracts of the 3 South American symposium on isotope geology, Pucón, 21–124 Oct 2001

Cornejo P, Matthews S, Pérez de Arce C (2003) The "K-T" compressive deformation event in northern Chile (24–27° S). In: Abstracts of the 10 Congreso Geológico Chileno, Concepción, 6–10 Oct 2003

Encinas A, Folguera A, Oliveros V et al (2015) Late Oligocene–early Miocene submarine volcanism and deep-marine sedimentation in an extensional basin of southern Chile: Implications for the tectonic development of the North Patagonian Andes. Geol Soc Am Bull 128(5–6):807–823

Fennell LM, Folguera A, Naipauer M, Gianni G, Rojas Vera EA, Bottesi G, Ramos VA (2015) Cretaceous deformation of the southern Central Andes: synorogenic growth strata in the Neuqu en Group (35 30'–37 S). Basin Res. https://doi.org/10.1111/bre.12135

Fennell LM, Iannelli SB, Folguera A et al (2017) Interrupciones extensionales en el desarrollo de la faja plegada y corrida de Malargüe (36° S). In: Abstracts of the 20 Congreso Geológico Argentino, San Miguel de Tucumán, 7–11 Aug 2017

Folguera A, Orts D, Spagnuolo M, Rojas Vera E et al (2011) A review of Late Cretaceous to Quaternary palaeogeography of the southern Andes. Biol J Linnean Soc 103:250–268

Franchini M, Meinert Ly, Montenegro T (2000) Skarns related to porphyry-style mineralization at Caicayén Hill Neuquén, Argentina: composition and evolution of hydrotermal fluids. Econ Geol 35:1197–1213

Franchini MB, Lopez Escobar L, Shalamuk IBA et al (2003) Paleocene, calc-alkaline subvolcanic rocks from Nevazón Hill area (NW Chos Malal fold belt), Neuquén, Argentina, and comparison with granitoids of the Neuquén Mendoza volcanic province. J S Am Earth Sci 16:399–422

Franzese JR, D'Elia L, Bilmes A et al (2011) Superposición de cuencas extensionales y contraccionales oligo-miocenas en el retroarco andino norpatagónico: la Cuenca de Aluminé, Neuquén, Argentina. Andean Geol 38:319–334

Gana P, Wall R (1997) Evidencias geocronológicas 40Ar/39Ar y K/Ar de un hiatos Cretácico superior-Eoceno en Chile central (33°–33° 30′ S). Andean Geol 24:145–163

García Morabito E, Ramos VA (2012) Andean evolution of the Aluminé fold and thrust belt, Northern Patagonian Andes (38° 30′–40° 30′ S). J S Am Earth Sci 38:13–30

Giambiagi L, Mescua J, Bechis F et al (2012) Thrust belts of the southern Central Andes: along-strike variations in shortening, topography, crustal geometry, and denudation. Geol Soc Am Bull 124(7/8):1339–1351

Gianni GM, Pesce A, Soler SR (2018) Transient plate contraction between two simultaneous slab windows: insights from Paleogene tectonics of the Patagonian Andes. J Geodyn 121:64–75

Godoy E, Yáñez G, Vera E (1999) Inversion of an Oligocene volcano-tectonic basin and uplifting of its superimposed Miocene magmatic arc in the Chilean Central Andes: first seismic and gravity evidences. Tectonophysics 306(2):217–236

Gulisano CA, Gutiérrez Pleimling AR (1994) The Jurassic of the Neuquén Basin: field guide. Asociación Geológica Argentina, Series E, Buenos Aires, pp 1–111

Horton BK, Fuentes F (2016) Sedimentary record of plate coupling and decoupling during growth of the Andes. Geology 44(8):647–650

Heredia N, Farías P, García-Sansegundo J et al (2012) The basement of the Andean frontal cordillera in the Cordón del Plata (Mendoza, Argentina): geodynamic evolution. Andean Geol 39:242–257

Iannelli SB, Litvak VD, Fernández Paz L et al (2016) Late Paleogene arc-related volcanism in the North Patagonian Andes (39–41° S). In: Abstracts of the 1 Simposio de Tectónica Latinoamericana, Santiago de Chile, 14–19 Nov 2016

Iannelli SB, Litvak VD, Fernández Paz L et al (2017a) Evolution of Eocene to Oligocene arc-related volcanism in the North Patagonian Andes (39-41° S), prior to the break-up of the Farallon plate. Tectonophysics 696–697:70–87

Iannelli SB, Fennell LM, Litvak VD et al (2017b) Volcanismo cretácico superior-paleoceno en la alta cordillera al sur de la provincia de Mendoza (35° 30′ S). In: Abstracts of the 20 Congreso Geológico Argentino, San Miguel de Tucumán, 7–11 Aug 2017

Iannelli SB, Fennell LM, Litvak VD et al (2018) Geochemical and tectonic evolution of Late Cretaceous to early Paleocene magmatism along the Southern Central Andes (35°–36° S). J S Am Earth Sci 87:139–156

Jordan T, Burns W, Veiga R, Pángaro F et al (2001) Extension and basin formation in the Southern Andes caused by increased convergence rate: Amid-Cenozoic trigger for the Andes. Tectonics 20:308–324

Kay SM, Godoy E, Kurtz A (2005) Episodic arc migration, crustal thickening, subduction erosion, and magmatism in the south-central Andes. Geol Soc Am 117:67–88

Kay SM, Burns M, Copeland P (2006) Upper Cretaceous to Holocene magmatism and evidence for transient Miocene shallowing of the Andean subduction zone under the northern Neuquén Basin. In: Kay SM, Ramos VA (eds) Evolution of an Andean margin: a tectonic and magmatic view from the Andes to the Neuquén Basin (35–39° S). Geological Society of America, SP 407, pp 19–60

Legarreta L, Uliana MA (1991) Jurassic-Cretaceous marine oscillations and geometry of back-arc basin fill, central Argentine Andes. Int As Sed SP 12:429–450

Litvak VD, Poma S, Jones RE et al (2018) The Late Paleogene to Neogene volcanic arc in the Southern Central Andes (28°–37° S). In: Folguera et al. (eds) The Making of the Chilean-Argentinean Andes, Springer, Berlin, pp 503–536

Litvak VD, Spagnuolo MG, Folguera A et al (2015) Late Cenozoic calc-alkaline volcanism over the Payenia shallow subduction zone, South-Central Andean back-arc (34° 30′–37° S), Argentina. J S Am Earth Sci 64:365–380

Llambías EJ, Rapela CW (1987) Las vulcanitas de Collipilli y sus relaciones con las provincias volcanicas del Terciario inferior de Neuquen-Mendoza y Patagonia. In: X Congreso Geol ogico Argentino. Tucumán, 249–251

Llambías EJ, Rapela CW (1989) Las vulcanitas de Collipilli, Neuqu en. y su relación con otras unidades paleogenas de la Cordilera. Rev Asoc Geol Argent 44:224–236

Llambías EJ, Mailvicini L (1978) Geología, petrología y metalogénesis del área de Colipilli, provincia del Neuquén, República Argentina. Rev Asoc Geol Argent 33:257–276

Llambías EJ, Quenardelle SY, Montenegro T (2003) The Choiyoi Group from central Argentina: a subalkaline transitional to alkaline association in the craton adjacent to the active margin of the Gondwana continent. J S Am Earth Sci 16:243–257

Llambías EJ, Aragón E (2011) Volcanismo Paleógeno. Geología y recursos naturales de la Provincia del Neuquén. Asociación geológica Argentina, Buenos Aires, pp 265–274

Lonsdale P (2005) Creation of the Cocos and Nazca plates by fission of the Farallón plate. Tectonophysics 404:237–264

Manceda R, Figueroa D (1995) Inversion of the Mesozoic Neuquén rift in the Malargüe fold and thrust belt, Mendoza, Argentina. In: Tankard AJ, Suárez R, Welsink HJ. (eds) Petroleum Basins of South America. AAPG Memoir 62:369–382

Mateo Fernández Caso MP, Montero DG et al (2011) Petrografía y geoquímica del magmatismo cretácico superior-eoceno en el área de Pichaihue. Provincia de Neuquén Rev Asoc Geol Argent 68:173–184

Maloney KT, Clarke GL, Klepeis KA et al (2013) The Late Jurassic to present evolution of the Andean margin: drivers and the geological record. Tectonics 32(5):1049–1065

Mazzoni MM, Kawashita K, Harrison S et al (1991) Edades radimétricas eocenas. Borde occidental del macizo Norpatagónico. Rev Asoc Geol Argent 46:150–158

Mosolf JG, Gans PB, Wyss AR et al (2018) Late Cretaceous to Miocene volcanism, sedimentation, and upper-crustal faulting and folding in the Principal Cordillera, central Chile: Field and

geochronological evidence for protracted arc volcanism and transpressive deformation. Geol Soc Am Bull 131(1–2):252–273
Mpodozis C, Allmendinger RW (1993) Extensional tectonics, Cretaceous Andes, northern Chile (27° S). Geol Soc Am Bull 105:1462–1477
Muñoz J, Troncoso R, Duhart P et al (2000) The relation of the mid-Tertiary coastal magmatic belt in south-central Chile to the late Oligocene increase in plate convergence rate. Andean Geol 27:177–203
Muñoz M, Fuentes F, Vergara M et al (2006) Abanico East Formation: petrology and geochemistry of volcanic rocks behind the Cenozoic arc front in the Andean Cordillera, central Chile (33° 50' S). Andean Geol 331:109–140
Muñoz M, Tapia F, Persico M et al (2018) Extensional tectonics during Late Cretaceous evolution of the Southern Central Andes: evidence from the Chilean main range at ~35° S. Tectonophysics 744:93–117
Müller RD, Seton M, Zahirovic S et al (2016) Ocean basin evolution and global-scale plate reorganization events since Pangea breakup. Annu Rev Earth Planet Sci 44:107–138
Pardo Casas F, Molnar P (1987) Relative motion of the Nazca (Farallón) and South America plates since Late Cretaceous time. Tectonics 6:233–248
Piquer J, Hollings P, Rivera O et al (2017) Along-strike segmentation of the Abanico basin, central Chile: New chronological, geochemical and structural constraints. Lithos 268–271:174–197
Radic JP, Rojas L, Carpinelli A et al (2002) Evolución tectónica de la cuenca terciaria de Cura-Mallín, región cordillerana chileno argentina (36° 30′–39° 00′ S). In: Abstracts of the 15 Congreso Geológico Argentino, Córdoba, 23–26 Aug 2002
Ramos VA, Folguera A (2005) Tectonic evolution of the Andes of Neuquén: constraints derived from the magmatic arc and foreland deformation. In: Veiga GD, Spalletti LA, Howell JA, Schwarz E (eds) The Neuquén Basin, Argentina: a case study in sequence stratigraphy and basin dynamics. Geol Soc, SP 252, London, pp 15–35
Ramos VA, Mosquera A, Folguera A et al (2011) Evolución tectónica de los Andes y del Engolfamiento Neuquino adyacente. In: Leanza HA, Arregui C, Carbone O et al (eds) Geología y Recursos Naturales de la Provincia del Neuquén. Asociación Geológica Argentina, Buenos Aires, pp 335–348
Ramos VA (2010) The tectonic regime along the Andes: Present-day and Mesozoic regimes. Geol J 45:2–25
Ramos ME, Folguera A, Fennell L et al (2014) Tectonic evolution of the North Patagonian Andes from field and gravity data (39–40° S). J S Am Earth Sci 51:59–75
Ramos VA, Folguera A (2005) Tectonic evolution of the Andes of Neuquén: constraints derived from the magmatic arc and foreland deformation. In: Veiga GD, Spalletti LA, Howell JA et al (eds) The Neuquén Basin, Argentina: a case study in sequence stratigraphy and basin dynamics. The Geological Society, London, SP 252, pp 15–35
Rapela CW, Spalletti LA, Merodio CJ (1983) Evolución magmática y geotectónica de la "Serie Andesítica" Andina (Paleoceno-Eoceno) en la Cordillera Nordpatagónica. Rev Asoc Geol Argent 38:469–484
Rapela C, Spalletti L, Merodio J et al (1988) Temporal evolution and spatial variation of early Terciary volcanism in the Patagonian Andes (40° S–42° 30′ S). J S Am Earth Sci 1:75–88
Rovere E (1998) Volcanismo Jurásico, Paleógeno y Neógeno en el Noroeste del Neuquén, Argentina. In: Abstracts of the 10 Congreso Latinoamericano de Geología y 6 Congreso Nacional de Geología Económica, Buenos Aires, 8–13 Nov 1998
Seton M, Muller RD, Zahirovic S et al (2012) Global continental and ocean basin reconstructions since 200 Ma. Earth Sci Rev 113:212–270
Somoza R, Ghidella ME (2005) Convergencia en el margen occidental de América del Sur durante el Cenozoico: subducción de las placas de Nazca, Farallón y Aluk. Rev Asoc Geol Argent 60:797–809
Somoza R, Ghidella ME (2012) Late Cretaceous to recent plate motions in western South America revisited. Earth Planet Sci Lett 331–332:152–163

Spagnuolo MG, Folguera A, Litvak VD et al (2012) Late Cretaceous arc rocks in the Andean retroarc region at 36.5° S: evidence supporting a Late Cretaceous slab shallowing. J S Am Earth Sci 38:44–56
Suárez M, Emparán C (1995) The stratigraphy, geochronology and paleophysiography of a Miocene freshwater interarc basin, Southern Chile. J S Am Earth Sci 8:17–31
Tapia F (2015) Evolución tectónica y configuración actual de los Andes Centrales del Sur (34° 45'–35° 30' S). Ph.D. thesis, Universidad de Chile
Thorkelson DJ (1996) Subduction of diverging plates and the principles of slab window formation. Tectonophysics 255:47–63
Utgé S, Folguera A, Litvak V et al (2009) Geología del sector norte de la cuenca de Cura Mallín en las lagunas de Epulaufquen, Neuquén. Rev Asoc Geol Argent 64:231–248
Zamora Valcarce G, Zapata T, del Pinto A et al (2006) Structural evolution and magmatic characteristics of the Agrio fold-and-thrust belt. In: Kay SM, Ramos VA (eds) Evolution of an Andean Margin: a Tectonic and Magmatic View from the Andes to the Neuquén Basin (35°–39° S). The Geological Society, London, SP 407, pp 125–146
Zamora Valcarce G (2007) Estructura y Cinemática de la faja plegada del Agrio, Cuenca Neuquina. Ph.D. thesis, University of Buenos Aires

The Oligo-Miocene Tectonic Mode Switch: From a Brief Period of Widespread Extension to the Final Closure of the Neuquén Basin

Lucas Fennell, Javier Quinteros and Andrés Folguera

Abstract Geological observations in the Neuquén Basin indicate a Late Oligocene to Early Miocene episode of extension followed by an abrupt shift towards regional compression. However, the reasons behind this brief extensional episode and the Oligo-Miocene tectonic mode switch are not fully understood. Through the aid of numerical modelling, it has been shown that after a period of limited subduction in Early Palaeogene times, the penetration of Nazca's slab tip into the mantle transition zone in Late Oligocene times resulted in the renewal of effective subduction due to the effect of the slab pull force. This renewed subduction consists of an initial stage of higher trench hinge retreat and steep slab dips, leading to extension and mantelic upwelling processes east of the trench. Then, the natural evolution of the slab produces a deceleration of roll-back and shallowing of the subduction angle once it reaches the lower mantle, resulting in horizontal shortening. These results indicate that the effect of the slab pull force is a potential responsible for the Oligo-Miocene tectonic mode switch, causing the opening of a series of intra-arc basins and widespread magmatism partially devoid of arc-like components followed by an increasing influence of the slab in the magmatic arc and the final closure of the Neuquén Basin.

Keywords Oligocene to Miocene sedimentation · Roll-back · Subduction · Numerical modelling

1 Introduction

The largest tectonic mode switch that affected the Neuquén Basin occurred in Late Cretaceous times, when the long-lived Late Triassic to Early Cretaceous extensional

L. Fennell (✉) · A. Folguera
CONICET—Universidad de Buenos Aires, Instituto de Estudios Andinos Don Pablo Groeber (IDEAN), Buenos Aires, Argentina
e-mail: lucasfennell90@gmail.com

J. Quinteros
GFZ Helmholtz Centre Potsdam, Potsdam, Germany

D. Kietzmann and A. Folguera (eds.), *Opening and Closure of the Neuquén Basin in the Southern Andes*, Springer Earth System Sciences,
https://doi.org/10.1007/978-3-030-29680-3_17

regime was replaced by a contractional one, resulting in the final disconnection from the Pacific Ocean and the initial uplift of the Andes, marking the transition of the basin into its foreland stage (Vergani et al. 1995) (see Chapter "The Late Cretaceous Orogenic System: Early Inversion of the Neuquén Basin and Associated Synorogenic Deposits, 35°-38° S").

Once established, the foreland stage became modulated by a complex interaction between structural, magmatic and sedimentary processes, primarily responding to geodynamic scale events taking place in the subduction zone (Ramos and Folguera 2009; Horton 2018a). However, after its early inversion in Late Cretaceous times, the Neuquén Basin experienced an Atlantic-derived marine flooding (Parras and Griffin 2013) during the occurrence of voluminous magmatism and localized extension in the core of the Late Cretaceous orogen (Fennell et al. 2017; Iannelli et al. 2018; Muñoz et al. 2018), which has been recently related to a shallow dipping subducting slab (Gianni et al. 2018). In the Palaeocene, the basin entered a stage signed by low accommodation rates in the foreland (Horton et al. 2016), limited evidences of active thrusting in the fold-thrust belt (Cobbold and Rosello 2003; Sagripanti et al. 2012; Álvarez Cerimedo et al. 2013) and slow accumulation of volcanic and volcaniclastic strata within an arc platform (Mosolf et al. 2018) (see Chapter "Evolution of the Andean foreland During Variable Tectonic Regimes of the Neuquén Basin"). This scenario became abruptly interrupted between late Oligocene and earliest Miocene times by an enigmatic phase of sediment bypass in the foreland (Horton and Fuentes 2016; Horton 2018b) coeval with the opening of a series of intra-arc basins and mafic volcanic outbursts recorded in the present fore-arc and retroarc areas (Fig. 1) (Jordan et al. 2001; Kay et al. 2006). These intra-arc basins can be both temporally and lithologically correlated throughout the entire Southern Central Andes (see Fennell et al. 2018 for a recent review), being the Abanico (Charrier et al. 2002; Kay et al. 2005), Cura Mallin (Jordan et al. 2001; Radic 2010) and northern sector of the Ventana basins (Ramos et al. 2014; Bechis et al. 2014) comprised within the Neuquén Basin domain (Fig. 1) (see Chapter "Late Cretaceous to Oligocene Magmatic Evolution of the Neuquén Basin").

A volcanic belt localized nowadays in Chile's Longitudinal Depression (Lopez-Escobar and Vergara 1997) characterized the western border of the Neuquén Basin, while the eastern sector was partially flooded by alkaline basalts devoid of arc-like components, located in the present retroarc zone (Fig. 1) (Kay and Copeland 2006; Dyhr et al. 2013a, b). These magmatic rocks are presently exposed in the fold-thrust belt due to their incorporation into the Andean wedge, product of the final structuration of the Andes during Miocene to present times (Giambiagi et al. 2012). This structuration event was accompanied by an eastward migration of the magmatic arc (Litvak et al. 2015) and the renewal of flexural subsidence in the foreland basin (Horton and Fuentes 2016), which have been linked to a second slab shallowing event in this area (Kay et al. 2006) (see Chapter "The Miocene Magmatism in the Malargüe and Chos Malal Fold and Thrust Belts").

While an extensional regime is pointed out as the responsible for the intra-arc basin generation and the widespread mafic volcanic outpouring, a contractional regime produced the inversion of these basins and the eastward propagation of deformation

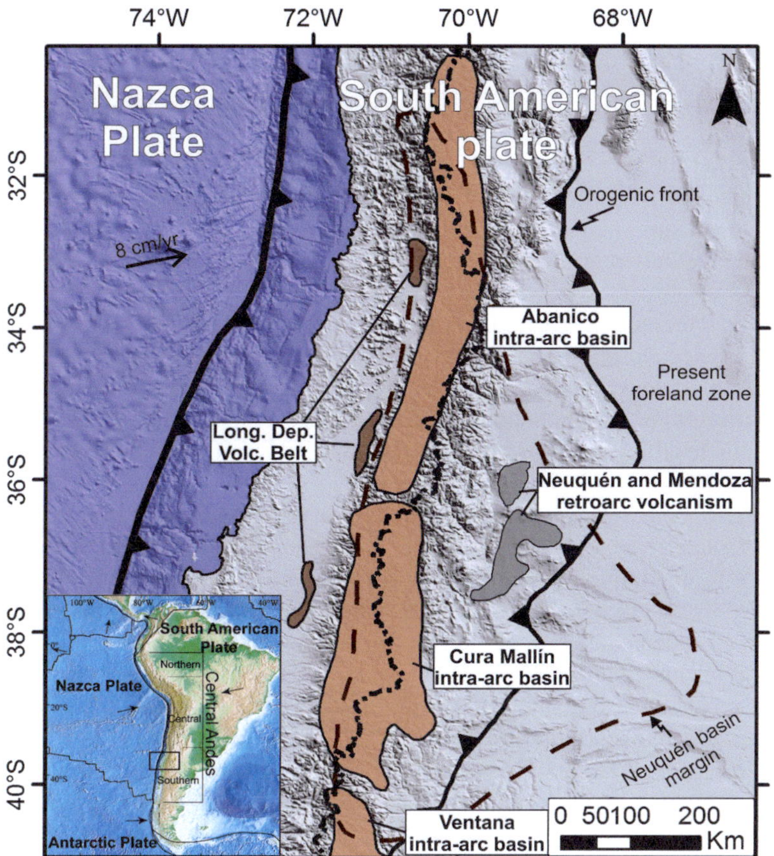

Fig. 1 Location of the intra-arc basins (light brown), magmatic belts (dark brown) and volcanic plateaux (dark grey) developed between the late Oligocene and earliest Miocene in the Neuquén Basin domain (Longitudinal Depression Volcanic Belt, Lopez-Escobar and Vergara 1997; Cura Mallin Basin, Jordan et al. 2001; Neuquén and Mendoza retroarc volcanism, Kay and Copeland 2006; Abanico Basin, Charrier et al. 2007; Ventana Basin, Ramos et al. 2014)

and magmatic loci (Godoy et al. 1999; Folguera and Ramos 2011). However, the reasons behind both regimes, as well as the transition between them, are not fully understood. The late Oligocene–earliest Miocene extensional regime has been linked to both slab steepening due to slab roll-back (Muñoz et al. 2000; Folguera and Ramos 2011) and diminished plate coupling (Horton and Fuentes 2016), while the Miocene contractional deformation has been associated with shallowing of the oceanic slab (Folguera and Ramos 2011) or the overriding of a mantelic plume (Gianni et al. 2017) (see Chapter "Plume Subduction Beneath the Neuquén Basin and the Last Mountain Building Stage of the Southern Central Andes"). Therefore, geodynamic events seem to be controlling the Oligo-Miocene tectonic mode switch.

2 Geodynamic Modelling

During the last years, numerical modelling has proven to be a powerful tool when used to deal with not-well-understood large-scale geodynamic processes (Sobolev and Babeyko 2005; Gerbault et al. 2009; Capitanio et al. 2011; Manea et al. 2012, among many others). The results of these models are validated by statistical analysis of modern oceanic subduction zone parameters, suggesting that the primary control behind the tectonic regime is the absolute motion of the overriding plate, although second-order processes may temporally dominate, resulting in minor perturbations in the overall tectonic scenario (Lallemand et al. 2005; Heuret and Lallemand 2005). One of these second-order processes corresponds to the slab pull force, a gravitational force acting on dense and negatively buoyant subducted slabs, which has shown to have an impact on the subduction rate relative to the convergence rate in subduction zones, thus resulting in horizontal extension or shortening in the overriding plate (Royden 1993).

Quinteros and Sobolev (2013), based on a thermomechanical self-consistent model of plate subduction, showed that changes in the convergence rate can be explained by the natural evolution of a slab, due to the activation of the slab pull force once its tip reaches the mantle transition zone. They show that after long-term slow and oblique subduction of the Farallón Plate below South America, an abrupt increment in convergence occurs during a change towards nearly perpendicular subduction, resulting in the splitting of the Farallón Plate into the Cocos and Nazca Plates in late Oligocene times. Afterwards, the strong reduction in the convergence velocity responds to the resistance offered by the highly viscous lower mantle to the oceanic slab penetration, regardless of the absolute motion of the overriding plate and the increasing load exerted by the Andes (Quinteros and Sobolev 2013). More recently, Fennell et al. (2018) analysed the role of the slab pull force during the late Oligocene to earliest Miocene stage of extensional deformation, finding a causal relation between both. However, its transition towards the contractional regime remains unsolved.

Therefore, in order to understand the tectonic mode switch that occurred in the Neuquén Basin during Oligo-Miocene times, we will analyse the results of the numerical model presented in Fennell et al. (2018), linking it to the tectonic scenario that led to its final closure.

3 Numerical Model and Set-up

The set-up of Fennell et al. (2018) is based on the model from Quinteros and Sobolev (2013), as well as the overriding plate layers, properties and thicknesses, which are considered suitable for pre-Miocene Andean settings (Fig. 2). For a detailed explanation of each of these properties, the reader is referred to the supplementary material of Quinteros and Sobolev (2013) and Fennell et al. (2018). A free surface is included on

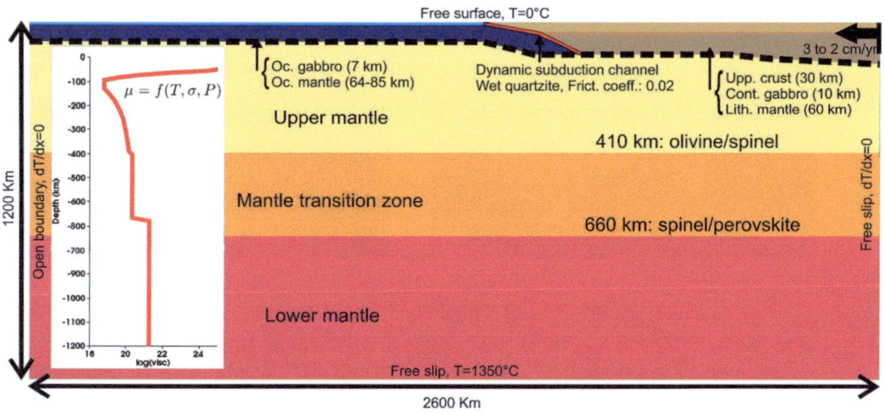

Fig. 2 Model set-up and parameters used in the experiment of Fennell et al. (2018), which are based on the Andean subduction model from Quinteros and Sobolev (2013). The inset shows the viscosity profile of the different layers, whose transitions correspond to major compositional boundaries responsible for the action of the slab pull force. Taken from Fennell et al. (2018)

the upper side of the model representing topography, which can be extracted from the model as the $T = 0$ °C isotherm, while the geometry of the slab can be tracked with the $T = 1300$ °C isotherm, which reflects the lithosphere/asthenosphere boundary (Fig. 2).

In order to mimic the long-lasting Palaeogene oblique and slow subduction, the beginning of subduction in the model was set at 30 Ma, which is supported by various seismic tomographic images showing the Nazca's subducting front still within the shallower lower mantle (Pesicek et al. 2012). Therefore, the experiment begins with an imposed initial overriding plate velocity of 3 cm/year and a subduction velocity of 2 cm/year, until the tip of the slab enters the mantle transition zone triggering the slab pull force. Afterwards, the slab is driven by gravitational instability, while the velocity of the upper plate suffers a reduction to 2 cm/year at 10 Ma, following plate reconstructions (Somoza and Ghidella 2012).

4 Modelling the Oligo-Miocene Geodynamic Setting

The results of the model presented by Fennell et al. (2018) show that an effective and geometrically equilibrated subduction setting is reached after 5 Myr (Fig. 3a), when the subduction of the oceanic slab is induced solely by the velocity imposed in the model, simulating the ridge push from the Pacific mid-ocean ridge (Fig. 3b). However, once the tip of the slab begins its interaction with the 410 km olivine/spinel phase transition at ca. 21.5 Ma, the slab subducts dynamically, increasing its velocity, as well as generating a viscous drag-induced circulation in the asthenospheric wedge (Fig. 3c). This initial increase in velocity marks the beginning of vigorous trench

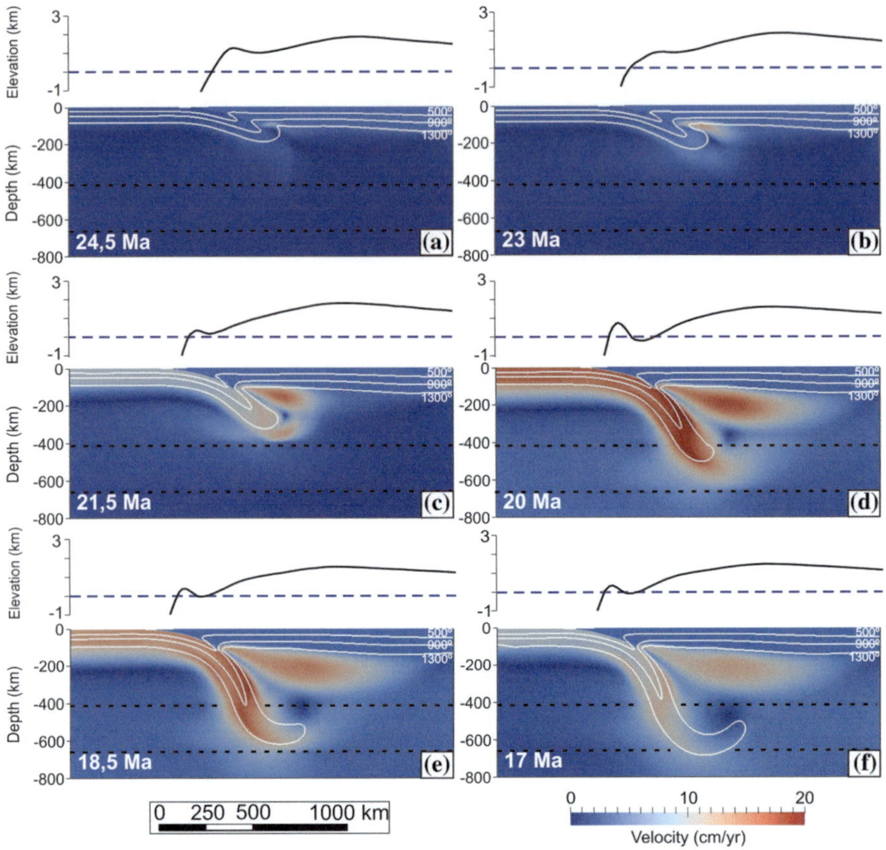

Fig. 3 2D geodynamic model of the Oligo-Miocene transition taken and modified from Fennell et al. (2018) showing changes in the geometry of the slab, velocity and topography as the subducted Nazca Plate interacts with the 410 and 660 km phase transitions (black-dashed lines). Noteworthy, the maximum influence of the slab pull force is registered at 20 Ma, evidenced by the formation of a topographic depression 200–300 km east of the trench, a steep subduction angle, a fast subduction velocity and an astenospheric influx towards the mantle wedge

hinge retreat and steepening of the slab, which translates in the creation of a basin-like geometry in the overriding plate located 200–300 km east of the trench (Fig. 3c). The influence of the slab pull force over the tip of the slab reaches its peak at 20 Ma, when subduction velocity suffers a sudden increase coeval to an important asthenospheric influx below an 800 km wide area east of the trench, due to the void left in the mantle wedge created by the steepening of the slab (Fig. 3d). At these times, the model predicts a fully developed basin subsiding below sea level, which is associated with a fast retreat of the trench hinge and major steepening in the angle of the oceanic slab (Fig. 3d). At 18.5 Ma, the tip of the slab encounters resistance from the viscous lower mantle as it reaches the 660 km spinel/perovskite phase transition, resulting in an upward bending and slowing down of the oceanic slab (Fig. 3e). This results in a waning of the astenospheric influx, along with both the abortion of basin formation and retreat of the subduction zone (Fig. 3e). An incipient slab shallowing event is observed at 17 Ma as the slab stagnates along the 660 km boundary and the subduction velocity continues to decrease, generating the stabilization of both the topography and the influx of hot material beneath the continental plate (Fig. 3f).

5 Linking the Oligo-Miocene Geologic and Geodynamic Scenarios

The relative length increment of the upper plate represented by the stretching factor, the depth of the basin located 200–300 km east of the trench and the convergence velocity were extracted from the model of Fennell et al. (2018), allowing their direct comparison with the Neuquén Basin's geologic framework and subduction parameter reconstructions (Fig. 4). The neutral regime that characterized the Neuquén Basin during most of the Palaeogene (Horton and Fuentes 2016) is reflected by an initial stability period that ended at ca. 23 Ma, when a rise in the stretching factor indicates an increase in the continental plate's length relative to its initial length (Fig. 4a). This event coincided with an acceleration in the convergence velocity and the creation of

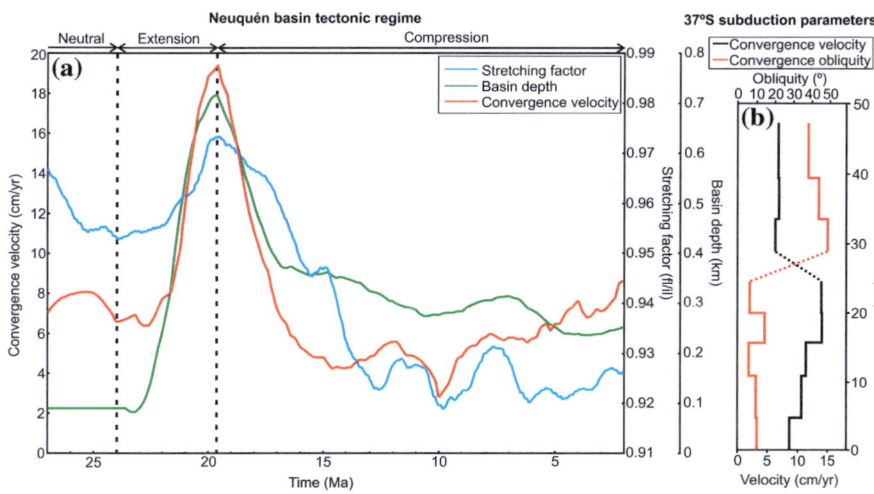

Fig. 4 **a** Variations in the upper plate's length relative to its initial length (illustrated as the stretching factor), basin depth and convergence velocity, extracted from the numerical model, follow an increasing trend overlapping with extensional deformation in the Neuquén Basin, while a decreasing trend during latter development of the compressional regime (modified from Fennell et al. 2018). **b** A major shift from oblique towards a more orthogonal subduction regime and a sudden increase in convergence velocity occurred in Late Oligocene times at 37° S (curves based on Somoza and Ghidella 2012), in coincidence with data extracted from the model and the onset of extension in the Neuquén Basin

a basin, reflecting the triggering of the slab pull force and the initiation of extension in the Neuquén Basin (Fig. 4a). This trend reaches its highest point at around 20 Ma (Fig. 4a), in agreement with the climax of magmatic production throughout the Neuquén Basin (e.g. Kay et al. 2005, 2006; Dyhr et al. 2013a, b) and the occurrence of marine sedimentation in the southern Ventana intra-arc basin (Bechis et al. 2014). Afterwards, the modelled basin experiences a gradual shallowing during a decrease in the stretching factor and convergence velocities, which reflects the beginning of the interaction of the Nazca Plate with the lower mantle. Both shallowing of the basin and shortening of the upper plate are in agreement with the shift towards a contractional regime recorded in the Neuquén Basin through the dating of synorogenic deposits (Fig. 4a) (Horton et al. 2016). This trend stabilizes after 15 Ma, remaining relatively undisturbed during the rest of the experiment, which supports the continuation of compression until present throughout the Neuquén Basin (Giambiagi et al. 2012; Horton and Fuentes 2016).

When comparing the convergence velocity obtained from the model and the reconstructed subduction parameters at 37° S (Fig. 4B) by Somoza and Ghidella (2012), these show a good correlation. An averaged convergence velocity of 13.9 cm/year during the 24-20 Ma period coincides with the sudden increase in velocity from 2 to 20 cm/year predicted by the model in the same time span (Fig. 4). Moreover, a major obliquity change also occurs during these times, shifting from 50° to less than 10° (Fig. 4b) (Somoza and Ghidella 2012), indicating a more orthogonal and effective subduction, in agreement with the numerical model and geologic observations.

6 The Oligo-Miocene Tectonic Mode Switch

The numerical model presented by Fennell et al. (2018) provides important insights regarding the sudden and enigmatic tectonic mode switch recorded in the Neuquén Basin during late Oligocene to early Miocene times. After a long-lived Palaeogene neutral regime insinuated by sedimentological and subsidence analyses carried out in the foreland basin (Horton and Fuentes 2016), the shift towards Miocene compression is preceded by a brief period of widespread extensional deformation (Fig. 5a). This is produced by a geodynamic scale process occurring at the mantle transition zone (i.e. the effect of the slab pull force), whose impact in the surface is reflected by the accelerated retreat of the trench hinge (Fig. 5a). Retreating subduction zones are characterized by back-arc extension and space creation in the overriding plate (Waschbusch and Beaumont 1996), in agreement with the basin shown in the model and the presence of normal faulting associated with synextensional deposition in the late Oligocene to early Miocene intra-arc basins (Jordan et al. 2001; Charrier et al. 2002; Folguera et al. 2010; Rojas Vera et al. 2010; Garcia Morabito and Ramos 2012; Ramos et al. 2014; Jara and Charrier 2014). Furthermore, the mafic flood devoid of subduction-related components recorded in the present retroarc zone (Fig. 5a) provides evidence for the presence of back-arc mantle being injected far east of the

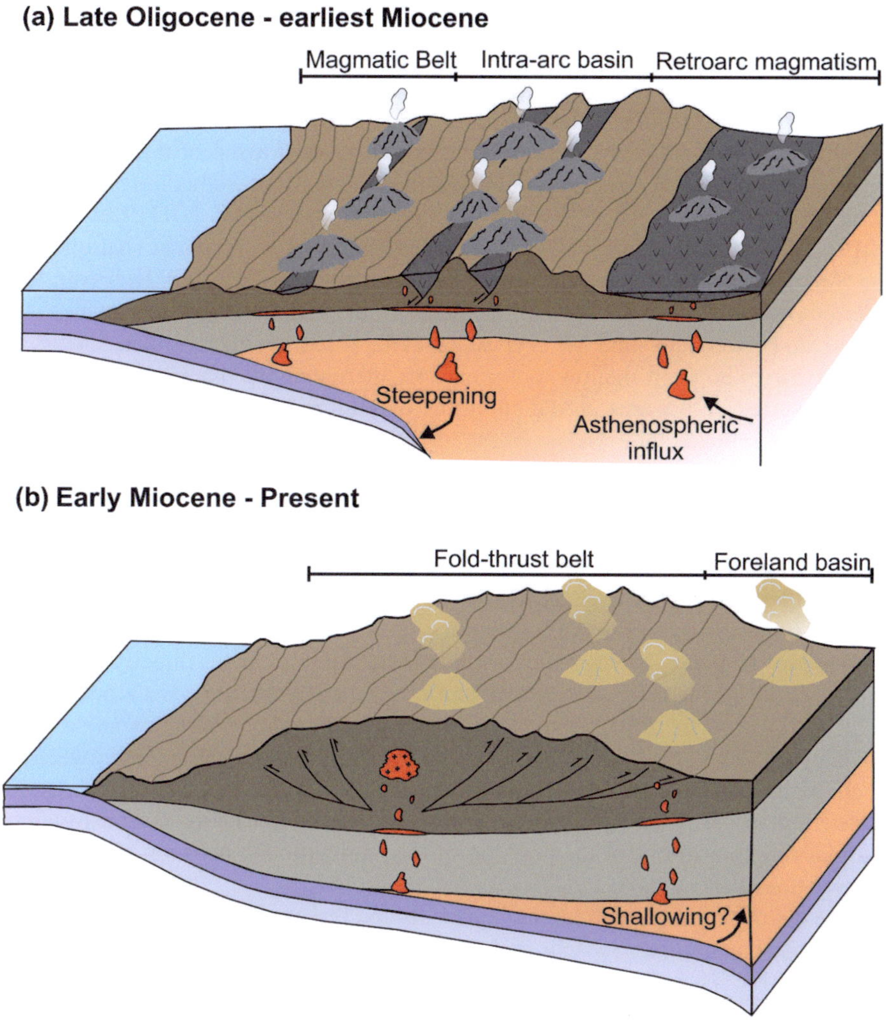

Fig. 5 Schematic representation of the Oligo-Miocene tectonic mode switch, with retreat of the trench hinge and a steep subduction angle associated with extension during late Oligocene and earliest Miocene times (**a**), and slowing down of slab roll-back causing the pushing of the upper plate against the trench hinge between early Miocene and present times, resulting in orogenesis and final closure of the Neuquén Basin (**b**). Magmatism is also affected by this geodynamic scenario, with an episode of widespread mafic volcanism influenced by mantelic processes (**a**), followed by an eastward migration of increasing slab-influenced magmatism (**b**) related to changes in slab dip and/or convergence rates

trench (Kay and Copeland 2006; Dyhr et al. 2013a, b), in correlation with what is observed in the model.

Afterwards, a contractional regime is installed, characterized by an eastward propagation of the thrust and magmatic arc fronts with geochemical evidences of increasing slab influence (Fig. 5b) (Litvak et al. 2015). A shallowing of the oceanic slab between 15 and 3.5 Ma has been pointed out as the main cause behind arc and orogenic front migrations (Kay et al. 2006; Folguera and Ramos 2011; Litvak et al. 2015), whose initial phase is suggested in the model by an incipient shallowing of the Nazca's slab after its tip reaches the 660 km boundary (Fig. 3F). However, seismic tomographies reveal a normal 30° angle in the slab subducting beneath South American between 35° and 39° S (Pesicek et al. 2012), giving less than 5 Myrs for the slab dip to go from shallow to normal. An alternative hypothesis involves a slab tear registered in this area produced by the overriding of a mantelic plume, which would have relieved the pressure in the overlying mantle, favouring the steepening of its angle (Pesicek et al. 2012; Burd et al. 2014; Gianni et al. 2017). However, given that the Miocene orogenic front (Sagripanti et al. 2011) coincides with the easternmost recognized evidences of Quaternary deformation (Folguera et al. 2009; Sagripanti et al. 2018) and that arc-like signatures are still recorded 400–500 km east of the trench (May et al. 2018), it becomes evident that a shallow subduction angle is unnecessary to explain migration of both the magmatic and orogenic fronts to their current position. Since the common belief that volcanoes directly overlie the places where the slabs reach a critical depth shows inconsistencies with modern volcanic settings, further hypothesis not requiring shallow dip subduction angles must be analysed (England et al. 2004). Considering the inverse correlation between the depth to the slab beneath the volcanic fronts and the descent speed of the subduction plate (England et al. 2004), the deceleration of Nazca's plate (Fig. 4) constitutes an alternative explanation for the position of the magmatic arc between Neogene and present times.

The history of major subduction zones around the world has shown that the tectonic mode can abruptly switch, even during uninterrupted convergence (Lister and Forster 2009). These switches are characterized by an initial retreat of the trench hinge creating a gravitational potential into which the orogen collapses, after which a slowing down of slab roll-back pushes the overriding plate against the adjacent subduction hinge promoting orogenesis (Lister and Forster 2009). The numerical model presented by Fennell et al. (2018) suggests that the natural evolution of a subducting slab after a period of limited subduction could be the responsible for the Oligo-Miocene tectonic mode switch, generating a brief period of widespread extension in the first place, which is later followed by horizontal shortening resulting in the final closure of the Neuquén Basin (Fig. 5).

Acknowledgements We acknowledge Marius Walter and the GFZ Geodynamic modelling section for their assistance with the numerical model. We are also grateful to Mark Brandon, whose valuable feedback improved the quality of this contribution. This is the R-317 contribution of the Instituto de Estudios Andinos "Don Pablo Groeber".

References

Álvarez Cerimedo J, Orts DL, Rojas Vera EA et al (2013) Mecanismos y fases de construcción orogénicos del frente oriental andino (36° S, Argentina). Andean Geol 40(3):504–520

Bechis F, Encinas A, Concheyro A et al (2014) New age constraints for the Cenozoic marine transgressions of northwestern Patagonia, Argentina (41°–43° S): paleogeographic and tectonic implications. J S Am Earth Sci 52:72–93

Burd AI, Booker JR, Mackie R, Favetto A, Pomposiello MC (2014) Three-dimensional electrical conductivity in the mantle beneath the Payún Matrú Volcanic Field in the Andean backarc of Argentina near 36.5°S: evidence for decapitation of a mantle plume by resurgent upper mantle shear during slab steepening. Geophys J Int 198:812–827

Capitanio FA, Faccenna C, Zlotnik S et al (2011) Subduction dynamics and the origin of Andean orogeny and the Bolivian orocline. Nature 480:83–86

Charrier R, Baeza O, Elgueta SE et al (2002) Evidence for Cenozoic extensional basin development and tectonic inversion south of the flat-slab segment, southern Central Andes, Chile (33–36 SL). J S Am Earth Sci 15(1):117–139

Charrier R, Pinto L, Rodriguez MP (2007) Tectonostratigraphic evolution of the Andean Orogen in Chile. In: Moreno T, Gibbons W (eds) The geology of Chile. The Geological Society, London, pp 21–114

Cobbold PR, Rosello EA (2003) Aptian to recent compressional deformation, foothills of the Neuquén Basin, Argentina. Mar Petrol Geol 20:429–443

Dyhr CT, Holm PM, Llambías EJ (2013a) Geochemical constraints on the relationship between the Miocene-Pliocene volcanism and tectonics in the Palaoco and Fortunoso volcanic fields, Mendoza Region, Argentina: New insights from 40Ar/39Ar dating, Sr–Nd–Pb isotopes and trace elements. J Volcanol Geoth Res 266:50–68

Dyhr CT, Holm PM, Llambías EJ, Scherstén A (2013b) Subduction controls on Miocene back-arc lavas from Sierra de Huantraico and La Matancilla and new $^{40}Ar/^{39}Ar$ dating from the Mendoza Region, Argentina. Lithos 179:67–83

England P, Engdahl R, Thatcher W (2004) Systematic variations in the depths of slabs beneath arc volcanoes. Geophys J Int 156:377–408

Fennell LM, Iannelli SB, Folguera A et al (2017) Interrupciones extensionales en el desarrollo de la faja plegada y corrida de Malargüe (36° S). In: Abstracts of the 20 Congreso Geológico Argentino, San Miguel de Tucumán, 7–11 Aug 2017

Fennell LM, Quinteros J, Iannelli SB et al (2018) The role of the slab pull force in the late Oligocene to early Miocene extension in the Southern Central Andes (27°-46° S): Insights from numerical modeling. J S Am Earth Sci 87:174–187

Folguera A, Ramos VA (2011) Repeated eastward shifts of arc magmatism in the Southern Andes: A revision to the long-term pattern of Andean uplift and magmatism. J S Am Earth Sci 32(4):531–546

Folguera A, Naranjo JA, Orihashi Y et al (2009) Retroarc volcanism in the northern San Rafael Block (34°-35°30'S), southern Central Andes: Occurrenge, age and tectonic setting. J Volcanol Geoth Res 186:169–185

Folguera A, Rojas Vera EA, Bottesi G et al (2010) The Loncopué Trough: a Cenozoic basin produced by extension in the southern Central Andes. J Geodyn 49(5):287–295

Garcia Morabito E, Ramos VA (2012) Andean evolution of the Aluminé fold and thrust belt, Northern Patagonian Andes (38 30′–40 30′ S). J S Am Earth Sci 38:13–30

Gerbault M, Cembrano J, Mpodozis C et al (2009) Continental margin deformation along the Andean subduction zone: Thermo-mechanical models. Phys Earth Planet Int 177(3):180–205

Giambiagi L, Mescua J, Bechis F et al (2012) Thrust belts of the southern Central Andes: Along-strike variations in shortening, topography, crustal geometry, and denudation. Geol Soc Am Bull 124(7–8):1339–1351

Gianni GM, Garcia HPA, Lupari M et al (2017) Plume overriding triggers shallow subduction and orogeny in the southern Central Andes. Gondwana Res 49:387–395

Gianni GM, Dávila FM, Echaurren A et al (2018) A geodynamic model linking Cretaceous orogeny, arc migration, foreland dynamic subsidence and marine ingression in southern South America. Earth-Sci Rev 185:437–462

Godoy E, Yáñez G, Vera E (1999) Inversion of an Oligocene volcano-tectonic basin and uplifting of its superimposed Miocene magmatic arc in the Chilean Central Andes: first seismic and gravity evidences. Tectonophysics 306(2):217–236

Heuret A, Lallemand S (2005) Plate motions, slab dynamics and back-arc deformation. Phys Earth Planet Int 149(1):31–51

Horton BK (2018a) Tectonic regimes of the Central and Southern Andes: Responses to variations in plate coupling during subduction. Tectonics 37. https://doi.org/10.1002/2017tc004624

Horton BK (2018b) Sedimentary record of Andean mountain building. Earth-Sci Rev 178:279–309

Horton BK, Fuentes F (2016) Sedimentary record of plate coupling and decoupling during growth of the Andes. Geol 44(8):647–650

Horton BK, Fuentes F, Boll A, Starck D, Ramírez SG, Stockli DF (2016) Andean stratigraphic record of the transition from backarc extension to orogenic shortening: a case study from the northern Neuquén Basin, Argentina. J S Am Earth Sci 71:17–40

Iannelli SB, Fennell LM, Litvak VDA et al (2018) Geochemical and tectonic evolution of Late Cretaceous to early Paleocene magmatism along the Southern Central Andes (35-36° S). J S Am Earth Sci 87:139–156

Jara P, Charrier R (2014) Nuevos antecedentes estratigráficos y geocronológicos para el Meso-Cenozoico de la Cordillera Principal de Chile entre 32° y 32° 30' S: Implicancias estructurales y paleogeográficas. Andean geol 41(1):174–209

Jordan TE, Burns WM, Veiga R et al (2001) Extension and basin formation in the southern Andes caused by increased convergence rate: A mid-Cenozoic trigger for the Andes. Tectonics 20(3):308–324

Kay SM, Copeland P (2006) Early to middle Miocene backarc magmas of the Neuquén Basin: Geochemical consequences of slab shallowing and the westward drift of South America. Geol Soc Am SP 407:185–213

Kay SM, Godoy E, Kurtz A (2005) Episodic arc migration, crustal thickening, subduction erosion, and magmatism in the south-central Andes. Geol Soc Am Bull 117(1–2):67–88

Kay SM, Burns M, Copeland P (2006) Upper Cretaceous to Holocene magmatism and evidence for transient Miocene shallowing of the Andean subduction zone under the northern Neuquén Basin. In: Kay SM, Ramos VA (eds) Evolution of an Andean Margin: a tectonic and magmatic view from the Andes to the Neuquén Basin (35–39° S). Geological Society of America SP 407, pp 19–60

Lallemand S, Heuret A, Boutellier D (2005) On the relationships between slab dip, back-arc stress, upper plate absolute motion, and crustal nature in subduction zones. Geochem Geophys Geosyst 6(9):Q09006. https://doi.org/10.1029/2005GC000917

Lister G, Forster M (2009) Tectonic mode switches and the nature of orogenesis. Lithos 113:274–291

Litvak VD, Spagnuolo MG, Folguera A et al (2015) Late Cenozoic calc-alkaline volcanism over the Payenia shallow subduction zone, South-Central Andean back-arc (34°30-37´S) Argentina. J S Am Earth Sci 64(2):365–380

Lopez-Escobar L, Vergara M (1997) Eocene-Miocene longitudinal depression and Quaternary volcanism in the Southern Andes, Chile (33–42.5° S): a geochemical comparison. Rev Geol Chile 24(2):227–244

Manea VC, Pérez-Gussinyé M, Manea M (2012) Chilean flat slab subduction controlled by overriding plate thickness and trench rollback. Geology 40(1):35–38

May VR, Chivas AR, Dosetto A et al (2018) Quaternary volcanic evolution in the continental back-arc of southern Mendoza, Argentina. J S Am Earth Sci 84:88–103

Mosolf JG, Gans PB, Wyss AR et al (2018) Late Cretaceous to Miocene volcanism, sedimentation, and upper-crustal faulting and folding in the Principal Cordillera, central Chile: field and geochronological evidence for protracted arc volcanism and transpressive deformation. Geol Soc Am Bull. https://doi.org/10.1130/B31998.1

Muñoz J, Troncoso R, Duhart P et al (2000) The relation of the mid-Tertiary coastal magmatic belt in south-central Chile to the late Oligocene increase in plate convergence rate. Rev Geol Chile 27(2):177–203

Muñoz M, Tapia F, Persico M et al (2018) Extensional tectonics during Late Cretaceous evolution of the Southern Central Andes: evidence from the Chilean main range at ~35° S. Tectonophysics 744:93–117

Parras A, Griffin M (2013) Late Cretaceous (Campanian/Maastrichtian) freshwater to restricted marine mollusk fauna from the Loncoche Formation, Neuquén Basin, west-central Argentina. Cretaceous Res 40:190–206

Pesicek JD, Engdahl ER, Thurber CH et al (2012) Mantle subducting slab structure in the region of the 2010 M8.8 Maule earthquake (30°–40° S). Chile. Geophys J Int 191:317–324

Quinteros J, Sobolev SV (2013) Why has the Nazca plate slowed since the Neogene? Geology 41(1):31–34

Radic JP (2010) Las cuencas cenozoicas y su control en el volcanismo de los Complejos Nevados de Chillán y Copahue-Callaqui (Andes del Sur, 36-39° S). Andean Geol 37(1):220–246

Ramos VA, Folguera A (2009) Andean flat-slab subduction through time. J Geol Soc London 327(1):31–54

Ramos ME, Folguera A, Fennell LM et al (2014) Tectonic evolution of the North Patagonian Andes from field and gravity data (39–40° S). J S Am Earth Sci 51:59–75

Rojas Vera EA, Folguera A, Zamora Valcarce G et al (2010) Neogene to Quaternary extensional reactivation of a fold and thrust belt: the Agrio belt in the Southern Central Andes and its relation to the Loncopué trough (38–39 S). Tectonophysics 492(1):279–294

Royden LH (1993) The tectonic expression slab pull at continental convergent boundaries. Tectonics 12(2):303–325

Sagripanti L, Bottesi G, Naipauer M et al (2011) U/Pb ages on detrital zircons in the southern central Andes Neogene foreland (36°–37° S): Constraints on Andean exhumation. J S Am Earth Sci 32(4):555–566

Sagripanti L, Bottesi G, Kietzmann D et al (2012) Mountain building processes at the orogenic front. A study of the unroofing in Neogene foreland sequence (37° S). Andean Geol 39(2):201–219

Sagripanti L, Colavitto B, Jagoe L et al (2018) A review about the Quaternary upper-plate deformation in the Southern Central Andes (36°–38° S): A plausible interaction between mantle dynamics and tectonics. J S Am Earth Sci 87:221–231

Sobolev SV, Babeyko AY (2005) What drives orogeny in the Andes? Geology 33(8):617–620

Somoza R, Ghidella ME (2012) Late Cretaceous to recent plate motions in western South America revisited. Earth Planet Sci Lett 331:152–163

Vergani GD, Tankard J, Belotti J et al (1995) Tectonic evolution and paleogeography of the Neuquén basin, Argentina. In: Tankard AJ, Suárez R, Welsink HJ (eds) Petroleum Basins of South America. AAPG Memoir 62:383–402

Waschbusch P, Beaumont C (1996) Effect of a retreating subduction zone on deformation on simple regions of plate convergence. J Geophys Res 101(B12):28133–28148

Tectonics Associated with the Late Oligocene to Early Miocene Units of the High Andes (Cura-Mallín Formation). A Review of the Geochronological, Thermochronological, and Geochemical Data

Eduardo Agustín Rosselot, María Hurley, Lucía Sagripanti, Lucas Fennell, Sofía B. Iannelli, Darío Orts, Alfonso Encinas, Vanesa D. Litvak and Andrés Folguera

Abstract The tectonic regime associated with the Oligo-Miocene Cura-Mallín Formation and equivalents in the Main Andean Cordillera between 35° and 40° S, the Ventana and Abanico formations to the south and north, respectively, is still matter of debate. While most authors have agreed in relating them to an extensional regime that could have interrupted Andean orogenesis, others have provided evidence of coexistence with an ongoing compressional regime at least for the upper part of their record. Available geochronological, structural, geochemical, thermochronological, and basin subsidence data between 33° and 43° S are compiled and analyzed in this chapter in order to provide a tectonic framework for these rocks. Based on this analysis, two different mechanisms seem to have succeeded through the accumulation of these sequences, a late Oligocene to earliest Miocene period characterized by syn-extensional deposits in an attenuated crust evidenced by low La/Yb ratios in contemporaneous magmatism, and a younger time period in the mid- to late Miocene that coincides with regional exhumation and compression revealed by thermochronological data and syncontractional sedimentation.

E. A. Rosselot (✉) · M. Hurley · L. Sagripanti · L. Fennell · S. B. Iannelli · V. D. Litvak · A. Folguera
Universidad de Buenos Aires, Facultad de Ciencias Exactas y Naturales, Departamento de Ciencias Geologicas, Buenos Aires, Argentina
e-mail: earosselot@gl.fcen.uba.ar

CONICET – Universidad de Buenos Aires, Instituto de Estudios Andinos "Don Pablo Groeber" (IDEAN), Buenos Aires, Argentina

D. Orts
Universidad Nacional de Río Negro, Pcia de Río Negro, Argentina

A. Encinas
Departamento de Ciencias de la Tierra, Universidad de Concepción, Concepción, Chile

D. Kietzmann and A. Folguera (eds.), *Opening and Closure of the Neuquén Basin in the Southern Andes*, Springer Earth System Sciences,
https://doi.org/10.1007/978-3-030-29680-3_18

Keywords Late Oligocene extension · Early Miocene contraction · Cura-Mallín Basin · Abanico Basin · Ventana Basin

1 Introduction

The highest Andes of Neuquén are characterized by the development of extensive volcanic and volcano-clastic successions with (Eocene) late Oligocene to early Miocene age commonly grouped in the Cura-Mallín and Abanico Formations (Niemeyer and Muñoz 1983; Suarez and Emparan 1995; Charrier et al. 2002; Vergara et al. 2004). These sequences are considered to have been deposited in a larger group of volcanogenic basins that were formed after the Incaic tectonic event through the southern Central and Patagonian Andes such as the Valle del Cura (in the southern Puna plateau), Ventana, Coastal magmatic belt (at the latitudes of the Neuquén Basin in Chile) and Traiguén basins (in Patagonia) (Godoy et al. 1999; Munoz et al. 2000; Charrier et al. 2002; Winocur et al. 2015; Encinas et al. 2016) (Fig. 1). The geochemistry of these rock associations indicates that they belonged to a voluminous arc front, although the tectonic controls associated with them are still matter of extensive debate. While under some hypotheses these rocks are related to a regional extensional regime that affected vast proportions of the arc and retroarc regions (see Chap. "The Oligo-Miocene Tectonic Mode Switch: From a Brief Period of Widespread Extension to the Final Closure of the Neuquén Basin") (Jordan et al. 2001; Charrier et al. 2002; Burns et al. 2006; Fennell et al. 2018), others have related them to contractional deformation that would have been continuous through the Andean orogenesis (Cobbold and Rossello 2003; Cobbold et al. 2008). A series of syndepositional sections have been described in several of these basins and associated with extensional and/or contractional tectonic settings, although without conclusive temporal constraints in most cases (Jordan et al. 2001; Pananont et al. 2004; Folguera et al. 2007; Winocur et al. 2015; Rubilar et al. 2017; Mosolf et al. 2018). No single tectonic model has been completely validated up to now in relation to the genesis of these rocks, basically because proposed models have been automatically reproduced without adding in some cases additional constraints. This chapter summarizes the temporal database associated with these rock associations, analyzing their distribution and their relation to the available fission track data as indicative of cooling ages linked with exhumation and enhanced erosional rates both in the forearc and in the retroarc through the Malargüe, Chos Malal and Agrio fold and thrust belts (FTB) (Spikings et al. 2008; Folguera et al. 2015; Rojas Vera et al. 2015; Sánchez et al. 2018). Additionally, this chapter compares the variable lifespan that characterizes these successions with available $(La/Yb)_n$ ratios that were calculated on them, which are sensitive to variable crustal thicknesses through time (Kay et al. 2005; Iannelli et al. 2017; Piquer et al. 2017). Then the successions that belong to the Cura-Mallín formation can be separated in two units with contrasting exhumation and geochemical patterns: One that evolved previously to the Neogene contractional period described in the Southern Central and Patagonian Andes that is constrained by thermochronological and

Fig. 1 Late Oligocene to earliest Miocene extensional basins (red polygons) developed in the Southern Central and Patagonian Andes, where the Cura-Mallín, Abanico, and Ventana basins, focus of this chapter, are located. DEM corresponds to Shuttle Radar Topographic Mission from NASA

synorogenic sedimentation data (see Chap. "The Andean Foreland Evolution of the Neuquén Basin: A Discussion), and a younger group that coexisted with it.

2 Geological Setting

The study area is localized in the Southern Andes between 36° and 39° S, where a series of N-trending morpho-structural units develop, listed below from west to east: the Coastal Cordillera, the Central Valley, the Main Cordillera, the Loncopué Trough and the Agrio, and Chos Malal FTB (Fig. 1). Mesozoic and Cenozoic sequences are exposed in the main cordillera deposited over a Permian–Triassic basement, the Choiyoi volcanic Province. These units have been contractionally deformed in two main deformational phases, one spanning from the late Early Cretaceous to the Eocene and another in Neogene times (Rojas Vera et al. 2015; Orts et al. 2012; Horton and Fuentes 2016). These compressional events were interrupted by a generalized extensional event that lasted from the (Eocene) Oligocene to earliest Miocene times, forming different basins along the Southern Central and Patagonian Andes (Fig. 1) (Jordan et al. 2001; Radic 2010). This extensional stage that interrupted Andean orogenesis has been linked to an increase in the velocity of the subducted slab that was associated with an accelerated roll back at a time with a nearly stationary behavior of the upper plate (see Chap. "The Oligo-Miocene Tectonic Mode Switch: From a Brief Period of Widespread Extension to the Final Closure of the Neuquén Basin") (Fennell et al. 2018).

Among the different basins associated with this stage, the northernmost Abanico Basin or Coya Machalí Basin, developed between 33° and 36° S, is filled up with over 2500 m of mainly acid to intermediate lavas and volcaniclastic deposits with continental sedimentary intercalations (Charrier et al. 2002). These deposits are divided into an extensional lower section, the Abanico Formation and equivalents, and a compressional-related upper section, corresponding to the (uppermost Abanico) Farellones Formation and equivalents.

Between 36° and 39° S and along the international border, the Oligo-Miocene deposits are grouped in the Cura-Mallín Formation. The Cura-Mallín Formation consists of pyroclastic deposits and volcanic flows interbedded with continental deposits (Niemeyer and Muñoz 1983; Muñoz and Niemeyer 1984; Suarez and Emparan 1995) which have been assigned to the Cura-Mallín Formation in Chile and the Lileo and Arroyo Palao formations in Argentina (Zanettini 2001; Rovere et al. 2004) (Fig. 2). Over these, the Trapa-Trapa, Cajón Negro, and Mitrauquén formations, characterized by volcanic and clastic rocks, conformably cover the Cura-Mallín Formation, with some few unconformities described in some works.

South of the Cura-Mallín Formation, between 39º and 43ºS, a series of Oligo-Miocene synextensional units have been gathered in the Ventana Formation (see Chap. "Late Cretaceous to Oligocene Magmatic Evolution of the Neuquén Basin") (Rapela et al. 1988; Iannelli et al. 2017). The Ventana Formation comprises volcanic, sedimentary, locally marine, and volcano sedimentary sequences (García Morabito

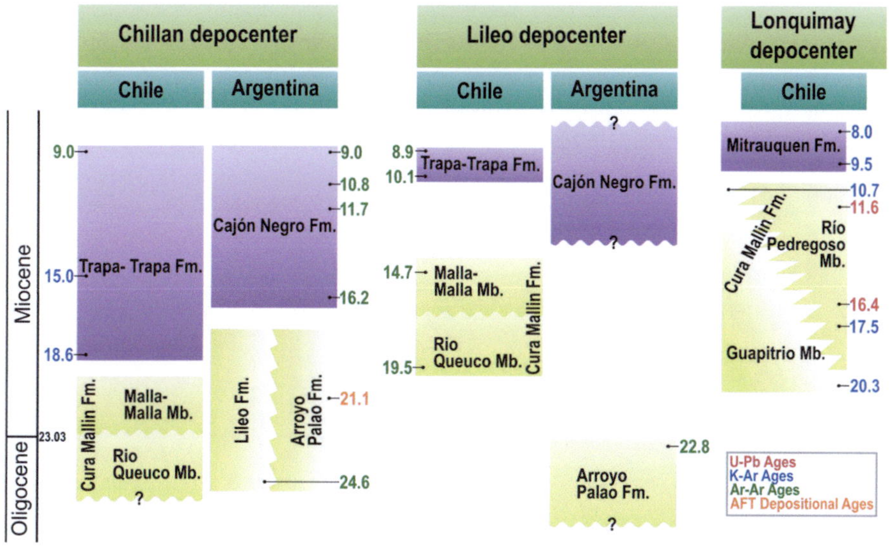

Fig. 2 Stratigraphic chart of the different zones of depocenters of the Cura-Mallín Formation outcrops in Argentina and Chile with the most representative age data (see text for specific references). Ages are in millions of years (Ma) (Muñoz and Niemeyer 1984; Munoz Bravo 1988; Suarez and Emparan 1995, 1997; Jordan et al. 2001; Burns et al. 2006; Herriott 2006; Flynn et al. 2008; Pedroza et al. 2017)

and Ramos 2012; Orts et al. 2012; Bechis et al. 2014; Fernández Paz et al. 2018) and is also controlled by extensional structures (Franzese et al. 2011) forming wedge-like depocenters in seismic and field profiles (Orts et al. 2015).

3 Cura-Mallín Formation

The Cura-Mallín Basin has been divided from north to south into three sub-basins which were interpreted as half-grabens of alternating polarities separated by NW and NE transfer zones (Radic 2010). These accommodation zones are inferred to have controlled long linear–fissural Quaternary volcanic complexes (Radic 2010) (Fig. 3). In this scheme, the northern compartment is the Chillán sub-basin between 36° and 37° S that is separated by the Chillán Volcanic Complex (37° S) from the Lileo sub-basin to the south. This sub-basin is limited in the south by the Callaqui-Caviahue-Copahue-Mandolegüe transfer zone that coincides with a volcanic belt formed by the amalgamation of the homonymous centers (38° S). South of this feature, the Lonquimay sub-basin is where the most comprehensive stratigraphic schemes have been proposed. In this work, we refer to these sub-basins as zones, considering the lack of consensus about the tectonic origin of the Cura-Mallín Formation rocks. In seismic profiles conducted east of these outcrops, the basal part of Cura-Mallín

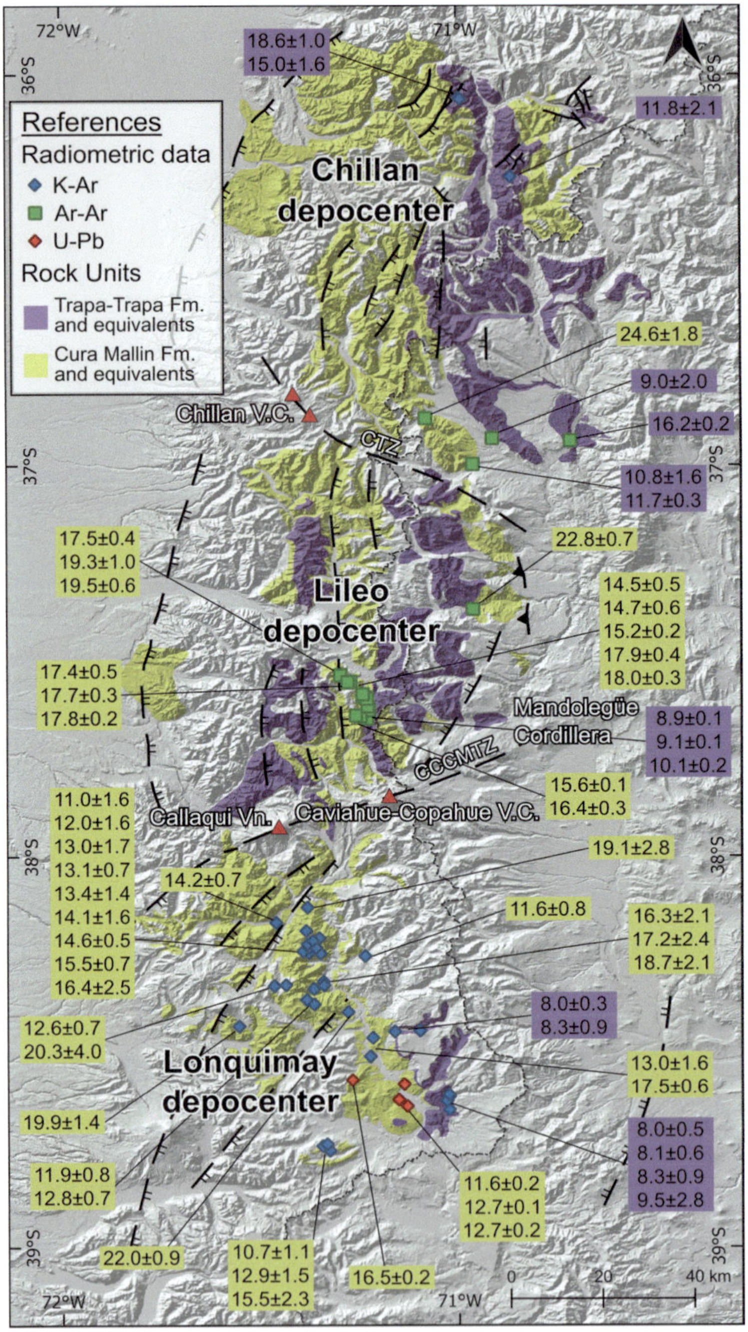
References
Radiometric data
K-Ar
Ar-Ar
U-Pb
Rock Units
Trapa-Trapa Fm. and equivalents
Cura Mallin Fm. and equivalents
Chillan depocenter
Lileo depocenter
Lonquimay depocenter
Chillan V.C.
CTZ
CCCMTZ
Mandolegüe Cordillera
Callaqui Vn.
Caviahue-Copahue V.C.
72°W
71°W
36°S
37°S
38°S
39°S
0 20 40 km
18.6±1.0
15.0±1.6
11.8±2.1
24.6±1.8
9.0±2.0
16.2±0.2
10.8±1.6
11.7±0.3
22.8±0.7
17.5±0.4
19.3±1.0
19.5±0.6
14.5±0.5
14.7±0.6
15.2±0.2
17.9±0.4
18.0±0.3
17.4±0.5
17.7±0.3
17.8±0.2
8.9±0.1
9.1±0.1
10.1±0.2
15.6±0.1
16.4±0.3
11.0±1.6
12.0±1.6
13.0±1.7
13.1±0.7
13.4±1.4
14.1±1.6
14.6±0.5
15.5±0.7
16.4±2.5
19.1±2.8
14.2±0.7
11.6±0.8
16.3±2.1
17.2±2.4
18.7±2.1
12.6±0.7
20.3±4.0
8.0±0.3
8.3±0.9
13.0±1.6
17.5±0.6
19.9±1.4
8.0±0.5
8.1±0.6
8.3±0.9
9.5±2.8
11.9±0.8
12.8±0.7
11.6±0.2
12.7±0.1
12.7±0.2
22.0±0.9
10.7±1.1
12.9±1.5
15.5±2.3
16.5±0.2

◀**Fig. 3** Radiometric database of the Cura-Mallín and Trapa-Trapa formations on the axial Andean zone above SRTM-30m DEM (Muñoz and Niemeyer 1984; Suarez and Emparan 1988, 1995, 1997; Jordan et al. 2001; Burns et al. 2006; Herriott 2006; Glodny et al. 2008; Pedroza et al. 2017). Mapping is based on Niemeyer and Muñoz (1983), Muñoz and Niemeyer (1984), Suarez and Emparan (1997), Zanettini (2001), and Rovere et al. (2004). Structure and zones after Radic (2010). Red-triangules represent active volcanoes. CCCMTZ: Callaqui-Caviahue-Copahue-Mandolegüe transfer zone. CTZ: Chillan Transfer Zone

Formation deposits shows wedge-like geometries (Jordan et al. 2001; Folguera et al. 2010; Rojas Vera et al. 2010), which partially support their extensional origin. It has also been proposed in these seismic lines that the Trapa-Trapa Formation shows onlap relations against inverted contractional structures (Folguera et al. 2010; Rojas Vera et al. 2010). Nevertheless, given the lack of wells attached to seismic lines or a stratigraphic surface control, these interpretations need further confirmation.

In the Chillan zone, the Cura-Mallín Formation has been divided into two members, the Río Queuco lower member and the Malla-Malla upper member (Muñoz and Niemeyer 1984). The Río Queuco Member contains mainly volcanic and volcano-clastic rocks and comprises the majority of the outcrops of this zone. On the other hand, the Malla-Malla Member is formed mostly by clastic rocks. Mid-Miocene volcanic and sedimentary rocks of the Trapa-Trapa Formation in Chile and Cajón Negro Formation in Argentina (Pesce 1981; Niemeyer and Muñoz 1983; Muñoz and Niemeyer 1984) overlay the Cura-Mallín Formation. The contact relationship between the Cura-Mallín Formation and Trapa-Trapa or Cajón Negro Formations is not clear; while in some sectors they are concordant to transitional (Niemeyer and Muñoz 1983; Muñoz and Niemeyer 1984), in other locations they are separated by an angular unconformity (Utgé et al. 2009) indicating contractional deformation of the underlying units previous to their deposition. The Chillan zone contains the poorest isotopic age coverage (Fig. 3) with only one Ar–Ar plateau age in the Cura-Mallín Formation of 24.6 ± 1.8 Ma near the top of the Rio Queuco member (Jordan et al. 2001). Nevertheless, this Ar–Ar age did not yield a solid plateau with an inverse isochron analysis with a high error (±23 Ma) and a high MSWD value (=30), reason by which it is considered highly doubtful. The overlying Trapa-Trapa Formation in this northern zone contains four K–Ar and three Ar–Ar isotopic ages that range between 18.6 ± 1.0 and 9.0 ± 2.0 Ma (Muñoz and Niemeyer 1984; Jordan et al. 2001; Burns et al. 2006) (Figs. 2 and 3).

The Lileo zone has a better development in the Argentinean Andean slope (Radic 2010), where its infill was subdivided into two units, the Lileo and Arroyo Palao formations (Rovere et al. 2004) (Figs. 2 and 3). These two formations are interpreted as proximal and distal facies, respectively, of contemporaneous volcanism, rather than two different age units (Burns et al. 2006). On top of them, the volcanic Trapa-Trapa Formation lies concordantly or pseudoconcordantly; and through angular unconformity the late Pliocene to Quaternary volcanic rocks (i.e., Invernada Vieja Formation, Cola de Zorro Formation) (Niemeyer and Muñoz 1983; Rovere et al. 2004). In this zone, Jordan et al. (2001) dated the lower Malla-Malla Member

(Arroyo Palao Formation), obtaining an Ar–Ar age of 22.8 ± 0.7 Ma. Nevertheless, this age is considered not trustworthy, because it was not steeply heated but analyzed after a single step at total fusion (1550 °C). Later, Herriott (2006) and Flynn et al. (2008) used Ar–Ar isotopes to date the Cura-Mallín and Trapa-Trapa Formations over the Chilean side of the Andes in the Lileo zone, obtaining younger ages for both units, ranging between 19.8 ± 0.4 and 14.5 ± 0.5 Ma for the older one and 10.1 ± 0.2 and 8.9 ± 0.1 Ma for the younger one. In this area, Burns et al. (2006) indicated that compressional tectonics should have begun post 11.7 Ma, based on an Ar–Ar age of an ash-flow tuff assigned to the Cajon Negro Formation cut by a reverse fault. In addition, the unconformity between the Cura-Mallín Formation and the Cola de Zorro Formation suggests that deformation must have taken place priorly to ~3 Ma, the oldest age of the subhorizontal Cola de Zorro andesites in this area. Furthermore, geochemical isotopic data available for this area indicate an increase in $(La/Yb)_n$ ratios at ~17 Ma, which would imply a contemporaneous crustal thickening (Kay et al. 2006) (Fig. 4).

In the Lonquimay zone to the south, the Cura-Mallín Formation crops out only in the Chilean Andean side and is overlain by the Mitrauquén Formation (Suarez and Emparan 1995) (Figs. 2 and 3). In this area, the Cura-Mallín Formation has been divided into two members, the mainly volcanic and volcaniclastic Guapitrío Member and the sedimentary Rio Pedregoso Member, being these units partially contemporaneous (Suarez and Emparan 1995). Yet, it is worthy to note that Salinas (1979) had previously considered the sedimentary rocks as part of the Guapitrio Member defined by Suarez and Emparan (1995), and therefore younger than the volcanic rocks of this area. On top of the Cura-Mallín Formation, the Mitrauquén Formation, composed of fluvial conglomerates, ignimbrites, and lavas, conformably overlies the Río Pedregoso Member, even though the contact is covered. However, it is worth mentioning that while the Cura-Mallín Formation is gently folded, the Mitrauquén Formation is subhorizontal, which could be interpreted as an angular unconformity between these two units or as a different response to deformation (Suarez and Emparan 1997). Suarez and Emparan (1995, 1997) performed several K–Ar ages on the Cura-Mallín and Mitrauquén Formations of the Lonquimay zone. They constrained the Guapitrío Member age between 19.9 ± 1.4 and 10.7 ± 1.1 Ma and the Rio Pedregoso Member between 17.5 ± 0.6 and 13.0 ± 1.6 Ma, proposing an interdigitation between both units. Additionally, they constrained the Mitrauquén Formation age between 9.5 ± 2.8 and 8.0 ± 0.5. Recently, Pedroza et al. (2017) obtained new U–Pb ages of 16.4 ± 0.2 and 11.6 ± 0.2 Ma for the Río Pedregoso Member. Based on sedimentary environmental analysis and a comparison with previous works (Jordan et al. 2001; Folguera et al. 2003), Pedroza et al. (2017) proposed two deformational phases: The first from ~13 to ~12 Ma, and a second beginning at 10 Ma.

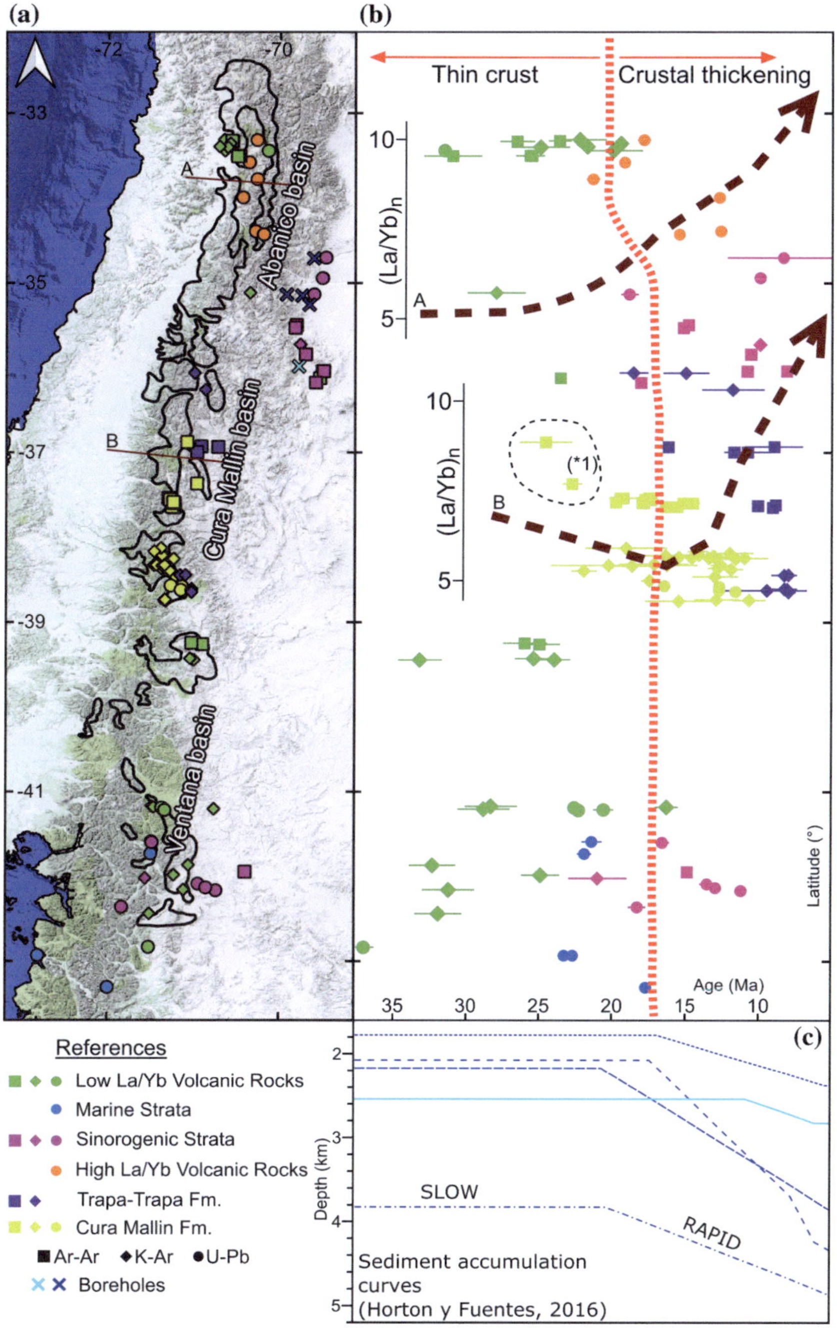
(a)
(b)
-72
-70
-33
-35
-37
-39
-41
A
B
Abanico basin
Cura Mallin basin
Ventana basin
Thin crust
Crustal thickening
(La/Yb)n
10
5
(*1)
Latitude (°)
Age (Ma)
35
30
25
20
15
10
(c)
Depth (km)
2
3
4
5
SLOW
RAPID
Sediment accumulation curves
(Horton y Fuentes, 2016)
References
Low La/Yb Volcanic Rocks
Marine Strata
Sinorogenic Strata
High La/Yb Volcanic Rocks
Trapa-Trapa Fm.
Cura Mallin Fm.
Ar-Ar
K-Ar
U-Pb
Boreholes

◂**Fig. 4** Available radiometric constraints of the sequences in the different depocenters developed between late Eocene and mid-Miocene times in the Southern Central and Northern Patagonian Andes compared with $(La/Yb)_n$ ratios as indicative of Andean crustal thickening. **a** Isotopic age data location, with indicated method over a Google Terrain Map. Black bold polygons represent basin outcrops. Dark-red lines (A and B) represent latitudes for compiled $(La/Yb)_n$ ratios (Kay et al. 2005; Profeta et al. 2015; Piquer et al. 2017). **b** Age versus Latitude diagram showing the approximate shift on tectonic regimes at each latitude (red dotted line) (Drake 1976; Thiele and Cubillos 1980; Muñoz and Niemeyer 1984; Munoz Bravo 1988; Rapela et al. 1988; Suarez and Emparan 1988, 1995, 1997; Cazau et al. 1989; Gana and Wall 1997; Sellés 1999; Vergara et al. 1999, 2004; Jordan et al. 2001; Kay et al. 2005; Silvestro and Kraemer 2005; Burns et al. 2006; Herriott 2006; Silvestro and Atencio 2009; Franzese et al. 2011; Orts et al. 2012; Bilmes et al. 2013; Encinas et al. 2014; Bechis et al. 2014; Ramos et al. 2015; Horton and Fuentes 2016; Pedroza et al. 2017; Piquer et al. 2017; Fernández Paz et al. 2018). Dark-red dashed lines are $(La/Yb)_n$ ratios interpreted as crustal thickness variations through time (Kay et al. 2005; Profeta et al. 2015; Piquer et al. 2017). All age error bars that are larger than the symbol are 1 sigma. (*1); ellipse encloses underestimated ages. **c** Stratigraphic age versus thickness diagram for five sites in the Andean foreland basin (borehole sites in A) (Horton and Fuentes 2016)

4 Oligocene to Miocene Volcanogenic Basins in the Southern Central and Patagonian Andes

4.1 Abanico Basin

Between 33 and 36° S, the Abanico Formation was deposited over the axial Andean zone during the Eocene to the early Miocene (Charrier et al. 2002, 2007) (Fig. 4). This unit is composed of volcaniclastic deposits and acid to intermediate lavas with continental sedimentary intercalations. Geochemical data in intercalated volcanic rocks show that magmas originated through relatively high percentages of partial melting and latter equilibrated and crystallized under low-pressure conditions (i.e., thin crust) (Vergara et al. 2004; Kay et al. 2005; Profeta et al. 2015; Piquer et al. 2017), which is consistent with structural data that point to an extensional setting at that time (Charrier et al. 2002; Piquer et al. 2017). Vergara et al. (2004) constrained this extensional event in Santiago de Chile area (33°–33° 30′ S) from 30.9 to 25.6 Ma, based on K–Ar and Ar–Ar ages. Recently, Piquer et al. (2017) performed an integrated geochronological, geochemical, thermochronological and structural work of the basin, dividing it into a northern and a southern segment, separated by the Piuquncillo fault system, a WNW-ESE structure at 34° S. They performed U–Pb LA-ICP-MS isotopic ages to constraint the Abanico Formation emplacement period between 34 and 21 Ma. In addition, they differentiated the Coya Machalí Formation from the Abanico Formation, being the former younger in age, richer in a volcano-clastic component, and mostly developed in the southern segment of the basin. Then the Coya Machalí Formation was inferred to have deposited between 23 and 12 Ma, with two new U–Pb ages that yielded 15.48 ± 0.29 Ma for the mid-part of the sequence and 12.62 ± 0.35 Ma for the top.

After that, the inversion time of the Abanico Basin has been divided into three main events, one that took place at ~22 Ma affecting the northern basin segment, with a second pulse beginning at ~12 Ma, and a third occurring at ~7 Ma, the last two registered all through the basin (Piquer et al. 2017). In this context, the increase in La/Yb ratios through the sequence is interpreted as related to the contemporaneous crustal thickening coupled with the basin inversion (Profeta et al. 2015) (Fig. 4). The Farellones Formation whose age span is constrained between 22 and 16 Ma, and lies unconformable over the Abanico Formation, is characterized by high $(La/Yb)_n$ ratios (Kay et al. 2005; Piquer et al. 2017). Contemporaneously, synorogenic basins were developed to the east in the foreland zone between 34° and 36° S, such as the Río Grande and Alvear basins and smaller depocenters such as Tralalhué and Puesto Burgos, whose growth strata yielded K–Ar (Silvestro and Kraemer 2005), Ar–Ar (Silvestro and Atencio 2009), and U–Pb (Horton and Fuentes 2016) ages comprehended between 18.5 and 7 Ma. In this sense, subsidence curves at the latitudes of the Abanico Basin indicate an accelerated subsidence starting by 20 Ma, while southern synorogenic deposits at the latitudes of the Cura-Mallín Basin show accelerated subsidence by 10 Ma (Horton and Fuentes 2016).

4.2 Ventana Basin

South of the Cura-Mallín Basin, late Eocene-early Miocene extensional arc-related volcanism is assigned to the Ventana Formation (González Bonorino 1973), as a part of the El Maitén Belt (Rapela et al. 1988) (40–43° S), and the Rancahue Formation (Turner 1965), that constitutes the older infill of the Aluminé Basin (39–40° S) (Folguera et al. 2003; García Morabito and Folguera 2005). These volcanic associations include a ~3 km thick succession of mafic to intermediate volcanic rocks, interbedded with marine volcaniclastic strata (Rapela et al. 1988). The volcanic sources varied from tholeiitic to calc-alkaline as the magmatism evolved trough time (Fernández Paz et al. 2018). The Ventana Formation volcanic rocks also show a marked arc signature, shown by trace elements concentrations and ratios (Fernández Paz et al. 2018). This arc association is characterized by low $(La/Yb)_n$ and Dy/Yb ratios that indicate low-pressure conditions and restricted slab interaction (Fernández Paz et al. 2018). In this context, the marine transgression associated with the volcanism is interpreted as related to crustal stretching (Bechis et al. 2014). The age of this crustal extension event at the latitudes of this basin system is constrained by several isotopic ages to between 37 and 20 Ma (Fig. 4). In particular, Rapela et al. (1988) indicated an Oligocene time span (~33–24 Ma) for the El Maitén Belt based on whole-rock K–Ar analyses. More recently, Ar–Ar ages from the Rancahue Formation ages support an Oligocene age (Franzese et al. 2011), yielding two ages of 26.0 ± 1.5 and 25.0 ± 1.4 Ma for the top of these basaltic sequences. In addition, Bechis et al. (2014) obtained three U–Pb ages between ~22 and 19 Ma in the northern section of the El Maitén Belt (41–42° S). In the same area, these authors dated the Troncoso Formation, obtaining two U–Pb ages of 21.4 ± 0.7 Ma for the basal part of this unit and one of 16.6 Ma for the top of the units. Growth strata allow interpreting them as synextensional (Bechis et al. 2014). Fernández Paz et al. (2018) informed U–Pb age of 37.0 ± 0.7 Ma for the lower section of the Ventana Formation near Esquel (43° S), identifying the beginning of the extensional event at least in the late Eocene.

The subsequent synorogenic sedimentation has been identified through growth strata and dated at 18.5 Ma by Cazau et al. (1989), 18 Ma by Orts et al. (2012), <16 Ma by Bechis et al. (2014) in the Andean hinterland, and at ~13.5 Ma by Ramos et al. (2015) in the foreland area (Fig. 4). Further east, Bilmes et al. (2013) constrained the beginning of the compressive event in the Gastre Basin (42° S), in the easternmost part of the broken foreland zone, to the 16.1–14.86 Ma time period based on new Ar–Ar age for undeformed strata from the Collón Cura Formation, and previous isotopic ages for the underlying deformed La Pava Formation. In the light of these data, some authors proposed that compression and synorogenic sedimentation began in the early Miocene (19–18 Ma) (Orts et al. 2012; Ramos et al. 2015), while others proposed that extension lasted until 16 Ma, whereas compression and synorogenic sedimentation began after that (Bechis et al. 2014).

5 Fission Track Analysis

Low-temperature thermochronological analyses have been carried out in the main Andes at the latitudes of the Neuquen Basin (Burns et al. 2006; Spikings et al. 2008; Glodny et al. 2008) (Fig. 5). Burns et al. (2006) performed apatite and zircon fission-tracks (AFT and ZFT) analyses around 37° S in sedimentary rocks of the Cura-Mallín Formation and intrusive bodies in the eastern slope of the Andes. Unfortunately, only one sample on a granite with ZFT and AFT analyses yielded useful data. Assuming a normal geothermal gradient of 30 °C/km, ~2–3.6 km of exhumation is inferred in the last 8.9 Ma, which is equivalent to an exhumation rate of 0.225–0.404 km/My. In a more regional trend, Glodny et al. (2008) and Spikings et al. (2008) carried out thermochronological analyses on the Cretaceous to Miocene North Patagonian Batholith in the western Main Andes between 35° and 40° S, and also in the Coastal Cordillera, constituting the basement of the Cura-Mallín Basin. For the Cretaceous granites of the Main Andes (K–Ar ages between 123 and 83 Ma), Glodny et al. (2008) obtained AFT ages of 34.3 ± 4.6 and 40.5 ± 6.2 Ma. This implies ~3 km of exhumation for the Cretaceous intrusives since the Paleogene with a normal average geothermal gradient (30°/km). Whether the exhumation was slow and constant since Paleogene times or rapid during a shorter period is not clear with the available data. In addition, Spikings et al. (2008) noticed a possible exhumation pulse around 18 Ma for the Miocene granites, although they were unable to discriminate among exhumation and sub-solidus thermal relaxation. In spite of these uncertainties, both works reported elevated exhumation rates since ~8 Ma, with an exhumation increase starting at 5 Ma and an N-S exhumation gradient between 35 and 39° S, reaching in the northern sector almost 5 km and in the southern sector less than 1 km (Fig. 5). Further

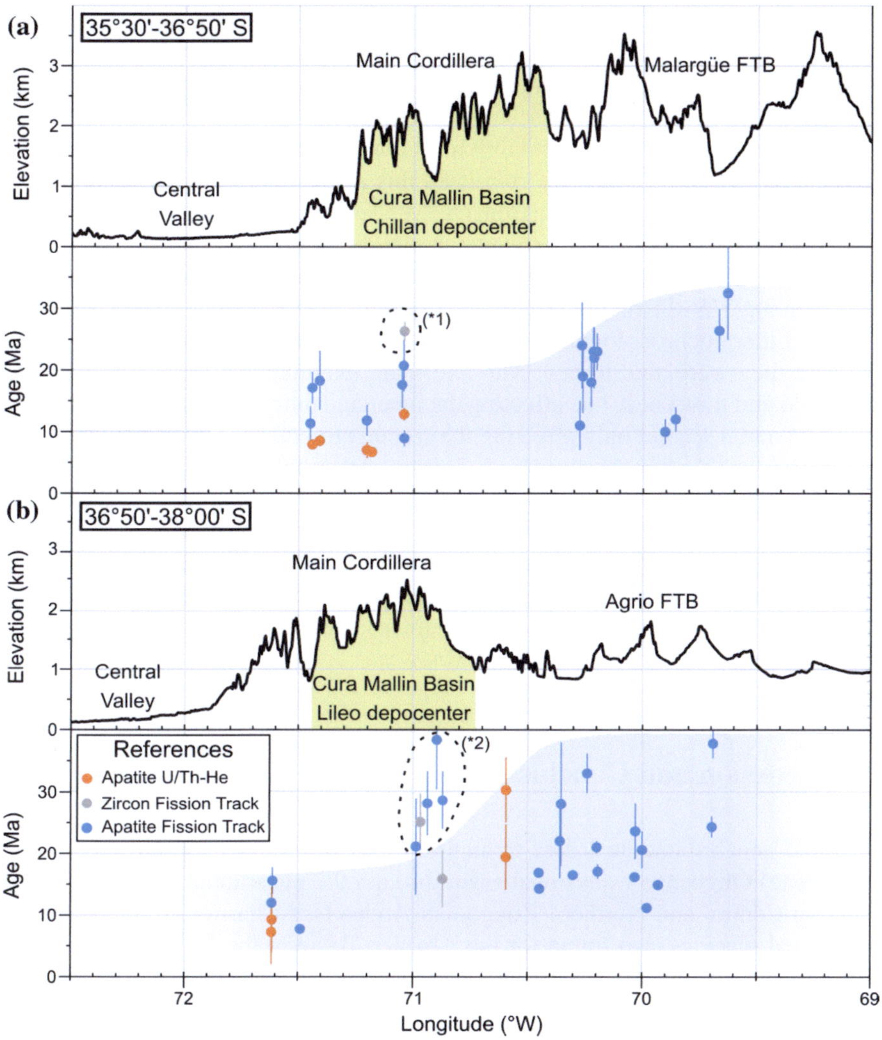

Fig. 5 East-west transects across the Main Andes and back-arc zone at different latitudes showing available thermochronological constraints and their spatial relationship with the Cura-Mallín Formation outcrops, CZ: Chillan zone, LZ: Lileo zone. Gray areas indicate enhanced exhumation periods; note the difference between the Main Cordillera and the back-arc area (Muñoz 1996; Burns et al. 2006; Spikings et al. 2008; Glodny et al. 2008; Folguera et al. 2015; Rojas Vera et al. 2015; Sánchez et al. 2018). The (*1) and (*2) ellipses indicate underestimated ages by Burns et al. (2006) (see Burns et al 2006, for further details). All age error bars that are larger than the symbol are one sigma. Topography has a 10:1 of vertical exaggeration.

east, in the Malargüe, Chos Malal and Agrio fold and thrust belts thermochronological analyses had been carried out in the last years (Muñoz 1996; Folguera et al. 2015; Rojas Vera et al. 2015; Sánchez et al. 2018) (Fig. 5). Folguera et al. (2015) analyzed five W-E structural transects between 36° and 37° 30′ S and presented AFT thermochronological data determining a regional exhumation event in the middle Miocene (15–10 Ma). Then they correlated this age with the angular unconformity that separates the Paleogene strata from late Miocene volcanic rocks. Similarly, Rojas Vera et al. (2015) reported high exhumation rates between 15 and 10 Ma for the Chos Malal and Agrio fold and thrust belts between 37° and 38° S, based on 9 AFT distributed along 3 W-E structural transects. Recently, Sánchez et al. (2018) performed a detailed thermochronological and structural survey on a W-E transect at 37°15′ S, where they were able to constrain two main exhumational stages on the Chos Malal fold and thrust belt, one affecting the inner and outer sectors of the belt during 15–7 Ma, and a second only affecting the frontal structures at ~9–7 Ma (see Chap. "Structural and Thermochronological Constraints on the Exhumation of the Chos Malal Fold and Thrust Belt ~37° S"). All the thermochronological data focused on the foreland FTBs in the Argentinean slope agreed in an exhumation event around 15 Ma, which is consistent with the age of the synorogenic sedimentation in the area; while data from the western Cretaceous to Miocene granites on the Chilean slope indicate a younger ~9–8 Ma age for enhanced exhumation, with a previous not differentiable phase most likely at ~18 Ma.

6 Discussion and Conclusions

Field and geochronological data through the last years have allowed recognizing an (Eocene) Oligocene extensional event through the transitional zone between the Southern Central and Northern Patagonian Andes (see Charrier et al. 2007). This extensional phase is also found for the Abanico Basin to the Ventana Basin infills between ~34/37 and ~21/16 Ma (Rapela et al. 1988; Kay et al. 2005; Charrier et al. 2007; Bechis et al. 2014; Piquer et al. 2017; Fernández Paz et al. 2018). The subsequent inversion of these basins started at around ~20 to ~17 Ma and lasted until ~10 to ~5 Ma, based on growth strata recognition (Silvestro and Kraemer 2005; Silvestro and Atencio 2009; Horton and Fuentes 2016), $(La/Yb)_n$ ratios (Piquer et al. 2017), thermochronological analysis (Orts et al. 2012; Bilmes et al. 2013; Bechis et al. 2014; Ramos et al. 2015; Folguera et al. 2015; Rojas Vera et al. 2015; Sánchez et al. 2018), and subsidence analysis (Horton and Fuentes 2016).

The strongest evidence supporting an extensional origin for the Cura-Mallín Formation rocks are the wedge-like geometries analyzed in seismic profiles (Jordan et al. 2001; Folguera et al. 2010; Rojas Vera et al. 2010). However, no age control exists on these buried sequences since there are no wells drilled in this specific area, located in the northern Cura-Mallín Basin. In addition, Fig. 4 shows that most of the Cura-Mallín Formation ages range between ~20 and ~11 Ma, suggesting that it is unlikely that these were entirely accumulated under extensional conditions, since at least part of them temporarily coincide with synorogenic strata (~18 to ~10 Ma) in nearby foreland basins and compressionally related volcanic rocks lying in the Andean axis (~20 to ~14 Ma). Moreover, in the Andacollo area, strata of the Lileo Formation were interpreted as growth strata (Cobbold et al. 2008), whose isotopic ratios suggested crustal thickening at the time of deposition (Kay et al. 2006; Profeta et al. 2015). Consequently, based on the available evidence, at least the upper part of the Cura-Mallín Formation and equivalents (Fig. 2) seem to be related to compressional, rather than extensional conditions. Contrastingly, wedge-like geometries most likely corresponding to the lower units described in seismic profiles are assigned to extensional conditions (Jordan et al. 2001; Folguera et al. 2010; Rojas Vera et al. 2010).

Acknowledgement This study has been funded by PICT-2016-12252, FONDECYT Project 1151146, PIP 11220150100426 and UBACYT 20020150100166BA. This is the R-315 contribution of the Instituto de Estudios Andinos "Don Pablo Groeber".

References

Bechis F, Encinas A, Concheyro A et al (2014) New age constraints for the Cenozoic marine transgressions of northwestern Patagonia, Argentina (41°–43° S): paleogeographic andtectonic implications. J S Am Earth Sci 52:72–93

Bilmes A, D'Elia L, Franzese JR et al (2013) Miocene block uplift and basin formation in the Patagonian foreland: The Gastre Basin, Argentina. Tectonophysics 601:98–111

Burns WM, Jordan TE, Copeland P, Kelley SA (2006) The case for extensional tectonics in the Oligocene-Miocene Southern Andes as recorded in the Cura-Mallín basin (36°–38°S). In: Ramos VA, Kay SM (eds) Evolution of an Andean margin: a tectonic and magmatic view from the Andes to the Neuquén Basin (35°-39°S lat). Geological Society of America, SP 407, pp 163–184

Cazau L, Mancini D, Cangini J, Spalletti LA (1989) Cuenca de Ñirihuau. Cuencas sedimentarias argentinas. Universidad de Tucumán, Serie Correlación Geológica 6:299–318

Charrier R, Baeza O, Elgueta S et al (2002) Evidence for Cenozoic extensional basin development and tectonic inversion south of the flat-slab segment, southern Central Andes, Chile (33°–36°S.L.). J S Am Earth Sci 15:117–139

Charrier R, Pinto L, Rodriguez PM (2007) Tectonostratigraphic evolution of the Andean Orogen in Chile. In: Moreno T, Gibbons W (eds) The geology of Chile. The Geological Society, London, pp 21–114

Cobbold PR, Rossello EA (2003) Aptian to recent compressional deformation, foothills of the Neuquén Basin, Argentina. Mar Pet Geol 20:429–443

Cobbold PR, Rossello AE, Marques OF et al (2008) Where is the evidence for Oligocene rifting in the Andes? Is it in the Loncopué Basin of Argentina? In: Abstracts of the 7 international symposium on Andean geodynamics, Nice

Drake RE (1976) Chronology of Cenozoic igneous and tectonic events in the central chilean Andes—Latitudes 35° 30′ to 36° S. J Volcanol Geotherm Res 1:265–284

Encinas A, Pérez F, Nielsen SN et al (2014) Geochronologic and paleontologic evidence for a pacific-atlantic connection during the late oligocene-early miocene in the patagonian andes (43–44° S). J S Am Earth Sci 55:1–18

Encinas A, Folguera A, Oliveros V et al (2016) Late Oligocene-early Miocene submarine volcanism and deep-marine sedimentation in an extensional basin of southern Chile: Implications for the tectonic development of the North Patagonian Andes. Geol Soc Am Bull 128:807–823

Fennell LM, Quinteros J, Iannelli SB et al (2018) The role of the slab pull force in the late Oligocene to early Miocene extension in the Southern Central Andes (27°–46° S): insights from numerical modeling. J S Am Earth Sci 87:174–187

Fernández Paz L, Litvak VD, Echaurren A et al (2018) Late Eocene volcanism in North Patagonia (42° 30′–43° S): arc resumption after a stage of within-plate magmatism. J Geodyn 113:13–31

Flynn JJ, Charrier R, Croft DA et al (2008) Chronologic implications of new Miocene mammals from the Cura-Mallín and Trapa Trapa formations, Laguna del Laja area, south central Chile. J S Am Earth Sci 26:412–423

Folguera A, Ramos VA, Melnick D (2003) Recurrencia en el desarrollo de cuencas de intraarco Cordillera Neuquina (37°30′–38°S). Rev Asoc Geol Argent 58:3–19

Folguera A, Ramos VA, Zapata T, Spagnuolo MG (2007) Andean evolution at the Guañacos and Chos Malal fold and thrust belts (36° 30′–37° s). J Geodyn 44:129–148

Folguera A, Rojas Vera EA, Bottesi G et al (2010) The Loncopué Trough: a Cenozoic basin produced by extension in the southern Central Andes. J Geodyn 49:287–295

Folguera A, Bottesi G, Duddy I et al (2015) Exhumation of the Neuquén Basin in the southern Central Andes (Malargüe fold and thrust belt) from field data and low-temperature thermochronology. J S Am Earth Sci 64:381–398

Franzese JR, D'Elia L, Bilmes A et al (2011) Superposición de cuencas extensionales y contraccionales oligo-miocenas en el retroarco andino norpatagónico: La cuenca de aluminé, neuquén, Argentina. Andean Geol 38:319–334

Gana P, Wall R (1997) Evidencias geocronológicas 40Ar/39 Ar y K-Ar de un hiatus cretácico superior-eoceno en Chile central (33–33°30′). Rev Geol Chile 24:145–163

García Morabito E, Folguera A (2005) El alto de Copahue-Pino Hachado y la fosa de Loncopué: un comportamiento tectónico episódico, Andes neuquinos (37 -39° S). Rev Asoc Geol Argent 60:742–761

García Morabito E, Ramos VA (2012) Andean evolution of the Aluminé fold and thrust belt, Northern Patagonian Andes (38° 30′–40° 30′ S). J S Am Earth Sci 38:13–30

Glodny J, Gräfe K, Echtler H, Rosenau M (2008) Mesozoic to Quaternary continental margin dynamics in South-Central Chile (36–42°S): the apatite and zircon fission track perspective. Int J Earth Sci 97:1271–1291

Godoy E, Yañez G, Vera E (1999) Inversion of an Oligocene volcano-tectonic basin and uplifting of its superimposed Miocene magmatic arc in the Chilean Central Andes: first seismic and gravity evidences. Tectonophysics 306:217–236

González Bonorino F (1973) Geología del área entre San Carlos de Bariloche y Llao-Llao, provincia de Río Negro. Publ Dep Recur Nat y Energía, Fund Bariloche, p 16

Herriott TM (2006) Stratigraphy, structure, and 40Ar/39Ar geochronology of the southeastern Laguna del Laja area: Implications for the mid-late Cenozoic evolution of the Andes near 37.5° S, Chile, University of California

Horton BK, Fuentes F (2016) Sedimentary record of plate coupling and decoupling during growth of the Andes. Geology 44:647–650

Iannelli SB, Litvak VD, Fernández Paz L et al (2017) Evolution of Eocene to Oligocene arc-related volcanism in the North Patagonian Andes (39–41° S), prior to the break-up of the Farallon plate. Tectonophysics 696–697:70–87

Jordan TE, Burns WM, Veiga R et al (2001) Extension and basin formation in the southern Andes caused by increased convergence rate. Tectonics 20:308–424

Kay SM, Godoy E, Kurtz A (2005) Episodic arc migration, crustal thickening, subduction erosion, and magmatism in the south-central Andes. Geol Soc Am Bull 117:67–88

Kay SM, Burns WM, Copeland P, Mancilla O (2006) Upper Cretaceous to Holocene magmatism and evidence for transient Miocene shallowing of the Andean subduction zone under the northern Neuquén Basin. In: Kay SM, Ramos VA (eds) Evolution of the Andean Margin. Geological Society of America, SP 407, pp 19–60

Mosolf JG, Gans PB, Wyss AR et al (2018) Late Cretaceous to Miocene volcanism, sedimentation, and upper-crustal faulting and folding in the Principal Cordillera, central Chile: Field and

geochronological evidence for protracted arc volcanism and transpressive deformation. Geol Soc Am Bull 131(1–2):252–273
Muñoz N (1996) The thermal evolution of Jurassic and Cretaceous source rocks in the Malargüe thrust belt, Argentina: implications for hydrocarbon exploration. Independent project report, Royal Hollow University London 98 p
Munoz Bravo JO (1988) Evolution of Pliocene and Quaternary volcanism in the segment of the southern Andes between 38 and 39 S, University of Colorado
Muñoz J, Niemeyer H (1984) Hoja Laguna del Maule: regiones del Maule y del BioBio: carta geológica de Chile 1: 250.000. Servicio Nacional de Geología y Minería, Santiago de Chile
Munoz J, Troncoso R, Duhart P et al (2000) The relation of the mid-Tertiary coastal magmatic belt in south-central Chile to the late Oligocene increase in plate convergence rate. Rev Geol Chile 27:177–203
Niemeyer H, Muñoz J (1983) Hoja Laguna de La Laja: región de Bio Bio: carta geológica de Chile 1: 250.000. Servicio Nacional de Geología y Minería, Santiago de Chile
Orts DL, Folguera A, Encinas A et al (2012) Tectonic development of the North Patagonian Andes and their related Miocene foreland basin (41°30'-43°S). Tectonics 31. https://doi.org/10.1029/2011tc003084
Orts DL, Folguera A, Giménez M et al (2015) Cenozoic building and deformational processes in the North Patagonian Andes. J Geodyn 86:26–41
Pananont P, Mpodozis C, Blanco N et al (2004) Cenozoic evolution of the northwestern Salar de Atacama Basin, northern Chile. Tectonics 23:1–19
Pedroza V, Le Roux JP, Gutiérrez NM, Vicencio VE (2017) Stratigraphy, sedimentology, and geothermal reservoir potential of the volcaniclastic Cura-Mallín succession at Lonquimay, Chile. J S Am Earth Sci 77:1–20
Pesce AH (1981) Estratigrafía de las nacientes del río Neuquén y Nahuever, Provincia del Neuquén. In: Abstracts of the 8 Congreso Geológico Argentino, San Luis, 20–26 Sept 1981
Piquer J, Hollings P, Rivera O et al (2017) Along-strike segmentation of the Abanico Basin, central Chile: new chronological, geochemical and structural constraints. Lithos 268–271:174–197
Profeta L, Ducea MN, Chapman JB et al (2015) Quantifying crustal thickness over time in magmatic arcs. Sci Rep 5. https://doi.org/10.1038/srep17786
Radic JP (2010) Las cuencas terciarias y su control en el volcanismo de los complejos chillan y copahue-callaqui (andes del sur 36°–39 s). Andean Geol 37:220–246
Ramos ME, Tobal JE, Sagripanti L et al (2015) The North Patagonian orogenic front and related foreland evolution during the Miocene, analyzed from synorogenic sedimentation and U/Pb dating (~42° S). J S Am Earth Sci 64:467–485
Rapela CW, Spalletti LA, Merodio JC, Aragón E (1988) Temporal evolution and spatial variation of early tertiary volcanism in the Patagonian Andes (40° S–42°30′ S). J South Am Earth Sci 1:75–88
Rojas Vera EA, Folguera A, Valcarce GZ et al (2010) Neogene to Quaternary extensional reactivation of a fold and thrust belt: The Agrio belt in the Southern Central Andes and its relation to the Loncopué trough (38°–39° S). Tectonophysics 492:279–294
Rojas Vera EA, Mescua J, Folguera A et al (2015) Evolution of the Chos Malal and Agrio fold and thrust belts, Andes of Neuquén: Insights from structural analysis and apatite fission track dating. J S Am Earth Sci 64:418–433
Rovere EI, Caselli A, Tourn S et al (2004) Hoja Geológica 3772-IV, Andacollo, provincia del Neuquén. Instituto de Geologia y Recursos Minerales. Servicio Geologico Minero Argentino, Buenos Aires
Rubilar J, Martínez F, Arriagada C et al (2017) Structure of the Cordillera de la Sal: A key tectonic element for the Oligocene-Neogene evolution of the Salar de Atacama basin, Central Andes, northern Chile. J S Am Earth Sci 87:200–210
Salinas P (1979) Geología del área Lolco-Lonquimay. Cordillera de los Andes, Alto Bio-Bio, IX Región, Chile. Memoria de título. Universidad de Chile. Departamento de Geología, 153 p

Sánchez NP, Coutand I, Turienzo M et al (2018) Tectonic evolution of the Chos Malal fold-and-thrust belt (Neuquén Basin, Argentina) from (U-Th)/He and fission track thermochronometry. Tectonics 37:1907–1929

Sellés D (1999) La Formación Abanico en el Cuadrángulo Santiago (33° 15′–33° 30′ S; 70°30′–70° 45′ O). PhD Thesis, Universidad de Chile

Silvestro J, Atencio M (2009) La cuenca cenozoica del Río Grande y Palauco: Edad, evolución y control estructural, faja plegada de Malargüe (36° S). Rev Asoc Geol Argent 65:154–169

Silvestro J, Kraemer P (2005) Evolución Tecto-Sedimentaria de la Cordillera Principal en el Sector Surmendocino a los 35°–30° S. Faja Plegada de Malargüe, Republica Argentina. Rev Asoc Geol Argent 60:627–643

Spikings R, Dungan M, Foeken J et al (2008) Tectonic response of the central Chilean margin (35–38 S) to the collision and subduction of heterogeneous oceanic crust: a thermochronological study. J Geol Soc London 165:941–953

Suarez M, Emparan C (1988) Geocronología y asociación de facies volcánicas y sedimentarias del Mioceno de Lonquimay, Chile (lat. 38–39 S). In: Abstracts of the 5 Congreso Geológico Chileno, Santiago de Chile, 8–12 Aug 1988

Suarez M, Emparan C (1997) Hoja Curacautin, Regiones de la Araucania y del Biobio. Carta Geológica Chile, No. 71, Servicio N. Servicio Nacional de Geología y Minería., Santiago de Chile

Suarez M, Emparan C (1995) The stratigraphy, geochronology and paleophysiography of a Miocene fresh-water interarc basin, southern Chile. J S Am Earth Sci 8:17–31

Thiele R, Cubillos E (1980) Hoja Santiago: región metropolitana: carta geológica de Chile escala 1: 250.000, Instituto de Investigaciones Geológicas

Turner JC (1965) Estratigrafía de Aluminé y adyacencias, provincia de Neuquén. Rev Asoc Geol Argent 20:153–184

Utgé S, Folguera A, Litvak V, Ramos VA (2009) Geología del sector norte de la cuenca de Cura-Mallín en las lagunas de epulaufquen, neuquén. Rev Asoc Geol Argent 64:231–248

Vergara M, Morata D, Villarroel R et al (1999) 40Ar/39Ar ages, very low grade meta morphism and geochemistry of the volcanic rocks from "Cerro El Abanico, Santiago Andean Cordillera (33°30′ S–70° 25′ W). In: Abstracts of the international symposium on Andean geodynamics

Vergara M, López-Escobar L, Palma JL et al (2004) Late tertiary volcanic episodes in the area of the city of Santiago de Chile: new geochronological and geochemical data. J S Am Earth Sci 17:227–238

Winocur DA, Litvak VD, Ramos VA (2015) Magmatic and tectonic evolution of the Oligocene Valle del Cura basin, main Andes of Argentina and Chile: evidence for generalized extension. The Geological Society, London SP 399:109–130

Zanettini JCM (2001) Hoja Geológica 3772-II, Las Ovejas, provincia del Neuquén. Instituto de Geología y Recursos Minerales, Servicio Geológico Minero Argentino, Buenos Aires

The Miocene Magmatism in the Malargüe and Chos Malal Fold and Thrust Belts

Vanesa D. Litvak, Sofía B. Iannelli, Lucía Fernández Paz and Andrés Folguera

Abstract Late Oligocene to Pliocene magmatism at the latitudes of Malargüe and Chos Malal fold and thrust belts, in the Neuquén Basin, is distributed from the main Andean axis to the retroarc zone. While arc magmatism maintained relatively similar compositional and geochemical features during late Oligocene to Pliocene times, major variations are seen in volcanic sequences developed in the retroarc zone, due to the development of a shallow subduction regime by mid-late Miocene times. Thus, late Oligocene-early Miocene period is characterized by an extensional regime that conditioned mainly tholeiitic magmas in the main Andean axis and alkaline intraplate magmas in the retroarc zone. Early-late Miocene marks a change to compressional tectonics in the Andean margin. Main arc magmatism showed a change to clearer arc, calc-alkaline signature, while retroarc magmatism showed the progressing input of slab-derived products as the shallow subduction regime triggered the eastward migration of the asthenospheric wedge. Thus, arc-derived lavas expanded into the mid to far retroarc zone (~69° 30″ to 68° 30″) with arc-like andesitic to dacitic compositions. With the progressive influence of the shallow subduction regime, arc-derived products reached almost 500 km away from the Chilean trench at ∽36° S by latest Miocene-Pliocene. By the middle Pliocene, re-steepening of the slabconditioned extensional tectonics that favored a widespread alkaline volcanism in the present-day Payenia retroarc and rhyolitic calderas in the main arc zone. After ∽3.5 Ma, retroarc magmatism at the latitudes of the Malargüe and Chos Malal fold and thrust belts lack of clear arc-related geochemical features.

Keywords Miocene magmatism · San Rafael Block · Volcanic arc · Payenia · Shallow subduction

---

V. D. Litvak (✉) · S. B. Iannelli · L. F. Paz · A. Folguera
Universidad de Buenos Aires, Facultad de Ciencias Exactas y Naturales, Departamento de Ciencias Geológicas, Buenos Aires, Argentina
e-mail: vane@gl.fcen.uba.ar

CONICET-Universidad de Buenos Aires, Instituto de Estudios Andinos "Don Pablo Groeber" (IDEAN), Buenos Aires, Argentina

D. Kietzmann and A. Folguera (eds.), *Opening and Closure of the Neuquén Basin in the Southern Andes*, Springer Earth System Sciences,
https://doi.org/10.1007/978-3-030-29680-3_19

1 Introduction

The main controls in the distribution of Neogene magmatism within the Neuquén Basin are related to particular changes in the subduction zone geometry through time. In this sense, arc-related magmas developed from the main Andean axis to the foreland zone, as a result of a shallow subduction regime that started by mid-late Miocene times at the latitudes of Malargüe and Chos Malal fold and thrust belts (FTBs) (~34°–37° S; Fig. 1) (Kay et al. 2006a, b; Dyhr et al. 2013a; Folguera et al. 2011; Ramos et al. 2014; Litvak et al. 2015; Spagnuolo et al. 2012). Over time, these magmas show a zonal distribution relative to the trench, as well as contrasting compositional features, which reflect the progressive variable input of slab-derived components registered from the arc to the far retroarc zone during late Miocene times.

These changes in Neogene arc magmatism correlate with distinct deformational stages, which include variable cycles of shallowing and steepening of the slab (Spagnuolo et al. 2012; Ramos et al. 2014; Ramos and Kay 2006). For example, late Oligocene-early Miocene stage is mostly controlled by extensional tectonics that conditioned arc magmatism in the main Andean axis and alkaline magmas in the retroarc zone (Kay et al. 2006b; Dyhr et al. 2013b; Ramos et al. 2014) (see Chapter "The Miocene Sequences Cropping Out in the High Andes of Neuquén and Their Relation to Andean Orogenesis"). However, by the latest early Miocene to late Miocene times, a stage of compressional tectonics promoted migration and expansion of the arc front and subsequent contractional deformation in the foreland zone, which was linked to the development of the late Miocene Payenia shallow subduction regime (Kay et al. 2006a, b; Ramos et al. 2014; Litvak et al. 2015). Finally, the re-steepening of the slab by early Pliocene times controlled the emplacement of rhyolitic calderas in the cordilleran axis, while large alkaline basaltic volcanic flows developed in the retroarc zone under extensional conditions (e.g., Llambías et al. 2010; Gudnason et al. 2012; Ramos et al. 2014; Søager et al. 2013; Hernando et al. 2014) (see Chapter "Plume Subduction Beneath the Neuquén Basin and the Last Mountain Building Stage of the Southern Central Andes").

This work summarizes the evolution of Miocene magmatism developed through the Neuquén Basin, strongly controlled by the variable geodynamic setting during that time. For this purpose, we compiled geochemical and geochronological data of late Oligocene to Pliocene arc-related magmas and associated alkaline magmatism developed from the main arc zone to the retroarc area between ~34° and 37° S. Variations in the geochemical signature of the arc and retroarc products, together with the contrasting distribution of the analyzed magmatism through time, are good indicators of the changing geodynamic conditions at the studied latitudes.

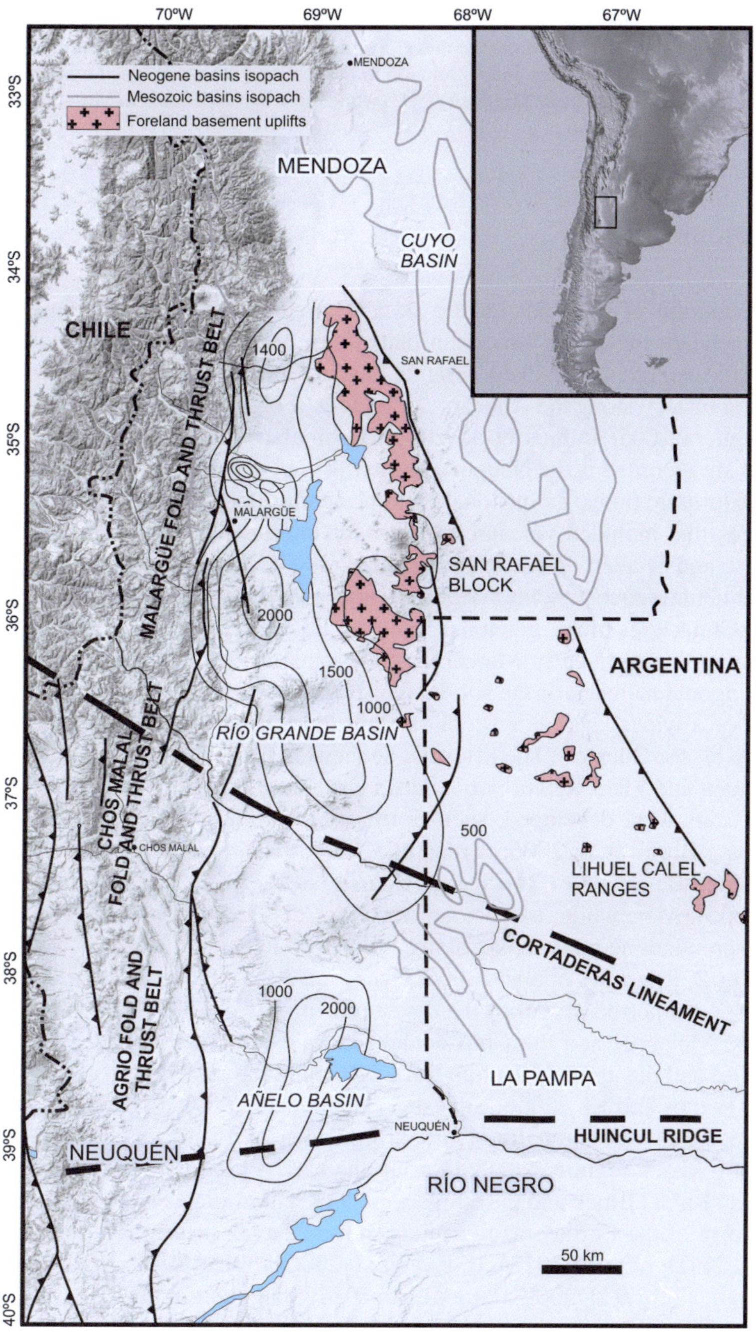
70°W
69°W
68°W
67°W
33°S
34°S
35°S
36°S
37°S
38°S
39°S
40°S
Neogene basins isopach
Mesozoic basins isopach
Foreland basement uplifts
MENDOZA
MENDOZA
CUYO BASIN
CHILE
MALARGÜE FOLD AND THRUST BELT
1400
SAN RAFAEL
MALARGÜE
SAN RAFAEL BLOCK
2000
1500
1000
ARGENTINA
RÍO GRANDE BASIN
CHOS MALAL FOLD AND THRUST BELT
CHOS MALAL
500
LIHUEL CALEL RANGES
CORTADERAS LINEAMENT
AGRIO FOLD AND THRUST BELT
1000
2000
LA PAMPA
AÑELO BASIN
NEUQUÉN
HUINCUL RIDGE
NEUQUÉN
RÍO NEGRO
50 km

◂**Fig. 1** Regional setting of the Malargüe and Chos Malal fold and thrust belts with location of the middle to late Miocene structure and its relation to the foreland basins. Isopach maps of Neogene synorogenic sequences are from Yrigoyen (1993) while those corresponding to Mesozoic basins are from Ramos (1998), Vergani et al. (1995), Alvarez and Ramos (1999) and Giambiagi et al. (2005, 2008)

2 Tectonic Setting

The subduction of different Pacific oceanic plates (e.g., Catequil, Farallon, and Nazca) beneath the South American plate since Mesozoic times, with contrasting ages and kinematics, provoked variable changes in the tectonic, sedimentary, and magmatic history along the Andean margin (e.g., Folguera and Ramos 2011; Ramos and Folguera 2005; Ramos et al. 2014). Particularly, a key area where all these changes are recorded is the Neuquén Basin (34° 30′–38° S). During Middle Triassic to Early Jurassic times, extensional tectonics controlled the development of synrift sequences that included volcanic products of intermediate to silicic compositions (Franzese and Spalletti 2001; Llambías et al. 2007). Since then, alternating marine and continental sequences were deposited until the beginning of the Andean orogeny by Late Cretaceous times, when the Neuquén Basin turned into a foreland basin.

By late Oligocene-early Miocene times, a major tectonic change triggered the steepening and rollback of the subducting Nazca plate and consequently the development of a widespread extensional setting (e.g., Cande and Leslie 1986; Jordan et al. 2001) (see Chapter "The Miocene Sequences Cropping Out in the High Andes of Neuquén and Their Relation to Andean Orogenesis"). Along the arc zone, extensional basins were developed, such as the intra-arc Cura Mallín Basin (~36°–38° 30′ S; e.g., Burns et al. 2006; Jordan et al. 2001; Suárez and Emparan 1995); and the Abanico Basin (~30°–34° S; e.g., Charrier et al. 2002; Godoy et al. 1999; Kay et al. 2005). Meanwhile, extensional tectonics also conditioned the emplacement of back-arc alkaline magmatism along the Sierra de Huantraico and La Matancilla areas (Fig. 2; Kay and Copeland 2006; Dyhr et al. 2013a, b). Later, a middle Miocene compressional period provoked the development of two east-verging mountain systems: the Malargüe and the Chos Malal fold and thrust belts (Fig. 1; Ramos and Kay 2006) and the uplift of the San Rafael Block in the eastern broken foreland area (Fig. 1; Ramos and Kay 2006; Silvestro and Atencio 2009). Toward the mid to late Miocene, a shallow subduction regime triggered an eastward arc expansion, while a renewed rollback regime installed during the early Pliocene provoked the collapse of the San Rafael Block and the emplacement of within-plate lava flows, constrained to the north of the Cortaderas lineament in the Payenia present-day retroarc region (Figs. 1, 2) (Dyhr et al. 2013a, b; Kay et al. 2006a, b; Litvak et al. 2015; Llambías et al. 2010; Nullo et al. 2002; Ramos et al. 2014; Spagnuolo et al. 2012) (see Chapter "Plume Subduction Beneath the Neuquén Basin and the Last Mountain Building Stage of the Southern Central Andes").

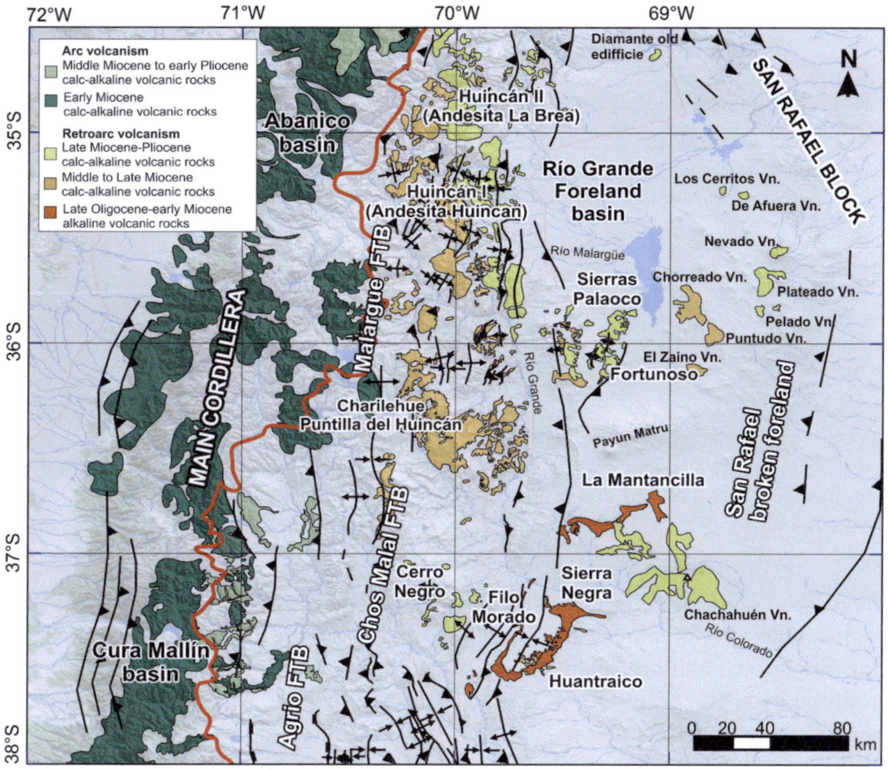

Fig. 2 Location of main arc and retroarc volcanism along the Malargüe and Chos Malal fold and thrust belts. Late Oligocene to early Miocene alkaline volcanism is distributed in the retroarc area; while from early Miocene to Pliocene, calc-alkaline series expanded from the near (~70° W) to the middle (~69° 30′ W) and into the far (~68° 30′ W) retroarc zones. Based on Charrier et al. (2002), Dyhr et al. (2013a, b), Folguera et al. (2009), Kay et al. (2005, 2006a, b), Kay and Copeland (2006), Litvak et al. (2015, 2018), Piquer et al. (2017), Ramos et al. (2014), and Spagnuolo et al. (2012)

3 Distribution and Age of the Neogene Magmatism in the Malargüe and Chos Malal FTBs

Late Oligocene to Pliocene magmatism along the Malargüe and Chos Malal fold and thrust belts is located from the present-day main arc to the far retroarc zone (Fig. 2). Main arc magmatism developed uninterruptedly in the Andean axis within the studied time span (∽25–2.5 Ma), while arc-related products expanded during middle Miocene to early Pliocene from the near (∽70° W), to the middle (∽69° 30 W) till the far (∽68° 30 W) retroarc zone, reaching the San Rafael Block area. In the westernmost retroarc zone, most of the ages are between 19 and 10 Ma; while to the east, they range from 14 to 4 Ma (Fig. 2).

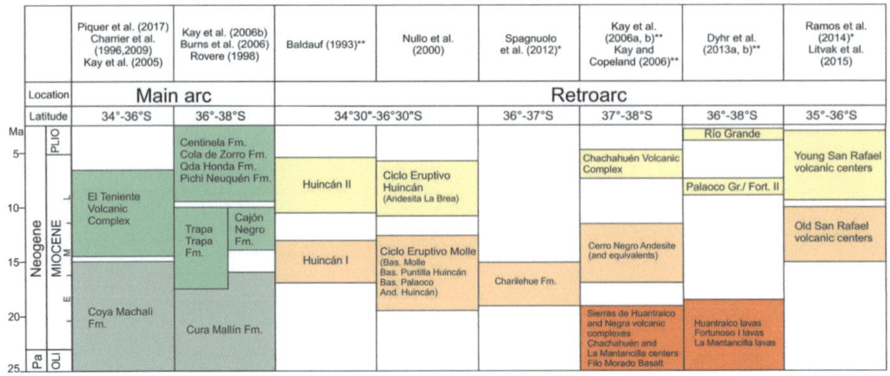

Fig. 3 Neogene stratigraphic scheme of main arc and retroarc magmatism along the Malargüe and Chos Malal fold and thrust belts. Retroarc volcanism is differentiated according the variable nomenclature and life span proposed from the mentioned authors along their studied locations. Age spans comprise a compilation of ages reported by each author; particularly, new reported ages correspond to * K/Ar and ** Ar/Ar.

Over the years, age constraints for arc magmatism were synthetized in a relatively homogenous stratigraphical scheme, while retroarc magmatism was grouped in several formational and informational units according to the study areas and life span proposed by different authors (Kay et al. 2006a, b; Kay and Copeland 2006; Baldauf 1993; Baldauf 1997; Combina and Nullo 2000; Nullo et al. 2002; Spagnuolo et al. 2012; Dyhr et al. 2013a, b; Ramos et al. 2014; Litvak et al. 2015). Figure 3 shows a compilation of the variable nomenclature and proposed life span both for the main arc and retroarc magmatism developed along the Malargüe and Chos Malal fold and thrust belts (∽34° 30′–37° 30′ S), based mostly on reported K/Ar ages and more recent Ar/Ar ones.

As mentioned, retroarc magmatism shows a more complex and variable nomenclature than the main arc; more importantly, the variable composition, distribution, and age of these retroarc sequences reflect major geodynamic changes in the Andean margin during the late Cenozoic. Therefore, in order to avoid nomenclature problems, the evolution of retroarc volcanism along the Malargüe and Chos Malal FTBs will be analyzed based on three volcanic stages: (1) late Oligocene-early Miocene (~25–18 Ma), (2) middle Miocene-late Miocene (~18–10 Ma), and (3) latest Miocene-Pliocene (~10–2.5 Ma) (Figs. 2, 3).

Along the Malargüe and Chos Malal FTBs latitudes, frontal arc magmas within the analyzed lifespan are particularly distributed in the southern part of the Abanico (∽34° 30′–35° 30′ S) and the northern part of Cura Mallín (∽35° 30′–37° 30′ S) basins (Fig. 2). In the Abanico Basin, south of 34° S, early Miocene to middle Miocene volcanic activity is grouped within the Coya Machalí Formation (∽23–15 Ma), which comprises volcanic and volcaniclastic deposits interbedded with continental sedimentary rocks (Kay and Kurtz 1995; Kay et al. 2005; Charrier et al. 2009, 1996; Piquer et al. 2017). Coeval main arc activity in the Cura Mallín Basin is formally included in the Cura Mallín Formation with an overall life span of ~25–15 Ma; particularly, deposits in the Argentinean Andean slope include basaltic to andesitic lava flows and silicic pyroclastic rocks interbedded with minor deltaic and lacustrine sequences with ages between ∽24 and 20 Ma (Suárez and Emparan 1995, Jordan et al. 2001, Burns et al. 2006; Utgé et al. 2009).

Middle Miocene to late Miocene main arc magmatic activity crops out mostly near the international limit and the Chilean Andean slope included in the Trapa-Trapa Formation (∼37°–37° 30′ S; 18–10 Ma), which comprises volcanic rocks and associated subvolcanic intrusives (Kay et al. 2006b; Rovere 1998). Partially coeval, the Cajón Negro Formation (∼36° 30′ S) includes basaltic andesitic to dacitic lavas, surge, and pyroclastic deposits, which also represent main arc magmatism ranging from 14 to 10 Ma (Pesce 1987, Burns et al. 2006). Frontal arc magmas from mid to late Miocene age (14.4–6.5 Ma) are also represented by the southernmost exposures of the El Teniente Volcanic Complex, cropping out in the Abanico Basin at ∼34° 30′ S (e.g., Kay et al. 2005). Finally, late Miocene to Pliocene frontal arc volcanism is represented by intermediate lavas and subvolcanic rocks from Quebrada Honda, Pichi Neuquén, and Centinela formations (∽7–2.5 Ma) cropping out in the main Andean axis at ∼36°–37° S (Kay et al. 2006b; Burns et al. 2006 and reference therein).

In the retroarc zone, the first volcanic stage (25–18 Ma) is represented by mafic volcanism cropping out at the easternmost sectors of the Chos Malal fold and thrust belt (∼36° 30″–37° 30″ S) in La Mantancilla and Huantraico hills region (Fig. 2). They comprise mostly alkaline olivine-, clinopyroxene-bearing basaltic flows, sills, and dykes that evolved to clinopyroxene-, hornblende-bearing basaltic to trachy-basaltic lavas, overall ranging from ∽25 to 17 Ma (Kay et al. 2006b; Dyhr et al. 2013b) (Fig. 1).

The second volcanic stage (18–10 Ma) is represented by arc-like rocks that expanded from the near (∽70° W) to the middle (∽69° 30′ W) retroarc region along the Malargüe and Chos Malal FTBs (Fig. 2). As mentioned, these intermediate volcanic rocks were grouped in variable formal and informal geological units, as new K–Ar and Ar–Ar ages were obtained over the past years (Fig. 3). This volcanic stage mostly reunites magmatic rocks included in the following units: Huincán I, Ciclo Eruptivo Molle, Cerro Negro Andesite, Charilehue Formation, and old volcanic centers along San Rafael Block area (Kay et al. 2006a, b; Kay and Copeland 2006; Baldauf 1993; Combina and Nullo 2000; Nullo et al. 2002; Spagnuolo et al. 2012; Dyhr et al. 2013a, b; Ramos et al. 2014; Litvak et al. 2015) (Fig. 3). They comprise pyroxene- and amphibole-bearing andesites to dacites, associated subvolcanic rocks, and minor pyroclastic deposits, which crop out as extended lava flows in Huincán I, Puntilla del Huincán (Ciclo Eruptivo Molle), and Charilehue regions, and as part of eroded stratovolcanoes, such as Chorreado, Puntudo, and El Zaino, located along the San Rafael broken foreland area (Fig. 2).

Finally, the third volcanic stage (10–2.5 Ma) comprises arc-related rocks that crop out from the middle (∽69° 30′ W) to the far (∽68° 30′ W) retroarc zone. They correspond to andesitic and dacitic rocks included in the following units: Huincán II, Ciclo Eruptivo Huincán, Andesita La Brea, Chachauén Volcanic Complex, Palaoco Group, and younger San Rafael volcanic centers (Kay et al. 2006a, b; Kay and

Copeland, 2006; Baldauf 1993, Baldauf et al. 1997; Combina and Nullo, 2000; Nullo et al. 2002; Dyhr et al. 2013a, b; Ramos et al. 2014; Litvak et al. 2015) (Fig. 3). Outcrops are located from the middle (Sierra de Palaoco, Cerro Negro, Huincán II area) to the far retroarc zone, the latter as isolated and eroded eruptive centers (Los Cerritos, De Afuera, Pelado) and as volcanic complexes (Nevado-Plateado and Chachahuén) (Fig. 2). These latter volcanic complexes constitute the easternmost exposures of the late Cenozoic calc-alkaline magmatism in the far retroarc zone. Overall, these retroarc volcanic sequences share similar petrographical characteristic than the rocks from the second volcanic stage, but differ in some key geochemical features as will be discussed in the following section.

4 Geochemical Evolution of the Miocene Magmas in the Malargüe and Chos Malal FTBs and the Changing Geodynamic Framework

4.1 Late Oligocene-Early Miocene (25–18 Ma): Alkaline Magmatism in the Retroarc Zone

Main arc magmatism during this period is represented by the Coya Machalí and Cura Mallín formations, which are characterized by basaltic to rhyolitic compositions ($\sim$47–74 SiO_2 wt%; Fig. 4a). Their La/Ta ratios over 25, as well as their Ba/La ratios and low Ta/Hf indicate an arc-related mantle source (Figs. 4b, c, d). Compositional signature of main arc magmas in this period ($\sim$25–16 Ma) is compatible with a tholeiitic source that can be explained as the result of magmas that were generated during an extensional tectonic setting developed along the Andean margin (Kay et al. 2006b). This event coincides with the initial stage of the arc after the breakup of the Cocos and Nazca plates into the Farallon Plate by late Oligocene (Kay et al. 2006b), which caused a change to an orthogonal subduction with increasing convergence rates (Cande and Leslie 1986; Lonsdale 2005; Pardo Casas and Molnar 1987; Somoza 1998; Somoza and Ghidella 2012). In this context, magmatism from the Cura Mallín and Abanico intra-arc basins was equilibrated within the crust with low-pressure mineral assemblages, as reflected in its La/Sm and Sm/Yb ratios (e.g., Kay et al. 2006b; Figs. 5a, b).

Meanwhile, in the retroarc zone, magmatic activity was represented by basaltic flows from La Mantancilla and Huantraico areas (Fig. 2). Contrarily to main arc magmatism, these basaltic to trachybasaltic lavas show an alkaline and intraplate signature, stated by their $Na_2O + K_2O$ versus SiO_2 values and low (<25) La/Ta ratio (Figs. 4a, b, c). Significantly, their high Ta/Hf ratios (>0.30) are also evidence of an enriched mantle source for this mafic magmatism (Fig. 4d), while Sm/Yb (Fig. 5a, b) ratios are indicative of a deep source that reach the garnet stability field (Dyhr et al. 2013a; Kay et al. 2006b).

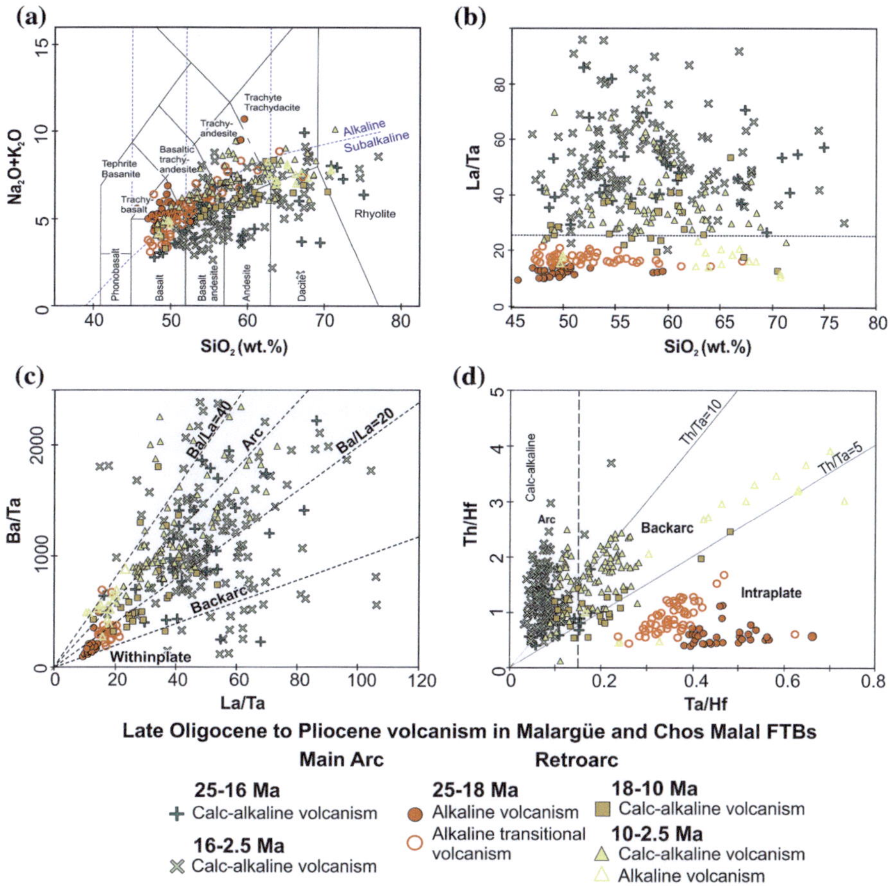

Fig. 4 Variable arc and alkaline-like geochemical signature of late Oligocene-Pliocene main arc and retroarc volcanism along the Malargüe and Chos Malal FTBs. **a** TAS diagram with alkaline versus subalkaline field (Le Maitre et al. 1989); **b** La/Ta versus SiO_2, and **c** Ba/Ta versus La/Ta diagrams that show the contrasting arc versus alkaline signature from the analyzed volcanic series; **d** Th/Hf versus Ta/Hf ratios as evidence from a subducted oceanic slab versus enriched mantle sources; Th/Ta ratios for intraplate, backarc, and arc are also shown (based on Kay et al. 2006b). Compiled data come from Dyhr et al. (2013a, b), Baldauf (1993), Kay et al. (2005, 2006a, b, 2005), Kay and Copeland (2006), Litvak et al. (2015), Nullo et al. (2002), Piquer et al. (2017), Spagnuolo et al. (2012), Utgé et al. (2009).

However, it is worth noticing a major change in the composition of mafic volcanism in the retroarc region by early Miocene. Lava flows erupted before 20 Ma lack subduction-related components, whereas those after 20 Ma start to show relatively higher La/Ta (15–26) and lower Ta/Hf (0.29–0.45) ratios (Figs. 4b, c, d). Depletions in HFSE, relative to fluid-mobile elements in these younger alkaline transitional lavas (20–18 Ma), indicate an hydrous component in the mantle source,

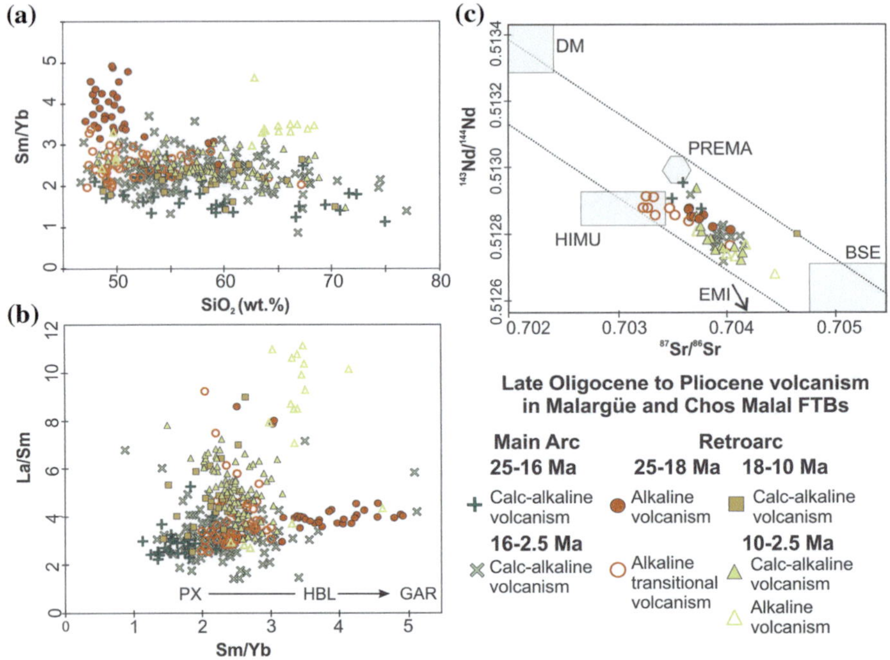

Fig. 5 **a** Sm/Yb versus SiO_2 and **b** La/Sm versus Sm/Yb as indicative of residual mineral assemblage at magma equilibration site; particularly, late Oligocene alkaline basaltic magmas show Sm/Yb ratios that indicate a garnet-bearing depth of melting. **c** $^{144}Nd/^{143}Nd$ versus $^{87}Sr/^{86}Sr$ ratios for analyzed volcanism. Compiled data come from Dyhr et al. (2013a, b), Baldauf (1993), Kay et al. (2005, 2006a, b, 2005), Kay and Copeland (2006), Kurtz et al. (1997), Litvak et al. 2015, Nullo et al. (2002), Piquer et al. (2017), Spagnuolo et al. (2012), Utgé et al. (2009)

which is interpreted as the first evidence of the introduction of slab-derived components beneath the retroarc zone as a result of the initiation of a shallow subduction regime at the latitudes of the present-day Payenia region (Dyhr et al. 2013b; Kay et al. 2006b). Consistently, Dyhr et al. (2013a, b) associated the isotopic trend of older alkaline volcanism in the retroarc zone (La Mantancilla lavas, ~25–20 Ma) with an OIB-like mantle component, whereas in younger lavas, a depleted source dominates (Huantraico lavas, ~20–18 Ma), with the influence of an arc end-member toward the youngermost expression of this retroarc alkaline mafic volcanism (Fig. 5c). It is worth noticing that the depth of melting of these transitional melts (~20–18 Ma) is relatively shallower than in the older alkaline lavas, as their Dy/Yb ratios are compatible with a spinel-bearing mantle source (Dyhr et al. 2013b).

4.2 Early Miocene-Late Miocene (18–10 Ma): Slab Input into the Retroarc Zone

The beginning of this stage marks the change to a compressional regime along the Andean margin, related to the initial development of the Payenia shallow subduction regime (Kay et al. 2006a, b; Dyhr et al. 2013a; Litvak et al. 2015). Main arc magmatism in the Cura Mallín Basin turned to a more basaltic to andesitic composition during the initial stage of this period (18–10 Ma; Trapa-Trapa and Cajón Negro formations, Fig. 3; Kay et al. 2006b), showing higher arc-like signature than lavas from the previous late Oligocene-early Miocene period. This is also reflected in the first expressions of the andesitic main arc volcanism in the southern sector of the Abanico Basin (El Teniente Volcanic Complex; Kay et al. 2005; Piquer et al. 2017). Overall, a clear arc-related signature in the main Andes comes from the increase in fluid mobile, incompatible elements (LILE), depletions in HFSE (e.g., higher La/Ta, Ba/Ta ratios) and increase in Th/Hf ratios as indicative of calc-alkaline sources (Fig. 4). Overall, geochemical features of this main arc volcanism along the Malargüe and Chos Malal FTBs remained through the following magmatic stage (late Miocene-Pliocene) considered in this work.

Major changes in the signature of arc-related magmas are seen throughout this period in the retroarc zone, as the shallow subduction regime progressed associated with the eastward migration of the asthenospheric wedge. Thus, an initial stage of arc-derived magmas developed in the near retroarc zone (~70° W) (Charilehue Formation, Huincán I lavas, Ciclo Eruptivo Molle, Cerro Negro Andesite, old San Rafael Block lavas; Figs. 2, 3), which presents mostly an andesitic to a dacitic composition (55–65 SiO_2 wt%) (Fig. 4a). Arc-like signatures come from La/Ta $\sim$20–40 and Ba/La $\sim$16 to 30 ratios, regardless their silica content (Figs. 4b, c), while Th/Ta ratios resemble backarc values (Fig. 4d). Particularly, the older expression of this stage, corresponding to the Charilehue Formation lavas (18–14 Ma; Fig. 3), shows a tholeiitic composition and the lowest Ba/La ratios representing the initial arc stage in the near retroarc zone (~70° W), (Spagnuolo et al. 2012; Litvak et al. 2015) (Fig. 6). Meanwhile, the younger expression of this stage, such as Huincán I lavas, Ciclo Eruptivo Molle, and Cerro Negro andesite, shows more typical arc signatures (increasing Ba/La, La/Ta; Fig. 4b, c). San Rafael volcanic rocks (Cerros Chorreado, Puntudo, and El Zaino, Figs. 2, 3) also follow the trend toward an increase in slab-derived components, being the easternmost rocks that register slab inputs during this volcanic stage (Fig. 6), and thus, the youngermost influence of the shallow subducting regime into the mid to far retroarc zone (~ 69° 30″ to 68° 30″) at ~15 Ma (Litvak et al. 2015).

The initiation of a shallow subduction configuration during this volcanic stage was linked to the increase of the westward shift of the South American plate since late Oligocene times (Spagnuolo et al. 2012). Moreover, expansion of arc-derived magmas from middle Miocene to early Pliocene in the retroarc zone across the Malargüe and Chos Malal FTBs shows a particular SW–NE projection into the South American plate that resembles a hot spot trail, as pointed out by Spagnuolo et al. (2012). Recent studies suggested, on the base of tectonic reconstructions, that the Payenia shallow subduction regime could in fact be caused by a plume-modified orogeny model (Gianni et al. 2018), which could explain the migration pattern and geometry of the magmatic arc focus toward the retroarc zone (see Chap. "Plume subduction beneath the Neuquén Basin and the last mountain building stage of the Southern Central Andes").

4.3 *Latest Miocene-Pliocene (10–2.5 Ma): Easternmost Arc Expansion into the Retroarc Zone*

As the shallow subduction regime progressed, arc-derived magmas that had expanded from the near (70° W) to the middle retroarc zone (69° 30′ W) reached the far retroarc area (68° 30) during this volcanic stage (10–2.5 Ma) (Fig. 2). These magmatic expressions comprise Huincán II lavas, Ciclo Eruptivo Huincán, Palaoco Group, Chachahuén Volcanic Complex, and young San Rafael rocks (Fig. 2; reference therein). They correspond mostly to mid to high-K andesitic and dacitic lavas with La/Ta, Ba/Ta ratios, and Th/Ta ratios that reflect an arc component with clearer calc-alkaline contributions into the mantle source (Figs. 4, 6). Besides, given the evolved nature of some of these volcanic rocks, fractional crystallization processes and potentially crustal assimilation during magma ascent also explain their geochemical signature, despite Nd and Sr isotopic ratios resemble scarce crustal components into the mantle source (Fig. 5c). Particularly, the dacitic lavas of San Rafael unit show low REE, Cs, Rb, Th, and U contents that were attributed to lower crustal contamination (Litvak et al. 2015). The same process would have affected the genesis of the late Miocene Palaoco volcanism, despite crustal contamination would have been a minor process in comparison with the influence of subduction related fluids to the mantle source (Dyhr et al. 2013a).

It is worth mentioning that part of the magmatism in the retroarc zone during the late Miocene shows a geochemical alkaline-like behavior. This is the case of Fortunoso II volcanism (8.6 Ma; Dyhr et al. 2013b) and the oldest lavas included in the Chachauhén Volcanic Complex (Vizcachas Group, 7.3–6.8 Ma; Kay et al. 2006a). Late Miocene Fortunoso II volcanism corresponds to basaltic rocks, which show an intraplate signature given by their low (<25) La/Ta and high (<0.2) Ta/Hf ratios; however, their intraplate enrichment is lower (Ta/Hf between 0.2 and 0.4 in Fig. 4d) when compared to the late Oligocene (25–20 Ma) alkaline lavas. Moreover, Dyhr et al. (2013a) also reported that Fortunoso II lavas show minor Ta–Nb negative anomalies and Ba/Nb and La/Nb values that resemble those of the early Miocene transitional alkaline lavas (20–18 Ma), as indicative of scarce arc component in the mantle source. For this reason, late Miocene Fortunoso II would suggest the prevalent influence of the early Miocene OIB-type mantle source still affecting magmatism developed in the retroarc zone by late Miocene (Dyhr et al. 2013a). On the other hand, intraplate lavas from Chachauhén Volcanic Complex corresponds to dacitic to rhyolitic rocks with lower La/Ta and higher Ta/Hf (La/Ta ∽10–14; Ta/Hf ∽0.7–1.1) ratios than coeval and relatively younger arc-related lavas in the retroarc zone (Figs. 4). Genesis of this silicic volcanism was explained by strong input of intraplate-like crustal components affecting magmas by the time of the formation of the late Miocene Vizcachas Group (Kay et al. 2006a).

Main arc magmas from this stage (10–2,5 Ma) reflect similar geochemical features as the ones observed in the previous volcanic period, suggesting that crustal and mantle composition beneath the main arc remained almost constant through the time (Kay et al. 2006b). The progressive input of slab-derived components into the retroarc area reached the easternmost zone around 3.5 Ma (Fig. 6), given by the location of the Chachahuén Volcanic Complex and the young San Rafael volcanic rocks, which show the highest slab-derived input with Ba/Nb > 100 and Ba/La between 20 and 50 Ma (Fig. 3b, c). Thus, slab influence in the retroarc zone reached almost 500 km away from the Chilean trench at 36° S, and after 3.5 Ma, retroarc magmatism at these latitudes lacks of clear arc-related geochemical features (Litvak et al. 2015, 2018) (Fig. 6).

The shallow subduction regime at the latitudes of the Malargüe and Chos Malal FTBs correlates with an increase on the compressional conditions and deformation across the foreland area during mid to late Miocene times. Magmas evolving within this setting show an increase in Sm/Yb ratios with time (Fig. 5a, b), which could be evidence for an increase of crustal thickness between 12 and 10 Ma (Nullo et al. 2002; Kay et al. 2006b). However, even though crustal thickness seems to have increased through this time, the youngermost volcanic units within the Malargüe and Chos Malal FTBs latitudes (Chachahuén Volcanic Complex and young San Rafael volcanic) show REE trends that suggest that magmas equilibrated at depth with pyroxene-amphibole as residual mineral assemblages, as evidence for low-pressure conditions, is still suggesting a crust of normal thickness (40 km) that resembles the present-day crustal thickness (Kay et al. 2006b and reference there in).

Shallow subduction configuration took place at the studied latitudes until the middle Pliocene, when the slab re-steepened and acquired its present configuration. This event would have caused extensional stress conditions in the upper plate that led to a widespread alkaline volcanism in the present-day Payenia retroarc and rhyolitic calderas and dome complexes in the main arc zone (e.g., Gudnason et al. 2012; Llambías et al. 2010; Ramos et al. 2014; Søager et al. 2013; Hernando et al. 2014). The alternative plume-modified model proposed to explain the shallow subduction regime also suggests that ponding of the Payenia plume at the base of the downgoing slab may have favored its tearing around 38° S at 5–3 Ma (Gianni et al. 2018) (see Chapters "Plume subduction beneath the Neuquén Basin and the last mountain building stage of the Southern Central Andes" and "Pliocene to Quaternary Retroarc Extension in the Neuquén Basin: Geophysical Characterization of the Loncopué Trough").

This would have promoted the ascent of the asthenospheric melts during the re-steepening of the slab and the development of alkaline volcanism in the retroarc zone by late Pliocene times. As suggested by Dyhr et al. (2013a), similar geochemical

Fig. 6 Age, La/Ta, and Th/Ta ratios versus longitude of the studied main arc and retroarc volcanic sequences along the Malargüe and Chos Malal FTBs. Late Oligocene-early Miocene alkaline volcanism is registered along the entire retroarc, whereas calc-alkanline volcanism expanded to the far retroarc area since middle Miocene to early Pliocene times

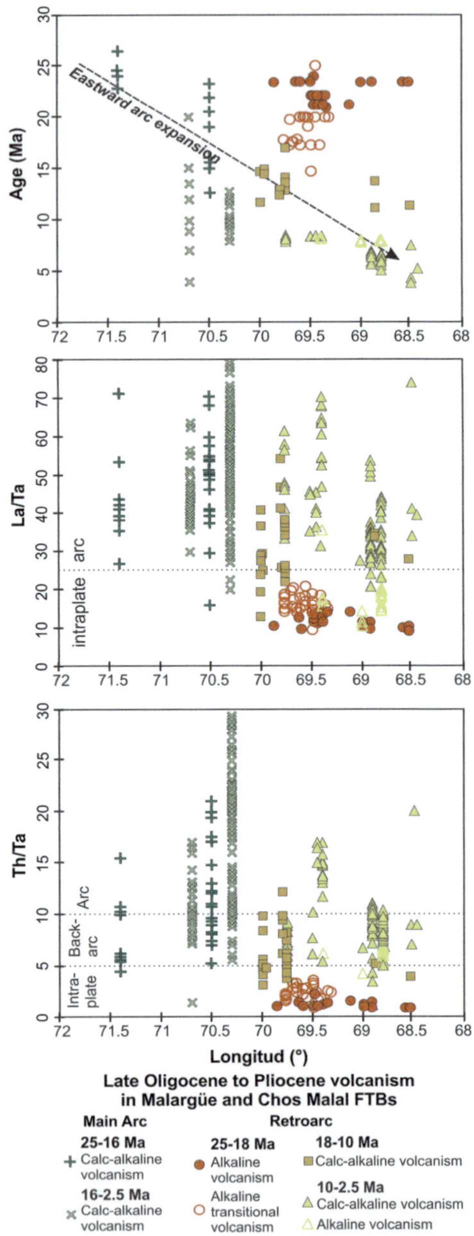

features of the early Miocene and the Pliocene alkaline basalts could be evidence from a long-lived nature of an OIB-type component in the retroarc mantle.

5 Concluding Remarks

The evolution of late Cenozoic volcanism along the Malargüe and Chos Malal fold and thrust belts is strongly controlled by changes in subduction slab geometry. While main arc magmatism shows similar compositional and geochemical features through the late Oligocene to Pliocene time period, remarkable changes are noticed in volcanic sequences developed in the retroarc zone. In this eastern region, alkaline magmatism dominated since late Oligocene times with a clear OIB-like mantle source, which prevailed till Pliocene times in the present-day retroarc. This late Oligocene-early Miocene mafic volcanism is developed in the context of a rollback regime and consequent extension.

Changes in volcanism along the studied latitudes are registered in the retroarc zone as the Payenia shallow subduction regime developed since middle Miocene and asthenospheric-derived magmas expanded toward the retroarc zone, together with a change to compressional conditions in the foreland region. Retroarc magmatism changed since early Miocene from an alkaline composition to arc-like series by middle Miocene. Particularly, as the shallow subduction progressed, arc-derived magmas developed in the retroarc zone changed from a tholeiitic signature to a calc-alkaline type. Thus, arc expanded from the main arc Andean axis ($\backsim$71° W) till the far retroarc area ($\backsim$68° 30′ W) from 15 to 3.5 Ma at the latitudes of the Malargue and Chos Malal fold and thrust belts ($\sim$34°–37° S). Then, as the slab re-steepened by 3.5 Ma, intraplate OIB-like volcanism was reinstalled in the present Payenia retroarc zone.

References

Alvarez PP, Ramos VA (1999) The Mercedario rift system in the Principal Cordillera of Argentina and Chile (32° SL). J S Am Earth Sci 12:17–31

Baldauf P (1993) Timing of deformation in the central Andean Foreland, Western Mendoza, Argentina, using $^{40}Ar/^{39}Ar$ dating techniques. Master thesis, The George Washington University

Baldauf P (1997) Timing of the uplift of the Cordillera Principal, Mendoza Province, Argentina. M.S. Thesis, George Washington University, pp. 356

Burns WM, Jordan T, Copeland P et al (2006) Extensional tectonics in the Oligo-Miocene Southern Andes as recorded in the Cura-Mallín basin. In: Kay SM, Ramos VA (eds) Evolution of an Andean margin: a tectonic and magmatic view from the Andes to the Neuquén basin (35–39° S). Geol Soc Am, SP 407, pp 163–184

Cande SC, Leslie RB (1986) Late Cenozoic tectonics of the southern Chile trench. J Geophys Res Solid Earth 91:471–496

Charrier R, Wyss AR, Flynn JJ et al (1996) New evidence for late Mesozoic-early Cenozoic evolution of the Chilean Andes in the upper Tinguiririca valley (35 S), central Chile. J S Am Earth Sci 9:393–422

Charrier R, Baeza O, Elgueta S et al (2002) Evidence for Cenozoic extensional basin development and tectonic inversion south of the flat-slab segment, southern Central Andes, Chile (33–36 S). J S Am Earth Sci 15:117–139

Charrier R, Farías M, Maksaev V (2009) Evolución tectónica, paleogeográfica y metalogénica durante el cenozoico en los Andes de Chile Norte y Central e implicaciones para las regiones adyacentes de Bolivia y Argentina. Rev Asoc Geol Argent 65(1):5–35

Combina A, Nullo F (2000) La Formación Loma Fiera (Mioceno medio) y su relación con el volcanismo y el tectonismo contemporáneo, Cuchilla de la Tristeza, Mendoza. Revista de la Asociación Geológica Argentina LV (3):201–210

Dyhr CT, Holm PM, Llambías EJ (2013a) Geochemical constraints on the relationship between the Miocene-Pliocene volcanism and tectonics in the Palaoco and Fortunoso volcanic fields, Mendoza Region, Argentina: New insights from 40Ar/39Ar dating, Sr–Nd–Pb isotopes and trace elements. J Volcanol Geoth Res 266:50–68

Dyhr CT, Holm PM, Llambías EJ et al (2013b) Subduction controls on Miocene back-arc lavas from Sierra de Huantraico and La Matancilla and new $^{40}Ar/^{39}Ar$ dating from the Mendoza Region, Argentina. Lithos 179:67–83

Folguera A, Ramos VA (2011) Repeated eastward shifts of arc magmatism in the Southern Andes: A revision to the long-term pattern of Andean uplift and magmatism. J S Am Earth Sci 32:531–546

Folguera A, Orts D, Spagnuolo M et al (2011) A review of Late Cretaceous to Quaternary paleogeography of the Southern Andes. Biol J Linn Soc 103:250–268

Franzese JR, Spalletti LA (2001) Late Triassic-Early Jurassic continental extension in southwestern Gondwana: tectonic segmentation and pre-break-up rifting. J S Am Earth Sci 14:257–270

Giambiagi LB, Bechis F, García V, Clark A (2005) Temporal and spatial relationship between thick- and thin-skinned deformation in the thrust front of the Malargüe fold and thrust belt, Southern Central Antes. In: VI° International Symposium on Andean Geodynamics, Extended Abstracts, pp. 315–318

Giambiagi L, Bechis F, García V, Clark A (2008) Temporal and spatial relationship between thick- and thin-skinned deformation in the Malargüe fold and thrust belt, Southern Central Andes. Tectonophysics 459:123–139

Gianni GM, Echaurren A, Fennell L et al (2018) Cretaceous Orogeny and Marine Transgression in the Southern Central and Northern Patagonian Andes: Aftermath of a Large-Scale Flat-Subduction Event? In: Folguera A, Contreras Reyes E et al (eds) The making of the Chilean-Argentinean Andes. Springer, Berlin, pp. 291–328

Godoy E, Yáñez G, Vera E (1999) Inversion of an Oligocene volcano-tectonic basin and uplifting of its superimposed Miocene magmatic arc in the Chilean Central Andes: first seismic and gravity evidences. Tectonophysics 306(2):217–236

Gudnason J, Holm PM, Søager N et al (2012) Geochronology of the late Pliocene to recent volcanic activity in the Payenia back-arc volcanic province, Mendoza Argentina. J S Am Earth Sci 37:191–201

Hernando IR, Aragón E, Frei R et al (2014) Constrains on the origin and evolution of the magmas in the PayúnMatrú Volcanic Field, Quaternary Andean Back-arc of Western Argentina. J Petrol 55(1):209–239

Jordan TE, Burns WM, Veiga R et al (2001) Mid-Cenozoic intra-arc basins in the southern Andes. Tectonics 20. https://doi.org/10.1029/1999tc001181

Kay SM, Copeland P (2006) Early to middle Miocene back-arc magmas of the Neuquén Basin: Geochemical consequences of slab shallowing and the westward drift of South America. In: Kay SM, Ramos VA (eds) Late Cretaceous to Recent magmatism and tectonism of the Southern Andean margin at the latitude of the Neuquén basin (36–39° S). Geological Society of America, SP 407, pp 185–214

Kay SM, Kurtz AC (1995) Magmatic and tectonic characterization of the El Teniente Region. CODELCO-Chile, Unpublished report, pp 180
Kay SM, Godoy E, Kurtz A (2005) Episodic arc migration, crustal thickening, subduction erosion and magmatism in the South-Central Andes. Geol Soc Am Bull 117:67–88
Kay SM, Mancilla O, Copeland P (2006a) Evolution of the Back-arc Chachahuén volcanic complex at 37° S latitude over a transient Miocene shallow subduction zone under the Neuquén Basin. In: Kay SM, Ramos VA (eds) Late Cretaceous to Recent magmatismand tectonism of the Southern Andean margin at the latitude of the Neuquén basin (36–39° S). Geological Society of America, SP 407, pp 215–246
Kay SM, Burns WM, Copeland P, Mancilla O (2006b) Upper Cretaceous to Holocene magmatism and evidence for transient Miocene shallowing of the Andean subduction zone under the northern Neuquén Basin. In: Kay SM, Ramos VA (eds) Late Cretaceous to Recent magmatism and tectonism of the Southern Andean margin at the latitude of the Neuquén basin (36–39° S). Geological Society of America, SP 407, pp 19–60
Kurtz A, Kay SM, Charrier R, Farrar E (1997) Geochronology of Miocene plutons and Andean uplift history in the El Teniente region, Central Chile (34–35 S). Rev Geol Chile 24:75–90
Le Maitre RW, Bateman P, Dudek A et al (1989) A Classification of Igneous Rocks and Glossary of Terms: recommendations of the International Union of Geological Sciences Subcommission on the Systematics of Igneous Rocks. Blackwell Scientific Publications, Oxford
Litvak VD, Spagnuolo MG, Folguera A et al (2015) Late Cenozoic calc-alkaline volcanism over the Payenia shallow subduction zone, South-Central Andean back-arc (34° 30′–37° S), Argentina. J S Am Earth Sci 64:365–380
Litvak VD, Poma S, Jones RE et al (2018) The Late Paleogene to Neogene volcanic arc in the Southern Central Andes (28°–37° S). In: Folguera A, Contreras Reyes E, Heredia N et al (eds) The making of the Chilean-Argentinean Andes. Springer, Berlin, pp 503–541
Llambías EJ, Bertotto GW, Risso C, Hernando I (2010) El volcanismo cuaternario en el retro-arco de Payenia: una revisión. Rev Asoc Geol Argent 67(2):278–300
Llambías EJ, Leanza HA, Carbone O (2007) Evolución tectono-magmática durante el Pérmico al Jurásico Temprano en la Cordillera del Viento (37º50'S-37º15'S): nuevas evidencias geológicas y geoquímicas del inicio de la cuenca Neuquina. Rev Asoc Geol Argent 62:217–235
Lonsdale P (2005) Creation of the Cocos and Nazca plates by fission of the Farallon plate. Tectonophysics 404:237–264
Nullo FE, Stephens G, Otamendi J et al (2002) El volcanismo del Terciario superior del sur de Mendoza. Rev Asoc Geol Argent 57(2):119–132
Pardo Casas F, Molnar P (1987) Relative motion of the Nazca (Farallon) and South American plates since late Cretaceous time. Tectonics 6:233–248
Pesce AH (1987) Evaluación geotermica del área Cerro Domuyo, Provincia del Neuquén, República Argentina. Rev Brasil Geofís 5:283–299
Piquer J, Hollings P, Rivera O et al (2017) Along-strike segmentation of the Abanico basin, central Chile: new chronological, geochemical and structural constraints. Lithos 268–271:174–197
Ramos VA (1998) Estructura del sector occidental de la faja plegada y corrida del Agrio, cuenca Neuquina, Argentina. In: X° Congreso Latinoamericano de Geología, Actas 2, pp. 105–110
Ramos VA, Folguera A (2005) Tectonic evolution of the Andes of Neuquén: Constraints derived from the magmatic arc and foreland deformation. In: Veiga G, Spalletti L, Howell J, Schwarz E (eds) The Neuquén Basin: a case study in sequence stratigraphy and basin dynamics. The Geological Society, London, SP 252, pp 15–35
Ramos VA, Kay SM (2006) Overview of the tectonic evolution of the southern Central Andes of Mendoza and Neuquén (35°–39° S latitude). In: Kay SM, Ramos VA (eds) Evolution of an Andean margin: a tectonic and magmatic view from the Andes to the Neuquén Basin (35°–39° S latitude). Geological Society of America, SP 407, pp 1–17
Ramos VA, Folguera A, Litvak VD, Spagnuolo M (2014) Andean tectonic cycle: From crustal thickening to extension in a thin crust (34–37 SL). Geosci Front 5:351–367

Rovere E (1998) Volcanismo Jurásico, Paleógeno y Neógeno en el Noroeste del Neuquén, Argentina. In: Abstracts of the 10 Congreso Latinoamericano de Geología y 6 Congreso Nacional de Geología Económica, Buenos Aires, 8–13 Nov 1998

Silvestro J, Atencio M (2009) La cuenca Cenozoica del Río Grande y Palauco: Edad, evolución y control estructural, faja plegada de Malargüe (36º S). Rev Asoc Geol Arg 65(1):154–169

Somoza R (1998) Updated Nazca (Farallon)—South America relative motions during the last 40My: implications for mountain building in the central Andean region. J S Am Earth Sci 11:211–215

Somoza R, Ghidella ME (2012) Late Cretaceous to recent plate motions in western South America revisited. Earth Planet Sci Let 331–332:152–163

Spagnuolo M, Litvak VD, Folguera A et al (2012) Neogene magmatic expansion and mountain building at the southern Central Andes, 36°–37° S, Argentina. J Geodyn 53:81–94

Suárez M, Emparan C (1995) The stratigraphy, geochronology and paleophysiography of a Miocene freshwater interarc basin, Southern Chile. J S Am Earth Sci 8:17–31

Søager N, Holm PM, Llambías EJ (2013) Payenia volcanic province, southern Mendoza, Argentina: OIB mantle upwelling in a back-arc environment. Chem Geol 349–350:36–53

Utgé S, Folguera A, Litvak V et al (2009) Geología del sector norte de la cuenca de Cura Mallín en las lagunas de Epulaufquen, Neuquén. Rev Asoc Geol Argent 64:231–248

Yrigoyen M (1993) Los depósitos sinorogénicos terciarios. In: Ramos VA (ed) Geología y Recursos minerales de Mendoza, Relatorio I(11):123–148

Plume Subduction Beneath the Neuquén Basin and the Last Mountain Building Stage of the Southern Central Andes

Guido M. Gianni, Agustina Pesce, Héctor P. A. García, Marianela Lupari, Sebastián Correa-Otto, Silvina Nacif and Andrés Folguera

Abstract The occurrence of a Neogene shallow subduction stage, as well as, a Pliocene slab-tearing, and steepening of the Nazca plate in the southern Central Andes are well established. However, a satisfactory explanation for the origin and connection between these complex processes is still elusive. In this contribution, we revise the late Cenozoic tectonic and magmatic evolution of the southern Central Andes between 35° and 38° S and discuss different proposals for the Miocene slab shallowing and its Pliocene destabilization. Recent plate kinematic reconstructions show that Neogene arc-front expansion linked to slab shallowing, fold belt reactivation in the main cordillera and intraplate contraction in the San Rafael Block correlates with the subduction of the ancient Payenia plume, a deep mantle anomaly potentially rooted in the lower mantle. Also, the Nazca slab tear determined from tomographic analyses and subsequent slab steepening may also be a direct consequence of this plume subduction process. Considering the westward drift of South America and the presence of several neighbor hotspots over the Nazca plate, the Payenia plume overriding could be the first of future episodes of plume–trench interaction in the Andes.

Keywords Plume-modified orogenesis · Flat slab · Fold and thrust belt · Plume–subduction zone interaction

1 Introduction

The Wilson cycle illustrates how former oceanic basins are consumed during plate convergence and continental collision. In a larger scale, this process is well recorded

---

G. M. Gianni (✉) · A. Pesce · H. P. A. García · M. Lupari · S. Correa-Otto · S. Nacif
IGSV, Instituto Geofísico Sismológico Ing. Volponi, Universidad Nacional de San Juan, San Juan, Argentina
e-mail: guidogianni22@gmail.com

G. M. Gianni · A. Folguera
CONICET, Instituto de Estudios Andinos Don Pablo Groeber (IDEAN), Universidad de Buenos Aires, Buenos Aires, Argentina

D. Kietzmann and A. Folguera (eds.), *Opening and Closure of the Neuquén Basin in the Southern Andes*, Springer Earth System Sciences,
https://doi.org/10.1007/978-3-030-29680-3_20

during episodic supercontinent assembly and breakup, which constitutes the supercontinent cycle (see Nance et al. 2014, for a review). In this context, if past oceanic basins held magmatic products associated with several mantle plumes as currently seen (e.g., French and Romanowicz 2015) and as indicated by worldwide evidence oceanic plateaux accretions in subduction settings (e.g., Betts et al. 2015), consumption of oceanic basins would necessarily have to involve direct interactions between plumes and active margins (Fig. 1).

In the last decades, the major question regarding the fate of mantle plumes during ocean basin closure defined the study of plume–subduction zone interactions (Murphy et al. 1998; Fletcher and Wyman 2015; Mériaux et al. 2016; Chang et al. 2016, among others). Although much remains to be learned about these complex geodynamic scenarios, conceptual, numerical, and analog modeling studies have revealed some first-order potential responses to this peculiar interaction. For instance, weak mantle plumes are expected to be suppressed during subduction (Druken et al. 2014; Kincaid et al. 2013; Steinberger and O'Connell 1998), while stronger plumes may resist subduction leaving particular tectonomagmatic imprints on the active margin. Only a few examples of plume survival after subduction have been identified so far

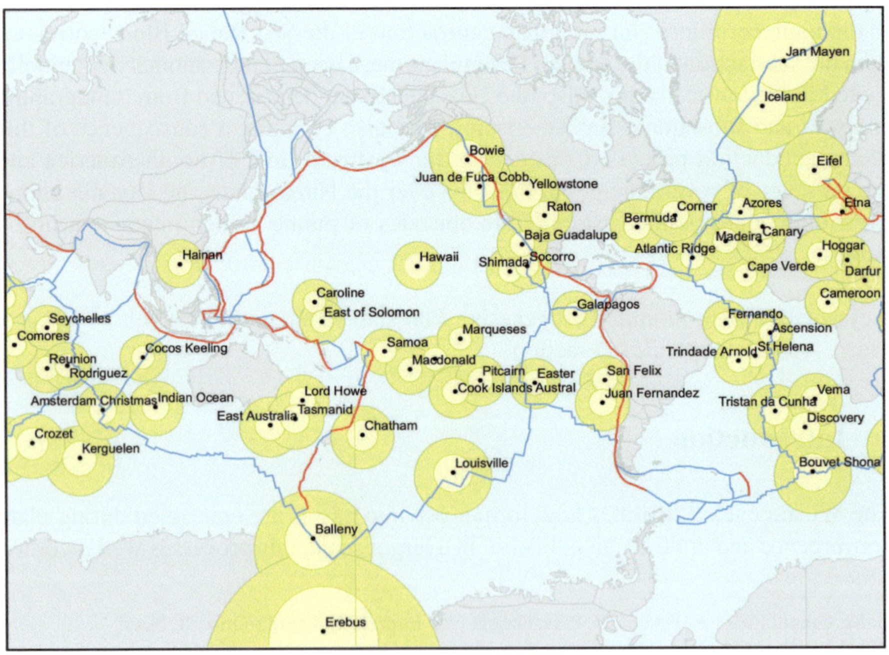

Fig. 1 World map showing mantle plumes with 1000 and 500 km diameter zones of potential plume–subduction interaction. In particular, for the 1000 km diameter, almost the entire northeast Pacific subduction is influenced by subduction zones. Thick red lines are subduction zones, and blue lines denote spreading ridges, transforms, terrane boundaries, and inferred paleoboundaries. Figure modified from Fletcher and Wyman (2015)

(Murphy et al. 1998, 1999; Dalziel et al. 2000; Betts et al. 2009). In these cases, the vigorous plume upwelling accompanied by a marked positive buoyancy changes the slab angle leading to a shallowing of the subducted plate (Murphy et al. 1998; Murphy and Keppie 2005; Betts et al. 2009, 2015). Such subduction configuration drives plate margin contraction, cratonward migration of orogenic and magmatic activity, and the eventual cessation of arc magmatism. Also, upper plate contraction may be aided by trench advance produced by plume–margin interaction (Betts et al. 2012). During subduction, the plume is less likely to undergo decompression melting inhibiting the typical eruption of large igneous provinces (Drunken et al. 2014; Murphy 2016).

The orogenic consequences of plume subduction have been denominated as "plume-modified orogeny" by Murphy et al. (1998). This process terminates when the subducting lithosphere breaks off due to the buoyancy contrast between the plume and the overlying slab (Murphy et al. 1998, 1999; Dalziel et al. 2000) or delaminates, which is assisted by the presence of mechanical weaknesses in the slab (Macera et al. 2008; Obrebski et al. 2010). The latter stages of this process are usually accompanied by slab steepening, subduction reactivation followed by rollback, and plume impact below the continental lithosphere (Betts et al. 2009; Dalziel et al. 2000; Murphy et al. 1998; Oppliger et al. 1997). Plume-modified orogeny has been proposed for the latest Cretaceous–Paleogene tectonic evolution of western North America (Murphy et al. 1998; Murphy 2016), the Paleozoic Acadian orogeny in the northern Appalachians (Murphy et al. 1999; Murphy and Keppie 2005), the Gondwanic orogeny in southwestern Gondwana (Dalziel et al. 2000), and the Mesoproterozoic evolution of Australia (Betts et al. 2009).

In this work, we review one of the most recent and complete examples of this process which is recorded by the final closure of Neuquén Basin and the building of the southern Central Andes (Gianni et al. 2017) (Fig. 2a). The extraordinary record of all the tectonic stages associated with the Payenia plume subduction makes the Neuquén Basin an ideal place to understand similar potential processes in ancient subduction settings.

2 Geological Setting

The southern Central Andes between 35° and 38° S mainly consists of two mostly thick-skinned fold and thrust belts, the main cordillera to the west and an intraplate belt to the east known as the San Rafael Block (Sagripanti et al. 2011; Giambiagi et al. 2012; Ramos et al. 2014) (Fig. 2a, b). These orogenic belts are separated by the Neogene Río Grande foreland basin (Yrigoyen 1993) (Figs. 2b and 3).

Andean orogenesis at studied latitudes took place through tectonic inversion of the Mesozoic Neuquén Basin during Late Cretaceous to Eocene (~98–55 Ma) and Neogene times (18–5 Ma) (Cobbold and Rossello 2003; Ramos et al. 2014; Folguera et al. 2015). The last deformational phase was related to a significant expansion of

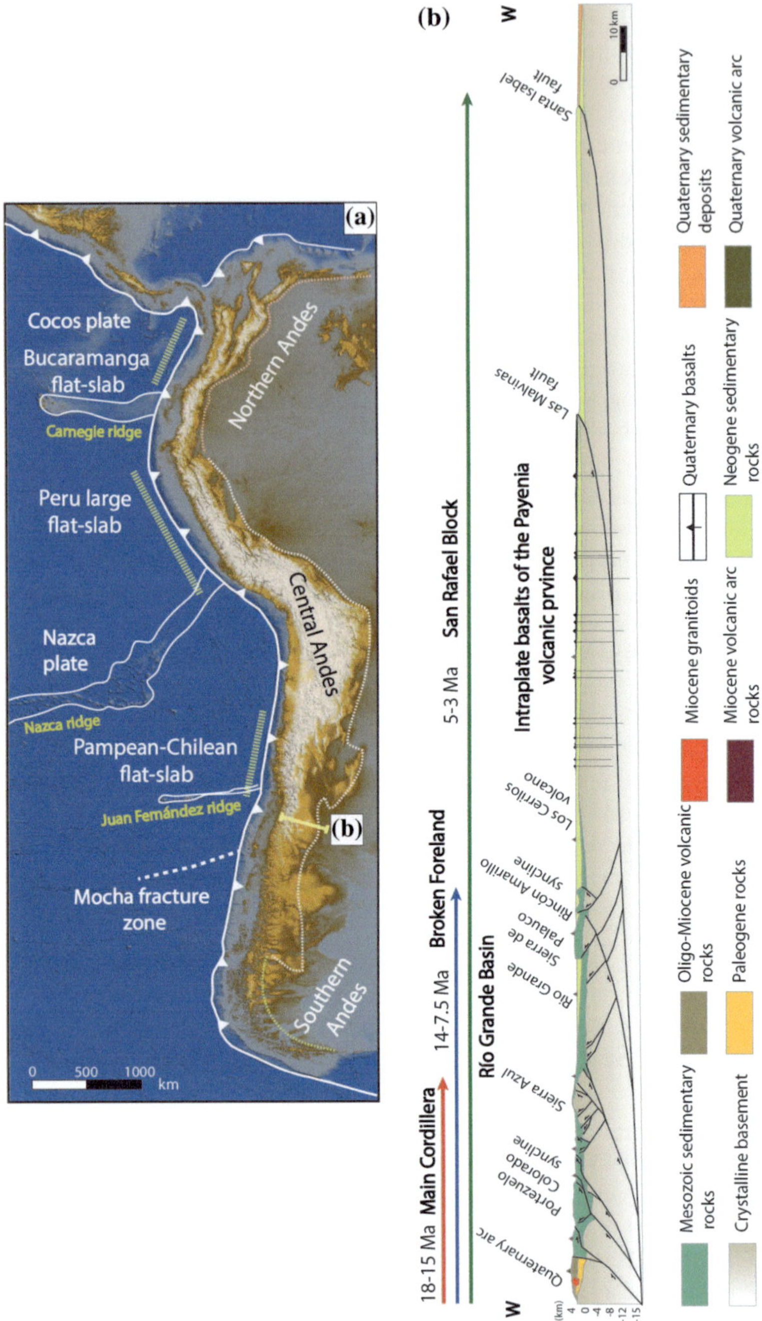

Fig. 2 **a** Tectonic setting of the study area. **b** Cross section depicting the general structure and fold–thrust belt propagation in the study area. Modified from Ramos et al. (2014)

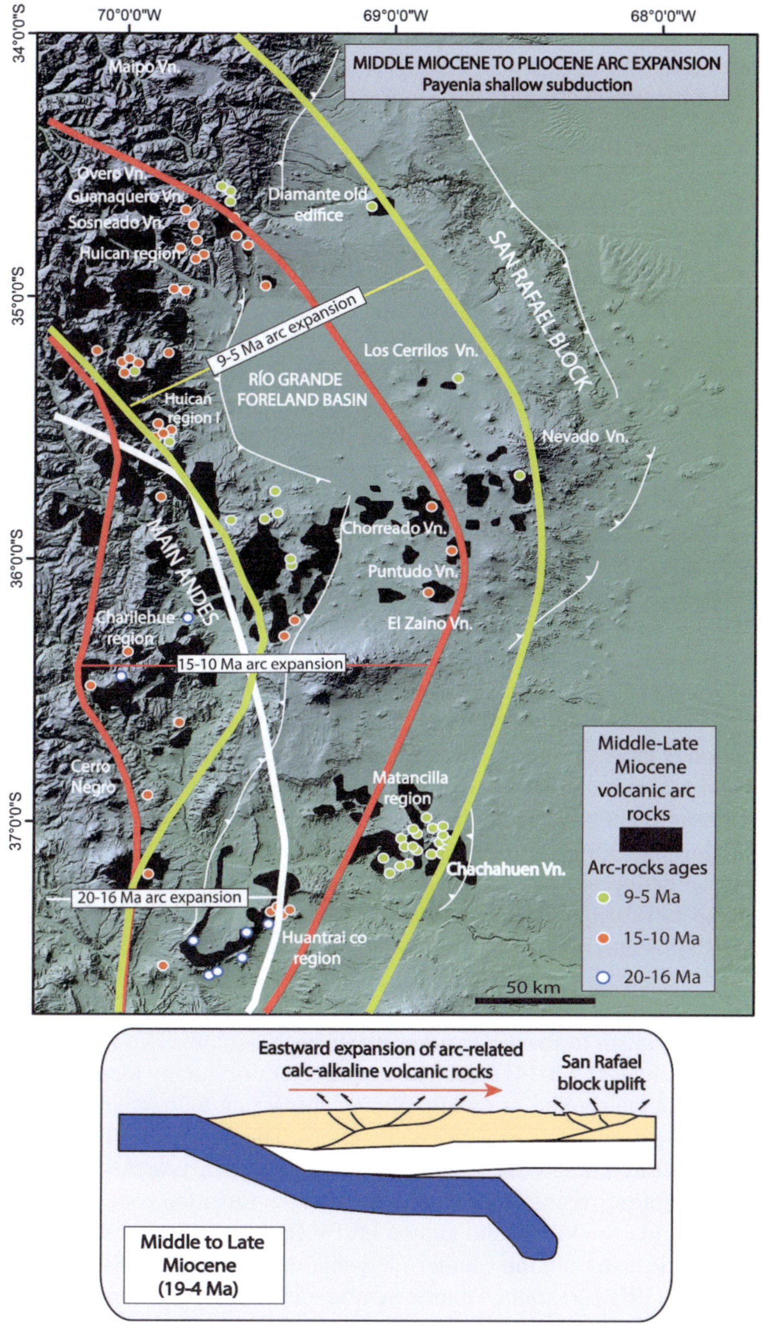

70°0'0"W
69°0'0"W
68°0'0"W
34°0'0"S
35°0'0"S
36°0'0"S
37°0'0"S
MIDDLE MIOCENE TO PLIOCENE ARC EXPANSION
Payenia shallow subduction
Maipo Vn.
Overo Vn.
Guanaquero Vn.
Sosneado Vn.
Huican region
Diamante old edifice
SAN RAFAEL BLOCK
9-5 Ma arc expansion
Los Cerrilos Vn.
RÍO GRANDE FORELAND BASIN
Huican region I
Nevado Vn.
MAIN ANDES
Chorreado Vn.
Puntudo Vn.
El Zaino Vn.
Charilehue region
15-10 Ma arc expansion
Cerro Negro
Matancilla region
Chachahuen Vn.
20-16 Ma arc expansion
Huantrai co region
50 km
Middle-Late Miocene volcanic arc rocks
Arc-rocks ages
9-5 Ma
15-10 Ma
20-16 Ma
Eastward expansion of arc-related calc-alkaline volcanic rocks
San Rafael block uplift
Middle to Late Miocene (19-4 Ma)

◀**Fig. 3** Above: spatial distribution of geochronological data of arc-related rocks between 34° 30′ and 37° 30′ S describing an eastward expansion of the middle Miocene to the early Pliocene arc. Modified from Folguera et al. (2015) and Litvak et al. (2015, 2017). Ages are taken from Nullo et al. (1993, 1999, 2002), Ramos and Barbieri (1989), Ostera et al. (1999), Cobbold and Rossello (2003), Giambiagi et al. (2005), Kay and Copeland (2006), Kay et al. (2006a, b), Folguera et al. (2009), Spagnuolo et al. (2012), Dyhr et al. (2013a, b) and Ramos et al. (2014). Below: proposed geodynamic model to explain anomalous eastward arc expansion between 20–19 and 4 Ma. Figure modified from Gianni et al. (2017)

the orogenic front (~550 km away from the Chilean trench), uplifting the intraplate San Rafael Block (Fig. 2b).

Several studies have noted that Andean reactivation and foreland deformation in Neogene times were concomitant with an eastward expansion of the magmatic arc reaching as far as ~300 km from the current arc front (Figs. 2b and 3) (Kay et al. 2006; Ramos et al. 2014; Sagripanti et al. 2015a) (see Chapter "The Miocene Magmatism in the Malargüe and Chos Malal Fold and Thrust Belts"). These two simultaneous processes have been interpreted as produced by a shallowing or flattening of the subducted Nazca plate (Kay et al. 2006a, b; Spagnuolo et al. 2012; Dyhr et al. 2013a, b; Ramos et al. 2014; Litvak et al. 2015) (Fig. 3).

A large geochronological dataset of arc-related rocks has allowed constraining this event between ~20–19 and ~5–4 Ma (see Litvak et al. 2015 for a synthesis) (Fig. 3). Noteworthy, the driving mechanisms behind the shallow subduction and in turn the Neogene widespread contraction remain enigmatic. The shallow slab configuration is thought to have destabilized in Pliocene times following a series of tectonic and magmatic events. The magmatic arc retracted between 3 and 0 Ma and a massive eruption of basaltic flood basalts (~40,000 km^2) took place in the back-arc area related to the extrusion of the Payenia volcanic province. These processes are thought to be connected and have been attributed to steepening of the Nazca slab triggering an asthenospheric upwelling after the previous shallow subduction stage (Fig. 4a) (Ramos and Folguera 2011). Slab steepening was probably a diachronous process as inferred by Gudnason et al. (2012) based on the spatial distribution of magmatic rocks from the Payenia volcanic province showing a progression of the intraplate magmatism to the west and north (Fig. 4b).

Recently, Burd et al. (2014) through a magnetotelluric survey identified two resistivity anomalies interpreted as mantle plumes impacting below the southern sector of the Payenia volcanic province (Figs. 5 and 6a). The mantle anomalies are a shallower plume located below the Payún Matrú and Tromen volcanoes called SWAP and a deeper anomaly rising at least from the mantle transition zone located beneath the Auca Mahuída shield volcano named DEEP (Fig. 6a). The presence of an EM1 OIB-type mantle source in the southern Payenia rocks where the SWAP anomaly is located and a MORB-like south Atlantic mantle source in the northern Payenia basalts reflects this complex mantle structure (Søager et al. 2013; Holm et al. 2016). In this study, we suggest a possible lower mantle origin for the Payenia plume as indicated

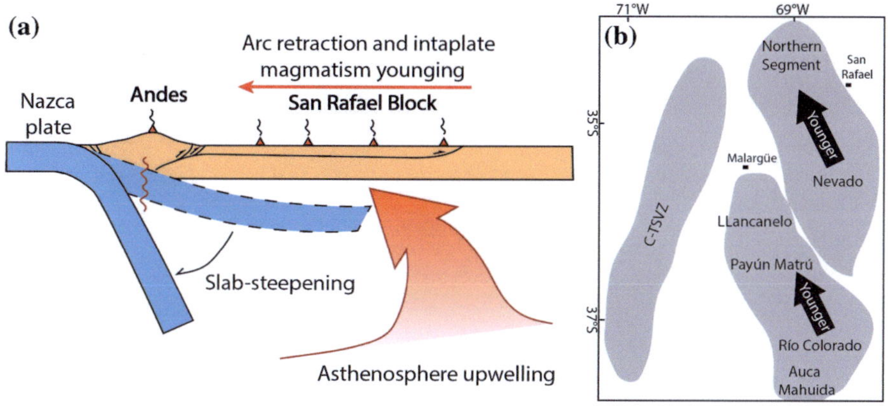

Fig. 4 **a** Tectonic model of Ramos and Folguera (2011) linking asthenospheric upwelling and eruption of the Payenia volcanic province to Plio–Pleistocene Nazca slab steepening. **b** Spatio-temporal distribution of volcanic products of the Payenia volcanic province showing a northwestward age progression of the magmatic activity. Modified from Holm et al. (2016)

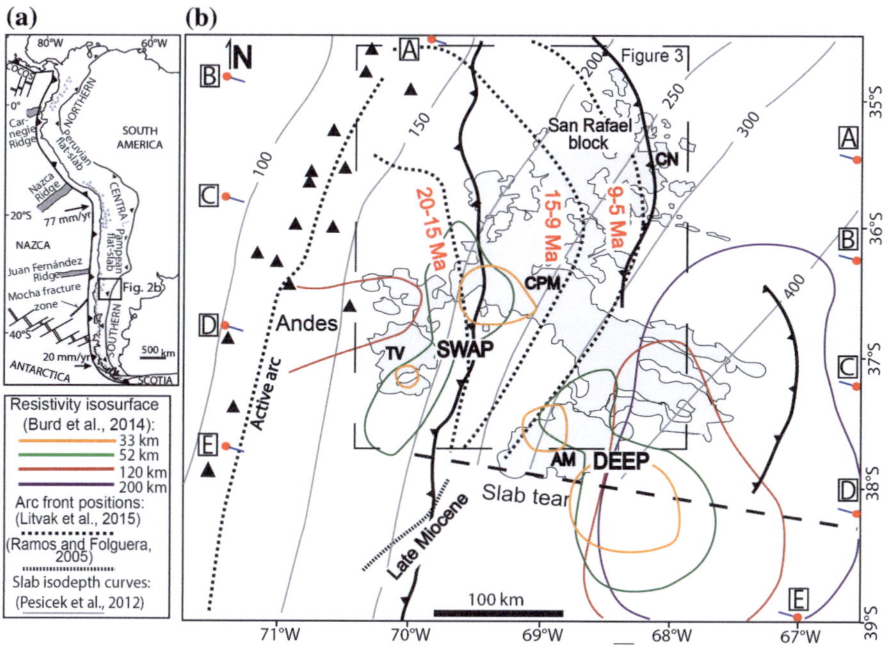

Fig. 5 **a** Tectonic setting of the Andes. **b** Sketch map showing the southern Central Andes between 35° S and 39° S, the Payenia volcanic complex (gray area) and the arc-front paleopositions during Neogene times (based from Ramos et al. 2014 and Litvak et al. 2015). Contours corresponding to the 35 Ω-m isosurface of upper part of DEEP and SWAP mantle plumes at 33, 52, 129, and 200 km are shown (modified from Burd et al. 2014). Abbreviations are TV: Tromen volcano, CPM: Payún Matrú Caldera, CN: Cerro Nevado volcano, AM: Auca Mahuída volcano. Figure modified from Gianni et al. (2017)

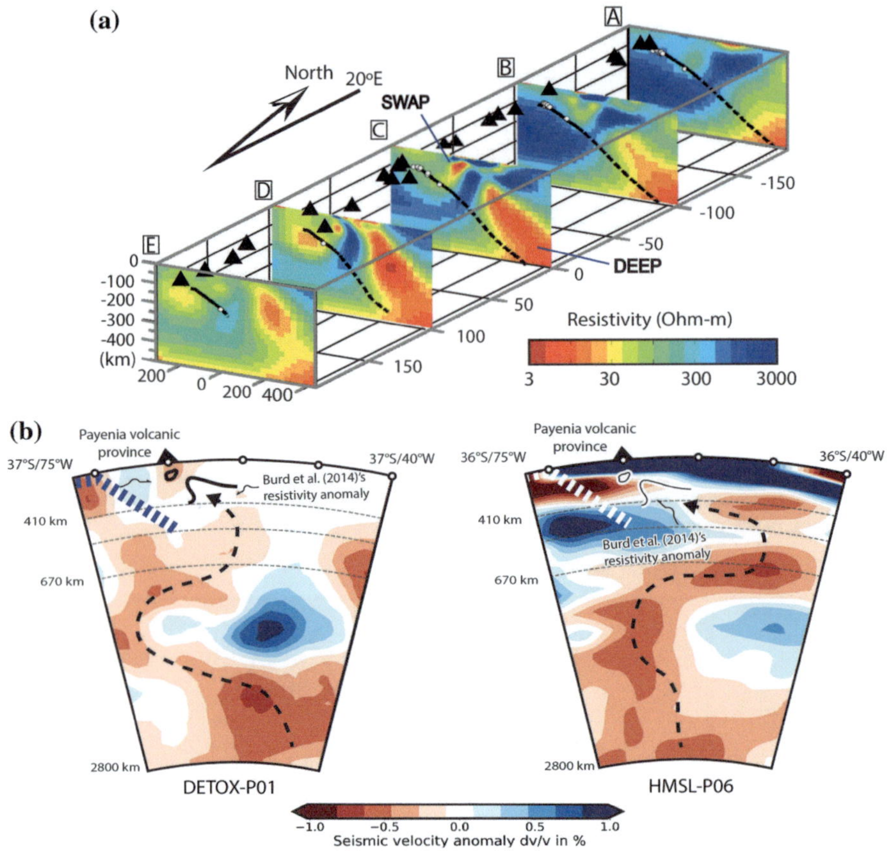

Fig. 6 **a** Resistivity model from Burd et al. (2014) displaying the geometries of SWAP and DEEP mantle anomalies. Solid and dashed black lines on profiles A to E indicate the subducted Nazca slab and white circles are slab seismicity. See profile locations in Fig. 5. **b** P-wave global tomography models DETOX-P01 (Hosseini 2016) and HMSL-P06 (Houser et al. 2008) showing a pronounced low-velocity anomaly extending from the lower to upper mantle connecting with the zone where the SWAP and DEEP mantle anomalies have been described beneath the study area

by recent global P-wave tomography models (DETOX-P01, Hosseini 2016; HMSL-P06, Houser et al. 2008). Although with slightly different geometries, both models show a contorted low-velocity anomaly that rises from the core–mantle boundary connecting with the DEEP anomaly of Burd et al. (2014) at the mantle transition zone (Fig. 6b).

According to Burd et al. (2014), the SWAP and DEEP anomalies were part of a single mantle plume in Miocene times that was decapitated to form the SWAP, when shallow northwestward mantle flow activated during late Pliocene steepening of the Nazca slab. Based on a regional seismic tomography survey, Pesicek et al. (2012) described a vertical and horizontal slab tearing in the Nazca plate at 38° S (Fig. 7). According to these authors, this process would have taken place between

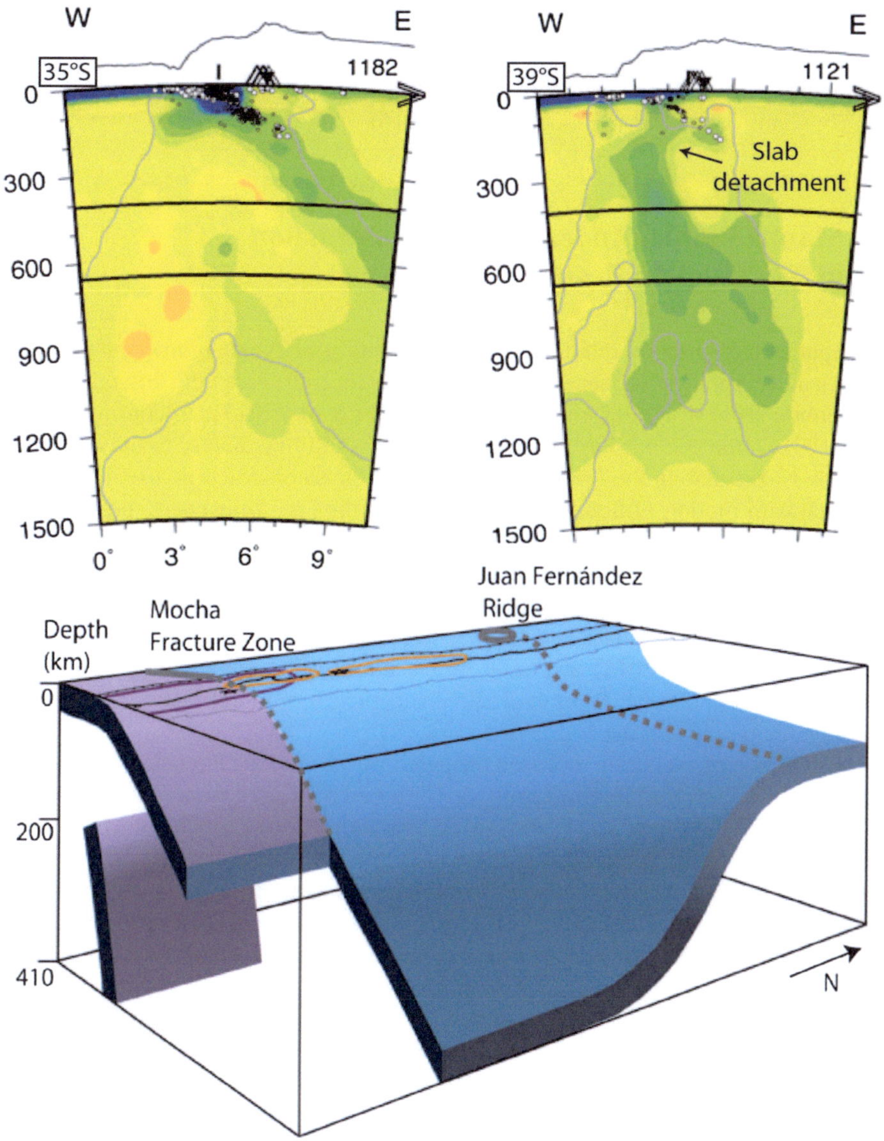

Fig. 7 Seismic tomography at the study area showing a continuous and a detached slab at 35° and 39° S, respectively, and the interpretation of the subducted Nazca plate morphology. Modified from Pesicek et al. (2012)

~ 5 and 3 Ma. Neotectonic activity in the Andean retroarc at studied latitudes has been documented in several studies that showed contractional structures deforming Quaternary volcanic and sedimentary strata (Galland et al. 2007; Sagripanti et al. 2015b; Gianni et al. 2014; Branellec et al. 2016).

3 Mantle Plume–Andean Margin Interaction in the Neuquén Basin

The spatial and temporal relations between the ancient Payenia plume and the South American margin in Neogene times have been recently assessed through a series of tectonic reconstructions carried out by Gianni et al. (2017). According to Burd et al. (2014), the ancient Payenia plume was located immediately to the east of the shallow Nazca slab in the Neogene. However, their reconstruction does not include the westward motion of South America between the Late Miocene to present times which is completely unlikely (see Fig. 14 in Burd et al. 2014). In this regard, a 2-D reconstruction at 37° S of the subduction system during flat-slab full development at 10 Ma including a slab flexural unfolding (Wu et al. 2016), plate kinematic parameters for this region (Pardo-Casas and Molnar 1987; Somoza 1998; DeMets et al. 2010; Somoza and Ghidella 2012), and taking into account different potential Neogene slab angles (Kay et al. 2006a; Dyhr et al. 2013b) depicts a different interaction between the Payenia plume and the Nazca plate (Fig. 8).

Also, this reconstruction included the paleo-position of the Payenia plume in Late Miocene times inferred by Burd et al. (2014) based on a restitution of the SWAP and DEEP anomalies (Fig. 8). This simple reconstruction shows that at 10 Ma the main body of the Payenia plume did not locate in the continental interior as suggested by Burd et al. (2014) but was likely positioned beneath the shallow dipping or flat Nazca slab (Fig. 8). The latter argues by a direct plume–subduction interaction at that time. Further details into this complex interaction are provided by a second plate kinematic reconstruction using a recent global plate reconstruction model (Müller et al. 2016). This reconstruction assumed a fixed mantle plume (Murphy et al. 1998) and introduced a restored position of the SWAP and DEEP anomalies and a time variable arc-front position from Oligocene–Miocene to Miocene–Pliocene (Folguera et al. 2015; Litvak et al. 2015) (Fig. 9a, b). This reconstruction shows that the ancient Payenia plume could have been completely introduced in the subduction system between ~20 and 16 Ma, when the first magmatic arc expansion is documented and at about the time of contraction in the southern Central Andes (Kay et al. 2006a, b; Litvak et al. 2015; Folguera et al. 2015). From 16 to 5 Ma further plume subduction is closely correlated with the arc expansion linked to the complete development of the shallow subduction and intraplate contraction in the San Rafael Block (Fig. 9b). Finally, this reconstruction shows a close relation between plume paleopositions and the place where the slab teared at 38° S between 5 and 3 Ma.

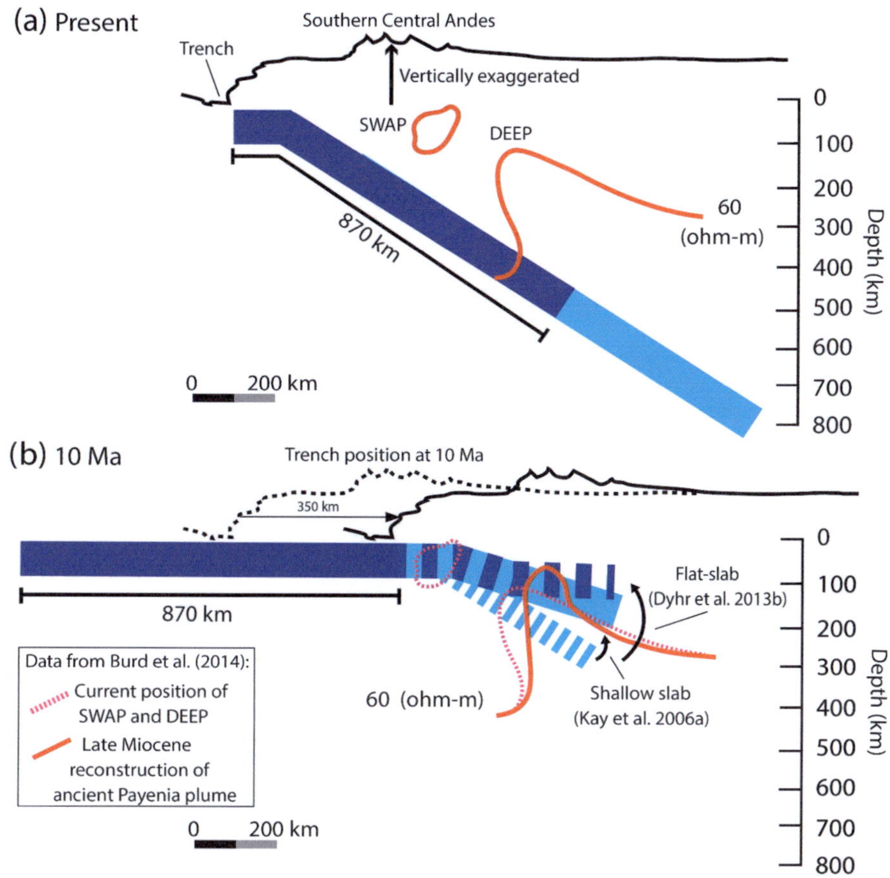

Fig. 8 **a** Current configuration of the convergent margin at 37° S showing the Nazca slab geometry at 36° S (Pesicek et al. 2012) and the 60, 30, and 15 Ω m isoresistivity contours for SWAP and DEEP mantle plumes (Burd et al. 2014). **b** Reconstruction of the convergent margin in Late Miocene times at 10 Ma. The mid-slab line in the Nazca plate was restored to the surface by flexural unfolding. The trench position in Late Miocene was reconstructed taking into account convergence rates and the absolute motion of South America. Plume reconstruction at that time was taken from Burd et al. (2014) (see Gianni et al. 2017 for further details on the reconstruction). Figure modified from Gianni et al. (2017)

4 Discussion and Conclusions

Presently, a satisfactory explanation for the Neogene shallow subduction and related orogenic activity in the Andes between 35° S and 38° S is lacking. In this sense, Ramos et al. (2014) suggested that slab shallowing was driven by the subduction of a buoyant aseismic ridge. Nevertheless, neither the latter authors nor subsequent studies have found evidence for the existence of this oceanic feature Litvak et al. (2017) linked the Payenia shallow subduction to the impingement of the Juan Fernandez

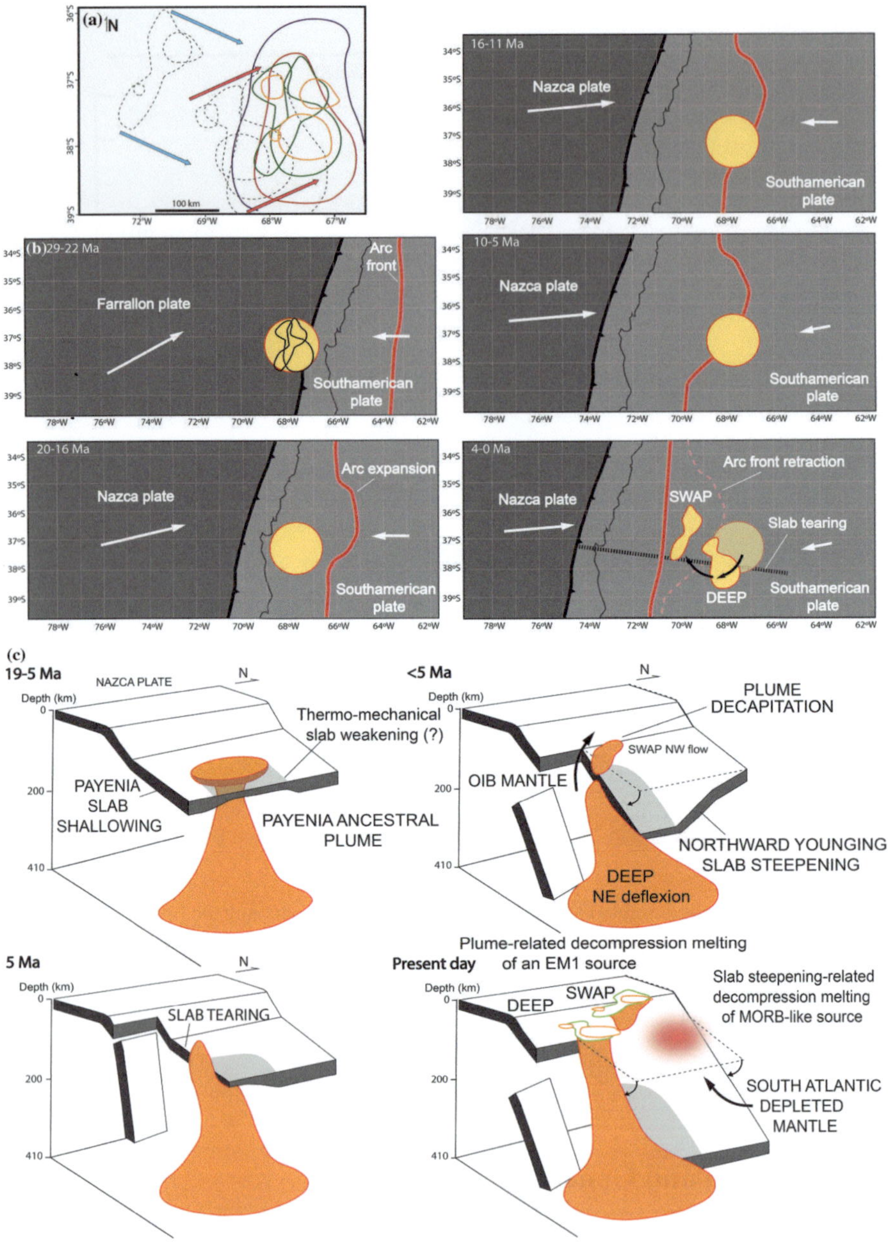
(a)
N
100 km
(b) 29-22 Ma
Farrallon plate
Arc front
Southamerican plate
16-11 Ma
Nazca plate
Southamerican plate
10-5 Ma
Nazca plate
Southamerican plate
20-16 Ma
Nazca plate
Arc expansion
Southamerican plate
4-0 Ma
Nazca plate
Arc front retraction
SWAP
Slab tearing
DEEP
Southamerican plate
(c)
19-5 Ma
NAZCA PLATE
Depth (km)
Thermo-mechanical slab weakening (?)
PAYENIA SLAB SHALLOWING
PAYENIA ANCESTRAL PLUME
<5 Ma
PLUME DECAPITATION
SWAP NW flow
OIB MANTLE
NORTHWARD YOUNGING SLAB STEEPENING
DEEP NE deflexion
5 Ma
SLAB TEARING
Present day
Plume-related decompression melting of an EM1 source
DEEP
SWAP
Slab steepening-related decompression melting of MORB-like source
SOUTH ATLANTIC DEPLETED MANTLE

◀**Fig. 9** **a** Restoration of the ancient Payenia plume position previous to the Pliocene separation of SWAP from DEEP introduced in the plate kinematic reconstruction of Müller et al. (2016). Blue and red arrows are the directions of restoration for SWAP and DEEP mantle anomalies, respectively. **b** Reconstruction of plume–Andean active margin interaction in Neogene times. **c** Gianni et al. (2017)'s tectonic model linking, Payenia plume subduction, slab shallowing, and later oceanic plate tearing, steepening, and plume decapitation. 19–5 Ma: Payenia plume subduction and shallowing of the Nazca plate. 5 Ma: tearing of the Nazca slab due to plume-related thermo-mechanical lithospheric weakening. <5 Ma: upward flow of the Payenia plume during plate steepening from beneath the slab through the teared area. During this process, the plume was separated into SWAP and DEEP mantle anomalies. Present day: normal subduction and the current location of both mantle anomalies in the Andean retroarc. Figure modified from Gianni et al. (2017)

aseismic ridge, often related to the origin of the Pampean flat-slab to the north (33° S–27° 30 S, Fig. 5a). According to these authors, the aseismic ridge influenced the slab buoyancy in the south between 35° and 38° S at the study area. However, most recent kinematic reconstructions have ruled out the effect of this ridge as a driving mechanism for the Chilean-Pampean flat-slab (Skinner and Clayton 2013) and its development has been instead explained by other causes (e.g., hydrodynamic suction, Manea et al. 2012). Thus, any possible connection between the development of the Payenia slab shallowing stage and the subduction of the Juan Fernandez ridge is unlikely.

According to the 2-D reconstruction of Gianni et al. (2017), the Payenia plume was potentially impinging beneath the shallow or flat slab in Late Miocene (Fig. 8). Additionally, the Neogene plate kinematic reconstruction indicates a strong correlation between progressive plume subduction and arc migrations linked to slab shallowing development (Fig. 9b). Thus, Neogene fold and thrust belt reactivation and intraplate deformation in the southern Central Andes could be conceived as a potential case of recent plume-modified orogeny (Gianni et al. 2017). In this regard, dynamic slab uplift produced by the Payenia plume impingement along with a thermal modification of the Nazca slab (Murphy et al. 1998) would have triggered the Neogene slab shallowing. Further slab flattening could have been enhanced by a fast continental overriding at that time. Noteworthy, this model explains several apparently unconnected tectonic events associated with the late evolution of the Neuquén Basin in Neogene times. For instance, the Pliocene slab tearing event at 38° S after shallow subduction and the subsequent slab steepening could also be a consequence of Payenia plume overriding as documented in other examples of plume-modified orogeny (e.g., Murphy et al. 1998; Betts et al. 2009) (Fig. 9c).

Pesicek et al. (2012) suggested that slab tearing localized in the Mocha fracture zone currently subducting at studied latitudes (Fig. 5a). Although this preexisting weakness may have played some role in localizing the slab tear, the observed ESE-WNW direction of the tearing in seismic the tomography is at odds with the NE—direction of the Mocha fracture zone (Fig. 5a) suggesting the action of additional factors influencing this process.

Alternatively, Gianni et al. (2017) noted that when the Nazca slab tearing took place (~5–3 Ma), the ancient Payenia plume partly overlapped the sector where vertical tearing developed (Fig. 9b). These authors suggested that Payenia plume material accumulating at the base of the Nazca slab (Betts et al. 2012) would have triggered thermo-mechanical weakening and/or assimilation (Macera et al. 2008; Obrebski et al. 2010), facilitating slab detachment. Indirect evidence for slab assimilation and erosion is indicated by the presence of eclogite bodies of possibly recycled MORB detected in the mantle source (SWAP and DEEP) of the southern Payenia basalts (Søager et al. 2013). Slab tearing created a window for the asthenosphere to flow during the slab steepening process (Ramos and Folguera 2011; Burd et al. 2014) detaching the upper part of the ancient Payenia plume (SWAP) toward the NW.

In this model, the slab window may have allowed the Payenia plume to escape from below the Nazca slab to locate at the base of the South American plate (Fig. 9c). Similar mantle dynamics have been proposed in other plume–slab interaction settings based on seismic tomography studies (e.g., Obrebski et al. 2010). This model also elegantly explains the diachronous character of slab steepening that started in the south (Gudnason et al. 2012), where the ancient Payenia plume was possibly located and the slab tearing took place and progressing to the north (Figs. 4b and 9c). The latter is consistent with the presence of an EM1 OIB-type mantle source in the south and a depleted MORB-like mantle in the north of the Payenia volcanic province (Søager et al. 2013; Holm et al. 2016).

The recent occurrence of plume subduction in the Neuquén Basin and the well-preserved stages of this interaction could help to identify similar evolutionary stages of plume-modified orogenesis in ancient convergent settings. Finally, considering the westward motion of South America and that the Nazca plate currently holds four plume-rooted hotspots (Juan Fernandez, Easter, Galápagos, and San Felix), it is reasonable to expect additional plume-modified orogenic events in the Andes in the future. In this regard, subduction of the Payenia plume may constitute the first of future plume-modified events.

Acknowledgements This study was supported by the Consejo Nacional de Investigaciones Científicas y Técnicas (CONICET).

References

Betts PG, Giles D, Foden J et al (2009) Mesoproterozoic plume-modified orogenesis in eastern Precambrian Australia. Tectonics 28:1–28

Betts PG, Mason WG, Moresi L (2012) The influence of a mantle plume head on the dynamics of a retreating subduction zone. Geology 40:739–742

Betts PG, Moresi L, Miller MS, Willis D (2015) Geodynamics of oceanic plateau and plume head accretion and their role in Phanerozoic orogenic systems of China. Focus Geosc Front 6:49–59

Branellec M, Niviere B, Callot JP et al (2016) Evidence of active shortening along the eastern border of the San Rafael basement block: characterization of the seismic source of the Villa Atuel earthquake (1929), Mendoza province, Argentina. Geol 153:911–925

Burd AI, Booker JR, Mackie R et al (2014) Three-dimensional electrical conductivity in the mantle beneath the Payún Matrú volcanic field in the Andean backarc of Argentina near 36.5° S: evidence for decapitation of a mantle plume by resurgent upper mantle shear during slab steepening. Geophys J Int 198:812–827

Chang SJ, Ferreira AM, Faccenda M (2016) Upper-and mid-mantle interaction between the Samoan plume and the Tonga-Kermadec slabs. Nat Commun 7

Cobbold PR, Rossello EA (2003) Aptian to recent compressional deformation of the Neuquén Basin, Argentina. Mar Petrol Geol 20:429–443

Dalziel IW, Lawver LA, Murphy JB (2000) Plumes, orogenesis, and supercontinental fragmentation. Earth Planetary Sci Let 178:1–11

DeMets C, Gordon RG, Argus DF (2010) Geologically current plate motions. Geophys J Int 181:1–80

Dyhr CT, Holm PM, Llambías EJ (2013a) Geochemical constraints on the relationship between the Miocene-Pliocene volcanism and tectonics in the Palaoco and Fortunoso volcanic fields, Mendoza Region, Argentina: New insights from 40Ar/39Ar dating, Sr–Nd–Pb isotopes and trace elements. J Volc Geotherm Res 266:50–68

Dyhr CT, Holm PM, Llambías EJ, Scherstén A (2013b) Subduction controls on Miocene back-arc lavas from Sierra de Huantraico and La Matancilla and new 40Ar/39Ar dating from the Mendoza Region, Argentina. Lithos 179:67–83

Druken K, Kincaid C, Griffiths R, Stegman D, Hart S (2014) Plume–slab interaction: the Samoa-Tonga system. Phys Earth Planet Interiors 232:1–14

Fletcher M, Wyman DA (2015) Mantle plume–subduction zone interactions over the past 60 Ma. Lithos 233:162–173

Folguera A, Naranjo JA, Orihashi Y et al (2009) Retroarc volcanism in the northern San Rafael Block (34°–35° 30′ S), southern Central Andes: Occurrence, age, and tectonic setting. J Volc Geotherm Res 186:169–185

Folguera A, Bottesi G, Duddy I (2015) Exhumation of the Neuquén Basin in the southern Central Andes (Malargüe fold and thrust belt) from field data and low-temperature thermochronology. J S Am Earth Sci 64:381–398

French SW, Romanowicz B (2015) Broad plumes rooted at the base of the Earth's mantle beneath major hotspots. Nat 525:95–99

Galland O, Hallot E, Cobbold PR et al (2007) Volcanism in a compressional Andean setting: A structural and geochronological study of Tromen volcano (Neuquén province, Argentina). Tectonics 26. https://doi.org/10.1029/2006tc002011

Giambiagi LB, Bechis F, García V, Clark A (2005) Temporal and spatial relationship between thick- and thin-skinned deformation in the thrust front of the Malargüe fold and thrust belt, Southern Central Andes. In: Abstracts of the 6 international symposium on Andean geodynamics, Barcelona, 12–14 Sept 2005

Giambiagi L, Mescua J, Bechis F, Tassara A, Hoke G (2012) Thrust belts of the southern Central Andes: along-strike variations in shortening, topography, crustal geometry, and denudation. Geol Soc Am Bull 124:339–1351

Gianni GM, Sagripanti L, Folguera A et al (2014) Tectónica cuaternaria en el retroarco Andino a la latitud del volcán Tromen (37° S). Rev Asoc Geol Argent 71:513–525

Gianni GM, Garcia HPA, Lupari M, Pesce A, Folguera A (2017) Plume overriding triggers shallow subduction and orogeny in the southern Central Andes. Gondwana Res 49:387–395. https://doi.org/10.1016/j.gr.2017.06.011

Gudnason J, Holm PM, Søager N, Llambías EJ (2012) Geochronology of the late Pliocene to recent volcanic activity in the Payenia back-arc volcanic province, Mendoza Argentina. J S Am Earth Sci 37:191–201

Hosseini K (2016) Global multiple-frequency seismic tomography using teleseismic and core-diffracted body waves. Dissertation, LMU München, Fakultät für Geowissenschaften

Holm PM, Søager N, Alfastsen M, Bertotto GW (2016) Subduction zone mantle enrichment by fluids and Zr–Hf-depleted crustal melts as indicated by backarc basalts of the Southern Volcanic Zone, Argentina. Lithos 262:135–152
Houser C, Masters G, Shearer P, Laske G (2008) Shear and compressional velocity models of the mantle from cluster analysis of long-period waveforms. Geophys J Int 174(1):195–212
Kay SM, Mancilla O, Copeland P (2006a) Evolution of the Backarc Chachahuén volcanic complex at 37° S latitude over a transient Miocene shallow subduction zone under the Neuquén Basin. In: Kay SM, Ramos VA (eds) Late Cretaceous to recent magmatism and tectonism of the Southern Andean margin at the latitude of the Neuquén basin (36–39° S). Geological Society of America, SP 407, pp 215–246
Kay SM, Burns WM, Copeland P, Mancilla O (2006b) Upper Cretaceous to Holocene magmatism and evidence for transient Miocene shallowing of the Andean subduction zone under the northern Neuquén Basin. In: Kay SM, Ramos VA (eds) Late Cretaceous to recent magmatism and tectonism of the Southern Andean margin at the latitude of the Neuquén basin (36–39° S). Geological Society of America, SP 407, pp 19–60
Kay SM, Copeland P (2006) Early to middle Miocene backarc magmas of the Neuquén Basin: Geochemical consequences of slab shallowing and the westward drift of South America. In: Kay SM, Ramos VA (eds) Late Cretaceous to Recent magmatism and tectonism of the Southern Andean margin at the latitude of the Neuquén basin (36–39° S). Geological Society of America, SP 407, pp 185–214
Kincaid C, Druken KA, Griffiths RW, Stegman DR (2013) Bifurcation of the Yellowstone plume driven by subduction-induced mantle flow. Nat Geosc 6:395–399
Litvak VD, Spagnuolo MG, Folguera A et al (2015) Late Cenozoic calc-alkaline volcanism over the Payenia shallow subduction zone, South-Central Andean back-arc (34° 30′–37° S), Argentina. J S Am Earth Sci 64:365–380
Litvak VD, Poma S, Jones RE et al (2017) The Late Paleogene to Neogene volcanic arc in the Southern Central Andes (28°–37° S), Argentina. In: Folguera et al (eds) The making of the Chilean and Argentinean Andes. Springer, Berlin
Macera P, Gasperini D, Ranalli G, Mahatsente R (2008) Slab detachment and mantle plume upwelling in subduction zones: an example from the Italian South-Eastern Alps. J Geod 45(1):32–48
Manea VC, Pérez-Gussinyé M, Manea M (2012) Chilean flat slab subduction controlled by overriding plate thickness and trench rollback. Geology 40(1):35–38
Mériaux CA, Mériaux AS, Schellart WP, Duarte JC, Duarte SS, Chen Z (2016) Mantle plumes in the vicinity of subduction zones. Eart Planet Sci Lett 454:166–177
Müller RD, Seton M, Zahirovic S et al (2016) Ocean basin evolution and global-scale plate reorganization events since Pangea breakup. Annl Rev Earth Planet Sci 44:107–138
Murphy JB, Oppliger GL, Brimhall GH, Hynes A (1998) Plume-modified orogeny: an example from the western United States. Geology 26:731–734
Murphy JB, van Staal CR, Keppie JD (1999) Middle to late Paleozoic Acadian orogeny in the northern Appalachians: a laramide-style plume-modified orogeny? Geology 27:653–656
Murphy JB, Keppie JD (2005) The Acadian orogeny in the northern Appalachians. Int Geol Rev 47:663–687
Murphy JB (2016) The role of the ancestral yellowstone plume in the tectonic evolution of the Western United States. Geosc Canada 43:231–250
Nance RD, Murphy JB, Santosh M (2014) The supercontinent cycle: a retrospective essay. Gondwana Res 25(1):4–29
Nullo FE, Franchi M, González P et al (1993) Mapa Geológico de la Provincia de Río Negro. Dirección Nacional del Servicio Geológico, Buenos Aires
Nullo FE, Panza JL, Blasco G (1999) El Jurásico y Cretácico de la Cuenca Austral. In: Caminos, R. (Ed.), Geología Argentina. Servicio Geológico Minero Argentino, pp 528–535
Nullo FE, Stephens G, Otamendi J, Baldauf P (2002) El volcanismo del Terciario superior del sur de Mendoza. Rev Asoc Geol Argent 57:119–132

Obrebski M, Allen RM, Xue M, Hung SH (2010) Slab-plume interaction beneath the Pacific Northwest. Geophys Res Let 37:L14305. https://doi.org/10.1029/2010GL043489
Oppliger GL, Murphy JB, Brimhall GH Jr (1997) Is the ancestral yellowstone hotspot responsible for the tertiary "Carlin" mineralization in the Great Basin of Nevada? Geology 25:627–630
Ostera H, Linares E, Haller M (1999) Paramillos Altos intrusive belt, Southern Mendoza, Argentina. Ages, chemical and isotopic constraints. In: Abstracts of the 2 South American symposium on isotope geology, Villa Carlos Paz, 10–15 Apr 1999
Pardo-Casas F, Molnar P (1987) Relative motion of the Nazca (Farallon) and South American plates since Late Cretaceous time. Tectonics 6:233–248
Pesicek JD, Engdahl ER, Thurber CH et al (2012) Mantle subducting slab structure in the region of the 2010 M8.8 Maule earthquake (30–40 S), Chile. Geophys J Int 191:317–324
Ramos VA, Barbieri M (1989) El volcanismo cenozoico de Huantraico: edad y relaciones isotópicas iniciales, provincia del Neuquén. Rev Asoc Geol Argent 43:210–223
Ramos VA, Folguera A (2011) Payenia volcanic province in the Southern Andes: an appraisal of an exceptional Quaternary tectonic setting. J Volcanol Geother Res 201:53–64
Ramos VA, Litvak V, Folguera A, Spagnuolo M (2014) An Andean tectonic cycle: from crustal thickening to extension in a thin crust (34–37° S). Focus Geosci Front 5:351–367
Sagripanti L, Naipauer M, Bottesi G et al (2011) U/Pb ages on detrital zircons in the Southern Central Andes Neogene foreland (36°–37° S): constraints on Andean exhumation. J S Am Earth Sci 32:554–565
Sagripanti L, Aguirre-Urreta B, Folguera A, Ramos VA (2015a) The Neocomian of Chachahuén (Mendoza, Argentina): evidence of a broken foreland associated with the Payenia flat-slab. Geol Soc London SP 399:203–219
Sagripanti L, Rojas Vera E, Gianni GM et al (2015b) Neotectonic reactivation of the western section of the Malargüe fold and thrust belt (Tromen volcanic plateau, Southern Central Andes). Geomorphology 232:164–181
Skinner SM, Clayton RW (2013) The lack of correlation between flat slabs and bathymetric impactors in South America. Earth Planet Sci Let 371:1–5
Søager N, Holm PM, Llambías EJ (2013) Payenia volcanic province, southern Mendoza, Argentina: OIB mantle upwelling in a backarc environment. Chem Geol 349–350:36–53
Somoza R (1998) Updated Nazca (Farallon)-South America relative motions during the last 40 my: implications for mountain building in the central Andean region. J S Am Earth Sci 11:211–215
Somoza R, Ghidella M (2012) Late Cretaceous to Recent plate motions in western South America revisited. Earth Planet Sci Let 331–332:152–163
Spagnuolo MG, Litvak VD, Folguera A, Bottesi G, Ramos VA (2012) Neogene magmatic expansion and mountain building processes in the southern Central Andes, 36–37°S, Argentina. J Geodyn 53:81–94
Wu J, Suppe J, Lu R, Kanda R (2016) Philippine Sea and East Asian plate tectonics since 52 Ma constrained by new subducted slab reconstruction methods. J Geophys Res Solid Earth 121:4670–4741
Yrigoyen M (1993) Los depósitos sinorogénicos terciarios. In: Ramos VA (ed) Geología y Recursos minerales de Mendoza. Asociación Geológica Argentina, Buenos Aires, pp 123–148

Quaternary Deformation in the Neuquén Basin, Explained by the Interaction Between Mantle Dynamics and Tectonics

Lucía Sagripanti, Bruno Colavitto, Ana Astort and Andrés Folguera

Abstract Quaternary deformations described in the retro-arc region at the latitudes of the Neuquén Basin can be divided into two main groups: a northern group characterized by Quaternary deformation zones concentrated near the main topographic breaks of different morphostructural systems, and a second group located in the southern Neuquén Basin distinguished by disconnected, sparced and noncontinuous Quaternary deformational zones. In the northern Neuquén Basin, evidence of active deformation is associated with the Frontal Cordillera (33°–34° S); while in the foreland area, young deformations concentrate in the San Rafael Block. In the southern Neuquén Basin, a western deformational belt constitutes the continuation to the north of an intra-arc fault system associated with dip- and strike-slip displacements in a strain-partitioned regime (Liquiñe–Ofqui fault system). At these latitudes, to the east, isolated evidence of Quaternary deformation, regional uplift and development of non-equilibrated fluvial profiles are recognized in the Tromen and Auca Mahuida volcanic plateaux and sierra de Cara Cura-sierra de Reyes area. These systems are short and unconnected and have been explained through an intricate pattern of asthenospheric anomalies evidenced from magnetotelluric data. These mantle anomalies could be related to the tearing of the subducted Nazca plate at depth evidenced by seismic-tomographic data. We therefore suggest that the thermally weakened crust at the southern Neuquén Basin latitudes could be the main control responsible for focalizing contractional, extensional and transpressional deformations in isolated mountain systems.

Keywords Quaternary deformation · Mantle dynamics · Slab-tear · Retro-arc

L. Sagripanti (✉) · B. Colavitto · A. Astort · A. Folguera
Facultad de Ciencias Exactas y Naturales, Departamento de Ciencias Geológicas, Instituto de Estudios Andinos "Don Pablo Groeber" (IDEAN), CONICET, Universidad de Buenos Aires, Buenos Aires, Argentina
e-mail: lusagripanti@gmail.com

D. Kietzmann and A. Folguera (eds.), *Opening and Closure of the Neuquén Basin in the Southern Andes*, Springer Earth System Sciences,
https://doi.org/10.1007/978-3-030-29680-3_21

1 Introduction

Most of the studies focused on the Quaternary deformation affecting the Southern Andes over the eastern Argentinean Andes are localized around the Chilean-Pampean flat-slab region (27°–33° S), where more than 90% of the Quaternary deformation concentrates (Costa et al. 2006a). There, a flat subduction system determines a high-amplitude orogen, where the foreland area is fragmented into a series of basement blocks such as the Sierras Pampeanas and southern Precordillera (Bastias et al. 1993; Costa and Vita-Finzi 1996; Cortes and Costa 1996; Costa et al. 1999, 2006b; Verges et al. 2006). To the south, the Neuquén and Cuyo basins are developed where the subduction of the Nazca Plate becomes gradually moderately steepened (30° E), and the broken foreland morphology is less defined (Fig. 1). In the northern sector of the Cuyo Basin (33° S–34° S), the Quaternary deformation is linked to the topographic

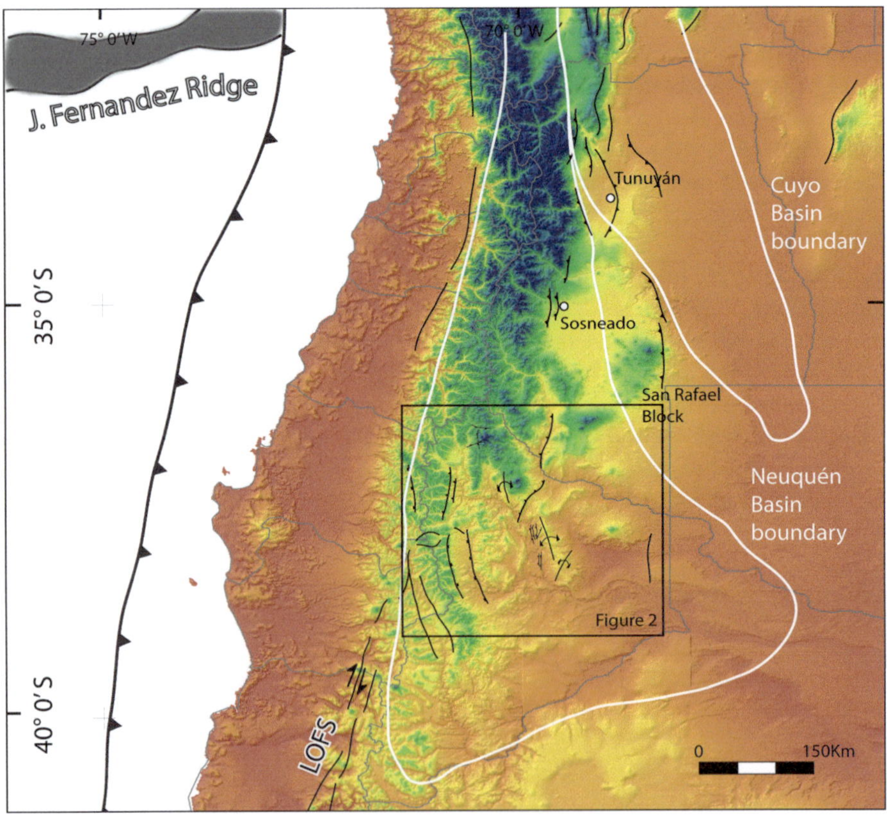

Fig. 1 Digital elevation model with main structures where Quaternary deformation has been described along the Southern Central Andes and particularly at the Neuquén Basin latitudes (black solid lines indicate described Quaternary deformation in map view). White lines represent the Neuquén and Cuyo basins boundaries

breaks, associated with the orogenic front of the Frontal Cordillera (García and Casas 2014; Cortes et al. 1999; Perucca et al. 2009). The Frontal Cordillera had a rapid uplift in late Miocene–Pliocene times through an east-verging basement-cored thrust (Giambiagi and Ramos 2002; Irigoyen et al. 2002; Giambiagi et al. 2003, 2014; Hoke et al. 2014; García and Casas 2014, among others). The progression of the deformation front toward the Frontal Cordillera piedmont during the late Pliocene – early Pleistocene has been registered by coarse conglomerates at the top of the growing anticlines and by the presence of progressive unconformities at their limbs (Yrigoyen 1993; Garcia et al. 2005). Further south (35° S), in the northern part of the Neuquén Basin near the Sosneado locality, scarce and localized evidence of Quaternary deformational activity place along the Andean topographic front (Baker et al. 2009; Turienzo 2010). To the east in the foreland area, young deformations also concentrate in the San Rafael Block (34°–36° S), where Quaternary deformations and prehistorical ruptures associated with seismic events have been recognized (Costa et al. 2006c; Folguera et al. 2009; Branellec et al. 2016) (Fig. 1). At 36°–39° S latitudes, crustal seismicity on the eastern Andes diminishes substantially, and Andean morphology becomes narrower and lower (Ramos et al. 2004; Nacif et al. 2017), in coincidence with strong changes in neotectonic patterns (Folguera et al. 2015a; Colavitto et al. 2019; Sagripanti et al. 2018) (Figs. 1 and 2). The purpose of this chapter is to characterize these contrasting neotectonic patterns at the latitudes of the Neuquén Basin. Then, we discuss possible sublithospheric controls on these

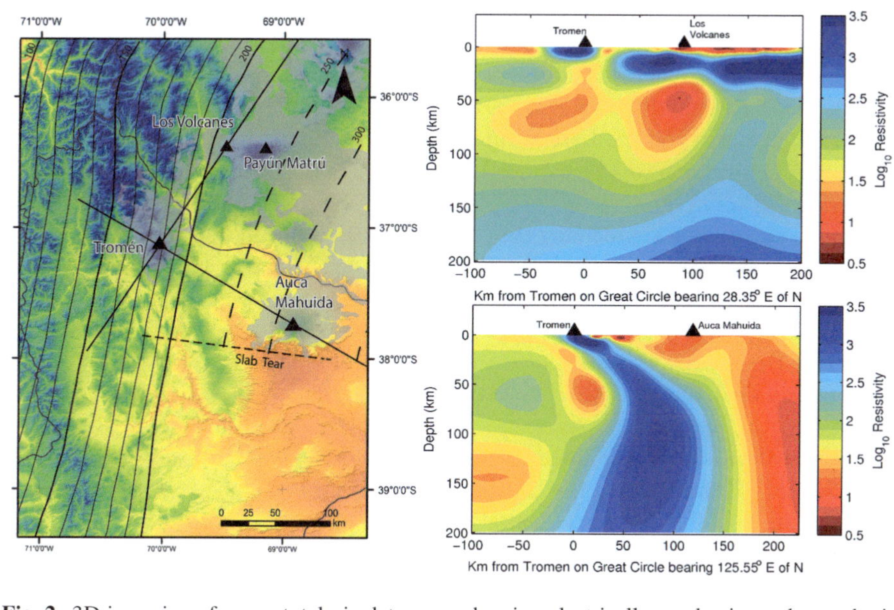

Fig. 2 3D inversion of magnetoteluric data array showing electrically conductive asthenospheric anomalies in 2D profiles (Burd et al. 2014). On the left, a digital elevation model is shown with reference to the indicated sections

changes from tomographic data (Pesicek et al. 2012) and three-dimensional electrical conductivity models (Burd et al. 2014).

2 Sublithospheric Structure of the Southern Central Andes (36°–39° S)

The Neuquén Basin infill between 36° and 39° S has been exhumed through different fold and thrust belts with contrasting morphology and kinematics of deformation. From north to south, the Malargüe, Chos Malal and Agrio fold and thrust belts (MFTB; CMFTB and AFTB) are the product of Cretaceous to Neogene deformations that partially inverted Late Triassic–Early Jurassic extensional depocenters (Kozlowski et al. 1993; Manceda and Figueroa 1995; Zamora Valcarce et al. 2006, 2009; Sagripanti et al. 2011; Folguera et al. 2015b; among others). However, locally these structures were reactivated during Pleistocene–Holocene times in isolated structures such as those recognized in the Loncopué Trough, the Cara Cura Range, the Tromen volcanic plateau and Cerro Domuyo (Folguera et al. 2010; Colavitto et al. 2019; Galetto et al. 2018; Rojas Vera et al. 2014, 2015; Sagripanti et al. 2015, 2018; Messager et al. 2010, 2013) (Fig. 2).

Recent magnetotelluric studies (MT) focused in the retro-arc area have shown the occurrence of an intricate pattern of electrically conductive asthenospheric anomalies feeding the volcanic fields of the Payenia volcanic plateau (Fig. 2) (Burd et al. 2014). These anomalies are interpreted as being fed from a deep source related to a plume-like feature located further to the east, branched into a series of shallower astenospheric anomalies (Burd et al. 2014) (see Chapter "Plume Subduction Beneath the Neuquén Basin and the Last Mountain Building Stage of the Southern Central Andes"). 3D inversion of MT data reveals an eastern conductive anomaly denominated deep eastern plume (DEEP), dipping to the east and located above the Nazca subducted slab with a source near the top of the 410 km deep transition zone (Burd et al. 2014), that is located directly beneath the Auca Mahida volcano (Figs. 2 and 3). The western conductive anomaly denominated shallow western asthenospheric plume (SWAP) developed to the west and rises from the subducted slab surface impacting the shallow crust under the Payun Matrú Caldera, the Tromen volcano and the Cara Cura Range. The SWAP and DEEP asthenospheric anomalies are interpreted as being connected in Miocene times when the uppermost part of the DEEP anomaly would have decapitated to form the SWAP anomaly (Burd et al. 2014) (see Chapter "Plume Subduction Beneath the Neuquén Basin and the Last Mountain Building Stage of the Southern Central Andes").

After the subduction-related Maule earthquake (2010), a complex geometry of the subducted Nazca plate at 38° S was imaged that was previously unknown (Pesicek et al. 2012). Thus, seismic-tomographic data showed a strong tearing affecting the subducted slab south of 38° S, steepening the older than 5 My Nazca plate to a nearly vertical position to the south of 38° S. This tearing coincides with the northern edge

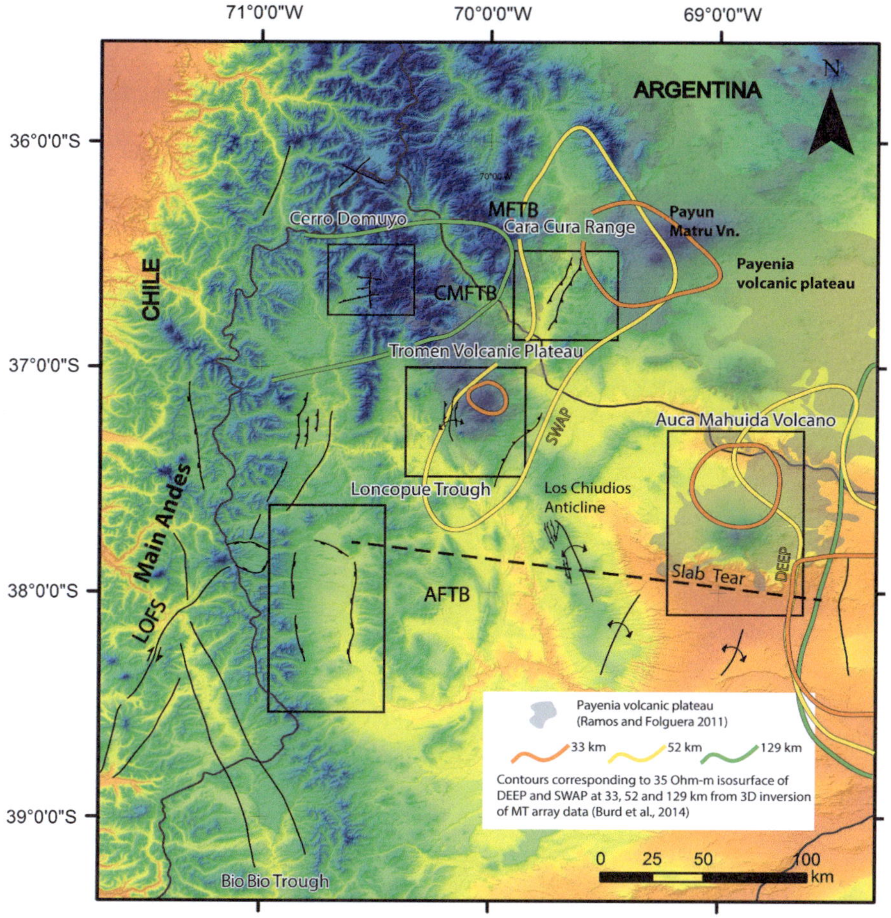

Fig. 3 Quaternary structures recognized in the Southern Central Andes at the latitudes of the Neuquén Basin (see inset in Fig. 1) (taken from Folguera et al. 2010; Galetto et al. 2018; Messager et al. 2010, 2013; Rojas Vera et al. 2014; Sagripanti et al. 2015, 2018; among others). Orange, yellow and green curves represent different isoresistivity values for the asthenospheric mantle in the model of Burd et al. (2014). Dashed line represents the horizontal projection of the tearing affecting the Nazca plate at depth, described by Pesicek et al. (2012) from seismic-tomographic data (MFTB: Malargüe fold and thrust belt, CMFTB: Chos Malal fold and thrust belt, AFTB: Agrio fold and thrust belt; LOFS: Liquiñe-Ofqui Fault System)

of the Loncopué Trough (Fig. 3) (see Chapter "Pliocene to Quaternary Retro-arc Extension in the Neuquén Basin: Geophysical Characterization of the Loncopué Trough").

3 Surface Uplift and Quaternary Deformation Evidence

Large shield-like volcanoes, such as Auca Mahuida, Payun Matru and Tromen volcanic plateaux (Fig. 3), emplaced over the Chos Malal and Malargüe fold and thrust belts as part of the western Payenia volcanic plateau (Gudnason et al. 2012). These centers have a back-arc (intraplate-like) geochemical signature and are separated from the volcanic arc to the west (Llambías et al. 1982; Litvak et al. 2015; Kay et al. 2006). In particular, three different areas through this volcanic plateau, the Cara Cura Range, the Tromen volcanic plateau and the Auca Mahuida volcano, show evidence of Quaternary deformation-exhumation.

The Tromen volcanic plateau consists of a basal section of 1.4–1.9 Ma unconformably emplaced over the Mesozoic sediments of the Neuquen Basin (Sagripanti et al. 2015; Galland et al. 2007). This basal volcanic plateau is deformed through an array of west-vergent faults (Figs. 3 and 4) (Galland et al. 2007; Folguera et al. 2008; Sagripanti et al. 2015). Steepness and concavity indexes calculated in river profiles show that the fluvial network is in a non-equilibrium state (Fig. 4). Drainage patterns show a correlation between the position of knickpoints and the neotectonic structures (Fig. 4) (Sagripanti et al. 2015, 2018).

Wind gaps, small ridges and linear depressions with an echelon distribution and river deflections reflect that the drainage is gradually deflected and defeated by the growth of Quaternary structures (Sagripanti et al. 2015) (Figs. 4 and 5). Through this reorganization associated with the Pleistocene orogenic uplift, the fluvial network changes its pattern from a radial fluvial network, typical of volcanic edifices, to a parallel—to the neotectonic structure fluvial network.

At these latitudes to the east, successive changes in the base level of the main drainage of the Colorado River have been described as well as anomalies in its terrace profiles. The Colorado River shows a peculiar behavior, with a nearly southward orthogonal deviation, producing a rotation of the channel section (Figs. 3 and 6). Nivière et al. (2013) inferred through a geomorphological analysis that this river would have started with an E-directed flow that gradually changed to SE-, S- and finally to a SW-directed flow (Fig. 6), leaving four paleo-channels defining a particular arrangement in which each of them has a progressive lower altitude (5-2; Fig. 6). Fan-shape morphology, river deviation and progressive deepening of subsequent channels were interpreted as being associated with changes in base level, presumably related to surface uplift, connected with the DEEP astenospheric anomaly (see Chapter "Plume Subduction Beneath the Neuquén Basin and the Last Mountain Building Stage of the Southern Central Andes").

To the west of the Auca Mahida volcano, Los Chihuidos Anticline develops as a long wavelength (30 km) positive structure between 37° 30′ and 38° S (Fig. 3). This feature has a steeper west limb, suggesting a west vergence of a basement-cored anticline, interpreted as the result of the inversion of Triassic half-grabens during Miocene times (Cobbold and Rossello 2003; Cristallini et al. 2006; Mosquera and Ramos 2006). Messager et al. (2010, 2013) analyzed Neuquén River remnant

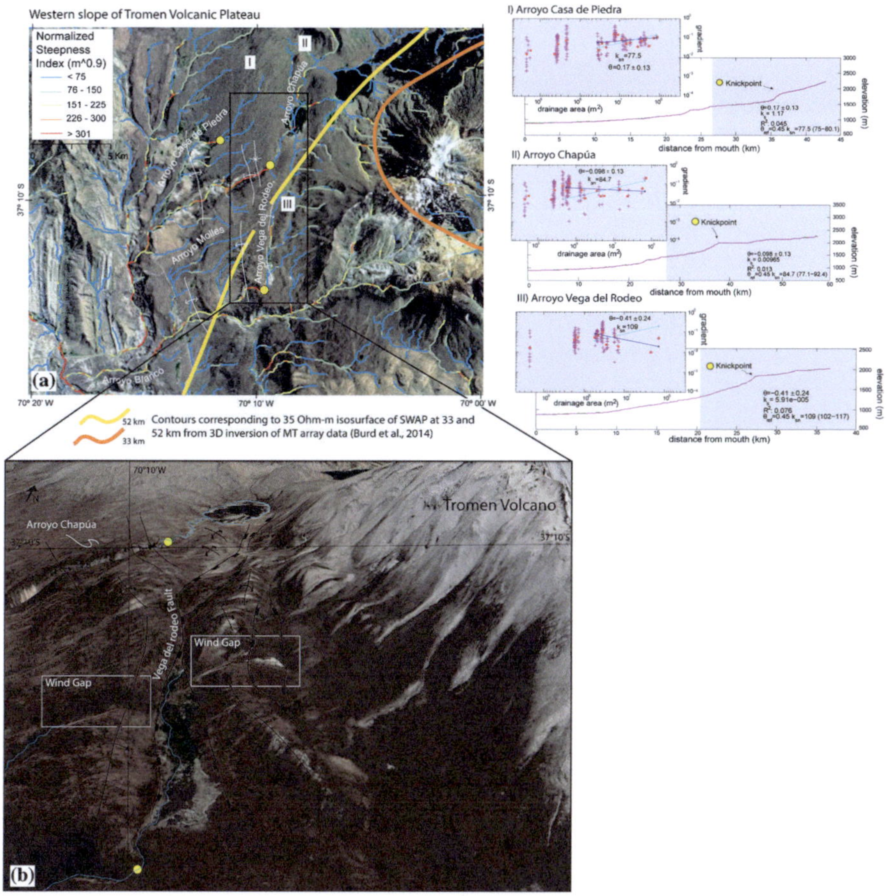

Fig. 4 **a** Drainage network at the western slope of the Tromen volcanic plateau represented by the normalized steepness index (ksn). Right, knickpoints calculated from longitudinal river profiles and (LogS-LogA) regression used to calculate concavity of selected streams (modified from Sagripanti et al. 2018). **b** Oblique ikonos image with faults associated with anticlines affecting Pleistocene basalts associated with wind gaps, river deflections and sang ponds. Yellow dots represent knickpoints location

Pleistocene fluvial terraces flanking this structure, showing that the terraces are tilted and folded at the southern end of the anticline, implying its Quaternary activity.

To the north, the Cara Cura Range is a basement structure composed of two N-trending anticlines associated with a back-thrust system (Fennell et al. 2015). To the west short-wavelength anticlines, affecting Upper Cretaceous rocks and Quaternary fluvial and volcanic deposits show evidence of young deformation (Fig. 3). Faulting and folding of young lava plateaux, water gaps, tilted and folded pediment surfaces and anomalous drainage patterns suggest landscape rejuvenation in response to Quaternary deformation (Fig. 7) (Colavitto et al. 2019; 2016).

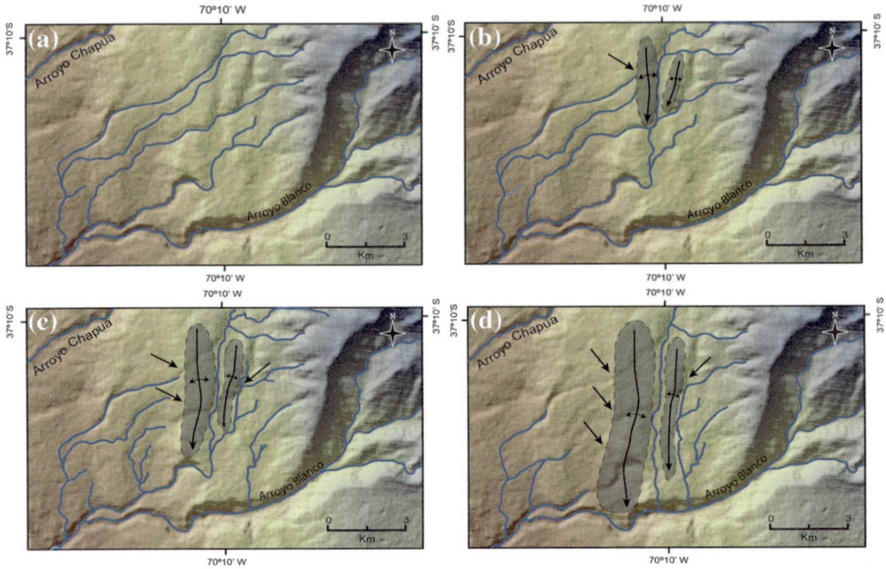

Fig. 5 Drainage evolutionary model of the western slope of the Tromen volcanic plateau. Black arrows indicate wind gaps, while gray-shaded ellipsoids represent the uplifted neotectonic anticlines. **a** and **b** represent the inferred drainage pattern before the inception of the neotectonic structures after correlating wind gaps with dissected channels. **c** Present drainage pattern with described wind gaps localized at the mid-part of both neotectonic anticlines. **d** Hypothetical future drainage pattern following the direction of propagation of the anticlines (taken from Sagripanti et al. 2015)

Quaternary deformation is described also in the highest Andes region to the west of the Payenia volcanic plateau. The Cerro Domuyo and the Loncopue Trough areas are located in a retro-arc position showing localized neotectonic deformations (Fig. 3). In particular, the Cerro Domuyo is a basement structure uplifted in Late Cretaceous and late Miocene times (Spagnuolo et al. 2012; Folguera et al. 2015b) from the inversion of Late Triassic structures (Carbone et al. 2011; Giambiagi et al. 2009; Bechis et al. 2010; Sagripanti et al. 2014, among others). From the upper Pliocene to the lower Pleistocene, dome complexes and mafic volcanism were controlled by previous structures (Miranda et al. 2006; Kay et al. 2006) that show evidence of Quaternary reactivation (Galetto et al. 2018). The neotectonic Domuyo fault system is located to the north of the Liquine-Ofqui and Antiñir-Copahue fault zones that run through the arc front (Sagripanti et al. 2018; Lavenu and Cembrano 1999; Folguera et al. 2004). Nevertheless, its development is also linked to the westernmost isosurface of the SWAP conductive anomalies described by Burd et al. (2014), implying some kind of connection with a thermally weakened lower crust as other neotectonic structures in the area (Fig. 3).

Further south, between 38° and 39° S, the Loncopue Trough is an elongate-topographic depression of nearly 200 km of extension and 50 km width (Fig. 3) (see

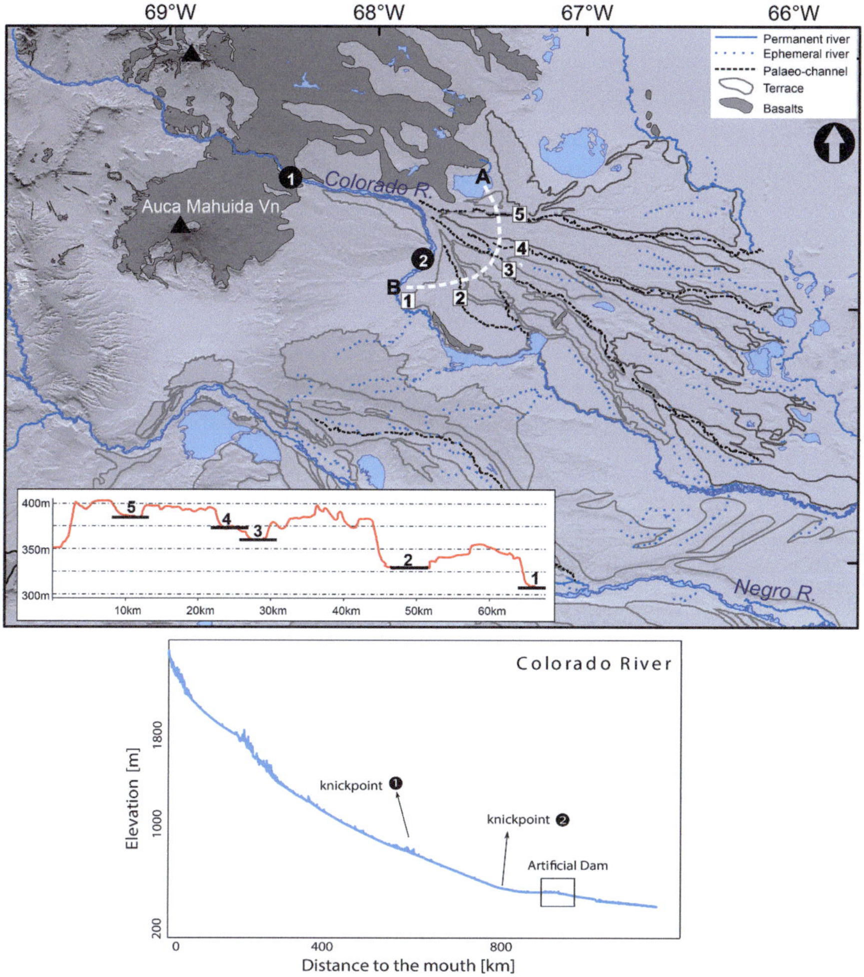

Fig. 6 Digital elevation model showing the proposed dynamics of the Colorado River through time (based on Nivière et al. 2013). Note the fan-shape arrangement defined by the different paleo-channels (5-1 in dashed white line). Topographic profile indicated with the white dashed line, crossing the different paleo-channels of the Colorado River that exhibits a decrease in the different terraces altitudes from location 5 to 1. The black dots correspond to knickpoints identified over the equilibrium profiles shown below

Chap. "Pliocene to Quaternary Retro-arc Extension in the Neuquén Basin: Geophysical Characterization of the Loncopué Trough"). The evolution of this depocenter started with an extensional regime during Oligocene–early Miocene times producing an asymmetric graben (Jordan et al. 2001). This was later extensionally reactivated during Pliocene–Pleistocene (Ramos and Folguera 2005), where Pliocene

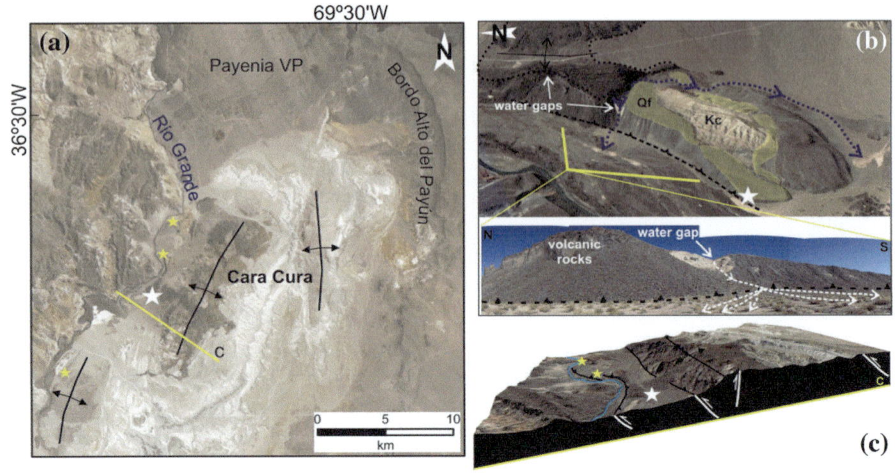

Fig. 7 **a** Ikonos image of the Cara Cura Range Anticline showing the location of Quaternary deformation sites (yellow and white stars indicate places where Quaternary deformation was described; Colavitto et al. 2019, 2016). **b** 3D model showing water gaps caved in the foothills of the Cara Cura Range; below: field picture of the same structure; **c** interpretation of the regional 3D structural model (modified from Sagripanti et al. 2018)

basandesitic sequences linked to wedge-like depocenters (Varekamp et al. 2010), followed by 2.6–2 Ma silicic volcanic successions associated with ignimbrites and distal pyroclastic flows related to a series of caldera collapses, east of the arc front (Rojas Vera et al. 2014). Finally, between 800 and 300 ka, volcanic eruptions coexisted with glacial sedimentation particularly to the west of the trough. Then, postglacial volcanic activity is described until historical times (Folguera et al. 2010; Rojas Vera et al. 2014).

Normal faults are limiting the trough, through which main rivers are incising over their channel slopes adjusting their equilibrium profiles (García Morabito 2004: Folguera et al. 2010). The geometry of the trough is visualized using swath profiles, where primary structures produce topographic breaks of nearly 800 m high (Fig. 8).

An attenuated Moho up to 33 km depth across the Loncopue Trough and western Agrio fold and thrust belt was determined from receiver function analyses and gravimetric data (Yuan et al. 2006; Folguera et al. 2007; Rojas Vera et al. 2014). This crustal attenuation produced an isostatically non-equilibrated relief south of the slab tearing that affects the Nazca plate at depth (Folguera et al. 2007, 2010).

4 Conclusions

Quaternary deformation at the latitudes of the Neuquén Basin is concentrated in the Loncopue Trough, Cerro Domuyo, Tromen volcanic plateau, Cara Cura Range and

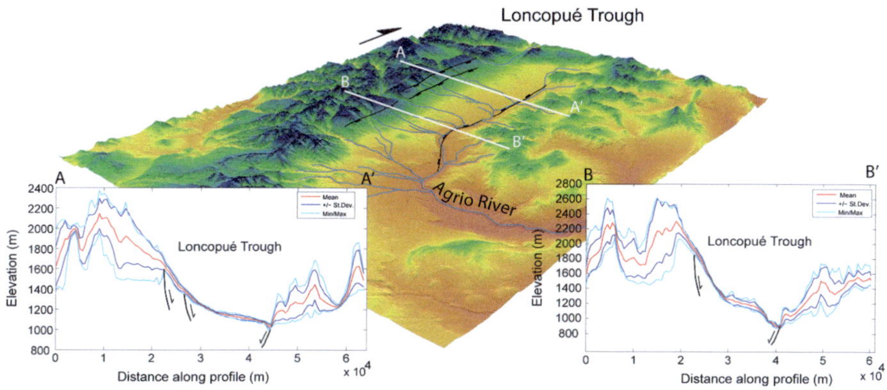

Fig. 8 Digital elevation model and drainage of the Agrio fluvial basin that drains through the Loncopué Trough (main structures flanking the Loncopué Trough are shown). Swath profiles developed through the northern and southern edges of the trough, where main neotectonic structures are highlighted

Auca Mahuida volcano surroundings. In particular, Quaternary structures that characterize the retro-arc have a spatial relationship with areas where asthenospheric anomalies are localized at the lower lithosphere (Burd et al. 2014). In particular, Quaternary deformations at the Tromen volcanic plateau and at the Cara Cura Range and possibly also at Cerro Domuyo correlate with the SWAP anomaly (shallow western asthenospheric Plume) described by Burd et al. (2014) (Fig. 3), while Quaternary regional uplift described in the Auca Mahuida volcano occurs above the DEEP anomaly (Deep Eastern Plume) (Burd et al. 2014). On the other hand, the extensional Loncopue Trough is explained in the light of seismic-tomographic data obtained after the Maule earthquake (2010), where Pesicek et al. (2012) describe a tearing in the subducted Nazca plate with a southern steepened sector that coincides with the longitudinal extent of the Quaternary extensional structures and related volcanism (Rojas Vera et al. 2014). Finally, the distribution of the Quaternary deformational systems across the retro-arc zone of the Southern Central Andes, characterized by short and disconnected fault arrays, could be related to the complex arrangement of asthenospheric anomalies.

Acknowledgement This work was supported by the National Council of Science of Argentina and funded by PIP 2015-2017 (11220150100426CO), UBACYT 20020110100019 and PICT - 2016-2252. We acknowledge Carlos Costa for sharing discussions that improved this work. This is the contribution R313 of the Instituto de Estudios Andinos Don Pablo Groeber (UBA-CONICET).

References

Bastias H, Tello G, Perucca L, Paredes J (1993) Peligro sísmico y neotectonica. In: Ramos V (ed) Geología de la provincia de Mendoza. Asociación Geológica Argentina, pp 645–658

Bechis F, Giambiagi L, Garcia V, Lanes S, Cristallini E, Tunik M (2010) Kinematic analysis of a transtensional fault system: The Atuel depocenter of the Neuquén basin, southern Central Andes, Argentina. J Struct Geol 32(7):886–899

Branellec M, Niviere B, Callot JP, Regard V, Ringenbach JC (2016) Evidence of active shortening along the eastern border of the San Rafael basement block: characterization of the seismic source of the Villa Atuel earthquake (1929), Mendoza province, Argentina. Geol Mag, 1–15

Burd AI, Booker JR, Mackie R, Favetto A, Pomposiello MC (2014) Threedimensional electrical conductivity in the mantle beneath the Payún Matrú Volcanic Field in the Andean back-arc of Argentina near 36.5° S: decapitation of a mantle plume by resurgent upper mantle shear during slab steepening. J Geophys Int 198(2):812–827

Carbone O, Franzese J, Limeres M, Delpino D, Martinez R (2011) El Ciclo Precuyano (Triasico Tardio – Jurasico Temprano) en la Cuenca Neuquina. In: Abstracts of the 18 Congreso Geologico Argentino, Neuquén, 2–6 May 2011

Cobbold PR, Rossello E (2003) Aptian to recent compressional deformation, foothills of the Neuquén Basin, Argentina. Mar Petroleum Geol 20:429–443

Colavitto B, Sagripanti L, Lupari M, Fennell L, Folguera A (2016) Quaternary deformation at the eastern Malargüe fold and thrust belt, Mendoza, Argentina. In: Abstracts of the 1 Simposio de Tectónica Sudamericana, Santiago, Santiago de Chile, 14–16 Nov 2016

Colavitto B, Sagripanti L, Fennell L, Folguera A, Costa C (2019) Evidence of Quaternary tectonics along Río Grande valley, southern Malargüe fold and thrust belt, Mendoza, Argentina. Geomorphology

Cortes J, Costa C (1996) Tectónica cuaternaria en la desembocadura del Río de las Peñas, borde oriental de la precordillera de Mendoza. In: Abstracts of the 13 Congreso Geológico Argentino y 3 Congreso Exploración de Hidrocarburos, Mendoza, 13–18 Oct 1996

Cortes J, Vinciguerra P, Yamín M, Pasini M (1999) Tectónica Cuaternaria en la región andina del nuevo Cuyo. In: Caminos R (ed) Geología Argentina, vol 29. SEGEMAR, Buenos Aires, pp 760–778

Costa C, Vita-Finzi C (1996) Late Holocene intraplate faulting in the SE Sierras Pampeanas, Argentina. Geology 24:1127–1130

Costa CH, Giaccardi AD, Díaz EFG (1999) Palaeolandsurfaces and neotectonic analysis in the southern Sierras Pampeanas, Argentina. Geol Soc, London, Special Publications, 162(1):229–238. https://doi.org/10.1144/GSL.SP.1999.162.01.18

Costa C, Audemard FA, Bezerra FHR, Lavenu A, Machette MN, París G (2006a) An overview of the main Quateranry deformation of South America. Rev Asoc Geol Argent 61(4):461–479

Costa C, Gardini C, Diederix H et al (2006b) The Active Andean Orogenic Front at the Southernmost Pampean Flat-slab. Backbone of the Americas. In: Abstracts of the GSA Annual Meeting, Mendoza, 3–7 Apr 2006

Costa C, Cisneros H, Salvarredi J, Galluci A (2006c) La neotectónica del margen oriental del bloque de San Rafael: Nuevas consideraciones. Asociación Geológica Argentina, Serie D SP 6:33–40

Cristallini E, Bottesi G, Gavarrino A, Rodriguez L, Tomezzoli R, Comeron R (2006) Synrift geometry of the Neuquén basin in northeastern Neuquén province, Argentina. In: Kay SM, Ramos VA (eds) Evolution of an Andean Margin: a Tectonic and Magmatic View from the Andes to the Neuquén Basin (35°–39° S lat). Geological Society of America, SP 407, pp 147–162

Fennell L, Folguera A, Naipauer M, Gianni G, Rojas Vera E, Bottesi G, Ramos VA (2015) Cretaceous deformation of the southern Central Andes: synorogenic growth strata in the Neuquén Group (35° 30′–37° S). Basin Res 29(1):51–72

Folguera A, Ramos VA, Hermanns RL, Naranjo J (2004) Neotectonics in the foothills of the southernmost central Andes (37°–38° S): evidence of strike-slip displacement along the Antiñir Copahue fault zone. Tectonics 23. https://doi.org/10.1029/2003tc001533

Folguera A, Introcaso A, Gimenez M, Ruiz F, Martínez P, Tunstall C, García Morabito E, Ramos VA (2007) Crustal attenuation in the Southern Andean retroarc (38′°–39° 30′ S) determined from tectonic and gravimetric studies: the Lonco-Luán asthenospheric anomaly. Tectonophysics 439:129–147

Folguera A, Bottesi G, Zapata T, Ramos VA (2008) Crustal collapse in the Andean back-arc since 2 Ma: Tromen volcanic plateau, Southern central Andes (36°40′–37°30′ S). Tectonophysics 459:140–160

Folguera A, Naranjo JA, Orihashi Y, Sumino H, Keisuke N, Polanco E, Ramos VA (2009) Retroarc volcanism in the northern San Rafael block (34°–35° 30′ S). J Volcanol Geoth Res 186:169–185

Folguera A, Rojas Vera E, Bottesi G, Zamora Valcarce G, Ramos VA (2010) The Loncopue trough: a cenozoic basin produced by extension in the southern central Andes. J Geodyn 49:287–295

Folguera A, Gianni G, Sagripanti L, Rojas Vera EA, Novara I, Alvarez O, Orts D, Tobal J, Gimenes M, Ruiz F, Martinez P, Ramos VA (2015a) A review about the mechanisms associated with active deformation, regional uplift and subsidence in southern South America. J S Am Earth Sci 64:511–529

Folguera A, Bottesi G, Duddy I, Martín-González F, Orts D, Sagripanti L, Rojas Vera E, Ramos VA (2015b) Exhumation of the Neuquén Basin in the southern Central Andes (Malargüe fold and thrust belt) from field data and low temperature thermochronology. J S Am Earth Sci 64:381–398

Galetto A, García V, Caselli A (2018) Structural controls of the Domuyo geothermal field, Southern Andes (36°–38′ S), Argentina. J Struct Geol 114:76–94

Galland O, Hallot E, Cobbold PR, Ruffet G, de Bremond d'Ars J (2007) Volcanism in a compressional Andean setting: a structural and geochronological study of Tromen volcano (Neuquén province, Argentina). Tectonics 26. https://doi.org/10.1029/2006tc002011

García VH, Casas AL (2014) Quaternary tectonics and seismic potential of the Andean retrowedge at 33°–34° S. In: Sepúlveda SA, Giambiagi LB, Moreiras SM, Pinto L, Tunik M, Hoke GD, Farías M (eds) Geodynamic Processes in the Andes of Central Chile and Argentina. The Geological Society, London, SP 399, pp 3–11

Garcia VH, Cristallini EO, Cortés JM, Rodríguez C (2005) Structure and neotectonics of the Jaboncillo and Del Peral anticlines: New evidences of Pleistocene to? Holocene deformation in the Andean Piedmont. In: Abstracts of the 6 International Symposium on Andean Geodynamics, Barcelona, 12–14 Sept 2005

García Morabito G (2004) Geología del sector occidental de la depresión de Loncopue, provincia del Neuquén. Degree thesis, Universidad de Buenos Aires

Giambiagi LB, Ramos VA (2002) Structural evolution of the Andes in a transitional zone between flat and normal subduction (33° 30′–33° 45′ S), Argentina and Chile. J South Am Earth Sci 15:101–116

Giambiagi LB, Ramos VA, Godoy E, Alvarez PP, Orts S (2003) Cenozoic deformation and tectonic style of the Andes, between 33° and 34° south latitude. Tectonics 22. https://doi.org/10.1029/2001tc001354

Giambiagi L, Tunik M, Barredo S, Bechis F, Ghiglione M, Alvarez P, Drosina M (2009) Cinematica de apertura del sector norte de la cuenca Neuquina. Rev Asoc Geol Argent 65(2):278–292

Gudnason J, Holm PM, Søager N, Llambías EJ (2012) Geochronology of the late Pliocene to recent volcanic activity in the Payenia back-arc volcanic province, Mendoza Argentina. J S Am Earth Sci 37:191–201

Hoke GD, Graber N R, Mescua J F., Giambiagi LB, Fitzgerald PG, Metcalf JR (2014) Nearpure surface uplift of the Argentine Frontal Cordillera: insights from (U/Th)/He thermochronometry and geomorphic analysis. In: Sepúlveda SA, Giambiagi LB, Moreiras SM, Pinto L et al (eds) Geodynamic processes in the Andes of Central Chile and Argentina. The Geological Society, London, SP 399. https://doi.org/10.1144/sp399.10

Irigoyen MV, Buchan KL, Villeneuve ME, Brown RL (2002) Cronología y significado tectónico de los estratos sinorogenicos neogenos aflorantes en la región de Cacheuta-Tupungato, Provincia de Mendoza. Rev Asoc Geol Argent 57:3–18

Jordan TE, Burns WM, Veiga R, Pangaro F, Copeland P, Kelley S, Mpodozis C (2001) Extension and basin formation in the southern Andes caused by increased convergence rate: a mid-Cenozoic trigger for the Andes. Tectonics 20:308–324

Kay SM, Burns WM, Copeland PC, Mancilla O (2006) Upper Cretaceous to Holocene magmatism and evidence for transient Miocene shallowing of the Andean subduction zone under the northern Neuquén Basin. In: Kay SM, Ramos VA (eds) Evolution of an Andean Margin: a tectonic and magmatic view from the Andes to the Neuquén Basin (35°–39° S Latitude). Geological Society of America, SP 407:19–60

Kozlowski E, Manceda R, Ramos VA (1993) Estructura. In: Ramos VA (ed) Geología y Recursos Naturales de Mendoza. Asociación Geológica Argentina, Buenos Aires, pp 235–256

Lavenu A, Cembrano J (1999) Estados de esfuerzo compresivo plioceno y compresivo-transpresivo pleistoceno, Andes del sur, Chile (38 y 42 30′ S). Rev Geol Chile 26(1):67–87

Litvak V, Spagnuolo M, Folguera A, Poma S, Jones R, Ramos VA (2015) Late Cenozoic calcalkaline volcanic rocks over the Payenia shallow subduction zone, South-Central Andean back-arc (34° 30′-37° S), Argentina. J S Am Earth Sci 64:365–380

Llambías EJ, Palacios M, Danderfer J (1982) Las erupciones Holocenas del Volcan Tromen (Provincia de Neuquén) y su significado en el perfil transversal E - O a la latitud 37° S. In: Abstracts of the 5 Congreso Latinoamericano de Geología, Buenos Aires, 17–22 Oct 1982

Manceda R, Figueroa D (1995) Inversion of the Mesozoic Neuquén rift in the Malargüe fold and thrust belt, Mendoza, Argentina. In: Tankard AJ, Súarez SR, Welsink HJ (eds) Petroleum Basins of South America. AAPG Memoir 62, pp 369–382

Messager G, Niviére B, Martinod J, Lacan P, Xavier JP (2010) Geomorphic evidence for PlioQuaternary compression in the Andean foothills of the southern Neuquén basin, Argentina. Tectonics 29. https://doi.org/10.1029/2009tc002609

Messager G, Nivière B, Martinod J, Lacan P, Hervouët Y, Xavier JP (2013) Plio-Quaternary thin-skinned tectonics along the crustal front flexure of the southern Central Andes: a record of the regional stress regime or of local tectonic-driven gravitational processes? Int J Earth Sci 103:929–951

Miranda FJ, Folguera A, Leal P, Naranjo J, Pesce A (2006) Upper Pliocene to Lower Pleistocene volcanic complexes and Upper Neogene deformation in the south-central Andes (36°30′-38°00′SL). In: Kay SM, Ramos VA (eds) Evolution of an Andean margin: A tectonic and magmatic view from the Andes to the Neuquén Basin (35°-39°SL). Geol Soc Am, SP 407:287–298

Mosquera A, Ramos VA (2006) Intraplate deformation in the Neuquén embayment. In: Kay SM, Ramos VA (eds) Evolution of an Andean Margin: a Tectonic and Magmatic View from the Andes to the Neuquén Basin (35°-39° S Lat), Geol Soc Am, SP 407: 97–123

Nacif S, Lupari M, Triep EG, Nacif A, Alvarez O, Folguera A, Gimenez M (2017) Change in the pattern of crustal seismicity at the Southern Central Andes from a local seismic network. Tectonophysics 708:56–69

Nivière B, Messager G, Carretier S, Lacan P (2013) Geomorphic expression of the Southern central Andes forebulge (37 S, Argentina). Terra Nova 25:361–367

Perucca L, Mehl AE, Zárate MA (2009) Neotectónica y sismicidad en el sector norte de la depresión de Tunuyán, provincia de Mendoza. Rev Asoc Geol Argent 64:263–274

Pesicek JD, Engdahl ER, Thurber CH, DeShon HR, Lange D (2012) Mantle subducting slab structure in the region of the 2010 M8.8 Maule earthquake (30–40 S). Chile. Geophys J Int 191:317–324

Ramos VA, Folguera A (2005) Tectonic evolution of the Andes of Neuquén: constraints derived from the magmatic arc and foreland deformation. In: Veiga GD, Spalletti LA, Howell JA, Schwarz E (eds) The Neuquén Basin, Argentina: a Case Study in Sequence Stratigraphy and Basin Dynamics. Geol Soc, London, SP 252:15–35

Ramos VA, Zapata T, Cristallini E, Introcaso A (2004) The Andean thrust systemlatitudinal variations in structural styles and orogenic shortening. In: McClay KR (ed) Thrust Tectonics and Hydrocarbon Systems. AAPG Memoir 82:30–50

Rojas Vera EA, Selles D, Folguera A, Gímenez M, Ruíz F, Orts D, Zamora Valcarce G, Martínez P, Bechis F, Ramos VA (2014) The origin of the Loncopué trough in the retro-arc of the Southern Central Andes from field, geophysical and geochemical data. Tectonophysics 637:1–19

Rojas Vera EA, Mescua J, Folguera A, Becker TB, Sagripanti L, Fennell L, Orts D, Ramos AV (2015) Evolution of the Chos Malal and Agrio fold and thrust belts, Andes of Neuquén: insights from structural analysis and apatite fission track dating. J S Am Earth Sci 64:418–433

Sagripanti L, Bottesi G, Naipauer M, Folguera A, Ramos VA (2011) U/Pb ages on detrital zircons in the southern central Andes Neogene foreland (36°-37° S): constraints on Andean exhumation. J S Am Earth Sci 32:555–566

Sagripanti L, Folguera A, Gimenez ME, Rojas Vera EA, Fabiano JJ, Molnar N, Fennell L, Ramos VA (2014) Geometry of Middle to Late Triassic extensional deformation pattern in the Cordillera del Viento (Southern Central Andes): A combined field and geophysical study. J Iber Geol 40(2):349–366. https://doi.org/10.5209/rev_JIGE.2014.v40.n2.45305

Sagripanti L, Rojas Vera E, Gianni G, Folguera A, Harvey J, Farías M, Ramos VA (2015) Neotectonic reactivation of the western section of the Malargüe fold and thrust belt (Tromen volcanic plateau, Southern Central Andes). Geomorphology 232:164–181

Sagripanti L, Colavitto B, Jagoe L, Folguera A, Costa C (2018) A review about the quaternary upper-plate deformation in the Southern Central Andes (36°-38°S): A plausible interaction between mantle dynamics and tectonics. J S Am Erath Sci 87:221–231

Spagnuolo MG, Litvak VD, Folguera A, Botessi G, Ramos VA (2012) Neogene magmatic expansion and mountain building processes in the southern Central Andes, 36–37°S, Argentina. J Geodyn 53:81–94

Turienzo MM (2010) Structural style of the Malargüe fold-and-thrust belt at the Diamante River area (34°30′ – 34°50′S) and its linkage with the Cordillera Frontal, Andes of Central Argentina. J S Am Earth Sci 29:537–556

Varekamp J, Hesse A, Mandeville C (2010) Back-arc basalts from the Loncopue graben (province of Neuquén, Argentina). J Volcanol Geoth Res 197:313–328

Verges J, Ramos V, Meigs A, Cristallini E, Bettini F, Cortés J (2006) Crustal wedging triggering recent deformation in the Andean thrust front between 31°S and 33°S: Sierras Pampeanas-Precordillera interaction. J Geophys Res 112, B03S15. https://doi.org/10.1029/2006jb004287

Yrigoyen MR (1993) Los depósitos sinorogénicos terciarios. In: Ramos VA (ed) Geología y recursos naturales de la provincia de Mendoza. Asociación Geologica Argentina, pp 123–148

Yuan X, Asch G, Bataille K, Bock G, Bohm M, Echtler H, Kind R, Oncken O, Wölbern I (2006) Deep seismic images of the Southern Andes. In: Kay SM, Ramos VA (eds) Evolution of an Andean Margin: a tectonic and magmatic view from the Andes to the Neuquén Basin (35°-39°S Lat). Geol Soc Am, SP 407, p 61–72

Zamora Valcarce G, Zapata T, Del Pino D, Ansa A (2006) Structural evolution and magmatic characteristics of the Agrio fold-and-thrust belt. In: Kay SM, Ramos VA (eds) Evolution of an Andean Margin: a tectonic and magmatic view from the Andes to the Neuquén Basin (35E 39S Lat). Geological Society of America, SP 407, pp 125–145

Zamora Valcarce G, Zapata T, Ramos VA, Rodríguez F, Bernardo ML (2009) Evolución tectónica del frente andino en Neuquén. Rev Asoc Geol Argent 65:192–203

Pliocene to Quaternary Retroarc Extension in the Neuquén Basin: Geophysical Characterization of the Loncopué Trough

Agustina Pesce, Guido M. Gianni, Mario E. Giménez and Andrés Folguera

Abstract The Loncopué Trough is an extensional basin produced by the extensional reactivation of the hinterland area of the Southern Central Andes. Neotectonic extensional structures in this basin bound a broad topographic low filled with volcanic and volcaniclastic rocks. The studies carried out in the area of the Loncopué Trough have concentrated on the study of its neotectonic activity, volcano-sedimentary infill and the surface structure. Less effort has been paid to characterize the magnetic properties of the crust and to unravel the deep geometry of this Pliocene to Quaternary extensional setting. Therefore, magnetic and gravimetric data were used to highlight the boundaries of the magnetic sources and to obtain a crustal-scale 2D density model at 38°S. To complement this work, an effective susceptibility model using the Magnetization Vector Inversion method was estimated, which takes into account the combined effects of remanence and induced magnetization. Additionally, the Curie depth points were calculated through the spectral analysis technique in order to determine the thermal structure of the retroarc area. From this analysis, we were able to characterize the main structures associated with this extensional trough. Based on this analysis, only the Loncopué eastern fault system is considered as having a crustal-scale hierarchy. Additionally, the susceptibility model revealed possible fluid (magmatic and or hydrothermal) reservoirs in the area of the Copahue volcano and the Codihue and Cajón de Almanza depocenters/volcanic fields. These spots coincide with shallower values of the calculated Curie depth point, implying higher heat flows. Finally, the 2D density model shows an area of lower crustal attenuation that is coincident with one of the described potential magmatic/hydrothermal reservoirs and is decoupled from the upper crust extensional structures immediately to the west in the Loncopué Trough. This crustal configuration could be explained

---

A. Pesce (✉) · G. M. Gianni · M. E. Giménez
IGSV, Instituto Geofísico Sismológico Ing. Volponi, Universidad Nacional de San Juan, San Juan, Argentina
e-mail: pesce.agustina@gmail.com

A. Folguera
CONICET—Universidad de Buenos Aires, Instituto de Estudios Andinos Don Pablo Groeber (IDEAN), Buenos Aires, Argentina

D. Kietzmann and A. Folguera (eds.), *Opening and Closure of the Neuquén Basin in the Southern Andes*, Springer Earth System Sciences,
https://doi.org/10.1007/978-3-030-29680-3_22

by a simple shear deformation model with a crustal-scale master fault dipping to the east.

Keywords Loncopué Trough · Extensional basin · Retroarc volcanism · Simple shear model · Effect susceptibility model · Southern Central Andes

1 Introduction

The Loncopué Trough is one of the few examples where extension is associated with the Andean deformational front (Ramos 1978; Folguera et al. 2006). This Pliocene to Quaternary extensional basin is located between the Agrio fold and thrust belt and the Main Andes (see Chap. "Quaternary Upper-Plate Deformation in the Neuquén Basin, Explained by the Interaction Between Mantle Dynamics and Tectonics"). This basin was defined as a 200-km-long topographic depression linked with broad basaltic and ignimbrite fields controlled by two tectonically active fault bounding systems (Ramos 1978) (Fig. 1a).

Most studies in the Loncopué Trough region have focused in the volcano-sedimentary infill characteristics, magma sources, surface structure and neotectonic activity (Folguera et al. 2006; Varekamp et al. 2010; Rojas Vera et al. 2010). Previous geophysical studies in this region used gravity and seismological data, aimed to receiver-function analysis and to produce 2D seismic reflection data. These studies provided a general idea of the crustal structure beneath the Loncopué Trough, which identified (1) an attenuated lower crust coinciding with the southernmost extreme of the basin at 38.9°S, where the Moho is locally uplifted to 30 km depth, presumably resulting from the back-arc spreading beginning in the Pliocene–Pleistocene (Yuan et al. 2006) and (2) synrift wedge geometries compatible with an extensional regime (Folguera et al. 2010; Rojas Vera et al. 2010). However, the spatial relation between surface extensional structures, retroarc volcanism and Moho morphology remains unclear. Noteworthy, the latter is crucial to better understanding the crustal deformation mechanism, which has particular consequences for thermal evolution, basin formation, subsidence history, local uplift and depositional patterns of basin sediments (Ruppel 1995).

In this chapter, we summarize the application of magnetic and gravimetric data used to analyze the spatial relation between the surficial structure, the development of volcanic fields and magnetic anomalies, with the aim of contributing to a poorly known aspect of the Loncopué Trough regarding its crustal-scale structure, particularly its deep geometry.

To asses these issues, the tilt angle derivative was calculated to enhance the edges of the magnetic sources. Then, we estimated the Curie depth point through the power spectrum analysis of the magnetic anomalies using the Tanaka proximation (Tanaka et al. 1999), which corresponds to the temperature at which the magnetic minerals lose their magnetic properties and give a general idea about the thermal gradients in the area. Additionally, in order to visualize susceptibilities contrasts and compare them

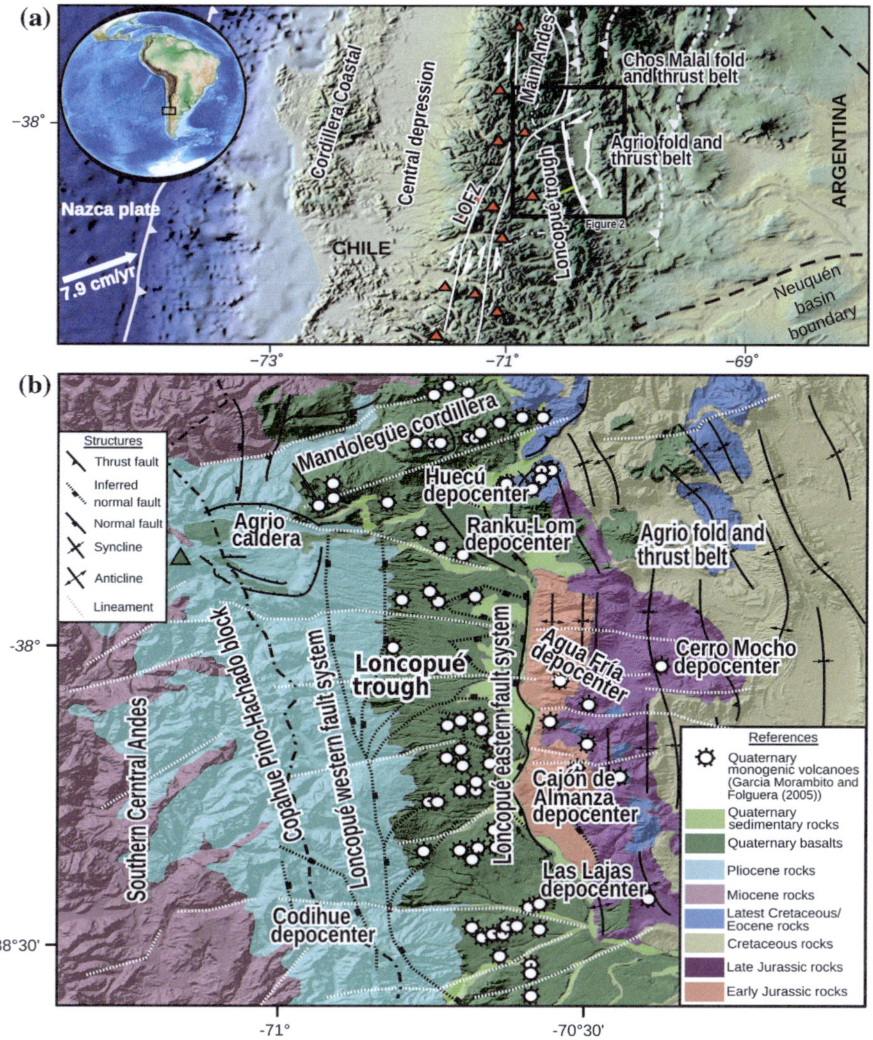

Fig. 1 **a** Tectonic configuration of the study area in the Southern Central Andes region and main units that characterize the Loncopué Trough. **b** Structural map of the study area (modified from Rojas Vera et al. 2010). LOFZ: Liquiñe-Ofqui fault zone

with surficial volcanic fields, a 3D model was computed using a magnetic inversion method. Finally, we obtained a 2D density model applying these new geophysical data and using the existing geological and geophysical information as constraints from previous works in the study area.

2 Geological Setting

The Southern Central Andes in the study region exposes marine to nonmaine sedimentary sequences deposited during the Mesozoic extension and sag stages in the Neuquén Basin. Nonmarine volcano-sedimentary deposition took place in (Eocene?) Oligocene to Miocene times during a second extensional stage (Radic et al. 2002) (see Chapter "The Miocene sequences cropping out in the High Andes of Neuquén and their relation to the Andean orogenesis"). The Mesozoic units are well exposed to the east of the Loncopué Trough in the Agrio and Malargüe fold and thrust belts, mostly deformed by thick-skinned and thin-skinned structures (Fig. 1b) (Valcarce et al. 2006; Rojas Vera et al. 2014). Contrastingly, Cenozoic units are mostly exposed over the Main Andes, west of the Loncopué Trough (e.g., Rojas Vera et al. 2014). This sector is interpreted as a thick-skinned west-vergent fold and thrust belt related to the inversion of Paleogene to Neogene extensional structures (Jordan et al. 2001; Rojas Vera et al. 2014). The Andes at the studied latitudes experienced two main contractional events, an older in late Early Cretaceous-Eocene and a younger in late Miocene times, respectively (e.g., Valcarce et al. 2006). These contractional stages took place in concert with significant eastward arc expansions which have been interpreted in terms of shallowing of the subducted plate (Folguera et al. 2006; Kay and Copeland 2006) (see Chapter "The Miocene magmatism in the Malargüe and Chos Malal fold and thrust belts"). Particularly, the youngest arc expansion has been recently related to the subduction of the Payenia mantle plume in Neogene times, which is currently impacting the Andean back-arc region to the east of the study area (Gianni et al. 2017) (see Chap. "Plume Subduction Beneath the Neuquén Basin and the Last Mountain Building Stage of the Southern Central Andes"). This process ended in Pliocene times with slab detachment and steepening of the Nazca plate (Pesicek et al. 2012), the development of extension in the hinterland region of the Andes (Rojas Vera et al. 2014) and localized contraction in the Andean foothills.

From Pliocene to Quaternary times, extensional activity related to the Loncopué Trough formed a narrow topographic depression between the volcanic arc to the west and the Agrio fold and thrust belt to the east. The 200–km-long trough is filled by 2 km of syn-extensional volcano-sedimentary successions of late Oligocene(?)–early Miocene to Quaternary age (Rojas Vera et al. 2014). The last syn-extensional volcanic stage started in the early Pliocene with wedge-like depocenters linked to extrusion of broad lava plateaux. This was followed by 2.6–2 Ma silicic volcanic sequences formed by ignimbrites and distal pyroclastic deposits associated with a series of caldera collapses. Two north–south main fault systems have controlled these volcanic centers and the early Pliocene to Quaternary monogenetic alkaline basaltic eruptions, one located to the east affecting the western sector of the Agrio fold and thrust belt and the other located at its western flank formed by east-facing scarps (Fig. 1b) (Ramos 1978; Rojas Vera et al. 2008, 2010). The Loncopué western fault system is inferred in map view and linked at the surface to small halfgrabens less than 100 m across associated with east-facing scarps affecting Quaternary lavas. The Loncopué eastern fault system represents a series of west- and east-facing normal fault scarps affecting Quaternary lavas and previously deformed Mesozoic strata in Cretaceous and Neogene times (Rojas Vera et al. 2014) (Fig. 1b).

3 Methods

3.1 Magnetic Data

A total magnetic field grid with a cell size of 500 m was obtained by Carson Aerogravity through the processing of aeromagnetic data collected at the end of 2004 (see Pesce et al. 2019, for further details). We calculated a magnetic anomaly map (Fig. 3a) by eliminating from the total magnetic intensity map the International Geomagnetic Reference Field (IGRF) at the acquisition date (Blakely 1996). Then, we also estimated a residual anomaly map (Fig. 3b) to study the density heterogeneities present in the shallower portions of the crust. It was computed by eliminating the long-wavelength components of the magnetic anomaly through a frequency filter, call Butterworth, of eighth order and a wavelength of 40 km.

To remove or minimize the dipole nature of the anomalies, which usually introduces complexity to the interpretation of the different geological structures, we transformed the residual anomaly map into a reduction to the pole map assuming that all the magnetization is parallel to the Earth's magnetic field (Baranov 1957; Phillips 2007). This transformation was calculated adopting inclination and declination values of -38° and 6°, respectively (Fig. 3c). Since remanence data on the Loncopué Trough are not available, the reduction to the pole map (Fig. 2c) constitutes an auxiliary map to analyze the correspondence between the anomalies, geological structures and Quaternary volcanic fields.

3.1.1 Application of the Tilt Angle Derivative

To enhance the edges of the bodies hosted in the upper crust that generate anomalous effects in the magnetic field, we applied the tilt angle derivative (TDR) from the reduced to pole anomaly map (Fig. 2c) (Hinze et al. 2013). On the TDR map, the zero contours (Fig. 3a) signal the edges of the structures.

3.1.2 Calculation of the Curie Point Depth

The Curie point depth (CPD) is related to the Curie isotherm (approximately 580 °C for magnetite), where the ferromagnetic rocks become paramagnetic, losing their magnetic ability. Therefore, the CPD is used to analyze the thermal structure of the Loncopué Trough.

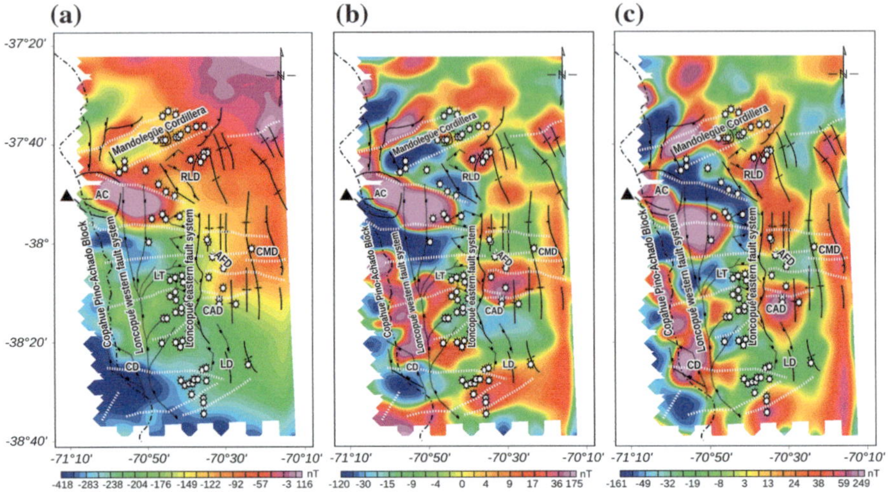

Fig. 2 Processed magnetic data. **a** Magnetic anomaly. **b** Residual anomaly. **c** Reduction to pole map. AC: Agrio caldera; RLD: Ranku-Lom depocenter; AFD: Agua Fría depocenter; CMD: Cerro Mocho depocenter; CAD: Cajón de Almanza depocenter; LD: Las Lajas depocenter; CD: Codihue depocenter; LT: Loncopué Trough

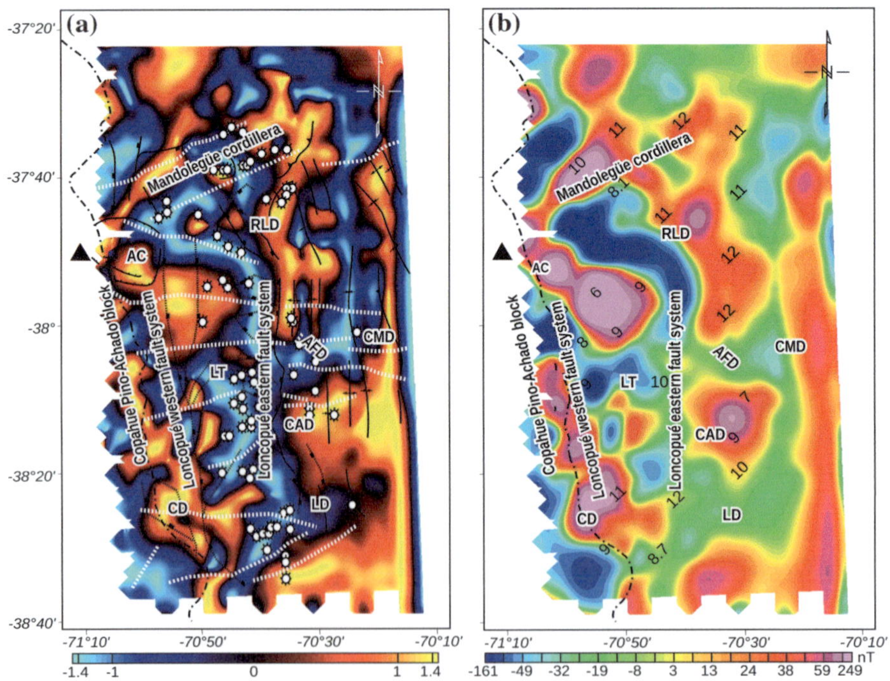

Fig. 3 **a** Tilt angle calculated from the reduction to pole map. **b** Curie point depth calculation (CPD) in kilometers over the reduction to pole map (black numbers indicate the center of the window to estimate the CPD)

The technique used to estimate the CPD is based on the statistical analysis of the magnetic anomalies in the frequency domain, following the Tanaka approximations (Tanaka et al. 1999). It models the magnetized basement as a horizontal semi-infinite plate with a random magnetization, where its bottom is assumed as the Curie point depth (CPD). Then, the CPD can be determined from the power density spectra of the magnetic anomaly (Blakely 1996).

Through the CuDePy program (Soler 2015), we could determine the Curie point depth (CPD) in the area of the Loncopué Trough (Fig. 3b). This program allows making an interactive selection of square sub-areas of the magnetic anomaly grid and to determine a CPD point in each sub-area. Therefore, we used variable windows of 30–40 km to estimate the Zb.

3.1.3 Inversion Model

An inverse modeling of the residual magnetic anomaly allows obtaining a model that represents the effective susceptibility contrasts present in the upper crust applying the VOXI Earth Modeling software from Geosoft, which implements the Magnetization Vector Inversion (MVI) method developed by Ellis et al. (2012). The algorithm defines the effective magnetization ($\mathbf{M}_{\text{eff}}$) as follows:

$$\mathbf{M_{eff}} = \mathbf{M} + \mathbf{M_{nrm}} = \chi|\mathbf{H}_\text{e}| + \chi_\text{nrm}|\mathbf{H}_\text{e}| = \chi_\text{eff}|\mathbf{H}_\text{e}|$$

where $\mathbf{M}$ and $\mathbf{M}_{\text{nrm}}$ are the magnetization and the normal remanent magnetization, χ and χ_{nrm} are the magnetic susceptibility and pseudo magnetic susceptibility, $\mathbf{H}_\text{e}$ is the external magnetic field, and χ_{eff} is the effective susceptibility (anisotropic + remanent).

The magnetic inversion model (Fig. 4) is obtained applying the Magnetization Vector Inversion method to the residual magnetic anomaly (Fig. 2b). Each cell in the model mesh is 500 × 500 × 250 m, and the inverted data were subject to a uniform uncertainty of 15 nT.

3.2 *Gravity Data Processing*

The data set used in this study is a compilation of different sources: gravity databases of YPF (former State Oil Company), stations measured by IGSV-UNSJ and field gravity measurements from IGM (Instituto Geográfico Nacional Argentino). In order to ensure the accuracy of the measurements and to homogenize all stations, each gravity station has been linked to the national altimetry network. They were referred to the IGSN 71 network (International Gravity Standardization Net 1971) and linked to the fundamental Miguelete station (Buenos Aires).

This methodology allows making a proper data reduction for anomaly calculation using the classical corrections (Blakely 1996).

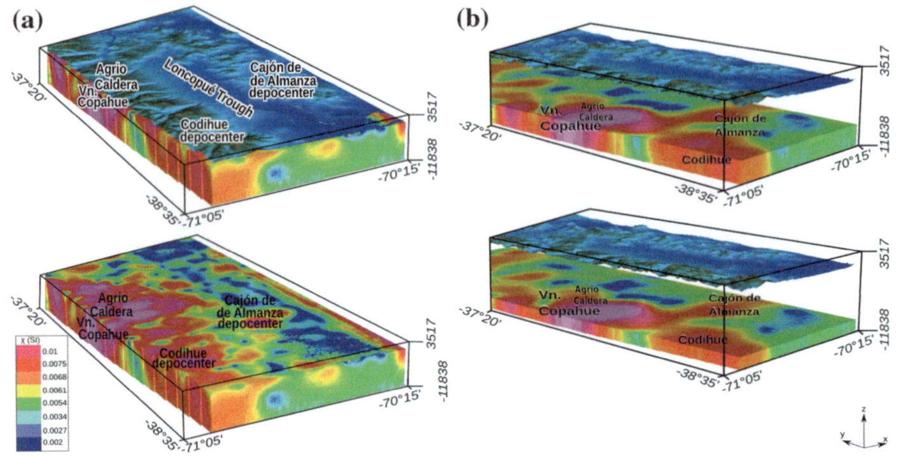

Fig. 4 Magnetic Inversion Model. **a** Model with and without the topography. **b** Lateral views of the model

Therefore, the anomalies were calculated adopting the normal gradient of 0.3086 mGal/m for the free-air correction and a density of 2670 kg/m^3 for the Bouguer correction (Hinze et al. 2005). To adjust the gravity effect produced by a mass excess (mountain) or deficit (valley) with respect to the elevation of the observation point, the topographic correction was applied to the anomalies. We used two digital elevation models obtained from the Shuttle Radar Topography Mission (SRTM) of the United States Geological Survey (USGS). One of them has a resolution of 1×1 km (regional aspect), which far exceeds the study area, and the other is a local grid of 90×90 m. We used the Oasis Montaj software to calculate this correction that combines the algorithms developed by Kane (1962) and Nagy (1966). The resulting maximum error for this correction was ± 1.5 mGal. Finally, the Bouguer anomaly values (Fig. 5a) were calculated on a regular grid cell size of 1231×1231 m, using the minimum curvature method (Briggs 1974).

3.2.1 Density Model of the Loncopué Trough and the Western Agrio Fold and Thrust Belt

In order to quantify the relationship between the gravity signals and the crustal structure, a 2D density model was constructed using a gravity profile (black line in Fig. 5a) and a direct modeling through the GM-SYS 2D software developed by Webring (1985), along the eastern slope of the Andes at around 38°S. Rojas Vera et al. (2010, 2014) constructed a structural model through this transect using field and borehole data, as well as previously published seismic lines, which had been analyzed in previous works as well (Valcarce et al. 2006; Zapata et al. 1999). Based on these studies, we used in our model a fold and thrust belt decollement located at

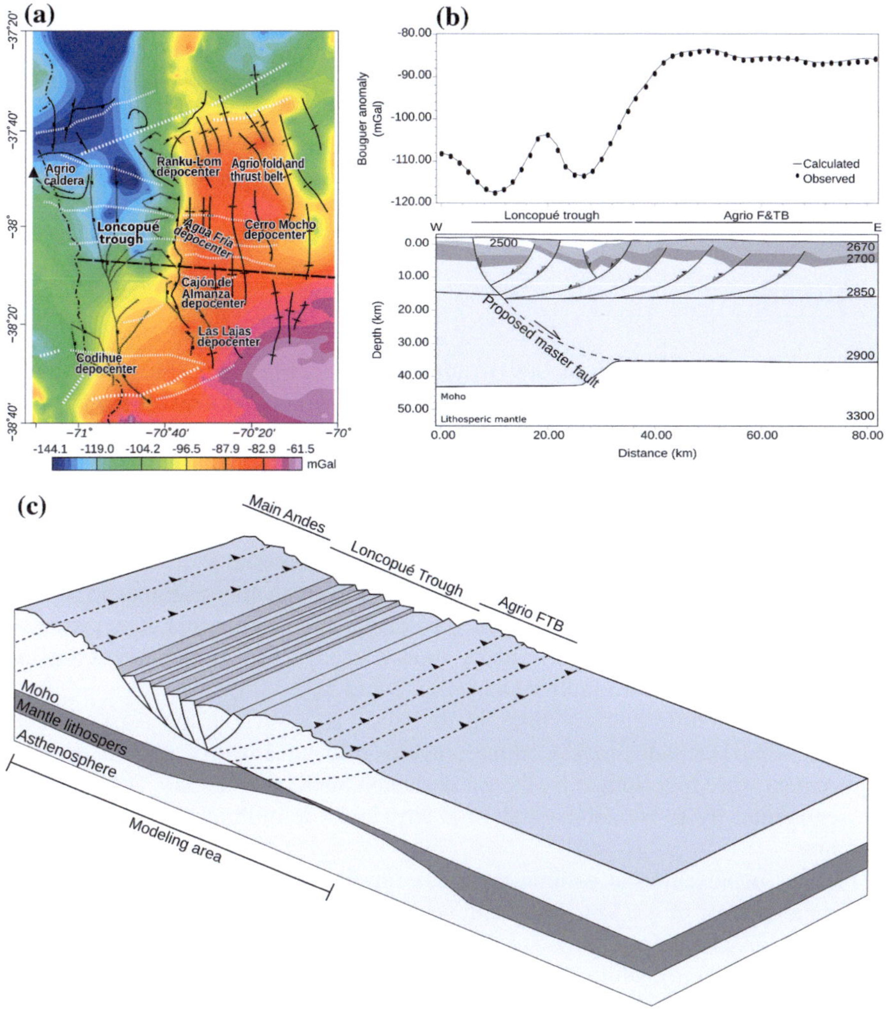

Fig. 5 **a** Bouguer anomaly map. The black-dashed line is the anomaly profile used to make the density model. **b** 2D density model. **c** Diagram of the geological model, which represents the extension of the continental crust using a single lithospheric dislocation. The detachment fault represents the upper levels of a shallow-dipping shear zone that crosses through the lithosphere (Lister and Davis 1989)

15–20 km depth. The modeled crust was represented by five layers with densities that correspond to (a) the densities converted from the P-wave (Lüth et al. 2003), (b) the global average density values (Christensen and Mooney 1995) and (c) the density values used in Rojas Vera et al. (2010) in their structural cross section, which were determined using borehole and surface data. Then, we assigned 2900 kg/m^3 for the lower crust, 2850 kg/m^3 for the mid crust, 2500–2700 kg/m^3 for the upper crust, while for the lithospheric mantle, we used a value of 3300 kg/m^3. To determine the geometry of the crust-mantle interface (continental Moho), we considered a receiver-function profile at 38.9°S (Yuan et al. 2006), where a crustal attenuation zone in the southernmost extreme of the Loncopué Trough and western Agrio fold and thrust belt is shown. We considered these data as a starting point to model the Moho in the study area, with our model located north of the receiver-function profile.

4 Results

The processed magnetic anomalies (residual anomaly, Fig. 2b and reduced to pole maps, Fig. 2c) are positive in the area of the Copahue–Pino Hachado block and Agrio caldera-Copahue volcano, adjacent to the Loncopué Trough. These can be directly associated with surficial volcanic and volcaniclastic deposits (Rojas Vera et al. 2010) that conform broad volcanic fields of monogenetic centers in these areas (Anci et al. 2016). Other positive effects are coincident with less known volcanic fields such as the Codihue and Cajón de Almanza regions that concentrate Pleistocene volcanic activity (Rojas Vera et al. 2010). On the other hand, a negative response in the area of the Loncopué Trough could be linked to the sedimentary infills. The processed magnetic data and the Bouguer anomaly show a strong E-W gradient that coincides with the Loncopué eastern fault system (Fig. 2 and 5a).

The edges of different structures in the study area such as the Agrio caldera, Copahue–Pino Hachado block, Codihue and Cajón de Almanza volcano-sedimentary depocenters were highlighted by the tilt derivative method (Fig. 3a). These features are segmented by E-W-, ENE- and ESE structures delimiting different magnetic domains.

Figure 3b presents the estimated values for the depths of the Curie isotherm with a mean error of 1.6 km. The shallowest depths are found in the area of Agrio caldera (approximately 8 km) and Codihue and Cajón de Almanza depocenters (approximately 9 km), while in the neighboring areas the mean depth is about 11 km.

The inversion model obtained is shown in Fig. 4 as a 3D affective susceptibility model. This model lacks of specific measurements of remnant magnetization within the area; however, it represents a general 3D representation of the different conductive bodies that can be correlated with different volcanic fields at the surface. In particular, this model shows that the highest effective susceptibility values are located bellow the region of Agrio caldera-Copahue volcano, Codihue and Cajón de Almanza volcano-sedimentary depocenters.

Bouguer anomaly values in the study area (Fig. 5a) go from a minimum, located in the Main Andes region and west of the northern axial part of the trough, of -144 mGal to a maximum of -61 mGal in the southeast part of the study area around the region of Las Lajas town. The anomalies become less negative to the south of the Agrio caldera (from −144 to −104 mGal). Other features from this map are constituted by a strong east to west gradient at the eastern boundary of the Loncopué Trough (−100 to −80 mGal) and a local maximum located over the area of the Agrio fold and thrust belt (−83 mGal). In order to adjust the anomalies, the zone of crust attenuation had to be displaced eastward with respect to the model of Yuan et al. (2006) (Fig. 5b). Then, an acceptable match was obtained between the calculated anomalies from the density model and the observed Bouguer anomalies, being the maximum discrepancy of ±1.3 mGal.

5 Discussion and Conclusions

Through the new geophysical data from the Loncopué Trough (magnetic and gravimetric data), we have analyzed (i) the structure of the upper crust in a region characterized by neotectonic and Quaternary magmatic activity and produced (ii) a crustal-scale 2D density model. Magnetic and gravimetric data (Figs. 2 and 5a) show that only the eastern boundary of this trough, the Loncopué eastern fault system, has a significant crustal hierarchy. Contrastingly, E-W-, ENE- and ESE structures are relatively minor structures segmenting the basement and neighboring areas (Fig. 2) that could be the result of reactivation of ancient fault zones linked to Jurassic depocenters of the Neuquén Basin (Rojas Vera et al. 2014).

The inversion model (Fig. 4) shows that the regions of the Agrio caldera (Copahue volcano), Codihue and Cajón de Almanza volcano-sedimentary depocenters have similar values of susceptibility with the highest determined values. The Agrio caldera, where the Copahue volcano is sited, constitutes an active volcanic field, while the Codihue and Cajón de Almanza depocenters concentrate only Pleistocene volcanic activity (Rojas Vera et al. 2008, 2010). These volcanic fields have similar values of susceptibility, positive magnetic response and contrastingly shallower Curie isotherm depths suggesting that they could be emplaced over active fluid (magmatic/hydrothermal) reservoirs.

According to our density model at 38°S (Fig. 5b), the lower crust attenuation identified by Yuan et al. (2006) to the south continues through the northern sector of the Loncopué Trough, indicating that the crust below this retroarc basin is about 10 km thinner than in the surrounding areas. Noteworthy, the zones of maximum lower crustal attenuation and higher susceptibility, such as the Cajón de Almanza volcano-sedimentary depocenter, are placed eastward of the area of extensional deformations (Fig. 5b). This decoupling between the attenuated lower crust, from the upper crust extension in the Loncopué Trough could be explained by a simple shear model of deformation (e.g., Lister and Davis 1989; Lister et al. 1991; Wernicke and Burchfiel 1982) (Fig. 5c). In this model, the extension is produced by asymmetric faulting of the upper crust that is accommodated in a detachment level inside the lower crust, hence producing an asymmetry between the ductile region of the lithosphere and the structures related to the rift in the upper crust (Fig. 5b and c). In the analyzed area, the crustal-scale simple shear deformation would be related to a master fault dipping to the east (Fig. 5b and c).

References

Anci S, Ruiz F, Lince Klinger F, Leiva F, García H, Acosta G (2016) Interpretación de la corteza superior a partir de meétodos potenciales en la región de la Payenia y faja plegada y corrida de Malargüe. Rev Asoc Geol Argent 73:159–172

Baranov V (1957) A new method for interpretation of aeromagnetic maps: pseudo-gravimetric anomalies. Geophysics 22:359–382

Blakely RJ (1996) Potential theory in gravity and magnetic applications. Cambridge University Press, Cambridge

Briggs IC (1974) Machine contouring using minimum curvature. Geophysics 39:39–48

Christensen NI, Mooney WD (1995) Seismic velocity structure and composition of the continental crust: a global view. J Geophys Res Solid Earth 100:9761–9788

Ellis RG, de Wet B, Macleod IN (2012) Inversion of magnetic data for remanent and induced sources. In: Abstracts of the ASEG 2012. https://doi.org/10.1071/aseg2012ab117

Folguera A, Ramos VA, Díaz EFG, Hermanns R (2006) Miocene to Quaternary deformation of the Guañacos fold and thrust belt in the Neuquén Andes between 37°S and 37°30′S. Geol Soc Am SP 407:247–266

Folguera A, Rojas Vera E, Bottesi G, Valcarce GZ, Ramos VA (2010) The Loncopué Trough: Cenozoic basin produced by extension in the southern Central Andes. J Geodyn 49:287–295

Gianni GM, García HPA, Lupari M, Pesce A, Folguera A (2017) Plume overriding triggers shallow subduction and orogeny in the southern Central Andes. Gondwana Res 49:387–395

Hinze WJ, Aiken C, Brozena J, Coakley B, Dater D, Flanagan G, Forsberg R, Hildenbrand T, Keller GR, Kellogg J et al (2005) New standards for reducing gravity data: The North American gravity database. Geophysics 70:J25–J32. https://doi.org/10.1190/1.1988183

Hinze WJ, Von Frese RRB, Saad AH (2013) Gravity and magnetic exploration: principles, practices, and applications. Cambridge University Press, Cambridge

Jordan TE, Schlunegger F, Cardozo N (2001) Unsteady and spatially variable evolution of the Neogene Andean Bermejo foreland basin, Argentina. J South Am Earth Sci 14(7):775–798

Kane MF (1962) A comprehensive system of terrain corrections using a digital computer. Geophysics 27:455–462

Kay SM, Copeland P (2006) Early to middle Miocene backarc magmas of the Neuquén Basin: geochemical consequences of slab shallowing and the westward drift of South America. Geol Soc Am SP 407:185–213

Lister GS, Davis GA (1989) The origin of metamorphic core complexes and detachment faults formed during Tertiary continental extension in the northern Colorado River region, U.S.A. J Struct Geol 11:65–94

Lister GS, Etheridge MA, Symonds PA (1991) Detachment models for the formation of passive continental margins. Tectonics 10:1038–1064

Lüth S, Wigger P, I.R. Group (2003) A crustal model along 39°S from a seismic refraction profile-ISSA 2000. Rev Geol Chile 30:83–101

Nagy D (1966) The gravitational attraction of a right rectangular prism. Geophysics 31:362–371

Pesce A, Gimenez ME, Gianni GM, Folguera A, Martinez P (2019) Magnetic characterization of a retroarc extensional basin: the Loncopué Trough. J S Am Earth Sci 89:55–62

Pesicek JD, Engdahl ER, Thurber CH, DeShon HR, Lange D (2012) Mantle subducting slab structure in the region of the 2010 M8.8 Maule earthquake (30–40°S), Chile. Geophys J Int 191:317–324

Phillips JD (2007) Geosoft eXecutables (GX's) developed by the US Geological Survey, version 2.0, with notes on GX development from Fortran code. US Geological Survey

Radic JP, Rojas L, Carpinelli A, Zurita E (2002) Evolución tectónica de la cuenca terciaria de Cura-Mallín, región cordillerana chileno argentina (36°30′–39°00′S). In: Abstracts of the 15 Congreso Geológico Argentino, Córdoba, 23–26 Aug 2002

Ramos VA (1978) Estructura. Geología y recursos naturales de la Provincia del Neuquén. Asociación Geológica Argentina, Buenos Aires, pp 99–118

Rojas Vera E, Folguera A, Valcarce GZ, Ramos VA (2008) The Loncopué Trough: a major orogenic collapse at the western Agrio fold and thrust belt (Andes of Neuquén, 36°40′–38°40′S). In: Abstracts of the 7 international symposium on Andean geodynamics, Nize

Rojas Vera EA, Folguera A, Valcarce GZ, Giménez M (2010) Neogene to Quaternary extensional reactivation of a fold and thrust belt: The Agrio belt in the Southern Central Andes and its relation to the Loncopué trough (38°–39°S). Tectonophysics 492:279–294

Rojas Vera EA, Folguera A, Zamora Valcarce G (2014) Structure and development of the Andean system between 36° and 39°S. J Geodyn 73:34–52

Ruppel C (1995) Extensional processes in continental lithosphere. J Geophys Res: Solid Earth 100:24187–24215

Soler SR (2015) Métodos Espectrales para la determinación de la Profundidad del Punto de Curie y Espesor Elástico de la corteza Terrestre

Tanaka A, Okubo Y, Matsubayashi O (1999) Curie point depth based on spectrum analysis of the magnetic anomaly data in East and Southeast Asia. Tectonophysics 306:461–470

Valcarce GZ, Zapata T, del Pino D, Ansa A (2006) Structural evolution and magmatic characteristics of the Agrio fold and thrust belt. Geol Soc Am SP 407:125–145

Varekamp JC, Hesse A, Mandeville CW (2010) Back-arc basalts from the Loncopué graben (Province of Neuquén, Argentina). J Volcanol Geoth Res 197:313–328

Webring M (1985) SAKI: a Fortran program for generalized linear inversion of gravity and magnetic profiles. Technical Report 85, 122 p

Wernicke B, Burchfiel BC (1982) Modes of extensional tectonics. J Struct Geol 4:105–115

Yuan X, Asch G, Bataille K et al (2006) Deep seismic images of the Southern Andes. Geol Soc Am SP 407:61–72

Zapata T, Brissón I, Dzelalija F (1999) The role of basement in the Andean fold and thrust belt of the Neuquén Basin. Thrust tectonics 99:122–124

MIX
Papier aus verantwortungsvollen Quellen
Paper from responsible sources
FSC® C105338

If you have any concerns about our products,
you can contact us on
ProductSafety@springernature.com

In case Publisher is established outside the EU,
the EU authorized representative is:
Springer Nature Customer Service Center GmbH
Europaplatz 3, 69115 Heidelberg, Germany

Printed by Libri Plureos GmbH
in Hamburg, Germany